4—
2016

CONI

THE MAN WHO NETWORKED
THE WORLD

MARC RABOY

OXFORD
UNIVERSITY PRESS

OXFORD
UNIVERSITY PRESS

Oxford University Press is a department of the University of Oxford.
It furthers the University's objective of excellence in research, scholarship,
and education by publishing worldwide. Oxford is a registered trade mark of
Oxford University Press in the UK and certain other countries.

Published in the United States of America by Oxford University Press
198 Madison Avenue, New York, NY 10016

Library of Congress Cataloging-in-Publication Data
Names: Raboy, Marc, 1948- author.
Title: Marconi : the man who networked the world / Marc Raboy.
Description: Oxford : Oxford University Press, [2016] | Includes
bibliographical references and index.
Identifiers: LCCN 2015042075 | ISBN 9780199313587
Subjects: LCSH: Marconi, Guglielmo, 1874-1937. | Electrical
engineers—Italy—Biography. | Inventors—Italy—Biography. | Telegraph,
Wireless—History. | Telegraph, Wireless—Marconi system. |
Radio—Italy—History.
Classification: LCC TK5739.M3 R33 2016 | DDC 621.384092—dc23 LC record
available at http://lccn.loc.gov/2015042075

The following photos courtesy of the Bodleian Libraries, the University of Oxford.

Insert 1
p. 3 top left   MS. Marconi 89, fol. 118 (119)
p. 3 top right   MS. Photogr. d. 50, Cuthbert Hall (HIS 247)
p. 3 bottom   MS. Photogr. c. 238, fol. 19 (top photograph)
p. 4   MS. Marconi 166, fol. 57r
p. 5   MS. Marconi 159, 1902 Autographed Banquet Menu, recto
p. 7 top left   MS. Marconi 41, fol. 65r (=p.47)
p. 7 top right   MS. Photogr. b. 62, 220/X (8016)
p. 7 bottom   MS. Marconi 249, Certified Track Chart of S.S. Philadelphia American Line (HIS 136)
p. 8 top   MS. Photogr. c. 235, fol. 81

Insert 2
p. 1 top   MS. Photogr. b. 60 (1), box 2, fols. 104 (105)
p. 3   MS. Marconi 21, fol. 101

1 3 5 7 9 8 6 4 2
Printed by Sheridan Books, Inc., United States of America

"... wires, wires everywhere. Up and down, and underground. Wires in your house I'd say. Wireless your heads one day.... Yup I'd say, world's gonna shrink one day.... We'll start with dots and dashes, codes and patches then it's double back slashes and faxed out hashes. Know what I mean?"

—"Hatter Pete's prophecy," in *Whispers in the Air,* a play for children
by Attila Clemann, 2012

Contents

Prologue: Marconi in His Time and Ours

Travellers to Bologna arrive at Guglielmo Marconi Airport. From there, it's a short ride to the Marconi quarter in the city centre. Tourists, if they are interested, can follow a Marconi itinerary of key locations in the early life of the legendary inventor. On a blistering July day in 2009, the city bus I was on left the centre by Guglielmo Marconi Avenue and meandered out through Bologna's suburbs before arriving at Villa Griffone, headquarters of the Fondazione Guglielmo Marconi, near the satellite town of Sasso Marconi.

The Fondazione is the world's main repository of artifacts related to the young Marconi, containing as well a small museum, an array of research materials, and an eclectic collection of books that once belonged to the Marconi family. Housed on the property where Marconi grew up, undertook his first groundbreaking experiments, and is buried, the Fondazione's library still holds many of the books Marconi was exposed to as a boy, including an early biography of Benjamin Franklin.

The place was bustling when I arrived. A film crew from Italian television was conducting interviews for an upcoming documentary on the centenary of Marconi's 1909 Nobel Prize in Physics. A group of engineers from Telecom Italia was having a look at a replica of the homemade transmitter Marconi used in 1895 to send signals from Villa Griffone to a neighbouring hillside. The small staff, headed by historian Barbara Valotti, was fielding phone calls, supervising student interns, and dealing with passing sightseers.

I had come here, to Marconi's childhood home, seeking answers to the question that had brought me to Bologna. Why had there never been a proper biography of Marconi? I took advantage of a serendipitous encounter with Elena Lamberti, a scholar of comparative literature at the University of Bologna who was waiting to be interviewed for the documentary. "Marconi is a myth—Marconi, the wizard of radio," she replied without hesitation. "He had his work, he was a womanizer, he had two families but his life was not that interesting. Look out this window," she said, pointing to the back of a bust of Marconi that faces away from the property. "Did you see the inscription on the mausoleum?" The inscription—*Diede con la sua scoperta il sigillo a un'epoca della storia umana* ("His discovery marked an era of human history")—is signed "Mussolini." Lamberti went on to point out that while the region saw fierce partisan fighting toward the end of the Second World War, Marconi's mausoleum had not been touched. "Marconi holds a special place in the Italian imaginary. People are not ready to look at his uncanny relationship with fascism. Marconi is part of the mythology of modern Italy." A few days earlier, Silvio Berlusconi's party had swept local elections and was at the height of its power. "We are not ready yet to deal with the past, and the past is tied up to the present."

Among the many things that compelled me to write this biography of Guglielmo Marconi, his "uncanny relationship with fascism," as Lamberti put it, was one of the most compelling. Even when acknowledged, it has been treated lightly. Marconi himself sugar-coated it with charm and guile. Cultivating an air of naive political neutrality, he was anything but naive or neutral. At the same time, toward the end of his life, Marconi became increasingly ambivalent about the degeneration of fascism and about Mussolini in particular. Mussolini's political police, who tracked Marconi's movements and pronouncements from 1927 to his death and beyond, were well aware of these contradictions and kept a file on this most visible of Italians. There is no evidence that Marconi ever knew of the existence of this file—which has also remained unexplored by researchers until now, although it is easily accessible if one knows to ask for it in the Italian state archives.

Marconi's relationship with fascism was not the only complication in his life. There was his relationship with the Catholic Church, the British government, the US telecommunications industry, German science, European colonialism, global modernist mythology, the international news media, and, lastly, the people in his life. They are all worth exploration and they all render his life, despite Lamberti's assessment, very interesting indeed. From our vantage point, what makes it especially interesting is that a century before the Internet, Marconi was living in a wireless global village of his own creation. Before the dawn of the twentieth century, Marconi envisioned a world without communication borders, and set out to create it using all the tools he could muster. Born with roots in Italy and Ireland, he planted others in the United Kingdom, the United States, and Canada, extended branches into Portugal, Japan, Germany, Argentina, South Africa, Russia, Turkey, and India, and left seeds in places on the imperial margins, like Montenegro and the Congo.

Marconi's story and its myths do not belong exclusively to Italy. From Signal Hill, Newfoundland, to Buenos Aires, from Melbourne to Saint Petersburg, one finds claims to Marconi. Mexico City, Montreal, and Hong Kong, among many others, have streets named after him. At Kahuku, on the north shore of Oahu, Hawaii, a move is underway to preserve and restore the Marconi Wireless Station, site of an early outpost. Portugal Telecom, that country's largest telecommunications provider, has a website dedicated to Marconi's role in the development of communication between Portugal and its former African colonies. In Clifden, a pretty seaside town in Connemara, on the west coast of Ireland, Marconi is remembered for having brought the area its first modern industry (although the Irish Republican Army seized the station, hastening its shutdown, in 1922).

There are many reasons why no one has been able to effectively capture the full scope of Marconi's story, despite all the earnest attempts that have been made to embrace him. He is diffused not only among countries, eras, and cultures, but also among ideas and technologies, as well as political currents. There has been no satisfying biography of Marconi because he was ceaselessly self-inventing and reinventing, his life a maddeningly elusive moving target with neither neatness nor coherent structure. Marconi was simultaneously, restlessly everywhere and connected to nowhere—like wireless, one might say.*

Nearly eighty years after his death, there are traces of Marconi in the least likely places. A blog called The Marconi Experiment offers lively commentary on "music, politics, pop culture and anything else that comes to mind." Marconi memorabilia turn up at auction houses from Dublin to Cape Town. A display at Oxford (which once gave him an honorary degree) includes a civilian uniform complete with sabre that Marconi wore to the coronation of King Edward VII. In Montreal's Little Italy, a church built by Italian immigrants a century ago and upgraded in the 1930s features a fresco depicting Mussolini astride a horse and, on foot beside him, Marconi.

---

* The most relevant existing Marconi biographies are discussed in the text and fully referenced in the Bibliography.

All of this points to the fact that Marconi was arguably the first truly global figure in modern communication. Not only was he the first to communicate globally, he was the first to *think* globally about communication. He was certainly not the greatest inventor of his time, but he brought about a fundamental paradigm shift in the way we communicate. A century before iconic figures like Bill Gates and Steve Jobs permeated our lives, sixty years before Marshall McLuhan proclaimed media to be "the extensions of man," there was Marconi. Today's communication explosion would have been inconceivable without him.

The globally networked media and communication system that we take for granted has its origins in the mid-nineteenth to early twentieth centuries, when, for the first time, messages were sent electronically across great distances. The telegraph, the telephone, and the radio were the obvious precursors of the Internet, iPods, and mobile phones. Yet many important aspects of early electronic communication and its impact on the way we live today remain unexplored. The connection between technical innovation and the corporate business model of contemporary capitalism is one such aspect. The role of government regulation as an arbiter of complex technologically mediated social interactions is another. The relationship between national sovereignty, colonialism, imperialism, and transnational governance mechanisms is yet another.

All of these can be traced back to Marconi, and his story serves as an entry point. Between 1896, when he applied for his first patent in England at the age of twenty-two, and his death in Italy in 1937, Marconi was at the centre of every major innovation in electronic communication. While Marconi is known in the popular imagination as "the inventor of radio," his contribution as an inventor was much more limited than is commonly thought—and has been arduously contested. While Marconi indeed authored some of the most significant technical advances in wireless radio communication, he was also, unlike his rivals, a skilled and sophisticated organizer. He was unquestionably more adept than his competitors as an *entrepreneurial* innovator, mastering the use of corporate strategy, media relations, government lobbying, international diplomacy, patents, and litigation.

Marconi was really interested in only one thing: the extension of mobile, personal, long-distance communication to the ends of the earth (and beyond, if we can believe some reports). Some like to refer to him as a genius, but if there was any genius to Marconi it was this vision. In 1895 he started by trying to send a signal beyond the hill, right here, in front of the Villa Griffone. Some say that was genius. In 1901 he succeeded in signalling across the Atlantic, from Cornwall to Newfoundland, despite the claims of science that it could not be done. If not genius, that took at least a lot of chutzpah. In 1924 he convinced the British government to scrap a plan to girdle the world with a chain of wireless stations, and replace it with the latest technology that he had devised, shortwave radio. In 1931 he created the world's first international broadcasting service for his friend, the pope, who didn't trust Marconi's other benefactor, Mussolini, to leave the Vatican free to spread unfiltered messages to the faithful. There are some who say Marconi lost his edge when commercial broadcasting came along; he didn't see that radio could or should be used to frivolous ends. In one of his last public speeches, a radio broadcast to the United States in March 1937, he deplored that broadcasting had become a one-way means of communication and foresaw it moving in another direction, toward communication as a means of *exchange*. That was prophetic genius.

What characterized Marconi's career was—as Malcolm Gladwell wrote in the *New Yorker* in 2011 about the origin of the computer mouse—"the evolution of a concept."[1] Marconi's concept was long-distance, point-to-point wireless communication, and his career was devoted to making it happen cheaply, efficiently, smoothly, and with an elegance that would appear to be intuitive and uncomplicated to the user—user-friendly, if you will. There is a direct connection from Marconi to today's social media, search engines, and program streaming that can best be summed up by an admittedly provocative exclamation: the twentieth century did not exist. In a sense, Marconi's vision leapfrogged from his time to our own.

I had other compelling reasons for writing this book. When I was five years old, my family moved into a new house in Montreal. From our kitchen window I could see a brick factory with a neon sign: "Marconi." Although I had never seen a factory before, I had a vague idea of what they did there; in our dining room we had a yellow plastic radio set with the same name moulded on to its front. Later, I learned that the station we used to listen to, CFCF (the oldest in Canada, they claimed), belonged to the Canadian Marconi company. Canadian Marconi was a wholly owned subsidiary of a British company by the same name, and when the Trudeau government brought in foreign ownership rules in 1969, Marconi was forced to divest. By that time, I had developed some strong views on media and politics, and

I thought that was a good thing (I still do). But I hadn't yet made much of a connection between these corporate, political, and media phenomena and the man whose name was on those factories, equipment, and companies.

In the late 1990s, having meandered through a number of careers before staking out some ground as a media policy scholar, I was doing research on the origins of radio regulation. There were a number of key markers in this history—a series of international conferences at which governments had negotiated a basic framework for use of the radio spectrum. As I read the documents, I noticed that the further back one went through this history, the more one kept coming across the name Marconi, until, at the first international radio conference in 1903, the man himself was at the centre of the discussions. At the turn of the twentieth century, Marconi's name was virtually synonymous with wireless.

Like most people who had at least heard of Marconi, I associated him with broadcast radio (those ubiquitous household boxes surely had something to do with that—even today, older folks in Britain still refer to what Americans call "radio" as "the wireless"). But that's not quite right. What Marconi invented was the idea of global communication—or, more prosaically, globally networked, mobile, wireless communication. In Marconi's initial vision, this was wireless Morse code telegraphy, an improvement on the principal communication technology of his day. Wireless telegraphy existed before Marconi, but he was the first to develop a practical system for doing it using the newly discovered electromagnetic radio spectrum. He certainly borrowed technical details from many sources, but what set him apart was a self-confident, unflinching vision of the paradigm-shifting power of communication technology, on the one hand, and, on the other, of the steps that needed to be taken to consolidate his own position as a player in that field. Tracing Marconi's lifeline leads us into the story of modern communication itself. There were other important figures, but Marconi towered over them all in reach, power, and influence, as well as in the grip he had on the popular imagination of his time. Marconi was quite simply the central figure in the emergence of a modern understanding of communication.

In his lifetime, Marconi foresaw the development of television and the fax machine, GPS, radar, and the portable hand-held telephone. Two months before he died, newspapers were reporting that he was working on a "death ray," and that he had "killed a rat with an intricate device at a distance of three feet." It seems to have been something like a taser: "I dropped the experiment after that," he said. "If you have to crawl up to within three feet of something to kill it with elaborate, costly and ungainly apparatus needing the most sensitive adjustment it's cheaper to use a gun."[2] This anecdote was typical Marconi. The story originated with an exclusive wire service interview he gave to a young journalist. The headlines were sensational, the substance much less so. The real story was Marconi's denial that his latest invention had any great use at all. But by 1937, anything Marconi said or did was newsworthy, and had been for more than forty years. Stock prices rose or sank according to his pronouncements. If Marconi said he thought it might rain, there was likely to be a run on umbrellas.

Marconi's biography is also a story about choices and the motivations behind them. At one level, Marconi could be fiercely autonomous and independent of the constraints and designs of his own social class. He mastered the exercise of power by association, the art of staying above the fray. On another scale, he was a perpetual outsider—the insiders' favourite outsider, one might say. Wherever he went, he was never "of" the group; he was always the "other," considered foreign in Britain, British in Italy, and "not American" in the United States. At the same time, he also suffered tremendously from a need for acceptance that drove, and sometimes tarnished, every one of his relationships.

So there are a number of reasons for a biography of Marconi, one that dismantles the myths that continue to surround the man. To be clear, this is not meant to be a book about technology or science. It is about a life and career that placed an indelible stamp on the way we live. Marconi not only "networked the world," he was himself the consummate networker. Marconi needs to be seen fully in his own time and for the reflection his story offers on our own, for there is a direct link between Marconi's pioneering work and today's technological environment. The French term *fil conducteur* roughly translates as "connecting thread." In Marconi's case, the thread is sometimes invisible; the connection is *wireless*.

I thought, when I started out, that this book might in some sense resemble A.J.A. Symons 1934 classic *The Quest for Corvo*, subtitled *An Experiment in Biography*, but I soon realized that any ambitions I might have had in this regard were eclipsed by Marconi's lifelong quest to control his own biography. Marconi the inventor was forever reinventing himself—in letters, interviews, public speeches, official testimony at parliamentary commissions, and patent litigation hearings. This capacity for reinvention was one of his greatest assets, and also a source of restlessness and alienation from those who thought they knew him. Marconi's greatest invention was himself.

To begin to understand Marconi, one needs to engage—as he did—with several cultures and historical contexts. Unlike other great figures such as, say, Bismarck, or Churchill, it is insufficient to look at a single place and time for a starting point. But a story needs to start somewhere, and for this story, central Italy in the mid-nineteenth century is as good a starting point as any. Not far from Bologna, in fact.

PART
I
The Prodigy

1

Bologna: Beginnings

Capugnano consists of a series of hamlets strung out along a provincial road, winding up into the lush Apennine hills outside the spa town of Porretta Terme, an important site of Roman thermal baths in north-central Italy.* About a mile from the Bologna highway is the seventeenth-century church of San Michele Arcangelo, standing alone in a field by the road. A plaque dating from the 1930s notes that this church was the baptismal site of Giuseppe, the father of the famous inventor Guglielmo Marconi; a second plaque to veterans of the 1915 to 1918 war commemorates the loss of three Marconis, among many others.

Another mile or so up the road lies Le Croci, a cluster of eighteenth- and nineteenth-century farmhouses, what locals call a *borgata*, or township. Today, the home of Domenico Marconi and Teresa Dalli, Marconi's paternal grandparents, has been beautifully restored. Although it is divided now into four separate apartments, it retains much of its original state, notably a weathered wooden door to a storage space at the back of the house—a space where Marconi may have played on family visits. The view from the property is magnificent, looking south toward Tuscany over the wide vistas of the Apennines.

The Marconis trace their ancestry back to the sixteenth century, but they were largely *montanari*—mountain people—before Domenico's time.[1] For centuries, they moved around the area between Granaglione, where the name Marconi originated, and Capugnano in the Porretta hinterland.

---

* Until 1931, the town was officially known as Bagni della Porretta, although it was often referred to locally as Bagni di Porretta or, simply, Porretta, which are the names that appear in most of the Marconi family correspondence. Where there is a discrepancy between historic and current place names, I will generally use the name that fits the context.

A sixteenth-century house in Granaglione was home to early generations of the numerous Marconis in the region. Eventually, they began to spread. A widow Marconi and her three sons are noted in the Capugnano census for 1700, the first sign of a Marconi presence in the area where Domenico, born in 1788, would eventually settle.

On August 28, 1817, the twenty-nine-year-old Domenico, a relative newcomer to Capugnano, signed a marriage contract with nineteen-year-old Teresa Dalli, daughter of a well-established landowning family from the nearby village of Castelluccio. The socio-economic ascent of Marconi's paternal line dates from that act; the couple soon had five children, all born at Le Croci di Capugnano and baptized at San Michele: Giovan Battista, Carolina, Giuseppe (born July 5, 1823), Arcangelo, and Luigia.

Domenico managed to buy the house at Le Croci and some of the surrounding land, and soon after the marriage, possibly with the support of the Dallis (or at least a substantial dowry brought to the marriage by Teresa), he began to extend the base of his small landholding, and also began making hemp canvases, in the age-old local tradition. He collected the raw material right outside his front door and first sold his product in the Saturday market in Porretta, then around Tuscany, and eventually in widespread European outlets and markets, reached by the port of Livorno. Bolognese canvas was a desirable consumer item in Europe's great cities, and Domenico Marconi was soon a wealthy man by the standards of the day.

Domenico also had a distinct advantage in that he knew how to read and write. As he expanded his business and acquired numerous properties in the region in the 1820s and '30s, he became increasingly active as a local notable among the largely illiterate population. Well regarded, and a valued intermediary in various deals, he was now ready to become a full-fledged member of the Porrettana bourgeoisie.

In November 1831, Domenico bought one of the most beautiful houses in Porretta, a four-storey, eighteen-room residence in the town's main square, Piazza Maggiore. He agreed to pay a princely sum, but the vendors were in difficulty and accepted a stable that he owned in town as part payment. Around this time he also purchased a property at Montechiaro, in the commune of Praduro e Sasso (now known as Sasso Marconi), an agrarian area south of Bologna. He was now a well-to-do figure in three neighbouring districts: Capugnano, Porretta, and Montechiaro.

In their teens, the two older Marconi boys, Giovan Battista and Giuseppe, were sent to school in Bologna, while the youngest, Arcangelo, was enrolled in the seminary. For the sons of a rising family like the Marconis, there were still only two occupational outlets aside from the family business: law and the church. Giovan Battista became a lawyer and Arcangelo a priest, but Giuseppe never completed his studies; in the 1848 census his occupation was listed as landowner living from rents, as he managed his family's properties. Not surprisingly for this place and time, nothing is known about the destinies of the daughters, Carolina and Luigia.

The area around Bologna was now an increasingly important regional centre for mercantile traffic, and both the feudal town of Porretta and rural communities like Granaglione and Capugnano now began to transform. In the 1830s and 1840s, the traditional power structure was challenged by the rise of newly prosperous former *montanari* like Domenico Marconi. As the family fortune continued to grow, Domenico transferred his base to Montechiaro, where he began raising silkworms in his home. When the democratic-republican insurrection and war of independence in the papal state of Bologna broke out in 1848, the Marconis astutely followed the prevailing winds and supported the republican forces. They were able to take advantage of the upheavals to acquire a substantial estate that had once belonged to the noble Griffone family, at Pontecchio, nine miles south of Bologna. The Villa Griffone was a significant site; combined with their other properties it made the Marconis an important presence in the area.

Domenico died in 1848, soon after the acquisition of Villa Griffone, and his three sons became co-owners of the estate. The Porretta house was now occupied only a few months of the year, during the summer spa season. Giuseppe, meanwhile, spent most of his time in Bologna, where he became involved in various business ventures. One of his acquaintances was a banker, Giovanni Battista de'Renoli, and Giuseppe was soon courting the Renolis' young daughter, Giulia. Giuseppe and Giulia de'Renoli were married in January 1855, and a son, Luigi, was born in October. In 1858, at the age of twenty-four, Giulia died, leaving her husband not only bereft but also overwhelmed by the burden of looking after a small child. Giuseppe remained close to his in-laws, and, according to family legend, it was in their home, a few years later, that he met a young Irishwoman, Annie Jameson, whose family, exporters of distilled liquor from Wexford, had asked their Italian banker de'Renoli to look after her while she studied bel canto at the conservatory in Bologna. Giuseppe and Annie were soon in love, and began making secret plans to marry.

Bologna in the late 1850s was at the centre of the revolutionary upheavals of the Risorgimento, the unification movement that defined and created the modern state of Italy. One of the prizes of papal power since the middle ages, the city changed hands frequently during the nineteenth century. Napoleon and the French army occupied Bologna in 1796, and for the next eighteen years, Bologna was a city marked by social, cultural, and economic change. Between 1796 and 1799 alone, more than forty religious institutions disappeared, their property and land redistributed to the benefit of a new bourgeoisie. In July 1815, however, following Napoleon's defeat at Waterloo, Bologna was restored to the papal states, and it was soon a cauldron of political activity and intrigue, giving rise to a host of oppositional secret societies that drew their membership from the city's numerous university students. Bologna took part in the European political turmoil of 1831 and 1848, and on June 12, 1859, Pope Pius IX finally gave up temporal power over the city. Less than a year later, in March 1860, Bologna (in fact, all of the province of Emilia-Romagna) joined the liberal constitutional monarchy of King Vittorio Emanuele II.[2]

There is no indication that Giuseppe Marconi was involved in politics, although he must certainly have been affected (his older brother, Giovan Battista, ran for public office). Giuseppe was not part of the urban, liberal elite; he was a more conventional man and closer to the old order, but like others of his social class, he would have pragmatically accepted change if not quite embraced it. By the time he met Annie Jameson, Bologna, and Italy, must have been in many ways unrecognizable to him. Giuseppe was nominally Catholic; however, in the climate of the Risorgimento he was at best indifferent to religion, if not outright anticlerical. This bode well for his romance with Annie, because the Jamesons were strongly Protestant, and Annie was determined that no children of hers would be educated by priests. When Annie returned to Ireland, the couple carried on a clandestine correspondence and in the spring of 1864 they met up at Boulogne-sur-Mer, France, where they were married on April 16. Giuseppe was forty; Annie, twenty-four or twenty-five.[†] Their sons would grow up in a very different society, one in which Enlightenment ideas were gaining currency and where science and industry were beginning, however gingerly, to supplant agriculture and the gentler arts.

Giuseppe and Annie's first child, Alfonso, was born at Pontecchio on November 22, 1865. The following year, the Marconis moved back to Bologna, where for the next few years they changed residence a number of times as Giuseppe prospered in various business ventures and their social

---

[†] For more on the uncertainty surrounding Annie's date of birth, see discussion on page 17.

status rose. By 1871, they were living in a palatial residence at the Palazzo Albergati (temporarily, according to the government census) and had three live-in servants: a housekeeper, a cook, and a stablehand. Their frequent moves around the centre of Bologna were unusual for the urban bourgeoisie of the time, as was their attachment to their country property at Pontecchio. Some have speculated that this signified a certain restlessness on Giuseppe's part, as he was always uncomfortable with city life; Annie too was attached to the country. Interestingly, neither one seemed uneasy about the frequent moves—a characteristic that their second son, Guglielmo, would share.

Giuseppe has been described in the standard Marconi literature as dour, severe, and uncompromisingly authoritarian, but this is not borne out by the documentary evidence—or by Marconi's own testimony. Marconi's friend and biographer Luigi Solari quotes Marconi as saying, "My father had a gruff appearance, but in fact he was a jovial Bolognese."[3] In an 1897 family photo, taken during Guglielmo's first triumphant visit home from England, Giuseppe looks like the country gentleman he was, relaxed, leaning back, in a dark coat and light high-waisted trousers, with a white felt hat in his lap. Guglielmo, on the other hand, looks grim and worried, with furrowed eyebrows, a little moustache centring his elongated Modigliani face, hands crossed demurely in front of him. Although the son's star was rising and the father's life winding down, the lines of familial authority are clear.

Marconi later tended to play down his father's considerable wealth, preferring to state simply that Giuseppe was "of independent means."[4] The lands around Villa Griffone were vast and productive, and Giuseppe attended to them assiduously. He also had a surprisingly broad array of business interests, extending as far as London where, according to one account (and despite the fact that he spoke no English), in the 1860s he was the silent partner in a Charing Cross restaurant and music hall operated by two expatriate brothers, Carlo and Giovanni Gatti.[5] In 1869, he travelled to Egypt to attend the opening of the Suez Canal.[6] Giuseppe has been portrayed as tight-fisted, but while he certainly insisted on precise justification and accountability for any expense he might be expected to pay, he was open to being convinced of a good investment. Marconi inherited his father's approach to business and financial practice, although, as we shall see, he had a more utilitarian attitude toward money and was not interested in accumulating it for its own sake. "Do you appreciate money?" Solari once asked him. "Yes," Marconi replied decisively. "Money is a unit of measure. If you are not paid, you can not measure the product of your work."[7]

Anne Fenwick Jameson was considerably younger than Giuseppe, at least fifteen or sixteen years. There were many similarities to their backgrounds, but Annie—as she was called—was from an even more conservative, and more established, family. Her father, Andrew Jameson, was descended on both sides from Church of Scotland (Scottish Presbyterian) whisky distillers. He was born in Clackmannan in 1783, the fifth son of John Jameson, the sheriff of Clackmannanshire, and Margaret Haig, eldest daughter of one of Scotland's most prominent whisky manufacturers, John Haig. The Jamesons had sixteen children; ten of them survived past childhood.

When Andrew Jameson was an infant, the family moved to Dublin where his father acquired an interest in a recently opened distillery in Bow Street. The Jamesons took over and renamed the distillery in 1805, and eventually John Jameson & Son was producing over a million gallons of whiskey a year, much of it stored in great vaults beneath the streets of the city. Jameson's fabulous Dublin parties earned him the nickname Glorious John, while Jameson's became, of course, the world's most famous and bestselling brand of Irish whiskey. The family motto, *Sine Metu* (without fear), is said to stem from a history of battling pirates on the high seas in the sixteenth century; the motto still adorns every bottle of Jameson whiskey to this day.[8]

John Jameson's older sons—John Junior, William, and James—were all prominent in the business, but little is known about Andrew; his name doesn't figure in any of the company biographies. As far as is known, he established himself in Wexford County around 1815, settling in a prominent manor house called Daphne Castle and opening a distillery in a former forgeland at Fairfield, near the town of Enniscorthy in 1818.[9]

The home to around ten thousand people on the River Slaney, Enniscorthy—or "island of the rocks"—dates its history to the sixth century and is said to be one of the longest continuously occupied sites in Ireland; it officially celebrated its fifteen-hundreth anniversary in 2010. Vinegar Hill, overlooking the town, was the headquarters of rebels who formed the short-lived Wexford Republic during the nationalist rebellion of 1798, and Enniscorthy played a small but significant part in the Easter Rising of 1916.[10] Enniscorthy is mostly Catholic, with an important Church of Ireland (Anglican) minority that dominated local industry in the nineteenth and early twentieth centuries—and at least one Quaker family, the Davises, who were united to the Jamesons through business and marriage, and would play an important role in Marconi's story.

Andrew's family situation prior to 1827 is also obscure. Records show an Andrew and Catherine Jameson married in Dublin in 1810, and six children were born between 1811 and 1819, but it is by no means clear if this was the same Andrew Jameson, and there is no trace of these Jamesons in Wexford (although Daphne Castle would certainly have been large enough to accommodate them, and why would Andrew have bought such a large house in 1815 if he were still a bachelor?). In 1827, however, Andrew Jameson "of the parish of Monart near Enniscorthy" unquestionably married Margaret Millar of Perth, Scotland. According to their granddaughter, Daisy Prescott, Andrew and Margaret were cousins.[11] They eventually had four daughters; the youngest, Annie, was said to possess remarkable musical talent.[12]

There is little aside from family stories and local memory to establish the facts about Annie's early life. There are no birth or baptismal records for her or any of her three sisters—which may be due to the destruction of the Church of Ireland registers for the parish of Monart (where Fairfield is located) during the Irish civil war in 1922. Or, as was often the custom at the time, Margaret may have returned to her own family in Scotland to give birth. References to Annie's age place her birth anywhere between 1837 and 1843. The closest to a reliable source is the burial register at London's Highgate Cemetery, which states she was in her eighty-first year when she died in June 1920, dating her birth at 1839 or 1840. Considering the interest in ancestry and the abundance of genealogical resources for nineteenth-century Ireland, the gaps in factual detail about Andrew and Annie Jameson are somewhat surprising.

Andrew ran his Fairfield distillery until 1840, when he was put out of business by a temperance crusade led by Father Theobald Mathew, who visited Wexford that year.[13] He then rented the property to a local merchant, Abraham Grubb Davis, who married his eldest daughter, Helen, in 1850. (Their second son, Henry Jameson Davis, born in 1854, would introduce Marconi to London's finance and engineering circles in the 1890s.) At some point in the 1850s, Andrew moved his family to Dublin, where he died sometime before 1859. (When Isabella Jameson married Edward Cookman, the son of her father's Fairfield neighbour, in Dublin in March 1859, the marriage announcement in the *Wexford Independent* said that Andrew was deceased.[14]) A third daughter, Elizabeth (Lizzie), married a British Indian army officer by the name of Prescott and eventually spent most of her time in Italy, where the Prescotts and the Marconis were close.

Though little is known about Annie Jameson's upbringing, she was, according to one account, a comely girl "with a glorious singing voice and a will of her own."[15] When just out of her teens, she was sent to study at the conservatory in Bologna—a long, long way from Enniscorthy—after her parents refused to allow her to accept the offer of a stage engagement at London's Covent Garden.

In Bologna, Annie met Giuseppe Marconi, a widower with a young child, "and, most heinous of all, a foreigner."[16] Family legend recalls a starstruck couple falling head over heels in love, and that when Annie returned home to ask for her parents' permission to marry it was briskly refused. The couple persisted, however, nurturing their love by smuggling messages to each other until they were able to elope. "Obstinacy was a feature of both spouses," according to historian Barbara Valotti. "This quality was undoubtedly transmitted to their second son, Guglielmo."[17] Marconi's eldest daughter, Degna, who published a perceptive memoir of her father 1962, remarked that her grandparents' romance had all the makings of a Victorian novel.

Was Annie Jameson born in Enniscorthy, Perth, or Dublin? In 1839, 1840, or 1843? How old was she when she met Giuseppe Marconi? Was her father still alive? Just how strongly did her parents protest the match? Genealogists, historians, and family descendants have been unable to answer these questions, and there are no remaining Jamesons in Enniscorthy (although there is no shortage of Cookmans and Davises).[18] Whatever the facts, the story has been told and retold down through the years, setting a backdrop of romance, passion, and determination that serves as an overture to the unconventional life and career of their son, Guglielmo. What is certain is this: the union of the Marconis of Porretta, Italy, and the Jamesons of Wexford, Ireland, yielded a partnership that turned out to be a remarkable success; its outstanding characteristic was family solidarity, which as much as anything else was one of the keys to Guglielmo Marconi's accomplishments.

❁   ❁   ❁

Until the 1990s, most of what was known of Marconi's early life was based on his own accounts and the memories of witnesses. Then, in 1993, a stunning discovery was made by an Italian archivist, Giovanni Paoloni, who came upon eight boxes of documents gathering mould in the basement of the Villa Farnesina in Rome, which had been the headquarters of the Royal Academy (Reale Accademia d'Italia) under Marconi's presidency in the 1930s. The boxes included dozens of family letters, records of household accounts, newspaper cuttings, and photographs, as well as Marconi's childhood report cards and adolescent notebooks. The documents had been put up for sale along with an assortment of furniture, paintings, and other material from Villa Griffone in an antique shop in the nearby town of Casalecchio di Reno, and were acquired in 1940 by the newly established Fondazione Guglielmo Marconi. They were put in storage in the Farnesina basement in the midst of the Second World War and forgotten until Paoloni stumbled over them more than fifty years later while looking for something else.[19]

These and other archival documents reveal that there are two versions to most of the stories about Marconi's early life: a heroic version, in which an unusually gifted genius overcomes incredible odds to prevail; and a more prosaic version, in which, by a combination of intelligence, determination, attention to wise counsel, class privilege, and plain good fortune, he achieves success. Marconi himself became a master at fostering ambiguity between these two versions of his biography; it was an ambiguity that served his interests.[20]

But about his beginning there is no ambiguity. Guglielmo Marconi was born on April 25, 1874, in his parents' home in the Palazzo Marescalchi, a seventeenth-century baroque palace decorated with magnificent frescoes and valuable works of art steps away from the city hall in the centre of Bologna. While Giuseppe, by now, much preferred the rustic elegance of Villa Griffone, he had rented a well-appointed apartment in the Marescalchi a few months before Guglielmo was born, possibly to ease Annie's pregnancy.[21] Guglielmo was baptized at the nearby cathedral of San Pietro.[22]

By the time Guglielmo was eighteen months old, his family had relocated to the Villa Griffone, also built in the 1600s, and situated on a hilltop set back from the Porrettana highway at Pontecchio. The villa was described in 1926 by an Italian journalist as "one of these dwellings, airy, serene and patriarchal, vast and resounding, in which living is sweet. Plane-trees and elms shade its entrance, around the garden; flourishing pines guard it on the opposite side, under which opens the wide valley and the train whistles joyfully on its course."[23] In Marconi's youth, the property included a stable, a barn, a chapel, and a large country mansion where a large room on the upper floor had once been dedicated to raising silkworms. Marconi did his experiments in this room, which is reconstituted today as it was in 1895.[‡]

‡ The house was occupied by the Nazis during the Second World War, and afterwards nothing remained of its original state. A bust of Marconi in front is scarred with a bullet hole, apparently made accidentally. After the war, the Fondazione Marconi received a letter from a German soldier who had been billeted there, apologizing for the damage (B. Valotti, interview, July 1, 2009).

Although Villa Griffone was Giuseppe's home for the next thirty years, Annie and her two sons lived there intermittently, mainly in the summers. When Guglielmo was two or three years old, Annie moved with her boys to Bedford, in the east of England.[24] It is not known why she chose Bedford (the birthplace of John Bunyan, author of *The Pilgrim's Progress*), although education may have been a factor—Alfonso was then eleven and Bedford was considered to have a good school system.[25] Alfonso attended Bedford School between 1876 and 1880, and Marconi himself later wrote that he received his first elementary education at a private school in Bedford for two years, when he was five and six years old. From the school registers, it appears that the Marconis left Bedford and returned to Italy in the spring of 1880.[26]

Giuseppe had evidently intended to follow Annie and the children to England, at least part-time. In 1877, the year he first visited his Irish in-laws, he renounced his Italian citizenship and began steps to obtain British naturalized status.[27] His precise motivation for taking this serious step has not been clearly established, but he was said to be uncomfortable with the increasing liberalization of Italian politics[28] and did not enjoy the long periods of separation from his family. The children's education was sure to be an ongoing issue; Annie insisted that she would not have them educated in schools run by the Catholic church,[29] and while Guglielmo was baptized a Catholic at birth, the children were raised as Protestants. There is no evidence that Giuseppe was at all opposed to this.

Giuseppe filed an application for British naturalization in April 1880 but was thwarted by an irregularity. A Bedford solicitor engaged by the family, William Stimson, reported that the application had not been successful because of a change in Home Office immigration rules. "I consider it an undoubted hardship that an application properly made in April 1880 should be made subject to regulations not in existence till 2 months later," Stimson wrote to Annie in April 1881. "Should Signor Marconi contemplate renewing the application I shall be pleased to receive instructions to prepare the necessary documents." By then Annie and the boys were back in Italy and the family's plans had changed.[30]

A rare and affectionate first-hand reminiscence by Daisy Prescott (daughter of Annie's sister Lizzie), describes Marconi as a child. Daisy, who was a few years older, recalled her cousin Guglielmo at age five as "a small boy dressed in a blue sailor suit... a remarkably pretty little boy... with a fringe of golden brown hair lying low on his forehead, under which a pair of wide open and wonderfully intelligent deep blue eyes looked question-ingly about him. He always struck me as being just what an artist would love to choose for his model, or a sculptor for his chisel, as his features were absolutely perfect in every detail."[31] Later, second-hand accounts describe the young Guglielmo variously as quick-witted, strong-willed, introspective, solitary, and taciturn, but the traits most commonly applied to him are focused, disciplined, and determined. Margherita Sarfatti, whose family vacationed with the Marconis at Porretta in the mid-1880s, became a life-long friend (as well as, notoriously, the mistress of Benito Mussolini in the 1920s); a few years younger than Marconi, she remembered the eleven-year-old Guglielmo carrying her on his shoulders and teaching her the names of the stellar constellations.[32]

After he became famous, although frequently asked to reflect back on his childhood, Marconi rarely agreed to do so in public. Exceptionally, in an interview published just a few months before he died, he said: "My most vivid memory from my childhood is the care I took to keep hidden from everyone my overwhelming feeling that I would one day do something new and great. By the age of 8 or 10 I was more than certain of it, and with this I consoled myself about the rebuffs from my teachers for not diligently preparing my lessons, which didn't interest me."[33] One of his first and most discerning biographers, Giuseppe Pession (who also knew him personally), wrote that as a child Marconi appeared "as though in his mind there was fluctuating something mysterious."[34]

Alfonso and Guglielmo were both fully bilingual in English and Italian, as well as in the local dialect of the Bologna countryside.[35] Like many children with two mother tongues who are raised in a bilingual environment from an early age, Guglielmo grew up speaking both languages fluently but with slight, yet perceptible, imperfections. In England, his Italianness gave him an exotic air, although he was in habit and mannerisms quite British, so much so that in Italy as a lad he was sometimes teasingly called l'inglesino—"the little Englishman."[36] When he became a public figure in the 1890s, the British and US press frequently marvelled at how well he spoke English, with only the hint of a vaguely "foreign" accent (sometimes taken to be Irish). He spoke both English and Italian with his mother and brother, Alfonso; English with his Jameson relations; and Italian with his father and the rest of the Marconi clan.

The details of Marconi's early life are sparse and sprinkled with gaps, but it is clear that as a youth, he had a difficult time engaging with formal studies. He did not do particularly well in school, but impressed a succession of private teachers with his enthusiasm and creativity regarding the subjects in which he was interested. The library at Villa Griffone was (and still is) well stocked, and Marconi read widely, everything from heathen mythology to tales of Captain Cook's South Sea voyages to the biography of Benjamin Franklin—giving rise to the perception that he was "self-taught," or in the more precise term of historian Barbara Valotti, "self-guided."[37] Certainly, his education was unconventional. Most of all, he was unusually, almost disturbingly, earnest—both in temperament and in focus. By age ten or eleven, he had started doing experiments that were consuming nearly all of his time.[38]

For roughly a decade beginning in 1882, Annie and her sons spent large parts of each year first in Florence and then Livorno, while Giuseppe stayed at Villa Griffone. Annie found the Bologna winters particularly hard. The Marconi archive at the Accademia Nazionale dei Lincei in Rome contains more than forty letters from Annie to Giuseppe—from Florence, Livorno, and Porretta—constituting the most complete existing source regarding the family's life in these years, including their various moves, financial issues, and the children's health and education.

The most frequently recurring theme in the letters is the fragile health of young Guglielmo, whom Annie refers to throughout as *il Bimbo*—the kid, or the baby—well into his teens. Guglielmo suffered a succession of minor complaints (mostly colds and stomach troubles), and the letters reveal some evident over-nurturing by his mother, which would continue throughout her life, even after he became a world-famous figure, married, and had children of his own. Guglielmo's adult correspondence with family and business associates is also littered with constant references to his health—partly characteristic of the time he lived in and clearly also a residue from his childhood. Luigi Solari, who knew him well for thirty-five years, referred to Marconi as a *salutista*, or "health fiend."[39] It is not entirely clear how many of these worries were real and how much was mild hypochondria or an inclination to bouts of depression, but Guglielmo's delicate nature certainly caused his parents, and himself, a good deal of anxiety.[40]

In Florence, Marconi briefly attended the Istituto Cavallero, a private school in Via delle Terme, a narrow medieval street close to the Ponte Santa Trinita. The family moved frequently at first, settling finally at 8 Via della Scala, a short distance away, near the Santa Maria Novella train station. Solari, who was a year older than Marconi and coincidentally just ahead of him at the Istituto Cavallero, describes meeting him as he arrived at school one day with his mother. Marconi, Solari wrote, "was slender and delicate, but had a hard and severe look that didn't match his age."[41]

Another recurring concern was the state of the household economy. Annie went to Giuseppe for every unanticipated expense, from *il Bimbo*'s need for a silk sun umbrella to his having outgrown his shoes. Florence in the 1880s was already a popular destination for British and American expatriates and certainly not cheap. (The pleasures of Florence in this period have been captured by Mark Twain, who lived for a time with his family in a villa in the Florentine suburb of Settignano in 1892.[42]) The Marconis' move from Florence to Livorno in mid-1885 was partly motivated by cost, as Giuseppe was evidently concerned about the expense of maintaining two households (although he must have been pleased at the marriage that year of his eldest son, Luigi, to Letizia Maiani, the daughter of a wealthy Bolognese family[43]).

Sometime in 1885, Annie and the boys moved to Livorno, only sixty miles away yet a distant world politically, economically, and culturally.[44] A relatively young city, founded in the late sixteenth century, it became (and still is) a major port on the Tyrrhenian Sea. Superficially, the city appears dull and nondescript compared to the grandeur of such neighbours as Pisa, Lucca, and Florence itself. It is a quirky place, at the origin of some interesting historical footnotes. Known since the late Middle Ages as a centre of tolerance, it became a haven for Jews, Protestants, and Orthodox Greeks fleeing religious persecution in other parts of Europe. Its most recognizable landmark is a seventeenth-century statue in the old port known as *I Quattro Mori* (The Four Moors), depicting the Medici Grand Duke Ferdinando I, in marble, looking down on the bronze figures of four black men in chains, trying to burst free. Livorno is also the birthplace of the modern artist Amedeo Modigliani (1884) as well as the Italian Communist Party (1921).

Known in English as "Leghorn," Livorno was in Marconi's day a popular summer colony for upper-middle-class Brits. Annie's sister, Lizzie Prescott, spent the winters there with her four daughters while her army officer husband was stationed in India. Livorno was also the site of the Italian Naval Academy, founded in 1881 by Admiral Benedetto Brin, the local member of parliament soon to become minister of the marine; and the birthplace of Costanzo Ciano, a future minister of communication in the Mussolini regime. These two key political figures would play a role in Marconi's relationship to Italy. The two periods Marconi spent in Livorno, between the ages of eleven and fifteen (1885–89) and eighteen and nineteen (1892–93), were critical to his formation, in every respect.

Annie had her boys attend the Waldensian Church (Chiesa Valdese di Livorno), in the city centre. This may well have been simply a convenient choice, as Annie herself was Anglican and the Waldensian Church was known locally as "the English church." However, it left a lifelong impression on Marconi. The Marconi brothers' names appear in the church registers, where it is noted that in 1892 Guglielmo passed an exam demonstrating sufficient knowledge and religious conviction to be received as a member.[45] He was confirmed as a Waldensian, and although never much of a church-goer, the link remained real enough that he had his own son, Giulio, born in 1910, baptized by a Waldensian minister.[46]

In Livorno, Marconi made the first significant non-family relationships that would mark his development. Between 1885 and 1889, he attended the Istituto Nazionale, a technical school in Via Cairoli, facing Piazza Cavour. His early studies there instilled in him a lively interest in physics and chemistry and especially anything to do with electricity, but he didn't finish his course, despite the fact that the school's director felt he was a good student, quiet, dutiful, and getting along well with his classmates.[47] During the summers, the family returned to Villa Griffone. "As I seemed to have a special taste or aptitude for mechanics, physics and chemistry," Marconi remembered, his father hired a young engineering graduate from the University of Bologna to teach him "the principles of physics, which were not taught at the school which I regularly attended in the winter."[48] His cousin Daisy Prescott wrote that he was always inventing something and told her, "'If you could only know, Daisy, what a lot of ideas I have got in my head.'" However, she added, "He never would talk about what he wanted to invent."[49]

In 1891, recognizing her son's interest in the sciences, Annie arranged for him to receive private lessons from Vincenzo Rosa, a high-school physics teacher who maintained a well-equipped laboratory at Livorno's Liceo Niccolini.[50] One morning, Rosa later recalled, Signora Marconi arrived accompanied by her son, who wanted some physics lessons. "You want to take an exam? You have to present yourself in a competition?" Rosa asked. "No," replied the youth. "I want to study science." Rosa was stupefied; he had never met anyone who wanted to study science in and of itself. He recalled Guglielmo as serious, pensive, and a loner. Marconi later warmly remembered and acknowledged Rosa's influence in lectures, memoirs, and interviews.[51]

Guglielmo also took private math lessons from a young teacher, Giotto Bizzarrini, introduced to the family by his brother Alfonso, who had been a classmate of Bizzarrini. They would meet daily, during Bizzarrini's lunch hour. According to Bizzarrini, Marconi demonstrated an "instinctive passion for the study of electrical applications and a mentality exceptionally geared to scientific specialization."[52]

Marconi soon became completely absorbed in his informal studies and, encouraged by Rosa, he began doing all sorts of experiments on his own. He set up a contraption on the roof of the house where the Marconis lived in Via del Passeggio, hooked it up to a receiver inside, and managed to make it set off an electric bell.[53] Solari, then a naval cadet, tells what is probably an apocryphal story, yet a colourful one nonetheless. "One day passing through Livorno with a classmate from the Academy, I saw some metal wires hanging from the roof of a house. 'What are they doing on the roof?' I asked my friend, to which he replied smiling: 'In that house lives Guglielmo Marconi who makes doorbells ring with electricity from the air.'"[54] Sanctifying though the story is, it doesn't change the fact that whatever Marconi was doing at this time was certainly not known beyond a tiny circle of intimates. But by the time he reached eighteen years of age, Marconi was telling close friends and family members that he wanted to invent something that would revolutionize the world.

❧　　❧　　❧

In Livorno, Marconi developed a single-minded, almost obsessive interest (or "passion," to use the heroic term) for electrical experiments that occupied him practically to the exclusion of all the other activities that normally fill the time of adolescent boys. The only exceptions were the piano, which he learned from his mother (and which he would play for relaxation during periods of stress throughout his life), and his other great passion: the sea. One of the earliest pieces of correspondence in the Italian Marconi archives is a note from Annie that reads simply, in English: "Darling Baby, Do not go for your bath until Alfonso comes home at 4 o'clock, and do not stay long in the sea for you have your master at 5. Your loving Mamma."[55]§ Guglielmo would often set out with his brother and some friends in small sailboats from the port of Livorno. He enjoyed the sea because it allowed him to get absorbed in reflection on his research, far from the racket of the city and the indiscreet curiosity of others.[56]

By the time he turned eighteen in 1892, Guglielmo was already thinking of how to pursue his interest in "my electricity," as he told Daisy Prescott. He subscribed to an Italian journal, *L'Elettricità*, a weekly devoted

§ Annie usually wrote to her sons in English, while they wrote to each other and all three wrote to Giuseppe in Italian.

to the popularization of scientific knowledge in the field of electricity, but he still had no clear idea what he wanted to do.[57] Records show that he often ordered specialized equipment with which to replicate or improve on experiments described in the magazine, and was planning to enter an international competition it was sponsoring for the development of a new type of electrical battery.[58] He had also done some experiments with electric engines (also a very recent invention). His father was bankrolling him.[59]

In fact, he was encouraged and supported by a sturdy network of immediate and extended family members—including his father, who was initially concerned that his youngest son was recklessly mortgaging his future by not pursuing formal studies or a more reliable career path, like the navy, but eventually came around. Daisy Prescott wrote that Marconi's father "had greatly opposed his son's progress in inventing on account of his inability to find out what the boy really wanted to do; but once he saw clearly that there was something tangible in it all, he was delighted and did his utmost to assist him." Annie, meanwhile, "was always the boy's right hand," fully devoted to him from the time he was a child to manhood.[60] Marconi himself wrote, toward the end of his life, "I owe what success I have had more than anything to the encouragement and inspiration of my mother."[61] So, while Annie's support was fundamental, and her Anglo-Irish business connections were crucial to her son's success, the documentary record shows that Giuseppe also contributed to Guglielmo's ambitions in numerous ways, providing substantial financial assistance and opening doors for him among his contacts in the Bologna establishment. Marconi was irritated by the false myth of his father's resistance, telling the Giornale d'Italia in 1903: "Not only has he not resisted my attempts, he encouraged them, so much so that my first experiments ... were made with the means he provided to me spontaneously."[62] He always felt that the story of his father's adversity to his work had "arisen from nowhere ... as a cautious businessman, he was not so enthusiastic at first, as my mother; but he was not in the slightest adverse to it," he wrote to a would-be biographer in 1937, a few months before he died.[63]

In a letter to his brother, Alfonso, written around this time, Marconi said he had been looking into requirements for a college degree in physics and was not sure he would be able to pass the entrance exam without further private lessons. "I will need a capable teacher at Griffone this summer [1892] in order to prepare for these exams. Perhaps it would not be hard to find a student for this purpose." He also reported to Alfonso on the success of his latest experiments—"I am sure that the last machine I have constructed has industrial merits"—revealing his early conviction in the importance of seeking commercial value in his work.[64]

The letter also referred to another figure who would play an important role in Marconi's rapid development in these early years: Professor Augusto Righi of the University of Bologna, an internationally known physicist who happened to summer in the same region as the Marconis and was introduced to the family by common friends.[65] Righi lived at Sabbiuno, just over a mile from Villa Griffone as the crow flies, but across the Reno river and hence several hours away by winding hilly roads. Marconi supposedly made the trip to visit Righi at his villa facing the Marconi estate riding on a donkey led by an illiterate Griffone gardener, Antonio Marchi, who told him stories to while away the time.[66]

The story of the donkey is the Marconi equivalent of the story of Abraham Lincoln studying his school lessons by candlelight, and Marconi himself later called it "mere fiction."[67] Still, the importance of Righi is undoubted, though exactly how important he was to Marconi's education is one of the minor controversies surrounding Marconi's development. Righi was more of an occasional mentor, providing some basic career advice and facilitating access to laboratory and library privileges at the university. It was Righi who suggested to Marconi that he look into following a formal university degree program; Righi also listened to Marconi describe the experiments he was doing and confirmed that they might have some commercial application. Furthermore, Righi allowed Marconi to audit a few of his classes and observe demonstrations related to his own research.[68] Marconi was certainly exposed to Righi and his work during his crucial formative years between 1892 and 1895.[69]

By 1893, the Marconis were back full-time at Villa Griffone, and it was around this time that Marconi read of the experiments of a German physicist, Heinrich Hertz, on the generation and propagation of electromagnetic waves. Hertz's breakthrough had attracted worldwide excitement when he published his results in 1888, but no one had yet found a practical application for the discovery. Righi himself was experimenting with the new so-called Hertzian waves and may have been the one who introduced Marconi to the phenomenon, and witness reports place Marconi at the scene of some of Righi's experiments. However, Righi was primarily a theoretician and never claimed to be *il maestro di Marconi*—Marconi's mentor—as he has often been portrayed.[70]

After studying receipts for purchases in the Lincei archives, Italian researcher Maurizio Bigazzi hypothesized that Marconi had begun his initial wave experiments by November 1893, soon after the family returned to Bologna from Livorno.[71] Marconi's own notebook notations show that he was already familiar with Morse code.[¶] In all likelihood, he was soon reading about Hertz and those working in his wake (including Righi and the British physicist Oliver Lodge) in the pages of *L'Elettricità*, which started reporting research on electromagnetic waves around the same time. Historian Barbara Valotti points out that an October 1893 issue of the journal includes an unsigned article in which one reads that "the slow vibrations of the ether would allow the marvelous concept of wireless telegraphy without underwater cables, without any of the expensive installations of our time."[72] Marconi's grandson Francesco Paresce, himself a distinguished research physicist, suggests that this may have been where Marconi first "hit upon the very idea of wireless communication"—something that he himself always denied.[73] In fact, nothing published at that time in any way suggested a connection between Hertz's work and the communication of intelligible signals. As Righi later recognized (and as we shall see in chapter 3), this was Marconi's idea, and he was certainly the first to put it into practice.

Soon after reading about Hertz's experiments, Marconi became fascinated with the possibility of developing an application of the principles demonstrated by Hertz, especially the idea that electromagnetic waves could be used as a medium of communication.[74] Marconi was thinking about "wireless telegraphy"—extending the current practice of telegraphy signalling by Morse code to telegraphy without wires. The practical value of telegraphy was unquestionable, and Marconi's initial contribution was to imagine that the use of Hertzian waves could eliminate the need to build a wired infrastructure. This would not only reduce the cost of electronic communication dramatically, it would allow it to reach where wires could not go, where they had been cut, or where there was simply no wired connection possible (between ships at sea, for example). Marconi was not the first to experiment with Hertzian waves, nor the first to try to send telegraph signals without wires, but by combining the two, he became the first person to effectively use what we now call the radio spectrum for communication.

¶ According to family lore, Marconi learned Morse code in Livorno, from a retired telegraph operator by the name of Marchetti. In some versions of this story, Marchetti is blind.

In the summer of 1894 or 1895—Marconi later used both dates almost interchangeably—he travelled with his much older half-brother Luigi (also known as Gigino) to join other family members on holiday at Andorno, in the Italian Alps. He planned to remain about a week and then return to Bologna, stopping in Milan "to get some electrical equipment I need for my experiments,"[75] as he wrote Daisy Prescott. Marconi always maintained that it was on this trip that he had the flash of insight that launched his career. The year is crucial to the story, because of Lodge's well-documented and publicly performed experiments of 1894 (which will be discussed in the next chapter). However, there continues to be an unresolved controversy about the exact date of Marconi's epiphany.[76] Marconi himself later described what happened in various ways. One was poetic:

> In the summer of 1894, contemplating the countryside around Biella from the high mountains of Oropa, the thought came to me that man might find new energies in space, new resources and new means of communication. The un-encumbered pathways of space for the transmission of human thought have, ever since then, enthralled me. They are inexhaustible springs of inspiration for fresh achievements to the benefit of mankind.[77]

This quote appears in both the original Italian and in English translation in the authoritative 1974 *Bibliografia marconiana*. However, in the document from which the quote is taken, a hand-written letter reproduced in the inaugural issue of the journal *Le Vie del mare e dell'aria* in 1918, Marconi has unmistakeably written "*Nell'estate del 1895 ...*," in the summer of *1895*.[78] This was also the date Marconi gave in his first press interview with the Rome daily *La Tribuna* on December 27, 1896, when the event was presumably still fresh in his memory.[79]

This is one of the central elements of the Marconi myth—"the thought came to me ..."—and it is impossible to validate the date; we must simply take it at face value, or not. Exactly what Marconi may have read at this time cannot be established. As we've seen, various sources have reported that Marconi became familiar with Hertz by reading descriptions of his experiments in *L'Elettricità*. Marconi himself said that he read of Hertz's experiments "in Italian and French publications."[80] However, he may have also read a report of Lodge's lecture and demonstration of Hertzian waves at London's Royal Institution on June 1, 1894, which was reported in two English journals, *Nature* and the *Electrician*.[81] Marconi, of course, read English. It is also possible, even quite likely, that in the University of Bologna library he had access to the publications of such scholars as Lodge and the Russian physicist Alexander Popov, who were very close to coming to the same conclusion as his.

Whenever it was that Marconi had his seminal "thought," he could not be certain that none of these reputed scholars had not had the same idea and were already acting to move it forward. It seemed so obvious to him that he was, in fact, convinced that someone more experienced than himself would turn Hertzian-wave communication into a practical application very shortly. Nevertheless, he realized that he was on to an idea that, at least, if not already explored had not been described in published results.

To determine what happened next we have only Marconi as a first-hand source, and—as at so many pivotal moments in his life—he told the story many different ways. The last time he wrote about his post-revelation activities at some length was in a memo to his first American biographer, Orrin Dunlap, shortly before his death in 1937:

> When I was twenty years of age, in 1894, I went with my family for a holiday in the Alps.... When I got back from the Alps I shut myself up in my attic laboratory, and got to work on my new theory. For months I lived the life of a hermit.... I found two local youths who were prepared to help me.... They did not always understand what I was doing but they were fired by my enthusiasm and stoutly defended me against the skepticism of the other young men of the neighbourhood.... It was in the spring of the following year that I made my first great experiment.... I had the transmitter near the attic window and the receiver a few hundred yards away on a small hill. I sat at the transmitter and Mignani, one of my assistants, watched the receiver. I tapped out the letter "s," and if there was any response Mignani waved a white handkerchief.... Would the waves overcome obstacles such as hills? There was only one way to solve the problem, and that was by experiment. I instructed Mignani to take the receiver to the other side of the hill out of sight of the house and watch the signals. Take this gun, I told him. I'll tap three times. If there are three clicks on the receiver, fire the gun. Mignani went off with the gun, and I called my mother into the room to watch the momentous experiment. And here is what happened. I waited to give Mignani time to get to his place. Then breathlessly I tapped the key three times. For what seemed an eternity I waited. Then from the other side of the hill came the sound of a shot.... That was the moment when wireless was born.[82]

Note that in this text, Marconi situates his Alpine brainwave in 1894 and the gun experiment in the spring of 1895; other sources, including the researchers at the Fondazione Marconi in Bologna, say the experiment may have taken place as late as November or December 1895. Marconi also wrote in this

1937 memo: "While I was in the Alps I got the idea that not merely signals but actual words and voices could be transmitted from one place to another without wires."[83] This is the only time he ever made this claim—very late in his life, when he was trying to establish his place as the inventor of "radio" and not merely "wireless." A more credible version of the Griffone experiments comes from Marconi's court testimony in a 1913 US patent suit, made under oath, where he stated:

> At my father's villa...during the fall of 1894 or early in 1895, I conceived the idea that by means of the invention of efficient telegraphic transmitters and receivers capable of sending and receiving Hertzian waves, it would be possible to transmit and receive messages over great distances without the necessity of using connecting or conducting wires between the transmitters and receivers. I commenced my experiments on this subject in the early summer of 1895 at the above mentioned country house.... Having ascertained and tested the practicability of my invention, and being also convinced that much greater distances could be obtained by further improvements in details.... I decided that some commercial use should be made of my invention.[84]

But this was in the summer of 1895, not 1894.

Whether the discovery occurred in 1894 or 1895, the standard narrative of what happened next, repeated by nearly all of Marconi's biographers, is that the family now wrote to the Italian Ministry of Posts and Telegraphs, offering the government the invention, and that receiving no reply they decided to take it to England. There is, however, no documentary evidence that the Marconis actually contacted the Italian government before Annie and Guglielmo left for England in February 1896, although they certainly considered doing so.[85] Asked by a journalist in 1903 whether it was true that the Italian government had refused to support him initially, Marconi was quoted as replying: "That's false; I asked nothing and they refused me nothing."[86] Research suggests Marconi himself made the claim only once, in a 1923 letter to Mussolini that was certainly politically motivated.[87]** The story became part of the mythology of fascism, used by both Marconi and Mussolini's regime as evidence of the incompetence and lack of patriotism of Italy's liberal governments of the 1890s and 1900s.[88]

On the contrary, it appears that after much discussion within the family and consultation with their British relatives, it was decided that Annie and

---

** In this letter Marconi declined the presidency of the newly created Italo Radio company, in protest of the Italian government's inclusion of his French and German competitors in the concession. See chapter 26.

Guglielmo would try their luck in England, where Jameson business contacts were in a position to help them exploit their interests. Marconi's own
clearest account of what happened, written in 1913, is this:"I was advised by
my mother's relatives that England was, amongst the European countries,
the country in which, owing to its large fleet, extensive coast and large shipping interests, my invention would be most readily employed." He added,
however, that he made his invention known to the Italian government through
General Annibale Ferrero, a "particular friend" of the Marconi family then
serving as the Italian ambassador in London.[89]

Marconi's 1913 testimony indicates that the decision to go to London
was made *before* the family contacted any government official, and even then,
the purpose of the contact was to *inform* the government, not to offer it the
invention. Coincidentally, before being posted to London in February 1895,
Ferrero, a geodesist by profession, had been commander of a military division in the Bologna region and was known locally. Yet another family friend,
their physician Gardini, wrote confidentially on their behalf to Ferrero, who
reportedly replied that young Marconi should apply for patents everywhere,
using the "right to freedom of action" provided by the Italian government,
and do so before the secret of his important invention got out.[90]

This, in effect, is exactly what Marconi did. Ferrero's subsequent advice further discouraged Marconi from placing too much faith in the Italian
government. One of the reasons the standard version of the story has received so much currency may be because it fit the view, prevalent in Italy in
the 1930s (and promoted by Marconi himself), that prior to fascism Italian
governments were not dynamic enough to take advantage of opportunities
such as the one he could have offered them.[91] While this view was consistent with much of Marconi's own experience in dealing with the Italian
government—and as we shall see again and again, he was quite adept at
asserting his position when he felt it would serve him and he was in a position of strength—he also always remained cautious not to ruffle any potentially friendly or useful feathers in his own accounts of events.

Degna Marconi writes that up until that point her father had been
"simple and provincial," but that moving to England to apply for the patent
marked the end of his childhood. A few months shy of his twenty-second
birthday, he set out on a course that was anything but simple and provincial,
moving to the world's most cosmopolitan city and marshalling a complex
set of legal, financial, entrepreneurial, and political tools to advance his goals.

In London, "the men he met were invariably surprised to find him so young but no one ever treated him like a child."[92]

As he travelled to England, Marconi was well aware that he was attempting to make history. He knew much about the history of technology and he also knew how it had been adopted and utilized by great powers in the past. Before carrying on with his personal story, it's important to survey this history. And since Marconi was also entering a competitive environment and seeking access to power, it's also important to look at those who had made, or were making, major strides in Marconi's own field. To understand Marconi's influence and power, one must look at the historical context, a context of which he was fully aware when he went to England in 1896.

2

Priority and Detractors

There is a historic link between communication and empire, going back to the accountants of the Egyptian pharaohs, whose records of tax payments, scratched onto sheets of papyrus, were a source of power and control five thousand years ago.[1] One can revert even further, to the earliest representations of real and imagined forms on cave walls and their place in maintaining the stable hierarchies of tribal life, or to the first petroglyph rock carvings. One of the tools that historians use to explain the evolution of human societies is chronological typology—reference to this or that "age" or "era"—which enables us to distinguish periods by their new or dominant characteristics; in the long sweep of history, scholars of communication speak of things like the age of morality, the invention of writing, and the introduction of machines.[2]

Gutenberg's invention of movable type, in the mid-fifteenth century, was arguably the most important single development in communication technology of the past thousand years, in terms of its impact on the struggles for unhindered human expression and the corresponding attempts to exercise social and political control over it. Coupled with the spread of literacy, the printing press enabled the Protestant Reformation, among many other revolutions of early modernity. But the thing about literacy, British cultural theorist Raymond Williams once wrote, is that you cannot teach someone to read the Bible without also, simultaneously and unintentionally, empowering them to read less holy tracts.[3]

From the sixteenth century onward, the printing press became the focus for all sorts of political struggles, but it was not, at least at first, primarily an instrument for the spread of either empire or capitalism. Transportation was much more critical to those designs, especially with the opening of new markets and vast uncharted territories with the European "discovery" of new worlds to the west and south, and competition over the ancient trade routes to the east. The "press"—the media derived from printing technology—was by its very nature oppositional and mobilizational, encouraging and enfranchising individuals, and their publishers, to act more effectively as political citizens. Governments became obsessed with a sense that they needed to control the press and, not surprisingly, the First Amendment to the United States Constitution, adopted in 1788, stated that Congress shall make no law interfering with freedom of the press. The press worked slowly and subtly, spreading ideas ranging from religious consolidation to political subversion—*ideologies* that came to be tied to interests as well as beliefs. By the early nineteenth century, the press was a tool not only for democrats but for all sorts of propagandists as well.

A representative example of the latter might be an anonymous pamphlet that appeared in London in 1828, entitled *On the Rise, Progress and Present State of Public Opinion in Great Britain and Other Parts of the World*. It later came out that the author was one William Alexander Mackinnon, a Cambridge-educated Scot who was in and out of Parliament between 1819 and 1865. This was one of the first treatises on the importance of public opinion as the touchstone of governance and power, and Mackinnon's idea of public opinion was profoundly based on notions of class and racial superiority. (The abolition of slavery in the United Kingdom was still a few years away.) The tract lauded the moral and political superiority of British society, explaining that Britain's position as a world power was attributable to four elements: information, proper religious feeling, ease of communication, and the material wealth of the individuals making up the community. Because these four elements were amply present—and more so in England than in any other country—Britain was in a position to impose its domination on the world, or so the argument went. Needless to say, the author thought it was a very good thing.[4]

This was possibly the first attempt, however primitive, to theorize the link between communication and globalization, and its 1828 date of publication is propitious, for it was just a few years before the introduction of the most important new communication technology since the Gutenberg printing press—electrical wired telegraphy. By the 1830s, large commercial press interests as well as a new type of company, the national news agency, started to emerge. Press technology could operate as well on a very small scale as on a large one. Getting one's hands on a small printing press and

using it to go into business or politics was not beyond the reach of entrepreneurs or activists. Telegraphy was another matter. Telegraphy was a complex technology, requiring huge capital investment; therefore, access to it
was regulated either by companies or governments, or, more typically, both.
With the telegraph, for the first time, there was a separation of means and
message, and the emergence of a belief that the tremendous power bestowed
by ownership and control of the means of communication had to be offset
by responsibilities.

Another new feature of telegraphy was that the messages sent along telegraph lines did not recognize national borders. (Neither did carrier pigeons,
which is one reason it took some time for telegraphy to catch on.[5]) Governments
were already putting in place regulations to deal with international postal service and customs tariffs. The first meeting to discuss international postal
issues was held in Paris in 1863, but it took another eleven years before an
agreement, the Treaty of Berne, established rules for cross-border postal delivery, and the General Postal Union was set up only in 1875 (becoming the
Universal Postal Union in 1878). The Treaty of Berne established "a single
postal territory" among the signing nations; prepaid postage became an international norm; and a "right of transit" obliged intermediary countries to allow
mail to pass through their territories on the same conditions as their own.[6]

The mail had to physically cross a border. Not so with the telegraph.
Once the lines were laid, the only thing that could be done to stop information from entering or leaving a sovereign country's space was to cut
those lines, a rather drastic step. So it was soon recognized that some type of
international (that is to say, intergovernmental) regulation was necessary. It
was, in fact, surprising that it took until 1865, some twenty years after
cross-border telegraphy began, that the first multilateral treaty organization,
the International Telegraph Union (ITU) was created. Known since 1932 as
the International Telecommunication Union, the ITU celebrated its 150th
anniversary in 2015 and is very much alive today. In fact, the questions of the
1860s prefigured issues, such as "net neutrality," that we are facing today,[7]
and it is not surprising that the ITU has been trying to convince governments
to charge it with setting international rules governing the Internet.

The international regulatory system that was being put in place in
Marconi's time was modelled on the postal system, the first of the great state
monopolies in communication, but it really began to take shape around the
telegraph.[8] With telegraphy, the need for international regulation was far
more urgent; the problems were greater, the stakes higher. The ITU's first

objective was therefore to establish a standardized system for international telegraph traffic. Unlike the post, telegraphy was not simply a question of message delivery but of an entire technological system of communication. (Anyone could write a letter, seal it, and arrange for its delivery, but to send a telegram meant consigning your message into the hands of a corporate structure that had absolute control over it, from composition to reception.)

Wired telegraphy had some significant limitations, however. It did not reach everywhere, and often needed to be combined with another, usually more primitive, form of communication. To send someone a "telegram," or "wire," one needed, first, to get the message to a local office. Then, at the other end, someone had to deliver it by hand to the intended receiver. There were issues of security and confidentiality. By and large, however, the new technology was so attractive and had so many potential applications that by the 1850s it was difficult to imagine life without it. War, banking, and journalism, among other things, would never be the same again.

Like building railroads, creating the infrastructure for telegraphy in the mid-eighteenth century led to unimaginable fortunes. In the United Kingdom, John Pender, a Manchester textile manufacturer, became managing director of the British and Irish Magnetic Telegraph Company, and one of the leading cable barons of the age. The German electrical entrepreneur Carl Wilhelm Siemens launched a global empire by building a network in Persia and then extending his lines throughout the world. In the United States, New York paper magnate Cyrus Field obtained monopolistic rights to operate along the Atlantic seaboard, western landing-point for the transatlantic cable system, leaving the integrated domestic US and Canadian telegraphy market to another burgeoning giant, Western Union. (Founded in 1851 as the New York and Mississippi Valley Printing Telegraph Company, the name was changed in 1856 to reflect the company's transcontinental expansion. Western Union had over a million miles of telegraph lines by 1900. It only ceased its telegram business in 2006 and is now largely a financial services company.)

After the first international underwater cable was laid between Dover, England, and Calais, France, in 1850, the idea of a transatlantic cable started to take shape. In 1854, capitalizing on a chance encounter with a British-Canadian engineer by the name of Frederic Gisborne who was then working in Newfoundland, Field formed the New York, Newfoundland and London Telegraph Company and was granted a fifty-year concession for telegraphic communication to and from Newfoundland (the closest North American land point to Europe).[9] In 1856, Field and a consortium of British investors formed the Atlantic Telegraph Company, and the first effort was made to lay the cable.[10] It failed. In England, a merger of Britain's five leading cable suppliers was formed in 1864 to spearhead a second transatlantic effort. The Telegraph Construction and Maintenance Company, presided over by John Pender, was also unsuccessful. Yet another new company was formed, the Anglo-American Telegraph Company, and in 1866, on the fourth attempt, a transatlantic cable was successfully laid from Valentia, Ireland, to Heart's Content, Newfoundland.[11] Cable manufacturing was now a capital-intensive high-tech industry. Submarine cables were of complex construction; they were thick and heavy and in order to work they had to be flawless. At £200 ($1,000) per half mile, cables were inaccessible to all but very large companies and wealthy governments.[12]

In 1868, the British government nationalized all domestic telegraph companies—enriching Britain's cable manufacturers (Pender's firm was the main beneficiary) and turning their interests outwards. Within two years, the number of international cables had multiplied, and in 1872, a four-company merger established the Eastern Telegraph Company. The dizzying cycle of mergers and acquisitions continued, and by 1875, companies controlled by British and American capital owned and operated a network of telegraph cables between England, the United States, India, Australia, and the Far East. By 1904, two-thirds (twenty-eight of forty-one) of the world's "cable ships"—ships designed to carry and lay down cable—were British.[13]

The cable-based global communication infrastructure expanded tenfold between 1870 and 1900, and doubled again in the next decade. The system provided the platform for the shaping of a global political economy and culture(s)—in short, it enabled the emergence of a new phase of what we now call globalization. The wired connection of vast parts of the world, over land and under the seas, was a key part of this process, a huge technical and economic accomplishment for the project of global capitalism (or globalization through capital investment). The new international communication system also sharpened the rivalry between the great colonial powers.[14] An ominous example: the question of sheltering the international cable system from wartime hostilities arose from time to time during this period. The cable companies were nervous about the possibility of attacks on their property in a time of war, and the United States advocated for an international declaration of the neutrality of cables in wartime. Britain, France, and Germany didn't like the idea.[15]

Meanwhile, the core principles governing telegraphy were enshrined in an international convention signed at Paris in 1865 (at the conference that launched the ITU). It included an article that is a landmark in the history of communication: "The High Contracting Parties recognise the right of all persons to correspond by means of the international telegraphs."[16] As scholar Ted Magder has pointed out, this was the first time that something approaching the idea of a "right to communicate" was embedded in an international legal document.[17] When the International Telegraph Convention was revised in St. Petersburg in 1875, the landmark clause was moved up from Article 4 to Article 1, an upgrade that highlighted the importance of the right to communicate in the architecture of international law.[18]

The St. Petersburg Convention also laid out principles regarding the privacy of communication that read ominously in light of issues we are dealing with today. In Article 2, the parties agreed "to adopt all the necessary measures to ensure the secrecy and prompt dispatch of telegrams"; but then, in Article 7, they reserved the right to intervene to control messages deemed dangerous to security. In other words, Article 7 *limited* the legal commitment to privacy spelled out in Article 2. In the tension between networked communication and national security, what the state giveth the state taketh away.[19]

This was the corporate and political context into which Marconi dropped in 1896, and the one in which he was to solidify his place. Marconi actually believed that extending the relatively new capacity to communicate instantaneously by removing the need for wires—"freeing" communication, in other words—would enhance the chances for world peace and social harmony. He also knew that having figured out how to communicate wirelessly was not enough. He needed a base from which to expand his reach. Given his family circumstances, as well as the fact that Britain was the centre of global communication at the end of the nineteenth century, London was the logical place for Marconi to patent his invention and establish his company. Ironically, and not for the last time, this meant tying his own trajectory to the interests of an imperial system.

As his fame and influence grew in the late 1890s, Marconi began to construct a narrative of the history of global communication that was meant to locate his own place in the progression of technological thought. In the first of several historical accounts he signed at different points in his life, written in 1899 when he was still trying to legitimize his recent rise to prominence, he started with the ancient Egyptians, whose practice of signalling with rockets was "telegraphing without wires, if you so choose to call it." The Greeks picked up the idea and took it further, inventing "a system of looking glasses (that) reflected the rays of the sun, reading messages from the reflections that were flashed across from mountain top to mountain top." This system was used to announce the fall of Troy, "a practical application of Wireless Telegraphy in war 3000 years ago." Since those early days, there had been few improvements in signalling methods: "Between the pre-Christian era and practically modern times—as we count historical dates—almost nothing was done in the way of practical improvement."[20]

Practically at the beginning of his career, Marconi was conscious of the need to tell his story on his own terms. After this relatively brief historical background, most of the unpublished 147-page manuscript describes Marconi's early experimental work much as we've seen in the preceding chapter.[21] What is interesting here is Marconi's view of wireless telegraphy as a means of communication rather than of electrical transmission, as it is typically considered by historians of science and technology. In the history of science, although the idea had been around for some time, the concrete possibility of wireless telegraphy only emerged in the late nineteenth century with the coming together of developments in human understanding of three naturally occurring phenomena: electricity, magnetism, and light.[22]

Magnetism was noticed untold millennia ago, as it occurs naturally in the mineral magnetite. Rocks containing magnetite, known as lodestones, were used by ancient Chinese navigators to create the first magnetic compasses. Some claim that the magnetic compass was used in battle as far back as the twenty-seventh century BCE by China's "Yellow Emperor," Huang-ti (who has also been portrayed as the originator of the centralized state). As Huang-ti's historical existence is itself a contested topic, it is hard to say whether this is legend or fact.

Around 600 BCE, the Greek philosopher Thales of Miletus discovered the properties of amber (*elektron*), a substance that mysteriously created what we now call an electrostatic charge, or spark, when rubbed on cat fur. Thales's discovery was transmitted orally and only recorded centuries later by Aristotle; as a consequence, it is not known whether Thales made the discovery himself or had witnessed it on his extensive travels, which had taken him to Egypt, among other destinations.

Little formal scientific attention was paid to these questions until 1600, when the English scientist and natural philosopher William Gilbert, physician to Queen Elizabeth I, published an opus, *De Magnete,* which among other things distinguished the lodestone effect from static electricity. Gilbert used the term *electricus*—like amber—to describe the phenomenon that would enter the English language as "electricity."

The popular understanding of electricity originated in the 1750s with the famous experiments of Benjamin Franklin, who demonstrated that a powerful charge would travel from an electrical source (in this case, lightning) along a line connecting two metal objects (in this case, a kite and a key). Most interesting—and dangerous—Franklin found that sparks leaped from the key to his hand, although he was holding the kite by a non-conducting silk string. In 1756 Franklin became one of the few Americans to be named to Britain's Royal Society, which had in fact been the site of the first recorded demonstration of conduction of electricity over wires by an amateur astronomer, Stephen Gray, in London on July 2, 1729. Gray died, unknown and destitute, a few years later, well before Franklin rediscovered electrical conduction.

Marconi questioned the originality of Franklin's experiment. In his 1899 manuscript, Marconi noted that a "poor Bohemian monk" had fashioned a lightning rod several years before Franklin, and the fact that "one man or a number of men shows the public that a certain thing can be done with certain instruments, proves nothing." There was clearly some self-reflection here, as in the same paragraph he also wrote: "Science is full of examples of misplaced credit, and it is almost a useless task to undertake to right the errors.... The question of originality in wireless telegraphy and of credit for actual developments is a difficult one to answer."

The idea that information could be transmitted over wires was suggested around the time Franklin did his famous test, and experiments were done on this problem beginning in the 1770s. In 1795, a Catalan physician, Francisco Salvà i Campillo, may have been the first to suggest the use of electricity for telegraphy, in an academic paper in which he also remarked, almost casually, that it could probably be done with or without wires.[23] Salvà is mainly remembered for the huge medical library he left to the Royal Academy of Medicine of Barcelona.

In 1789, an assistant working in the lab of Bologna physician Luigi Galvani accidentally discovered that a dissected frog's leg twitched when its nerve was touched by a metal scalpel. Galvani concluded that an electro-static spark had created the convulsion. Had he gone further down this path, Galvani could have become the inventor of wireless, but, as a medical doctor and anatomist, he was more interested in what was happening to the frog than in the spark. Galvani is considered the father of bioelectricity. After Marconi's success was publicized, Galvani's experiment was repeated in Paris by a French physiologist by the name of Lefeuvre, who set up a spark transmitter on the Eiffel Tower and sent a signal to a receiver connected to a frog leg two hundred miles away in Rennes; a needle attached to the frog leg recorded the signal on a rotating cylinder covered with soot.[24]

Another Italian, Alessandro Volta, a physicist and professor of natural philosophy at the University of Pavia, realized after studying Galvani's discovery that the frog's leg in this case had been the detector of an electric current passing between the scalpel and the metal plate on which the frog had been lying. This led him to look for a chemical source for electricity and to develop what became known as the "Volta pile," or electric battery, which he unveiled in 1800.[25] This type of device may have been known in the Middle East more than two thousand years earlier. In 1938, in Baghdad, the German archaeologist Wilhelm König, then director of the National Museum of Iraq, unearthed something resembling an electrochemical cell made of copper and iron immersed in wine vinegar—the so-called Baghdad Battery, which has been dated to 250 BCE.[26]

It was not uncommon, in the early nineteenth century, for researchers on both sides of the Atlantic to be working on similar problems at the same time. Typically, only one would emerge as the leading figure. The American physicist Joseph Henry, of Albany, New York (later the first secretary of the Smithsonian Institution), was searching for a way to convert magnetism into electricity when, in 1831, a British scientist, Michael Faraday described the relationship between electricity, magnetism, and light. Consequently, Faraday is considered the foundational figure in the study of electromagnetism.

Faraday in fact *anticipated* the experimental work of those such as Hertz, Righi, Lodge, and Marconi, who all acknowledged his role as an influential precursor. Faraday wrote in 1832 that "I cannot but think that the action of electricity and magnetism is propagated through space in some form of vibration." The letter in which Faraday wrote these words lay sealed in the custody of the Royal Society for more than a century; it was only opened in 1937 and its contents revealed at a memorial event for Marconi a few months after his death.[27]

For Marconi, Faraday's discovery—"the seed from which it can be truly said that wireless has sprung"—was "that it was not necessary for two electrical circuits to be in actual physical contact in order that electric energy might pass across a small space between them."[28] Faraday, Marconi wrote in a 1931 article in the London *Times*, drew attention to "the medium" separating the two electric circuits, which was essential for communication to take place.[29] This space became known as the "spark gap," because the bit of electrical current generated by the "transmitting" circuit created a spark as it leaped across the space to the "receiving" circuit. However, no one yet understood the nature of the medium through which this took place.

The debates and discoveries of the 1830s paved the way for a monumental invention: the electric telegraph, the first means of long-distance electronic communication—or telecommunication. Practical prototypes emerged almost simultaneously in the United States and Britain. In England, William Cooke and Charles Wheatstone patented and successfully demonstrated a telegraph system using magnetic needles in 1837; their system went into commercial service alongside the newly opened Great Western Railway line between London's Paddington Station and the town of West Drayton, thirteen miles away, in 1839.

Also in 1837, Samuel Morse patented and publicly demonstrated an electric telegraph at New York University, sending recorded signals over a distance of about a quarter of a mile. It was not until 1844 that Morse's electric telegraph connected Washington, DC, and Baltimore, a distance of about forty miles. Morse's telegraph had a distinct advantage over previous devices: a unique code for transmitting information that he had developed along with an associate, Alfred Vail. The code consisted of dots and dashes representing the letters of the alphabet, thus making Morse's telegraph particularly well-suited for sending text messages. Because of the elegant simplicity of the Morse code, as well as its binary basis, some have argued that the telegraph can be considered the forerunner of digital communication.[30] Morse's claim was contested—as Marconi's would later be; his patent application was refused in Britain and challenged no less than fifteen times in the United States within a few years. With his entrepreneurial drive and the backing of the US Congress, however, Morse has generally come to be known as the inventor of the telegraph.

A raft of related communication apparatus soon followed. In 1849, an Italian immigrant to the United States, Antonio Meucci, demonstrated the first telephone in Havana, Cuba. He was incapable of navigating the US patent system, which would have enabled him to claim the invention as his own and benefit from its further development. Meucci's experience was instructive to Marconi, who fought, unsuccessfully, to rehabilitate Meucci's memory internationally in the 1930s.[31] But there is at least a nod to Meucci in contemporary popular culture. In the 1990 Hollywood film *The Godfather, Part III*, mobster Joey Zasa presents Michael Corleone with the Italian-American Society Meucci Award. Michael asks: "Who was Meucci?" Zasa replies incredulously: "He was the *paesan* who invented the telephone!"

So the world's first telephone patent was awarded in 1876 to an elocu-tion teacher, Alexander Graham Bell, for an electromagnetic voice trans-mitter. A physicist, Elisha Gray, had arrived at the patent office with a similar application just two hours after Bell, late enough to settle the issue in Bell's favour. Interestingly, Bell's patent application was for "Improvements in Tele-graphy." It became the most lucrative—and most contested—patent in his-tory, and was challenged in some six hundred cases.[32] The Bell Telephone Company, created in 1877, won every one of them. An International Bell Telephone Company was established in Brussels in 1880, and through a series of agreements with other countries the company became a global corporate power. By 1886 there were more than 150,000 telephones in operation in the United States alone. Bell himself soon resigned from the company, ex-hausted by the constant court appearances needed to defend his patent and interested in pursuing new inventions.[33]

While Morse's telegraph diminished distance, Bell's invention added something more liberating and influential—the use of the human voice. Telephony was indeed a transformative technology, and Bell had a vision of a universal network that would reach into homes and workplaces every-where.[34] Marconi would extend this vision a few decades later to include places where wires could not reach and by adding an additional dimension: mobility.

❖   ❖   ❖

Mark Twain once observed that "Truth is stranger than fiction, because fic-tion is obliged to stick to possibilities; truth isn't." The world in which wire-less communication developed reveals how accurate that assessment is. In his 1899 historical survey, Marconi mentions Jules Allix, a journalist and member of the Paris Commune who died in 1897. Known in his lifetime as a feminist, socialist, and eccentric inventor, Allix had been associated with something called the "snail telegraph," a contraption that took advantage of the "sympathetic bond" that occurs when one puts two snails in contact with one another. Allix claimed to have witnessed the inventor of the snail telegraph, Jacques Toussaint Benoit, use it to send a message across Paris in

1851, and he publicized Benoit's claim that the snail telegraph could be used to communicate across any distance. Benoit was trying to establish a "snail link" to America when he mysteriously disappeared; Marconi (who was sloppy about citing sources) wrote that Benoit went insane. Despite the fact that it purported to create a physical link, the snail telegraph was considered in its day a form of telepathy.[35]

Marconi's version of history, as expressed in that 1899 work, reveals a decidedly orientalist vein. He liked to quote odd instances of "unexplained" intuitive communication, such as when, during the mutiny in India, "British officers were continually troubled because their proposed movements became known to the natives many miles away almost before they could be communicated to their own men." Tourists who visit the fortress of Golkonda in Hyderabad today are shown an example of how this may have occurred, a remarkable acoustic effect whereby someone standing on the rampart of the fortress can, by clapping hands, send a message to someone else nearly a mile away. There are surely similar and less mystical explanations for the other exotic examples cited by Marconi, such as when an English officer related that people on the street in Cairo had heard of the death of General Gordon of Khartoum on the same day he was killed, more than a thousand miles away—"indeed a most remarkable occurrence." Or when a political assassination in the Andaman Islands was known in Calcutta "by the natives" hours before the arrival of the ship bearing the news. Marconi reminds us that North American Indians used smoke signals, and tribes in central Africa drums. All these means proved valuable, Marconi noted, until they were displaced in the nineteenth century by the arrival of the electric current. Marconi's exploits too were often considered magical.[36]

"The mid-nineteenth century was a wonderful time to be alive for a young man with a quick brain and endless curiosity," Charlotte Gray wrote in her biography of Alexander Graham Bell. "It soon appeared as though every young go-getter had a blueprint for a new gadget in his back pocket."[37] Sure enough, and not only young go-getters. The Marconi Archives in Oxford have a thick file of newspaper and magazine cuttings about claimants to the mantle of "first" to accomplish the feat of wireless telegraphy, some more serious than others.[38] In the welter of discovery, finding a neat narrative line that can unequivocally justify a claim to "priority" is far from evident. The genealogy of wireless communication is embedded in a tangled family tree.

An American dentist named Mahlon Loomis began attempting to communicate without wires in the 1850s and received what is considered to

be the world's first patent for wireless telegraphy in 1872.[39] Loomis had succeeded in signalling across some thirteen miles by "aerial telegraph," using a
wire kite. Loomis's patent called for employing an aerial "to radiate or
receive pulsations caused by producing a disturbance in the electrical equi-
librium of the atmosphere." This disturbance, he wrote in his 1872 patent
application, "would cause electric waves to travel through the atmosphere
and ground."[40] Although the US Congress had voted the then-colossal sum
of $50,000 to develop such a system in 1869, Loomis was not able to get his
hands on any of this development money, and his system went no further.
He is more happily remembered for having invented the first successful por-
celain dentures, on which he obtained patents in the United States, England,
and France.

   Loomis has a good claim to priority, but there were many others
working on the verge of wireless who, for one reason or another, failed to
turn their work into a viable *system* of communication. In England, credit is
often given to David Edward Hughes, who invented a device for amplifying
sound that he called a "microphone." Hughes's invention—which he reported to the Royal Society on May 8, 1878, but idealistically chose not to
patent—was soon incorporated into Bell's new telephones, vastly improving
the quality of telephone voice transmission.[41] (Hughes had already made a
fortune with an 1855 patent for a letter-printing telegraph receiver, or "tele-
printer.") Hughes found that his microphone was sensitive enough to pick
up what he called "aerial electric waves," and, pursuing his research, he
produced a wireless hand-held receiver that was able to pick up sounds from
a few feet away. However, Hughes himself was evidently more sensitive than
his microphone, and fearing a mocking reception from the British scientific
establishment (correctly, it seems), he never published his work, and eventually abandoned it.[42] For decades, it was known only to a small circle of
insiders.[43] One of these, the engineer A.A. Campbell Swinton, described
Hughes walking down a street with a working telephone in his pocket in
1879. However, Swinton wrote, "Hughes seems to have had no conception
that he was dealing with electromagnetic waves."[44]

   One more inventor worth mentioning is Nathan B. Stubblefield, a
melon farmer from Murray, Kentucky. The facts of his accomplishments are
obscured by folklore, but it is said that Stubblefield started experimenting
with acoustics in the 1880s and, in 1892, demonstrated something akin to a
wireless telephone.[45] He later staged an exhibition in front of more than
a thousand people in Murray, where conversations were carried on at a

distance of 250 feet between telephones that were not connected by wires. Stubblefield took out a patent for his invention in 1907, but shied away when backers approached him about developing it, out of fear they would steal it from him. The man was eccentric to a T; his home was said to be surrounded by wireless lighting suspended from trees and poles. Stubblefield became a recluse and died in 1928 of starvation. His gravestone says he was "the father of radio broadcasting," one of many to claim that mantle, and Murray at one time billed itself as "the birthplace of radio."[46]*

❀   ❀   ❀

In countless speeches and other public statements after his fame and importance were established, Marconi always acknowledged what scientists and historians of science generally agree: that the immediate pre-history of wireless began in 1864, when the Cambridge physicist James Clerk Maxwell formulated the electromagnetic theory of light, predicting—but, crucially, not demonstrating—the existence of what we now call radio waves.[47] Maxwell laid the foundation for the theory of electromagnetics, which became the basis for early wireless research. Building on Faraday, Maxwell amalgamated existing theories on magnetism, electricity, and light, and developed a series of mathematical formulas with which he was able to predict electrical phenomena with mathematical precision. The formulas indicated that, aside from light (which was already known to be based on electrical currents in the air), there must be other, as yet unknown, types of waves operating at lower frequencies. Physicists today still consider Maxwell's original theory to be valid.

Maxwell's theory excited European theoretical physicists to the point that his followers came to be known as "Maxwellians." Many aspects of the theory were so obscure that even the most engaged Maxwellians found it difficult to work with, and some, like the British physicist Oliver Heaviside, built careers on clarifying and interpreting it. In the 1870s and 1880s many others were looking for experimental applications. One of the most prominent of the Maxwellians was Oliver Lodge, a physicist at the University of Liverpool. Lodge began looking into the possibility of producing electromagnetic waves in his lab shortly before Maxwell's death in 1879.[48] Following

---

* The term *wireless telephony* caught on in the early 1900s when researchers began looking for ways to extend Marconi's method to include voice and sound transmission. The term was then replaced by *radio broadcasting* in the 1920s, with the development of the activity we know by that name today.

Maxwell, Lodge suspected the existence of electromagnetic waves, but he had no idea how to generate or detect them. This was to be accomplished by Heinrich Hertz, a young professor of physics at the University of Karlsruhe, in Germany.

Hertz's mentor, the distinguished professor Hermann von Helmholtz at the University of Berlin, had encouraged him to test Maxwell's theory. Hertz later wrote that he was not quite certain he'd fully grasped the theory, but working in his laboratory in Karlsruhe, in 1886 and 1887, he observed that "the oscillatory discharge of a Leyden jar or induction coil through a wire loop caused sparks to jump a gap in a similar loop a short distance away."[49] Published in 1888, this was the first physical description of low-frequency electromagnetic waves—the oscillating currents that allow a spark to leap across a gap. There was now a direct line from insight to theory to practical demonstration—from Faraday to Maxwell to Hertz—and Hertz is recognized, undisputedly, as the discoverer of those waves which to this day are commonly known by his name in most modern languages. (In addition, the unit of frequency that describes the number of "cycles per second" of electromagnetic wave oscillations is known as the Hertz, or Hz.)

Others before Hertz, including Thomas Alva Edison, had observed electromagnetic waves but had not recognized them as such, as they were unaware of the connection to Maxwell's theory. Hertz was the first to knowingly generate, transmit, and receive (or, in the jargon, "radiate and detect") electromagnetic waves.[50] After Hertz's results were published, he was embraced by the Maxwellians, who set about to replicate and extend his experiments in their own laboratories. Hertz's discovery thus focused and defined the new field of electromagnetics. Hertz continued his work but died prematurely, at age thirty-seven, in 1894. That was, of course, the year (or at least one of the two possible years) that Marconi claimed he had his revealing "thought." No one before Marconi had managed to *use* Hertzian waves to send intelligible signals—to communicate (in these early days, by Morse code)—over significant distances. Marconi's detractors, including Lodge, pointed out that he was not the first to communicate wirelessly, or even to patent a system for wireless communication, but the crucial distinction is that Marconi actually employed Hertzian waves to do so, for this is what enabled him to later transmit and receive across ever-increasing distances, eventually eliminating distance altogether while also reducing the necessary power and cost enough to make it a *practical* system. Simply put, Marconi discovered and developed a practical use for the electromagnetic spectrum. It was at least another decade before other experimentalists learned how to transmit voice by wireless—and also discovered that they had to deal with Marconi's ironclad patents in order to commercialize it.

To step back for a moment: physicists distinguish between three forms of what we normally call telegraphy—conduction, induction, and electromagnetic.[51]

The breakthrough in what came to be known as "conduction telegraphy"—or simply, telegraphy—came with the wired system developed by Morse in the 1840s, which proved to be faster, more practical, and, in fact, superior in every way to all others known at that time. Morse also experimented with using water as a conductor, but with much less successful results. He was able to send signals wirelessly across the mile-wide Susquehanna River in Pennsylvania, but the system was so cumbersome that it was not considered practical. Others experimented, equally unsuccessfully, with conductive wireless until Marconi's system made the idea obsolete.

"Induction telegraphy" seemed more promising for reaching where wires could not go. Marconi, in his 1899 manuscript, acknowledged the role of Karl von Steinheil, "a noted electrician in Munich," who, in 1838, set the stage for wireless induction by demonstrating that earth or water could be a conducting medium.[52†] This method was used in 1880 by a Tufts University professor named Amos E. Dolbear to invent an electrostatic telephone receiver, different from Bell's, that worked even when the line was broken. Dolbear then did experiments with wireless telegraphy and patented a device in 1882 (one that would lead to a famous lawsuit involving Marconi). Dolbear's device resembled the one Marconi would patent in 1896, with the difference that Marconi's was based on electromagnetic waves, which were unknown to Dolbear when he filed his patent.

In 1885, Edison, too, filed a patent application for transmitting signals wirelessly, but his was also based on induction rather than electromagnetic waves. Except for that caveat, Edison's description of his discovery is remarkably similar to what Marconi did ten years later: "I have discovered that if sufficient elevation be obtained to overcome the curvature of the earth's surface and to reduce to the minimum the earth's absorption, electric telegraphing or signaling between distant points can be carried on by induction

---

† The term *electrician* was used in Marconi's day to describe a practitioner we would probably refer to today as an electrical engineer. Marconi frequently gave his own occupation as "electrician."

without the use of wires connecting such distant points," Edison wrote in his application.[53] A much-handled, dog-eared copy of the patent, falling apart and held together by adhesive tape, is one of the oldest documents in the Marconi Archives at the Bodleian Library in Oxford.

Edison had been trying to send messages without wires since the 1870s (even before he patented the phonograph; his two eldest children were nicknamed "Dot" and "Dash") and had, without realizing it, come very close to discovering electromagnetic waves. He had observed the sparks created across the gap between two distant conducting objects, and was puzzled because they were unlike anything with which he was familiar. He concluded that the sparks were due to "a radiant force, somewhere between light and heat on the one hand and magnetism and electricity on the other," which he called the "etheric force," and built an "etheroscope" for detecting the spark gap. Busy and juggling many different things, Edison soon dropped this line of research.[54]

The patent for Edison's system—known colloquially as the "grass-hopper telegraph"—was granted only in 1891, but by then Edison had moved on and he never used it. (Edison amassed more than a thousand patents during his career, and always focused on ones he felt could be profitable in the moment.) Edison also became an early American booster of Marconi and sold his wireless patent to the American Marconi company in 1903, joining the company's board as a technical advisor at the same time. The announcement was made on Wall Street.[55]

In England, where telegraphy had taken on commercial as well as strategic military importance, a Royal Commission on Electrical Communication with Lighthouses and Light-Vessels was appointed in 1892.[56] The chief engineer of the British Post Office (officially known as the General Post Office or GPO), William H. Preece, was charged with developing a system for communicating with offshore lighthouses and ships sailing within British territorial waters. Preece began doing induction experiments using loops of wire to transmit across the Bristol Channel, but this proved to be terribly unreliable and cumbersome; the wires needed on either side to send a signal across a relatively narrow distance had to be several miles long. When Preece met Marconi in 1896, as we shall see, he immediately recognized that Marconi's method was far more promising, as it showed that Hertzian waves could connect transmitters and receivers much more easily.

Preece had another, more mundane, motive for embracing Marconi. He had for some years been involved in an acrimonious controversy with the Maxwellians, whose fixation on Maxwell's theory he saw as a hindrance to finding a practical approach to wireless telegraphy. In Preece's view, the Maxwellians, in fact, were not interested in communication at all, except in a purely experimental sense. The Maxwellians, meanwhile, saw Preece as a vulgar "practician," poorly versed in theoretical physics, and simply looking to solve a banal administrative problem—how to send telegrams where there was no wired connection (they would soon view Marconi the same way). The debate pitting "practice against theory" was not merely academic, but was followed and reported on in the popular and technical London press. The news of Hertz's discovery gave new impetus to the Maxwellians and seemed to isolate Preece, who had a very concrete job to do as Britain's chief public servant responsible for improving the country's domestic communication capacity. But the Maxwellians, despite the relevant work of Lodge in particular, were not really interested in developing a practical system.

The key that tied together this wide range of ideas and experimentation was the notion that electromagnetic waves, far from being an abstraction of merely theoretical interest, actually constituted a medium of communication. The properties of magnetism had been understood since Faraday, as we've seen, although before Maxwell there had been no simple law to explain it (and Maxwell's law was far from simple[‡]).

Maxwell, as we have seen, predicted the existence of electromagnetic waves in 1864, and Hertz generated and detected them in 1887. Hertz showed that electromagnetic waves were similar to light except for the wavelength—an important precision in that the Maxwellians were primarily interested in studying light. Like Edison, Hertz used a spark gap to detect his waves, but he was not thinking at all about telegraphy. The first to suggest, even hypothetically, that Hertzian waves could be used for telegraphy was William Crookes, one of Britain's most imaginative physicists, in an article in the *Fortnightly Review* in 1892. After Marconi burst onto the scene four years later, his critics pointed to Crookes's article as counter-evidence to Marconi's claim to priority.[57] Sungook Hong, the historian of technology, writes that "before Marconi, Crookes was sneered at or neglected; after Marconi, Crookes was considered a visionary."[58] (Crookes's vision for Hertzian waves included such possible uses as "improving harvests,

---

[‡] Hertz himself had reformulated Maxwell's mathematical equations in simpler form, and some later scholars, including Albert Einstein, took to referring to them as the Hertz-Maxwell equations (Sarkar et al 2006, 544).

killing parasites, purifying sewage, eliminating diseases and controlling weather."[59]) Marconi himself always insisted that he never read the article before filing his patent application in 1896.[60]

What set Marconi apart from the rest was that he saw wireless communication, in his mind's eye, quite literally as telegraphy without wires. Thus, in his experiments he set out to reproduce as closely as possible the physical conditions of telegraphy, and use them to the same ends. The "homology"[61] Marconi set out to create between wire and wireless, coupled with the fact that he was not weighed down by the baggage of theoretical physics and the academic controversies swirling in that field, greatly played to his advantage. Experimentalists influenced by Maxwell's theory were hampered by their understanding of electromagnetic waves as light, while practitioners like Preece and Edison were looking completely elsewhere for the key to wireless communication. It was Marconi who made the leap from Hertz's lab experiments to practical wireless telegraphy using electromagnetic waves as the medium of communication. This was his original contribution, and what caused him to be labelled a genius in the heroic version of his life story.

Marconi's entire career was built on this one idea. He saw intuitively that the development of wireless technology using the radio spectrum (although he never called it that) would free communication from the constraints of wires—which were not only physical constraints but required heavy corporate, financial, and physical infrastructures to install and maintain. Wires were also symbolic constraints, corseting the idea of freedom intrinsic to the very notion of mobile communication. This also fit Marconi's vision—to maximize human beings' capacity to use technology to communicate freely— and translated, in his mind, to the idea of communication across greater and greater distances at lower and lower cost (importantly, not for free). From 1895 onward, everything Marconi did was meant to serve this vision; his career was a progression in this direction, and at the end of his life he took great pride and found serenity in being personally able to travel wherever he liked and to communicate with anyone, anywhere, at any time. Marconi not only invented mobile communication, it shaped and defined his world.

❧   ❧   ❧

Hertz's 1887 discovery was the basis of Marconi's 1896 patent; between them a dizzying array of experimental developments took place in Russia, Germany, Italy, England, and France, as well as the United States and as far afield as India. Some of the period's most accomplished physicists, such as the future Nobel laureate Ernest Rutherford, immediately began working

to improve the apparatus Hertz had used. Many of these distinguished scientists would soon be rejecting Marconi's claims of originality as spurious
or invented.

One of the most important new inventions in this critical decade was
a wave detector that quickly replaced the resonator Hertz had used as a receiver. In 1890, both Lodge in England and Edouard Branly in France, separately, invented a new type of receiver. Branly's model, consisting of fine
copper filings in a glass tube whose conductivity increased exponentially
when a spark was generated nearby, became the prototype for experimental
communication with electromagnetic waves; however, Lodge's term for the
new invention, the "coherer," caught on and he was consequently often
credited with its development, if not invention. Marconi always, correctly,
said he had used "a Branly coherer" in his early work, irritating Lodge, who
claimed, quite simply, that Marconi had done nothing that he himself had
not already done before.

Marconi's Bologna neighbour, Augusto Righi, meanwhile, devised a
new type of transmitter he called a "spark oscillator," an improvement on
what Hertz had called a "radiator." Righi began working on Hertzian wave
research as early as 1889,[62] and as we've seen, he had an important, if indirect
and somewhat obscure, influence on Marconi. Marconi acknowledged,
however, that the transmitter he used in his groundbreaking experiments
was a "Righi oscillator."[63]

Another scholar whose work attracted international attention in the
early years after Hertz's death was the Indian physicist Jagadish Chunder
Bose. Bose was one of the first to experiment with what later came to
be called "microwaves," generating and detecting wireless signals of six-
millimetre wavelength. In 1895—a full year before Marconi applied for his
patent—Bose set off a charge of gunpowder and rang a bell using microwaves
at a public demonstration at Calcutta's town hall. However, Bose, like Lodge,
was not trying to "communicate." Bose's work was well known in Britain
and he actually met Marconi, probably through Preece, during a London
lecture tour in 1896. Modest and idealistic, Bose was not interested in commercializing his scientific discoveries and felt that others should just freely
use his work. In fact, he found the idea of commercialization of science so
repugnant that when he founded his own research institute in India in 1915,
he forbade its members from applying for patents. Bose is considered one of
the pioneers of microwave optics—as well as the father of Bengali science
fiction literature.[64]

One of the strongest claims to priority over Marconi was that of the Russian physicist Alexander Stepanovich Popov. Popov presented a wireless communication system using Hertzian waves at a meeting of the Russian Physical and Chemical Society in St. Petersburg on May 7, 1895, and published a paper, in Russian, in the society's journal in December of the same year.[65] The dates and venue are important, given the claims Marconi would make in his 1896 patent specification. Popov himself never directly challenged Marconi, but his 1895 experiment and publication were frequently cited in legal arguments by others to discredit Marconi's priority. Preparing a patent defence in 1899, Marconi's scientific advisor James Ambrose Fleming established that Popov's publication was placed on the shelves of the library of the Chemical Society in London in March 1896, adding, however, that as it was written in Russian it didn't constitute "prior publication" in England.[66] In the United States, Popov's paper was received at the library of the American Academy of Sciences in Boston on April 4, 1896, eight months before Marconi filed his US patent. However, Marconi was able to successfully argue that he had conceived and put into practice all the points on which Popov's paper anticipated his claims *before* Popov's own publication of December 1895.[67] So was Marconi the first to develop wireless telegraphy using Hertzian waves or just the first to *patent* it?

When Marconi first came to public attention, it was obvious that established figures in the field would consider him an interloper, and Popov, although far more gently than others, was one of those who questioned his priority. In a letter to the British journal the *Electrician* in December 1897, Popov, like Lodge, claimed that Marconi's receiver was a "reproduction" of his own. In later years, as Marconi's leadership (if not priority) in the field was established, Popov became, in some respects, his friendliest rival; in 1902, he called Marconi the "father of wireless," and when Marconi married in 1905, Popov sent him a seal skin and a silver samovar as gifts. Marconi, on his part, was always generous in publicly acknowledging Popov's work, which he simply claimed he was unaware of until after he filed his patent application. Popov was far enough away from the heart of the action and not a threat to Marconi. That said, in Russia—under the tsar, during the Soviet period, and to this day—Popov is considered the true inventor of radio.[68]

Perhaps the most compelling of Marconi's detractors—and possibly the only one who could have presented a real challenge to him on practical grounds—was Nikola Tesla, a brilliant, cranky figure who may well have been the journalistic prototype of the eccentric inventor. Tesla, a Serbian immigrant to the United States and thus, like Marconi, an outsider working in a not always welcoming host environment (England, in Marconi's case), resented comparisons to others—especially Marconi, whom he considered his nemesis.[69] Older, and a loner, he was also closest to Marconi in global vision.

Born in Smiljan (then part of Austria, now Croatia), Tesla lived in Budapest and worked for the Edison company in Paris before immigrating to the United States in 1884. By 1889 he had opened a lab in New York City and quickly made a name for himself with a series of odd inventions, flamboyant stunts, and outrageous assertions. In 1892 he visited London and presented his work to the Royal Society, impressing such luminaries as Preece, Heaviside, Fleming, Lodge, and Crookes by firing up a great coil that emitted erupting thunderbolts, speaking like an oratorical sorcerer, and speculating that transatlantic cable telephony would soon be possible.[70] In February 1893, in Philadelphia, he laid out a scheme for wireless communication, before an audience familiar with the experiments of Henry, Loomis, and Edison.[71] It was patented later that year.

Although he was a bona fide inventor with many practical patents to his credit, Tesla was seen as a wizard, an iconoclast, a dreamer, and a madman, walking a tight line between recluse and showman. His 1893 patent was for a grandiose system of wireless communication that effectively proposed to use a layer of the atmosphere as transmitter and the whole planet as a receiver (technically not as far-fetched as it sounds, although impossible to realize in practice). By early 1895 he had lined up backing from Wall Street, formed a company, and started to build an expensive transmitter. But Tesla had an insatiable ego, and rather than focus on a clear goal he kept himself busy with a diverse array of high-profile ventures, like the construction of a huge electrical power plant at the 1893 Chicago World's Fair. Unable to develop his ideas into practical applications, he eventually lost the financial support he needed to exploit his wireless patent when Marconi appeared headed down the same track with a less costly plan that seemed more likely to succeed.[72]

Tesla was already famous when he heard about Marconi from Preece, in 1896.[73] He was now in a race with a newcomer. Marconi also had the tacit backing of another of Tesla's rivals, Edison (who was threatened by Tesla's experiments with alternating current). Between 1897 and 1902, the press pitted Tesla and Marconi against each other as sparring boxers, reporting their respective moves and speculating on their every word and movement. The press treated Tesla as nasty and churlish, Marconi as humble and charming; they could understand Marconi's experiments more easily than Tesla's—Marconi often demonstrated for journalists while Tesla tended only to talk. Among Marconi's supporters, Tesla's brilliance was recognized and feared. Correspondence from 1900 and 1901, when Marconi was preparing his first transatlantic connection, shows company officials and Marconi himself concerned, even terrified, that Tesla might beat them to it.

Finally, we return to Oliver Lodge, the most obstinate of Marconi's detractors. As we have seen, Lodge was one of the most prominent British Maxwellians and an early experimenter with Hertzian waves. Marconi had unwittingly invaded Lodge's bailiwick, England, and been turned into a poster boy by Lodge's rival, Preece of the GPO. Like the Player Queen in *Hamlet*, it may be that Lodge did protest too much. He certainly had an important following, but as Marconi went from strength to strength, Lodge's star faded.

Lodge's claims were based on two lectures he gave in 1894, one in London in June, and the other in Oxford in August, thus at least a full year before anything Marconi claimed to have done. The problem was, it was difficult even at the time to ascertain clearly what Lodge was trying to achieve. Heinrich Hertz died on January 1, 1894, and the Royal Institution asked Lodge to commemorate his work at one of its celebrated Friday evening lectures, on June 1. There is no existing text or transcript of the lecture, and no report at the time or subsequently refers to the use of Hertzian waves for telegraphy. What Lodge said and did during his lecture to the British Association for the Advancement of Science, the UK's main learned society, in Oxford on August 14 was even less clear-cut. Again, there is no verbatim record of the event. Lodge later claimed that in Oxford, using a Morse telegraph key, he performed the first public demonstration of transmission and reception of a signal by Hertzian waves. Marconi's supporters always insisted that this was not at all what Lodge had done, and that despite his successful demonstration—of something—it was in no way anything to do with wireless telegraphy. Interestingly enough, Lodge made no reference to such an accomplishment in the first edition of *The Work of Hertz*, a book he published a few months later, yet he included the claim in later editions. He also famously stated, on many occasions, that the greatest distance that could be covered by Hertzian wave transmission would be around seventy yards.

Marconi himself, in his typical ironic fashion, wrote in his 1899 manuscript:

In the beginning I had the whole scientific world against me, saying and doing
everything they could to upset me, but I want to state that I feel very grateful
to them for this opposition. Particularly do I wish to thank Professor Lodge
who has continually opposed me.... The fact that a great man like Professor
Lodge should take notice of a young man never heard of before who was not
saying anything, only working, was significant.... Their criticism had a great
effect on me in the feeling of future satisfaction which I anticipated when
I would have the privilege of showing them they were wrong.[74]

The debate over the legitimate inheritors of Heinrich Hertz, and particularly the debate about whether Lodge or Marconi should have priority, has
gone on for more than a century, and continues to be an important controversy among specialists in the history of technology. Lodge may well have
demonstrated the use of Hertzian waves for signalling. On the other hand,
he had no idea that this could have a practical application. Certainly, the fact
that Lodge did not try to patent his discovery, if that's what it was, supports
this view. After his Oxford lecture, despite the encouragement of colleagues,
Lodge did not pursue his work on Hertzian waves; he only rekindled his
interest in wireless later, retrospectively, after Marconi put the subject on
the map.[75]

In any case, Lodge was not an outlier. His criticism of Marconi was
shared by other members of the British scientific establishment who
deplored Marconi's commercialism and keenly scrutinized his every move
to take advantage of any flaws in his patent claims. But after seriously considering legal action, Lodge abandoned his opposition to Marconi. In 1897
he patented a method for synchronizing transmitters and receivers that became the fundamental method for wireless "tuning," and promptly tried to
sell it to the Marconi company. Lodge eventually came to respect Marconi
and the two maintained cordial relations for the rest of their lives.[76]

In sum, Marconi's most credible rivals—such as Popov, Tesla, and
Lodge—all in their own ways rallied to Marconi as he unstoppably rose above
the fray. Marconi's youth, his lack of theoretical baggage, his vision, his outsider status, all became assets. The others, with their friends, enemies, and
constricting theories, were unable to think or act imaginatively. The Maxwellians looked at electromagnetic waves and saw optical applications, not
communication technology. A theoretical physicist such as Lodge looked at
electromagnetic waves and saw a light source and a human eye as, respectively, transmitter and receiver. Marconi looked at the same electromagnetic
waves and saw telegraphy without wires.

Sungook Hong sums up Marconi's accomplishment this way: working by trial and error over several months in 1895, Marconi perfected the coherer, invented a stable tapper, increased the efficiency of the induction coil, connected a Morse inker and telegraphic relay to the transmitter and receiver, and controlled the resulting electrical sparks. "Although most of these components had been invented by others, they were unstable or unconnected with one another before Marconi," writes Hong. The whole thing fit into a small black box. "When Marconi 'opened' this 'black-box' by publicizing his first patent in 1897, people were amazed and intrigued by its simplicity. The solutions appeared so simple and so obvious that many began to wonder why no one else had come up with them."[77]

Others said they had, however, and the hullabaloo surrounding claims and counterclaims continued well into the twentieth century. It didn't matter. Once he had his patent in hand, all Marconi needed was an argument to defend it, and aided by his advisors, he soon found one. The argument was nicely expressed by Fleming in an internal memo as the company was preparing for a crucial US lawsuit in 1904: "Marconi's invention was not any particular element, but the combination of the elements, some new and some old, which made Hertzian wave telegraphy possible...prior to 1896 there was no wireless telegraphy and subsequent to 1897 there has been no wireless telegraphy except that initiated by Marconi."[78]

3

London: Start-up

Marconi, Annie, and a dubious-looking box of electrical equipment left Italy for England in February 1896.[1] They left behind a country in the throes of a disastrous colonial war in Ethiopia and wracked by violent political unrest at home. The first Italo-Ethiopian War lasted barely a year and ended with an Ethiopian military victory over Italy. Under the October 1896 Treaty of Addis Ababa, Italy recognized Ethiopia as an independent country—becoming the first European country to be defeated in an African colonial war.

Compared to Italy, England was enjoying both economic prosperity and social peace. London was booming, its population increasing by nine hundred thousand during the 1890s. The British Empire was at its height, Queen Victoria was preparing to celebrate her diamond jubilee, and two of her grandchildren sat on imperial European thrones (the German Emperor Wilhelm II, and Empress Alexandra of Russia, wife of the Russian tsar).* The Britain that greeted Marconi was the centre of the world's political economy. However, soon enough the placid end of the nineteenth century would give way to an atmosphere of international rivalry and mutual suspicion. Communication would figure large in the scenarios of power.

In England, the Victorian era was precursor to a new age, in which, it was said, a self-made man could "make it" to the highest level of society in one generation. The English had become an urban people, and the basis for accumulating wealth had definitively shifted from ownership of the land to

---

* As Tsar Nicholas II was also a nephew by marriage of Victoria's successor Edward VII (whose queen was a sister of Nicholas's mother), Germany's Wilhelm, Russia's Nicholas, and Victoria's grandson, King George V, were first cousins. Wilhelm and Nicholas would both be the last of their lines to rule.

industry. This transitional period between the Victorian and Edwardian eras was marked by a series of innovations that would have as profound an effect on British social and economic life as had those of the Industrial Revolution decades earlier.[2]

A revolution in popular culture was also under way. Theatre provides an example. In London, Herbert Beerbohm Tree was about to move into Her Majesty's Theatre and transform the foundations of the imperial stage. George Edwardes's Gaiety in the Strand (next door to where Marconi would establish his corporate headquarters in 1912) was a place where, according to historian Robert Cecil, "so many young peers of the realm lost their hearts and, in their parents' eyes, debased their coronets" as the musical moved to the centre of popular entertainment.[3] The D'Oyly Carte played the operettas of Gilbert and Sullivan at the Savoy (later Marconi's favourite London hotel), where the liquor flowed to lubricate assignations of every sort. Two weeks after Marconi's arrival in London, the Lumière brothers began projecting their moving pictures at the Empire Theatre in Leicester Square.

The period also saw the emergence of the first wave of a phenomenon we now call "media convergence." The impact of the age's new media—shaped by the connections between the press, new forms of mass entertainment, and the birth of electronic communication—was to have tremendous influence. The empire was held together by the telegraph (Britain controlled the world's telegraph cables, as we've seen). Telegraphy brought the day's international news to the breakfast tables of London's elite, sometimes even before it had been "officially confirmed."[4] Marconi was one of the first to recognize that, before long, wireless would bring about another great leap forward, intuitively acknowledging the liberating power of communication technology.

Marconi and his mother settled into a flat at 71 Hereford Road, a short walk from the Bayswater underground station. Soon they moved around the corner to slightly more upscale 67 Talbot Road in Westbourne Park. The area at the time was made up of fine Georgian row houses (and still is). A London County Council plaque marks Marconi's Hereford Road residence; today it is the shabbiest house on the block. Sixty-seven Talbot Road no longer exists as a separate address, having been combined some time ago with the building next door, number 69, to create a brace of gentrified apartments facing a lovely neighbourhood park, Shrewsbury Gardens. It is close by St. Stephen's Anglican Church—which may have been an important factor for the pious Annie.[5]

Annie's nephew, Henry Jameson Davis, a well-connected Irish milling engineer practising in London, helped the Marconis get settled and took his young cousin (not yet twenty-two) under his wing. Davis was the son of a prominent merchant/industrialist, Abraham Grubb Davis, who was married to Annie's sister, Helen Jameson. After marrying his cousin Emma in 1898, Davis made his official home at Killabeg, an estate just outside Enniscorthy, in County Wexford. He spent most of his time in London, however, where his business was conducted from 82 Mark Lane, next to the Corn Exchange in London's financial district. Twenty years older than his new protégé, Davis had some experience in both finance and—importantly—patents. He had already had some success with two patents of his own. Echoing the advice of Italian Ambassador Annibale Ferrero, Davis advised Marconi to file a patent application as soon as possible to protect his invention. In the meantime, he covered Marconi's expenses and turned his own flat into a showroom where Marconi's odd-looking apparatus could be shown to potentially influential people.[6]

Almost immediately upon Marconi's arrival Davis set up a meeting with one of London's top patent agents, Carpmael & Company, to begin establishing his rights. Marconi later wrote that without a secure patent, "it would be highly impolitic, to say the least, to try to bring [the invention] to the attention of likely commercial or Government users."[7] Before going ahead, however, Marconi took one last stab at verifying the Italian government's possible interest. On February 29, 1896, he met with Ferrero, who told him that as far as he was concerned, "my find can be of great use to the Italian government though he advises me as a friend not to reveal any secrets until I have received the patent." The ambassador warned the young inventor that the Italian government was "imbecile," and was filled with unscrupulous types who would steal his find. Marconi was a little taken aback by Ferrero's criticisms of their country.[8]

Now convinced, Marconi met on March 2 with Edward Carpmael, who advised him to file a patent application and test the waters for interest in his discovery. A letter to his father two days later shows Marconi as a quick study on a steep learning curve, systematically collecting the information he needed to protect his interests, and acting decisively. He was also trying to impress Giuseppe that he was the one in charge, telling him that "Yesterday this lawyer [Carpmael] wrote the description of my discovery according to my instructions"—perhaps unaware that this was highly unusual according to the prevailing practices of London patent agents in the 1890s.[9]

By the late nineteenth century, patent agents in the United Kingdom had established themselves and gained institutional recognition as a profession, bridging the fields of law and technology. These agents worked closely with lawyers and consulting engineers specializing in patent litigation and industrial property rights. The role of the patent agent was to determine the rights a client might have with regard to an invention and accompany the inventor at every stage of the patent process; this might involve helping to write the patent application.[10]

Be that as it may, barely a month after his arrival in London, Marconi filed a provisional specification for his patent on March 5, 1896.[11] He had written a twelve-page draft, dated "*Febbraio 1896*," in longhand, in Italian, in a schoolboy's exercise book that he had purchased in London—perhaps his first English acquisition.[12] The specification as filed was a crude description of his apparatus and what he had done and was hardly the basis for a sound application—this would have been immediately apparent to someone like Carpmael.[13] Nonetheless, Marconi was determined to push ahead with his case.

Under British patent law, a provisional specification was a placeholder, a way of staking a claim while the details of the complete specification were worked out. The document remained secret and an applicant had up to nine months in which to perfect the description of his invention. Unlike American and German patent law at the time, British law protected the *claim,* rather than the actual invention, as property.[14] In other words, it was not necessary to have invented the means for, say, wireless telegraphy—meaning the apparatus for generating and detecting electromagnetic waves—in order to be the first to claim those means for that purpose. The date of the first filing was crucial in establishing priority among competing claims. Therefore, no changes could be made in the scope of the invention between the two stages.

Marconi realized (or was told) that his provisional specification was weak, and on June 2, 1896, he replaced it with a new one, incorporating the results of more recent experiments. He then spent the next nine months—his maximum court-allotted time—revising and fine-tuning the patent papers, consulting widely, and engaging Britain's leading patent experts.[15] The result was ten pages of precision and technical clarity, the tone set by its opening words: "My invention relates to the transmission of signals by means of electrical oscillations of high frequency ..."[16] The patent went on to make nineteen claims based on this method of transmission. A contemporary scholar wrote that the patent was "a marvel of completeness."[17] However, Marconi had little to do with the final draft.

The patent drafting process was the first instance of a working method that Marconi would use for the rest of his life. This method involved collaboration, led and driven by Marconi's vision, and involving the best available expert support. Marconi's patent effort was perhaps the best-known example of the new enabling conditions of technological innovation at the end of the nineteenth century. An entire industry had developed around the protection of intellectual property, involving lawyers, technical consultants, and highly specialized experts in the arcane practices surrounding patents. Although he found the process "laborious and difficult" Marconi would become one of the most astute and effective users of patents.[18]

Once the initial patent application was out of the way, Davis began introducing Marconi to British engineering circles. On March 30, 1896, one of London's most prominent electrical engineers, A.A. Campbell Swinton (whom we encountered briefly in chapter 2), wrote on Marconi's behalf to William Preece. At sixty-two, the chief engineer of the General Post Office was a venerable and well-known public figure, and the senior authority responsible for developing public communication in Great Britain. Preece's own experimental work trying to establish wireless communication from ship to shore had been widely reported in the British press, but he had never thought of using Hertzian waves for the purpose.[19] By 1896, his research had reached a dead end.[20]

"Dear Mr Preece," began Swinton's letter, "I am taking the liberty of sending to you with this note a young Italian of the name of Marconi, who has come over to this country with the idea of getting taken up a new system of telegraphy without wires, at which he has been working...."[21] Marconi evidently met Preece almost immediately, for he mentions the meeting in a letter to his father on April 1, 1896.[22] Preece's response was generous and enthusiastic and he eagerly received Marconi at the GPO headquarters in London. Preece was ready to swallow his pride and cut his losses when Marconi came along;[23] it may have suited him to embrace this unlikely newcomer rather than concede failure to his own adversaries.

Some forty-five years later, P.R. Mullis, a junior assistant in Preece's office, wrote a detailed memoir of that first meeting.[24] Marconi, "a young dark looking foreigner," had arrived carrying two large leather bags. While Preece cleaned his gold-rimmed spectacles, the contents of the bags were placed on a table. They seemed to consist of a number of brass knobs fitted to rods, a large spark coil, some odd terminals, and "most fascinating of all a rather large sized tubular bottle from which extruded two rods, which as far as could be seen terminated inside the bottle, on two bright discs very close together and between which could be seen some bright filings or metal particles." This was Marconi's version of the famous "coherer" invented by Branly and perfected by Lodge, which Marconi recognized as the key to transmitting an intelligible signal by means of Hertzian waves.

According to Mullis, the apparatus "immediately took the Chief's eye." Preece asked Mullis to go and get a telegraph key and some lengths of wire. "Some batteries were also obtained, and returning with these to the Chief's room, the pieces of apparatus were then carefully joined together.... The curious glass tube or bottle was placed on a small table...." At this point, Preece looked at his watch; it was noon, so he told his assistant to take Marconi out for lunch and come back at two. "I had come to like this quiet young foreigner with his quaint English and dexterous manner by which he connected up the pieces of apparatus," remembered Mullis, "bending the obedient wire just where he wanted it, this way and that." Over lunch, Marconi told Mullis about Italy and then they strolled around the neighbourhood, while Marconi inspected the curbside stalls and their displays of fruit, junk, and old books. They were back in Preece's office at two sharp.

Everything was as they had left it, and after Marconi made a few adjustments he depressed the key "and immediately the bell on the adjacent table commenced and continued to ring." Marconi then went over and sharply tapped the coherer, causing the ringing to stop. After Marconi repeated the procedure a few times, Mullis said, "I knew by the Chief's quiet manner and smile that something unusual had been effected." Mullis was particularly impressed with Marconi's modesty. "Marconi's invariable expression 'we' will try this and this and that or we must do so and so...never 'I', connected us together in a most friendly manner." The experiments continued for the rest of the week.

Preece was soon advising Marconi on financial issues as well. Marconi had received a proposal for a "suggested arrangement" from two potential investors, Frank Wynne and David Urquhart, who proposed to form a company and purchase his invention in exchange for just over 40 percent of the shares.[25] Marconi's letter to his father on April 1, 1896, indicates that he was leaning toward accepting the offer, if he could get Wynne and Urquhart to up his allotment. However, his meeting with Preece had left him wondering whether the invention was more important than he had thought. He was puzzled over what to do and had not yet come to a conclusion. On April 8, 1896, he wrote to his father that Wynne and Urquhart were willing to wait one more week and that "I do not consider it wise to refuse their proposals."[26] He managed to keep them at bay for another month and then—following Preece's advice—he let the offer lapse.

During this month Preece provided personnel and space for Marconi to continue his experiments, although no direct funding. This meant that Marconi had to continue returning to his father for funds to bear the costs of equipment as well as the expensive patent filings. Giuseppe offered to "lend" him the necessary funds, but Marconi was testy in his response: What do you mean by "lend" me? he asked. "You will surely understand that in the hypothetical case of a failure, I could not take responsibility for the money spent." Giuseppe was nevertheless bankrolling him, as Marconi acknowledged with gratitude. When his father sent him £100 in May 1896, Guglielmo responded: "I will try to use the money in the best possible way and take great care not to spend it unnecessarily."[27] Giuseppe's financial support for his son had expanded exponentially, from a few tens of lire in 1892 and 1893, to hundreds and then thousands after Guglielmo moved to England.[28] According to Giuseppe's accounts in the family archives, Marconi received around 45,000 Italian lire, roughly equivalent to £1,800 or $9,000, from his father during 1896 and 1897, his first year in England. In comparison, Preece's salary that year was £1,500.[†]

Preece pressed Marconi to demonstrate the invention publicly, but his Irish relatives were advising caution on that front until the patent application was approved. On June 2, as we've seen, he replaced his original provisional specification with a new one, starting the clock ticking again. At the same time, he began hedging his bets. On May 30, 1896, he wrote to the War Office, informing them that he had discovered "electrical devices which enable me to guide or steer a self propelled boat or torpedo from the shore or from a vessel without any person being on the said boat or vessel."[29] This was more than wireless telegraphy. Marconi was now talking about remote control. He was beginning to imagine more elaborate applications, with military consequences. Two weeks later the inspector general of ordnance wrote back making an appointment.[30]

The collaborative nature of Marconi's research is also evident from the "experimental room notebooks" kept by George S. Kemp, an older technician

---

[†] Calculated at twenty-five lire to the pound and five lire to the dollar. Anna Guagnini (2006) totalled Giuseppe's contribution to be £888 plus 23,100 lire in 1896 and 1897.

who joined Preece's staff around the time Marconi arrived at the GPO
with his two leather bags.[31] Close to forty years old (he was born in 1857),
Kemp had served twenty-five years in the Royal Navy, eventually as an electrical and torpedo instructor. Kemp's first encounter with Marconi was on
July 27, 1896, when Marconi demonstrated his apparatus to Post Office
officials with a rooftop transmission between two GPO buildings in central
London.[32]

Kemp's notebooks provide an exhilarating view of how the GPO saw
Marconi. They also illustrate the collaborative nature of the research even at
this early experimental stage:

> July 27th. Monday. Experiment with new Instrument for communicating
> across space. Invented by G. Marconi Esqr.
> July 28th. Tuesday, Experiments with new Instrument for communicating
> across space.
> July 29th. . . . Experimenting with Mr. Marconi's Telegraph Apparatus . . . Found
> that when working morse it was necessary to revolve the contact or you
> would get breakdown in your signal. . . .
> July 30th. Preparing motor experiment for ringing bells and noticed they did
> not act as well as previous experiment. . . .[33]

The experiments continued through August 1896 as Marconi and the GPO
technicians tinkered with the apparatus, combined it with other equipment,
and tried to improve the distance and quality of the transmissions. On
August 11, they began experimenting with a so-called Tesla coil, an electrical transformer used for producing high-frequency current in spark gap
generators, patented by Nikola Tesla in 1890.[34]

On Tuesday, September 1, Marconi and the GPO team decamped to a
military training ground at Salisbury Plain, about eighty miles southwest of
London, for several days of experiments and demonstrations, some of them
in the presence of army and navy officials. Here they were able to signal
over a distance of three-quarters of a mile, the greatest achieved thus far.
Marconi's work had by now attracted the attention of Royal Navy Captain
Henry B. Jackson, a wireless experimenter in his own right who would become another powerful Marconi booster as he rose through the ranks to
become admiral of the fleet. On September 15, after Marconi informed him
of the Salisbury experiments, Jackson wrote from his ship HMS *Defiance* to
say that if the apparatus could produce signals that travelled over three miles
it would be of real use. Two days later the War Office requested that Marconi
indicate how much he would want "to demonstrate the possibility of

working either of two independent receivers, enclosed in the same steel box, in the sea, 1 mile from the shore where the transmitters are set up."[35] Soon Marconi was receiving enquiries from agents for foreign companies and governments, seeking to know whether his apparatus was for sale commercially.

Preece, for his part, was amused by the sudden interest. Following the developments from his country retreat in Wales, he wrote to a colleague that at long last "the War Department [is] waking up to a system of signaling without wires which we have been working at for nearly ten years!"[36] In this letter, Preece also expressed some skepticism about the success of the Salisbury experiments. A report on the trials by a member of his staff was "very interesting but not encouraging. I think our plan of using electro-magnetic waves is cheaper and more practical. I do not at present propose going further with it...." But he didn't act on that thought. To the contrary, his support for Marconi increased.

Kemp was soon seconded to work exclusively with Marconi and eventually left the GPO to become Marconi's chief assistant in November 1897.[37] From 1897 to 1932, Kemp kept a diary recording his work with Marconi. The diary is a precious resource for understanding Marconi's research in the early days, when Kemp was by his side during his most important experiments. Throughout his life Marconi considered Kemp his first collaborator and a valued friend—in some ways it was as close a relationship as he was capable of having, the type of relationship he was most at ease with and liked best. Marconi always felt closest to people with whom he could discuss technical matters without cluttering the conversation with emotional issues—invariably men of a more modest social class than himself. Kemp—described as looking like "a typical sea-faring man with ruddy complexion, and closely-cropped grizzled hair and heavy moustache,"[38] and speaking in stentorian tones—was one of those; and while the relationship remained clearly professional and hierarchical, Kemp was one of the few who had complete access and could speak his mind to Marconi. Kemp was still employed by the Marconi company when he died on January 2, 1933, at the age of seventy-five. Marconi was one of the witnesses to his will.[39]

❋ ❋ ❋

Marconi had literally come out of nowhere when he arrived in England and attracted the attention of William Preece in 1896. He was so unassuming that even a young office boy like Mullis could enjoy taking him to lunch. The older, more experienced Kemp quickly became completely devoted to him. Seasoned military men like Jackson wanted to work with him. Before long, word of Marconi's experiments was making the rounds of London's scientific circles. "The calm of my life ended then," Marconi later told his friend Luigi Solari.[40]

Preece reported Marconi's discovery at the annual meeting of the British Association for the Advancement of Science in Liverpool on September 22, 1896. The news was buried among some twenty-five other items in a long, dry article in the *Times* the following day, and went almost unnoticed, except among scientists.[41] That quickly changed. Preece was scheduled to give a public lecture at Toynbee Hall, a universities settlement house in East London, on Saturday, December 12, and decided he would illustrate the possibilities presented by Marconi's system. Marconi agreed to rig up an apparatus capable of transmitting signals across the lecture room. However, he was not ready to disclose his method, as he told Preece, "until my whole study can be laid before some scientific society."[42] He was also anxious about the working condition of his equipment.

The meeting had the air of an illusionist's performance (if not a snake oil demonstration)—the gawky youth in his undoubtedly cheap suit running about the room setting off bells—saved only by the earnest reputation of Preece. The press reports of Marconi's debut were sensational:

> Mr. W.H. Preece, the telegraphic expert of the Post Office, had a surprise in store for his audience at Toynbee Hall on Saturday night, when he lectured on "Telegraphy without Wires." There is, of course, nothing new in the fact of being able to communicate without wires, but towards the close of his lecture Mr. Preece announced that a Mr. Marconi, a young Italian electrician, came to him recently with a system of telegraphy without wires.... Mr. Marconi was present that night, and this was the first occasion on which the apparatus had been shown, except to the Government officials....

> The apparatus was then exhibited. What appeared to be just two ordinary boxes were stationed at each end of the room, the current was set in motion at one and a bell was immediately rung in the other. "To show that there was no deception" Mr. Marconi held the receiver and carried it about, the bell ringing whenever the vibrations at the other box were set up....[43]

Despite his brief moment of doubt following the Salisbury experiments, Preece now announced that the Post Office had decided "to spare no expense in experimenting with the apparatus," in which, he added, he had "the greatest faith."[44] Preece was smiling to himself in vindication of his rivalry with Oliver Lodge and the Maxwellian theoreticians. Also, his championing of Marconi's cause was by no means disinterested; he had to be concerned about the appearance of a new technology that could change everything. It was much better for him to keep Marconi on a short leash than have him offering his invention to private entrepreneurs on the open market.

The day after the Toynbee Hall demonstration Marconi wrote thanking Preece for the kind reference he had made to his work and asking him to recommend a patent expert as he was still afraid "the specifications and claims have not been drawn up with sufficient accuracy and skill." He was worried about making a serious mistake in his application.[45]

Nonetheless, by early January 1897, Marconi seems to have been convinced that his application was ready for scrutiny, for he had applied for patent rights in Russia, France, Germany, Austria, Hungary, Spain, India, and the United States, as well as, of course, Great Britain. He had been unveiled to the public and was beginning to receive the attention of important people—beginning with Ambassador Ferrero, who invited him to dinner at the Italian embassy where, Marconi wrote his father, he was subjected to another diatribe about the hopelessness of the Italian government.[46] Marconi was learning about self-promotion; he had written to Ferrero on December 20, 1896, updating him about his latest experiments and reiterating his view that "the system could be of great use to the Italian army."[47] The Austro-Hungarian embassy had also asked for a demonstration.

Patents notwithstanding, he wrote to Giuseppe that he still expected litigation. He was now thinking of selling the patents piecemeal and ridding himself of the worry. He told his father that "two American gentlemen" were offering £10,000 for his US patent rights. He had also not forgotten Wynne and Urquhart, and was keeping them informed of his progress. "I believe that it may be better for me to accept one of these early offers since I could benefit from a tangible sum of money and thereafter, obtain the patent rights in other nations such as Belgium, Switzerland, Portugal, Denmark, Brazil, etc etc, where for the time being I have not yet applied for any patent." If he accepted a deal at least he would come out with something. He asked his father to obtain legal advice.[48]

The following week Marconi went to see "one of the best local lawyers," whom Preece had recommended. Preece had told Marconi that there might be "some danger dealing with patents," especially in the United States. Marconi was anxious to act, but not before obtaining his parents' approval for any business deal. Still, he was worried about the vagaries of the

patent system: "Just think that some ninety per cent of the patents that are
granted over here, are made invalid within ten years from the date of their
licence," he told Giuseppe. He added that he did not have much hope for a
timely decision from the British government as to whether or not they
would make an offer for the rights and even if they did, he did not expect
that they would pay a great deal for them. "Please think carefully about all
this and let me know your opinion."[49] There is no record of Giuseppe's response, but within days Marconi and Davis had their lawyers draw up a
contract proposal to submit to a group of unnamed purchasers.

One of the most remarkable documents in the Oxford Marconi
Archives is a twenty-six-page itemized account from the London legal firm
of Morten, Cutler & Co.,[50] that, read alongside the Marconi family correspondence, paints a rich portrait of the machinations of the next six months,
leading to the incorporation of the company in July 1897. The handwritten
sheaf of bound ledger sheets begins with a meeting on January 25, 1897:
"Attending Mr Davis on his calling instructing us that in this matter on behalf of Mr Marconi to prepare the necessary contracts in order to enable the
sale of your Patents to be carried through." Over the next five days, the
counsellors met several times with Marconi and Davis and hammered out a
draft contract that was sent to the purchasers' solicitors.[51]

Morten et al advised Marconi to wait and see the money before dis-
closing any "particulars" of his invention. Meanwhile, the purchasers asked
for a demonstration before lawyer J. Fletcher Moulton "so that he might see
results attained." Marconi's counsel advised, however, that the prospective
buyers would have to wait. Eventually, Moulton was scheduled to see a
demonstration; however, after he failed to attend three appointments the
lawyers wrote the purchasers to say that unless the contract was returned
approved and signed within twenty-four hours "the matter would be at an
end." Marconi instructed them to take a less aggressive approach and withdraw the ultimatum, but the deal fell through. On February 13, the purchasers "declined the business and returned (the) draft proposed contract."

The solicitors advised Marconi and Davis "to go on with the registration of the patents and to place the matters in other likely channels at once,"
cautioning Marconi to be careful about publicizing details of his invention
before the patent application was accepted. On February 19, they were
instructed by Marconi and Davis "to draft a prospectus for a proposed
Company"—a radically different approach than what they had been considering until then. As they began drafting the prospectus, they received a

boost in the form of a consultant's report to Preece, stating that Marconi's instrument was not only "entirely novel" but also "the first really successful application of wireless telegraphy." The invention was "exceedingly simple and reliable" and "capable of immediate industrial application."[52]

The solicitors then met with John Cameron Graham—the lawyer recommended by Preece—who suggested that they retain the services of…Moulton. Along with Carpmael, Graham and Moulton both expressed the opinion that Marconi's patent, when granted, would be valid. Marconi filed his complete specification on March 2, 1897—the last possible date under the nine-month limit since the filing of the provisional version. The final draft was Moulton's. Marconi's invention was now in play.[53]

In the rarefied atmosphere of patent lawyers, industrial speculators, venture capitalists, and technical experts in late nineteenth-century London, two new figures now entered the picture. On February 27, Morten et al reported that a "Mr. Fleming" had come to their office to discuss Marconi's invention. James Ambrose Fleming was a professor of physics with a lucrative sideline as a consultant and trial expert.[54] He was acting as agent for Colonel Sir Henry Hozier, secretary of Lloyd's insurance firm. On March 1, 1897, Marconi's solicitors met with Fleming and Hozier, explained the invention, and agreed to do a demonstration for Lloyd's.[55]

On March 3, Morten et al met with Marconi to apprise him of the new developments. Marconi told them to put the plans to form a company on hold, and the next day the solicitors met with Davis to give him this latest news. Marconi still entertained some hope that the GPO might come through with an acceptable offer for his patent. (He was aware of the recent enthusiastic consultant's report to Preece.) By March 9, the solicitors and Davis thought they had convinced him that the company underwriting should go ahead, however, two days later he met separately with the lawyers, informing them of "his wish that Mr Davis should not proceed with the matter at present."

While the standard Marconi literature canonizes Henry Jameson Davis as the wise and trusted mentor leading Marconi to corporate success, the archives, as we see, tell a more nuanced story. On March 13, 1897, Morten et al wrote to Davis "that Mr Marconi could not sell his invention until after the completion of the experiments and thereon." Could not or would not? Davis's reply is not recorded. Three days later, the lawyers wrote to him again "informing him Mr Marconi did not wish to throw him over but that his father did not wish him to bind himself in any way until the experiments had been completed."[56]

This was quintessential Marconi—firm and unequivocal, yet not closing any doors. There was also a backstory of the Marconi family discussions during those days that does not appear in the lawyers' account. A letter from Annie to her son on March 20, while he was off doing another round of tests at Salisbury Plain, cautioned him "not to write to H. Davies [sic] or have anything more to say to him until you return home.... You must not on any account write to him or promise him anything. He is a dangerous person to have anything to do with."[57]

Clearly relations between Marconi and his cousin were not quite as cordial as they seemed. In fact, the Marconis and their *consigliere* were close to a rupture. On March 29, Morten et al met with Marconi "with reference to correspondence between himself and Mr Davis and advising him not to break with Mr Davis but to suggest a meeting." A meeting took place on March 31. Davis had already raised some subscriptions for the prospective company and Marconi "expressed a wish that the money subscribed should still be kept but that he should have a free hand." They met again the following day, and the lawyers brokered a compromise, in which Marconi still did not pledge to sell while it was understood that Davis should have the first offer.

The solicitors kept meeting separately with the two cousins, discussing the financial prospects of the future company with Davis and informing Marconi of the bona fides of Davis's proposed investors. By April 8, Marconi seemed to be back on board with Davis's plan, although he and Davis were still meeting the lawyers separately. Marconi was still uneasy about taking the entrepreneurial route rather than staying with a public agency like the Post Office. The new company was offering to pay him £15,000 in cash and give him a large proportion of the shares; Marconi, however, was hesitant lest it appear that he was ungrateful to Preece. "I understand from him that he is under a great debt of gratitude to you in more than one respect," J.C. Graham wrote to Preece. "The matter is evidently weighing heavily on his mind," Graham added, and "he appears to have only one object in view, viz. to do the right thing."[58] Marconi himself wrote to Preece on April 10. The company's backers were pressing to know whether Marconi intended to accept their offer or not. "I beg to state, however, that I have never sought these offers, or given encouragement to the promoters."[59]

Marconi was playing a high-stakes game, trying to keep all bases covered and seeking to alienate no one, and looking after himself first and foremost as he tried to figure out where his interests lay—as a private entrepreneur or a partner of a great public agency. The Post Office solved his dilemma by failing to move quickly enough. The pressure applied by Davis eventually helped Marconi make his decision. But not without a lot more tension.

On April 14, Davis met with Marconi and then wrote to the lawyers that there could be no question of anything less than that Marconi "complete with the Company." The money was there and Marconi "so unbusinesslike" as not to be trusted to decide on his own best interests. "I am now going to see Mr Marconi, and if possible will come along to you later, but it will be no use proposing anything weaker than this from the Company's point of view."[60]

Marconi may well have been "unbusinesslike," but he was trying to be honourable with respect to the Post Office and, more important, he was struggling to determine where his own best interests really lay. From the very beginning, his preference, as we've seen, seemed to be for a partnership with a public or government body, first in Italy and now in England, and it was Davis pushing him toward a private business solution. Davis, however patronizing, genuinely believed it was in Marconi's best interests to create a commercial concern.

On April 19, Graham wrote to Marconi, telling him that he had no legal obligation whatsoever toward Preece or the government, "except for assistance received in carrying out a scientific investigation." He advised Marconi to send Preece a formal letter telling him that he had received an offer to buy the patents and asking whether the British government had any objections (thus informing them as he had previously done with the Italians). The crux of Graham's legal advice involved the money. "The main point in my opinion is to make certain that the £15000 is in the form of cash in the Bank and ready to be handed over. Promises to pay are not worth the paper they are written on."[61]

While Marconi and Davis continued haggling in their lawyers' office, Davis's investors were becoming impatient. One of them wrote on April 26 that unless Marconi signed the agreement at once he would ask for his £1,500 back. "If you get a <u>written</u> agreement with Mr Marconi…I would be content to wait, but even then there should be a time limit fixed within which all was to be completed."[62]

Marconi was indeed stalling. He didn't want to do anything final before completing tests that he was about to begin with the Post Office at Penarth, Wales, on the Bristol Channel.[63] He was also waiting for approval of the plan from his father, who had become his chief confidante and advisor. Giuseppe's advice was consistent and sound. He insisted that Marconi retain all rights to his patent, especially for Italy. He also strongly advised him to place his name on the company's title. Marconi was being pressed from all sides but kept his focus. On April 27, Davis wrote to him that the time had come to decide. He could not keep the investors or their money waiting indefinitely. "Dear Guglielmo, I have several thousand pounds in the Bank and at this office, and am quite prepared to pay the deposit and sign the formal agreement, and we can easily get the balance of the money required within a few days. This we would have had long ago, but as you know we suspended the matter for the time being at your request." He signed the letter, "Your affectionate cousin, Henry."[64]

Davis then wrote to the lawyers on April 29, 1897, asking that the subscription money be returned or the company floated, and Marconi instructed them to inform Davis that he agreed to proceed with the formation at once. On April 30, Marconi agreed to Davis's terms and the deal was made. A company would be formed within thirty days of conclusion of the Penarth experiments. Marconi would assign his patents to the company for £15,000 in cash plus sixty thousand of the total one hundred thousand share offering (at a par value of £1 each), out of which he was to give Davis ten thousand shares (five thousand as "commission for your trouble" and five thousand "to pay all expenses in connection with the formation of the company"). The company would have a working capital of £25,000 and Marconi would receive a salary of £500 a year for the next three years.[65] Appended to Marconi's letter was a list of eighteen patents "to be taken out at once" and eight more to be taken out "later."[66]

More meetings followed as the lawyers drew and redrew successive versions of the draft agreement. An unaddressed note, evidently a memo to file on Davis's personal stationery and signed by him, dated May 4, 1897, records Marconi's verbal agreement, in a meeting with Graham, "that, under no circumstances, would he accept an offer from the Government through Mr Preece or from any other party for his invention."[67] (Davis and his investors' nervousness was understandable as Marconi was still in contact—and publicly—with other suitors.[68])

While this turned out essentially to be the deal, negotiations over details dragged on while Marconi went about his research. The Bristol Channel tests were his priority. These were Marconi's first experiments across water, and Kemp's diary indicates their fragility:

Mr Marconi's apparatus was set up on the cliff at Lavernock Point, which is about twenty yards above sea-level…On the 11th and 12th [May 1897] his experiments were unsatisfactory—worse still, they were failures—and the fate of his new system trembled in the balance. An inspiration saved it. On the 13th May the apparatus was carried down to the beach at the foot of the cliff, and connected by another 20 yards of wire to the pole above, thus making an aerial height of 50 yards in all. Result: The instruments which for two days failed to record anything intelligible, now rang out the signals clear and unmistakable, and all by the addition of a few yards of wire![69]

After this transmission, from Lavernock to Flat Holm, a small island in the Channel just over three miles away, the receiver was relocated to Brean Down—a distance of more than nine miles from Lavernock. Again, the communication was successful. These distances over water were the greatest yet achieved, far greater than anything accomplished on land (where the record was still the three-quarters of a mile covered on Salisbury Plain). Marconi later referred to Lavernock as "the world's first seaside wireless laboratory."[70]

The Bristol Channel tests took place in the presence of an important observer: Professor Adolf Slaby, scientific advisor to the German kaiser and a renowned researcher at the technical college in Charlottenburg. Slaby, in Berlin, had heard of Marconi's work and asked Preece if he could come to observe the tests. Preece sent an invitation, against Marconi's better judgment. Preece, meanwhile, continued publicizing Marconi's work in high-profile lectures at such prestigious London venues as the Royal Institution (June 4, 1897) and Royal Society (June 16, 1897), as well as in local meeting halls, where Marconi sometimes appeared on the platform by his side.[71]

Marconi was also now planning a trip to Italy. The Italian naval attaché in London had written to Naval Minister Benedetto Brin, informing him of Marconi's experiments. Brin immediately telegraphed Marconi, inviting him to Italy to demonstrate his invention to the Italian navy.[72] In London, the parties to the company deal hoped to have an agreement signed before his departure; however, there were still a number of sticking points. Marconi was to be appointed "Engineer" of the company, and would be a director, with the right to name two additional directors to the seven-member board. But the investors wanted him to commit to residing in England for the next three years, and to that he would not agree. Everyone was anxious about the fact that the English patent had not yet been approved. They were also concerned about the eventual legal status of the invention and sought a legal opinion from Moulton, who advised that it would come under the Telegraph

Act of 1869, just as though it were using wires.[73] Thus, as Marconi always maintained, his invention was just that: an extension of conventional telegraphy—telegraphy without wires.

On June 23, 1897, Marconi told the solicitors that he was leaving for Italy and asked them to clear up the outstanding issues at once. This was impossible, so a power of attorney was arranged for another of Marconi's lawyers to act on his behalf. The power of attorney described what was about to happen—namely, a contract to sell to the new company "my invention for improvements in transmitting electrical impulses and signals...and my patents and patent rights in connection therewith," except with respect to the Kingdom of Italy, its colonies and dependencies.[74]

The new entity was described by Marconi as "a company which my relative Mr. H. Jameson Davis is promoting for the purpose of purchasing my invention" on terms essentially as spelled out in the exchange of notes dated April 30, 1897.[75] In the original draft prospectus, the company was to have been called Marconi's Patent Telegraphs Limited—the title was overwritten in Marconi's own hand and the parties settled on Wireless Telegraph & Signal Company Limited (WTSC). Giuseppe Marconi was particularly displeased as he felt it was important that the family's name appear in the title, but Marconi accepted it; he was to receive his cash payment immediately upon signing.

Just as Marconi was preparing to leave for Italy, his persistent rival Oliver Lodge, frustrated by all the attention Marconi was getting, published a letter in the *Times* reiterating his view that Marconi had done nothing really new.[76] Lodge was also corresponding with Augusto Righi, who was somewhat ambivalent about his own role with regard to Marconi's claims and was resisting being drawn into a public polemic. Righi had recently given an interview to a Bologna newspaper about what he referred to (to Lodge) as "the so-called invention of young G. Marconi. This young man is very intelligent but has little training in physics. I advised him to follow some university courses. I would be very curious to know more about his apparatus, but it seems that it strongly resembles my oscillator and your coherer." Lodge promptly replied (getting in a dig at Preece as well): "I must tell you that your protégé Marconi has obtained the ear of the British Post Office officials, some of whom are, like him, not well versed in physics." Righi, however, was far less excited than Lodge, and despite some skepticism quite prepared to acknowledge Marconi's contribution. "I have not seen him in two years and it is precisely in this period and without my knowledge that he has made his alleged invention...He is the first, I believe, to have had the idea to use electric waves for long-distance signals."[77] The rest of Marconi's career would flow from this type of authoritative recognition.

Marconi arrived in Rome on June 28 and was received the following morning by Brin, who was "very interested in my discoveries," as he wrote to his father.[78] Despite the stifling heat, a shock after sixteen months in England, he went on to meet with a number of admirals and naval commanders, as well as officials of the war and postal ministries. Everyone was thoughtful and kind to him, he told Giuseppe. Some of his equipment had been broken during the trip and took a few days to repair, but by July 2—the very day his British patent was awarded—he was able to transmit and receive messages successfully between two floors of the ministry offices in Via della Scrofa (substantially less than he had been doing with ease in England yet more than enough to impress his Italian hosts).[79] In the parliamentary Chamber of Deputies, the minister was asked to report on the results, and in the coming days there were banquets and more presentations to government members as well as the press, culminating on July 6 when Marconi demonstrated his invention before King Umberto and Queen Margherita at the royal palace, the Quirinal. Marconi's command performance made headlines back in England, where it did not go unnoticed by his increasingly impatient backers.[80]

With the awarding of his patent, Marconi arguably owned the exclusive right to use what is recognized today as a global public resource: the electromagnetic or radio spectrum. At least that is what the embryonic Marconi company would claim. It would be a few years before anyone realized that Marconi's patent could mean an effective monopoly on use of the spectrum, and then it would be aggressively contested by competing interests, including powerful governments. That astounding victory aside, an obscure squabble between the various lawyers handling the incorporation talks led to further delays—while Marconi continued his Italian tour, triumphantly.[81]

Davis pressed him to return to London as soon as possible. "The thing has been left far too long in this unsettled state. . . . [The subscribers] are very dissatisfied, and think they have not been well treated." A day or two later he wrote again: "If you can make arrangements to come to London for two or three days the matter can be arranged, but I think it exceedingly unwise in the present state of affairs to be away." Davis reminded his cousin that he, Marconi, would control the new company; under the terms of the agreement he had a seat on the seven-member board and got to name two of the other directors, and, Davis added just a bit innocently, "of course my vote can always be counted on." He then again urged Marconi to come back "while there is still time…if I had this chance I should not let anything prevent me coming and settling at once."[82]

Marconi, laying the foundation of a lifelong relationship with the Italian authorities, was in no hurry to return to England. On July 10 he went on to the naval base at La Spezia, located between Pisa and Genoa on the Ligurian Sea, where he did more detailed experiments over a period of a week. At La Spezia, again, the officials he met were astonished by his youth as well as his earnest demeanour. The results of his tests were a great success. Setting up what he later described as "the first floating wireless laboratory" on the battleship *San Martino*, he achieved clear signals over a range of about ten miles.[83]

While he worked at La Spezia, Marconi's new notoriety was being closely followed from Villa Griffone, where Annie had joined Giuseppe. Annie kept up a small stream of letters to Guglielmo in Rome, congratulating him on his successes but deploring that his name was not mentioned more frequently in the press (although it is hard to imagine how it could have been more), and echoing Giuseppe's anxiety about the company negotiations back in England. They both wrote to him on July 7, proudly reporting that they had read of his visit to the Quirinal. More important, they buttressed his resolve to hold out for exactly what he was seeking in his talks with Davis. They also raised the issue of the company's name. "The Italian Government would be very pleased if your invention could be kept in your name," wrote Giuseppe, recommending that Guglielmo instruct his London lawyer to put a hold on the contract until they had a chance to discuss it in person. "Do not pay too much attention to those people who say that the signatories will lose their patience to wait any longer. These are all but tricks of theirs in order to push you to accept conditions that are far too much beneficial to them." Giuseppe went on to commend his son for giving his patent "as a present" to the Italian government, something that had brought much honour to Guglielmo. Giuseppe's parental pride was palpable.[84]

Meanwhile, another issue had come up: Marconi's passport, issued in February 1896, was valid only for one year, after which he was supposed to return to Italy to do military service.[85] He now had three alternatives: suspend his experimental work, become a British subject, or try to negotiate a deal with the Italian authorities. The first two were unacceptable; the third was the obvious course. Ambassador Ferrero prompted Naval Minister Brin to intervene.[86] Marconi was enrolled in the Italian Royal Navy as an ordinary seaman and immediately detached to serve with the naval attaché of the Italian embassy in London for a period of three years. It was understood that this was an administrative arrangement; he would pursue his research and other duties under the auspices of his company, and his embassy duties would be less than nominal. He was even to be paid a stipend. Again, Marconi set a pattern and learned an important lesson: with the right connections, anything was possible.[87]

Marconi joined the family in Pontecchio on July 20—the day his company was launched in London. He wired Davis that he would be arriving in England the following week; the cable was upbeat, as Marconi reported "splendid results from experiments, 12 miles small apparatus. Must decide as large French Company wants patents." Twelve miles was the greatest distance yet achieved and Davis replied that he was very pleased. "It will be most desirable to get to work without delay and try and make some money." From this point, the pressure would be on to produce revenue-generating results.[88]

Marconi stayed on longer still in Italy, spending days meeting with old friends and associates (he had a "long and learned" conversation with Augusto Righi, as it was reported in the press) and generally basking in the attention he was getting. Before his departure from Bologna his father gave a sumptuous lunch at one of the city's finest hotel restaurants, the Albergo Italia, for a large number of family, friends, and political and business personalities. A local member of parliament, Enrico Pini, gave a moving speech to which Marconi replied with emotion. Such affairs would become a ritualistic routine for Marconi. Two days later, accompanied by his mother, he left Bologna by train for England.[89]

The Wireless Telegraph & Signal Company was incorporated on July 20, 1897, and the board held its first meeting two days later, with Marconi absent.[90] Marconi was informed by telegram on July 24 that the sum of £15,000 had been deposited to his account with the London Westminster Bank. He also received 60 percent of the company's shares, and virtually controlled its board of directors. Just a few months past his twenty-third birthday, Marconi was now a wealthy, powerful man.‡ Morten, Cutler's bill came to £95, or $475, quite a bargain considering their efforts.

❀   ❀   ❀

Marconi must have been supremely confident to leave England at the critical moment leading to the creation of his company. His next challenge would be to convince his new business colleagues to continue funding his

‡ £15,000 was equivalent to US$75,000 in 1897. Based on the UK retail price index, this translated into £1.5 million in 2014; or, according to the US consumer price index, roughly $2.2 million. However, based on relative share of gross domestic product—a measure of economic power, or how rich someone would be in relation to the overall economy—Marconi was

experimental work, anxious as they were to begin seeing a return on their investment after months of financial idleness.[91] As Davis reminded the investors in a note informing them of the signature of the contract and updating them on Marconi's Italian success, the company could now "deal with the invention as it thinks fit."[92]

First, however, Marconi felt compelled to inform Preece of the news. Writing from Villa Griffone just before leaving Italy on July 27, he laid out the motives for his decision. The first was that the business had become "too large for myself alone, as all the governments in Europe want experiments carried out." The apparatus, which was still cumbersome, needed to be made more practical, and many new tests would need to be done. Second, the expense of patenting was too great for him to bear alone. In addition to protecting the invention everywhere, he needed to patent some further improvements. Marconi then expressed a wish and a pledge: "Hoping that you will continue in your benevolence towards me. . . . I shall also do my best to keep the company on amicable terms with the British Government."[93]

Whether Marconi was being disingenuous or merely naive, Preece's reply was unequivocal. In a handwritten note on August 6, 1897, Preece wrote: "I am very sorry to get your letter. You have taken a step that I fear is very inimical to your personal interests. I regret to say that I must stop all experiments and all action until I learn the conditions that are to determine the relations between your Company and the Government Departments who have encouraged and helped you so much."[94] It was not the end of Marconi's relationship with the Post Office or the government by any means—far from it—but he would no longer enjoy the unequivocal support of one of his most powerful and generous benefactors. Marconi may well have felt he no longer needed that support.

much wealthier; according to that index, his deal placed him in the same category as someone worth £16.7 million in the United Kingdom, or $83.5 million in the United States in 2014. Marconi's sixty thousand shares, at a par value of £1 each, raised his "economic power" to the equivalent of more than US$400 million at the time of the company's launch. In comparison, Mark Zuckerberg pocketed an estimated $17 billion from Facebook's initial public share offering in 2012, making him around forty times as "powerful" as Marconi at an equivalent stage of career (based on www.measuringworth.com.).

4

The Magician

It didn't take long for the European press to pick up the news of Marconi's success. The first public presentation of his invention using the platform provided by Preece was reported in a Viennese newspaper, the *Neue Freie Presse*, on December 17, 1896.[1] The news was relayed with a proud local twist by Bologna's *Il Resto del Carlino* on December 22, which noted that the son of a distinguished Pontecchio family, the Marconis, had unveiled a new invention of wireless telegraphy to clamorous acclaim in London.[2] Marconi's first press interview appeared a few days later in Rome's *La Tribuna*, then considered the city's most authoritative newspaper. The most notable thing about the interview was the modest simplicity with which Marconi explained his invention; he told the paper's London correspondent: "My discovery does not contain any new principle, but applications and extensions of principles already known."[3]

Within weeks, the major US newspapers began exploring the impact of Marconi's discovery, elevating him instantly to the same plane as established, famous figures like Edison and Tesla. Marconi's system was "fraught with immense possibilities which scientists are studying," waxed the *Chicago Daily Tribune*, one of the early US purveyors of Marconi's aura.[4] In March 1897, barely three months after Marconi was unveiled to the public, H.J.W. Dam did a feature-length story on him in London's *Strand Magazine*.[5] The much-published iconic photograph of Marconi sitting reflectively in front of his apparatus was taken at Marconi's home in Westbourne Park by the *Strand* photographer and first published with this article, which circulated widely in English, Italian, and French versions.

Dam's article sang Marconi's praises, comparing his discovery to German physicist Wilhelm Roentgen's X-rays, only "more wonderful, more important and more revolutionary." (In 1895, Roentgen discovered "an invisible form of radiant energy" he called X-rays, for which he was awarded the first Nobel Prize in Physics, in 1901.) Strange as it may seem, wrote Dam, considering all the brilliant scientists working on the problem, "it has been left to a young Italian, Guglielmo Marconi, to conceive what might be done with electric waves, and to invent instruments for doing it." Marconi's story "sounds like a fairy tale," yet he "makes no claims whatever as a scientist and simply says that he has observed certain facts and invented instruments to meet them."

Dam gave credit to Preece's "enterprise and shrewdness" in bringing Marconi to the public, and reiterated Preece's Toynbee Hall lecture, creating the impression that the British government was not only supporting but overseeing Marconi's work. Marconi's tone was casual and, again, modest. Dam asked a lot of technical questions, and Marconi often replied that he was not a scientist and didn't know the answer. Actually, a scientifically knowledgeable reader would have noticed some embarrassing errors, or at least gaps, in Marconi's tentative answers to some of Dam's technical questions; an unkind one (Oliver Lodge, for example) might have said Marconi didn't know what he was talking about.

The interview was revealing in light of Marconi's later reconstruction of his early experiments. He maintained the waves he was using were not Hertzian waves, but some other type of wave that he could not quite describe. "My receiver will not work with the Hertz transmitter, and my transmitter will not work with the Hertz receiver. It is a new apparatus entirely." Marconi was trying to distinguish his originality. Still, as Preece had recognized, Marconi's instrument worked. "Do you think that sitting in this room you could send a dispatch across London to the GPO [four and a half miles away]?" Dam asked. Although he had yet to achieve more than three-quarters of a mile, Marconi replied: "With instruments of the proper size and power, I have no doubt about it." Marconi believed that eventually his system would be used for lighthouses and lightships, and placed the practical range limit at twenty miles. Dam asked why he couldn't send a dispatch to New York, and Marconi replied, "I do not say that it could not be done." Then, despite what he had just said: "I do not wish to be recorded as saying that anything can actually be done beyond what I have already been able to do." Later that would change, and Marconi would often make predictions, partly as a commercial strategy, and partly as he came to feel that his achievements entitled him to believe in future ones.

The *Strand* article and its offshoots—it was picked up by *McClure's Magazine* in New York, and was the basis for pieces in newspapers across the United States—were stock examples of popular science and technology reporting in the late nineteenth-century British and American press, reporting that emphasized and celebrated the role of the inventor-hero,[6] men such as Morse, Edison, Bell, and now Marconi. These stories always focused on picking *winners*; they eschewed the Hughes, the Meuccis, the Stubblefields, except to deplore their shortcomings, and even the contenders like Tesla or Lodge were invariably portrayed as jealous, self-destructive, or narrow-minded. The press also brought out the role of a new class of technology brokers, like Preece, who had the knowledge and wisdom to recognize valuable new commodities and the power to promote them with apparent disinterest.

The remarkable thing about the press's embrace of Marconi was how quickly, totally, and uncritically it occurred. If anything, Marconi himself appeared to be more prudent, more humble, and more circumspect than the press in his claims. Portraits like Dam's created the image of a young man one could trust. Marconi's lack of pretention, his reflectiveness, and the fact that he did not claim to have all the answers served him well. At the same time, he quickly became an adept, lifelong practitioner of image control. A copy of the Dam article in the Lincei archives in Rome indicates that Marconi himself had corrected the proofs of the article.[7]

In Britain, press fascination with Marconi seemed to reach a new peak during his Italian trip in July 1897. A full description of his system appeared in the *Electrical Review*. London's *Daily Chronicle* observed that "There is no lack of spirit and versatility in the Marconi boom." The *Globe* put his work on the same plane as H.G. Wells's recent science fiction novel, *War of the Worlds*. *The Morning* reported: "Signor Marconi, the famous young Italian electrical engineer, whose name in the popular mind is fast becoming affixed to the system of telegraphing without the instrumentality of wires, is being boomed in a manner truly American." The *Daily Mail* asked the question it said was on everyone's lips: "Is Signor Marconi's wireless telegraphy a new discovery, or merely an adaptation by ingenious methods of that which other scientists had found out before him?"[8] Some of the articles noted that Marconi's own claims were typically more restrained than those of his boosters.

The major US papers, too, were now fully part of the act. Preece's early reports had already attracted some transatlantic attention and the *New York*

Times took sober notice of the interest in Marconi's system in May 1897. By July he was a news phenomenon. The *Herald* presented Marconi as some kind of boy wonder. The *New York World* hailed him as "a new wizard." In Philadelphia, the *Press* reported that the inventor of the "gramophone," Emile Berliner, considered Marconi's wireless telegraph "the invention of the year." (Berliner's gramophone was on its way to becoming the standard record player, replacing Thomas Edison's more awkward "phonograph," patented in 1878.) The Wizard of Menlo Park, Edison himself, was quoted by the *New York Herald*, approvingly: "That lad is smart." Rarely, some of the coverage was more skeptical. In a dispatch from London, the *New York Sun* wrote: "It is almost impossible to pick up a newspaper or magazine without finding a puff…and the people who are running the Marconi boom are not the sort of folks to give something for nothing."[9]

Marconi was not the first nor the only technological figure of his time to be referred to as a wizard (as well as, most famously, Edison, Nikola Tesla too attracted this label).[10] As his fame grew, in press accounts of his exploits he was soon being called a magician. Before long, Marconi's name was being mentioned in the same breath as that of another famous illusionist and showman, Harry Houdini.

Houdini, whose real name was Ehrich Weiss, was a performance artist from New York. Born into an orthodox Jewish family in Budapest (his father was a rabbi) in March 1874, exactly one month before Marconi, he immigrated to the United States with his parents and four brothers in 1878. By the time Marconi began attracting media attention in the late 1890s, Houdini was already a huge celebrity and had mastered the strategy of using the popular press to cultivate an image.[11] There is no evidence that Marconi knowingly borrowed this strategy from Houdini but he had to have been aware of Houdini and his success. Marconi and Houdini both developed the practice of inviting selected members of the press to witness their feats and performances. By 1897 they were both international media stars, constantly photographed and recognized wherever they went.

Marconi and Houdini also shared a place in the popular imagination of the time. They were both considered conjurers, even charlatans by some. The "magic" label seemed to stick. When Marconi first demonstrated his machinery, people would look for the hidden wire. When it became clear that Marconi *could* make bells ring or send signals across rooms and through walls without connecting wires, in some minds he came to be associated with the idea of "telepathy."[12] With the discovery of "communication via waves," even serious journalists like H.J.W. Dam made a link between Marconi's work and psychic phenomena, such as séances, concluding that underlying it all was "the mystery of the ether," and evoking a connection to "thought waves," impulses from one person's brain affecting others over distance.[13]

This was a respectable notion in the 1890s. The possibilities of "mental telepathy" intrigued personalities as diverse as Alexander Graham Bell, Mark Twain, and Arthur Conan Doyle—not to mention established academic scientists such as Sir William Crookes and Oliver Lodge—but no one had thought that electrical impulses, carried on "waves" in the ether, could connect minds. Now a link was made between telepathy, telegraphy, and spiritualism—the idea that people could communicate with the dead. In 1897 there were said to be more than eight million followers of the spiritualist movement in Europe and America.[14]

Marconi, "the wizard of the air waves," never bought into this. Although constantly confronted with the most outlandish hypotheses, he always remained a skeptical rationalist. Here again, Marconi found a strong ally in Houdini, who, despite his showmanship, knew he was creating illusions, not representing the supernatural. They both had many admirers among the millions of believers in spiritualism, notably Doyle, who in addition to being the creator of Sherlock Holmes was the leading proponent of spiritualism in Great Britain.[15]

Marconi's work was also of interest to the Society for Psychical Research, established in England in 1882, with a US branch opened in 1885. The society's official aim was to examine "in a scientific spirit those faculties of man, real or supposed, which appear to be inexplicable."[16] One of the attractions of psychical research (or what we now call parapsychology) was that it offered a secular alternative to religious explanations of scientifically unexplainable phenomena. When Marconi said he couldn't explain the scientific basis for how his airwaves worked, psychical research had an answer: telepathy.

Far from a meeting place for cranks, the Society for Psychical Research attracted prominent members of the social mainstream (and even avantgarde) such as the future British prime minister Arthur Balfour, American psychologist William James, and French philosopher Henri Bergson. Significantly, its ranks also included some of the leading figures associated with the early research on wireless communication. From 1896 to 1899, precisely when Marconi was establishing his reputation, the president of the society was none other than Sir William Crookes, who, as we've seen, had been the first scientist to suggest the possibility of wireless communication by Hertzian waves in 1892. In 1898, when Crookes was also president of the British Association (making him Britain's most authoritative academic), the *Times* of London editorialized that it was a sign of the times that a declared believer in the "fundamental law that thoughts and images may be transferred from one mind to another without the agency of the recognized organs of senses" was president of both associations.[17]

In Crookes's presidential address to the British Association that year, he recognized Marconi's role in building on Lodge's experimental research to make wireless telegraphy a practical reality; at the end of the long lecture, which treated many different subjects, he touched on current issues in parapsychology, suggesting that some aspects of telepathy fell within the scope of physical science.[18] When Crookes raised the idea that telepathy could be considered a matter of "brain waves," a writer following the conference made the link between wireless telegraphy and mental telepathy, suggesting that perhaps the brain sets up a mechanism that acts like a Marconi transmitter.[19]

A few years later, from 1901 to 1903, the president of the Society for Psychical Research was Oliver Lodge.

❉   ❉   ❉

Marconi now spent the rest of 1897 and 1898 developing working relations with his business colleagues, setting up experimental stations in the south of England, and managing his new-found fame and fortune.

Despite William Preece's initial reaction to the formation of Marconi's private entity, negotiations were soon going on between the company and the GPO with a view to making some arrangement for the government to use Marconi's patent. Soon after returning to England, Marconi wrote his father saying, "It seems likely that [Preece] will remain friends with me, even though I have sold the patent rights." A few days later, Marconi met with Preece, who assured him that the British government would come to an agreement with the company.[20] Preece had his own reasons for maintaining good relations with Marconi; in a note to a colleague about Marconi's upcoming tests, he wrote: "It is well for us to keep hold of him and know exactly what he is doing."[21] From now on, the company's relations with the Post Office, and particularly Marconi's relations with Preece, would run hot and cold.[22]

In August 1897, the Wireless Telegraphy & Signal Company went about setting up shop. Henry Jameson Davis was named managing director for a period of two years (at his insistence, as he was eager to get back to his own career), and an important appointment was made: Henry W. Allen, formerly Davis's clerk, was named company secretary; he would hold the position until 1930, becoming one of Marconi's closest associates. The company rented a part of Davis's existing premises at 82 Mark Lane.[23]

Marconi now had a modest but effective corporate machine at his disposal. One of the first things Davis did was order up a set of precise technical drawings of Marconi's instruments.[24] At the first working meeting following the company's incorporation, the board of directors approved nineteen new patent applications and instructed its agents to apply for patents in British North Borneo, the Straits Settlements, Hong Kong, Gibraltar, Victoria (Australia), Ceylon, and Canada. The board also considered various proposals that Marconi had received directly from a number of foreign governments and decided "that it would be advantageous to the Company for Mr Marconi to make experiments in Russia and Austria, at an early date."[25] The German fleet was said to be testing Marconi's system at sea, between Kiel and Kronstadt.

Meanwhile, the Italian navy announced that it was adopting Marconi's system and the board "decided to accept the Italian Government as customers for their machines at a fair profit." Marconi may have given them the use of the patent at no charge, but in order to exploit the rights effectively, they would still have to deal with the company. Another pattern was established: the company's clients would be locked into long-term deals involving multiple revenue streams. With the exception of its deal with the Italian government, the company would also keep control over its technology. Italy thus became the first country to adopt Marconi's system; it also became the company's first client, and the first to commit itself to long-distance wireless communication.[26]

All this interest encouraged WTSC to pursue its relations with the British government as well, and one of the board's first acts was to send a message to Preece expressing its appreciation of his "friendly assistance in experimenting with Mr Marconi's system" and its desire to continue their collaboration. For obvious reasons of proximity, England would be Marconi's principal testing-ground and the board decided that experiments should be carried out at Salisbury, and then between Dover and Calais, "for which Government assistance is to be invited." Preece, meanwhile, saw the new company as a potential competitor and even challenger for the Post Office's monopoly in communications; he intended to stay in touch, but Marconi still had a way to go before he could do concrete business with Her Majesty's government.

The company now also began cultivating business with the private sector. The first major firm to recognize the commercial potential of Marconi's invention was Lloyd's, the world's leading provider of maritime insurance and, hence, dealer in shipping information. On August 20, 1897, the secretary of Lloyd's, Sir Henry Hozier (one of the first businessmen to show an interest in Marconi's invention, as we saw in chapter 3) wrote to Marconi stating that he thought wireless could be of considerable assistance to Lloyd's as it might enable them to communicate with points they might not be able to reach using cables. Lloyd's was thus the first private company to recognize the worth of wireless and would become the Marconi company's first commercial client, as the Post Office had been its first public benefactor, and, shortly, the Admiralty its first military contractor.[27]

The company continued receiving queries from far and wide. In September the board heard of "various proposals received from continental firms" but resolved to wait until after its next round of experiments at Salisbury before making any deals. There was a problem, however. The Post Office was now planning its own trials, to take place in Dover, and Marconi was not invited. The board asked Marconi to query Preece regarding the government's intentions.[28] Marconi duly wrote to Preece, regretting that "certain improved details of which I recognized the importance" in Italy were not being applied by the Post Office, although the Italian navy was now using the system experimentally every day. He also dropped two important names: Army Captain Baden Baden-Powell, who had loaned the company five kites and was coming down to observe their use at Salisbury; and Captain Henry Jackson of the Royal Navy, who had been instructed to attend "my experiments" at Dover on behalf of the Admiralty.[29] (The Royal Navy was the naval branch of the British armed forces; the Admiralty was the authority that ruled it.)

The company seemed to be on a roll, although it had yet to earn a penny. After a few months it moved to larger offices, still in Mark Lane, and hired a group of assistants willing to put in long hours. Most of them—like Marconi—were well under the age of thirty, and Davis, at forty-three, was by far the oldest. "Our youth was probably one reason why progress was so fast," Allen said some years later.[30] In fact, driven by Marconi's enthusiastic confidence of ultimate success and Davis's business skills, it was stunningly fast.

The board minutes for this period show, with precision, both the breadth of activity the company had quickly become involved with and the unique place occupied by its most precious asset, Marconi himself. At the insistence of some of the company's financial backers, provisions had been written into its regulations to ensure that Marconi did not become too powerful. The company's statutes placed a cap on the number of shares that could be voted by any individual shareholder, 40 percent of the issued total, regardless of the number of shares that individual actually owned. This was not a hypothetical check; Marconi started out with 60 percent of the company's shares (although that was immediately reduced to 50 percent once he gave Davis his promised allotment), and in the first year of operations he continued to increase his holding. So, for that matter, did other members of his family. Marconi's letters home have him reporting on the fluctuation in share values and advising his father to keep buying. Annie and Alfonso also owned shares. But although pushed by Giuseppe to occupy every inch of available space, Marconi himself had no problem taking a more prudent course. He went along with keeping his name from the company's title and in September 1897, when the board resolved to omit the words "Marconi's System" from the share certificates, it was with his explicit consent.[31]

The company was now attracting queries and business proposals from all over the world, and yet, the investment market was wary. An article in *Investors Review* reminded readers that 75 percent of the company's initial share offering had gone toward the payment for patent rights (sixty thousand shares plus £15,000 to Marconi) and it still had no earnings: "The public will be well advised to keep clear of this concern," the journal wrote, pending the development of a serious commercial outlet for "Signor Marconi's ingenious ideas."[32]

The company's financial prospects were, however, aided by reports that government agencies as diverse as the US Navy and the British lighthouse authority, Trinity House, were looking at the new technology. The Post Office was interested in establishing a wireless connection between the isles of Guernsey and Sark. In October 1897, Marconi's American patent was granted—an event treated with irony by the US press, which speculated that the patent would face the same fate as the telephone, the light bulb, and similar universally useful inventions, and end up in the courts.[33]

Less than a year after he first came into view, Marconi had already become a common popular reference point. Feeling that "all sorts of wild claims were being reported about him and his work," and with an eye toward the prospects of his company (and undermining the supposedly mystical or telepathic properties of his invention), Marconi decided to give an "authoritative interview" to the *Daily Chronicle*. Here, Marconi asserted that before his trials, no one had signalled with Hertzian waves—not Lodge nor anyone else. (In the interview with H.J.W. Dam a few months earlier, he had said that *he* had not used Hertzian waves.) He proposed to set the record straight as to what he had done (sent a message across twelve miles) and not done (blown up magazines on distant ships), could do (transmit and receive thanks to the use of vertical conductors), and could not do (regulate watches in people's pockets). Despite the efforts of those who felt he was not entitled to it, he had a patent, and "that is the great thing."[34]

Marconi's priority was now to find a site for an experimental station. He settled on Dorset, a region that was relatively accessible by rail from London; well equipped in both infrastructure and social networks; offered a range of harbours, inlets, islands, and seashores that could be used for staging experiments at various distances; was close to passing sea traffic; and, when the time came (which he expected to be soon), could provide the British end of his first international links, to continental Europe.

In November, Marconi and Davis told the board that after a visit to the area they had found a suitable site for a station at Alum Bay, at the western tip of the Isle of Wight.[35] Here, amid a range of stunning chalk cliffs, overlooking a beach of multicoloured sands favoured as pigments by Victorian painters, Marconi set up what he claimed to be the world's first fixed wireless station at the Royal Needles Hotel. The station opened officially on December 6, 1897.[36] "The wireless room was a private sitting room on the ground floor," Marconi later wrote. "It had a carpet on the floor, and a large bay window leading on to the lawn. A hole was drilled in one of the window-panes and the aerial led through to the mast outside." The first experiments done here were with different types of aerials, and Marconi's goal was now to establish communication with ships at sea. In no time, he had succeeded in sending readable signals to a tug in Alum Bay, then to Bournemouth (fourteen miles), Poole (eighteen miles), and eventually to ships passing on the seaward side of the Isle of Wight.[37]

Writing in *McClure's* a bit later, journalist Cleveland Moffett described the "amazing business" at the Needles station this way:

> Looking down from the high ground, a furlong beyond the last railway station, I saw at my feet the horseshoe cavern of Alum Bay, a steep semicircle, bitten out of the chalk cliffs, one might fancy, by some fierce sea-monster, whose teeth had snapped in the effort and been strewn there in the jagged line of Needles....On the right were low-lying reddish forts, waiting for some enemy

to dare their guns. On the left, rising bare and solitary from the highest hill of all, stood the granite cross of Alfred Tennyson [Britain's poet laureate, who had lived on the Isle of Wight until his death in 1892]....[38]

Next, Marconi opened a mainland station in the Madeira House, a solid Victorian hotel on the seafront at Bournemouth, then a popular weekend and holiday destination for the middle class. The station's 115-foot mast flagged the site to the genteel couples, from all parts of the country, who strolled along the Bournemouth pier with parasols on their arms. Marconi considered the work he did here critical to the early development of his system, even if the accommodation was limited and the apparatus simple. The wireless room was barely eight feet by eight feet, situated in the base-ment, and lit by one small window. As at the equally spartan Needles, the aerial was passed through the windowpane to a spark coil standing on a small table outside. The only other furniture was some shelving.[39]

Sending messages between these two stations, and communicating with passing ships, Marconi and his staff kept improving the system.[40] On Christmas Eve of 1897 he reported to Giuseppe that the work was going extremely well, even unexpectedly so. After spending two weeks on a steamship in horrible weather, he had managed to communicate across six-teen miles through fog and clouds. "With the help of God everything went well," he wrote to his father.[41]

Despite being as busy as could be, Marconi kept up his correspond-ence with Giuseppe. Having advised his son on the crucial negotiations while setting up the company, and having provided critical financial support in the early stages of Marconi's research in England, Giuseppe was now in-creasingly a business partner.[42] Giuseppe continued to propose investments, mostly potential land purchases in the Bologna area; he didn't seem to re-alize that his son's future was now tied to actively running his company from England. (In fact, Marconi would not visit Italy again until September 1902.) Giuseppe suggested that Guglielmo might want to purchase a villa in Italy with his new fortune, but Guglielmo was clearly far more interested in pursuing his research.[43]

Marconi proudly reported to his father on the success of the business and the honours he was beginning to garner. The company was closing in on its deal with Lloyd's and shares were increasing in value as trials suc-ceeded over greater distances. Not yet twenty-four years old, he had re-ceived two Italian honours: the *Croce da Ufficiale* (Officer's Cross) and the *Regio Decreto* (Royal Decree), and expected soon to be entitled to use the

title *Commendatore* (Commander). He had also heard that the German kaiser was contemplating giving him a title. Honours were important to
Marconi, and he would accumulate them incessantly throughout his lifetime, a mark of recognition more important to him than media attention or
money.[44]

Marconi's interests were still being looked after by his cousin, Henry
Jameson Davis, who remained in the thick of things while Marconi ran
about. Marconi had not taken advantage of the provision in the company's
incorporation papers that allowed him to name an additional director of his
choosing before February 1, 1898, and time was running out. A few days before the deadline, Davis tracked him down. He suggested William Goodbody.[45]

The Goodbodys were an established Irish Quaker family (like the
Davises) that operated one of Ireland's oldest brokerage houses. One of the
original subscribers to the company launch was corn factor Marcus T.
Goodbody. Another Goodbody, Robert, represented the family (and the
Marconi company) in New York. Marconi took Davis's advice and nominated William Woodcock Goodbody, a Dublin banker, as a director.[46] It was
a wise move. When the company needed to restructure its finances in
August 1901, no fewer than nine Goodbodys were among the fifteen lend-
ers who formed the new financial syndicate.[47]

In January 1898, an incident occurred that showed Marconi's flair for taking
advantage of an opportunity. The ailing former prime minister William
Gladstone had gone to Bournemouth for his health, followed by a flock of
newspaper reporters, and took a turn for the worse just as a heavy snow-
storm knocked out the regular telegraph lines. Marconi improvised a transmission link from Bournemouth to the Needles and from there to the nearest
operating wire station, from which the journalists' reports on Gladstone's
health were retransmitted to London. The incident gave Marconi some good
publicity and made him newspaper friends—another pattern he would develop and exploit throughout his life.[48]

In April, Alfonso joined his brother in Bournemouth, and soon reported to Giuseppe: "I have a lot to do to keep all of Guglielmo's things in
order."[49] Alfonso now took over the task of keeping their father informed
of Marconi's affairs. The big news was that Marconi was considering forming a new company. Another of Annie's Irish business relatives, John Eustace
Jameson, had approached WTSC with the idea of a new venture involving
one of Britain's best-known scientists, Sir William Thomson (Lord Kelvin),

and Marconi's old nemesis Oliver Lodge. Kelvin was the inventor of the "mirror galvanometer," an essential component of cable reception, and had been chief engineer on the transatlantic cable project.[50] He had also been an early doubter of Marconi. "Wireless," he once sniffed, "is all very well but I'd rather send a message by a boy on a pony."[51] Now, however, he had become a fan. Jameson felt they could raise as much as £1 million by combining the prestige of Marconi, Kelvin, and Lodge under a single corporate roof. Alfonso assured Giuseppe that this time Marconi would make it *sine qua non* that his name be in the title. Marconi would be the head of this new company and his stipend would quadruple to £2,000 per year, but he was wondering if he should accept less in order to give the company a better chance of success. Alfonso asked his father what he thought of all this. But the venture never materialized.[52]

Travelling frequently between Bournemouth and London, Marconi now received a stream of important visitors: journalists, politicians, business people, and scientists. In early June, Lord Kelvin himself called at the Needles, accompanied by Hallam Tennyson, son of the late poet laureate. By way of Bournemouth, Kelvin was able to send a greeting to a colleague in Glasgow; he was so impressed that he insisted on paying a small fee for this service—thus, just as Preece had feared, flouting the Post Office monopoly on inland telegraphy, and allowing Marconi to forever claim that he had sent the world's first commercial wireless message from the Needles Hotel to Bournemouth on June 3, 1898.[53] Ambassador Ferrero and an entourage also spent two contented days at the Needles around this time, sending a telegram to Rome as well.[54]

These were publicity stunts more than anything else. All Marconi was really doing was sending a wireless message across a short distance, to then be relayed by conventional wire telegram to its final destination. There was still little prospect for a commercial wireless telegraph service, and little added value to be had. But the publicity was priceless and in Marconi's mind a revolutionary idea was starting to take shape. If he could indeed conquer distance—as he deeply believed he could—he would be able to do away with the cable altogether. If wireless could, say, bridge the Atlantic, it would become a serious, credible, and possibly devastating competitor to the cable companies. (The thought soon occurred to shareholders in the cable companies as well. In July 1899 the Eastern Telegraph Company attributed a reported fall in share value to the news of Marconi's experiments.[55]) He was careful not to state this so baldly until it had actually been accomplished, but that now became his goal.

Marconi's contacts with naval authorities, in Italy and Britain (and soon the United States), were critical to this plan. Marconi himself dreamed of the beauty of communicating while at sea, and his company saw the commercial possibilities of engaging with firms involved in the shipping industry, like Lloyd's, but no one had a greater interest in the development of wireless than the world's great navies. In May 1898 Davis wrote to the First Lord of the Admiralty suggesting they test the system on its war ships.[56] By December the Admiralty was proposing to adopt the system and negotiations with the company were under way. One of the issues concerned who would have the rights to the Marconi patents. Another concerned the ownership of the apparatus that would be custom-built for the Navy. Trials and talks would continue during 1899, and the Admiralty contract, concluded in July 1900, would be the company's first concrete deal with an arm of the British government.

❁   ❁   ❁

Still less than two years after first being noticed, Marconi now became an active media presence. In July 1898, he received a compelling invitation—to cover the Irish Kingstown Regatta (in modern Dún Laoghaire) for two sister Dublin newspapers, the *Daily Express* and the *Evening Mail*. Marconi would oversee the operations himself, demonstrating another practical application of wireless, this time to journalism. Once again, enthusiastic accounts appeared in major world newspapers. It was also one of the company's first paid engagements.

The experiment took place on July 20 and 21, 1898, in Dublin Bay. Marconi and his equipment were installed on a steamboat, the *Flying Huntress*, hired by the newspapers. This was Marconi's first public exposure in Ireland and the press focused hard on him. One report described him as "A tall, athletic figure, dark hair, steady grey blue eyes, a resolute mouth, and an open forehead," his manner "at once unassuming to a degree, and yet confident." While he was at work at his transmitter, his face showed "a suppressed enthusiasm....He stood there with a certain simple dignity, a quiet pride in his own control of a powerful force." Another reporter wrote that he felt as though he was attending an illusionist's performance—again attributing magical talents to Marconi. "How is it done? Well, Marconi himself can do little more than guess." The report referred to Marconi's invention as "a creation."[57]

The Kingstown regatta was also the setting for another Marconi first, when company director William Goodbody placed a securities purchase order to the Dublin exchange—the first ever financial transaction by wireless. The order was executed and reported back to the boat, which received the message at sea, miraculously out of sight of the shore station.[58]

While Marconi was in Kingstown, the Prince of Wales suffered a knee accident while sailing on the royal yacht *Osborne* in Cowes Bay on the Isle of Wight. The prince, who had already shown some interest in wireless telegraphy, asked Marconi if he could establish a link between the yacht and Osborne House, Queen Victoria's summer residence on the Isle of Wight, so that he could stay in touch with his mother while recuperating. Marconi was delighted at the opportunity and easily set up the connection. He stayed on the yacht for sixteen days, sending daily bulletins to the Queen—and, for a rare occasion, openly enjoying himself.[59] Observers—not least the royal family—were impressed with the smoothness of the system.[60] The prince, in particular, entertained Marconi "with his usual bonhomie." Marconi proudly reported to his father that the prince gave him "a very nice and valuable tie-pin" as a gift.[61]

In September, after a disagreement with the manager of the Madeira House (likely about money), Marconi had to find a new location for his mainland operations. He moved to the Haven Hotel, at Sandbanks, seven miles west of Bournemouth at the entrance to Poole Harbour—one of the attractions was that he would be able to make the trip by bicycle. The Haven was slightly farther from the Needles but still, at eighteen miles away, a reasonable transmission distance. The Haven became Marconi's principal English domestic research station, and often his home, for the next twenty-seven years. He would happily spend weeks at a time at the Haven, which developed into a popular resort run by a French couple, the Poulains, who looked after Marconi, his staff, and their frequent visitors well. In the early years, Marconi would often be accompanied there by Annie or Alfonso. It took just under four hours to get to the Haven from London's Waterloo station.[62] One could put in a full day's work in London, take the 4:10 train to Bournemouth, and be at the Haven in time for Madame Poulain's dinner.

Well into the early nineteenth century, the sandbanks of Poole Harbour were notorious for catering to smugglers (hence, evidently, a "haven"). An inn was in operation on the southern side as early as 1813, and in 1838, the North Haven Inn was built on the site still occupied today by the hotel that housed Marconi. By the time Marconi arrived, the area was beginning to be developed for tourism: a new steamer pier was under construction, and the sandbanks were being subdivided for high-end residential real estate. Today, it remains one of the most exclusive destinations in the south of England, favoured by yachting folk and well-off retirees. The Poulains sold the Haven in the 1920s, a few years before Marconi closed his station in 1926. It underwent a total reconstruction and is now a four-star hotel owned and operated by a small regional luxury chain, FJB Hotels.[63]

Marconi considered his quarters at the Haven more comfortable than any he had had up to that time. The main operations room was more than twice the size of that in the Madeira House and had two windows, one of them drilled for the aerial to run through, as at the Needles and in Bournemouth.[64] In the crucial period between 1898 and 1900, the Haven became a serious experimental research station with a semi-resident staff. Marconi spent as much time there as he could, driving his men hard. A large room on the ground floor served as the main laboratory, with other work carried on in a cluster of out-building huts as well as directly on the beach. The team worked long hours but the atmosphere was highly sociable, everyone sharing a common table at dinner, which was often followed by homemade musical entertainment (there was no other kind). One of Marconi's colleagues would play the cello; Alfonso, when present, the violin. Annie might sing, and Marconi, always, would play the piano.[65]

Work at the Haven was an exercise in improvisation—what historian Hugh Aitken has called Marconi's "determined empiricism."[66] Marconi and his assistants were continually developing new improvements to Marconi's basic apparatus; for example, a new type of transformer they called a "jigger," which was essentially a wire coil wound around a paper tube. Every one of these innovations incrementally enhanced the quality of wireless transmission and reception; the use of "jiggers," in this case, facilitated the development of tuned circuits, a crucial stage in the perfection of wireless technology.[67]

Marconi's over-riding goal was distance, and finding ways to increase it was the main job that continued to occupy his thoughts. He was convinced that there was theoretically no limit to the extent that wireless signals could travel, and that the key was to find the right combination of technical equipment, electrical power, and natural wavelengths. He had discovered, essentially by trial and error, that working with longer wavelengths at the low end of the frequency spectrum and increasing the height of his aerial antennas resulted in greater communication distance. He was now working with ever larger aerials and had not entirely abandoned the possibilities of using the shorter wavelengths at the higher end of the spectrum (where the original experiments of Hertz and researchers like Righi and

Lodge, as well as Marconi himself, had taken place). At the Haven he also began experimenting with shortwaves again, as in his early work, "using as an aerial a metal can elevated a few feet from the ground."[68] He would return to shortwaves in the 1920s; for now, he focused on greater height, greater power, greater wavelength.

❊   ❊   ❊

Marconi's correspondence with his parents indicates how his character was developing in this frenetic period of his life.

Giuseppe had recently asked his son how the company's stock was doing, and Marconi replied that the shares had dropped a bit and were now selling at roughly £4 (still a fourfold increase in just over a year). He advised his father not to sell, as that could cause the price to drop further, but instead to buy more of the new issue they would be putting out shortly to raise capital. Their roles had reversed: Marconi was now counselling Giuseppe on financial matters. He was also politely distancing himself from his father's advice: "Regarding those investments you told me about, I believe that, until I have accumulated a sensible sum, it would be preferable that my capital remains here, especially since I now earn 2% in interest." Furthermore, even bigger profits were in the offing as it was very likely they would soon sell the American rights to his invention. Marconi signed off by warmly inviting his father to join Annie the next time she came to visit him in England. He was now clearly in charge of their relationship.[69]

In his letters to his mother from the same period, Marconi addresses her informally, particularly compared to his greater formality with his father (Marconi would often sign letters to his father with his full name). His letters to Annie are conversational and chatty, and touch on family as well as business and financial matters. In one example, he urges her to insist that Giuseppe "get good clothes" if he comes to England. "Know that some very grand people come to see me and it would not do if he had not nice clothes."[70] Most of all, Marconi's letters to both his parents display a respectful and unshakeable confidence.

Annie's letters to Guglielmo are also revealing of the influence she exercised over her son, as well as of ongoing tensions within the family. She had an interesting way of not so subtly conveying her message, of taking a side while claiming not to take sides. She wrote three times while in Italy for a visit in October 1898. In the first of this set, after the usual niceties she got to the main point, which was that she was surprised that Guglielmo was offering to sell his father two thousand shares at a cut rate. She suggested it might be better for estate planning to sell them to his brother Alfonso instead, for "your papa changes his mind so often that one cannot be sure what he intends to do. He is not pleased at the way your Company affairs are arranged." Annie was also playing the stock prices herself; Alfonso had told her the price had gone down and she said that she would like to buy thirty-five more shares. A week later, she wondered whether Guglielmo could let her have "70 or 80 shares at 1 pound each of your last allotment."[71]

Annie and Alfonso also kept up a steady stream of breezy, newsy letters to Giuseppe from Bournemouth during 1898 and early 1899.[72] Alfonso's letters, as we've seen, were full of detail about the business; Annie's were equally newsy but also motherly. She worried that Guglielmo was so busy that he hardly had time to eat. She planned to go to Enniscorthy for Easter with the Davises (she was now evidently reconciled with her nephew) and hoped Guglielmo would be able to join them. Guglielmo had, she said, a great need of her to help put some order in his life.

Most of the time Marconi was too busy to write, and these letters constitute one of the best sources we have for his activities during this critical year, as he travelled back and forth between Dorset, the Needles, and London almost constantly, when he was not in Salisbury, Dover, Ireland, or France.

❦   ❦   ❦

On October 7, 1898, in London, the company held its first full-fledged annual shareholders meeting; the Liverpool-based *Journal of Commerce* termed it "in many respects the most interesting company meeting of the year."[73] A resolution was passed, unopposed, to increase share capital to £200,000, and another confirmed the statute capping the voting strength of any individual at 40 percent, regardless of the number of shares owned. Marconi, who had been buying shares regularly since the company's formation, now held well over 50 percent of the issued capital and was the only shareholder to which this measure could apply. The resolution carried with his consent.

Chairing the meeting, Henry Jameson Davis reported that Marconi had now managed to transmit and receive signals across a distance of twenty-four miles over land (from Salisbury to Bath) and twenty-five miles from ship to shore—wireless was the only means by which two moving objects could remain in communication, Davis recalled, and its importance for naval and merchant marine work was inestimable. The company now owned twenty-two foreign patents, and seven more were pending; it had ten electrical assistants, some of whom had been sent to do demonstrations in foreign countries where it was required to establish a physical presence in order to retain patent rights (for example, Malta). Expenses were growing, Davis said, and there was still no significant revenue stream; but the commercial business with Lloyd's was continuing to develop and this would help get the system adopted by merchant ships.[74]

One thing Davis did not mention was that the company had turned down an offer to acquire Oliver Lodge's 1897 patent for synchronization, or "tuning" of wireless transmitters and receivers.[75] In July 1898, after abandoning his thought of legal action against Marconi (as we saw in chapter 3), Lodge offered his patent to the Marconi company for the enormous sum of £30,000 (twice what Marconi had received for his patent). The offer was turned down after Lodge refused to demonstrate his invention to the company's satisfaction.[76] Lodge then formed a company of his own, with an associate, Alexander Muirhead. In 1911, the Lodge-Muirhead syndicate sold the tuning patent to Marconi, and Lodge himself became a scientific advisor to the Marconi company.[77]

❀   ❀   ❀

Of all the countries in Marconi's global design at this time the most uncertain was Germany, which saw wireless as an important instrument in its effort to challenge Britain's naval supremacy. The company was still waiting to hear about its German patent application, the most important of the seven foreign patents still pending. Germany was the only European power with the resources and interest to develop a competing system and Marconi was anxious about its intentions. The key to the German puzzle seemed to be held by Adolf Slaby, the gentlemanly Berlin professor who had visited Marconi in Wales in May 1897 to observe his tests at Lavernock, on the Bristol Channel (discussed in chapter 3). Slaby would become alternately a rival and a booster of Marconi in Germany and internationally.[78]

In the fall of 1897, reports from Berlin had begun referring to Slaby as "a new Marconi," and some German scholars to this day claim Slaby as the inventor of wireless.[79] Slaby had been trying for some years to develop a wireless system, with little success, and gave Marconi full credit for what he had achieved. After observing Marconi's tests at Lavernock, he wrote: "What I saw was something new: Marconi had made a discovery."[80]

Slaby was clearly ambivalent about how to treat Marconi. In April 1898, he published an important article in *The Century* magazine, simultaneously crediting Marconi's innovative discovery while establishing his own claims to distinction.[81] Marconi, he wrote, was the first to achieve "the transfer of forces through space" using electric rays, and in a form approaching practical application. Marconi's "electric eye" (a popular term for the coherer/receiver) was "a clever improvement" on Hertz's resonator. Still, argued Slaby, Lodge appeared to have preceded Marconi in the use of such equipment for studying Hertzian waves, and Lodge could fairly be considered "the father of the idea of telegraphing with electric rays." But Lodge had argued that a distance of about half a mile was the outer limit, "without ever having given any practical proof of the theory." Marconi always maintained that there was no absolute limit so long as the equipment could keep being improved. In a sense, Marconi claimed that "the sky's the limit."

In January of 1897, after hearing of Marconi's first successes, Slaby had done his own experiments and not been able to telegraph more than a few hundred feet. "It was at once clear to me that Marconi must have added something else—something new—to what was already known, whereby he had been able to attain to lengths measured by kilometers." Returning home from England, Slaby repeated his previous experiments, adapting his instruments slightly following his observations in England. "Success was instant." Slaby thus laid the basis for claiming his own original improvements on Marconi.

Marconi, meanwhile, kept up correspondence with Slaby, who became his point of connection to Berlin. Slaby offered to help open doors for Marconi in Germany—doors that the German high authorities were determined to keep shut, fearful that the British would use Marconi to establish their supremacy in wireless as they had in cable telegraphy. In May 1898, Slaby wrote to Marconi that he regretted very much that Marconi's German patent application was meeting with difficulties, and offered to help—as a former member of the German Patent Office, he knew very well how to overcome difficulties in such matters, he said. He invited Marconi to Germany.[82]

Slaby was clearly following Marconi's progress closely. In August 1898, he congratulated him on establishing service between Alum Bay and Bournemouth, noted Lord Kelvin's visit, and added, "I hope all English adversaries will now be put to silence and will confess freely, that you have done something more than Mr Lodge and consorts and that your invention is worthy of general appreciation." Then he got to his main point: the German kaiser had become "deeply interested" in Marconi's invention. Slaby asked Marconi for "some details." As he saw the kaiser nearly every month, Slaby maintained that he would have "good occasion, to interest him furthermore."[83]

Slaby was also aware that Marconi was in contact with the German firm Siemens & Halske (S&H), with whom they were both hoping to do business. In an undated 1898 letter to his mother cited above, Marconi mentioned that he was negotiating with "a large German Company," asking £200,000 for the rights to his patent.[84] S&H (as Slaby referred to the firm) would be "a very powerful associate" for Marconi in naval and military matters. Slaby reported that his own research was not achieving as much as Marconi's and repeatedly requested details on Marconi's recent practical "applications"—was Slaby holding back or was this really the case? All of this was intriguing: international politics, commerce, and military research were tied together; Slaby was playing an enigmatic role as he was now the bridge to Germany for Marconi and his invention.[85]

If one were inclined to use an espionage metaphor (at least industrial espionage), one could say that Slaby was trying to "recruit" Marconi—or at least his invention—to German commercial and political service. Slaby seemed to be acting as an agent for Siemens as well as the German navy, with the blessing and urging of the kaiser; he also had his own scientific and financial interests to promote. Exactly what he was doing was never fully clear. We don't have Marconi's side of the correspondence, but it is clear from Slaby's letters that Marconi was on the hook.[86]

The Marconi company held firm. In November, Slaby wrote that he had been told that Marconi's proposals were "so exorbitant" that S&H could not accept the business. S&H was prepared to pay "as high a royalty as your company would like to have"; Marconi, however, was insisting on cash up front and this seemed to be a deal-breaker. Slaby warned that Marconi's obstinacy would delay the application of his system in Germany and "your company will not earn a farthing at all from the German business."[87]

Slaby now introduced a new figure, Professor Ferdinand Braun of Strasbourg (then part of Germany), who was opposing Marconi's German patent claims. While Slaby was dismissive of Braun, he worried that Marconi's main claim was too narrow to withstand an inevitable Braun infringement. "The claim protects only a <u>combination</u> of a transmitter with a narrowly defined receiver. If anybody alters <u>one</u> of them your patent is not infringed." Slaby argued that Marconi needed to make a larger claim. Popov's 1895 article, which Slaby had heard of but not seen, could also be a problem; if the German patent authorities felt that Popov had alleged that his apparatus could be used for signalling purposes, Marconi's patent would not stand up in Germany.[88]

By January 1899, Marconi's German relations were deteriorating. Siemens & Halske were waiting for the German patent ruling, after which they intended to go into business without him; the military was pursuing its research, in conditions so secret that even the kaiser only heard of it after it was a *fait accompli*. Slaby began to distance himself: "I myself have no connexion at all with the navy authorities and must decline to offer my services in order to remain independent....As far as I know, it is excluded that foreigners will be admitted to the experiments. The only way for you to get into business with our navy is by a German firm."[89]

When Marconi received this letter, he replied to Slaby immediately, hinting that he had his own direct channels to Siemens & Halske and that Slaby must have been misinformed.[90] However, he was wrong and the Marconi company did not establish a foothold in Germany (with the exception of supplying some of the major German shipping lines—in fact, North German Lloyd was the first commercial shipping company to adopt Marconi's system. Its SS *Kaiser Wilhelm der Grosse* was equipped with Marconi wireless in February 1900, days after the Marconi company opened the first German wireless station on Borkum Island.[91])*

There would soon be an independent German system, under the patronage of the kaiser. After Marconi's failure to establish a commercial arrangement with Siemens & Halske, S&H began working with Ferdinand Braun, while Slaby and an associate, Count Georg von Arco, joined forces to develop a system based on Marconi's apparatus, under the aegis of the German electrical company Allgemeine Elektricitäts Gesellschaft. The Slaby-Arco and Braun interests merged in May 1903, at the urging of the kaiser, to form the Gesellschaft für drahtlose Telegraphie—known to this day as Telefunken.[92] Telefunken quickly became a potent international alternative to Marconi's company. The rivalry became tied up in great-power politics and would lead to the introduction, at German insistence, of a global regulatory regime for wireless in the first decade of the twentieth century.

❊   ❊   ❊

Marconi's international plans had begun to take shape almost immediately after the company was formed. At one of the very first board meetings on September 30, 1897, board member William Smith expressed the view that

---

* The German shipowner North German Lloyd had no association with Lloyd's insurance company; the latter was so predominant in marine insurance that "Lloyd" was actually a generic term for shipping company in the late nineteenth century.

they should set "a practical value" on the invention and market the patent to other European countries. He thought they would be able to get £100,000 and a small royalty in the largest countries. This soon evolved into the view that the company should create national subsidiaries and keep control of the invention itself.[93]

Expressions of interest came in from various quarters. In March 1898 Davis informed the board that he had received a visit from the Austrian naval attaché and had arranged for Marconi to do a personal demonstration on their ships. Marconi would receive expenses and fifteen pounds a week, as well as the expenses of an assistant, but the demonstration doesn't seem to have ever taken place. By this time, at any rate, he was certainly far too busy.[94] Foreign governments were closely monitoring Marconi's work; the Portuguese naval military club, for one, had published the results of his trials as early as 1898, and the Portuguese media were following his affairs closely.[95]

By December 1898, the company had reached a comfortable cruising speed. That month it set up the world's first wireless factory in Chelmsford, Essex, to manufacture equipment—in anticipation of contracts yet to come. No one seems to know why they set up shop in Chelmsford; the only indication in the Oxford archives is a note from the 1960s suggesting that maybe it was because "Colonel Crompton knew Marconi."[96] Also in December 1898, a few days before Christmas, the company installed a first wireless set on a lightship—the *East Goodwin*—off Dover, with a shore station at the South Foreland lighthouse. Marconi now decided he was ready to try to get a message across the English Channel.[97]

On March 2, 1899, Marconi graduated, so to speak, to a new level of credibility when he presented his first scholarly paper, to the Institution of Electrical Engineers, in London. The hall was packed to overflowing and Marconi's paper was unadorned if also full of technical detail. He was asked to present the talk a second time to satisfy those who had been turned away—a rarity indeed for an academic lecture.[98] The lecture ended with "quite a dramatic little incident," as Marconi, at the lectern, was handed a telegram informing him that the French government had granted permission to make the first attempt to send a wireless message across the English Channel from France to England.[99]

Marconi immediately went to work, building a station with a 150-foot mast on the beach at Wimereux, a coastal village near Boulogne-sur-Mer (where his parents had married in 1864). On March 27, 1899, a message was sent across the thirty-two-mile channel to Folkestone, with Marconi himself at the transmitter and a London *Times* reporter present. Public interest in the event was immense and the press was beside itself with high praise. The *New York Times* reported that the wireless channel crossing had "astounded the world," and now anointed Marconi as history's pick to be considered the inventor of wireless, despite the recognition due to precursors like Tesla, Righi, and Lodge. Men who do something practical are usually the ones who eventually get the credit, the *Times* philosophized.[100]

Crossing the English Channel had tremendous symbolic value and Marconi's supporters were also ecstatic. "It is a splendid triumph notwithstanding the presence of the group of French skeptics," wrote Agnes Baden-Powell, whom Marconi had befriended after her brother's visit to Salisbury in 1897.[101] Of all the accolades, the one that must have pleased Marconi most was the letter he received from William Preece, then in Paris: "My dear Marconi," Preece wrote, "I congratulate you very much on your success between Folkestone and Boulogne. I never had a doubt of it."[102] Preece was now so confident, he wrote, that he had recently promised the governor of Gibraltar that Marconi would one day establish communication between the Rock and Tangier, a formidable thirty-four miles away. A month later, a French government ship equipped with the Marconi apparatus reported that it had successfully communicated by wireless over forty-two miles.[103] Distances were getting longer, doubling nearly every month. Wireless had come a long way since Oliver Lodge declared that it had a maximum limit of half a mile. The world was getting smaller.

A few months later, in July 1899, Marconi took part in the Admiralty's annual manoeuvres, working his apparatus on three navy vessels, including the *Juno*, under the command of Captain Henry Jackson. The exercise took the form of an encounter between two fleets, one equipped with wireless and the other not. Marconi's system worked to everyone's satisfaction; messages were sent consistently over a distance of eighty miles—nearly twice the previous record—and on one occasion, 126 miles was reached.[104]

That summer, Marconi was also working frequently in Dover, doing transmission experiments between stations separated by mountains or other obstacles. In August, James Ambrose Fleming wrote to Davis suggesting he (Fleming) use the tribune of an upcoming British Association meeting in Dover to demonstrate Marconi's latest findings: "It will effectually dispose of those people (like Lodge) who have been saying that the magnetic deduction telegraphy is the only one which will work through obstacles and I know the experiment will excite great interest."[105]

While determined to keep his plans secret, Marconi was also mindful to remain in public view. He liked Fleming's idea, especially as the British Association meeting in Dover was set to coincide with a conference of the Association Scientifique de France in Boulogne and an academic congress in Como, Italy, marking the centenary of Volta's invention of the electrical battery. Marconi orchestrated an exchange of messages between the three groups; scientists of three countries were impressed and, of course, the press made much of the event.[106]

❦   ❦   ❦

In June 1899, *McClure's* magazine published Cleveland Moffett's long article on Marconi and his invention.[107] *McClure's*, a monthly founded by S.S. (Samuel Sidney) McClure and John Sanborn Phillips in 1893, was one of America's bestselling periodicals; it would make journalism history in the early 1900s with a new style of critical investigative reporting that came to be known by a term coined by President Theodore Roosevelt in 1905—"muckraking"—after publishing exposés of US cities and the steel industry, among other things. It also pioneered the new print technology of photo-engraving, and Moffett's article was illustrated by a striking portrait of Marconi, looking like a cross between Mark Twain and a prairie politician—more like young Abe Lincoln than a budding businessman. (Marconi actually met Mark Twain in June 1899, at a dinner at London's Savage Club.[108])

In the late 1890s, *McClure's* had not yet turned toward muckraking exposés, specializing instead in articles on people who were shaping the world, like Alexander Graham Bell and Thomas Edison. Moffett and his colleague Ray Stannard Baker, who joined the magazine in 1897, would author some of the earliest and most important articles introducing Marconi to the American public. Moffett's piece was the first not to be distracted by Marconi's youthful demeanour and unusual character traits, and to focus instead on his surroundings, equipment, staff, and prospects. The article was a total, uncritical boost of Marconi, speculative but remarkable in its foresight.

Moffett provided one of the most detailed descriptions to date of Marconi's apparatus: "…the relay and the tapper and the twin silver plugs in the neat vacuum tube, all essential parts of Marconi's instrument for catching these swift pulsations in the ether. The tube is made of glass, about the thickness of a thermometer tube and about two inches long. It seems absurd that so tiny and simple an affair can come as a boon to ships and armies and a benefit to all mankind; yet the chief virtue of Marconi's invention lies here in this fragile coherer."

As other journalists had done, Moffett noted Marconi's practical approach and his near indifference to the nature of the waves he used; it was enough to say they worked, and he was far more engaged with the results: "My system allows messages to be sent from one moving train to another moving train or to a fixed point by the tracks; to be sent from one moving vessel to another vessel or to the shore, and from lighthouses or signal stations to vessels in fog or distress," Marconi told Moffett, describing what we now call mobile communication. He emphasized the advantages of wireless: over water, it was less costly and easier to maintain than cable; over land, it eliminated the need to get the right of way to put up wires and poles; and in war, it would provide "for quick communication, and no chance that the enemy may cut your wires." The system might even replace newspapers: "The news might be ticked off tapes every hour right into the houses of all subscribers who...would have merely to glance over their tapes to learn what was happening in the world." As we do today when accessing online news on our telephones or computers, of course.

Marconi was now one of the bellwethers of a coming new era. Not only was he shaping the future with his wireless, he was also coming to symbolize it. A popular Australian period author, Max Rittenberg, captured this retrospectively a few years later in his novel *Every Man His Price*. Set at the turn of the century, the story opens in the London bachelor flat of the fictional financier Sir Wilmer Paradine, sharing cigars after a dinner à deux with the young scientist Hilary Warde (a barely disguised stand-in for Marconi). Warde, talking about the prospects of the company he works for, tells the older man: "Wireless telephony will blanket the earth in ten years' time. Just now it's only in its infancy. We're pioneers. We're making history. We hold the inner secrets...."[†] The conversation continues:

PARADINE: You dream of a world monopoly, don't you?
WARDE: Yes....I don't want it only for myself or my firm. That would be a poor ambition. I want it for England. I want to see England mistress of the ether! ...There are experimenters all over the world working at the practical problems of wireless telephony. It's excessively complex. I know we've got a long start...

[†] Within the time frame of Rittenberg's novel, his characters are actually talking about wireless telegraphy. The term *wireless telephony* only came into use a few years later, to describe the early practice of voice transmission which eventually evolved into *radio*.

And the narrator concludes: "[Paradine] had long since decided that the young scientist was essential to his financial schemes. He believed in brains coupled with high purpose—such a man would be far more useful to him than brains alone."[109]

This was fiction, but not far off the mark of where Marconi stood as the Victorian era was coming to its end. He was now moving in English social circles at a level suitable for his life and ambition, certainly within the social mainstream but also at the leading edge of change. As someone rising in social standing, radical in some respects while conservative in others, Marconi was ideally suited to the dawn of Edwardian England; his budding acquaintanceship with the Prince of Wales was a concrete sign. The future Edward VII consorted with many whom the old establishment considered upstarts—his nephew the German kaiser, for example, on hearing that his uncle was sailing with self-made tea mogul Sir Thomas Lipton, snorted that Edward was "boating with his grocer."[110] Wilhelm II did not think much of Marconi either, but Marconi and Lipton got along well. Marconi did not disdain upstarts; after all, he was one himself.

Marconi had been on a non-stop experimental blitz for just over two years, since returning to England from Italy in July 1897. He was gradually increasing the distance over which he could communicate, illustrating each incremental gain with dramatic moves like crossing the English Channel. As his transmission capacity expanded internationally, so did the corporate empire he was building. The two went hand in hand. He had enjoyed a string of successes, but it was time to focus on consolidation.

Communication with America was Marconi's next big goal. In August 1899, Fleming informed him: "I have not the slightest doubt I can at once put up two masts 300 feet high and it is only a question of expense getting high enough to signal to America." Marconi was now confident enough to tell a journalist he was "not unhopeful" that he would one day be able to send signals between England and the United States.[111] But first he would go there himself. On September 13, 1899, Marconi sailed from Liverpool on the steamship *Aurania*, with William Goodbody and three assistants. They arrived in New York eight days later.

5

New York: New Frontiers

Despite a harrowing work schedule, when the *New York Herald* invited him to cover the America's Cup yacht races off Long Island, Marconi crossed the Atlantic in the fall of 1899. He was fully expected. "I arrived in New York on September 21st and had to run the gauntlet, as soon as I had descended the gangway, of the numerous reporters and photographers who were awaiting me," read a memoir later ghostwritten for him.* The newspapers were surprised to learn that he spoke English fluently, "in fact 'with quite a London accent,' as one paper phrased it," and that he did not in the slightest resemble the popular notion of an inventor, "that is to say a rather wild-haired and eccentrically-costumed person."[1]

New York was in a technological boom. A few years earlier, Thomas Edison had installed electric lighting in the downtown core, as well as the first central power plant in the United States at his Pearl Street station near the Brooklyn Bridge. Edison was aware of Marconi and his impending visit; in May, he wrote to a friend that he understood Marconi was coming to America and that he would "probably produce something practicable here."[2] Marconi established his American subsidiary's headquarters on Broad Street in Lower Manhattan just a few months later.

New York was teeming with newly arrived immigrants, but it is not likely that Marconi encountered many steerage passengers on the crossing.

---

* In 1919, the company commissioned Marconi's private secretary, Leon de Sousa, to write Marconi's "autobiography" based largely on documents available in its files. Although never approved for publication, the text was written under Marconi's direct guidance and the company considered it to be "factually accurate" (G.G. Hopkins, cover note bound with the de Sousa manuscript, dated January 1959, in OX 55.) While de Sousa took many liberties with the facts and the manuscript is often not in Marconi's voice, this extract is particularly evocative of the excitement Marconi may well have felt upon arriving in New York City, an international media celebrity at the age of twenty-five.

His eye was more attracted to refined young ladies. Marconi, sporting a fashionably rakish moustache, cut a dashing figure in the bustle of midtown New York so well captured in Alfred Stieglitz's 1900 photograph, *The Street, Fifth Avenue*, which shows a city of elegant thoroughfares lined with horse-drawn carriages, their drivers in cloaks, and a thin layer of snow on the ground.[3]

Marconi and his entourage stayed at the stylish Hoffman House, on Broadway and Twenty-Fourth Street, just across from Madison Square. (A few years later, in 1902, the area was forever changed with the construction of the wedge-shaped twenty-two-storey Flatiron Building at Broadway and Twenty-Third.) Built in 1864, the Hoffman House was one of Manhattan's liveliest hotels, playing host to political, show business, and business people; Grover Cleveland lived there during his second presidential election campaign in 1892. The hotel bar was "one of the sights of New York which no visitor missed," according to the *New York Times*.[4] One of the bar's attractions was the valuable paintings on its walls, among them Bouguereau's *Nymphs and Satyr*, which was displayed there until 1901.[5][†] Said to typify the atmosphere of New York in the Gilded Age, the hotel went bankrupt in 1910, but its creditors kept it open and its rooms fully occupied until the building was sold to developers in 1915.[6] Marconi stayed there again in 1902, and later at the eight-storey Holland House a few blocks north, on Fifth Avenue at Thirtieth Street, preferring these smaller, slightly edgy hostels to grander, flashier hotels such as the Waldorf.

Marconi liked the Madison Square area, lunching on the hotel rooftop (or downtown at Delmonico's), dining at nearby Muschenheim's Arena, taking in a vaudeville show at Koster and Bial's Music Hall with one or another of his business associates or staff. One of the three assistants accompanying him, William Densham, kept a detailed diary that gives some idea how the group worked as well as of some of Marconi's movements and activities during his first New York sojourn.[7]

He was there as, of all things, a newspaperman—to report on the yacht races off Sandy Hook, south of the city in Lower New York Bay. As a result of his coverage of the Kingstown regatta, Marconi was approached by James Gordon Bennett Jr., the publisher of the *New York Herald* and an avid

---

[†] The painting was sold in 1901 to a buyer who found it "offensive" and kept it from public view; it resurfaced in 1943 and is now owned by the Clark Art Institute in Williamstown, Massachusetts.

yachtsman, and offered $5,000 to report the best-of-five series between the New York Yacht Club's *Columbia*, defender of the Cup, and Sir Thomas Lipton's challenger, *Shamrock*, of the Royal Ulster Yacht Club. At first Marconi demurred, as he feared it would take his focus away from his research. But once he had bridged the English Channel in March 1899, he became more interested in working in the United States and struck another of his trademark symbiotic relationships with powerful people. Bennett was notorious for making news happen. He had sent the explorer Henry M. Stanley on his legendary expedition to find David Livingstone, the Scottish missionary who had disappeared somewhere in central Africa. Bennett's paper had been the first to report on the inventions of the telegraph and the light bulb and the *Herald* had been a pioneer in using the transatlantic cable for publishing news. Bennett also understood the importance of closing the gap between transportation, travel, and communication. Marconi shared these views.[8]

The yacht races were due to begin on Wednesday, October 3, and there was a lot to prepare—the SS *Ponce* had to be fitted with 150-foot masts in order to do the transmission. While Marconi and his staff were working, word came that the ship carrying Admiral George Dewey, returning hero of the battle of Manila Bay in the Spanish-American War, was about to enter New York Harbor. Marconi had been persuaded to go out in a hired tugboat to board Dewey's ship and report its arrival but the plan was thwarted when the ship arrived two days earlier than expected. Marconi nonetheless got to send a message on behalf of a US Navy officer from ship to shore during Dewey's naval parade on September 30, 1899, said to be the first paid ship-to-shore telegram.[9] "As we went up the river cheers were hurled from every ship we passed," he later wrote.[10] This is how the *Herald* reported Marconi's first US performance:

> From the chart room of the steamship Ponce messages were flashed by wireless telegraphy yesterday afternoon across the city and over New York harbor and bay to the station at Navesink Highlands....
>
> [Signor Marconi] was greatly hampered in his work, but never complained. Questions were always answered and few of the women passengers went ashore without first having had the system of wireless telegraphy explained to them in all its intricacy by its inventor. Many of them were more interested in the man than in his work. They met a young man with an erect, athletic figure, an Irish cast of countenance, and a speech which was filled with English mannerisms. His voice was low, well modulated, and filled with earnestness....

Signor Marconi was well pleased with his day's work, but had little to say about it. "You may announce", he said to me, "that the tests made to-day on the Ponce were as fully successful as I had expected."[11]

The experience was a turning point in the US Navy's view of Marconi's system. A navy observer said: "This is no experiment. Signor Marconi shows by his every move that he has passed the stage of uncertainty....In the near future wireless telegraphy will be in general use by the navies of the world. Its value cannot be too highly estimated."[12]

The day the races began, Densham picked up Marconi at the Hoffman House and they set out for the docks. It was a fine, warm day and the New York crowds were excited. Two days earlier, two million people had lined Broadway to cheer Dewey. Now *le tout New York* watched the screen in Herald Square for minute-by-minute news of the first big race between the *Shamrock* and the *Columbia*.

Throughout the day bulletins written by *Herald* reporters were sent by Marconi and his assistants from the *Ponce* to a shore terminal at Navesink, New Jersey, to then be transmitted over telegraph wires to the *Herald* offices in midtown Manhattan. This "relay" took unnecessary time, the *New York Times* said grumpily, but results were still reported within one minute of when they were sent. The wireless transmitted twenty-five hundred words in all during the day. In comparison, the "regular" cable system sent twenty-five thousand words per day, but only Marconi's were sent directly from the scene. In addition to race coverage, private messages were sent and stock market quotes received. The *New York Times* framed Marconi's coverage as a "test," but even they reported that the results were "eminently satisfactory."[13]

Again, the *Herald* outdid the competition in lauding the wonders of its new content provider: the successful operation of Marconi's wireless telegraphy system was "far and away the greatest achievement connected with the yacht race," the paper reported. Several military officers were on board the *Ponce*; one of them, Navy Lieutenant Commander Edward F. Qualtrough, "became an enthusiast early in the day, and before nightfall was declaring his belief that the US government would do well to persuade Signor Marconi to install his system in the Philippines at the earliest date. 'If we could only have had this last year [during the Spanish-American War],' said Qualtrough, 'what a great thing it would have been.'" The *Herald* headlined: "Wireless bulletins worked like magic."[14]

That evening Densham and some cronies celebrated with a tour of New York's tenderloin joints.[15] Marconi stayed in; he still had work to do.

On October 9, Marconi's team dismantled the station on the *Ponce* and fitted up a new one on the excursion steamer *La Grande Duchesse*. The next ten days were more or less the same, mostly testing, fitting, testing, fitting, as Densham reported in his diary. The American scene was less stiff and secretive than the British, more open and popular than the Italian, and Marconi found he was exposed, for the first time, to rival systems that he was not even fully aware of. At one point he was in communication for a full hour with a nearby vessel fitted with unfamiliar equipment.[16] This was a *faux pas*, as the company would soon be insisting that "intercommunication" between transmitters and receivers based on different technical systems could not work. This sketchy claim would be critical to the company's efforts to establish a global monopoly, and its opponents (both government and commercial) would invest much energy disproving it.

The publicity Marconi received was priceless, and also precisely what he wanted. Already known to specialists in the United States, his name now became a household word. He was in the press nearly every day for over a month, often in the company of national heroes like Admiral Dewey. He was "entertained at lunch" by the director of the Metropolitan Museum of Art.[17] The Italian Chamber of Commerce feted him with a banquet where speakers declared Marconi to be one of Italy's gifts to the world, putting him on a par with Michelangelo and Columbus. Possibly perplexed, but getting used to this type of hyperbole, he appeared nonplussed and gave what was by now his stump speech on the history of wireless telegraphy.[18]

Marconi's efforts were closely followed by American scientific and trade periodicals, and four agencies of the US government were now considering the possibilities of using wireless as an extension of their cable and telegraph facilities: the Lighthouse Board, the Weather Bureau, the army, and the navy. Each agency had its own agenda and was doing its own investigations of wireless. The Weather Bureau, for example, hired Reginald Fessenden, a professor at the University of Western Pennsylvania in Pittsburgh and soon to become one of Marconi's most persistent rivals, to help them develop their own system.[19]

However, what really captured Marconi's interest was the attention he was suddenly getting from the US military. Both the army and the navy had been closely following the development of wireless in Europe and now declared that they wanted to take advantage of Marconi's presence in the United States to have him do tests for them as well. In a substantial and generally very sober report to the US secretary of war, the army's chief signal

officer, brigadier-general A.W. Greely, wrote that Marconi's "inventive genius and persistent application have demonstrated the practicability of this
system to the civilized world." Greely foresaw wireless telegraphy replacing
the use of cables, "with equal advantage to commerce and the Army," for
connecting harbour fortifications, communication with islands, as well as to
and from lighthouses, ships, and signal stations, both on coasts and at sea.
Greely's report stated that arrangements had been made for Marconi to su-
pervise some army wireless experiments, but curiously these never took
place.[20] A few days before Marconi left New York, Greely told the *New York
Times* the army had decided not to use his system.[21]

As in Britain and Italy, the US Navy was the government department
most interested in Marconi's work. Following the British naval manoeuvres
of July 1899 (discussed in chapter 4), the American naval attaché in London
was instructed to find out from the Marconi company how much it would
cost to have a wireless demonstration on shipboard, and the purchase price
for two sets of equipment. The company's reply, on September 2, 1899, indicated its policy—a demonstration could be done for cost but the equipment was not for sale.[22]

On October 23, 1899, Secretary of the Navy John Davis Long appointed a special board for the purpose of investigating Marconi's system.[23]
On October 26, the first full day of tests, a fifteen-hundred-word newspaper
article was transmitted across five hundred yards. Again, the press was called
in to witness and chronicle the event. Critics flagged the short distance and
problems with speed and accuracy, but Marconi pointed out that these were
due to the limitations of the operators, not the system. The *New York Times*
remarked that "there was a trace of deserved sarcasm in the inventor's tone"
when he said this.[24] There were a lot of questions and Marconi had an answer for everything. But his best answer was to show what he could do. By
November 2, he had managed to communicate between ships thirty miles
apart.[25] This was significantly less than what he had achieved in the British
naval manoeuvres a few months previously (and perhaps this accounted for
Marconi's impatience), but the US authorities needed to witness it for
themselves. Indeed, a Marconi company circular put out for potential US
investors at this time announced that wireless messages could now be sent
"with certainty" up to one hundred miles on water and at least fifty miles
on land.[26]

Although the navy was impressed with the potential value of the system, Marconi's elusiveness when questioned blocked them from striking a

deal. One of the naval experts assigned to follow the tests reported to his superiors that his relations with Marconi and his staff had been "most pleasant, but at no time did any of them describe or explain the apparatus more fully than had already been done in public print, and their answers to questions never gave any information beyond this." When they didn't want to answer a question, they would say they had to withhold the information because a patent was still pending but "frequently the answers given by different members of the party were conflicting, and sometimes it appeared that their answers were intended to mislead."[27]

Marconi's insistence on secrecy, and the way he justified it, looked like a shell masking the limited state of development of his system. The navy's official historian, Linwood S. Howeth, cites a letter from Marconi, dated October 29, 1899, in which he explained why he was unable to demonstrate how he could prevent interference, or tune and synchronize his system's instruments. "The reasons why I cannot give such demonstrations are: (a) the means employed are not yet completely patented and protected, (b) insufficient material and instruments here with me to give a full demonstration, and (c) no detailed information from the United States Navy Department was received by my Company, prior to my departure from England, as to the extent of the demonstrations required."[28] The instruments he had with him were strictly what he needed to carry out his contract with the *Herald*; they were not sufficient for the large-scale test the navy was seeking. This was Marconi being typically disingenuous—as we saw, the company had had detailed correspondence with the navy before the trip—and it was a bad omen for the possibility of any eventual business between the two parties. The navy was left with a lasting impression that Marconi would be difficult to deal with.

Despite this disappointment, the navy's Radio Telegraph Board recommended on November 4, 1899, that the navy try Marconi's system. However, the trials were pre-empted by his abrupt departure, when he was ostensibly called back to England because of the looming Anglo-Boer War in South Africa. At least, that is what the navy was told; there is no documentation to support this claim, but it was Marconi's stated reason for leaving America when he did. Marconi also turned down an invitation to visit Alexander Graham Bell at his home in Cape Breton, Nova Scotia, stating the same reason. Bell had gotten in touch with Marconi soon after he arrived in the United States, but the letter of invitation apparently went astray and Marconi only acknowledged it on the eve of his departure, politely declining the invitation because, he said, he had to return to England due to the British government decision to use his system in its South African war.[29]

Marconi had spent two and a half years making improvements to his system, publicizing it with a string of media events in collaboration with a sensation-hungry press, consolidating his patents commercially, and cultivating interest among several military powers. While he was in the United States, he was quoted as saying, "I'd like to try the system in war."[30] He was about to get his chance.

❁   ❁   ❁

As the nineteenth century was ending, the extent and ambitions of the British Empire appeared to be limitless. In southern Africa, the British fought two wars against the Dutch Afrikaner, or Boer, settlers of two independent republics, the Orange Free State and the Republic of the Transvaal. The two Anglo-Boer wars—or "freedom wars," as they were known to the Boers—took place against a backdrop of growing imperial rivalry in southern Africa, where the British (with strong footholds in the Cape Colony, Natal, Rhodesia, and Bechuanaland) feared a developing alliance between the Boer republics and German Southwest Africa. The Anglo-Boer wars were also overlain with the resistance of indigenous South African peoples to European colonialism. Britain abolished slavery in the Cape in 1834, a move regarded as unnatural by many Boers.

The first Anglo-Boer War, in 1880–81, was precipitated by the British annexation of the Transvaal in 1877 and ended with the re-establishment of self-government for the Transvaal (which now became known as the South African Republic, or SAR). The second, from 1899 to 1902, followed a massive British military buildup in the cape, which was seen by the Boers as an evident prelude to a new move to take over their republics. War is by definition hell, and the South African wars left their own uniquely terrible legacies, most memorably the British "scorched earth" policy of destroying farms and the internment of civilians in concentration camps. The wars ended with the demise of the Boer republics and their absorption into the British colonial system.

The Marconi company's approach to the second Anglo-Boer War tells us much about the uneasy coexistence of patriotism and the profit motive in colonial business practices at the turn of the twentieth century.[31] This would be the first time that wireless communication was used in a military conflict.

The SAR government began looking into wireless in the winter of 1898, when its general manager of telegraphs, a well-informed and memorable figure by the name of C.K. van Trotsenburg, wrote to the London office of Siemens Brothers (a branch of Siemens & Halske) asking if they would be able to supply instruments for communicating "telegraphically without wires" between a given spot in a valley and different points in the surrounding hills. There was no mention of the intended purpose for the communication. It is notable that van Trotsenburg should write to Siemens in London, as he could have addressed the query to the parent German company's agent in Johannesburg. However, he had a reason for going through London first: "Of course we require the best known instruments of this class, with all the improvements which have since been introduced in the instruments of Marconi."[32]

While he waited for Siemens to reply, van Trotsenburg wrote to his superior, SAR State Secretary L.W.J. Leyds, with a proposal regarding telegraphic communication between military camps. Van Trotsenburg favoured "the erection of an overhead line, to be worked with an ordinary telegraph or telephone instrument or perhaps with both," but added that communication "without wire (emphasis in original)" could be desirable for distances of six miles or so. Van Trotsenburg pointed out that large-scale wireless experiments were being conducted by the European powers, and that lately important improvements had been made. "I would suggest that I communicate with manufacturers and in case of satisfactory information being received, to order one set of instruments for trial," he wrote. The costs would be comparatively low, compared with cable.[33]

Meanwhile, Siemens's London director, Alexander Siemens, replied on March 26, 1898, that the exploitation of the Marconi system in England was in the hands of a company that owned the patents. There should be no great difficulty doing the installation the South Africans desired. However, Siemens noted, the Marconi company refused to sell the equipment outright but proposed a leasing arrangement instead, consisting of an initial payment and an annual fee; they also insisted on sending their own staff to install the equipment at the customer's expense. This, as we've already seen in the US case, would become the company's preferred model for government contracts. Siemens added that the company was anxious to know who was expressing interest, "but as you had marked your letter 'confidential' we did not feel at liberty to satisfy them on that point."[34] The Marconi company might be willing to do business with anyone, but they wanted to know with whom they were dealing.

Leyds replied to van Trotsenburg on April 20, requesting that he correspond with his suppliers in Europe. Van Trotsenburg then wrote directly to Siemens & Halske in Berlin, asking whether they could provide SAR with the necessary equipment. S&H referred the letter to their London office, and on June 11, 1898, Siemens London replied on Berlin's behalf. They had again met with the Marconi company but regretted to say "we can make no progress with them." It seemed the job could be quite easily done, but "The Company decline to name the sum to be paid as annual royalty unless they are made fully acquainted with the purpose for which the instruments are to be used." Siemens reiterated that the company would not sell the apparatus, only the right to use it; he had pointed out to them that this was not common practice, but they said they saw no other way to make sufficient profit. The Marconi company was obstinately sticking to their business model, which would allow them to retain full proprietary rights over their equipment and the underlying technology. This was outrageous, Siemens said. As had previously been the case with the US Navy, for similar reasons, Siemens was piqued: the company's insistence on retaining control over the apparatus seemed to be "inconsistent with the free use you would naturally expect to obtain for your outlay." They had begun making enquiries about using Lodge's system, although it was not yet commercially available, but clearly, they still saw Marconi's system with all its flaws as the leader in the new industry.[35]

Nothing substantial happened for over a year. S&H referred van Trotsenburg's letter to its South African agents in Johannesburg, who on June 21, 1898, offered to sell the SAR government some units. Van Trotsenburg also approached a French company, La Société industrielle des téléphones, which offered a quote for equipment.[36] Then there were no further exchanges until van Trotsenburg visited London, Berlin, and Paris in June 1899.

On June 30, 1899, van Trotsenburg met in London with Henry Jameson Davis. Davis wrote to him the following day, confirming that the company "would be willing to supply instruments for one or more installations of wireless telegraphy to your Government," on a royalty basis. The interesting thing here is that this offer came in the context of the brewing storm and rumblings of impending war between Britain and the Boer republics. If the company had any patriotic concerns, the least one can say is that these were not expressed nearly as forcefully as their business conditions.[37]

On August 24, 1899, van Trotsenburg wrote to his superiors explaining that he was opting for German-made Siemens equipment. The French installations he saw were too slow and not likely to be improved; the Marconi system "is still a novelty"—specified tasks could not be done; sometimes it worked and sometimes it didn't. The German equipment was likely to work best. Although this may have been an error in judgment on his part, it at least shows the prevailing ambivalence toward Marconi's system at that time in military circles. Siemens's conditions were much more favourable, about the same for a one-time purchase price as Marconi proposed to charge for an annual licence fee. The South Africans placed an order with Siemens for six sets of instruments. The need for secrecy was underscored: the instruments were "not to be shipped as if for the S.A.R. but as for the Agency of Messrs Siemens Ltd Johannesburg. Should, however, in case of war with England these instruments be confiscated or destroyed by the enemy, then the Agency will receive from us a reasonable refund for same."[38] This is considered to be the world's first military order for commercial radio equipment.

War between South Africa and Britain was now imminent. There is no record of the Marconi company's sentiments on having lost the Transvaal contract to Siemens & Halske; they may well have been relieved. At any rate, Marconi's comments in New York certainly left his American hosts with the impression that he was eager to serve the British war effort. While Marconi was reporting the yacht races off the coast of New Jersey, the South Africans issued an ultimatum, accusing the British of breaking the London Convention of 1884 (which had ended the first Anglo-Boer War) and interfering with South Africa's internal affairs. The ultimatum was ignored and on October 11, 1899, the governments of South Africa and the Orange Free State declared war on Great Britain. When the equipment ordered by van Trotsenburg finally arrived in Cape Town, it was promptly confiscated by British customs. Had SAR waited a bit before declaring war, it could well have become the first country in the world to use wireless in wartime—and it would have been Siemens, not Marconi equipment.[39]

As it turned out, the Marconi company seized the opportunity. Three months after bidding to supply wireless equipment to SAR, the company received its first actual contract from a government agency, the British War Office. Six Marconi engineers and five portable wireless sets left Southampton on a troop ship on November 2 (while Marconi was still in the United States) and arrived in Cape Town three weeks later, on November 24.[40] Soon after Marconi returned from the United States on November 16, 1899, it was reported that he would be going to South Africa himself, but, as we've seen, he was not interested in making the trip, fearing that it would distract his focus from his research.[41]

The equipment was designed to be used on ships, but with the help of army personnel from the Corps of Royal Engineers (known as the "Sappers") and—once again—kites supplied by Marconi's friend Baden Baden-Powell, the sets were modified for mobile land use at the battle front. Royal Engineers Captain J.N.C. Kennedy, who had been at Salisbury and was familiar with the Marconi system, cannibalized the S&H sets destined for the Boers which had been confiscated on arrival in Cape Town, and on December 4, 1899, successful demonstrations were performed under his oversight. However, it soon proved impossible to successfully deploy the equipment. While contact was established over fifty miles, about as far as could then be covered over land, further attempts were thwarted by dust and lightning storms, and the machines were too bulky for use in the field. Marconi's chief engineer, George L. Bullocke, reported back to London that the men were working in temperatures of 104 degrees Fahrenheit in the shade.[42] After six weeks, the army felt that the system wasn't functioning.

Marconi went on the offensive. In a lecture delivered before a large and enthusiastic audience at London's Royal Institution on February 2, 1900, he defended his engineers and his equipment and blamed the failure on the military authorities.[43] The lecture hall was "literally packed from floor to ceiling," he later wrote, and was so crowded he had a hard time getting in the door himself.[44] Marconi was in fighting form. First, he qualified the War Office's request for his company's assistance as "tardy." Then he implied that the army didn't know how to operate the equipment; had he himself been in South Africa, he said, he would have refused to try to operate the system without the help of Baden-Powell's kites. Finally he commented caustically on erroneous reports that the confiscated Siemens apparatus intended for the Boers had been supplied by his company. "I need hardly add that as no apparatus has been supplied by us to any one, the Boers cannot possibly have obtained any of our instruments."[45]

Marconi may not have been as diplomatic as he was later made out to be; here (as in some of his remarks during the US Navy exercises), he appeared brash and abrasive as well as impatient and inexperienced. He was also being disingenuous—another of his trademark character traits: while he was correct in stating that his company had not supplied SAR, he elided the fact that that was not by its own choice but the decision of SAR itself to order Siemens equipment instead of Marconi's. At any rate, the British director of army telegraphs was miffed by Marconi's comments and ordered the instruments withdrawn from service. In hindsight, Marconi considered the army withdrawal a fortunate disaster, as the British navy promptly asked to take over the equipment the army had rejected. The five Marconi sets in South Africa were installed on cruisers blockading Delagoa Bay (at present-day Maputo, Mozambique; then Lourenço Marques) in March 1900, and their performance in those circumstances was much more satisfactory.

As the war entered a new phase a few months later, there was no further need for a naval blockade and the Marconi equipment was withdrawn from service and put into storage. But the British experience in South Africa undoubtedly contributed to the Admiralty's decision to equip twenty-eight warships and four land stations with Marconi equipment a few months later, in July 1900.[46] The South African war continued until May 1902, and the Siemens equipment was eventually sold at auction, to a collector. It is now housed in the War Museum of the Boer Republics in Bloemfontein, South Africa.[47]

❧   ❧   ❧

Back in the United States, the chief of the Bureau of Equipment reported to the secretary of the navy on December 1, 1899, that the Marconi system appeared to be successful and well adapted for navy use; the main objection to it was the "interference" that occurred when a third station tried to transmit within the range of two communicating stations. Notwithstanding this problem, the bureau opined that the system "promises to be very useful in the future for the naval service." It was already being used by the Italian, French, and British navies, as well as by the British army in South Africa and at several lighthouse and shore stations in England and France. No one had been able to successfully duplicate Marconi's apparatus, and any attempts to do so would undoubtedly be "problematical"; they would also involve a risk of an injunction or patent suit. "Although the validity of the Marconi patents have not yet been tested, he is the recognized successful inventor and practitioner of wireless telegraphy," the report concluded.[48]

While the conditions outlined in the Marconi company's letter of September 2, 1899, were unacceptable to the US Navy, it did offer to purchase twenty sets of equipment. The company maintained its no-sale position, replying with a counter-offer of a lease/royalty agreement instead. The navy wasn't interested and there was no deal. Linwood Howeth believes that Marconi was sure that with his US patent and the new American company that he was about to create, the navy or some other US government department would soon want his business. But it did not happen. Marconi's insistence on leasing rather than sale—which was working in other situations (in

Italy, for example, and soon in Britain)—turned out to be a blessing in disguise for the US government. Unhampered by Marconi's restrictions, the navy ended up with a free hand to develop wireless and, later, radio in the United States.[49]

At the end of the day, Marconi's first American experience had a paradoxical result: it laid the basis for creating a popular image of Marconi the hero-inventor that would persist in the United States throughout his life, while revealing just how difficult he could be. In his relations with the American government he came across as either not well informed or not willing to communicate appropriately for someone apparently interested in doing business; at the same time, despite doubts about the paternity of his system (which would eventually be confirmed by the US courts), his patents mitigated against others simply duplicating his system without a legal agreement, while his financial and proprietary demands made such an agreement impossible. While Marconi always basked in the adulation and recognition he received in America, it was, at the same time, the place where (along with Germany) he would meet the most resistance from government.

❦   ❦   ❦

When he wasn't working on a ship, Marconi spent much of his time in New York with company director William Goodbody, meeting with potential investors. Goodbody had come over with Marconi to float the US and Canadian rights to the company's patents, which the company was willing to sell outright (£300,000 for both countries, £200,000 for the United States alone, or £150,000 just for Canada).[50]

The circular Goodbody prepared was anything but modest. "No doubt is felt in regard to sending messages hundreds if not thousands of miles on both land and water," it said. "Mr. Marconi fully expects to send messages across the Atlantic."[51] Prospective buyers were reminded that stock in the British company was currently worth more than six times the issue price of barely two years earlier. In the end the company decided to set up an American branch operation. Marconi's booster Lord Kelvin had provided a letter of introduction to his nephew John Bottomley, a prominent New York attorney with no aversion to risky ventures, and Bottomley and his law partner, Edward H. Moeran, became the key figures in establishing the Marconi Wireless Telegraph Company of America.[52]

The American company was incorporated under the laws of the State of New Jersey on November 22, 1899, with exclusive rights to exploit Marconi's patents in the United States, its possessions (except for Hawaii, where a newly established inter-island system was retained by the parent British company), and Cuba. The $10 million company (two million shares were issued with a par value of $5 per share) became sole owner of all rights to Marconi's patents in the United States.[53] Like a 1990s dot.com start-up—and like its British parent—the American Marconi company had no foreseeable revenue stream. Marconi was nonetheless systematically putting the building blocks of a global enterprise in place. The deal also increased his personal wealth substantially.

By early November 1899, Marconi and his party were ready to return home; they sailed two weeks before the American company was formed. On November 8 they boarded the SS *St. Paul*, one of the most luxurious liners plying the transatlantic route. It would become Marconi's favourite, another one of his temporary homes.

The crossing was momentous for Marconi. He decided to compound his US public relations success by communicating with his station at the Needles, as the ship passed within forty-five nautical miles of the Isle of Wight on its way to dock at Southampton, on November 15, 1899. Receiving the latest news of the South African war and other items, he published a single edition of a small newspaper he called the *Transatlantic Times,* which was sold to passengers for one dollar in the final hours of the trip. Lest there be any doubt, the promotional blurb read: "Through the courtesy of Mr G. Marconi, the passengers on board the 'St. Paul' are accorded a rare privilege, that of receiving news several hours before landing."[54]

Henry Jameson Davis and Samuel Flood-Page, the company's new managing director, were in Southampton to meet the *St. Paul* when it docked. They had a lot to discuss with Marconi and were eager to see him. They didn't realize—and he certainly didn't tell them—that on the transatlantic crossing he had fallen in love.

6

Love and Imperialism

Josephine Bowen Holman came from a distinguished Indianapolis family. On her late father John A. Holman's side they were judges and politicians, while Josephine's widowed mother, Helen Bowen Holman, was the civically active daughter of Silas T. Bowen, founder of the Bowen-Merrill publishing house (later Bobbs-Merrill) and president of the local board of trade. Josephine attended the Indianapolis Classical School for Girls, a modern school founded in 1882 by the feminist educator, suffragist, and peace activist May Wright Sewall, and graduated from Bryn Mawr (class of 1896), where she had been involved in amateur theatre. According to a family history published in 1909: "So great was the attraction and personal charm of this member of our kindred, that it was said of her, that no one could be in her company without being thereby made better."[1]

Holman was twenty-four or twenty-five years old when she met Guglielmo Marconi (her exact year of birth, 1874 or 1875, is unclear[2]). Either way, they were nearly the same age—he was either six or eighteen months older than she—though some published accounts had her somewhat younger. She had recently moved to New York with her mother and younger sister (also Helen) and was invariably described in press reports as unusually pretty, popular, vivacious, intelligent, well informed, and interested in the world. One newspaper article, announcing her engagement to Marconi in the spring of 1901, said, "She may be called a typical, wide-awake American girl. . . . A discriminating reader of current literature, she is much better informed of current events and questions of state than most young women are." The report went on to say admiringly that she became interested in wireless telegraphy after reading some articles in magazines.[3] It was also said that she had career ambitions.

An unlikely and probably apocryphal story highlights the serendipity of their meeting. Marconi had booked a passage to Liverpool but decided at the last minute to take an earlier ship, so the story has it. Holman had gone to the pier to see off some friends leaving for a month in England on the SS *St. Paul*; they persuaded her to remain on board and join them on the trip and she did, her trunk following her on the next steamer. A lively presence on board, she met Marconi on the third day and soon they were inseparable.[4]

Also on board the *St. Paul* was a journalist and publisher by the name of Harry McClure, cousin of S.S. McClure, the founder of *McClure's Magazine*. McClure was a family friend of the Holmans and may have been the one who introduced Guglielmo and Josephine. Holman knew all about wireless technology, she even knew Morse code, and Marconi revelled in her attention. Throughout his life, he would always be happiest and most relaxed when he was at sea, and in the company of a woman he found attractive. The masthead of the *Transatlantic Times*, the little newspaper Marconi published as the *St. Paul* approached England, listed McClure as managing editor and Josephine Holman as treasurer.

Holman was Marconi's first serious love interest. He had discovered girls as an adolescent in Livorno (where he had once had eyes for Sita Camperio, the older sister of his buddy, Giulio Camperio[5]), had partied with the Prince of Wales on the royal yacht *Osborne*, and in 1899 was possibly the world's most eligible bachelor. Except for one oblique comment to Luigi Solari about having had his first amorous adventure—"or rather to be more precise, a first caprice"—in Bournemouth around 1898,[6] there is nothing in any of his letters, memoirs, or secondary accounts by people close to him to indicate that he had ever engaged in so much as a flirtation. Now he was head over heels in love.

Marconi's relationship with Holman preoccupied him totally during the next two years, which were also the busiest of his life. Four long letters (and several fragments) from Holman to Marconi and more than fifty letters from Marconi to Holman have survived to tell the story. Along with Holman's diary and the scrapbook she kept during their courtship, these documents constitute the record of a dreamlike romance between two ambitious young people on the cusp of two continents, two centuries, and two eras.[7] The courtship recalls Annie's and Giuseppe's: mainly epistolary, long-distance, and discreet.

After they landed in England, Marconi and Holman met a few times in London, where, coincidentally or by design, both were staying at the Grand Hotel in Piccadilly. Guglielmo took Josephine to the British Museum, introduced her to his mother, and escorted them both to the theatre in Drury Lane. Then he returned to work while she went on to visit friends and family in Liverpool. Marconi's first letter to Holman, written at his Haven station barely ten days after the *St. Paul* docked, set the tone for the first period of their relationship—and also typified the style he would adopt in letters to intimates throughout his life—a combination of breezy chit-chat, awkward attempts at humour, minor health complaints, hints as to his sentimental state of mind, and significant details about his work; he had already evidently confided in her about the "great thing" he was working on, getting a wireless signal across the Atlantic Ocean. The letter resounds with Victorian propriety, addressing a variety of topics, but ends: "I would be very disappointed if I would not meet you again before you return to America. Any chance of your coming to this part of England?"[8]

They exchanged a few more letters as Holman completed her English visit and returned to the United States in mid-December. On New Year's Eve of the new century, she wrote a long letter to Marconi, laying out her hopes and dreams for herself, for him, and, rather immodestly, for humanity. The letter highlights Holman's frustration with the conventions that constrained educated young women in America in 1900. It offers a rich self-portrait and demonstrates the strength of character that Marconi found attractive. This is the first of Holman's letters that have survived.

December the thirty first, 1899.

Dear Mr. Marconi,

Today the Old Year gives place to the New. Her passing is like that of all great souls—quiet and without pomp. We are standing upon the threshold of a new century and one cannot help wondering what it has in store for our world, our country and ourselves. So conversely, one realizes that in greater or less degree each human soul will, through 1900 as in 1800, react upon the life of his country, and ultimately upon the world. There will be great men in 1900 as in all the other past centuries, who will make history, and better the human race. How wonderful the plan of the universe is—even the little we know of it, how inspiring and full of meaning. It does one good to stop to think, once in a while—one almost wishes that the birth of a century came oftener than once in a life-time. It makes one long to do one's best. If only we could be always living upon the heights! But then we might lose the power of sympathy which we learn from our moments spent upon the plains....[9]

Marconi wrote back, echoing Holman's idea of "progress," if with less eloquence: "The good old century has produced very great things, but what shall we expect from the new one? We are young and shall probably be permitted to have a peep of the further progress of mankind during the 20th century." He also gave her (and us) a glimpse at how he was working: she shouldn't believe everything in the papers about him or his discovery, he warned her. All he could say was that he was "preparing things for some long distance tasks." He and his staff were going to work quietly, and keep the press at bay.[10]

❁   ❁   ❁

By 1900 Marconi and his companies had reached cruising speed, though they still lacked one absolutely essential element: profitability. Marconi had surged ahead of every one of his rivals. He was personally one of the most recognizable figures in the world, certainly in the domain of science and technology. His patents were being confirmed and his company was attracting interest from governments and corporations everywhere. Still, it was far from being commercially viable. Marconi decided that this last elusive quality could come only from one source: a transatlantic connection. If he could achieve communication across the Atlantic he would gain the confidence of the last of the doubters and also, not least, be able to confront the cable companies on their most profitable terrain, the sea.

The company's shareholders confirmed their recognition of Marconi as their most valuable asset by changing its name to Marconi's Wireless Telegraph Company (MWTC) at a special general meeting on February 23, 1900. Self-effacing to a fault—or wanting to be sure to be seen that way—Marconi asked that the minutes record that he remained neutral on the name change. His entire family—Annie, Giuseppe, and Alfonso—were visiting England and were ecstatic.[11] From now on, Marconi's name would be on every letterhead, share certificate, and piece of equipment manufactured by the company. It would be mentioned in every press release and stock quotation. Already a household word thanks to three years of daily media exposure, Marconi's name now became one of the most recognizable brands in the world.

The Marconi system still had a few hurdles to clear: distance, the lack of privacy (anyone equipped with a receiver could receive any transmission), and interference. Marconi was convinced that the first of these obstacles would soon be overcome by sheer perseverance; to address the other two he began working on a new innovation: tuned circuits, or what his rival Oliver Lodge called "syntony." As we've seen, Lodge took out a patent for "syntonic wireless" in 1897. Marconi followed him with a first patent for

tuned circuits in June 1898 (claiming to skeptical reviews that it was different from Lodge's), another in 1899, and a third in 1900. The provisional
specification for a "master patent," with a conveniently memorable number,
7777, was filed on April 26,1900;[12] it provided for the tuning of transmitters
and receivers to the same frequency, thus ensuring secure communication
and a regulatory mechanism for controlling interference between stations
operating in the same vicinity. Marconi's colleagues claimed that the so-
called four-sevens patent made wireless telegraphy fully competitive with
the conventional wired telegraph, and his successful exploitation of the
patent worldwide illustrated his increasing sophistication in the intricacies
of strategy, secrecy, and tactical manoeuvring. Marconi considered it to be
one of his most important accomplishments in the quest to establish his supremacy in the field, and the patent's validity was eventually upheld by
courts in the United Kingdom, the United States, and France.[13]

A pattern was set by a patent suit launched in the United States by an
assignee for Amos Dolbear in October 1899, which was still not resolved when
Marconi filed his "four-sevens" application.[14] Dolbear was the first Marconi
rival to challenge him in court. His patents dated from the early 1880s, more
than a full decade before Marconi's—but also a few years before Hertz published his discovery of electromagnetic waves. The company was confident that
Marconi's original 1896 patent was strong and decided that its best course of
defence was to put Marconi on the stand. "No one can explain it so fully as Mr
Marconi himself," managing director Samuel Flood-Page wrote to Edward
Moeran in New York.[15] Marconi's personal touch was required—and would be
sufficient—to distinguish the invention from its competitors.

Fully prepped by Fleming and the Carpmaels, Marconi arrived in
New York in June 1900 to be examined, his presence duly noted in the
press. Flood-Page was right; Marconi's testimony was brilliant. He conceded
that he sometimes used a telegraph transmitter and receiver connected to
the ground, but argued that this was in no way Dolbear's invention; it had
been suggested by Francisco Salvà as far back as 1795 and had been used by
Humphry Davy, James Bowman Lindsay, and Alexander Graham Bell as
well.[16] He recounted the string of successes he had enjoyed since setting up
his system in England, describing his work between Bournemouth, Poole,
and the Isle of Wight; his Lloyd's installations at Ballycastle and Rathlin
Island, in the north of Ireland; the Kingstown regatta; as well as the foreign
countries in which his system was starting to be used.[17] Marconi was in his
element under questioning. Here again, he was his own strongest advocate.

In patent litigation, parliamentary testimony, or senate hearings he was a formidable presence, always impeccably well prepared, confident, and convincing. Over the years, some of his most important self-assessments and renderings of his story would be made from the witness stand. The Dolbear suit was dismissed in March 1901 and the company could claim that this recognized Marconi "as the first and true inventor of practical Wireless Telegraphy."[17]

Marconi's mastery of patent litigation was so formidable that only the most stubborn or naive competitors dared confront him. Some landmark cases still lay ahead, but generally, after the Dolbear suit, other inventors sought to offer their discoveries, or their services, to the Marconi companies—and when they wouldn't, he sued them. The actual solidity of Marconi's claims to 7777, for example, were never airtight, but they were close enough that many of his competitors (including, eventually, Oliver Lodge) found it more expedient—and profitable—to join him than fight him.[18]

❂   ❂   ❂

As Marconi's fame and influence grew, he recognized and willingly embraced his usefulness to the power structure of European colonialism. Just before Christmas 1899, he spent a pleasant day with the writer Rudyard Kipling in Brighton. One of Kipling's interests was technology, and he told Marconi he was going to write a poem about wireless.[19] He didn't address the topic until 1902, and then it was not a poem but a short story in *Scribner's,* set amid the development of wireless in the south of England. The story began: "It's a funny thing, this Marconi business, isn't it?"[20]

Kipling had recently coined the term "white man's burden," publishing a seminal poem by that name in *McClure's* in February 1899.[21] Coming in the wake of the US intervention in the Philippines in the Spanish-American War, Kipling thus popularized the idea that colonialism had a civilizing mission—an idea that became central to Marconi's own political philosophy. In the twentieth century, the notion of "the white man's burden" came to symbolize the racist ideology underlying Western imperialism; a few years after Kipling, the British journalist Edward Morel, one of the first Europeans to expose Belgian brutality in the Congo, coined the counter-notion "the black man's burden."[22]*

---

* Marconi and Kipling crossed paths again a number of times at high-class social functions in England; Kipling moved increasingly to the right during the 1920s and was, like Marconi, an early admirer of Mussolini before denouncing him as a dangerous dictator when he grew too close to Hitler.

Marconi seemed to be inattentive to such nuances. In March 1900, he responded with innocent and childlike excitement when King Leopold II invited him to Brussels to explore the possibilities of adopting his system both for commercial purposes in Belgium and also by his army in the Belgian Congo.[23] He travelled to Brussels on short notice to demonstrate his system to the King of the Belgians. A letter to Josephine from Brussels shows him in awe in the presence of the king, whom he had met in 1898 while on the Prince of Wales's yacht at Cowes: "I went to the Palace at 12 today. The King is a very kindly old gentleman and speaks English perfectly. The Queen is also very nice. . . . I have to give them a private lecture tomorrow. I believe the King is keen on having my system in the Congo Colony."[24]

Marconi's deferential attitude toward Europe's imperial rulers was unmistakeable. He had now "performed" for the royal families of Italy, the United Kingdom, and Belgium, and attracted the attention of the German and Russian emperors, without the slightest hint of discomfort about the system they represented. By 1900, the "kindly old gentleman" Marconi met in Brussels was already known to many as the "butcher of the Congo." Leopold had founded the Congo Free State and exploited it as his own personal property over the dead and often mutilated bodies of at least 10 million Congolese, until international pressure forced its takeover by the Belgian government a few years later.[25]

Marconi was either oblivious to what Leopold was doing in the Congo or he didn't care; most likely, it just didn't enter his frame of reference. What interested him was that Leopold was a head of state who could promote his standing. The Congo rubber boom had begun in the mid-1890s, and for a long time Leopold successfully masked the extent of the atrocity-driven pillage conducted under his rule, and to his huge benefit. Today it is seen by many historians as the most excessively brutal of the European colonial projects in Africa. Leopold was able to maintain a distance from published eyewitness reports of gross atrocities, which he generally received with feigned incredulity and promises to investigate. He was able to neutralize his critics with his formidable charm, encouraging them to trust him rather than take their cases to the press. But this approach soon wore thin.[26][†]

[†] Marconi was not alone in being taken in—initially—by Leopold's facade of benevolent innocence. The African-American journalist George Washington Williams described him, in 1889, as "a pleasant and entertaining conversationalist"; however, a year later Williams published one of the first and most scathing accounts of the "moral inferno" that was the Congo (Hochschild 1998, 106).

Certainly by the time Marconi emerged as a public figure, the situation in the Congo was known to readers of the *Times*—and Marconi was certainly a reader of the *Times*. An article in May 1897 reported on a London meeting of the Aborigines' Protection Society, one of the world's first international human rights organizations (it was founded in 1837), where a returned missionary told of routine cruelty inflicted on forced labourers in the rubber trade, ranging from random shootings to mutilation for refusal to work. Whole villages were being spoliated and destroyed. In one case the witness had seen a child with its hand cut off lying on its dead mother. The slave trade was being practised by the state; those who refused to work were shot and their hands cut off as proof to show the authorities.[27]

Marconi could be excused for overlooking a newspaper report of an obscure community meeting while he had his own busy life to attend to, but by the time he was invited to lunch with the King of the Belgians the situation in the Congo was difficult to ignore. Joseph Conrad's *Heart of Darkness*, a fictional account of colonial cruelty in the Congo, based on Conrad's own experience, was published in London in 1899 to sensational acclaim.[‡] By February 1900 alarming reports from the Congo were regularly reaching Europe. A mere fortnight before Marconi's Brussels visit, the *Times* reported a pogrom-like wave of terror by tribal mercenaries on native villages, sanctioned by Leopold's Congolese officials. According to an American missionary, "Almost daily slaves are brought here and exposed for sale by the Zappo Zaps, a cruel cannibalistic people who are armed and kept by the State . . . as plunderers, tribute collectors and slave raiders. . . . The iniquitous tribute system in vogue throughout the State, by which rubber and ivory in great quantities are obtained, is the primary cause of and often furnishes a plea for these crimes."[28] In the House of Commons a few days later, the government was asked whether it had received representations about these atrocities committed "under the Congo Free State flag, and with the alleged connivance of state officials. . . . " The government spokesman replied that no representations had been made.[29]

Remarkably, Leopold continued to enjoy popular favour at this time.[30] He may well have convinced Marconi that he needed his system in order to instill some European order in the unruly colony; or the question may not have come up. Marconi's view of his own role of service to the imperialist

---

[‡] *Heart of Darkness* was first published in serial form in *Blackwood's Magazine*, London, over three months beginning in February 1899.

power structure was one of his benchmarks for success. One of Leopold's defences to his critics was to say that what he was doing was certainly no worse, perhaps more benign, than what the great powers had been doing in Africa, Asia, and the Americas for centuries. This argument would have resonated with Marconi as the citizen of a "small" power, Italy, with comparable colonial designs. Although he never put it so baldly, he did not see why different rules should apply to lesser colonial powers like Belgium or Italy than they did to the big ones like France or England. The astonishing thing is that even in private communication he could be so unreflexive and not the least bit ironic about going to lunch with the King of the Belgians.

After the Brussels demonstration, Leopold immediately, and publicly, promised to adopt Marconi's system for his army.[31] Belgium became another pole of Marconi's research and business activities; he soon demonstrated the practicability of using his system at sea on the Belgian mail boat *Princess Clementine* running between Dover and Ostend; one of the first messages sent from the ship offered greetings to the king.[32] Leopold became one of Marconi's strongest European supporters, and the company set up its continental base in Brussels: the Marconi International Marine Communication Company (MIMCC), a new subsidiary to look after all of the company's maritime activities, was incorporated in Brussels in April 1900.[33]

By June 1901 the company had an important agreement to supply equipment and trained operators to King Leopold's Congo colony and began organizing a team to go down and carry it out. Henry M. Dowsett, who had already been to South Africa for the company, was put in charge since, it was understood, "he has no objections to going to tropical countries."[34] But more men would be needed and the company had other African business in its sights as well, in the Sudan, for one. The Congo project was important commercially, experimentally, and diplomatically—for Marconi and, certainly, for Leopold.[35]

Despite the intensity of his research work, Marconi insisted on approving every detail of the Congo agreement, especially those that had anything to do with technical issues. He was pleased with the agreement, which was crucial to the development of his European plans.[36] However, "the Congo question" was soon on the public agenda in Britain and elsewhere, taken up by progressive intellectuals like Mark Twain and more conservative figures like Arthur Conan Doyle. The Congo was an easy target as it did not involve powerful countries such as the United States, Britain, or Germany. By 1903, the great powers were calling for an end to Belgian exploitation of the Congo, citing the arbitrary nature of its government, corruption of its covetous officials, and abuses related to slavery. Leopold replied aggressively that it was none of their business.[37] By 1904, the King of the Belgians was reportedly so incensed with the British press systematically reporting his abuses in the Congo that he cancelled his subscription to the *Times*. As Adam Hochschild wrote in *King Leopold's Ghost*: "What happened in the Congo was indeed mass murder on a vast scale, but the sad truth is that the men who carried it out for Leopold were no more murderous than many Europeans then at work or at war elsewhere in Africa."[38] Marconi may have unconsciously sniffed this out.

❊   ❊   ❊

On the other side of the world, Marconi's company was having another edifying encounter with colonialism. In Hawaii, the company signed on with a small start-up seeking to connect the Hawaiian (or Sandwich) Islands, then a recently acquired US territory.[39]

Frederic J. Cross was an electrical engineer from Buffalo, New York, seeking opportunities in Hawaii. He had obtained a franchise to connect the islands for telegraphy, presumably by laying cables, but having read about Marconi's new system, he wrote to Marconi from Honolulu in July 1899. "I think the conditions existing here are almost ideal for the perfect operation of your system between the Islands of this group."[40]

It wasn't long before the project was public knowledge. Even before a contract was signed, the *New York Times* was reporting that Marconi's system would be used to connect the various Hawaiian islands. An organization had been formed, capital subscribed, and equipment was on its way, the newspaper reported, noting that except for the channel between Oahu and Kauai, the distances were smaller than the English Channel, and that had already been crossed.[41]

Company secretary Henry Allen replied to Cross on Marconi's behalf a few weeks later. It would be no problem. There would be no difficulty connecting the islands, and wireless would be less expensive than cable. The company would charge a royalty of £100 per annum for every ten miles covered, "and whenever you wish to give up the use of our system, we are prepared to take back the instruments we sell you, allowing for fair wear and tear."[42] At this stage they were still prepared to sell the equipment to nongovernmental clients; within months they would be refusing such sales, and only licensing the use of the Marconi system across the board.

Cross wrote back ordering two sets for tests. He wanted to hire local operators but Allen replied that it would be useless until "intelligent people of yours" could be trained to use the instruments as well as to set up the stations. He suggested that London send out one or two assistants "to put up the first installations and to instruct your men in the working of the same."

The Marconi company's business plan was to provide the system's operators, thus keeping even closer control. On October 31, 1899, an agreement was signed, by which the company licensed the use of the Marconi system to Cross and agreed to provide five sets of equipment and three "experts"; the rest of the staff would be hired locally. Cross formed a company, Inter-Island Telegraph Company Limited, to conduct the business; a note in the archives indicates that the Marconi company considered Cross's venture to be heavily underfunded.

The Marconi man sent to Hawaii reported back a few months later that the sites chosen for stations were pretty good; more interesting, an auction house was going to sell the first message for upwards of $1,000, a huge sum validating the cult status of wireless even at this early date. The first transmission took place from Kaimuki to Honolulu on June 17, 1900, in the presence of a large crowd, and was deemed a great success.

There were problems, and in September the company decided to send out its top engineer, Andrew Gray, to take charge of the operations. Gray met with Marconi at the Haven on September 20, 1900, and was told: risk nothing, take no chances. Gray left Liverpool on September 22 and arrived in Honolulu six weeks later, on November 2. Gray kept a logbook of his activity in Hawaii. On November 27, he attended a meeting of the directors of the Inter-Island Telegraph Company. "Proceedings stormy," he recorded. The same day he wrote to Cross: "I must assure you that I am as anxious as yourself that this commercial wireless should be a success... from Marconi's point of view the *system must work*" (emphasis in original).[43]

Marconi's international prestige was now at stake—Hawaii was more than just another location; it was remote and exotic and if wireless could be useful there it would succeed anywhere. But relations between the two companies were strained and the work continued to proceed poorly. In Gray's view, this was due to the quality of the operators. He wrote again to Cross on December 29: "Your co-directors want demonstrations and yet they fail to provide the operators necessary to give demonstrations. The instruments and stations are there."[44]

Gray cabled the news to London and meanwhile set up day-long classes for Cross's operators; but he found them not up to the task. Their attention span was short, they were undisciplined, and they wouldn't follow rules. Things got a bit better when Cross raised the operators' pay, but not much. Some of the men, Japanese migrant workers (whom Gray referred to in his logbook as "Japs"), were sent home; others deserted. Increasingly frustrated, Gray wrote to Cross on March 28, 1901, attributing the failure of the system to have been caused "partly by the blowing over of the poles which you constructed and partly by the class of operators which you have sent out."[45]

Matters kept getting worse. Cross reneged on paying the second half instalment for the first year of the contract and denied his personal liability, claiming he had assigned it to the Inter-Island Telegraph Company. On June 27, 1901, the Marconi company sued for payment in Hawaiian Circuit Court and its assistants were called home. Gray reported that the company had fulfilled its end of the contract, fitting six stations that worked satisfactorily, and having trained fourteen operators of the Inter-Island Telegraph Company in adjustment, sending, and receiving.

The Marconi company also considered taking over the system and running it itself but this option was fraught. In an undated memo to Flood-Page, Gray's second-in-command wrote that the majority of the operators were absolutely incapable of managing the instruments and were unable to speak English properly; they would have to be replaced. The existing accommodations would have to be rebuilt "as no white man would live in the huts which have been constructed," and salaries would have to be nearly doubled. Introducing even a single "white operator" would bring the cost of working the system up to about ten pounds a day.[46]

An interesting aspect, unmentioned in the archival correspondence but noticed by the press, was that four of the fourteen operators trained by Gray's team were women—possibly the first women wireless operators anywhere in the world. In fact, a reporter for the *Los Angeles Times* invited to observe the training program was favourably impressed. The women operators "all seemed eager to learn the new order of things and very quick to pick up the suggestions of their superiors in relation to the details of their work. They stood about the mysterious-looking instruments with their pencils in their hands and interpreted the messages as they came across the lines." According to this report, one of the women was slated to be the chief operator of the system.[47]

But the Hawaiian operation failed to take off and was seen in business circles as a stain on the company's global expansion plans. It confirmed to Marconi that it was better to lease equipment than to sell it, and to insist as part of the contract at least on training operators if not providing them out-right. Taking a longer view, the results were far more damaging, as Marconi, called to account for the "fiasco" at the company's 1902 annual shareholders meeting, made a stunning statement, publicly blaming the failure on "the inferior class of the operators—frequently ignorant half-breeds and negroes—whom the Hawaii company was ill-advised enough to employ for reasons of very false economy."[48]

Even in the context of the time, this was appalling. Some of Marconi's early writing, as we've seen, was mildly orientalist in tone, and as we saw in his correspondence with Josephine and his relations with European rulers like Leopold of Belgium, he was naive at best in his understanding of colonialism. But here he was vulgarly racist, shamelessly supremacist—and in an official corporate statement aimed at impressing shareholders, no less. This was the other side of the "white man's burden": "half-breeds and negroes" were an unreliable, inferior class of people, not worthy of employment.

Marconi's views and attitudes on race, gender, and politics were squarely camped in the mainstream of his time and place, but this moved him to an extreme. In the same speech, he went on to envision the benefits "not only to this Empire, but to the whole of the civilized globe . . . by the placing of new means of communication at the disposal and within the reach of an indefinitely wider public for social as well as for commercial ends." However, Marconi's civilized globe only included white people.

7

The Upstart Technology

Marconi left Liverpool on the *St. Paul* on May 26, 1900, arrived in New York on June 2, and set up again at the Hoffman House. It was to be a brief visit, mainly to make a deposition in the Dolbear patent suit,[1] but Josephine had invited him to visit the Holmans' country cottage in Cragsmoor, a tiny rustic hamlet in the Hudson Valley, to meet her mother and sister, and he hoped to come up to see her. Typically, time was tight and his priority was business. "I came over here to make a few preliminary arrangements for carrying out the 'great experiment' which will not be tried for some months yet as my patents (for new improvements) are not yet complete," he told her.[2] He was more cryptic in his public comments, telling the press, "I'm here on entirely personal and private business which relates to the securing of capital. I may make a few experiments in wireless telegraphy while here."[3]

He turned out to have some free time, but Josephine had family obligations and Marconi ended up making short trips out of town, including a climbing excursion with friends in New Jersey, before he finally managed to get to Cragsmoor at the very end of his trip.[4] It was only for a day or two—but the relationship now took a radically more romantic turn. On the eighteenth—just back from Cragsmoor on the eve of his departure for England—he wrote to say goodbye: "The pleasant time I have spent at Cragsmoor is a thing of the past, and the more I think of it, the more it seems like the memory of a very happy dream. I hope I shall dream of it again soon..."[5]

On June 20, 1900, heartsick (possibly for the first time in his life), Marconi wrote just as his ship sailed. "As the ship speeds away leaving behind it in the mist, the great American continent, a feeling of sadness such as I have never experienced comes over me." Then, at the end, he added—in Morse code: "My dearest Jo / I feel I cannot write much I feel so sad and yet so happy / I now have something to work for which I value far more than riches honour or glory or anything this world can give /"[6]

Marconi wrote every day during the crossing, compiling a rambling letter, written across several days, that he would only post a few days after his return to England. The letter continued to have a formal external facade, but each instalment ended with a few lines in Morse code where he opened his heart, exposing himself just a little bit more, and a little bit more: "I find it rather difficult to write for reasons which you know, but hope that soon I will have no difficulty. [in Morse code] Always yours Guglielmo."[7] There is a melancholy, self-pitying quality to the letter, as Marconi vacillates between his unmoored emotions and his work, which seems to ground him:

2 22:* I have begun to write out some specifications for instruments to be ordered when I get back to England, and that is keeping me more busy than I would have been otherwise.... [in Morse code] God bless you Jo;

3 23ᵈ: It is raining today and everything looks dismal.... I wonder what you have been doing and thinking the last few days.... I think you can guess of what and of whom I have been mostly thinking of.... [in Morse code] Don't forget me I hope to be yours for ever;

24th: It is very difficult to write on this ship, as the writing room is so very small and every one seems to want to occupy the tables for reading or writing.... [in Morse code] Your Guglielmo;

25th: We are now getting near to Ireland. I sat up till rather late last night watching the gray waves illuminated by the stars and the ships lights. I was thinking of how so much these waves mean to me. They will have to conduct the messages if my great experiment is a success, and they separate my country and the land I am going to from another great country which I have left. I hope that the electric waves that God is teaching us to understand and to handle, may be the means of bringing about a true union of these great countries, and other unions also. [in Morse code] You understand;

26th: Weather always dull lots of people asking for my autograph etc. I now know every one on board.... [in Morse code] Love to good sweet Jo;

27th: 11 p.m. We are back in Liverpool again and all is light and activity around the ship. We called at Queenstown this morning. Some reporters came on board to interview me and asked me the same old questions.... [in Morse code] I miss you very much but I hope it will come right.... It is the happiest thing to know that one is loved. God bless you. Guglielmo.[8]

Marconi's shipboard letter reflects the trip itself, dull and repetitive. He had time on his hands and one can almost feel the boat rocking to the tedium

---

* Marconi is using his own code: "2 22" means second letter in the series, written on the twenty-second.

of the chit-chat and card games. The Morse messages are unimaginative and reveal none of the thrill someone might feel writing—free of constraints— to his inamorata.

Back in England, Marconi recaptured his energy and his subsequent correspondence was more dynamic—just as his missing her became more urgent. In a letter from London, his Morse code postscript read: "My Jo, I am constantly thinking of you. I wish you were nearer. I hope I shall soon see you again. Goodbye, my true love, my ideal. Your Guglielmo."[9] He had found a soulmate.

    ❁   ❁   ❁

On the evening of July 29, 1900, Marconi's benefactor King Umberto of Italy officiated at an athletic prize-giving ceremony near one of his palaces at Monza, just north of Milan. As the king returned to his open-air carriage, a silk weaver by the name of Gaetano Bresci burst from the crowd and fired three bullets into his chest, killing him instantly. Bresci had recently returned from the United States, where he had been involved in anarchist politics, with the express intention of assassinating Umberto. He was tried, found guilty, sentenced to life in prison, and found dead in his cell less than a year later.

From London, Marconi wrote to Josephine: "You will have heard of the horrible murder of the King of Italy. It is hard to realize what terrible wickedness there is in this world. It has made me very very sad." He went on to extol the virtues of the monarch to whom he owed a "very grand deal." Umberto had lent Marconi use of the Italian Royal Navy's battleships for his early trials. His government was also "the first to adopt wireless and to pay for its use."[10] Perhaps most importantly, Umberto had helped Marconi get out of his military service by appointing him an attaché at the embassy in London.

In Italy, the Marconis reacted to the assassination with a mixture of indignation and opportunism. Annie wrote to Guglielmo: "I am sure you were dreadfully shocked and sorry to hear how the poor King was murdered by one of those horrid anarchists." The entire family was grateful to Umberto for settling Marconi's *volontariato* (military service) business. She wrote her son that his father thought it would be a "good thing" if Guglielmo sent the queen a telegram of *condoglienza*. "Perhaps you have done it. . . . Your Papà says if you were to write a letter of condolence to the Queen now it would do just as well."[11]

The assassination provided Annie an opportunity to comment on the politics of church and state in Italy: "I send you a paper by which you will see the bad jesuitical way the Priests and the Pope are going on, forbidding the Queen's prayers for the poor King to be read in the Churches." In the conflict between church and state, the Marconis were on the side of the state. Guglielmo, later, would become adept at playing both sides to his advantage. Meanwhile, without skipping a beat, Annie returned to the most mundane of secular concerns in the very next sentence: "I am glad to hear you have bought a Bicycle, but don't go fast down the hills it is very dangerous."[12]

※　　　※　　　※

The pace of Marconi and Holman's correspondence quickened during the following months, but, as we've seen, few of her letters have survived. The next trace we have is from the end of October 1900, when she wrote a long, newsy letter:

> My dear Mr. Marconi,
>
> It was very nice to hear from you again, and learn that everything is going well with "W.T." and the "great experiment." I am constantly seeing paragraphs in the papers about the success of wireless, and its rosy future—all of which I believe, you know. But I shall wait eagerly to hear from you the result of the first experiments with the great thing—which I do not read about in the papers. You must have a magical way of managing the press—to have succeeded in keeping even a line of it from the public ear. . . .
>
> I am a bit melancholy today. . . . There are some things which are troubling me, annoyances and embarrassments, which I can neither run away from nor improve, I fear. Life usually seems joyous to me, but sometimes, well, I promise not to bore you again, and I'll write you a jollier letter next time. . . .[13]

Holman ended her letter, in Morse code, with a revelation that shook Marconi to the core: "Then there has been another man and most embarrassing."

Marconi of course replied immediately: "What is the <u>other man</u> like? [his emphasis] I am rather curious."[14] Actually, this came at the end of a long letter from the Haven, in which he treated a wide variety of subjects before coming to that point. For the first time, he abandoned his Morse code and addressed her directly:

> My dear Josephine,
>
> I think I can now write to you in plain English . . .
> . . . the fact is my health is not as it should be, it may be overwork. I have always hoped it would be all right, but still as I say—I do not feel happy—I am not sure I am doing right in tying you up. You may meet someone who is better

than myself in many ways, and although the thought pains me very much it is possible that I may by keeping you engaged at this time of your life, remove the possibility of your liking some one who as I have said is better than me.

...it may not be so black as I imagine after all, but for the last two months I must say I have been discouraged.

It is a great happiness to me to have met such a good clever dear and sweet girl as you are, one who seems to raise the standard of humanity...

His ambivalence was pervasive and surprising. Did he want to marry her or not? Holman's next letter is equally serious but less auspicious.

My dear Mr. Marconi,

It is a long time since my last letter—and I have a lot of things to tell you...

I have a collection of "wireless" items which I read with great interest—Oh, how much you are doing for the world—and I am doing nothing...

I have been hoping for a letter in answer to my last. Perhaps I am a selfish friend, but I would like to hear from you more often—although I can fancy in what a busy rush you are at this time. I know how important these months are to your whole life's work. It is the part of good friends to mutually help one another—and not to be a hindrance. One owes to one's friends as much as one can give, in justice to oneself. When I wrote from Cragsmoor, that I would not expect letters at this busy time, I hope that you did not infer that I liked the scarcity. I only thought that I was a bit of a sacrifice to the cause of "wireless"...

[in Morse code] Your long silences have made me very unhappy.

[in English] I tell you this because I think you like to be told the truth.

[in Morse code] I pray for your success.[15]

Holman's letters are far more heartfelt than Marconi's, which are often breezy and self-centred. His letters are almost adolescent and childlike; hers are mature, thoughtful, and worldly. Holman was perhaps too strong and self-assured a woman for Marconi. She spoke her mind and her heart, knew what she wanted and how to express it. She was supportive and at the same time frustrated: "Oh, how much you are doing for the world—and I am doing nothing." Except for one brief allusion in an undated fragment, there is no further mention of "the other man" in any of the letters, his or hers, that have survived.[16] It is quite possible the "other man" was Holman's way of getting Marconi to be more direct about his feelings. For all that he was diminishing distance by wireless, he was not doing it emotionally.

Marconi was touched by Holman's candour and attentiveness, and he evidently enjoyed her admiration. But he had become monomaniacally focused on his own success. He had done little else but work during the first year of the new century and his next letters to her complained of minor health issues and overwork, partly to excuse his emotional remoteness, partly to cover up his growing melancholy. This was the most exciting period of his life and yet he seemed depressed. Photographs of Marconi during this time show him looking earnest and unsmiling. "I think my life is very different from that led by most people," he wrote in a letter to Holman.[17]

He also introduced a new element that would be significant in the unfolding of their relationship: "My dear Mother has arrived here from Italy. She often asks of you. I have not yet told her everything. I am afraid she would be pained if she knew. She told me plainly last spring that she thought I ought not to think of marrying for the present until my health was better, and I fear she knows I am not better." Annie had arrived in England in November, staying with Guglielmo at the Haven, fussing over him and fretting about his health. The previous Christmas, she had cancelled a plan to spend the holidays in Ireland with Henry and Emma Davis because she had caught a cold and Guglielmo was not yet recovered from an unspecified ailment. At her urging, Guglielmo now took a few pleasant days respite at Christmas, joining her at board member James Fitzgerald Bannatyne's country estate near Exeter, which Annie described in a letter to Giuseppe as "a magnificent *palazzo*." He also received some good personal news: in early December the Italian government officially confirmed that it was letting him out of the military call-up to which he was susceptible. Characteristically, he had Annie alert Giuseppe not to speak of this in Italy.[18]

Whether by design or not, Annie's observations indicate that Guglielmo was not quite pining for Josephine as much as his letters to her let on. In addition to his research, and love, he also had business, politics, and patents to attend to.

❀   ❀   ❀

Henry Jameson Davis's contract as managing director of the Wireless Telegraph and Signal Company had expired in August 1899. As he had announced when he first took the job, he was anxious to return to his own affairs. Although he would no longer be responsible for the company's management, he told his fellow directors, he would always take the warmest interest in its success—not least since, next to Marconi, he was the largest shareholder.[19] He remained a member of the board.

Davis was replaced as managing director by Samuel Flood-Page, a seasoned business administrator and engineer with an impressive background in overseeing electrical innovation. As the manager of London's Crystal

Palace, Flood-Page had staged a major exhibition in 1882, at which Edison's new electric light bulbs were introduced to the British public. As a result of that success, he was hired on as secretary-manager of the Edison and Swan United Electric Lighting Company, a position he held from 1883 to 1892. Under his stewardship, electric lamps were installed in London's Holborn Viaduct, the offices of the *New York Herald*, the colonial Legislative Assembly in Cape Town, and office buildings in Australia.[20] Marconi's scientific advisor, James Ambrose Fleming, had also been associated with Edison and Swan, and he was the one who introduced Marconi and Flood-Page at Dover in 1898. At sixty-five, Flood-Page was not an obvious choice to direct the company's day-to-day affairs, but he was well connected and, importantly, temperamentally suited to working with Marconi.

Flood-Page joined the company on September 23, 1899, while Marconi was in the United States.[21] His first task was to go to Berlin to try to unravel the intricacies surrounding the company's German patent applications. On October 6, after spending a week in the German capital, he wrote a long and confidential report to Marconi, who was still in New York.[22] The report gives a good idea of the sort of international industrial, political, and financial intrigue in which Marconi was now engaged. Marconi, as we've seen, had applied for a patent in Germany in December 1896, in his very first round of foreign patent applications, and the issue was still being actively opposed and was far from resolved. Whatever the intrinsic value of a German patent, Flood-Page wrote, there could be no doubt of its importance because of the careful examination that would be made by the German patent office. In fact, Germany was absolutely critical to the company's global strategy; everyone who came calling about negotiating foreign rights "invariably asked whether the German patent had been issued." It was a sign of the real strength and prestige of Marconi's claims.

Marconi's British patent agent, Carpmael & Company, felt that things were moving slowly yet in the right direction, but Fleming and others thought it was necessary to secure the co-operation of a German expert to lobby the patent office. The problem was, the leading German experts were all potential rivals. It was decided to approach Adolf Slaby, who had appeared sympathetic to Marconi and, in fact, had already been of some help, as we've seen.[23] Slaby was undoubtedly the most influential ally they might have in Berlin, but on October 2, 1899, the German sub-agent they had hired through Carpmael received a short note from Slaby regretting that he could not help out as he himself had an application for a wireless patent pending "and was not therefore any longer impartial."[24] The company decided it had to find out what Slaby was doing, what his application was about, and whether they could buy it or work with him.

The company was aware that the kaiser was personally very much interested in wireless and anxious to introduce it as soon as possible into his Imperial Navy, and that the German naval authorities were in close touch with Slaby on the subject. Marconi's Berlin agent went to see Slaby and learned that the German naval authorities were determined to have wireless, or, as Slaby called it, "sparks-telegraphy." As Marconi had not yet made any arrangement in Germany, Slaby himself had applied for a patent, based on a method he argued was different from Marconi's. Slaby had, in fact, tried his plan on German warships at Kiel, where they had been placed at his disposal for that purpose. Slaby told Marconi's agent that he had no direct financial interest in the matter, that since his efforts to arrange an agreement between Marconi and Siemens & Halske had failed because of Marconi's exorbitant demands, he had decided "to find a substitute" for Marconi's invention. Slaby said he was open to negotiating with Marconi if his own patent application was successful; Flood-Page felt this would be highly desirable.[25]

The company had to immediately adopt a strong, proactive policy, Flood-Page wrote to Marconi, advising that they make a deal with Siemens & Halske, which he considered the leading manufacturer of electrical equipment in the world. With such a partner, Flood-Page felt they could be the first to occupy the German market. It was abundantly clear, Flood-Page wrote, that if Marconi's didn't make arrangements for the immediate supply of wireless telegraphy to Germany, someone else would do so: "It must be realized that there are others besides Dr Slaby in Germany who are following on in Marconi's steps, and are trying to get round our patents."[26] One of these was Ferdinand Braun, who was already opposing Marconi at the German patent office.

Any of us can handle the business end, Marconi was told, "But only you can invent, you alone have the deep insight necessary for all the improvements which must be made."[27] However, they still didn't have German patent protection, and the approach Flood-Page was suggesting went against Marconi's basic strategy. After the German patent was awarded, in December 1900,[28] Flood-Page proposed that the company try to absorb its rivals. "I am confident that if we bring into one central organisation Braun and Popoff and perhaps [French researcher Eugène] Ducretet we shall cover the world," he wrote to Marconi on January 3, 1901.[29][†] But this approach was not followed by the company, which sought instead to use litigation and high-level lobbying to try to crush the competition. Sometimes it worked, but when it didn't, it made formidable enemies. Braun, for example, was willing to sell everything to Marconi, but the Company felt—wrongly, as it turned out—that it had nothing to fear from Braun.[30]

By the time Marconi received his German patent, the world's six most powerful governments (Britain, the United States, Germany, France, Russia, and Italy), and many lesser ones, were actively and even avidly interested in the military potential of wireless communication. While they were each trying to develop their own system, Marconi's was the gold standard, and he was to varying degrees involved with all of them. He was now working simultaneously in London, New York, and Berlin, the three most important centres of politics, technology, and finance at the turn of the twentieth century. For someone with global aspirations, the three capitals each had its strengths and weaknesses: London's strengths were politics and finance, New York's finance and technology, and Berlin's technology and politics.[31] Without necessarily seeing the situation in these terms, Marconi played these strengths and weaknesses against each other. In London, he was seen mainly as a technologist and attracted the interest of politicians and financiers; in New York, it was his international political connections that impressed; and Berlin was interested in his commercial dynamism. Not yet twenty-seven years of age, he was becoming a pivotal figure in the emerging global political economy.

Marconi's company saw its bread and butter as providing equipment, services, and expertise to military and merchant marine organizations, exemplified by its contract for the supply and equipment of wireless apparatus to the British Admiralty on an annual royalty basis, as well as its South African war activity. The new Brussels-based Marconi International Marine Communication Company, launched in April 1900, pioneered the business model of hiring out to a shipping company a full system—apparatus, operators, use of Marconi shore stations—while, from London, the parent company manufactured the equipment and developed contract relationships with governments.

[†] Ducretet is considered the first person in France to transmit signals using radio waves, employing Marconi's method in a transmission from the Eiffel Tower to the Panthéon in Paris on November 5, 1898.

The purpose of MIMCC was to exploit an exclusive licence of the Marconi patents "for all maritime purposes" throughout the world except for countries where the parent company had already made or was in the process of making deals. This was no small matter, as it excluded Italy, the United Kingdom, and the United States, as well as smaller jurisdictions such as Chile and Hawaii. (Chile, like Hawaii, was one of the first countries with which MWTC began developing prospects; the company had a representative working in Valparaiso as early as July 1899.[32]) The new maritime company's structure mirrored that of the parent company. It would be run by the same managing director and have the same technical advisor, Marconi. Its pitch was added safety and security for the world's passenger and trading vessels, and it was anticipated that a large maritime telegraphy business would ensue. MIMCC would work "hand in hand with all Governments, whose interests seem for the most part identical with those of the Company." By the time its maritime subsidiary was created, the Marconi company had feelers out in places as far afield as Portugal, Egypt, India, and Australia.[33] A launch prospectus proudly announced the internationalism of the MIMCC board, which had members from the United Kingdom, Germany, France, Spain, Italy, and especially Belgium. The Marconi maritime system would operate in English, French, German, Italian, Spanish, Portuguese, Dutch, and Latin; in other words, in all of the major European languages.[34]

Notwithstanding his increasingly global reach, Marconi's most important relations were still with the British, rightly perceived as his principal patrons. In Britain, the company was able to circumvent the existing government monopoly on domestic public telegraph communication—which dated from 1868 but did not apply on the high seas or for communication from outside territorial waters. The legislation also applied only to *commercial* telegraphy. Marconi's business was based on providing a service to other private companies; as it was not charging for messages, it claimed to be outside the scope of the law. This was a textbook case of public policy lagging behind a new communication technology—the British law-makers clearly intended telegraphy to be in the public domain; the Marconi company successfully privatized the newest part of it. MWTC would develop an ambivalent symbiotic relationship with the British government, frustrating it by persistently remaining a step ahead of the law while seducing it with schemes that served its policy goals for communication.

A prelude of things to come took place when the British House of Commons attacked the question of telegraphic communication in a long debate on May 22, 1900.[35] The debate addressed fundamentals: Was telegraphy essentially a commercial or a strategic activity? Should it be left to the private or the public sector? If public, what was the appropriate role for private companies? Should there be subsidies? Regulation? The debate was provoked by Sir Edward Sassoon, a reform-minded MP, who called for the establishment of "a system of federated cables between all branches of this Empire" and for a committee of inquiry "into the commercial and strategic defects of our Imperial telegraphic communication." Sassoon had been elected to Parliament with the Liberal Unionist Party in March 1899, slightly more than a year before this historic debate, and was one of a small group of politicians pushing for cheaper telegraph rates, especially in overseas communication.[36]

Sassoon reviewed the great profits the cable companies had made in both domestic and colonial communication, and took a decidedly public service stance: these private companies were serving public interests, and public considerations must prevail. Sassoon pleaded for a comprehensive state policy, "embracing the very wide interests of the constituent portions of our Empire," providing for the establishment of a scale of charges, and, above all, "a system by which the State could acquire the cables if it should think right to do so."

The debate touched on all the great themes of the day: imperialism, capitalism, and social welfare. Seconding the motion, Sir Charles Dilke, one of the British Parliament's most outspoken advocates of human rights, underlined the importance of an all-British cable route to prepare for time of war, with alternate routes in case the cable is cut or friendly states turn unfriendly. Nationalization was the preferable course to take, with assurance that only British nationals were employed in the industry. Liberal MP Sydney Buxton (a future postmaster general) said it was time that the cable companies returned some of the benefits they had reaped to the public that had so generously subsidized them.

Member of Parliament A.S.T. Griffith pointed out that it was up to the state to look after the "growth of community of feeling between the various portions of the Empire" by ensuring cheap communication between the mother country and the colonies. The excessive rates charged by the cable companies were detrimental not only to commercial interests and to strategic interests, but also to the sense of empire. "The Government ought to make every effort to break down this monopoly."

Speaking for the government, Financial Secretary to the Treasury Robert William Hanbury agreed that "these submarine lines are the true nerves of the Empire," but warned that any inquiry should be made by the government itself, not a public committee that would have to air strategic issues in full view of foreign nations. The first lord of the treasury, Arthur Balfour, agreed with his colleague; publicity would be most inexpedient. Persuaded "on patriotic grounds," Sir Edward Sassoon withdrew his resolution, ending the debate.[37]

It was notable that not a word had been said about wireless. However, a seed had been planted. Within a few years critics of the cable companies like Sir Edward Sassoon discovered an alternative. Wireless, the upstart technology, could challenge the cable cartel, provide substantially cheaper service, and work just as efficiently if not more so with the public sector. With almost no effort of his own, Marconi was beginning to make friends in Parliament.

Marconi, meanwhile, had his own sights set on the cable companies, and particularly their long-distance operations. This was where wireless could offer the greatest added value in terms of ease of operation, cost, and, ultimately, profitability. Wireless also had the tremendous advantage of being unregulated, and it stood to capture the popular imagination. Transatlantic wireless communication—"the great thing" he referred to in conversations with intimates—would be the jewel in his crown. The first thing he had to do was convince his board that the effort was worth the investment. They were still not making any money and Marconi's scheme involved building super-powerful stations of unprecedented size, complexity, and cost. Marconi put the full force of his personality into persuading the board to back his plan. As company historian W.J. Baker put it, "It was rather like proposing to build a cathedral in a world which had never seen anything more grandiose than a log hut."[38]

8

"The Great Thing"

Marconi was now a global media celebrity. He was secretly working on a project he referred to as "the great thing"—an attempt to signal across the Atlantic Ocean. Theoretical physicists said it couldn't be done because electric waves (they said) radiated in a straight line out into space and would not follow the curvature of the earth. Holders of this view included the French mathematician Henri Poincaré, who did indeed understand the properties of Hertzian waves, but not those of the earth's atmosphere.[1]

Once again, Marconi was convinced that the theoreticians were wrong; electromagnetic waves would bend to follow the curvature of the earth, he believed. He was also convinced that transatlantic wireless telegraphy would be commercially viable,[2] though he no longer had only his friends and family to convince. He was now under intense pressure from his financial backers to prove this to them. His every move was scrutinized by the press, and with the deep skepticism of a substantial part of the scientific community.

While the long-distance trials were now his number-one priority, multiple distractions kept getting in the way. Nearly every letter he wrote to Josephine Holman during their courtship refers to the preparations and the delays.[3] In the summer of 1900, the company began looking for an appropriate site for its long-distance experimental station in the west of England: they needed about an acre of land, with a regular water supply, on which they could erect a solid single-storied electrical plant surrounded by a huge field of masts and poles. In August, Samuel Flood-Page, the company's managing director, and Marconi found a suitable location adjoining a Coast Guard lookout on Angrouse Cliff, above a spot known as Poldhu Cove near the village of Mullion on the Lizard Peninsula in Cornwall. There was even a small hotel nearby. Flood-Page wrote to the owner of the property, Viscount Clifden, a lease was signed, and construction began a month later.[4]

Flood-Page, inexplicably—given the premium the company placed on secrecy—gave Clifden far more information than necessary. "We shall endeavour to communicate with Cape Finisterre, perhaps with the Azores, and with Newfoundland or America," he wrote in his lease offer.[5] The company also leased a second piece of land from Clifden, six miles away in a wheat field adjoining the nearby Housel Bay Hotel, for a smaller test station that they called "the Lizard."[6] The surroundings were more forgiving than Poldhu, and the hotel there more comfortable, so Marconi and his entourage stayed at the Lizard whenever they could. He expected it wouldn't be long before trials would begin.[7]

The Lizard Peninsula is the southernmost point of the British mainland, and Poldhu was about as close as one could get to America and still be in England—at the tip of the toe at the bottom of the map. It was also far from inquisitive minds and prying eyes. "Today there are a good hotel there and other buildings besides the station buildings themselves," Marconi's ghostwriter wrote in that 1919 memoir, "but in those early days there was little or nothing beyond the landscape which, in spite of its hardness and bleakness, possessed an inexpressible charm because of the soft airs, pungent with salt, that blew over it and because of the black rocks jutting far out into the ocean and lashed by sea foam of dazzling whiteness."[8]*

Marconi travelled constantly between London, Dorsetshire, and Cornwall during construction of the Poldhu station in the fall of 1900. It was by no means a simple journey, especially as Marconi often combined a trip to Poldhu with a stop at the Haven for a quick meeting with his technical staff, and would then leave for Cornwall the following morning. From Poole station, two and a half miles from the Haven, it was a long train ride with three changes, and then seven miles from the station at Helston to the Poldhu hotel by horse-drawn carriage.[9] Alternatively, he might go back to London from the Haven, and take the train for Penzance from Paddington. Sometimes he managed to squeeze in a whistle stop at his station on the Isle of Wight as well. The trip from London to Poldhu took eleven hours in all. Marconi never complained.[10]

---

* Marconi and his associates used Poldhu as a research station until 1935; in 1957 the company gifted the cliff land to the National Trust, and the area behind it (where the station and mast were located) in 1960. Today there is a memorial column to mark the spot.

While Marconi pursued his other work at the Haven, preparation of the new station at Poldhu was overseen by two key figures: the company's scientific advisor, James Ambrose Fleming, looked after the design, and Richard Norman Vyvyan, a recently hired engineer, the construction.[11] Fleming kept up a vigorous correspondence with Marconi, describing every step of the work and looking to him for both large and sometimes miniscule decisions.[12] On December 1, 1900, Fleming was officially named scientific advisor to the company at a salary of £500 a year, after some delicate finessing by Marconi. (It was the same amount that Marconi was then receiving, about $40,000 in today's money.) Fleming had first been appointed in May 1899 at a lower rate (£300), but he felt the work at Poldhu was demanding too much of his time.[13] Fleming was now insisting on an increase and Marconi agreed, but he had to convince his cost-conscious board. With Marconi's assent, the board approved the increase, and Flood-Page gave Fleming the news—with a crucial proviso: "If we get across the Atlantic, the main credit will be and must forever be Mr. Marconi's."[14]

This condition was quintessential Marconi. He believed unquestioningly that it would be possible to signal across the Atlantic, but he had no idea how to actually make it happen. Fleming could figure out how to do it, but only a corporate structure like the company Marconi had put in place could support such a venture. (The cost of Poldhu and the transatlantic research eventually came to £50,000—more than three times the cash payment Marconi had received for his patent four years earlier.) The company's biggest asset was Marconi, who was, indeed, its brand, and he wanted that to be clear to Fleming. Historians have been nearly unanimous in attributing a determining role to Fleming in laying the foundation for the success at Poldhu; by Fleming's own accord, however, the main credit would forever be Marconi's.[15]

Marconi himself wrote to Fleming on December 10, 1900, to forestall any ruffled feathers. Praising Fleming's contribution, he promised, in addition to the company stipend, to personally transfer to him five hundred shares in MWTC "in the event of our being able to signal across the Atlantic." The gesture worked. Fleming—who appreciated money and attention— did not even seem to notice the constriction of Flood-Page's letter; in fact, he deferred to Marconi as the master of the moment. "You have made such stupendous advances in four years that I have the strongest conviction you are destined to solve this transatlantic problem if not at once yet in time. We may have great difficulties but there is certainly in my opinion no impossibility about it and it will revolutionize everything."[16]

Marconi's relationship with Fleming offers a fine example of how Marconi co-opted talent. Fleming would remain on contract with the company until 1931. Under the terms of his agreement, Fleming gave the company rights to all his patents for improvements in wireless telegraphy, one of which, the 1904 invention of the thermionic valve—or vacuum tube—laid the foundation for the broadcast radio receiver.[17] In later years, as Marconi's fame and the trappings that came with it continued to grow, Fleming would occasionally resent his treatment at the hands of the company. In a lecture he gave on behalf of the Royal Society after Marconi's death, late in his long life (he died in 1945), he revised the view he had always championed and claimed that priority for the conception of wireless rightly belonged to Oliver Lodge.[18] In the same lecture, however, he paid tribute to Marconi's talents: "In the first place he was eminently utilitarian.... He had remarkable gifts of invention and ready insight into the causes of failure and means of remedy. He was also of equable temperament and never seemed to give way to impatience or anger, but he did not suffer fools gladly or continue to employ incompetent men. He also owed a good deal to the loyal and efficient work of those who assisted him."[19]

Tests at Poldhu began in earnest in January 1901. Although he made light of it, Marconi was concerned about rumours that Nikola Tesla was preparing a similar move. Indeed, Tesla was making even more sensational noises about his plans, as Marconi wrote to Josephine: "I think the long distance experiment will be a success, but I am afraid Mr Tesla is very visionary about his messages from Mars. I think it is necessary to learn how to send messages over distances in this little world of ours before thinking of sending them across the millions and millions of miles which separate us from the other planets." He was also now planning an imminent trip to the United States to select a site and set up the American side of the "great experiment."[20]

The Victorian era ended with the Queen's death on January 22, 1901, at Osborne House on the Isle of Wight. Although her health had been failing for months, it was still a shock to a monarchist nation that had known no other sovereign for more than sixty years. Victoria's eldest son, Albert, the Prince of Wales (now to be King Edward VII), and her grandson Wilhelm (the German kaiser) were among the family members at her bedside. "What a sad calamity the death of the poor old Queen has been to England," Marconi wrote to Holman from London. "I know she was a very good woman. The war in South Africa grieved her very much."[21]

Marconi spent most of February at the Haven while Fleming and the staff at Poldhu worked feverishly to move the long-distance trials along. By the end of February, Fleming was urging him to set up the US receiving site. He included a newspaper cutting about Tesla setting up a station in the United States, but, he said, "I have no faith in its genuineness." Fleming also urged the greatest secrecy. As he wrote in a confidential memo, "It is obviously essential that Mr Marconi's name should not appear in the matter, or otherwise the price of desirable situations may be raised against him or difficulties may occur in view of Mr Tesla's recent sensational announcements."[22] Even machinery should be purchased through an intermediary, he advised the company, which closely monitored Tesla's activities throughout 1901.[23]

Early in 1901, Marconi was planning a trip to America to scout sites for his US station, but he kept postponing it for various reasons, primarily related to his health and the demands of work. Finally, on March 6, 1901, he set sail on the White Star Line's *Majestic* out of Liverpool, bound for New York, along with Vyvyan. He was multi-tasking as usual. Just before the ship left port he received a cable from board member James Fitzgerald Bannatyne about some issues with the Admiralty contract, and spent the first hours of the crossing composing a long, confident memo to Flood-Page crafted to show the Admiralty that he had the tuning problem under control.[24]

He was also hoping to reunite with his sweetheart, whom he hadn't seen in nine months. "I shall call at 292 W 92 St [the Holmans' residence on New York's Upper West Side] the day I arrive if possible and see my love again," he wrote her. Once again, though, his health got in the way. On arriving at the quarantine station on Staten Island on March 14: "My dear Josephine. . . . I have got rather a nasty cold which I think I caught through having passed too often from the hot engine room to the deck. . . . I fear I shall not be able to come today if I take the Doctor's advice but hope to come tomorrow."[25]

The press was aware that Marconi had arrived in America, though he had kept a low profile as to the purpose of his visit. The scope of the preparations for his transatlantic work was deeply secret, as was his relationship with Holman. The public face of his visit to America, as reflected in the press, was thus, as one journalist put it, "veiled in mystery."[26] It is unclear exactly when and how Marconi and Holman finally did rendezvous (his appointments diary for 1901 is silent on the matter, although an unsigned telegram dated March 20, 1901, indicates that they probably finally met that evening),[27] but his letters show a clear shift in the intensity of their rekin-dled relationship.

> My own darling Jo,
>
> I feel I cannot go to bed without writing a line to my love....I feel so very happy and proud to know you are my very own....
>
> You are in every way my ideal—the girl angel I dreamed of but which I never thought I would meet....
>
> Yours for always and only yours.
>
> Guglielmo[28]

His priority was clearly business, however. Accompanied by Vyvyan and John Bottomley, now vice-president of the American company, Marconi visited a large number of sites on the coasts of New England. All but one of them were deemed unsuitable for various reasons (absence of roads and water, distance from railways or towns, impossibility of obtaining provisions, "and in some cases inability of the supposed owners in producing satisfactory deeds" to prove their title of ownership of the land). The one exception was at South Wellfleet, Cape Cod, in Massachusetts. The spot was less than a mile from the New York, New Haven, and Hartford railway stop, it was eight acres, and they were able to obtain the freehold for only $250. There was plenty of water, Marconi reported to his board on his return, and "a very bad inn is situated about 3 miles away; there is, however, a residential house which we can rent on very moderate terms within 200 yards of the site." It was a *fait accompli*; all the necessary machinery had already been ordered.[29]

Marconi had arrived by boat from Boston at Provincetown, on the tip of Cape Cod, where he was met by Ed Cook, a local businessman who knew the Cape inside out.[30] Cook had interests in lumber, shipping, fishing, salvage, and wrecking, and various real estate holdings; he immediately saw the importance of Marconi's scheme for the area—as well as for himself.[31] He first took Marconi to Barnstable, then the most developed part of the Cape, with relatively good railroad access and accommodations. Marconi, though, wanted direct access to the sea and Barnstable was too far inland. Cook showed him a piece of land next to Highland Light, but the light-house operators would not agree to let Marconi set up his station there (a wireless post would have undermined their major function of spotting arriving ships and cabling the news to Boston). Finally, Cook offered Marconi a piece of his own land, on a high bluff overlooking the beach, at South Wellfleet. Marconi paid Cook the $250, an astronomical sum for what was considered locally to be worthless land. The site offered a completely unobstructed line of contact with Poldhu and was reasonably close to both sea and rail supply lines. For Marconi, it was priceless.[32]

South Wellfleet is located about halfway between the elbow and the fist on what Henry David Thoreau is said to have called the "bare and extended arm of Massachusetts, where a man may stand and put all of America behind him."[33] The Marconi site was at the narrowest point of Cape Cod, barely a mile across (and getting narrower every year as more of the dune succumbs to the sea). Marconi set up his headquarters at the Holbrook House in Wellfleet, then the only place to stay in the area; ever resourceful, according to local lore, he discovered there was a householder in nearby Truro who cooked authentic Italian dishes and he borrowed a horse to ride over for a home-cooked meal. But he didn't stay long on the Cape, leaving the construction in the hands of Vyvyan. Marconi's ghostwritten 1919 memoir described South Wellfleet as "a very dreary spot," the latest in a string of outposts on both sides of the Atlantic that Marconi described in more or less the same way.[34]

There were some delays getting started. In mid-April Marconi was ready to return to Europe but had not yet acquired clear title to the land.[35] By early May, however, the *Barnstable Patriot* was able to report that "a sea wireless telegraph system will soon be installed at South Wellfleet." Its real purpose, though, was camouflaged: "The plant will operate a Marconi system by which signals can be exchanged with passing ships."[36] There was no hint at all that it had anything to do with "the great thing."

Vyvyan proved to be an able organizer and he was soon able to report to Josephine Holman (with whom he often commiserated about Marconi's aloofness) that not only was the station construction going well, he was also building a house "and there will be room for you if you live here at all after your marriage!"[37] Eventually, a twelve-hundred-square-foot bungalow was built to house the station crew; the house was generously furnished, and a piano installed for Marconi's use. Marconi himself, though, was not inclined to come to Cape Cod until he was sure he would be able to make the transatlantic leap. He wasn't good at keeping Vyvyan informed of his plans; for months they were constantly expecting his arrival from one week to the next.

There was no electricity on the outer Cape, so Vyvyan and his team had to generate their own. Building supplies were brought in from Boston by train, and thence on to the site by horse and cart. As it tended to be wherever Marconi installed a station, the activity created a commotion among local residents, and a barbed-wire fence had to be put up to keep bystanders away. The station was modelled on the one at Poldhu, with twenty two-hundred-foot masts supporting a circular inverted cone of antenna wires and one hundred wires suspended from a stay between each pair of masts. The masts were planted in sandy soil about 165 feet from the edge of the cliff.[38] The local people warned that they would not survive the next strong gale and they were right. The South Wellfleet towers blew down on November 25, 1901; fortunately for Marconi's transatlantic plans, he was by then no longer counting on Cape Cod.

The report in which Marconi recounted the establishment of the South Wellfleet station also covered all aspects of the company's US business, including the good prospects for his outstanding patent applications, and a promising lunch with railway magnate Cornelius Vanderbilt, who expressed interest in backing the development of the system in the United States. The report was vintage Marconi: a sense of irony and sophistication, attention to detail, self-confidence, determination to get only the best, and an appetite for success. He was in full control, progressive (in a business sense)—for example, recognizing the importance of having his system developed in the United States by Americans—and honest about his annoyance at the London office's omission to keep him fully informed of home developments in a timely fashion. Indeed, his American experience had him looking down on London's somewhat lackadaisical, laid-back approach.

One thing Marconi did *not* mention, oddly, was a meeting he and Bottomley had with Thomas Edison at Edison's facilities in West Orange, New Jersey, on April 16, 1901. The story of the meeting was told in some detail years later in two letters from a fellow inventor, Ralph Hamilton Beach, to Francis Jehl, a close associate of Edison. Beach had been asked to arrange for Marconi to meet Edison. "I spoke to [Edison] about it and he at once said sure bring him over." Beach did so and spent about four hours listening to a fascinating conversation between Marconi, who "seemed to be a fine young fellow," and Edison. "You know how the old man was in things of this kind. . . . he cross examined Marconi and told him more things about wireless, apparently, than Marconi had ever heard of." Edison brought out a copy of his own 1885 wireless patent. Then he said something like this, according to Beach: "You know Marconi there are sounds going all around us right at this moment right in this room. . . . If we only had the means to detect these tiny vibrations we would be able to select out of all of these the ones we desired to hear. The time will come when we shall be able to just push a button and listen to the finest music, coming from any place on this earth, to converse clearly without the use of any wires."[39] Edison at this time was no longer interested in developing wireless himself, but Marconi was paying close attention.

After about a month in the United States, Marconi sailed for Europe on the *Campania* on April 20, having agreed with Holman that they would announce their betrothal as soon as he had had a chance to tell his mother. He wrote to her immediately from the ship, and without recourse to Morse code: "My own dearest Jo, I got on board all right last night and was pleasantly awakened this morning by a steward who handed me your letter. And what a letter! Your love for me is the one great happiness of my life, and God has been kinder to me than I deserve by giving me your love...."[40]

As he did during his return voyage the previous year, Marconi kept a daily journal for Holman while crossing on the *Campania*.[41] Again, it was largely uninspired; he begins each daily instalment with a weather report: "The sun is shining.... The sea is a bit rough.... Rather foggy this morning...." But there are also loverly caresses: "I dreamed of you twice dear.... If you were on this ship the journey would seem to me very short.... Jo I love you. Sweet love good night.... A thousand kisses from your Guglielmo who does and will always love his Jo...."

He celebrated his twenty-seventh birthday during the crossing. Josephine had thoughtfully sent him a gift (although he had forgotten hers a few months earlier) with instructions not to open it until the day. It was an umbrella, a useful item for a man who lived in London, with his initials on the handle. He was pleased, especially as it had come from her and showed her practical side. He would spend the day thinking of her, he said, and on his next birthday "it will be different.... You will be with me."

A rare offhand remark showed him thinking of his faith, a topic he usually kept to himself. "Although today is Sunday there has been no church service. I think it very wrong. Why should there not be at least something to remind the forgetful of him who holds the 'waters in the hollow of his hand.' Although as you know I am by no means a Church goer I think that on ships in the ocean an informal service should be held."

Most important, though, he shared with her his thoughts about "the great thing." Although the longest distance he had achieved to date was only around one hundred miles, he was absolutely confident that in a matter of months he would send a signal across the Atlantic. Aside from his closest colleagues, his parents, and his brother, Josephine was the only person privy to his plan.

Finally, on the twenty-sixth, the day after his birthday, he sighted land as the *Campania* approached Queenstown (today, Cobh). Soon, he was back in London, and within hours of reuniting with Annie, he cabled Josephine: "Mother sends her love. You may tell. Guglielmo."[42] At last, their engagement could be announced. In fact, Jo had confirmed the news to the *New York Times* two days earlier.[43] Before long it was in the British press as well.

Despite his annoyance at the premature announcement in the press, Marconi was pleased—and relieved—at his mother's initial reaction to his engagement,[44] although she was not quite as enthusiastic as he seemed to think. Annie saw it rather soberly as a sound match, as she wrote her husband: "I have news that will perhaps surprise you. Guglielmo is engaged to an American young lady, and from what he tells me she is a very good girl. I saw her a year and a half ago at the Grand Hotel, she came with us to the Theatre one evening. She is nice, and very pleasant, and has a small dowry. Her father was a judge and her grandfather was President [sic, she means governor] of one of the united states. I have no other information yet. Guglielmo hopes that this will not displease you."[45]

However, displease Giuseppe it did, especially the timing. Before long both of Guglielmo's parents were pressuring him to postpone his marriage plans—which they viewed as frivolous, if not a complete misalliance—until he had completed his transatlantic research, at least. Annie wrote to Giuseppe again: "I am also unhappy that Guglielmo is decided to marry before he has completed his invention, but now things have gone so far that it would be useless to persuade him to do otherwise. The American newspapers are making much praise of this lady." Later she wrote that she had "no more news to give you of Guglielmo's marriage plan which may perhaps take place this autumn if all his American business goes well."[46]

It is hard to tell whether Annie shared her husband's concerns or was trying to smooth the way for her son. Guglielmo hoped it was the latter, and wrote to Josephine on May 21, 1901: "I am sorry to say my father is very angry with me because of our engagement but he gives no reason for his objections. My Mother will I hope put things right for us." On her side, Annie wrote to Holman: "I cannot help feeling anxious about my son who naturally I love beyond expression and who I have always had with me, and cared for up to the present." Nonetheless, she added that he had been very reassuring, telling her how "good and nice you are in every way."[47]

Marconi was receiving congratulations from friends and acquaintances as a result of the published news, but oddly, the fact that he was getting married didn't seem to register with his associates at work.[48] Indeed, while the published news was unequivocal, and widely acknowledged in Marconi's American circle, it was less clear in England, and even less so in Italy, where an impression lasts to this day that the engagement was not more than a rumour fuelled by the press.[49]

Much of this was due to ambivalence on Marconi's part, to his reticence to display signs of his emotional life, to his preoccupations with control and secrecy, which tended to colour all his relations, and to his concern about his parents. At any rate, we see here the emergence of one of Marconi's strengths, a gift for compartmentalization. However much in love he might have been, he would not allow his sentiments to interfere with his professional life. Still, he had discovered a need and a desire for intimacy that would continue for the rest of his life. Marconi's relationship with Holman became the template for those that followed.

As we've seen, Marconi and Holman kept up an intense correspondence between 1899 and 1901. As his side of the correspondence has not been accessible until now, the relationship has never before been fully explored, although Holman's side of the relationship is richly exposed in those four of her letters that remain in the Bodleian archive at Oxford. Separated by an ocean, the intensity of Marconi's work, and the restrictions of the late Victorian era, Holman's letters nonetheless ripple with excitement about Marconi's future. The last of these letters, written while Marconi was struggling with his family's opposition,[50] is the most openly revealing about this remarkable woman. She wrote it at Cragsmoor, in June 1901:

My own dearest love,

My heart felt a dozen great thrills of pride when I found yesterday a whole page of more wireless wonders in the "Herald."...As I read, dearest, I had a kind of indefinable vision of what this century [will] bring forth from science—and the whole part that you will play in this wonderful age we are living in....

....I know that these are the busiest and most strenuous days of your life, dear—and so it must not worry you if you can't write to your Jo even a line, for she isn't entirely uncomprehending even so far away. I know that you love me— this is not a holiday time but the day for unselfish loving. So, my darling, I shall be happy in the imagination of what you think and dream of me, and when the happy time comes when we are together for ever, you can tell me everything....

   Your own true and loving
   Jo[51]

Back in London, things were changing at the company's head office. For some time, Flood-Page—now sixty-eight years old—had been asking for more support; he was finding his position stressful and demanding as affairs continued to expand in every direction. He was feeling isolated and needed the help of a "youngish, active, experienced assistant manager." In March 1901 they found just the man.[52] H. (Henry) Cuthbert Hall was an electrical engineer with a forceful personality and a razor-sharp tongue, well versed in the affairs of London, ambitious, hard-working, and boundlessly energetic. A few years older than Marconi, Hall was an associate of the Institution of Electrical Engineers, an arbitrator in the electrical section of the London Chamber of Commerce, and had attracted some brief notoriety by successfully suing a prominent journal, the *Electrical Review*, for libel in January 1896—just days before Marconi arrived in England.[53]

Hall was appointed manager (Flood-Page remained nominally in charge as managing director) of MWTC on March 21, 1901,[54] coming on board while Marconi was still in the United States. The tone of the company's voice changed immediately. There was a powerful chemistry between Marconi and the new manager. Marconi wanted nothing more than to be free of the company's day-to-day operations, to concentrate on his experimental work—and, at the same time, to have the final say on every major decision. Hall was a perfect sparring partner. Marconi gave him room to move but Hall never overstepped the boundaries. He soon established himself as the crucial gatekeeper between Marconi and the growing array of supplicants who were forever trying to get a piece of him. They even looked alike: both tall, lean, and sober-looking, a bit like two fraternal vicars.[55]

Marconi resumed his travels between London, the Haven, and Poldhu, taking time to deliver an important lecture on tuning to the Royal Society of Arts on May 15, 1901.[56] On May 23, the company celebrated the departure from Liverpool of the first ocean vessel equipped with Marconi instruments for commercial wireless telegraphy, the Elder Dempster Lines' *Lake Champlain*. The same day, Hall sent Marconi the plans for fitting the Cunard liner *Lucania* with similar apparatus.[57] Company business was, literally, steaming ahead.

Marconi was not pleased with the work on the *Lake Champlain*, whose signals were lost about five hours out at sea, and wrote an uncharacteristically abrasive note to his new manager to say so. "I consider the performance very unsatisfactory. Kindly let me have the names of the assistants who installed and worked the apparatus." Hall was tough. Marconi was tougher.

And when the *Lucania* set sail in June its successful transmissions were reported in newspapers from Manchester to New York.[58]

By 1901, the company had eight stations in England and Ireland and was operating on thirty-two ships and stations belonging to the British government; it had stations in France, Corsica, Belgium, and Germany and was operating on ships belonging to several private steamship lines (Cunard, Elder Dempster, North German Lloyd) as well as the Belgian government and the Italian navy.[59] It had dozens of irons in the fire and its sights were set far beyond Europe. In British North America it had deals brewing in places stretching from Newfoundland to British Columbia. In the United States it was engaged in high-profile activities from covering the yacht races to reporting news to and from ships arriving in Boston and New York. An Indian firm had applied to be the company's agents in the subcontinent; it was negotiating with Lloyd's about building installations on the Upper Nile between Khartoum and Fashoda, and had applied to put up two stations on the Suez Canal. In Brazil, the Amazon Telegraph Company had inquired about equipping stations along the Amazon River.

Marconi kept his hand in every project, flitting from one to another like a juggler trying to keep so many plates spinning on top of sticks. Busy as he was conducting and overseeing his research, Marconi was a key figure in every business development. As the company's largest shareholder, he had the biggest direct stake. He was also the company's most important asset as well as its brand, and people wanted to know they would be doing business directly with him. Increasingly, both he and his companies saw the world as a single unit and were developing a global corporate strategy.[60]

Europe was, of course, the trickiest, especially as communication was so fundamental to the imperial rivalries of the time, and several countries were grooming possible competitors to Marconi. The MIMCC board, meeting in Brussels on January 7, 1901, decided that the best approach (also considering Marconi's personal connection to King Leopold) would be to create a Belgian company that would then form its own national subsidiaries in the various other countries. It would simply be impossible to obtain concessions in places like France, Spain, or Portugal—never mind Germany—if it was known that an English company was behind the scheme. It is hard to imagine governments failing to see through this ploy, but, be that as it may, that was the approach the company took, and in mid-1901 a committee consisting of Marconi, Flood-Page, and director Albert Ochs was charged with forming a Belgian company, with MIMCC as the sole shareholder.

The Belgian company, to be known as la Compagnie de télégraphie sans fil, would, in its own turn, create and own the national companies.[61]

In order to keep its international patents active, the company had to send an assistant to each country where it had a patent once a year to work the system for a brief period, even where there was no real activity going on. This was a costly process, and that alone was inducement to try to set up installations everywhere, and hopefully soon turn a profit. Typically, they would try to first get a concession from a small country and then leverage that using the incentive of a communication link with a larger one. In September 1900, for example, one of Marconi's chief assistants, C.E. Rickard, travelled to Malta to demonstrate the system and try to convince authorities to set up a Marconi station "for the purpose of communicating with passing ships and with Sicily." If the Maltese bought in, the Italians would have to follow suit with a similar station in Sicily.[62]

The further away they went, the more difficult it was to keep effective control over developments. In Valparaiso, Chile, a company representative learned that an agency had applied for a wireless patent on behalf of Marconi's German archrivals, Adolf Slaby and Georg von Arco. Fortunately, the Chilean naval minister wanted Marconi's system and none other.[63] All this frenetic activity meant that Marconi had to try and effectively establish a monopoly. Patent protection could keep pretenders from entering Marconi's markets, but the company had to occupy the space with active operations, and to do that they needed contracts or concessions.

They also had to deal with industrial espionage. In November 1901, the company learned that there seemed to be "a traitor" inside the Belgian government who knew the terms the government had arranged with the company's Belgian associates, and had alerted the Slaby-Arco group so that they could put forward a cheaper proposal. "We must not take it that we are going to lose this contract, but it brings a very sharp reminder that the Germans are very active," Flood-Page wrote to Marconi.[64]

While work on "the great thing" proceeded at Poldhu, Marconi realized it would be wise to test his new high-power station over a shorter distance before attempting to send a signal across the Atlantic. The opportunity arose in June 1901, when the company opened a new station at Crookhaven, in West Cork. The most southerly port in Ireland, Crookhaven had long been a supply point for vessels setting out for America and the West Indies, the last stop for mail and provisions before heading into the open Atlantic, and the first port of call for ships arriving from the other side. At the height of its importance as a shipping hub in the mid-nineteenth century, it was said that you could cross the harbour on the decks of boats.

Crookhaven's eighteenth-century church is dedicated to St. Brendan the Navigator, a sixth-century Irish monk from Tralee who, according to myth, set out to cross the Atlantic Ocean in search of the Garden of Eden; some even believe that Brendan was the first European to reach the shores of North America. More recently (and historically verifiable), in the nineteenth century, Paul Julius Reuter's news agency based staff in Crookhaven to go out and meet incoming ships that would drop canisters with news that Reuter's staff would retrieve and then telegraph to Cork, Dublin, and London.[65] Crookhaven was thus used to playing an important role in transatlantic communication, and this symbolic history was not lost on Marconi, who may have seen himself repeating St. Brendan's voyage electronically or, at least, more prosaically, modernizing Reuter's methods.

Crookhaven was the sort of place that attracted Marconi and sparked his imagination; a complicated seven-hundred-mile journey by rail and boat from Poldhu and twenty miles from the nearest train station, it was tiny, mountainous, wild, remote, and hard to reach, except by sea.[66] According to one of his staff, "The first impression on arriving at the village of Crookhaven is that 'the end of everywhere' has been reached."[67] After spending three weeks there in June and July 1901, Marconi described it in a letter to Holman: "This is a very wild place. The people are very wild but very poor. I hope to take you here some day (I don't expect however you would stay long!) It is nearly as bad although not quite so bad as Cape Cod."[68]

Crookhaven was considered extremely well situated for short-distance sea telegraphy,[69] a potentially important revenue stream for Marconi's company—the further out to sea that ship-to-shore signalling could take place, the more commercial wireless apparatus the company could lease. Marconi initially established a station in the village but soon moved it to the top of Brow Head, two miles outside town. Here, three hundred feet above the sea, he shared the site with a Lloyd's signal post that used semaphore and flashing lights to communicate with passing ships. Before Marconi's time, the Lloyd's station was part of a network that ringed the south of Ireland, recognizable in the distance by characteristic block towers. A cluster of buildings may have been used as operators' residences. Two overgrown tracks, the width of a car, lead from the road up to the remote hilltop, begging the question, How did they get up there every day?

In October 1901, Marconi succeeded in signalling 225 miles across the Celtic Sea from Poldhu to Crookhaven, then the longest distance ever achieved, and this convinced him that he was at long last ready to try a transatlantic signal.[70] His purpose fulfilled, he was gone; his last documented visit to Crookhaven was that same month.[71] As in many of the other places he passed through, Marconi didn't stay long, he didn't mix much, and he left when his work was done. The station continued to operate until 1922, when it was closed in the wake of the Anglo-Irish conflict.[72]

The building blocks of our modern communication system lie strewn about at forgotten places like this. There is no marker whatsoever at Brow Head, but Crookhaven today is a lively tourist stop, and a supply point for yachting regattas, with souvenir shops and pubs serving deep-fried seafood (and little else). One of the main spots, The Welcome Inn, was run for many years by a former Marconi operator, Arthur "Daddy" Nottage, who came out from England to work for Marconi, married a local, and stayed for sixty-five years. The nearby larger town of Schull has fine dining, comfortable B&Bs, including at least one run by Swiss expats, and an ATM. But aside from a billboard advertising rental apartments in a complex known as Marconi House, the only nod to Marconi in the area is in a display at Mizen Head, six miles away. It is as though the local chamber of commerce doesn't want people trekking out to a remote hilltop when they could be better spending their time and money in the souvenir shops and pubs in town.

❊   ❊   ❊

Marconi was now establishing the collaborative network on which he would count for the rest of his career: scientific or technical experts like Fleming and Vyvyan, lawyers and financial promoters like Moeran and Bottomley, managers like Flood-Page and Hall. His affairs were well on their way in the United Kingdom, the United States, and northern Europe. Next, he made a crucial contact that would strengthen his already substantial ties with Italy.

Marconi had had little direct contact with Italy since his experiments at La Spezia in 1897, and in fact felt that the Italians were trying to pursue the development of wireless on their own.[73] Now, in the summer of 1901, they sent a young naval lieutenant to England to find out what was up with Marconi's system. Marconi had met the officer, Luigi Solari, during the Spezia demonstrations. They had attended the same middle school in Florence in the mid-1880s; Solari had been a year or two ahead of Marconi and claimed to actually remember him.[74] Solari had been put in charge of the Italian navy's department of radio-telegraphy and was receiving reports of Marconi's progress from the naval attaché at the Italian embassy in London (to whom Marconi was nominally attached as the condition of exemption from his military service). In August 1901, Solari came to England to look up Marconi and see for himself.

Solari was the hereditary *marchese* (or marquess, a noble rank above a count and below a prince) of Loreto, a hill town near Ancona in Italy's Marche region, on the Adriatic Sea to the east of Tuscany. Born into a merchant family in Turin in May 1873, he graduated in electrical engineering from Turin Polytechnic and studied at the newly opened Naval Academy in Livorno before being commissioned as an officer in the Italian navy in 1890. From 1901, the rest of his career would be tied to Marconi's. Eventually, he would become the anchor of Marconi's interests in Italy, the manager of his Italian affairs, the antenna that kept him informed of the ins and outs of Italian politics, the most influential of his associates, and also his first biographer.[75]

Their relationship started auspiciously. According to his own account, Solari turned up at company headquarters in London's financial district requesting a meeting and was told that Marconi was too busy to meet him. After hanging around London for a month, he managed to cajole a doorman into revealing the three places Marconi was likely to be found—the Haven, Poldhu, or the Lizard. Solari cabled all three places, saying he was on his way; when he received a reply from the Haven asking him to wait in London, he guessed Marconi was there and sped immediately to Poole, where Marconi agreed to see him.[76]

In later years, Solari added various embellishments to this story, but the bottom line was that Marconi had almost written off Italy as a source of support, and felt rather aggrieved by the Italian Royal Navy's lack of acknowledgement after its initial interest.[77] A man of great charm and aplomb, Solari was able to restore the relationship; it helped that he was armed with the news that Marconi was to be awarded the gold medal of the Italian Science Society at a gala dinner in London. More important, Solari also had with him an adapted version of Marconi's coherer—small and transportable—that he and his colleagues at La Spezia had developed; this piece of equipment came to be known as "the Italian navy coherer" and would be used in the successful transatlantic signal a few months later.[78] Solari stayed on for another month, observing Marconi's experiments at the Haven and Lizard stations and eventually being allowed to visit the highly secret installation at Poldhu as well.

Marconi was now getting close to completing "the great thing," as well as planning his next trip to America to carry it out. As with the previous trip, though, this one was postponed again and again. By mid-August he didn't expect to come before the end of October. He had a great deal of work, he wrote Holman, and was not in the best of health. His mother was in Italy and he missed her. Marconi was also probably worried about his father's stubbornness in withholding approval of his marriage. As the months passed, his letters became more focused on his work and less on romance.

Everything was now geared towards the transatlantic trial. The 225-mile connection between Crookhaven and Poldhu had absolutely convinced Marconi that he could communicate over ten times that distance. He still had to secure financial backing. In the face of investor uncertainty and reticence over the cost, and pressure to concentrate on revenue-producing activity, Marconi now had a powerful new advocate in the head office: Hall. As Marconi continued moving about, the new manager kept an eye on all the pieces from London and reassured his boss. Still, the preparations were hit with delay upon delay.[79]

On September 17, 1901, a heavy squall struck the Poldhu towers and brought down the entire structure with its twenty two-hundred-foot masts. All that was left was a pile of splintered timber and tangles of wire.[80] Marconi immediately ordered the rebuilding of the masts, but he now got into a disagreement with Hall over the role of his trusted assistant George Kemp. "Am I to understand that you consider Kemp the most competent man for the work?" Hall asked. He had been informed that the staying of the masts had been insufficient, and if Kemp was responsible for their disposition and number (which he was), "it is conclusive evidence to my mind that he is not competent to plan the work."[81] To Hall, Kemp appeared to be "a very good man for carrying into execution other people's arrangements," not planning them himself. But however exacting, Marconi had "his men," from whom he expected everything and to whom he remained loyal. Kemp remained in charge of the masts.

Marconi estimated that the Poldhu disaster would set back "the great thing" by at least three months.[82] To speed things up, he decided to put up two temporary masts, only 160 feet high, with a much smaller aerial supported beneath them.[83] He also made another fortuitous decision, to try to make the transatlantic connection with Newfoundland, the British colony that was more than nine hundred miles closer to Poldhu than Cape Cod. Cape Cod was nearly thirty-one hundred miles from Poldhu, while the

distance to Newfoundland was just over twenty-one hundred miles—
significantly, less than ten times the distance that had just been achieved
from Poldhu to Crookhaven.[†]

The board was appalled by the collapse at Poldhu and on the verge of
abandoning the scheme. But Marconi, bursting with energy and supported
by Hall, forged ahead. A ten-acre site was found at Cape Race, on the Avalon
Peninsula at the southeastern tip of Newfoundland. Cape Race was often
the first sight of land for westbound transatlantic travellers; the site chosen
for Marconi was on a flat plateau 120 feet above sea level; half a mile from a
lighthouse; two minutes' walk from a telegraph station; well supplied with
fresh water; and only fifty hours from New York by steamboat, a relative
breeze. According to Hall it was "very satisfactory... [and] absurdly cheap—
about 30 cents per acre."[84] The key directors were brought on side and on
November 7, 1901, the company signed an agreement with the government
of Newfoundland "to install telegraph stations as required in Newfoundland
and on the Labrador" for which they would be paid a royalty of £100 per
station per year for ten years, after which the stations would become the
property of the government. This became the announced pretext for
Marconi's trip.[85]

There was, however, an apparently insurmountable obstacle: after the
successful opening of the transatlantic submarine cable in 1866, the Anglo-
American Telegraph Company had acquired Cyrus Field's 1854 monopoly
concession for telegraphic communication to and from Newfoundland.[86]
Marconi's New York representatives had checked the Anglo-American
charter and informed London that without the consent of the Anglo-
American it was doubtful that Marconi could operate in Newfoundland.
The Anglo-American charter covered not only cable but also "other means
of telegraphic communication" from any place whatsoever.[87]

With typical chutzpah, Marconi paid no attention to this information
(later feigning astonishment and claiming that he didn't believe the concession

---

[†] Marconi always said the distance he had signalled between Poldhu and St. John's was "1700
miles" (which is also the figure given on the official Newfoundland Heritage website). According
to the online distance calculator at http://www.globefeed.com/, the distance from Poldhu to
St. John's is actually 3,430 kilometres, which equals 2,131 miles, or 1,851 *nautical* miles. Poldhu
to Cape Cod is 4,924 kilometres, equivalent to 3,060 miles or 2,657 nautical miles. According to
the same calculator, Poldhu to Crookhaven is 353 kilometres, or 220 miles (190 nautical miles).
Marconi always said it was "225 miles," so he was being inconsistent, sometimes referring to the
distance "as the crow flies," and sometimes in nautical miles. Here (and elsewhere), except when
citing Marconi, I am using the figure for miles.

applied to tests); he did, however, modify his plans once again. Rather than build a new station at Cape Race (more time, more money), his scheme was now to set up a temporary installation, using kites and balloons instead of fixed masts, and try to receive a single signal from Poldhu. He would worry about the rest later.‡ Supported by his most important directors, Henry Jameson Davis and William Goodbody, as well as Hall, Marconi now began planning the details of the trip. Rather than wait passively in London, Goodbody and Hall set off for New York to move the company's North American business along, in anticipation of good news from Newfoundland.[88]

Marconi's decision to go to Newfoundland had enormous consequences for his relationship with Josephine. In addition to being closer to Poldhu than Cape Cod, Newfoundland was also more than a thousand miles from New York. Two new complications had already entered the mix: Helen Holman was becoming anxious about the length of her daughter's engagement, which had people beginning to talk. And there was something Marconi was reluctant to reveal. On October 15, 1901, in the course of a long letter about the stalled progress of the transatlantic preparations, he wrote that he thought "it would be better for you to make your plans for the winter, and trust that things may come right soon and the clouds lift." He promised to tell her shortly about the "one thing" that had been making him unhappy.[89]

It took another few weeks to put everything in place for the new plan. On November 4, Marconi cabled Kemp at Poldhu: "Please hold yourself in readiness to accompany me to Newfoundland on the 16th inst. If you desire holidays you can have them now."[90] By the sixteenth, though, they were still not ready to leave. The Marconi Archives in Oxford have a thick file of correspondence between company secretary Henry Allen and agents for the various shipping lines, as well as "correspondence and papers about obtaining balloons and hydrogen to fill them, and shipping these to Newfoundland, including sample of material for the balloons."[91]

Word inevitably leaked out. The New York Herald heard that Marconi was planning to go to Newfoundland to try experiments with balloons. But no one imagined quite what he actually had in mind. Marconi put out word that he was hoping to communicate with Grand Banks fishing vessels three hundred miles off the coast at Cape Race, and that in itself was news enough.

‡ A Marconi wireless station was eventually built at Cape Race in 1904 and played a key role in relaying news of the sinking of the Titanic in 1912.

"My mission is to first establish stations along the south coast of New-foundland, and thus to enable incoming vessels to get more quickly into communication with the mainland," he told the Liverpool *Journal of Commerce*. Marconi had the press in his hand at this time, and they would report whatever he told them.[92]

While plans were being finalized, Marconi wrote to Holman on November 18. His tone was cool and formal ("Dearest Josephine" rather than "Dearest Jo"); the flames of passion were flickering if not altogether extinguished. He was now only days from his departure. The pressures of work were, he admitted, excruciating. He hadn't heard from her and was not even sure where she was. And he still had an unspecified concern, which he raised again. "I am not at all happy about a thing I may write you about soon but don't be unhappy too."[93]

On November 22, Marconi provided detailed instructions to the Poldhu staff as to how to work the plant: "When I wish you to start sending across, I shall telegraph the date to the Office in London; this date will be immediately wired on to you and from this advice you are to understand that on the date named in the telegram you are to start sending the Dot programme from 3 p.m. to 6 p.m., Greenwich time, and are to continue sending same programme during the same hours every day (Sundays excepted) until you are instructed to stop."[94] Marconi left a set of even more elaborate instructions with Allen:

> If I cable the word "Accepted", it will mean that the signals are being received....
> If I cable the word "Press" it will mean that you are to communicate the information to the Press.
>
> All messages preceded by the word "Riviera" are to be re-transmitted by telegraph to Poldhu, and special care must be taken to obtain proper acknowledgements of these messages in order to remove any possibility of mistake in the transmission between London and Poldhu....

And a certainly unnecessary afterthought: "Please keep this letter entirely private."[95]

Place was booked for Marconi and his entourage—which included Kemp, another assistant, P.W. Paget, two tons of iron tunings, and several vats of sulphuric acid—to sail from Liverpool to St. John's on the Allan Line's *Carthaginian*. The ship was due to sail on November 23, but Marconi couldn't complete the preparations in time and re-booked on the *Sardinian* out of Liverpool three days later.[96]

Finally, on Tuesday, November 26, 1901, Kemp recorded in his diary: "Took Mr Marconi's baggage on board *Sardinia* [sic] after breakfast & then returned to Allan Line office . . . to enquire about Acid . . . (had lunch) . . . then went on board the *Sardinia* [sic] & had dinner at 7.30 pm. At 9 pm we started to leave the dock."[97] Only hours before sailing, they received word that the Cape Cod station had been damaged by a storm similar to the one at Poldhu, vindicating the choice of Newfoundland and providing yet another example of Marconi's Irish luck.

9

Newfoundland: The World Shrinks

Guglielmo Marconi's time in Newfoundland is perhaps the central piece of his saga, or at least that is how it has been interpreted over the years. It has been chronicled, fictionalized, dramatized, and commemorated in stage and radio performances, screenplays, puppet shows, websites, TV documentaries, and videos on YouTube. In Newfoundland, it became one of the foundational stories of the colony (and later, Canadian province), despite the fact that Marconi spent less than three weeks there. Nearly every minute of those nineteen days has been documented in press reports, diaries, government minutes, and legal briefs. The story of the first transatlantic signal is the keystone to the Marconi mythography.

Newfoundland, the pre-Columbian home to the Beothuk people, began receiving European visitors around a thousand years ago, starting with a Norse settlement in the eleventh century; followed by seasonal migrant cod fishermen from just about every Atlantic seafaring country; the Venetian sea captain Zuan Chabotto (also known as Giovanni Caboto, or John Cabot), who sailed on behalf of King Henry VII of England and is said to have landed at Cape Bonavista in 1497; and Sir Humphrey Gilbert, who claimed the island for Queen Elizabeth as England's first North American colony in 1583. The Beothuk did not survive the European encounter and were extinct a half century before Marconi's time (although some contemporary Mi'kmaq in eastern Canada today claim Beothuk ancestry).

Newfoundland changed hands several times during the French-English wars of the so-called long eighteenth century (1697–1815), then settled into British colonial status, nonetheless developing a distinct culture and identity that set it apart from the rest of British North America. (Among many other points of distinction, it is one of the rare places in the world that lies between two standard time zones, three and a half hours earlier than Greenwich Mean Time, and one and a half hours ahead of New York.) A legislative assembly met for the first time in 1832 and Newfoundland obtained "responsible government" in 1855, opting to stay out of Confederation when three other British North American colonies united to form the Dominion of Canada in 1867.[1*] Whether or not it was a good idea to become part of Canada had been a topic of controversy for some time when Marconi arrived.

Ironically, Marconi nearly didn't get to Newfoundland at all. The passage took longer than usual, ten days, and Marconi found it "awfully rough."[2] On the morning of Friday, December 6, 1901, George Kemp noted in his diary: "Sighted Block House St-Johns at daybreak Saw Iceberg to the north & whales spouting south of the Harbour. Landed at Shea Wharf & put up at Cochrane House...where Premier [Sir Robert] Bond had apartments."[3] Events then moved at amazing speed. That evening the party met with Newfoundland government officials and the following morning they moved all their instruments and equipment into an abandoned diphtheria hospital on Signal Hill, overlooking St. John's Harbour.

Originally called "the Lookout," Signal Hill got its name in 1762, when General William Amherst targeted the promontory in what became the final battle of the seven-year war between Britain and France. The name was first recorded in Amherst's journal, where he referred generically to "the signal hill" overlooking the village of Quidi Vidi on the inlet of Cuckold's Cove, and it stuck. The Battle of Signal Hill sealed the fate of Newfoundland; the irony of the place name was certainly not lost on Marconi.[4]

Marconi's 1919 ghostwritten memoir includes an account that describes how he picked the location: "After taking a look at the various sites which might prove suitable, I considered that the best one was to be found on Signal Hill, a lofty eminence overlooking the port and forming the natural bulwark which protects it from the fury of the Atlantic gales...." There was a small plateau on top of the hill, and a crag on this plateau with a new signal station,

---

* The three British colonies became four provinces, as the existing United Province of Canada was divided into Ontario and Quebec, and joined with Nova Scotia and New Brunswick. Three more provinces, Manitoba, British Columbia, and Prince Edward Island, had entered Confederation by the time Marconi came to Newfoundland, and two more, Saskatchewan and Alberta, joined a few years later. Newfoundland became a British Dominion in 1907, and, after three referendums, became the tenth Canadian province in 1949. In addition to the ten provinces, which have distinct constitutional powers in areas such as education and health care, Canada also has three "territories," the Northwest Territories, Yukon, and Nunavut, whose powers are delegated by the federal government.

the Cabot Tower (what Kemp had referred to in his diary as the "block house"). A few feet away from the tower was a small structure that had served to quarantine fevered diphtheria sufferers. "It was in a room in this building that I set up my apparatus and made preparations for the great experiment."[5]

Marconi was exhilarated to be in action at last. He enjoyed the attention he was getting and did not even mind the bitterly cold and blustery weather. A letter to a friend identified only as "Jumbo" shows him in a relaxed and light-hearted mood: "When I come to the City in the evening I have a d—good time. I know practically every one here and they are all very 'nice' and do their best to entertain me etc. I am going to a reception at the Governor's tonight. He has told me he is going to ask some charming ladies which is something to look forward to."[6]

The only cloud on the horizon was his lingering relationship with Josephine. He wrote soon after he got to St. John's. He had not heard from her in more than a month and was still clearly anxious about the "serious matter" he had mentioned in his last letter. Having previously described preparations for "the great thing" excitedly and in minute detail, all of it was now reduced to a single line: "I expect to try some experiments here in about a week."[7]

And that was the last of Marconi's letters to Josephine Holman. The difference in mood between this and his letter to Jumbo couldn't have been greater: "Having a very ripping time here.... Every one is exceedingly nice to me but I have to carry out experiments on the highest hill on this side of Newfoundland and am there all day long."[8] Who Jumbo was we haven't a clue. One would think it was most likely a male chum; "Jumbo" sounds like an English boys-school nickname and is certainly not a term of endearment one can imagine a Victorian gentleman using for a lady friend. On the other hand, as some have speculated, "Jumbo" could have been the "serious matter" that caused the breakup.[9] Either way, the breezy tone was a far cry from the earnest notes he had been sending to Holman. After that, all she got from him was a series of truncated telegrams.

On the work front his adrenalin was flowing. A letter waiting for him when he arrived in St. John's, from Cuthbert Hall in New York, informed him that Hall had met with Marconi's archrival Nikola Tesla, who had boasted that he was working on a one-hundred-foot "spark," much more power than Marconi had at Poldhu and easily enough to send a signal across the ocean. Tesla seemed to have no idea how far Marconi was ahead of him and was so confident that he had no qualms about providing Hall with details of his Long Island installation, although he said he had constructed no receiving station in Europe as yet. Hall noted that this may have been untrue, but he pressed Marconi to keep things moving quickly so that they could be first. Marconi had, in Hall's mind, done exactly the right thing in going to Newfoundland.[10]

Marconi and his assistants spent Monday and Tuesday (December 9 and 10) making preparations, setting up, and testing the improvised aerial system of balloons and Baden-Powell kites they had brought from England. Characteristically, they had little to do with the people of St. John's, keeping to themselves and staying quiet about what they were up to.[11] The barrack near Cabot Tower proved a good location, offering enough room inside for Marconi's equipment and enough room outside to fly the kites.[12]

According to Kemp, Marconi, excited as he was, also fretted, mentioning that one of the things they had to do when they got back to England was work on a more sensitive reception system. The first trials were attempted on Wednesday, December 11, and Kemp recorded: "Got Balloon up in a strong breeze & lost it at 3.9 m when it was blown away. Mr Marconi trying to get sigs during the time Balloon was up on various receivers calling me in at interval."[13] Marconi himself later wrote that a strong gust had broken the rope and that the balloon and four copper wires were lost, "nothing more having been heard of it since."[14]

What happened next has been the subject of so many reconstructions that it remains difficult to render a factual account separate from the various stories that have been told. As we have seen, Kemp's diary entries are usually clear, if laconic; but on Thursday, December 12 he was more, well, telegraphic. At the top of the page—in text possibly added later—he has written: "Cut Sigs 3 Dots." And then: "Lost first kite with two wires each 510 ft long after being up for 1 hour then got-up another kite with one wire 500 ft long & kept it up 3 hours which appeared to give Sigs good." The extended version of Kemp's diary, where he elaborated on events (relying on long-term memory and with some thirty years of hindsight), says: "We received the three dots or the S signal repeated."[15]

Marconi also carried around a small pocket diary in those days that he used occasionally for recording experimental notations. The notations are often stripped of any context and not necessarily placed on the pages bearing the dates when they were made. But under December 12, 1901, partly obscured by other notations that he may have made earlier or later, he has written: "Sigs at 12.30 1.10X and 2.20," and on December 13, 1901:

"Sigs at 1.38." These entries are the only ones in ink; the others are in pencil. In later years, Marconi frequently referred to these notations, as well as Kemp's diary, as evidence of the time and date the signals were received.[16]

What is perhaps most unusual is that neither Kemp's nor Marconi's diary indicate they felt that anything extraordinary had taken place. The entries are matter-of-fact and unadorned. In later years, they both embellished the story, turning it into drama. In Marconi's 1919 memoir, for example, the only thing missing is a pirate ship on the horizon:

> It was a bluff, raw day; at the base of the cliff, three hundred feet below us thundered a cold sea. Oceanward, through the mist I could discern dimly the outlines of Cape Spear, the easternmost reach of the North American continent while beyond that rolled the unbroken ocean nearly two thousand miles of which stretched between me and the British coast. Across the harbour the city of St. John's lay on its hillside, wrapped in fog. The critical moment had come for which the way had been prepared by six years of hard and unremitting work in the face of all kinds of criticisms and of numerous attempts to discourage me and turn me aside from my ultimate purpose....[17]

For this "critical moment," Marconi was using an adapted version of Bell's telephone receiver, which he expected to be more sensitive and reliable than the Morse inker he usually used. "Suddenly, about halfpast twelve there sounded the sharp click of the 'tapper' as it struck the coherer, showing me that something was coming and I listened intently. Unmistakeably, the three sharp little clicks corresponding to three dots, sounded several times in my ear." He had it corroborated by Kemp, "and I knew then that I had been absolutely right in my calculations." What he was hearing were signals being sent out from Poldhu across the Atlantic, "serenely ignoring the curvature of the earth which so many doubters considered would be a fatal obstacle."[18]

In an exhaustive report to his company directors on December 23, 1901, Marconi put it more prosaically: "On Thursday the 12th, a successful attempt was made to rise one aerial by means of a kite, and at 12.30, 1.10, and 2.20, the pre-arranged signals from Poldhu were received in a manner which left no room to doubt their authenticity by myself and Mr. Kemp, on the telephone receiver, but not on the ordinary receiver. Signals were also received, but less distinctly, on Friday the 13th. On Saturday the 14th, a strong gale from the northwest made it impossible to elevate either kites or balloons." As this was freshly written, for Marconi's closest associates, and presumably confidential, it can be considered the most authoritative of Marconi's many accounts of the events of those days.[19]

The distance was nearly ten times the length of the previous record, and confirmed Marconi's conviction that there was no limit to how far wireless communication could travel.[20]† Yet rather than announce the remarkable news immediately, he decided to try for more and clearer signals. However, the weather got worse and nothing further was received from Poldhu. Finally, on Saturday, December 14, after cabling London for advice as to the wisdom of making the news public, Marconi got the go-ahead and announced the news to the press (first informing the Italian government as well).[21]

There was great excitement in St. John's, and a media frenzy ensued. Newspapers around the world echoed the *New York Times*, which trumpeted: "Guglielmo Marconi announced tonight the most wonderful scientific development of recent times."[22] Still, the press also picked up on the question of whether Marconi had actually heard the signal or imagined it. No one was suggesting fraud, or anything like it—Marconi was in some ways above suspicion by this point—but there was a certain amount of confusion. Thomas Edison, for example, at first declared he thought the feat was impossible, then reversed his position and stated, "If Marconi says he did it, it must be so." (Edison had his own reasons for embracing Marconi, as Tesla was his nemesis as well.) A grateful Marconi telegrammed thanks to Edison.[23]

The endorsements signalled that the days of doubting Marconi were over. The physicist Michael Pupin, a prominent professor of electrical mechanics at Columbia, was the first American authority to anoint the moment, declaring that "Marconi deserves great credit for pushing this great work so persistently and intelligently, and it is only to be regretted that there are so many so-called scientists and electricians who are trying to get around Marconi's patent." As though on cue, Amos Dolbear declared from Boston that he accepted Marconi's announcement as fact.[24]

Aside from Marconi's and Kemp's recollections and diary notations, there is only one claim to an eyewitness account of what happened that day (Marconi's second assistant, Paget, was ill and not present). In 1951, for the fiftieth anniversary of the event, the company's in-house newsletter, the *Marconi Review*, published reminiscence by Captain H.J. Powys, who said he had been present on Signal Hill in his official capacity as aide-de-camp to the governor of Newfoundland. According to Powys, Premier Sir Robert Bond, in top hat

---

† Regarding the distance between St. John's and Poldhu, see footnote on page 166.

and frock coat, Governor Sir Cavendish Boyle, other government officials, local reporters, and a crowd of ordinary onlookers all witnessed the "great experiment." Given Marconi's own accounts, his penchant for secrecy, and the fact that no announcement of the news was made until December 14, this is more than unlikely; it is impossible. There were no witnesses, aside from Marconi and Kemp. Nonetheless, Powys wrote fifty years later: "I can still distinctly remember the excitement of that moment and can still visualize the scene as Marconi was congratulated by those present." Powys's tale speaks to the failings of memory and the desire to be at the centre of history that are so much a part of this story.[25] What he described was evidently the re-enactment that Marconi staged five days after the event, on December 17, immortalized in a remarkable series of photographs by James Vey that have been reproduced countless times and often represented as a record of the event itself.[26]

At any rate, the acceptance of the transatlantic connection by Edison, Dolbear, Pupin, and others sealed Marconi's reputation. Oliver Lodge later described it as "an epoch in human history."[27] Technically, scientifically—as with his previous breakthroughs—Marconi's accomplishment changed understanding of the properties of the earth's atmosphere. In 1901, science insisted that it was impossible to communicate across the Atlantic because Hertzian waves were thought to travel in straight lines, like light. Prompted to try to explain how Marconi had been able to receive a Hertzian wave signal nearly two thousand miles away, theoretical physicists Arthur Kennelly and Oliver Heaviside hypothesized a year later "that there might exist an ionized layer in the upper atmosphere capable of reflecting or refracting radio waves of certain frequencies back to earth."[28]

This, it turns out, is how Marconi's signal travelled from Poldhu to St. John's—by reflecting the signal-carrying waves off the "ionized layer in the upper atmosphere" so that they *did,* in effect, appear to follow the curvature of the earth. Like Columbus stumbling onto the New World, Marconi thus discovered the ionosphere while looking for something else. But Marconi wasn't interested in natural phenomena like the ionosphere; he was interested in connecting people and places—in *communication.* His tinkerer's experience told him it was logical that wave communication should be possible over greater and greater distances, and he made it happen by following hunch and intuition rather than scientific theory.

On Monday, December 16, after spending the day scouting possible locations for a permanent station near St. John's, Marconi was having dinner at his hotel with William Smith, a senior official of the Canadian Post Office who was also staying at the Cochrane House and with whom he had struck up a friendly relationship.‡ Marconi was telling Smith he had decided to build a permanent station at Cape Spear (just outside the city) when a messenger entered the dining room and handed him a letter from lawyers for the Anglo-American Telegraph Company, threatening legal action unless he ceased doing telegraphic communication in Newfoundland.[29] According to Smith, Marconi "showed much distress." After a moment or two of silence, Smith had an idea, and said to Marconi: "Drop those people and come to Canada."[30]

Marconi later said that he was absolutely astounded to receive the order to cease and desist—despite, as we've seen, having been forewarned that the Anglo-American had a monopoly and could block him from working in Newfoundland. Marconi immediately replied that he had no intention of infringing any rights the Anglo-American might have in Newfoundland and had already decided to remove his instruments and discontinue his tests.[31] The Anglo-American's legal threats added another dimension to Marconi's aura, pegging him as an underdog. The press was overwhelmingly in his favour. The *New York Times* called the Anglo-American gesture "an act of folly," carried out by "short-sighted, narrow-minded, unprogressive persons"; it was "an outrage against common sense...supremely ridiculous," huffed the *Times* of London. From London, Samuel Flood-Page cabled: "Anglo American action shows they afraid Cable shares down...all directors many others hearty congratulations." The publicity was priceless: "a very big advertisement," wrote Henry Jameson Davis.[32]

Marconi now had to think about his long-term North American plans. He was already considering (and had just about decided not to accept) an invitation from Alexander Graham Bell to use Bell's estate at Baddeck in Cape Breton, Nova Scotia, as a temporary experimental station.[33] But now he faced an unpleasant option: a costly and possibly lengthy legal case, which at the very least would keep him from concentrating on his scientific and technical research, or leaving Newfoundland—where, after all, he had no ties. Looked at this way, Smith's offer should have been an easy choice, yet Marconi was cautious. He told Smith that he would go to Canada only if he received an official invitation from the Canadian government.

❊   ❊   ❊

Oddly and unexpectedly, Canada now suddenly occupied a pivotal place in Marconi's corporate destiny. Canada in many ways was virgin territory, an

‡ Not to be confused with the William Smith who was one of the company's original directors.

ideal testing-ground for long-distance communication—domestic and international, transcontinental and intercontinental, over land and across seas. Canada's communication needs were as vast as the land itself. Most important, Canada was politically welcoming to Marconi. Nominally self-governing in internal matters but still dependent on the British Empire in decisive areas like foreign affairs, the young country was aching to shape an independent course among nations as it began to emerge from the shadow of British colonial domination while trying to resist the most excessive attractions of its southern neighbour, the United States of America. Developing wireless communication fit well with the quiet but ambitious nation-building vision of Canada's prime minister, Sir Wilfrid Laurier.

Laurier's Liberal Party had been elected in 1896 and was returned with an increased majority in 1900. Canada's first French-Canadian prime minister, born in the Quebec Laurentian village of Saint-Lin, Laurier was a law graduate of Montreal's McGill University. He had built the Liberal Party on the basis of conciliation between Canada's two main linguistic groups, French and English, which also mirrored the strong religious cleavage separating largely Catholic Quebec and Protestant Ontario. Laurier promoted a clear separation between church and state while guaranteeing, among other things, religious education rights for minorities. He also presided over the first wave of massive immigration that would transform the country, and famously proclaimed that the twentieth century would belong to Canada.[34]

One of the conditions of Confederation had been that the new Dominion government would build a railway connecting the provinces.[35] Railway construction continued through the end of the nineteenth century, and with the telegraph lines running alongside the railroad, a vast transcontinental communication system was fashioned.[36] By 1900, communication technology had become one of the central building blocks of Canadian sovereignty (and soon, identity).[37] But wires and railways lines couldn't go everywhere. Ships travelling the St. Lawrence River and Gulf of St. Lawrence in the 1890s communicated with land stations by visual semaphore. Attempts to link the telegraph system to important Gulf islands like Anticosti and Belle Isle by underwater cable were exorbitantly costly and unreliable; hence, there was an interest in wireless. A modern, sophisticated communication system was not only essential for national development, it would also be a valuable asset in the emerging North American economy: the east coast of Canada was two days closer to Europe than the great American ports of Boston and New York.

It was thus inevitable that Canada should be one of the first countries to embrace Marconi.[38] On March 19, 1900, a member of Parliament from Prince Edward Island raised the question in the House of Commons, asking the government whether it had given any consideration to adopting the Marconi system for its seacoast telegraphic service. The postmaster general replied that indeed, they were studying it.[39] An underwater cable was laid between the coast of Labrador and Belle Isle, Quebec, twenty-two miles away, in July 1901, but the risk of it being disabled by icebergs was great. The Department of Public Works recommended the adoption of the Marconi system as an alternative in case of a disruption of service, and an order was placed with the Marconi company for two sets of wireless instruments to connect Belle Isle with the shore station at Chateau Bay, Labrador. The Canadian government claimed that this was the first practical use of the system on the North American continent, a claim borne out by the company's own archival records.[40]

❀   ❀   ❀

William Smith prepared to leave for Ottawa on the next train, determined "to save Marconi to British North America." First he went to see Newfoundland premier Sir Robert Bond, to let him know what he was doing. Relations between Canada and Newfoundland were sensitive, and Bond, a Newfoundland nationalist, was annoyed; this looked to him like nothing but a crass attempt to poach Marconi. The Anglo-American monopoly would, after all, expire in 1904, only a few years away; meanwhile, perhaps Marconi could continue his work on the French islands of Saint Pierre and Miquelon, just offshore. Smith pointed out that if the idea was "to keep Marconi within the Empire," a French solution wouldn't help; he might as well be in the United States—where he would certainly end up if they didn't succeed in getting him to Canada.[41]

Smith left St. John's for Ottawa on Tuesday evening, December 17, wiring ahead—in code, to keep the plans from the cable company—to Postmaster General William Mulock, who immediately took the news to Prime Minister Laurier and Finance Minister William Stevens Fielding.[42] Fielding, a former premier of Nova Scotia, recognized that this was an important opportunity for his province and telegrammed Marconi assuring him of the full co-operation of the Canadian government.[43] Unlike in Newfoundland, Fielding said, there would be "no difficulties whatever" for him to carry on operations in Canada. Fielding invited Marconi to come to Cape Breton as soon as he could get there.[44] Marconi accepted and would later state repeatedly—and quite accurately—that never before, or since, had he received such encouragement from a government.[45] (This must have stuck in the craw of some of Marconi's erstwhile governmental supporters like William Preece, who was indeed one of those who expressed skepticism about Marconi's claim to have vanquished the Atlantic.[46])

Arriving in Ottawa, Smith was instructed to turn straight around and head back to meet Marconi in Cape Breton. By chance, Laurier was on the same train, as far as Montreal, and Smith had a good two hours to brief him further. Laurier had already warmed to the opportunity and had discussed it with his ministers. Now he was convinced. The good of the empire aside, Laurier immediately recognized what Marconi meant for placing Canada at the forefront of global communication technology development while building a domestic communications network.

Switching trains in Montreal, Smith found that he had some more helpful travelling companions, and by the time he arrived in North Sydney, he had put together a list of key contacts for facilitating a Marconi installation in Cape Breton, including the general managers of the Dominion Iron and Steel Company and the Dominion Coal Company, which controlled important railway lines and steamshipping on the island. Smith reached Sydney shortly after midnight on December 24, while Marconi was still in St. John's.

The "poaching" of Marconi did indeed have historic consequences on relations between Ottawa and St. John's. One of the purposes of Smith's mission to Newfoundland had been to prepare a confidential memorandum outlining Newfoundland's needs and expectations in the event of a union with Canada. After filing his report with Ottawa, Smith was asked to set up a secret meeting in Montreal between Bond and Laurier to discuss the project. He was unable to accomplish this. Bond was known for his political volatility and vacillation, especially on the question of Confederation, and when Smith called on him upon returning to St. John's in January 1902, "he was angry with me about Marconi, and in a most unreasonable mood." It was not the only reason, to be sure, but Newfoundland did not become a Canadian province until 1949. As at so many other important junctures in his life, Marconi influenced history as he passed through a place on his own particular journey.[47]

While planning his next move, Marconi spent these days mostly handling the press and local politicians, receiving and replying to congratulations, writing up a detailed report to his business colleagues in London—and long-distance managing his relationship with Holman. In St. John's, Marconi confirmed his marriage plans to the press but asked to be excused from stating when the wedding would take place. In Indianapolis, meanwhile, Holman told friends it was postponed indefinitely, until Marconi's experiments were completed. "Cupid waits on wireless telegraph," read one headline. In New York, Helen Holman waited as well, growing impatient as her putative son-in-law peppered her with updates about his arrival delays: "Going Ottawa tomorrow and hope return via New York... Marconi."[48]

The disconnect between Marconi's emotional life and the public attention he was getting was more than he could handle. Aside from the sheer wonder of transatlantic wireless communication, much was being made of the commercial, political, and social benefits of Marconi's achievement. Speaking at a luncheon in his honour hosted by Governor Boyle (the local pols were quick to demonstrate their support for Marconi over the cable monopolists), Marconi emphasized the importance of ease of communication between countries. The wireless rate to England could be less than a penny a word, he said, as compared with the twenty-five cents charged by the cable companies.[49] This was not just idle speculation. The cost of a transatlantic submarine cable was about $4 million; a Marconi station could be built and equipped for $60,000.[50]

After recounting all the details he got to the main point: "I may say that this test confirms my belief in the fact that given sufficient power, and properly erected permanent stations it is possible to work wireless telegraphy across any distance which exists between any countries in the world," Marconi wrote in his report to the company directors in London on December 23.[51] He had no doubt that with an appropriate permanent station, *commercial* communication between North America and England would be possible, and he began to make a case for increasing the power used at Poldhu. Finally, he asked the board to postpone the company's general meeting until he returned so that he could address the shareholders directly. He clearly had the wind in his sails.

On Christmas Eve day, Kemp and Marconi hired a sleigh and went around to say goodbye to the Newfoundland officials, then left St. John's in a private railway car provided for them by Robert G. Reid, owner of the Newfoundland Railway and one of the colony's most powerful businessmen. (Reid "practically controls the island," a Marconi company internal memo stated.[52]) As they crossed the forest on a 550-mile journey to take the ferry SS *Bruce* (also owned by Reid) at Port aux Basques, they spent the evening drinking to the health of Kemp's wife and children and dined the following day on a traditional turkey dinner washed down with

champagne—again, all provided by Reid. The Newfoundlanders still entertained some hope of doing business with Marconi.[53]

After a rough night crossing the Cabot Strait in a driving blizzard, they
arrived in North Sydney, Cape Breton, on Boxing Day. There Marconi was
welcomed like royalty by the local political elite (as well as by the leaders of
the Italian diaspora in what was then one of the most ethnically diverse regions of the New World[54]) before going on to Sydney in yet another private
railway car, this one laid on by the Canadian government, who were also
sparing no effort in courting him. That same afternoon, under the guidance
of the premier of Nova Scotia, George Henry Murray, the party continued
by train to the mining town of Glace Bay and then spent the next day
looking at possible station sites.[55] Early in the morning of December 28,
after barely two days in Cape Breton, Smith, Marconi, and Kemp set out for
Ottawa. As soon as their train crossed the Strait of Canso they were joined
on board by all sorts of promoters, many of them American, trying to make
deals with Marconi.

It was still snowing hard when they arrived in Montreal in the evening
of Sunday, December 29. They descended at the Windsor Hotel, right next
to the train station. Wherever he went, Marconi stayed in the most fashionable hotels, and Montreal was no exception. An opulent hotel that billed
itself as "the best in all the Dominion," the Windsor had opened in 1878 and
its early guests included Sarah Bernhardt, Mark Twain, and Rudyard Kipling.
At the Windsor, another horde of reporters was waiting. The following day,
December 30, Montreal's most important daily, *La Presse*, devoted its entire
front page to the story of Marconi's accomplishment, under a banner headline christening him *"l'homme du siècle"*—the man of the century.[56]

Marconi stayed in Montreal just long enough for a morning sleigh ride
and lunch with local industrialists at the exclusive St. James Club before
moving on to Ottawa, where he and Kemp put up at the Russell House,
again the city's leading hotel. Here they were met by still more clamorous
reporters. While Marconi gave interviews, Smith buttonholed Fielding, the
minister of finance, and told him the only way to keep Marconi in Canada
would be by setting up some kind of partnership. "I told him my idea was
that we should provide the money for the station and that Marconi should
run it." The cost would be around $80,000 (this was the figure Marconi had
given Smith). "But it's a pure gamble," Fielding told Smith. "Yes, a pure
gamble," Smith replied, "but I had seen several persons on the train who
would risk much more than that sum to have the same chance."[57] Smith

may have had an inflated sense of his own importance but the sum he mentioned became the basic kernel of the deal.

Marconi was treated like a visiting head of state, lunching with Sir Wilfrid Laurier and flirting with debutantes at the governor general's New Year's Day toboggan party—while his fiancée waited for him in New York. But the press were also following his escapades. Filed with the cables in Holman's papers is a cutting, probably from a New York newspaper, that reported how Marconi was being entertained in Ottawa. The *Herald* had him "whistling down the Government House slide on a toboggan accompanied by three or four pretty girls. Marconi liked the experience so much that he asked for more." His regal celebrations were sadly juxtaposed to Josephine's melancholy. The stress was taking a toll on her. By the time she arrived in New York, she was bedridden. Marconi cabled her mother from Ottawa at New Year's: "Hope Josephine is better. Marconi."[58]

Marconi's own diary during this heady time is uncharacteristically loquacious (usually it is limited to bare-bone prompts, reminders, or cryptic notations). After the austerity of St. John's, he was indeed having a happy time in Ottawa, socializing with young ladies when not in high-level meetings with politicians and senior officials.[59] One of the people who sought Marconi out in Ottawa was Sir Sandford Fleming, the Scottish-Canadian engineer and reformer who is considered the inventor of standard time. Fleming, an advocate of cheap telegraph rates, was pushing for the nationalization of Canada's wire lines and state control of the transatlantic cables. He was excited by the possibilities afforded by wireless in making telegraphy a widely accessible "common convenience" and lauded the Canadian government's encouragement of Marconi.[60]

On his own, with none of his seasoned London advisors at hand, Marconi held a round of government meetings and drafted a six-page letter to the prime minister in which he spelled out his wishes and expectations. Aside from the financial and logistical details (mainly the $80,000 subsidy for the Cape Breton station), his main point was that "it might be of reciprocal advantage if the Government could become in some way interested in or associated with the undertaking which I hope to carry to a successful issue."[61] He wanted to have the Government of Canada as a partner. Laurier and his ministers were interested. After delivering his letter on January 7, Marconi met with the governor general, the Earl of Minto, and told him the Canadian government might subsidize some of his undertakings. They had a long chat that Minto later summarized in a memorandum—emphasizing that Marconi cautioned that his negotiations with the government had to be kept entirely secret as there might be great disadvantage if word got out to the press.[62]

The main details of the contract were agreed before Marconi left Ottawa. It provided that Canada pay a flat fee of C$80,000 (equivalent to around US$2 million today[63]) for construction of the Marconi station in Cape Breton. With this, Canada became the first country to fully welcome Marconi and grant him a subsidized monopoly on wireless communication—a move that set a tone for Canada's ambivalent relations with monopolistic multinational communication companies that persists to this day.[64] Canada was also pioneering an economic development model based on joint public-private enterprise, in which infrastructure was laid at public expense by private companies who were then allowed to reap the profits.[65] The Canadian government's contract with the Marconi company would fit perfectly—and indeed was one of the prototypes—of this model.[66]

Why did Laurier and his colleagues—these seasoned and sophisticated politicians, bastions of the British Empire—line up with a twenty-seven-year-old youngster (and one still considered a foreigner by many in Britain) against a mighty company, the Anglo-American and its equally powerful friends? Perhaps it had something to do with the effort to break away from colonialism. Sir Wilfrid Laurier, for one, may have looked at Marconi and seen a spunky pioneer. As far as he himself was concerned, Marconi was temperamentally much better suited to working with the Canadian public officials, who shared his penchant for big public-private partnerships under monopoly conditions rather than speculative ventures funded by private capital, like the ones he was offered by the businessmen he met on the train from Cape Breton to Montreal. He was also susceptible to being courted, and the Canadians did that very well. "The Canadian Government gave splendid proofs not only of initiative but of generosity," Marconi's 1919 memoir recorded.[67]

Marconi was now ready to present the Canadian agreement for the approval of his associates in London. But first he had one more stop to make. It had now been several weeks that Marconi was expected in New York, to straighten out his engagement plans. Everything was different now. Josephine Holman had waited two years for him to complete "the great thing," but he was nowhere near ready to settle down, even if he could possibly imagine what married life with Holman might be like.

In New York, press speculation about whether the wedding was on or off fuelled the suspense, while even Holman's closest friends and family members weren't sure. (Harry McClure, for one, sent a note of congratulations on New Year's Eve, with the comment: "I am anxious to see Marconi and to shake his hand—for several reasons."[68]) Josephine herself was seeking advice from mentors like her old school principal, May Wright Sewall, now president of the International Council of Women, the first transnational organization of advocacy for human rights for women. Sewall wrote to her (on council letterhead), choosing her words carefully: "It is good of you to tell me of Mr Marconi's good work.... I shall be anxious to hear your decision in regard to your future plans. You will have my sympathy if they involve a still longer separation, and yet, I think it is much better that you should be separated longer now before marriage, than that long separation should follow after marriage."[69]

Marconi and Kemp left Ottawa and returned to Montreal on January 9, 1902, for a meeting with Sir Thomas Shaughnessy, president of the Canadian Pacific Railway. Whether or not he was thinking about Josephine as he packed his bags, a telegram from Hall must have shaken him back to reality: "Herald told me Holmans would serve writ on your entering United States. Hall."[70] Helen Holman, furious at Marconi's stalling, was intending to sue him for breach of promise. However this might have affected him, he took the overnight train from Montreal on January 11, with Kemp, and checked in at the Hoffman House in New York early Sunday morning, January 12, 1902.[71] He was received with a hero's welcome, and hailed for having signalled across the Atlantic, a feat believed to be impossible. But the press was equally hungry for news of his marital intentions, on which he was noticeably circumspect.[72]

Marconi's marriage plans were now commanding almost as much interest as his latest wireless accomplishment. A dozen visitors were waiting for him at the Hoffman House when he came down to breakfast the morning after his arrival. Queries were met "with a smile and a shrug of the shoulders."[73] He told the *New York Times* he had had "a royal time" in Canada and would have little time for anything other than business matters while in New York, beginning with a gala banquet set to take place in his honour that very night.[74] Acting with forty-eight hours' notice, the American Institute of Electrical Engineers (AIEE) had arranged a three-hundred-plate dinner at the Waldorf-Astoria. What he didn't report was that he had asked Josephine to accompany him to the banquet and she had refused.[75]

The cream of the US scientific community, fans and rivals of the guest of honour, were invited to the banquet on January 13. Alexander Graham

Bell was there, approvingly, at the head table and signed Marconi's menu, along with physicists Elihu Thomson and Michael Pupin.[76] Edison sent a telegram of congratulations and was represented by his wife, Mina. Nikola Tesla sent regrets. He told the *Herald* he felt that he could not rise to the occasion but wished to join in congratulating Marconi on his splendid results. "He is a splendid worker and a deep thinker," Tesla said.[77]§

    The event was a bit over the top. According to a trade journal:

> The menu, printed in Italian olive ink on sea foam green card, bore outside a sketch representing an ocean scene, with the two coasts of the Atlantic, and a Marconi staff on two tall lighthouses signaling "s"s' in three dots all the way across. In the middle was a medallion with Mr. Marconi's portrait, draped with the Italian flag....
>
>   At the two ends of the room were large tablets, one reading "Poldhu" in white lamps and the other "St. John's," in letters about a foot long. Immediately opposite the speaker's table was a similar tablet bearing the name of "Marconi." Between the three signs were strung strands of the conductor, into which were inserted clusters of three lamps at frequent intervals, to represent the three dots, or "s," sent across the Atlantic from the Cornish coast to New-foundland. At fitting times these were flashed or allowed to stay illuminated, and the success was great....[78]

Marconi was moved by the reception. After the years of buildup to his trans-atlantic signal crossing, the legal and political tensions of St. John's and Ottawa, and the pressure from London and the technical press, it was as though he was finally, unequivocally appreciated by the greatest of his peers. When he rose to speak it was "with a modesty almost amounting to diffi-dence" as he told of the long series of failures leading to his recent triumph, acknowledging that he had only built on the discoveries of other scien-tists.[79] The astonishing thing about these accolades was that the story of his triumph in Newfoundland was based exclusively on his word. "There have been few great facts in science thus accepted with unquestioning confi-dence on the authority of one known to be anything but disinterested," the *New York Times* wrote in an editorial." Marconi was now one of the most trusted names in the world.[80]

    But not in his private life. Holman's diary tells how the following days unfolded.[81]

---

§ Tesla was then living at the Waldorf-Astoria, and apparently snuck out in order to avoid the Marconi festivities; the dinner, organized by one-time Tesla booster Thomas Commerford Martin, featured "a gaggle of Tesla adversaries" on the dais (Seifer 1996, 277).

[Tuesday, January 14, 1902] G. came in afternoon—talked with Mamma and Helen [Jo's younger sister]—then he takes me to dine at the Endicott [an Upper West Side hotel, now a condo complex]—unhappy talk as we walk home—But very happy afterward—hopes everything would be right—more like last spring.

[Wednesday, January 15, 1902] G. to come at three—waited seven hours for him—but he did not come. [He pleaded illness but somehow couldn't manage to get a message to her.]

[Thursday, January 16, 1902] G. came at noon—everything changed—no hope— I agree to see Mr. McClure at five—talk with him—find him most kind—see G. alone afterwards and our engagement is at an end—I promise to wear ring for friendship—Mamma and Helen come in—most unhappy scene....

On January 21, Holman sailed for England on the *Kaiser Wilhelm der Grosse*. As the ship left Hoboken, McClure issued a statement to the press on behalf of the Holmans, simply stating: "Josephine has asked Mr Marconi to release her from her engagement to marry him, and...he has complied with her request." Confronted at the Hoffman House by a reporter, Marconi confirmed the break. "I am sorry, very sorry, that it should be so," he said, "but Miss Holman wrote to me asking me to release her from her promise to be my wife, and I had no alternative but to obey her wish. I have not one word to say regarding Miss Holman's action, except that she has done what she considers best for her future happiness." The reporter added: "Mr Marconi seemed greatly depressed."[82]

Marconi never revealed the unexplained cryptic topic he had mentioned in his last letters to Josephine. Holman's mother was quoted in the *Herald* as stating: "There have been disasters on both sides." But she did not elaborate. The *Herald* report went about as far as a newspaper could go in accounting for the outcome of the relationship, alluding to reports that Marconi's family was not pleased with the match, in light of his "conspicuous social position" both in Italy and in England; his friends and relatives, the paper said, "had greater ambitions for him in a social way than that he should choose an American girl, with neither a notable distinction in society nor an impressive fortune. It was a love match pure and simple, born of the dangerous propinquity of an ocean steamship acquaintanceship."[83] To a direct question from the *New York Times*, Marconi denied that the breakup was connected with his reported presence at a dinner in Montreal with a famous actress, adding: "There was a reason, but Miss Holman and I agreed not to disclose it. That is all there is to say."[84] He then sailed for England on the *Philadelphia* the following day, January 22, 1902.

An undated, unsigned typescript in Holman's papers, probably the transcript of a news or wire service dispatch, describes the public consternation that followed: "The announcement by Mrs. H.B. Holman that her daughter Josephine has broken her engagement to marry Mr. Marconi came as the biggest sort of surprise, not only to New Yorkers but to Americans in general." The *New York Times* was prompted to comment: "Naturally enough, the inventor was an unsatisfactory fiancé.... Geniuses make proverbially bad husbands."[85]

Holman rebounded quickly. Her diary reveals an uneventful ocean crossing, mostly spent struggling with seasickness. On disembarking at Cherbourg, she reported, the Holmans met a "courteous man" who looked after her. She spent an enjoyable train ride to Paris engrossed in interesting conversation with this Mr. Boross of Budapest. By the time they arrived in Paris, she had a new friend. Settling in to her room at the Hotel de l'Athenée—"nice rooms," full of "queer French things"[86]—one of the first things she noticed was an article in the Paris edition of the *Herald* about her and Marconi.

Eugen Boross called the next day and a romantic courtship blossomed, fuelled by the sheer exhilaration of being in Paris. Eugen took Josephine for tea at the Ritz, and the ballet at the Olympia, to the Louvre and the Folies Bergère. He sent her violets and listened to her talk about her breakup with Marconi—the topic was unavoidable; it was in the Paris papers every day. Eventually he went on to Budapest but cabled and wrote to her. The Holmans left Paris eventually as well, continuing their grand tour: Holman's diary has entries from Avignon, Hyères, St. Moritz, Genoa, Pisa, Rome, and Florence. The diary ends on April 5, 1902, but their travels apparently continued as far as Budapest, and a month later the *New York Times* announced the coming marriage of Miss Josephine Holman, "who was once engaged to marry Signor Marconi of wireless telegraphy fame," to Eugen Boross of Budapest, Hungary. The marriage took place in London, at St. Mary Abbots Church, Kensington, on May 22, 1902, in the presence of a handful of friends.[87]

The Borosses divided their time between Europe and New York and became prosperous enough through various business ventures to amass a major collection of museum-quality European art. They had two daughters, Eugenia (born 1903) and Alys (born 1905), who both graduated from Bryn Mawr in 1925 and married Episcopal church ministers at a double wedding in June 1930. The sisters were both active in the Oxford Group, the 1930s

Christian organization which later evolved into the Moral Re-armament Movement.[88]

Josephine Holman Boross died in October 1941 and is buried in Indianapolis. The burial notice reported that her date of birth was unknown. A gentle but firm feminist before the time, she has been effectively air-brushed from the standard Marconi biographies. Marconi's most candid biographer, his daughter Degna, reports that growing up she was only vaguely aware that her father had had a broken engagement in his youth, and didn't even know the woman's name until she began research for her book. Marconi himself, who later remained in touch with exes, does not seem to have had any further contact with Holman after their engagement ended. In the 1930s, he implored his American biographer, Orrin Dunlap, not to mention her in his book, ostensibly to respect her privacy, and Dunlap complied. Yet, for the two most intense years of Marconi's life, leading up to the success that sealed his reputation, Josephine Bowen Holman was his friend, lover, confidante, and occasional distraction; she was by far the person closest to him, possibly closer than anyone had ever been.

Alys Boross Smith inherited her mother's papers, did some work classifying and annotating them, and toward the end of her own life tried to interest an institution in taking the papers, with no success. In August 2005, Alys's son—Josephine's grandson—Peter Schermerhorn Smith, a retired California state judge, gifted the collection to the Huntington Library; it has never before been scrutinized by researchers. The finding aid[89] can be read online, a fitting denouement to the two-year romance between the inventor of long-distance wireless communication and the woman who captured his heart and then let him go.

PART II

The Player

10

Corralling the Brand

Ray Stannard Baker rushed to Newfoundland the moment he heard Marconi's announcement. Baker, thirty-one years old, was an associate editor of *McClure's*, the muckraking magazine that had helped put Marconi on the world map a few years earlier. Baker was a liberal, intellectual journalist, specializing in philosophical essays as well as articles exposing corruption and calling for social reform. He would soon be one of the most prominent American journalists of his generation, gaining renown for his coverage of labour strife and working conditions. He would also serve as director of the American press bureau at the Paris Peace Conference in 1919, and eventually wrote an eight-volume biography of Woodrow Wilson, for which he was awarded a Pulitzer Prize in 1940.

Baker's feature article on Marconi appeared in the February 1902 edition of *McClure's* and established Marconi's reputation.[1] Most important, it established clearly just what Marconi had done. "Mr. Marconi makes no claim to being the first to experiment along the lines which led to wireless telegraphy, or the first to signal for short distances without wires. He is prompt with his acknowledgement to other workers in his field, and to his assistants." What Marconi did, according to Baker, was to solve a practical, mechanical problem, devising an instrument that would produce a certain kind of electric wave, and receiving this wave in a second machine some distance away from the first. His apparatus was adapted from those invented by others, introducing features without which there could be no wireless telegraphy. This made the achievement all the more impressive. Marconi's kite-borne "aerial," for example, made it possible to do the unthinkable, pull a signal out of the air from more than seventeen hundred miles away.*

---

* See footnote about distance on page 166.

In Newfoundland, Marconi obliterated the obstacle of distance from wireless communication.

Baker accompanied Marconi to Nova Scotia, a trip that afforded him a glimpse of Marconi's discreet charisma. On the way to Cape Breton, "it seemed as if every fisher and farmer in that wild country had heard of him, for when the train stopped they came crowding to look in at the window...they wondered most at the inventor's youthful appearance." Baker provided a portrait of Marconi's bearing and character, describing his height ("somewhat above medium"), his temperament ("highly strung"), and his movements ("deliberate"). When Marconi is interested, or excited, he gets "a peculiar luster" in his eyes. He conveys a sense of "intense nervous activity and mental absorption.... He talks little, is straightforward and unassuming... he has accepted his success with calmness, almost unconcern; he certainly expected it."

Baker also gives us an insight into what drove Marconi. "The only elation I saw him express was over the attack of the cable monopoly in Newfoundland, which he regarded as the greatest tribute that could have been paid his achievement. During all his life, opposition has been his keenest spur to greater effort." Indeed, at twenty-seven years of age, Marconi had overcome the unforgiving atmosphere and isolation of his father's attic in Pontecchio, the indifference of the Italian authorities, the skepticism of the scientific establishment, the conservatism of London's financial markets, the anxieties of his Irish investors, the scorn of the technical press, and the impatience of military bureaucrats in a half dozen countries.

While stimulated by adversity, Marconi's talent was that he was able to marshal and subtly command support: from his immediate family circle, important well-placed backers who saw an interest in promoting him, the popular press, and political figures, from the King of the Belgians to the prime minister of Canada, who saw that they could use his technology to leverage their own positions in the global power game. Part of Marconi's attraction to the powerful was that he was certainly of their social class, yet still an underdog. Marconi had exceptional access to the powerful. At the same time, he himself was attracted to power, uncritically and oblivious to how it was used.

Marconi was also ubiquitous. He had become a universal brand. He now had more than 130 patents, in just about every country in which patenting was possible, and was continuing to perfect his core invention, attacking new obstacles, such as mountain and desert terrain, and, of course, ever greater distances. His system was now in use by the world's greatest navies and commercial steamship companies for an expanding array of purposes, from person-to-person greetings to signalling from ship to shore in time of distress. His company was beginning to show some revenue but was pouring profits back into new research and development as quickly as they could be earned. This appeared to be sound practice—Marconi was challenging the cable companies, which had literally sunk an estimated $400 million in investments. Was wireless a "new" industry, or a spinoff of the old? It was too early to tell and perhaps the question itself was premature or unnecessary.

Even if it turned out to be a great success, wireless would be a complement, not a competitor, the cable industry said. The reporting of Marconi's activities was having an adverse effect on the price of their shares. The cable companies were huge, impersonal entities; the names of their leaders were almost never mentioned in the press, while wireless—at least in the United States, the United Kingdom, Canada, and Italy—was always referred to as "Signor Marconi's system."[2] Marconi's most valuable asset was his quiet but boundless enthusiasm, which enthralled and infected everyone he met. His work was now also the measure by which technological innovation in the early 1900s was judged.

As his global stature grew, so too did the hurdles he faced. He was now travelling constantly, trying to keep control over a vast series of activities on several continents. His life seemed to be made up of separate threads that threatened to unravel or at least degenerate into loose ends. During the next few years his main challenge was just managing to stay on top of things and remain in charge of his affairs, even when he was physically away from the action, on the other side of the world.

❈　　❈　　❈

Flushed with his success in Newfoundland, Canada, and New York, yet disappointed in the breakdown of his relationship with Josephine Holman, Marconi sailed on the SS *Philadelphia* on Wednesday, January 22, 1902, arriving in London eight days later.[3] He still had no fixed address, and "home" at this time oscillated between Garlant's Hotel and the Bath Club, in Pall Mall.[4] (The Bath Club, also favoured by Mark Twain, was one of London's rare gentlemen's clubs that was also open to women.) But he had no time to contemplate his personal situation. A cable from the Canadian finance minister was waiting for him: "Government accept your proposal." On February 6, 1902, Marconi's company board, too, approved the draft agreement he had negotiated in Ottawa, and Marconi cabled back that he would come to finalize the deal at the end of the month. The agreement was mentioned enthusiastically in the Canadian government's Speech from the Throne a week later.[5]

Marconi spent the next days preparing for the company's upcoming annual general meeting, its fifth. The company was able to report to its shareholders that it now possessed "a complete international organization," aiming to achieve "uniform adoption of the Marconi System in other countries on the same lines" as in the United Kingdom—that is to say, as government-backed monopolies.[6] It was about to appoint agents in Chile, India, Argentina, Holland, and Brazil. It had already sold its US patent rights to the Marconi Wireless Telegraph Company of America in 1899, of course, and following the negotiation of the groundbreaking Canadian contract, a Canadian company would soon be formed as well.[7]

On February 20, Marconi made a triumphant address to the company's AGM.[8] Having been almost universally acclaimed for his exploit in New-foundland, his main concern now was to convince his shareholders of the commercial possibilities of wireless. There was still a lot of skepticism, and one comment stung him particularly hard: Sir William Preece (he was knighted in 1899) had been quoted at a meeting of the Anglo-American Telegraph Company as stating that the underwater cable industry had nothing to fear, commercially, from wireless.

Marconi minced no words rebutting his former mentor. Preece, Marconi said, may be "a gentleman with various claims to scientific distinction," but he had no competence to make such a statement. Preece's knowledge of his system, Marconi said, was at least three years old and out of date. "Of the conditions under which the system is now worked, Sir W.H. Preece is, in fact, wholly ignorant."[9] He also had a few well-placed barbs for Oliver Lodge who, despite his acknowledgement of the importance of Marconi's feat, had also questioned its commercial viability. As generous and gracious as he could be when he didn't feel under attack, Marconi on the defensive was a sight to behold.

The speech captured Marconi's sense of self and identity at this point in his life, when he had just had his greatest success and felt he was entitled to articulate an upbeat vision for the future. Although he was increasingly a global figure, he still situated himself—at least when he was in Britain—as a British imperialist, albeit one conscious of how the boundaries of the empire were changing. Marconi's transatlantic connection was nothing if not instrumental. "By the cheapening in the method of communication between them, I think we may also fairly claim that we are doing something to strengthen the bond that unites the two great English-speaking communities of the world."[10]

This comment, widely reported in the mainstream and specialist press, was not lost on Marconi's principal rivals in Germany. On February 22, 1902, Marconi set sail again on the *Philadelphia,* leaving Southampton bound for New York. While he was at sea, a crucial event illustrated the geopolitical stakes that were now at play. Crown Prince Henry of Prussia, the kaiser's younger brother, was due to make a "goodwill" visit to the United States, on behalf of the House of Hohenzollern. Prince Henry bore a remarkable resemblance to his first cousins Tsar Nicholas of Russia and the future British monarch, George V. As a young man, he was known as "Prince Charming." The trip was preceded by much fanfare and media attention and was an important event in German-American relations.[11]

Prince Henry sailed for the United States on the *Kronprinz Wilhelm,* a liner belonging to the Hamburg-American Line, out of the German North Sea port of Cuxhaven. It was equipped with Marconi's system. In New York, Marconi engineers and operators put in place elaborate plans to communicate with the *Kronprinz Wilhelm* in mid-ocean, as it approached US shores on its way to the dock on the Hudson River. Stormy weather delayed the ship's arrival in New York, and there was briefly some concern for the prince's safety. The *Kronprinz Wilhelm* arrived safely enough, if two days late, on February 24, 1902, and Prince Henry was feted in true regal fashion by the city's elite. Alexander Graham Bell, Thomas Edison, and Nikola Tesla were among the luminaries attending a banquet hosted by John Pierpont Morgan and a clutch of financiers in the prince's honour a couple of days after he arrived.[12]

One of the highlights of Prince Henry's crossing was an exchange between Marconi operators on the *Kronprinz Wilhelm* and the eastbound *Lucania* as the ships passed each other in mid-ocean. The *Kronprinz Wilhelm* was in contact with the German lightship at Borkum (Germany) for three hours after leaving Cuxhaven; next, with the Marconi station at North Foreland (England) near Dover; then with the Lizard (at distances ranging from one hundred to three hundred miles); then with the *Lucania*; and, finally, with Nantucket, where the prince sent a message of greeting to President Theodore Roosevelt. Needless to say, this was all priceless publicity for Marconi.[13]

Prince Henry had been so impressed that he thought he might send a message or two himself on his return to Germany on the *Deutschland* a few weeks later. There was only one very big problem: the *Deutschland*, although belonging to the same shipping company, was equipped with the rival Slaby-Arco apparatus, and ships and stations using Marconi's system were contractually forbidden from communicating with other systems except in case of emergency. There was a rash of conflicting reports as to what actually happened on board the *Deutschland*: according to the *New York Times,* the prince had an amusing and uneventful trip, but the European press reported that when the ship docked in Cuxhaven on March 18, a furious Prince Henry fulminated to his brother the kaiser, who was waiting on the dock to greet him.[14]

In fact, what happened was this: Marconi had been for some time aggressively cementing his monopolistic position by prohibiting "intercommunication" between ships or shore stations equipped with his system and any others (which were all, at the time, essentially German ones). Indeed, he claimed there were no competing systems, only usurpers—such as Slaby and von Arco—who were marketing illegal clones. Marconi claimed his patent protected him from any alleged competitors, and he had an ironclad clause put into all his leasing contracts prohibiting intercommunication. He also contended that intercommunication was technically impossible, which was not the case.

A forceful letter from Marconi to Cuthbert Hall a few months earlier, after the *Deutschland* had inadvertently been allowed to communicate with the Marconi-equipped *New York Herald* station at Nantucket, highlighted the importance Marconi attached to these conditions: "In view of this it will be your duty to communicate immediately with the *Herald* authorities pointing out to them in strong terms that this is a breach of their agreement with us. I should call your attention furthermore, to the fact that the very possibility of such communication being allowed by our Assistant-in-Charge at Nantucket points once more to inadequacy of instruction in his duties, or, less probably, to wilful breach of them." Hall immediately took steps to ensure that the terms of the agreement would be enforced.[15]

Marconi's 1919 memoir described what followed Prince Henry's ocean crossing as "a state of berserk fury which might almost have been called Marconiphobia."[16] The European press ran several articles on the righteous indignation of the insulted German royals but Marconi publicly defended his position in a long interview with the *New York Times* on March 30, 1902 (he was still in New York at the time). He confirmed his policy of refusing intercommunication in the name of patent rights and contractual freedom, expressed doubt that the *Deutschland* would have had the capacity to communicate with his land stations anyway, and dismissed as "absurd" the claim that his company had deliberately interfered with the working of the *Deutschland's* system. He stated, somewhat tongue-in-cheek, that he had the greatest respect for the German kaiser and considered him one of the world's great men, while recalling that, at the kaiser's request, he had years ago accommodated Adolf Slaby, who was now copying his system. (Slaby wasn't staying on the sidelines; he told the press that it was Tesla, in his 1893 St. Louis lecture, who had first described the idea that Marconi later put into practice.[17])

In London, the company moved into spin mode. At the insistence of its agent in Berlin—who virtually dictated the letter and urged that it be sent without delay—Flood-Page wrote an apology intended for the kaiser's naval adjutant and éminence grise Admiral Gustav von Senden-Bibran, who would be sure to pass it on to Wilhelm during one of their weekly meetings: "It is a matter of very great regret that His Imperial Highness Prince Henry should have suffered any inconvenience owing to his having been unable to communicate with our station at the Lizard. I need hardly say there was no intention of lack of courtesy on our part as we should have only been too glad to make any arrangement within our power to meet the convenience of His Imperial Highness."[18]

Flood-Page's salvage operation was too little and too late. Wireless was now a strategic piece of the German effort to build a navy capable of challenging British supremacy. Kaiser Wilhelm, unfazed, decided to call the big powers to an international conference in Berlin, aimed at agreeing on regulations to guarantee open communication between wireless systems, regardless of who owned them. It would take the next ten years to play out, but it was the beginning of the end of Marconi's embryonic monopoly.

The Germans seized the pretext of the Prince Henry incident to broach an issue that had been concerning them for some time, and which reflects how the new communication technologies were being governed at the beginning of the twentieth century. Britain, as we've seen, had established effective control of the international cable system, and Wilhelm was determined that this situation would not be repeated with wireless—especially as the Marconi company was perceived, accurately, to be a British company. In this sense, wireless became a chess piece in the game of inter-Imperial rivalry at a moment when that competition was perhaps most fierce.

At the same time, wireless was at the heart of conflicting ideas about the role of government and the state in commerce, social development, and, especially, communication. The British Post Office, in asserting a public utility monopoly on domestic telegraphy that went back to 1869, pioneered a model that would be replicated several decades later with radio broadcasting. In the United States, in contrast (albeit also with the sole, notable exception of the Post Office), the development of communication was driven almost entirely by private investment and conglomerates sanctioned by government regulation. Again, the governance structures of telegraphy and, later, broadcasting became world models for the development of communication as private enterprise.

Canada and the other British Dominions offered a middle ground, where colonial governments seeking to stake out increased independence and autonomy from Mother England sought to use communication in the building of their national economies and, indeed, identities. Marconi, as a well-connected outlier, was a perfect partner for these incipient nation-states. By 1902 he was making customized corporate arrangements in each of the countries where he operated.

It was inevitable that, probably sooner rather than later, some form of international regulation would be put in place to govern the practice of wireless communication. As we saw earlier, in the latter half of the nineteenth century the two main systems of international communication, the post and the telegraph, had each spawned an intergovernmental organization (the General Postal Union and the International Telegraph Union) to set rules for their respective fields, by international agreement. By 1902, the 1875 St. Petersburg convention on telegraphy begged to be updated to encompass wireless.

The idea of "an international agreement to regulate the wireless transmission of messages" in the interests of the greatest good was first publicly evoked by none other than Adolf Slaby, in March 1902, with direct reference to Marconi's monopolistic intentions as evidenced by the Prince Henry incident.[19] Within days, the press was reporting that German authorities were preparing a note to be presented to the United States, Britain, and France, proposing a conference for the purpose of agreeing on means to prevent a monopoly of wireless telegraphy on the high seas. This was framed as "a direct consequence of the reported refusal of the wireless station at

Nantucket to receive a message from Prince Henry on board the Hamburg-American Line steamer *Deutschland*."[20] It was a pretext, but a good one, symptomatic of the tensions provoked by the young industry.

The Berlin meeting would be the world's first international conference on governing the use of wireless, and the ensuing treaty paved the way for radio regulation in the twentieth century. Marconi argued that the technical specifications of his company's products prevented "intercommunication" between his equipment and that of rival companies (an argument that would be repeated a century later by companies like Microsoft and Apple). He also argued that it was simply unfair, in a business sense, for other companies to be allowed to profit from the technology and expertise that his company had developed. However, Germany, and, to a lesser extent, France and the United States, were not content to see a British enterprise establish a global monopoly in wireless communication. The Germans managed to convince the other major powers of the day (except for the United Kingdom and Italy, who were already locked into contracts with Marconi) that wireless should be open to alternative technical standards—and regulated by international agreements. Further diplomatic conferences in 1906, 1912, 1927, and 1932 then laid the foundation for the international regulatory framework for radio that still prevails today. Marconi and his company were key players in all of these negotiations, which established the basis for the global governance mechanisms that characterize the modern worldwide electronic media system, as well as the Internet.

Marconi's arguments would become the prototype for the corporate ideology of deregulation. First, he insisted that his company was entitled to enjoy the benefits of the system it had developed. Marconi maintained that the Germans' real design was to promote their own national wireless firms. The Germans, however, did not want to see a Marconi monopoly, notwithstanding his argument that a monopoly might be the legitimate spoils of a successful invention. They made the case that competition would lead to lower rates and improved service—in other words, would be good for consumers and hence a legitimate object of an international convention. The heart of the matter was an understanding of regulation: Was it right for governments to intervene where market forces and case law created paradoxical situations? Marconi and his supporters argued that legal property rights and corporate agreements were the only legitimate forms of regulation. The governments of the world, despite their differences, claimed the right to intervene in the name of a broader general interest. They didn't use the term, but they were in effect saying that communication was a public good.

      ❁   ❁   ❁

Marconi took advantage of his crossing on the *Philadelphia* to further extend his signal reach, this time with witnesses. As the ship neared America, he received signals from Poldhu at distances up to twenty-one hundred miles, the greatest yet, and in the presence of the ship's captain. He was now convinced that any distance could be overcome as long as he had sufficient electrical power to generate a bigger and bigger spark. The object of his new permanent station in Nova Scotia would be to perfect the system so that he could send complete messages regularly and reliably, and—soon, he hoped—set up a commercial service that could compete with the cable companies.

After a week in New York, Marconi went on to Canada, first stopping in Montreal and then arriving on March 10, 1902, in Ottawa, where some important contract issues remained to be resolved. Marconi showed himself to be a thorough, tough, and consistent negotiator, resisting the Canadians' efforts to insert a number of control points to the agreement. Laurier himself handled the detailed negotiations with Marconi. Finally, after some minor amendments, everyone was happy, and Marconi and Laurier signed the agreement on March 18, 1902. Again, the symbolism was clear: Marconi was on the same political plane as the prime minister. At the suggestion of the minister of justice, Charles Fitzpatrick—like Marconi, an Irishman—it was agreed to backdate the C$80,000 contract to the seventeenth—St. Patrick's Day—"for luck," and that is indeed the date that it bears, March 17, 1902.

The essence of the agreement was that the Marconi companies would erect two stations, one in the United Kingdom and one in Cape Breton, to conduct commercial communication between the two countries using "the Marconi system"; the company further undertook to provide commercial service at 60 percent below the rate being charged by the cable companies (ten cents a word rather than twenty-five cents) and to use, as far as possible, Canadian machinery, material, and labour in construction of the Nova Scotia station (the latter point an important concession to Canada's protectionist "National Policy").[21]

Marconi then left immediately for Cape Breton, arriving on March 21, 1902, in Sydney, late at night in a persistent rain, where he was met and accompanied to his hotel by torchlight by what seemed to him like "the whole of the local Italian colony."[22] After inspecting several locations, he selected a ten-acre site belonging to the US-controlled Dominion Coal Company, on the promontory of Table Head, just above one of the area's major coal mines and overlooking the ocean atop sixty-foot cliffs in the mining town of Glace Bay on the island's eastern shore. Marconi quickly settled the lease, left Richard Vyvyan in charge, and returned to New York before sailing back to England on April 9.

The agreement placed Marconi, Canada, and the relatively remote region of Cape Breton, Nova Scotia, in the forefront of world communication development; it also secured some urgently needed cash for the company, and stood in sharp contrast with the experience Marconi had been having with the British, Italian, and US governments. The Cape Breton station—again, a model of public-private collaboration—would figure prominently in Marconi's research, commercial, and personal activities over the next fifteen years.

Marconi was now spending every available moment at one or another of his experimental stations, working to broaden the scope of his system and establish its monopolistic position in the global market. From London, his company managers were trying to develop a viable business, extending the company's tentacles through an increasingly complex web of subsidiaries and intercorporate connections. As chairman and principal shareholder, Marconi had the final word on every important decision—as well as the ability to micromanage countless minor ones. There were some fundamental conflicts in business philosophy. One of the most important pitted Marconi against Hall, who was an aggressively competitive manager, whereas Marconi believed in disarming his interlocutors through charm and diplomacy.

Hall was named managing director of the company on July 25, 1902, succeeding Samuel Flood-Page.[23] The conflict in styles between Marconi and Hall was most evident in the company's dealings with the British Post Office, an agency that was at once a direct competitor, potential partner, and political advocate. Although steeped in new-found prestige and constant media attention, the Marconi company had yet to establish itself on solid financial footing. It had a small contract with the British Admiralty, but relations with the Post Office, a far more important potential source of revenue, were shaky. Hall was of the school that saw anything to do with government—any government—as backwards, inefficient, bureaucratic, and fraught with the tarnish of politics. Marconi saw government as a partner with deep pockets, extensive connections, and political clout; he understood the value to be gained from a strong alliance with the British Crown in his quest for global dominance.

For the next five or six years, Marconi and Hall made a dynamic corporate duo. Hall ferociously protected every square inch of the company's turf while Marconi wooed investors, politicians, and the press. When Marconi was at the Haven or Poldhu (as he was at least half the time), Hall would write to him several times a day about developments in London or at the company's scattered offices, his letters often running to a dozen pages or more. Marconi was more laconic and to the point, agreeing, disagreeing, or adding a twist to Hall's strategic suggestions. Marconi would come up to London as necessary—for a directors meeting, to speak to shareholders, or to take tea with a minister. Hall's tenacity was critical to keeping the wheels of business turning. When Marconi was not convinced about a project, rather than say no, he would frequently procrastinate. Hall was the only one who would dare shake him into action.

By March 1902, Marconi was tiring of British pokiness. "England is not the country in which I may hope to accomplish much," he admitted to a reporter from the *New York Times*. "They are a little old-fashioned over there you know."[24] Marconi now had twenty-five stations operating on the coast of England, but the Post Office monopoly on domestic telegraphy prevented them from communicating inland or with each other. For commercial purposes, wireless was limited to communicating with ships outside the three-mile limit, where British sovereignty, and hence regulatory authority over communication, ended. Marconi suggested that this constraint could lead him to move his base of operations away from England. The Republican, pro-business *New York Times* saw Marconi's problems as emblematic of the problem of government ownership of public utilities and their ruthless suppression of competition. "If the British officials make trouble for Mr Marconi and drive him to France or Spain, it will be unutterably stupid."[25] The British, for their part, recognized this and were prepared to engage in much negotiation to keep him in England and happy.

In addition to the fact that his main research installations and world corporate headquarters were based there, Marconi had concluded one commercial deal in Britain that was both lucrative and strategically important. On September 26, 1901, after years of negotiation, the Marconi International Marine Communication Company signed a fourteen-year agreement with Lloyd's, establishing a single system of mercantile wireless telegraphy for "maritime signalling" (worldwide except for certain territories, notably the United States, Hawaii, and Chile, where Marconi already had agreements).[†] Lloyd's was to have sole right to use the Marconi system and agreed to use only the Marconi system.[26]

Lloyd's had, of course, been the first commercial enterprise to take an interest in Marconi's invention, in the very early days when he was setting up his company. Marconi's initial arrangements with Lloyd's were small but crucial to establishing the company on a commercial footing, and also served to reveal the tensions between the research and business dimensions of Marconi's personal interests. However, it was not until September 1901 that a comprehensive deal was struck. By then, both Marconi and Lloyd's saw a possible partnership as a way of forestalling anticipated government intervention to take over wireless telegraphy.[27] The *Electrician* reported that the 1901 Lloyd's agreement placed the Marconi system "upon a secure and enduring footing for mercantile marine signaling and intercommunication between ships and coast signal stations." The "effective and practical monopoly" granted to Lloyd's was far more tangible and certain than the theoretical one of a master patent, even if supported by the courts. Lloyd's agreed not to receive or transmit with a vessel equipped with apparatus not supplied by the Marconi company.[28]

It was the sort of contract business people love: good for both parties.[29] At least at first. As new applications of wireless developed, disputes arose over the interpretation of rights under the agreement. Marconi was frequently called to intermediate between Henry Hozier of Lloyd's and Hall, and the companies were eventually involved in a lawsuit that continued to simmer even after a court settlement in 1906.[30] But the deal with Lloyd's was crucial to the monopolistic business model Marconi was developing. It demonstrated the effectiveness of keeping end-to-end control of wireless operations in the hands of a single entity and hence supported Marconi's argument against "intercommunication." It also became a concrete legal obstacle to the German effort to break Marconi's monopoly.

[†] Maritime signalling was defined as "the collection, publication and diffusion of all such intelligence and information with respect to shipping, cargoes, freights and insurance as it is at present collected or diffused by Lloyd's." It was crucially distinct from "sea-telegraphy," which was defined as "any messages transmitted to or from vessels which could not reasonably be described by the expression 'maritime signalling'," which the Marconi company reserved for itself. Marconi later claimed that the agreement gave Lloyd's no rights to transoceanic telegraphy (Memorandum, attributed to SFP, February 25, 1903, OX 248).

The Lloyd's contract gave Marconi an important leg up in his dealings with the British Post Office.[31] Marconi's own stance at this time was to try to establish a commercial wireless telegraph service owned by his company and grafted on to the Post Office monopoly (which, at this point, only explicitly covered conventional cable telegraphy); that is to say, a privately owned public service—and he went so far as to characterize it as such.

The tone of the relationship in this period before Marconi's sensational transatlantic leap was evident in a series of long and detailed exchanges regarding the company's efforts to establish its position. The Post Office view was outlined in a letter from a senior official, J.C. Lamb, commenting on the company's claim that the Telegraph Act of 1869 allowed it to connect its coast stations with the country's telegraph system. The Post Office did not agree, Lamb wrote; transmitting messages between Marconi's signal stations and ships within British territorial waters would infringe on the postmaster general's monopoly, and, consequently, "his license will be required if such operations are to be permitted." Most important, any arrangements that might be made between the company and the Post Office "must be taken in no way to prejudice any questions which may arise in regard to the validity or scope of the Marconi patents."[32]

The company's view was laid out in a confidential letter from Flood-Page to an unnamed official, possibly the postmaster general, Lord Londonderry, himself, in which the company sought a licence "to organise land telegraphy for the benefit of the public...in co-operation with the Post Office."[33] The Marconi company had been started with British capital and was regulated by British law. It was pioneering sea telegraphy (which, importantly, was unregulated) and had applied for a licence to work in British territorial waters; now it was asking to extend that licence to land telegraphy as well. It was all about revenue. The company argued that its proposal was not only fair but also beneficial to both parties—developing Marconi's discovery in such a way that "while increasing the revenue of the Post Office, we may receive adequate return for the large capital outlay we have made, and are still making."

On the surface, this approach, combining public and private enterprise, suited Marconi, who was by nature firm but conciliatory; but Marconi could also be ruthless in planning his moves. Even Hall, who loved nothing better than a fight, sometimes felt he had to rein in his boss. A letter from Hall to Marconi in July 1901 characterized the situation around this time, as Marconi was considering holding back his personal services from the company's arrangements with the Post Office: "It seems to be perfectly clear that the Government has absolutely no claim on this. It might, however, be very desirable to demonstrate to them the value of it, apart and in addition to the patents—to emphasize the fact that even if they are prepared to pirate the inventions, there is a great deal which they cannot get without your co-operation. The suggestion you make means war—which of course might result in a peace advantageous to the Company."[34]

Marconi persisted, trying various tactics to break the monopoly in practical ways. In May 1902 (from his sickbed at the Haven, where he was suffering from influenza), he instructed Hall to request to link Poldhu to London via the Post Office's land telegraph system. When that didn't work, he took a more direct approach. In October, Hall wrote to the new post-master general, Austen Chamberlain, reiterating the company's request for a licence. From Glace Bay, learning that the licence was not likely to be granted, Marconi instructed Hall to play his ace in the hole: "I hope you will point out that the result of further unpleasantness may quite probably be to shift the European centre of wireless telegraphy to Italy." Then he added, enigmatically, "This, of course, is what the Italian Government is now working for."[35]

In this context of jousting between the company and the Post Office, one issue was now emerging as predominant: the brewing conflict with Berlin. The interests of the British Empire were at stake, Hall argued, and this was the reason the government should support Marconi. Hall summed it up in a pitch to a senior official:

> As you are aware the Kaiser last year sent a diplomatic note to all the Powers calling attention to the question of wireless telegraphy and proposing a confer-ence. The avowed objects were to render Wireless Telegraphy in use for ships on sea of whatever system available for everybody...the attack is nothing but a veiled attack upon a British Industry...as a British Company, the owners of the original patent and the organizers of a commercial system of wireless teleg-raphy, which ultimately must have great value and benefit for Great Britain, we say it is important that the Government should give us every support.[36]

At the bottom of the memo someone added by hand: "Hostility of perma-nent post officials." Perhaps it was a self-fulfilling prophecy. On the last day of 1902, Lamb wrote to Hall informing him that the Post Office would not grant the Marconi company the licence it had requested.[37] As Hall intimated in his memo, the stakes were now much higher.

11

Regulation

Marconi now had an established North American beachhead in Canada. He was playing cat-and-mouse with the British authorities, but they needed him at least as much as he needed them. As he prepared to face the challenge from Germany—and as he had hinted in his letter to Hall from Glace Bay—he also held a wild card: Italy.

In the months leading up to his transatlantic success, Marconi had felt that he was being mildly denigrated in the Italian press. Luigi Solari had come to his defence and was "admonished...rather strongly" by a senior admiral, Carlo Mirabello, commander of an important naval division (and soon to be named minister of the marine).[1] The Italian government was following Marconi's developments with close interest. On February 18, 1902, the Italian consul general in Montreal noted two items worth mentioning from the recent opening of the Canadian Parliament: proposed measures to counter anarchist political activity (today we might call them anti-terrorist measures) and the Marconi agreement.[2]

Marconi decided to make new overtures toward Italy. On April 12, 1902, from aboard the *Majestic* en route from New York to England, he wrote a long, confidential letter to Solari, detailing his recent results on the *Philadelphia* and expressing confidence that he could establish a connection between Italy and its East African colony in Eritrea, as well as with South America, where there was a large Italian diaspora. He also reported on the Canadian contract as evidence that at least one government was taking him very seriously.[3]

Solari focused his lobbying efforts directly on the young king, Vittorio Emanuele III, who had taken over after his father's assassination in 1900, and in June 1902 the king ordered the navy to place his personal cruiser, the seven-thousand-ton warship *Carlo Alberto*, at Marconi's disposal—ostensibly to support his research, but also, from the navy's point of view, to see just how far his system had been perfected. Marconi was energized by this new opportunity and he threw himself into making preparations.

The *Carlo Alberto* sailed from Naples on June 10, 1902, with Solari and a skeptical Admiral Mirabello on board. The plan was to pick up Marconi in England while Vittorio Emanuele attended the coronation of Edward VII, but Edward had an attack of appendicitis and the coronation was postponed. The delay played in Marconi's favour. He joined the *Carlo Alberto* briefly off Poole on June 27, and was soon getting along well with Mirabello; he was able to write to Hall on June 29 that "They also want to negotiate terms for a transatlantic station to be erected on the coast of Italy. The Admiral and officers are very nice to me and are prepared to do anything possible on this ship to facilitate the tests."[4] Then he returned to the Haven and London where he made a statement to the board "as to the method preferred by the Italian Government of conducting our business in Italy" (that is, through himself—a different model than in the United Kingdom, Canada, and everywhere else that the company was doing business) and was mandated to take charge. Hall reported that Marconi had personally secured an order for twenty sets of apparatus from the Italian navy and that the Italian government wanted to erect a long-distance station "on the same terms as the Canadian station." Finally, Marconi informed the board to hearty approval that he had received an invitation to join the king of Italy on board the *Carlo Alberto* for a cruise to the Russian port of Kronstadt.[5]

The invitation to visit Kronstadt, Russia's naval headquarters, which also guards access to St. Petersburg at the eastern end of the Gulf of Finland, was another sign of Marconi's growing stature in European politics. The *Carlo Alberto* left Dover on July 6, 1902. Marconi joined the ship at Poldhu the following day and on July 16 he met Tsar Nicholas and reviewed the troops at St. Petersburg (then Petrograd). With Marconi's precursor, Alexander Popov, looking on, the tsar conferred the Order of Saint Anne on Marconi. Popov also presented Marconi with a photograph dedicated to "Guglielmo Marconi the father of wireless telegraphy"—an impressive acknowledgement of his position.[6] The *Carlo Alberto* left Kronstadt on July 22 and Marconi was back in England on August 1, in time to get in several days' work before the rescheduled coronation of Edward VII on August 9, 1902.

The *Carlo Alberto* remained at Marconi's disposal, and he sailed out again on August 23, 1902,[7] remaining in daily wireless contact with Poldhu

as the ship headed for Italy by way of Spain and Gibraltar. Solari kept a detailed journal of the voyage, which demonstrated to Marconi's satisfaction
that high land masses were no barrier to communication, that daylight signalling required more energy than nighttime, and that his newly patented
magnetic detector was a practical addition to his system. He was also convinced that he would soon be able to send not only signals but also complete
messages across any distance, and began planning to go to Cape Breton.[8]

Returning to Italian soil for the first time in five years, Marconi was
leaving nothing to chance. On September 3, he wrote to Hall from shipboard at Cadiz, Spain, that Mirabello suggested the company blaze a trail by
sending greetings to the king—it would be the first wireless transmission
between England and Italy and was part of a comprehensive strategy Marconi
was putting in place to re-establish his relations with Italy.[9] Having followed
the progress of the last few months, Mirabello had changed his tune, writing
a long, flattering, and enthusiastic letter to the minister of the marine lauding
"our great countryman . . . the young scientist who is as modest as he is great."[10]

On September 10, the *Carlo Alberto* arrived at La Spezia where, according to Solari, the first person to greet the ship was the Japanese naval
attaché, who had travelled from Rome expressly to extract a promise from
Marconi that his company would sell to the Imperial Japanese Navy its best
naval equipment and would train Japanese officials in their use.[11] Marconi
cabled Hall that the Italian Navy was making the *Carlo Alberto* available to
take him to "Cape Breton, Cape Cod and wherever necessary. Publicity in
Times desirable." Hall made sure the *Times* received the news.[12]

After a day or two in La Spezia, Marconi and Solari moved on to visit
the king at his summer residence in Piedmont. In Turin, a few days later,
Marconi met Tancredo Galimberti, the minister of posts and telegraphs, to
begin discussing a Canadian-style contract. He then went on to Bologna
and Pontecchio, for a reunion with his father, Giuseppe, and his first visit to
Villa Griffone in more than five years.[13] Finally, on September 24, 1902,
Marconi left Bologna for London by train, while the *Carlo Alberto* sailed to
England.

The German government had now invited England, France, Russia,
Italy, the United States, Austria, and Hungary to meet in Berlin in August
1903 to draw up an international convention settling the conditions for
wireless telegraphy. The *Times* reported the German view that Marconi's
arrangements with Lloyd's in particular threatened to create an absolute
monopoly that was both commercially and politically objectionable, but

that the suggestion of an international convention had been received "in a friendly spirit" by the governments concerned.[14]

Back in London, Marconi informed his board that the Italian government believed that only one system, his, should be used for international wireless communication, and that it would support that position in Berlin. Furthermore, only Marconi stations would have the right to communicate with Italian stations and ships.[15] It was important to complete the Italian contract as soon as possible, and certainly before the powers convened at the kaiser's conference.[16]

Marconi was dealing directly with the Italians, in his own name (as provided by the original patent agreement with his own company), and his directors expressed the wish that the company's name appear in the agreements he was negotiating with the Italian government. They also made it clear that they wanted Marconi himself to represent them in those negotiations. A strong deal with the Italians would go a long way to protecting the company's, and Marconi's, global interests. But Marconi played the Italian contract card close to his chest. On October 7, 1902, he wrote to Hall that "it would be most inadvisable for the Company to approach the Italian Government on the subject at all."[17]

Marconi then left for Poldhu, whence he wrote the first of several letters to Hall about strategy for Berlin. Now that the conference had been called, he wanted a thorough statement of the company's case that could be presented to friendly governments as well as to the conference itself. "I need hardly tell you that it should be based rather upon what we take to be the interests of the shipping community in general than upon our own special interests," he instructed Hall.[18]

Meanwhile, the Canadians were anxiously waiting for results from Glace Bay, where the staff at the new station was working on perfecting the system for transatlantic communication. On September 26, 1902, Hall wrote to Prime Minister Sir Wilfrid Laurier that progress was good, if a bit slower than anticipated; and Marconi would soon be arriving in Cape Breton on the *Carlo Alberto*. Upbeat, Hall reported to Laurier on Marconi's visit to Italy seeking support against Germany's diplomatic efforts.[19] Canada was also strategically important to Marconi's plans: his work there further cemented his relationship to the British Empire and strengthened his argument for British support at Berlin; at the same time, the latitude he was receiving from the Canadians stood in strong contrast to his treatment in Britain. Again, the arrangement served both parties: by dealing separately with

Marconi, Canada affirmed its independence from both England and the United States, while Marconi developed a practical model of co-operation with government that he would try to reproduce elsewhere—it suited him.

Hall was urging Marconi on the need for "absolute secrecy" regarding the transatlantic preparations. Marconi wrote nonetheless on October 10, 1902, to Charles Moberly Bell, editor of the London *Times*, informing him that he expected to shortly conduct a series of experiments between Cape Breton and England and that "I am not unhopeful that the results of these may be to establish wireless telegraphic communication between the two continents." He offered Bell the opportunity to receive the first wireless press message from America to England, "a short one, not exceeding, say, 20 or 30 words, since, naturally, there must be some possibility of something not working quite smoothly in these early tests." Hall then followed up with an exclusive arrangement for the *Times* to publish the message—when Marconi was ready.[20]

The Marconi station was by this point a major new feature of the Cape Breton landscape. The region's economy was still based on staple activities like farming and fishing, as well as a bewildering array of small concerns servicing the burgeoning lumber and shipbuilding industries—"forges, tanneries, carriage factories, cooperages, grist mills, sawmills and woollen mills," according to local historian Mary MacLeod. The island was a crossroads of traffic, as ships laden with fish left for Great Britain, Europe, the West Indies, and the United States, and returned with products like Jamaican rum, molasses, teas, and spices for onward shipment to the urban centres of North America. North Atlantic fishermen had been visiting Cape Breton since the sixteenth century, and "the owners, brokers and other businessmen who dealt in transatlantic trade were as familiar with the cities of Europe as they were with Halifax."[21]

The steel and coal industries arrived just as the government's so-called National Policy was beginning to direct trade and commerce to central and western Canada and ignore the traditional economic patterns of the Maritime provinces. Sydney became the centre of the Cape Breton steel industry, while towns like Glace Bay developed around the mines. The island was drained of its farmers, fishers, and loggers as the Boston-based Dominion Coal Company brought in workers from abroad, and Cape Breton became the most cosmopolitan centre in Canada east of Montreal (layered onto its basic Scottish Highland stock, of course). Often working in appalling conditions, the Cape Breton miners were soon among the most radical industrial workers in North America, organizing themselves into militant trade unions and calling for the overthrow of capitalism. In this climate, a job with the Marconi company was considered a prize.

Glace Bay, incorporated as a town in 1901, bent over backwards to accommodate Marconi. To help assuage his fears that electric flashes from a nearby trolley wire might interfere with his system, the municipality prohibited electric trolleys from passing within six hundred yards of the station. Given a free hand, Richard Vyvyan went about building the Table Head station with gusto, employing a "curious mixture" of one hundred to two hundred men of diverse backgrounds and languages. "An affable man with a handlebar moustache and a smoking pipe always at hand," as MacLeod described him, Vyvyan would remain in Glace Bay for eight years. He built a large residence on the station grounds, overlooking the sea, where Marconi would also stay when he was in town. The twelve-room, twenty-five-hundred-square-foot wooden structure was finished in natural wood and included "bedrooms and bathrooms, a large parlour, sitting room, dining room and kitchen. A wide hall ran end to end through the centre of the house with the rooms opening from either side of the hallway." As at any of the homes or stations where Marconi was likely to spend any length of time, there was a piano.[22]

On October 20, 1902, Marconi and George Kemp, his chief assistant, joined Solari on the *Carlo Alberto* at Plymouth. They arrived in Cape Breton on October 31 and Marconi remained at Glace Bay for most of the next two months. From Cape Breton, via his constant, daily contact with London, Marconi consolidated his British, US, and Italian activities and made preparations for Berlin (even though the conference was still nine months away). On November 26, 1902, he wrote to Hall that work was progressing but not as rapidly as he had hoped. He enclosed drafts of the Italian agreement and instructed Hall to write to him with company assent to the terms as agreed between Marconi and the Italian government. He mentioned that he also required a similar letter from the American company but would look after that himself.[23] Marconi later wrote in a similar vein to the American company, pointing out that they all had a stake in the Italian contract, which would secure that one European power at least would be committed to supporting Marconi interests at the Berlin conference.[24] Marconi was micromanaging, multi-tasking, and fully in charge, going so far as to dictate the wording of the letters of assent—all the while heading up the daily experimenting on transatlantic messaging.

On December 18, 1902, Hall reported to the board that Marconi had sent him a copy of his proposed Italian agreement and wanted to commit the company to the contract; the agreement was excellent in both form and substance and would go a long way toward protecting the company's interests at Berlin, he said. The board approved the agreement and gave Hall the authority to amend it at his discretion. There was also some interesting news about an apparent breach in solidarity among Marconi's German rivals: a proposal had been received from Ferdinand Braun in Strasbourg, providing for fusion of the Braun and Marconi interests.[25] Braun had been making overtures to the company for some time. He was frustrated by the "amalgamating tendencies" of Slaby and von Arco and thought that he could more easily work with Marconi.[26] (Braun soon joined with Slaby and von Arco but, ironically, he would share a Nobel Prize with Marconi in 1909, although they never worked together at all.)

Marconi was nearly ready to send a full wireless message across the Atlantic. Typically attentive to the publicity value of news regarding rulers as well as the benefits of flattering them, his plan was to send a message of congratulations from the Canadian governor general, Lord Minto, to the recently crowned King Edward VII of England. He orchestrated every aspect of the plan, instructing Hall to "take steps to ascertain privately what will be the most correct and acceptable method of sending to the King of England a message from Cape Breton. I should be very unwilling to expose myself to the possibility of not receiving a prompt reply."[27] Marconi had Minto's message in hand as early as November 7, but the system was not yet working perfectly and he insisted on waiting until he was absolutely certain the transatlantic messaging would work before announcing anything to the press.[28] He said as much to his old friend Harry McClure, writing to him that "it would be quite useless for your purpose of getting a magazine 'story' that you should come up here at present." Marconi was keeping the details of his work absolutely secret, and in fact was so dissatisfied with much that had appeared in the press, including the magazine press, in the past year that he had made up his mind "to communicate nothing whatever until after the completion of my programme in its entirety." But as usual, word got out. In London, the *Mail and Empire* reported on November 24, 1902, that "Marconi has triumphed," and that King Edward would be receiving an overseas wireless from the Lord Minto within a few hours.[29] However, that would not happen for another month.

The connection was finally made on December 21, 1902, in the presence of a representative of the *Times*, explicitly invited by Marconi to witness the event.[30] Sir George Parkin, a well-known Canadian educator, professor at Upper Canada College (Toronto), and until recently the *Times'* principal Canadian correspondent, described how it went.[31] Arrangements had been made with Poldhu to send the message around 1:00 a.m. Cape Breton time. "A little after midnight our whole party sat down to a light supper. Behind the cheerful table talk of the young men on the staff, one could feel the tension of an unusual anxiety as the moment approached. . . . All put cotton wool in their ears to lessen the force of the electric concussion." Marconi himself was at the transmitter.

Parkin recalled being struck by the "instant change from nervousness to complete confidence" in Marconi's face the moment he put his hand on the transmitting apparatus. After telling the party that it would be necessary to transmit the letter *S* to catch the attention of the operators at Poldhu, and to get them to adjust their instruments, Marconi then began to transmit a continuous sentence, the words spelled out "in short flashes of lightning." The message travelled to Poldhu at the speed of light—about one-ninetieth of a second. "The first message has been sent through the ether across the Atlantic. What that means for mankind no one can even guess," wrote Parkin."*

❈   ❈   ❈

Marconi was happy during the two months he spent at Glace Bay. Despite the remoteness of Cape Breton from the cosmopolitan capitals he was used to, he had a self-contained support system and was surrounded by both comfort and adulation. Not many inventors or commercial entrepreneurs had at their disposal the flagship of the Italian navy, the resources of the Government of Canada, the spotlight of the *Times* of London, and the attention of an admiring local population. He took advantage of his latest success to redouble his efforts to recruit the Canadians to help improve his fraught relations with the British government, and continued to lobby both Minto and Laurier directly, but despite their courteous responses, it was to no avail.[32]

At the same time, he had to deal with his company's chronic cash shortage. Remarkably, given the innovations Marconi was stewarding, the company's

---

* The Glace Bay station was relocated in 1904 (see below). The successors of the Dominion Coal Company continued to own Table Head well into the 1980s, until it was taken over by Parks Canada for a modest interpretation centre, the Marconi National Historic Site. The only remnants of the original structures are some traces of the station's stone foundation.

finances were in desperate straits. It was still in debt after taking out a £30,000 loan to finance Marconi's latest transatlantic adventure. While Marconi was in Cape Breton, a series of telegrams crossed the Atlantic in both directions, between Marconi, Hall, Davis, company director James Fitzgerald Bannatyne, and the London Westminster Bank regarding a financial syndicate that had been set up a few months earlier to support the company's costs. Bannatyne proposed that a core group of backers inject more funds. On January 5, 1903, Marconi cabled Davis: "Wish no new financial arrangements entered upon before my return.... If immediate borrowing necessary propose you Bannatyne and self provide funds for interim." He then had his London bank forward £2,000 to Bannatyne and left for the United States.[33]

Marconi was expected in Montreal on January 13, 1903, for the launch of the $5 million Canadian Marconi company but cabled from Glace Bay at the last minute to say he had to go directly to Cape Cod.[34] He had arranged for President Theodore Roosevelt to send a wireless message to King Edward via the new Marconi station at South Wellfleet. Marconi left Glace Bay on January 14 and took the express train from St. John, New Brunswick, to Boston the following day.[35] He arrived in Cape Cod on January 16, 1903, "with ten reporters in his wake," reported the *Barnstable Patriot*.[36] The South Wellfleet station, blown down in November 1901, had been completely rebuilt, modelled after the one at Glace Bay. Fifteen carloads of Oregon pine were shipped by rail to South Wellfleet. Four wooden towers were raised 210 feet high, in a two-hundred-foot square; each barn-red tower was twenty feet square at the base and eight feet square at the top, and was set in a four-foot-thick bed of concrete. The corner posts were twelve-by-twelve timbers and the towers were guyed by twelve one-inch steel cables per tower, anchored in twelve-by-twelve timbers buried under ten feet of sand.[37] The station was ready to transmit, but Marconi wanted to be there when it happened. Before leaving Glace Bay, he cabled South Wellfleet with precise instructions: "Do not spark before I arrive."[38]

It was a grand event. On January 19, 1903, Marconi entertained dignitaries from Massachusetts's Italian community before transmitting a message from Roosevelt to the King of England, the first transatlantic message ever sent from the United States (and though it was weeks later than the one from Canada, it is usually reported in the United States to this day as having been the first from "America"). Initially, Marconi's goal was to connect South Wellfleet and Glace Bay and then retransmit from Glace Bay to

England; he was as surprised as anyone to discover that the Cape Cod station was powerful enough to send a message across the Atlantic.[39]

With a firm connection established from Glace Bay to Poldhu, Roosevelt's message was purely symbolic, but Marconi recognized the publicity value of such a feat. He had a messenger standing by to relay the news of the successful transmission to the land telegraph station at the Wellfleet railway station four miles away. According to local lore, he told the messenger, Charlie Payne, to "drive like the wind and if you kill your horse I'll get you another one." Marconi always knew that control of a story went to the one who was first out of the gate. Sure enough, the next day's *New York Times* reported on its front page: "President sends wireless message to King Edward. England's ruler returns the compliment by means of the Marconi system."[40] The King's reply was sent by cable, not wireless, but Marconi's publicity triumph was no less impressive.

Marconi returned to South Wellfleet later that year, accompanied by six assistants, and spent four days overseeing improvements to the plant.[41] But the South Wellfleet station proved less reliable for transatlantic communication than Glace Bay, and was used mainly for ship-to-shore contact. Aside from a twenty-four-hour whistle stop with Henry Jameson Davis and John Bottomley in September 1908, Marconi never visited Cape Cod again.[42] The US Navy took possession of the station when the United States entered the First World War in April 1917 and closed it down in February 1918.[43] By 2013, the original site was only thirty-five feet from the sea; erosion had claimed the rest. All that remains today is a motley pile of red brick and a scrum of rusted chains.

❋   ❋   ❋

After Cape Cod, Marconi went to New York for a few days, then briefly back to Montreal, before returning to London. His attention now went to completing the Italian agreement, which he recognized as a crucial pole for developing his global activity. The contract gave the Italian government the rights to Marconi's inventions in Italy and its dependencies, for free, but tied them to exclusive use of Marconi equipment in wireless operations for fourteen years; it also provided for the erection, by Marconi but at Italy's expense, of a long-range, high-power wireless station in Italy that would be the most powerful station in the world.

Considering that Marconi wanted to make a gesture of real generosity to his beloved homeland, it was a hugely advantageous deal for him. He would have a second government ally in the upcoming battle at Berlin and a fourth high-power wireless station, in as many countries, at his disposal (along with Poldhu, Cape Breton, and Cape Cod). Thanks to the Italian contract, Marconi would soon be able to develop new links with Argentina and the Balkans, where Italy had privileged connections. He was now beginning to think of a worldwide network and told Solari his next goal was to build a station that would be able to connect to India and Australia.[44] The visionary idea that would crystallize a few years later as the British "imperial chain" was taking shape in his mind. Meanwhile, the Italian contract became a model for the company's expansion into the lucrative markets of continental Europe—where many small countries, such as Belgium, Spain, and Portugal, recognized the advantages of a communication system that could keep them in touch with their far-off colonies, and would be impervious to being shut down by enemies in the event of war.

Marconi signed the Italian agreement in London on February 12, 1903, and handed it to Solari, who delivered it immediately to Rome. It was then signed in turn by Minister of Posts and Telegraphs Tancredo Galimberti, and promptly presented to parliament. The new station, Galimberti announced, would be the first government-owned high-power radio-telegraphy station in the world, and finally meant some redemption for Italy from the "great blow" suffered by the fact that Marconi's "first experiments of really practical interest...took place in a foreign country."[45]

A lively debate, or rather a string of accolades, ensued.[46] In the Chamber of Deputies, the Italian lower house, Deputy Silvio Crespi, secretary of the committee studying the bill, reported on February 17: "Marconi has always freed his relations with Italy from all bonds of a commercial nature and has, without any remuneration, handed over to our Government all the results of his studies." In this agreement, Marconi was asking for no more than reimbursement of expenses—plus exclusivity. Crespi continued: "Rome will soon be able to communicate direct with any point of the globe wherever a Marconi installation may be made. . . . This reminds us of Rome's historical past, and the awakening of this memory will not be the least of the merits of Guglielmo Marconi." Marconi thus enlisted support for his key argument about intercommunication, and tied himself to Italy's imperial designs.

Veteran Deputy Luigi Rava, who hailed from the Appenines, the homeland of so many emigrants to America (as well as the Marconis), marvelled at what quick and easy affordable communication would mean for the four million Italians "who live in remote parts of America and are now experiencing such difficulties in rapidly communicating with their remote relations and their beloved homes." Science, he said, would thus make possible a great social and democratic reform.

Not all parliamentarians took such a charitable view. In discussion of the bill on February 20, Deputy Angelo Battelli did a detailed analysis of the contract and pointed out that it meant Italy would be cutting itself off from communication with countries that had or would adopt the German system; Marconi could surely cancel or modify his stipulation of exclusivity. But Battelli's more cautious tone was drowned out by those who saw only the wonders of Marconi's system. Deputy Arturo Galletti, for one, described the agreement as "a proof of great confidence in this Italian genius.... Let Marconi erect the station wherever and in whatever manner he may deem fit." Crespi was even more direct: "Marconi has the right to claim that, at least in the beginning, his country should exclusively make use of his system and use every endeavour to have it preferred in Europe."

On March 31, 1903, the agreement was adopted in the Italian Chamber of Deputies by a huge majority (216 to 14) and received unanimous consent in the Senate—a rare distinction that Marconi claimed had occurred only twice before in the history of that body, with regard to pensions for the father of the nation, Garibaldi, and the widow of the martyred King Umberto, Queen Margherita.[47] The station was to be built at a cost of 800,000 lire—equivalent to £32,000 or US$160,000, more than double the cost of one of Marconi's British or American stations—at Coltano, near Pisa, on land donated by the Italian Crown. Coming on the heels of the establishment of transatlantic communication between Europe and America, the Coltano station would be an important link between the home country and the growing Italian diaspora in the Americas. It was also meant to be an important strategic installation for Italian colonial expansion in Africa. Even more important to Marconi than the agreement itself was the renewed recognition it brought him in Italy. Municipalities across the country, including Rome and his native Bologna, offered him the keys to the city. He was hailed as Italy's greatest living genius, feted and coddled by the press, and, finally, completely vindicated in the eyes of his now ailing father, Giuseppe.

A few weeks after the agreement was signed, Marconi was in Rome to receive new honours and to speak in the shadow of Roman emperor Marcus Aurelius at the Campidoglio. On April 30, 1903, a crowd of several thousand Romans cheered him enthusiastically on his arrival at the train station, where he was greeted by the mayor of Rome, Prince Prospero Colonna, and several members of the municipal council. Marconi's ghostwriter Leon de

Sousa described the event in that 1919 manuscript: "Bands of students with banners, representatives of patriotic and scientific associations and the general public invaded the station and spread along the platforms and railway tracks...we got into the first cab we saw. But a large number of students and others had followed us and instantly detaching the horse from the shafts they insisted upon dragging me as far as the Grand Hotel where I intended staying and which was only some three or four hundred yards from the station."[48]

On May 4, 1903, the king and queen gave a dinner for Marconi at the Quirinal where, by coincidence, another guest was Marconi's nemesis, the visiting German emperor (whom the Italian press labelled for the occasion—and as an expression of Marconi's fame and status in his own country—"the other Guglielmo"). The dinner was the scene for a legendary exchange between Marconi and the kaiser that has been reported in various versions in all of the standard Marconi biographies. This was the first time the two had met. When they were introduced, the kaiser, trying to be cordial, said: "Signor Marconi, you must not think that I have any animosity against yourself. It is the policy of your Company I object to." To which Marconi replied: "Your Imperial Majesty, I should be overwhelmed if I thought you had any personal animosity against me. However, it is I who decided the policy of my Company."[49]

The Campidoglio speech was Marconi's first Italian account of his "triumph," beginning with the technical aspects of his initial discovery and subsequent experiments up to that time, but there was, remarkably, very little rhetoric or vision. Except for the introduction and closing passages where he thanked his Italian supporters, including Solari, and said a few patriotic words, it was like so many other speeches he had made, and would make, to audiences around the world. But in Italy it was considered a particularly important historical document, and the British War Office thought it important enough to have it translated into English.[50]

Marconi was joined in Rome by his mother and brother. Annie was received almost as warmly as her son.[51] Alfonso kept up a stream of letters to Giuseppe back at Villa Griffone, informing him of all the grand events, and also the more mundane news of Guglielmo's apparently wavering health. After the intensity of the previous months, Marconi found the pace of the trip exhausting. He was out all day and returned to the hotel each night worn out. One day his pulse was racing and he called for the doctor. Alfonso informed their father that Guglielmo's health would not allow him to come to Bologna after Rome as planned. He would instead go to Pisa, to visit the nearby site of the future Coltano station, and then back to London.[52]

Guglielmo had one important message for his father, which he communicated via Alfonso. Milan's *Corriere delle sera* had published an article in its issue of May 9, 1903, repeating the myth that Giuseppe had been a miserly father who had stood in his son's way. Guglielmo asked that his father write to the editor to set the record straight. The article, entitled "*Il maestro di Marconi*" ("Marconi's teacher"), was principally a tribute to Vincenzo Rosa, who had taught Marconi as an adolescent in Livorno. But at one point it digressed and charged that Giuseppe "did not want to spend a penny" on his son's gadgetry but insisted that he go to London and get into the import-export business. There is no mention of any such scheme anywhere in the literature, and Marconi's disclaimer to Giuseppe speaks for itself. Marconi was forever trying to correct the press. Alfonso enclosed a cutting of the offending article, which still sits in a file of Giuseppe's papers in the Marconi archive at the Accademia dei Lincei in Rome, embellished by neither comment nor context. There is no record of a letter to the editor from Giuseppe, but he kept the cutting for the rest of his life.[53]

❄   ❄   ❄

As soon as he got back to London, while still shaky from his Italian trip, Marconi set about immediately to catch up on work at his English research stations. As usual, he also took the opportunity to present his latest technical improvements to an audience. Unexpectedly, however, he suffered an embarrassing experience that shook his confidence in one of the stock tools in his repertoire: the public performance. The sensational blow was struck by, of all people, a professional stage magician.

Nevil Maskelyne was, like Marconi, a self-educated British inventor who had attracted some local attention in the late 1890s by publicly exploding gunpowder at a distance without any visible contact.[54] In 1900, he applied for a UK patent "to provide an improved means for exciting and detecting Hertzian oscillations, and for applying the transmitted impulses to the general purposes of signaling or telegraphy."[55] The patent application included an interesting historical interpretation of the origins of "these peculiar oscillations," crediting the contributions of Hertz, Branly, Lodge, and even Crookes, but conspicuously omitted any mention of Marconi.

Maskelyne had collaborated with Henry Hozier, developing a wireless system for use on Lloyd's ships and shore stations, but the shipping firm of course signed a contract with the Marconi company in September 1901, adopting the Marconi system (along with the company's restrictive conditions). While Hozier recognized that Maskelyne's system was different from Marconi's, Lloyd's went with Marconi, either because they felt his system was superior, or because they feared a patent challenge, or more likely because of some combination of both.[56]

An earlier Nevil Maskelyne had been the royal astronomer in the court of George III, but the contemporary Maskelyne was also a member of a famous family of stage magicians. His father, John Nevil Maskelyne (best known as the inventor of the pay toilet), owned and ran a popular exhibition venue in Piccadilly known as the Egyptian Hall, which he had turned into a showcase for magicians and illusionists. (In 1898, a young Harry Houdini had written to Maskelyne senior asking to join his show, and received a curt refusal.[57])

Nevil Maskelyne succeeded his father at the Egyptian Hall, but he wanted to be recognized for his work in wireless. His own son, Jasper—also a stage magician—wrote that while his father taught himself to be a master illusionist, "his heart never thrilled at the glitter of the footlights or the clatter of appreciative applause." Instead, the older Maskelyne had, in part because he lacked the confidence, "kept one foot in the theatre and one in the laboratory."[58] When it became clear that he could not compete with his rival, however, Maskelyne became one of Marconi's public detractors.[59]

By 1903, Maskelyne was one of a handful of more or less credible critics who were constantly tracking Marconi's claims, pointing out inconsistencies and inadequacies in the development of his system. On the other side were Marconi's high-profile supporters, such as Fleming, who—albeit on Marconi's payroll—had an impeccable academic pedigree and a long track record of association with important practical innovation based on scientific discovery. Detractors and supporters sparred constantly in the technical and popular press, but, as Marconi well knew, nothing could take the place of a flamboyant public performance.

In the early months of 1903, Marconi, with Fleming's support, was trying to demonstrate that he had solved the problem of "syntony" or tuning. He had an important patent to that effect (the so-called four-sevens patent number 7777 of 1900), but there was persistent doubt as to its practical effectiveness. On March 31, Marconi told the sixth annual general meeting of his company that recent experiments by Fleming at Poldhu had convinced them that they could eliminate the risk of interference between transmitters and receivers equipped with Marconi apparatus.[60] It remained to show this publicly.

A demonstration was arranged to take place at London's prestigious Royal Institution on June 4, to be preceded by scientific lectures by Fleming on May 28 and June 4, immediately before the planned demonstration. A sixty-foot receiving antenna was set up on the roof of the Royal Institution building on Albemarle Street in Mayfair, and Marconi himself was at the transmitter in Poldhu. (The signal was actually relayed by the Marconi factory works in Chelmsford, Essex, as the distance from Poldhu to London was too great to cover with a sixty-foot antenna.[61])

Maskelyne later recollected that he thought "the opportunity was too good to be missed." He set up his own transmitter on the roof of the Egyptian Hall, only a few blocks away from the Royal Institution. A few minutes after Fleming started his lecture, but before Marconi's signals were due to arrive, Maskelyne began transmitting. Fleming, who was nearly totally deaf, carried on with his lecture obliviously, but the incoming messages were clear to everyone in the room who could understand Morse code. First, the word *rats*—a term meant to suggest empty boasting, and popularized during the Anglo-Boer War to refer to the hubris of the British Army— and then, after a short pause, a few lines of doggerel which essentially went like this: "There was a young fellow of Italy, who diddled the public quite prettily. . . . "[62]

The event was followed by a chain of indignant exchanges between Fleming and Maskelyne in the pages of the *Times* and elsewhere. Most of the press dismissed the stunt as "scientific hooliganism" but some serious questions were raised, particularly regarding Fleming's role as a hired gun for the Marconi company. "When the philosopher stoops to commerce he must accept the conditions of commerce," commented the *Electrical Review*. On his end, Fleming denounced Maskelyne's attack as "dastardly."[63]

Maskelyne continued to be an occasional thorn in Marconi's side. He introduced a Danish wireless system known as the "Poulsen arc"[†] to Britain and helped the American inventor Lee de Forest to develop his British

---

[†] Danish engineer Valdemar Poulsen's "arc transmitter," developed in the first decade of the century, became a competitive alternative to spark telegraphy as the field of wireless systems proliferated. During the second Berlin conference, in 1906, Cuthbert Hall wrote to Marconi: "There is a great deal of talk here about the Poulsen Arc," and the *Times* reported that the Poulsen system "promises to revolutionize wireless telegraphy by varying with the mere turn of a handle the length of the electric waves"—prefiguring the mechanism that would one day be known as the radio dial (HCH to GM, October 14, 1906, OX 205; "The Wireless Telegraphy Conference," *LT*, October 8, 1906). Poulsen is also remembered today as the inventor of magnetic tape recording.

operations; he also formed a rival company with himself as scientific advisor.[64] The Maskelyne event has sometimes been considered the first recorded incident of deliberate wireless message jamming, or what the British military later called "electronic counter-measures." Today we would be more likely to call it hacking, or a denial of service attack. At any rate, it was another first for Marconi.

❀   ❀   ❀

Cuthbert Hall kept sparring with the British Post Office in the early months of 1903. The Post Office conceded that it would be good to concentrate the business of wireless telegraphy in as few hands as possible, to which Hall responded that this could be achieved by giving the Marconi company an exclusive licence. Marconi's was the only wireless manufacturing and supply company to grasp "the true nature of the business... that it is a telegraph business and that it should be organized on the lines of a telegraph company." The company's global strategy was based on that assumption, and treated the entire world as a single system; the crux of this policy was control of all international wireless operations under the Marconi corporate umbrella. Hall didn't hesitate to emphasize that the company was British. In other words, the interests of the Marconi company and those of the British Empire coincided. Hall's efforts were nothing if not a pitch for British support at Berlin.[65]

Marconi then got into the discussion himself, with a series of exchanges with Postmaster General Austen Chamberlain. The two met in Chamberlain's room at the House of Commons in late February, and Marconi followed up with a note. Chamberlain was interested in an arrangement where the Marconi company did commercial business between England and North America, while the Post Office kept control of inland communication. Marconi had deals pending with other UK government departments and also, of course, with Canada, Italy, and the US Post Office. Now he wanted the same rights as the UK's legacy cable companies: to distribute inland the messages they had transmitted overseas. He asked for a licence for commercial communication within the three-mile limit. "I may point out that at present it would be impossible for passengers on board a ship in distress in British territorial waters to communicate with their friends by Wireless Telegraphy, except in breach of the existing monopoly of His Majesty's Postmaster General."[66]

Chamberlain replied promptly and in a spirit that must have encouraged Marconi, for the essence of the reply was to suggest an "amicable settlement." A series of meetings and further exchanges ensued,[67] but nothing concrete emerged and Marconi's efforts to obtain a licence were soon over-shadowed by his concerns to get the strongest possible support from the British delegation to the conference in Berlin. He was lobbying in all direc-tions. On July 18, 1903, he received the Prince of Wales and a royal party for lunch at Poldhu. The prince was already aware of Marconi's difficulties with the Post Office "and shewed very plainly that he did not altogether approve of their methods." To be sure the prince got the point, Marconi had the sta-tion decorated with the Union Jack, the Italian naval ensign, the Irish flag, and the Canadian flag, "each of which had a significance of its own." Evidently a believer in flags (another a form of wireless communication), Marconi also asked to fly the Prince of Wales's standard.[68]

The Foreign Office meanwhile confirmed the official British position: the conference was only a preliminary discussion and its delegates had no power to commit the government to any particular course of action.[69] A long memo in the Marconi Archives, almost certainly written by Hall and addressed only "Sir," outlined the essence of the company's position. The "art of wireless telegraphy" was not yet ready for rules and regulations; the German proposals, however desirable in theory, were simply not practicable. The company had fundamental patents controlling wireless telegraphy in the United States, England, Italy (added by hand), and Germany, and had spent over £250,000 acquiring and developing Marconi's inventions. Its German competitors had "pirated" Marconi's invention, the company claimed.[70]

The Preliminary Conference on Wireless Telegraphy convened in Berlin on August 4, 1903. On the eve of the conference, Cuthbert Hall published a long letter in the *Times* in which he laid out the issues for a British reader-ship, equating Marconi's corporate interests and the British national inter-ests. The Germans were seeking international regulation of wireless between ships at sea and ship-to-shore stations, in order to require Marconi to accept signals from German apparatus. This was, first of all, impractical for technical reasons, Hall stated. The field was too young for rules and regulations, which would merely reduce the utility of the most advanced systems. The Germans say they are seeking to prevent a Marconi monopoly but they are really seeking to gain advantages for a German company that had failed to achieve them through fair business competition. Marconi had organized a system of shore stations for communicating with ships equipped with its apparatus; it had contracts with foreign and colonial governments as well as private corporations to this end; the conditions were always the same: only Marconi apparatus were to be employed, except for communicating with ships in distress. Far from levelling the playing field, any other approach would reduce the best to a lowest common denominator. The Marconi company hoped the British government would resist proposals that tended to deprive a British company of advantages it had secured in open competition.[71]

The German company Hall mentioned was Telefunken, newly formed (in May 1903) by the amalgamation of the Slaby-Arco and Braun concerns, as we've seen. Telefunken, meanwhile, issued a statement affirming its capacity to compete with Marconi and applauding the fight to block him from getting a world monopoly.[72] The British representatives were in a bind. Although they typically liked Marconi the man, there was no love lost between them and his company, and Hall's aggressive attitude had personally irritated many of the politicians and functionaries who made up or were behind the British delegation. On the other hand, they didn't mind the fact that an emerging wireless monopoly would be British. Plus they had their own contract (via the Admiralty, as did the British Lloyd's) forbidding intercommunication. The Foreign Office, however, was sensitive to German pressure. The Italian government, as a result of its four-month-old agreement with Marconi, was in a similar legal position, and it too had strong patriotic reasons for supporting Marconi. In fact, Marconi's man in Italy, Solari, still nominally an Italian naval officer with expertise in radiotelegraphy, was an official Italian delegate to Berlin.

The stated purpose of the conference was "to establish a basis for the international regulation of the radiotelegraph service." The Germans' starting proposal could not have been any clearer—or more clearly directed at stopping Marconi: "Radio-telegrams originating from and destined for ships shall be received and forwarded without regard to the system employed."[73] If Germany had its way, wireless concessions would be granted only to persons accepting this condition, and companies in non-adhering countries that refused to communicate with ships equipped with a different system would be excluded.

The opening remarks by German postmaster general Reinhold Kraetke set the tone.[74] Kraetke reviewed the history of wireless, naming as pioneers Faraday, Maxwell, Hertz, Kelvin, Hughes, Branly, Lodge, Popov (who was a Russian delegate to the conference), and Marconi, and concurrently those who had contributed to its perfection: Braun, Ducretet, de Forest, Fessenden,

Righi, Slaby, von Arco, and Tesla. It was just about the most thorough list of names ever assembled on the topic and Kraetke's point was unmistakeable: "We owe radiography to the intellectual collaboration of most of the great nations." The service was already so important that there was a need for some form of regulation, not only within the limits of individual countries, but internationally.

The German undersecretary of state and deputy minister for postal development, Reinhold von Sydow, was even more candid, addressing the issue with a directness rare if not unknown in diplomatic discourse. The object of the proposed protocol was first of all "to prevent the creation of a monopoly in favour of a single system." But wireless communication was developing in the direction of such a monopoly. Referring to Marconi's exclusive contract with Lloyd's, von Sydow said the German government believed that "the interests of navigation, as well as technical interests, imperatively demand that communication between shore stations and ships should be facilitated as much as possible, without regard to existing systems."

The British delegation responded with an inconclusive stance. Britain was in an awkward position because of its own contract with Marconi, as well as the fact that both Marconi's and Lloyd's were British companies. It was difficult to express a definite opinion on the questions formulated by the German government, the chief British delegate, J.C. Lamb, stated. The invention was not yet fully developed, its possibilities and limitations not yet fully known. British governmental authority did not apply to communication outside territorial waters (as Marconi, playing the system perfectly, was well aware); it would be necessary to ask Parliament for power before seeking to impose international control, and it would be difficult to get such powers "if the control contemplated were such as to prejudice those who had established a working service under the facilities which the present law grants them" (that is, Marconi). Britain remained attentive, but would not make a decision at the present conference.

Jean-Jacques Bordelongue, the French delegate, associated his country with the German view, agreeing with the need for regulating the practical applications of wireless telegraphy, which he characterized as "a work of common progress" that must be organized to meet the general interest and international interests. Adalbert von Stibral, the Austrian delegate, agreed with his French colleague. The Russian delegate, Ship's Captain Salewsky, assented as well. Toward lunchtime of the first day, the head of the Italian delegation, Admiral Charles Grillo, made an ambivalent statement, accepting the general view of the need for regulation, with reservation as to the proposed provisions, which he said might injure particular interests; some kind of intermediary solution might be necessary. But he avoided the crux of the matter: Should wireless communication be "open" or "closed"? And with that the conference's first sitting adjourned.

When the conference reconvened the following morning, the head of the US delegation, General A. W. Greely, outlined the scope and extent of international regulation the United States deemed advisable. The "inchoate state" of the science of wireless telegraphy made detailed regulation impracticable. The conference should confine itself to general resolutions, embodying principles that "ensure the greatest advantages to the commerce and trade of the world" while taking care "to avoid interference with the development of wireless telegraphy." This was brilliant diplomatese, echoing the Marconi company's rhetoric while effectively supporting the German position. (Formerly the US Army's chief signals officer, Greely had been a skeptic since his observations of Marconi's first US trials in 1899, as we saw in chapter 5.)

Solari then made a nuanced appeal for the delegates to consider the technical issues. Communication between apparatus of different systems would always be dubious, and regular connection could not be guaranteed (this was somewhat milder than what Hall had been writing in the press: that intercommunication was technically impossible). "I have no intention of saying that a given system is better than others," Solari stated, but "it was not until the Italian Navy ... adopted a single system, Marconi's" that it found an efficient service. There were also commercial concerns, the need to respect existing contracts, and the legitimate rights of companies that had "proven" the superior worth of their product. The adoption of a single, common international system would facilitate improved communication for all. The German proposition, he said, was not in the public interest.

The stakes were high and, aside from within the Italy delegation, the mood in Berlin was not favourable to a Marconi monopoly. Solari therefore proposed "the temporary adoption of a single system"—taking care to add that "I have not an incontrovertible preference for a given system." This was disingenuous, to say the least. Solari had provided Marconi with the receiver for his first transatlantic signal (the so-called Italian navy coherer), had arranged to put a cruiser at Marconi's disposal (the *Carlo Alberto*), had brokered Marconi's recent contract with the Italian government, and would soon be on Marconi's payroll. He now conceded that he had been closely following Marconi's "magnificent experiments" and doubted the possibility of obtaining similar results with any other system, but if another system offered superior advantages he would accept to substitute it instead.

Admiral Grillo was more prosaic: the Italian delegation could not adhere to a proposition that would give advantages to companies whose apparatus was not perfected to the detriment of those who had brought the technology to where it was today. It was clear, however, that with the exception of Italy and, to a lesser extent Great Britain, all the delegations favoured free competition based on the regulated obligation for all systems to communicate with one another. France went so far as to want to make it obligatory for each system to make its technology transparent. The delegates agreed that Italy found itself in an uncomfortable legal position, having agreed by contract that its stations would communicate only with Marconi equipped vessels, but for all other cases, there was no injury being done to Marconi by imposing free competition between systems; the Germans were blunt about it—the only thing Marconi was losing was the hope of creating a monopoly.

By the end of the third day, it was clear that the majority was moving to a decision. With the exception of the British, who continued to maintain that they could not commit themselves because they had no power over international communication, every delegation present claimed that their government had a monopoly on telegraphy and, hence, on wireless as well. Basically, they all agreed, they could do whatever they pleased. At the start of the fourth session, Solari made another impassioned plea, explicitly for Marconi. Having taken the pulse of the conference, he now argued for indemnifying Marconi to compensate him for his investment in the event it was decided to allow other companies to enter the international wireless business. "The same treatment should not be meted out to him who has carried out a work of genius with boldness, in defiance of dangers and sacrifices, at an expense of several million francs, as to those who have sought to profit by it while only contributing a small share to the progress of the invention."

Faced with this bombardment, Kraetke responded that the German government believed "that all inventors' rights arising from patents should remain reserved to the inventors and companies working the patents." That said, the Marconi company had already made a large profit from wireless telegraphy (Kraetke said, although this was far from the case) and stood to make even more from the increased traffic that would come from free competition. Bordelongue, the French delegate, was not long to note that Solari was no longer asking the conference to endorse a Marconi monopoly, but rather to indemnify Marconi for the losses that he would suffer as a result of competition. Italy's Grillo reiterated that the Marconi company already had "a kind of monopoly" and could not be compelled to submit to conditions that would restrict it without compensation. "Its right of making conditions cannot, then, be contested." One of the US delegates, Navy Commander Francis M. Barber, seemed to speak for the others when he said he saw no reason for granting an indemnity to existing systems. This was indeed code for a stance taken by Barber and other influential figures within the US Navy, since 1899, of supporting "anyone but Marconi."[75] The session was then adjourned.

The Italian delegation formalized its singular position at the opening of the conference's fifth session. Captain Quintino Bonomo del Casale submitted a memorandum stating that "in the existing state of wireless telegraphy, it is not possible to think of laying down rules with the object of seriously ensuring exchange of communications with the apparatus of different systems." He then added verbally that while the Italian government favoured adoption of a single system, it could not at present agree to the formulation of rules of any kind relating to international radio communication, "because of the engagements which Italy has entered into with Mr. Marconi."[76]

Marconi was following the conference closely from England, thanks to regular reports from Solari. On August 12, 1903, the eve of the conference close, he was sanguine enough to write to Hall that "the English delegates are sticking up for us."[77] But the outcome of the conference was troublesome for Marconi, in that the world's leading powers, except for Italy and with some reservation on the part of the United Kingdom, were in agreement that wireless communication should not be left in the hands of a single privately owned company. (What was even more exceptional about the Italian position was that its contract was not even with the Marconi company but with Marconi himself.)

Ultimately, a final protocol, dated August 13, 1903, was signed by the delegates of Germany, Austria, Spain, France, Hungary, Russia, and the United States of America, followed by separate declarations by the British and Italian delegations, recording their reservations. (The British and Italian delegations undertook to submit the protocol to their respective governments, with a "general reserve" on the part of the British, and a set of explicit reservations by the Italians. The Italians reiterated that their agreements with Marconi bound them to keep secret the details of their installations, and that they could not, without his agreement, compel him to accept the proposed provisions.)

The protocol proposed a set of regulatory principles governing "the exchange of correspondence between ships at sea and wireless telegraph coast stations," subject to a future international convention. The most important—and most contentious—point was the first: that "coast stations are bound to receive and transmit telegrams originating from or destined for ships at sea without distinction as to the systems of wireless telegraphy used by the latter." All the delegations agreed that priority be given to requests for assistance from ships in distress. The only other item agreed to by all was the stipulation that the 1875 International Telegraph Convention of St. Petersburg was applicable to the transmission of messages by wireless telegraphy.[78] With this unanimous diplomatic gesture, national governments asserted their right to regulate the radio spectrum.

This may have been the first and only time an international conference was held to discuss what to do about the growing power of a single individual—something no modern media mogul, not Bill Gates nor Steve Jobs nor Mark Zuckerberg, can claim. In the debate that persists to this day between "open" and "closed" communication systems, Marconi was the inventor of the "closed" model; to make a current analogy, he sought to eliminate his competitors by controlling how both hardware and software are used. Today's moguls and their companies, whether aware of it or not, are direct descendants of Marconi. Apple's products emulate Marconi's approach to "intercommunication"—they don't do it. Microsoft's Windows operating system, to this day, is licensed, not sold.[79]

12

Marriage

Within days of the end of the Berlin conference, Marconi, joined by
Luigi Solari, was off on another trip to America. It was less than two
years since the collapse of his relationship with Josephine Holman in January
1902. Surrounded by constant adulation, Marconi was nonetheless desperately lonely. Since adolescence, he craved female attention. As a young man,
famous as he was so early, he attracted it easily but superficially; when the
attention was serious, as in the case of Holman, he couldn't reciprocate.

Marconi's diary for 1902—more like the proverbial little black book
(though in this case it was actually brown)—is crammed with women's
names and addresses, often written in hands other than his own. Miss Susie
Edwards, Detroit...Miss Grigsby, London...Ethel Christie, Vancouver...
Maisie Goodday, London...Elsie R. Bouliwell, Boston...Winifred Dick,
New Brunswick...Adele Lasalle, Farnham, Quebec. In Cape Breton, newspaper reports linked him romantically to a Miss Nina MacGilivray of
Sydney, and there were rumours that he had fathered a child (some even
suggesting unkindly that Richard Vyvyan's newborn daughter was actually
Marconi's, a rumour that is still alive in Cape Breton to this day).[1] In Sydney,
he told Solari, the women were easygoing and not too clinging.[2]

Marconi's life at this time, marked by high tension and perpetual motion, didn't allow for much more than passing flirtations. There was often a
coterie of women around him in Dorset or Cornwall—mothers with mar-
riageable daughters on offer, or often the daughters themselves. Solari, who
seemed to enjoy portraying his friend as a ladies' man, wrote that Marconi's
mere presence in these spots was enough to attract onlookers, "in the majority of the female sex." He described an afternoon at Poldhu in September
1901 when Marconi abruptly interrupted what they were doing and said to
him, "Enough talk of technical matters. Let's go back to the hotel where I

will introduce you to some beautiful women. I am sure they will please you as much as they please me," adding that after a long day's work the company of the right woman was good for one's health.[3] But despite all the publicity and rumours, we have few clues about his sexuality and there is little concrete evidence that he had any actual sexual experience.

According to Solari, Marconi had a physique "that responded to the ideals of British taste"—lean and graceful, healthy, slim, and strong. He was tall (just under six feet) and well proportioned, with an oblong head and a high forehead, light brown hair, slightly pink-faced and clean-shaven, with gray-blue eyes from which he looked at you now vaguely now sharply. He had a prominent but normal nose, a subtle mouth, and "the air of someone who was part ancient Roman and part Anglo-saxon of refined heritage."[4] (This description was written in 1940, when Solari was undoubtedly influenced by fascist physical ideals.) Solari also evokes a characteristic that had often been remarked upon by journalists writing about Marconi: he was a cosmopolitan.

For all that he pretended to an intimacy with Marconi, Solari seems to have been completely unaware of Marconi's long-distance romance with Holman. At any rate, he never mentioned it—while he was not particularly laconic about Marconi's relationships in general. When Solari accompanied Marconi to the Haven in 1901, he said, their arrival was celebrated by three gracious sisters, the Skimmings, frequent bicycling companions of the inventor. Solari did note, however, that in the period leading up to Newfoundland Marconi had no time for anything but work, little time for politics, sports, art, or "worldly distractions." In August 1903, when they set out for America together after the Berlin conference, Solari—who probably knew him as well as anyone at this time—wrote that Marconi was leading "a rather austere life."[5] All of that was about to change.

They sailed on the *Lucania* on August 22, 1903. On the way over, Marconi was in simultaneous communication with Poldhu and Glace Bay, the first time a ship had been in contact with both sides from the middle of the Atlantic.[6] When they arrived in New York, a letter was waiting from Thomas Edison inviting Marconi to visit him in West Orange, New Jersey, that weekend. Marconi had received a priceless boost a few months earlier when Edison ceded several of his patents to the American Marconi company in exchange for shares and membership on the company's board of technical engineers.[7]

Marconi and Solari took the train from New York, then walked to Edison's house from the station. The meeting yielded one of Marconi's favourite anecdotes, as their absent-minded host had made no provision for lunch—an essential moment of the day for Italians, as they gently reminded him. They returned to New York in the evening, presumably had a good meal, and Marconi remained in close touch with Edison throughout his stay.[8]

He then spent a full week with lawyers in connection with the two pending lawsuits that had brought him to New York.[9] But he was anxious to get on with more interesting business and spent as much time as he could for the rest of the month shuttling back and forth between New York and Ottawa, Cape Breton, Cape Cod, and St. Louis, where he had been invited to install a model wireless station at the forthcoming 1904 World's Fair.[10] When they boarded the *Lucania*, the same ship that had brought them to America six weeks earlier, on October 3, 1903, Marconi told Solari: "I need rest."[11]

Marconi's only time for relaxation in these days was when he was at sea, en route between destinations where there was always too much to do, too many people waiting to see him, too many unwanted distractions. On this trip he met and began a romance with another young American woman, Inez Milholland, who was travelling with her mother and two younger siblings, returning home to the family's townhouse on the edge of London's Kensington Gardens. The Milhollands, a newly prosperous, politically progressive family based in New York City, had been living in London since 1899.

While she was all of seventeen years old when she met Marconi, everything in Inez Milholland's background was preparing her for the life she was going to lead. Her father, John Elmer Milholland, born in 1860, was the son of immigrants from Ulster; he grew up in the Adirondacks, where he bought a newspaper, the *Ticonderoga Sentinel*, and was a strong and early advocate for causes ranging from civil rights and environmental protection to severe penalties for abusive husbands. Inez's mother, Jean Torry, was from Scottish stock; she grew up in Boston and Jersey City and had a reputation for quietly challenging conventional views on the proper place of women in the workplace, in politics, and in family life. The Milhollands moved to Manhattan soon after they married, in 1884, and John (having tripled his investment on the sale of the *Sentinel*) became a political writer and editorialist for the *New York Tribune*. After a brief attempt to enter politics, he went into the business of pneumatic tubes, the new technology that was expected to revolutionize postal communication. Pneumatic tube postal delivery in New York began in 1897, and by the turn of the century the Milhollands found themselves economically part of America's one percent, yet socially on the cutting edge of change.[12]

Solari described Milholland as gracious, beautiful, and "exceptionally-cultured for her age." According to her biographer, Linda Lumsden, "Inez undoubtedly inherited her parents' passion; she enjoyed sex and the power of the role of seductress. As soon as she hit adolescence, men began buzzing around her like bees around honeysuckle." One of the earliest of these was Marconi. Inez became a high-profile first-wave feminist, one of the most visible stars of the American women's movement in the 1910s. A 1913 image of her on a white horse was one of the movement's iconic symbols. She was a suffragist, a militant campaigner for women's rights, a crusading journalist, and "one of the most sexually radical of the New Women."[13]

On the *Lucania*, the Milhollands were fascinated observers of Marconi's experiments. Inez wrote to her father that they had felt like the ship's wireless was their own "special property," and he in turn was impressed. "By his [Marconi's] wonderful invention and courtesy I was able to send them a message of cheer when they were away out in mid-ocean, at least a thousand miles from land," Milholland wrote in his diary, in one of the first firsthand accounts by a personal wireless user.[14]

In London, Marconi became a frequent visitor to Kensington Gardens, explaining his latest inventions while swearing the Milhollands to secrecy, as he always did. They practically adopted him, calling him "Billy." Marconi longed for family life and proposed marriage; the precocious seventeen-year-old Inez accepted. The engagement was whimsical rather than serious, although it lasted more than a year. (John Milholland, an avid diarist, never mentions his daughter's engagement when discussing Marconi's visits to their London home.) It is difficult to gauge how strong-felt were Marconi's sentiments. Solari wrote, "Marconi admired her very much, but given their different temperaments, after a short-lived engagement, their feelings remained limited to a pure and idealistic friendship."[15] The incompatibility may have been more than temperamental; Marconi later told friends he'd felt Milholland needed someone more masculine, or "stronger," than himself.[16] Marconi's daughter Degna, one of the sharpest observers of her father's personality, wrote: "Except for her beauty, Inez Milholland embodied everything he basically disapproved."[17] (As a child, Degna probably met Inez when she visited the Marconis at their Sussex mansion, Eaglehurst.) But Inez was clearly a type of woman to which Marconi was attracted throughout his life, strong-willed, independent of spirit, and on her way to great accomplishment. There would be many women like Inez in Marconi's life, but they were typically not the ones he married. Far from disapproving, Marconi always supported and encouraged Inez's endeavours.

When he was in London, Marconi lived an Edwardian bachelor's life, having traded the placidity of his mother's stuffy middle-class house in Westbourne Park for the hotels and gentlemen's clubs of central London. Early in 1903 he rented a suite of rooms on the top floor of 90 Piccadilly, a stately four-storey Victorian building near Green Park that became his first personal address. Correspondence with Annie Marconi indicates that he shared the space with his private secretary, Herbert Kershaw, who seems to have doubled as valet.[18]

Marconi loved living in the very centre of the throbbing metropolis. The city was being overtaken by the new urban means of transport. By 1903, horse-drawn carriages jostled with motor cars, hansoms with taxicabs. Between the new Underground railway—the Piccadilly line opened in 1906—the electric trams, and the steamboats on the Thames, London seemed to be in perpetual motion.[19] It was a good hour's walk, or twenty minutes by tram, from Marconi's flat to the company's offices in Finch Lane. The good thing was that the route along the Strand passed right by the Savoy, where Marconi was soon having lunch most days. Later, the Savoy also became his favourite London hotel.

Marconi was now one of the most recognizable figures in the capital. In March 1902, a wax likeness of him went on display at Madame Tussauds, where according to the *Times* it was one of the year's most popular exhibits,[20] and he was one of the first people in London to own a motor car—a Napier roadster, the car to own in Britain when Marconi acquired his in 1903. (An entire box of documents in the Marconi Archives in Oxford is dedicated to Marconi as auto aficionado.[21])

Marconi was in Italy to receive an honorary degree from the University of Bologna when his father, Giuseppe, died unexpectedly on March 26, 1904, of a combination of chronic lung and heart problems.[22] In his last few years, the elder Marconi had been spending winters at the Palazzo Albergati, a luxurious residential compound in Bologna. He was eighty years old and had been ill and bedridden for some time, but his death came as a shock nonetheless. Guglielmo was swept up in the outpouring of family grief; deeply shaken, he was soon overtaken by fever. He was, however, present at the reading of Giuseppe's will on March 29, 1904, where the family learned to its surprise that the patriarch had left his prized estate, Villa Griffone, to his youngest son, Guglielmo. It was a final break with tradition for a man considered by all around him to be the very definition of the word. Giuseppe left a property at Buonconvento, near Siena, to his eldest, Luigi; Alfonso received the cash in Giuseppe's bank accounts as well as a substantial sixteenth-century palazzo in the centre of Bologna, the Palazzo Orsi, that his father had purchased as a rental property in 1900. Alfonso was named administrator of the estate. But Villa Griffone, along with all its livestock, furniture, linen, and silver, comprised the bulk of it.[23]* Giuseppe had led an interesting life, of his own confection.

Marconi left immediately after the funeral for Pisa and then England. He was exhausted, as well as thoroughly disillusioned by his endless struggles with the British Post Office, US patent challenges, continuing German diplomatic hostility, and his company's failure to take off financially. He was also shaken by the death of his father and the new responsibilities it meant with regard to Annie and the rest of his family in Italy. He was enjoying his amusing flirtation with Inez but it was also obvious that she was not what he was looking for in a wife. He had the attention of other women as well, "so numerous that it disturbs me," he told Solari, but he felt completely unfulfilled: "I would like to have one woman fully to my taste and a well-organized home." Marconi told his friend he felt that he had reached the middle of his life (he was not far off) and that he had not yet found intimate happiness.[24]

In the summer of 1904 Marconi was in his usual frenetic state, having just returned from another trip to give patent case testimony in the United States, picked up an honorary degree at Oxford (in the speech presenting him for the DSc degree, he was again referred to as a "magician"),[25] presided at a board meeting in London, and finally headed to the Haven in early July. Always a prize for the social set, he had made friends with a wealthy Dutch family, the van Raaltes, who had recently bought historic Brownsea Island, five miles off the coast of Poole Harbour, and luxuriously renovated its old castle.

---

* Giuseppe also provided a yearly allowance of 6,000 lire (around £240, or $1,200) to Annie, to be paid in equal parts by each of his three sons.

Brownsea Island had been settled as early as 500 BCE. A blockhouse on the site of the first castle was built with the encouragement of King Henry VIII in 1545, as part of a string of forts designed to defend the south coast of England against invasion. The island was privatized in the nineteenth century, and went through a number of hands until Charles and Florence van Raalte bought it and settled there in 1901. Van Raalte had made a fortune in tobacco trading and stock investments. Brownsea became a focal point of Edwardian high society as the van Raaltes entertained lavishly, with pheasant shoots, golf, and music played by a resident band. The island was a short hop from the ferry dock at the Haven Hotel.[26]

Marconi used to visit Brownsea when he wanted a few hours' relaxation from work at the Haven. In July 1904, the van Raaltes had a house guest, a young woman from one of the oldest families of the Irish peerage, the O'Briens of Inchiquin Castle, Dromoland (pronounced "drum-OH-land"), in County Clare. Beatrice O'Brien knew of Marconi, of course—who didn't?—but she imagined him to be too famous and grand to think of as a potential suitor. After all, she thought, she was not yet twenty-two years of age, and he was probably thirty (which he indeed was). They noticed each other as soon as he got off his boat and was climbing the steps of Brownsea pier.[27]

Beatrice described her first impressions of Marconi in a memoir she wrote in the 1950s:

> I was more than pleasantly surprised to see a young and quite good-looking young man and what struck me more than anything else was his broad intellectual forehead, the expression in his blue eyes which as he addressed you looked above you as if into infinity. He had a very serious rather cold manner which however broke into great cordiality and a bright smile. I noted a great sense of humour. Immediately I felt at ease with him and we quickly made great friends. He seemed to have taken a great fancy to me immediately.... He showed immediately that he cared for me but I was shy of him.[28]

Marconi began visiting Brownsea whenever he could; it seemed that all he wanted to do was spend every minute with Bea, as she was known. A few days later they both returned to London (she by train, refused permission to travel with him unchaperoned in his rakish sports car), where Marconi attended a charity ball organized by her mother, Lady Inchiquin, at the Albert Hall.[29] There, at the top of the long lobby staircase, he proposed. Beatrice was torn; she was embarrassed and confused, and told him she would give him an answer in a few days. She was pleased, but he wasn't like the other young men who were seeking her attention. In fact, Beatrice thought Marconi seemed strange, and convinced herself that she didn't care for him "that way"—and also that his foreignness was an insurmountable obstacle to someone of her station. Days passed. Marconi, frantically waiting, began to run a high fever, as he often did in times of great stress. Finally Beatrice invited him to tea, told him of her doubts and fears, and gave him her answer: it was no.[30]

Degna Marconi's version is that her father reacted "like the jilted suitor in a romantic Victorian novel," leaving on an exotic trip "wearing his broken heart on his well pressed sleeve." Degna constructs a romantic narrative of boy meets girl, boy courts girl, girl rejects boy, boy pursues girl, boy gets girl. Marconi's letters—to Hall, to his mother, and to others—and activities, travels, and press accounts tell a rather different story, one in which the courting of Beatrice was one of several concurrent pursuits. First, he left for a long-planned family reunion in Italy and a high-profile visit to the Balkans, where he was feted by royalty while inaugurating the first connection between Bari (Italy) and Antivari (Montenegro)—and also contracted a case of malaria that would plague him for the rest of his life.[31] Then, after returning briefly to England, he was off again to America, for more patent hearings, to attend the St. Louis World's Fair, to do business in New York and Washington, and to put in a visit to Montreal and Glace Bay. But before sailing he visited Dromoland to meet the O'Briens.[32]

On September 8, 1904, Marconi wrote to his mother—mentioning Beatrice, characteristically, only at the end of a letter announcing that he was leaving for New York the following day, reporting his latest business news, reassuring her about his health, and swearing her to secrecy: "It will certainly interest you and perhaps amuse you if I tell you that I am rather in love with a very nice Irish girl. She is a Miss O'Brien, daughter of Lord Inchiquin of Dromoland Castle, County Clare. Her family like me very much but I am not certain whether she likes me or not."[33]

With this exception, the people around him were completely unaware of any emotional turmoil. In fact, if anything, it was business as usual, romantic and otherwise. While he was in the United States in October, the press reported an incident where a chauffeur-driven vehicle carrying Marconi and an unidentified young woman was stopped by police for speeding along Lafayette Street in Lower Manhattan (the car was doing a shocking twenty-two miles per hour and the police gave chase for five blocks before it stopped). The *Halifax Herald* reported that "Marconi expressed concern lest his woman companion be inconvenienced. He gave $500 cash bail for his chauffeur."[34]

The press also made a fuss about Marconi being seen around with Alice Roosevelt, the president's daughter known for her unconventional behaviour. Checking in with Annie on his return, he wrote: "My things here and in America are going well, there is no truth about that affair with Miss Roosevelt as published in the papers. I met her often in America and she is a very nice girl and of course I kept very friendly with her and with the family of the President as that is very good for the affairs of the Company in the United States." He wrote a similar letter to Cuthbert Hall from Washington, reporting on his meetings with the president and adding, a bit tongue-in-cheek: "Always in the interest of the Company of course, I have been meeting Miss Alice Roosevelt the President's daughter rather often. She takes a great interest in Wireless but I got into the papers over it."[35]

In December, Beatrice returned to Brownsea after Florence van Raalte assured her that Marconi would not be informed that she was there. But Beatrice's host betrayed her, alerted Marconi, and invited him to visit Brownsea as well. This time, Beatrice found herself not only flattered but also growing fond of him. He reported to Hall that he was visiting a friend on the island rather frequently; he had caught a cold from her and she was helping him look after it.[36] When he proposed again, she accepted—provided her favourite sister Lilah approved.

Lilah O'Brien, who was two years younger than Beatrice, was at finishing school in Dresden. Bea wrote to her on December 21, 1904; she was not in love with Marconi, she wrote, but she liked him well enough to marry him: "I think he really does love me and would try to make me happy... and to think I never meant to marry! I had always arranged to be an old maid."[37] Next she told her brother Barnaby (Barney), who was delighted and urged her to tell the rest of the family—she was fearful of her mother's reaction, and had to get the consent of her half-brother Lucius, who had succeeded their father as Lord Inchiquin.

Marconi wrote to Annie the same day; he would not be coming to Italy for Christmas as previously planned but would spend the holiday at Brownsea with the van Raaltes. "There is a large party stopping there, and I know many of them, and there is also my friend—the daughter of Lady Inchiquin—the Irish girl I wrote you about some time ago."[38]

Predictably, Lady Inchiquin was not encouraging when Marconi came around a few days later to ask for Beatrice's hand. "One could not get away from it, famous though Guglielmo was, he was undeniably a foreigner," Degna later wrote.[39] Lucius agreed with his stepmother, and Beatrice was told to break her engagement and send back the ring Marconi had given her. But Bea showed she had more backbone than any of them had thought and—reminiscent of her future mother-in-law, Annie Jameson—alone, she decided to go ahead.

Marconi, meanwhile, was off again to Rome, where he went hunting wild boar with the king.[40] Before long he was at the centre of another media firestorm, looking more like a cad than a jilted lover. It was reported in the Italian papers that he was engaged to Princess Giacinta Ruspoli, twenty-one-year-old daughter of Prince Francesco Ruspoli, master of the Holy Hospice, a high hereditary Vatican position, after they were seen sharing a box at the opera. The press reported the very next day that the two were not engaged at all, but not before the story had travelled around the world—thanks in part to the marvels of wireless.[41]

Marconi wisely rushed back to London. Mustering his considerable charm, he not only calmed Beatrice, he won over her family as well, and the real engagement was announced.[42] ("My heart isn't broken," Inez Milholland said when Marconi broke their own engagement, although she couldn't resist snipping to her sister Vida that she thought Beatrice looked like a "dinky doo."[43]) Marconi may not have been English, or titled, but he was a remarkable catch; and while he was unquestionably a foreigner, at least he wasn't Catholic (Marconi, like the O'Briens, was Church of Ireland Protestant). He was also, incorrectly, as it turned out, believed to be rich—certainly his lifestyle allowed observers to believe that, even though his assets were largely sunk in the leaky coffers of the Marconi company.[44] Beatrice's sister Lilah believed that Bea's main reason for marrying Marconi was to be a support to her family, in view of its shaky and unpredictable finances.[45] From his viewpoint, he was finally making it into high society, while also hoping to introduce some playfulness to his life; he gave his bride two wedding gifts, a diamond tiara and a bicycle.

The courtship was straight out of that episode of *Downton Abbey* where the Dowager Countess sniffs that if the romance between Matthew Crawley and Lady Mary doesn't work out, they can always marry her off to "some Italian who is not too picky." This is not too far from the Inchiquins' reaction to the match between Beatrice and Marconi. However, Marconi became a family favourite and many of the O'Brien-Inchiquins remained close to him for life.

✽   ✽   ✽

Dromoland, near the village of Newmarket-on-Fergus on the west coast of Ireland, was the ancestral home of the O'Briens.[46] The Barons of Inchiquin were direct descendants of the fabled eleventh-century High King of Ireland Brian Boroimhe (a.k.a. Boru), who was slain in his tent by Viking warriors at the battle of Clontarf, near Dublin, in 1014. The O'Briens claimed to have lived at Dromoland for over nine hundred years, since the time of the founder of the clan, a son of Brian Boru (hence "O'Brien"). A stone hermit's cave on the castle grounds bears the date "AD 1041." In 1543, clan chieftain Morrough O'Brien was forced to swear obeisance to Henry VIII, who made him the first Lord Inchiquin in exchange for his loyalty. The original castle was built in the sixteenth century, while the present one dates from 1736, with its main building completed in 1826. The surrounding estate covered fifteen hundred acres.

The O'Briens were at one time reputed to be the wealthiest family in Ireland, thanks to the rents paid by their tenant farmers. They kept a dairy farm and herds of beef cattle, raised fruit trees and vegetables, and manufactured various delicacies, while Europe's nobility flocked to Dromoland year-round for pheasant shooting, horse races, festive dinner parties, and fishing for pike and perch in the estate's lake. In Marconi's time, Dromoland had a glassed-in heated patio and a staff of four working around the clock just to wash and iron the household's laundry.

Dromoland was also the birthplace of William Smith O'Brien, a revolutionary hero and leader of the Irish nationalist uprising of 1848. A plaque on the property attests that "after 20 years as an MP and in the third year of the Great Famine this Protestant patriot asserted the independence of his country for which he and six companions were banished to Tasmania." A less noble traitor would have hanged, but O'Brien was pardoned and returned to Ireland in 1856. His role in Ireland's struggle for independence is commemorated with a statue in Dublin's main thoroughfare, O'Connell Street.

Some of the O'Brien forebears were thus good models for Marconi. Converts from Catholicism, the O'Briens professed loyalty to the British monarchy was a pragmatic one. They skilfully navigated the political waters of Anglo-Irish relations as well as Irish class politics, making alliances to maintain their own privileged position while supporting a large constituency of family and community dependents. If he did not actually study their history, Marconi may well have unconsciously adopted some of their patterns.

The O'Brien-Inchiquins' wealth began to dwindle after the Land Acts of the 1880s compelled them to begin selling their tenant farm lands, stripping them of their main source of income. Beatrice's half-brother Lucius succeeded their father to become the fifteenth baron in 1900 and presided over the further decline of the family fortune during the struggle against England that led to Ireland's independence in 1921. However, the O'Briens were proud of the fact that, unlike other great Irish landholdings, Dromoland survived the turbulence without being burned to the ground. It was twice targeted for destruction by the Irish Republican Army (IRA), which secretly sent gasoline shipments to County Clare from Dublin for that purpose; but Dublin's orders were countermanded both times by local IRA leaders who argued that the Inchiquins were "fair and benevolent" landlords. Sir Lucius, the thirteenth baron, was remembered for his relief work in the famine years of the 1840s, as was, of course, William Smith O'Brien.

Dromoland was increasingly difficult to maintain, and after the death of Lucius in 1929, the estate was supported mainly by the personal wealth of his widow, Lady Ethel. Lady Ethel still managed to keep a staff of twenty household servants (as well as eighteen gardeners and seventy-five farm labourers), but in the 1940s, her son Donough, the sixteenth baron, had to start taking in tourists as paying guests, and in 1962 the castle and all but some four hundred acres were sold and the castle was renovated and repurposed as a luxury hotel. Donough O'Brien later built a home, Thomond House, on the remaining land where he lived until 1968, and his brother Phaedrig, the seventeenth baron, managed the estate into the 1980s. The current and eighteenth Lord Inchiquin, Conor Myles John O'Brien (Beatrice's great-great-nephew), and his family are still living at Thomond House, until recently farming the land and running a sporting and leisure sideline on their property. Locals talk in hushed tones about a bitter dispute between the current baron and the new corporate proprietors of Dromoland over ownership of its original artwork—ironically, the name "Dromoland" translates from old Gaelic as "hill of litigation." Dromoland, meanwhile, has become one of Europe's most exclusive resorts. It has been owned and operated by a consortium of Irish and American investors since 1987, and its guests have included Bill Clinton, Nelson Mandela, Muhammad Ali, and Johnny Cash. One of its outstanding landmarks is an exquisite eighteenth-century Temple of Mercury—honouring the Roman god of communication and messages.

Beatrice was the tenth of fourteen children of the fourteenth Baron Inchiquin, Edward Donough O'Brien, a Cambridge graduate and Irish peer in the British House of Lords. According to Degna Marconi, Lord Inchiquin spoke out against Home Rule "but he was an Irishman through and through."[47] Beatrice's mother, Ellen White, also of royal lineage, was her father's second wife. Beatrice, born in 1882, had four half- and nine full siblings. She outlived them all.

When she became engaged to Marconi, the press reported that Beatrice was only eighteen or nineteen years of age but in fact she was nearly twenty-two. It is almost impossible to correct published misinformation, especially where celebrities are concerned, and Beatrice was invariably reported to be younger than she was throughout her life. Even Degna erroneously writes that Beatrice was only nineteen years old when her parents met—not a huge difference, but significant.[48] This was another theme that runs through Marconi's lifelong jousting match with the press. The women he was associated with were invariably young, but usually not as young as they were said to be. When he married for the second time in 1927, at the age of fifty-three, his bride's age was reported as twenty, although she was twenty-five.

Beatrice was totally unlike any of Marconi's previous love interests. Yet he was as ambivalent about Beatrice as he had been about Josephine and, to a lesser extent (because the relationship was much less serious), about Inez. His ghostwritten autobiography of 1919 devotes less than a full sentence to her, and as in his letter to his mother, what reference there is is bracketed with other details: "A fortnight after this lecture, on March 16th, 1905, I was married at St. George's [Church], Hanover Square, to the Hon. Beatrice O'Brien, a daughter of Lord Inchiquin and by the end of the month I was again on my way to New York, this time accompanied by my wife."[49]

Annie and Alfonso arrived in London in early February 1905, to meet the Inchiquins. Marconi was typically fussy, anxious that they should stay in a good hotel and that their clothes should be all right. "Beatrice's relations are so very high up in society that it's rather important," he wrote Annie. Solicitous as always, he tried to orchestrate the first encounter between his mother and his future wife. "She is such a sweet and good girl that I think you will like her and love her very much. If you think it's right, write her and say you are glad we are engaged, but if you think you should not write first, don't."[50]

In the tradition of the Edwardian upper class, Marconi made a financial settlement in Beatrice's favour a few days before the marriage. The settlement created a trust fund to which he transferred some £11,500 in stocks and securities, the income to be paid to Beatrice throughout her life, and a further twelve thousand shares in Marconi's Wireless Telegraph Company, on which he would receive any dividends payable during his own lifetime. The trustees were two of his future in-laws (Beatrice's uncle Charlie White, her mother's brother; and Noel Armar Lowry-Corry, husband of Beatrice's eldest sister, Clare) and his colleague, Cuthbert Hall. He also wrote a holograph will leaving half of everything he owned to Beatrice and the other half for the education of his issue to be divided among them equally when they became of age. Alfonso was best man at his wedding, but otherwise the Italian side of Marconi's life was remarkably absent from his plans.[51]

Guglielmo and Beatrice were married on March 16, 1905, at St. George's Church in Hanover Square, Mayfair, one of the most fashionable churches in London, amidst a cyclone of unwanted publicity, a gawking crowd, and a worrisome death threat that turned out to be bluster. If there was anything that Marconi and the Inchiquins agreed on it was that they all detested the press poking into their private affairs. But their efforts were hopeless. London's tabloid sandwich boards that day read "Marconi to wed daughter of Irish peer," and the *New York Times* reported that "the crowds outside the church were so dense that traffic had to be stopped."[52]

Immediately after the wedding, Guglielmo and Beatrice left for a honeymoon at Dromoland—where they were joined for part of the time by Annie and Alfonso.[53] In the five years since her father's death, Beatrice, her mother, and two of her sisters had been living in London, in a large comfortable house near Marble Arch. Beatrice now found Dromoland a bit dreary, the Inchiquin glories a thing of the past, but she put the best spin on things, writing to her mother on March 18: "We are both happy and it's <u>too</u> heavenly being back here again everything just the same as ever not changed one little bit...."[54]

Degna tells a different story. "At this stage the castle was virtually empty and Bea herself found it quite depressing.... [Marconi] was almost frozen with the cold and draughts and the whole stay was an unmitigated disaster." Marconi insisted on going for long walks alone; after a week he was called back to London on business, to both of their satisfaction. Beatrice only returned to visit Dromoland two or three times during her long life, and never again with Marconi.[55]

After a brief pit stop in London, the Marconis left Liverpool on the Cunard liner *Campania* on March 25, 1905. It wasn't long before Beatrice had a sense that her new husband had eccentric qualities. He hung clocks all around the cabin and had a habit of throwing his worn shirts and socks out the porthole windows to avoid bothering with laundry. He was also, and already, obsessively jealous, inquiring in detail about anyone and everyone she had contact with on deck. Beatrice, who didn't have a jealous bone in her body, felt that it was an "absolutely incomprehensible lack of trust."[56]

In many respects, Marconi and Beatrice were opposites. To paraphrase Degna, Beatrice was innocent, extravagant, and impractical, while Marconi was worldly, organized, and meticulous. But she also fussed over him, as only his mother had done before. When he was away working she would write to his associates, as on one occasion when she wrote to George Kemp at the Haven: "Will you take <u>great</u> care of him and let me know by wire if he isn't quite so well don't ask him, just let me know—<u>Promise</u> as really I am rather worried about him [emphasis in original]."[57]

Beatrice would be Marconi's confidante and epistolary foil for the next twenty-five years. When he was away, he wrote to her constantly; when he was under stress, he sought her counsel. Only after his second marriage did he distance himself from her, undoubtedly under the influence of his new wife. The various Marconi archives contain nearly two hundred letters from Guglielmo to Beatrice, the last one dated June 14, 1927, the very eve of his second marriage. They tell only one side of the story, however. Nearly all of Beatrice's letters to Guglielmo have disappeared, lost during one of their many moves[58] or buried among the articles he had in his possession when he died.

13

A Life in Litigation

On April 1, 1905, the Marconis arrived in New York, where their gilt-edged social calendar included Fifth Avenue visits with the likes of Grace Vanderbilt as well as lunch with President Roosevelt and his daughter Alice (whom the press had romantically linked with Marconi less than a year before) at Roosevelt's summer home at Oyster Bay.[1] Ten days later, on April 11, they were in Montreal, staying at the Windsor Hotel, when Marconi received a telegram from his patent attorneys in New York: "Marconi case decided. Court holds Marconi exercised high degree inventive ability... Congratulations."[2] After three years of litigation, Judge William Kneeland Townsend of the US federal circuit court ruled in Marconi's favour in a patent infringement case against the American inventor Lee de Forest. The Townsend decision was the most important and lasting patent ruling for Marconi and anointed him practically unassailable. It established Marconi's original patent as the foundation of wireless communication.[3]

Marconi's system was now pervasive. On ships it was sometimes suggested that wireless had ruined "the delights of complete repose which have hitherto, more or less incorrectly, been associated with the idea of a long ocean voyage,"[4] but this notion was discounted by the benefits it brought for minimizing danger at sea. It was also good for business travellers who could for the first time remain in touch with their offices as they crossed the Atlantic. With cheap long-distance telegraphy within reach, emigration took on a less onerous meaning; it would soon be easier for members of diasporic communities to keep in touch with their families back home. At the same time, ambitious corporations and military establishments everywhere vied for ways to use the new technology as an instrument for their grand designs. Indeed, the sentiments for and against Marconi's invention were not unlike those we hear today about the good and evil of the constant connectedness that comes with modern communication technology.

There was full agreement, however, on the basic point: wireless communication had changed people's relationship with time, distance, and mobility.

Litigation was never far from the surface. As Marconi sought to consolidate his company's global position, patent litigation became a standard piece in his repertoire. Marconi's original British patent of 1896 was replicated on behalf of the company and its subsidiaries around the world, and Marconi vigorously attacked any attempt by others to build on it. Marconi had a small army of legal and technical experts, including James Ambrose Fleming, who was on permanent retainer as a scientific advisor providing advice and testimony as to the primacy of Marconi's inventions, and keeping potential rivals at bay. In fact, Marconi was fond of saying, "A patent is merely a title to a law suit."[5]

As we've seen, there were early pretenders to the invention of wireless telegraphy, but none of them were able to channel their ideas into a practical *technology*, and indeed, these discoveries were well before their time. It was only after 1887, when Heinrich Hertz demonstrated the existence of electromagnetic waves in the "air," that scientists as well as tinkerers (as some still dismissively considered Marconi) were able to imagine a vehicle for the transmission of electric sparks. No one before Marconi had shown that these waves could actually transport intelligible signals.

This innovation became the legal basis for Marconi's claim to ownership of all manner of wireless. Marconi's original patent, dated June 2, 1896, was deceptively simple. The summary stated briefly that he, Guglielmo Marconi, was "in possession of an invention for Improvements in transmitting electrical impulses and signals and in apparatus therefor, that he is the true and first inventor thereof, and that the same is not in use by any other person, to the best of his knowledge and belief."[6] Within months, others such as Sir Oliver Lodge—who had, as noted, publicly demonstrated spark, but not signal, transmission a full year before Marconi—had submitted patent applications. But Lodge didn't file until May 1897.

German, French, and Russian competitors soon joined the crowded field, but the first serious challenges came from the United States, where inventor-entrepreneurs such as Samuel Morse, Thomas Edison, and Alexander Graham Bell had established a pattern for using patents to build huge and lucrative corporate empires. Giant companies such as General Electric (GE) and Westinghouse bought up patents by the cartload, just to take them off the market. At the turn of the century, Nikola Tesla became a one-man patent generator, selling dozens of schemes to the Morgans, Astors, and Westinghouses of the day. Tesla and Marconi would spar in court, in the press, and in salons in Europe and New York.

But the real threat to Marconi's plans came from two ambitious young inventors, Reginald Fessenden and Lee de Forest. Reginald Fessenden was the older of the two.[7] Born at East Bolton, on the shore of Lake Memphremagog in the bucolic Eastern Townships of Quebec in 1866, he attended Bishop's University in nearby Lennoxville while still a teenager, but did not finish (although he was brilliant enough to be taken on to teach math to younger students at the feeder Bishop's College School). He moved to Bermuda in 1883 and worked briefly as a school principal. After reading of Thomas Edison's lighting up a section of Manhattan, he moved to New York in 1885 and approached Edison for work. As he told it in a 1925 magazine article, his pitch was bold; in his application Fessenden wrote, "Do not know anything about electricity, but can learn pretty quick," to which Edison apparently replied, "Have enough men now who do not know about electricity."[8] He persisted, however, and eventually was hired as a tester with Edison Machine Works, the company laying the wiring mains under Manhattan. Soon, he was working on various projects in Edison's laboratory in West Orange, New Jersey.

Working for Edison was crucial to shaping Fessenden's approach to invention, but he left the company in 1892, the year Edison merged with Thomson-Houston to form General Electric.[9] After a brief stint teaching electrical engineering at Purdue University, he was spotted by George Westinghouse, who recommended him to be the chair of the electrical engineering department at Western University of Pennsylvania (later renamed the University of Pittsburgh). He remained at Pittsburgh from 1893 to 1900.

Fessenden is considered by many historians to be the first significant figure in US wireless development. He was soon making a name for himself. Dynamic, argumentative, and possessing a razor-sharp mind, Fessenden was notoriously impatient, and described by contemporaries as "choleric, demanding, vain, pompous, egotistic, arrogant, bombastic, irascible, combative, [and] domineering."[10] Not surprisingly, his career was tumultuous.

Fessenden became aware of Marconi as soon as news of his initial experiments with the British Post Office reached the United States, and he was one of the first North Americans to make note of Marconi in print. In August 1897, Fessenden mentioned Marconi's system in an article on a planned telegraph link between Alaska and Russia.[11] (Fessenden appears in this article to be a more mature thinker and writer than Marconi but makes no scientific claims of his own).

Fessenden wrote to Marconi, inviting him to lecture at the Carnegie Library in Pittsburgh, during Marconi's first visit to the United States in 1899. His letter was ambitious, self-promoting, and fawning, and seemed as much intended to introduce himself to Marconi as to convey an invitation (he also offered to put Marconi up). Among other things, he claimed it was he who had suggested to the *Herald* that they try to get Marconi to report on the yacht races. The letter was signed "Sincerely & admiringly yours...." Marconi may not have replied, as Fessenden wrote again two weeks later asking Marconi if he had received the initial letter.[12]

Fessenden presented himself to Marconi as the inventor of "multiplex signalling" and of something he referred to as "the 'sine wave' method of telegraphy."[13] It was probably not clear to Marconi what this meant, but Fessenden was soon known in the United States as a sharp critic of Marconi's system. Fessenden realized that if Marconi's spark transmitter could be replaced by one that gave off a continuous wave, it would be possible to transmit *voice* by wireless.[14] This was the technical breakthrough that enabled what would eventually be known as broadcasting, and for this reason, Fessenden is often claimed to be the inventor of radio. The spark–reliant in-termittent wave transmission that Marconi pioneered could transmit dots and dashes but not speech and music (hence the distinction between "wire-less telegraphy" and "broadcasting"). However, both methods relied on the medium of electromagnetic waves, and Marconi was unquestionably the first to use the wave spectrum for communication.

Fessenden went to work for the US Weather Bureau in 1900, doing experimental wireless research on the Atlantic coast. The Marconi organi-zation was soon tracking his work. On July 12, 1901, Cuthbert Hall sent Marconi a letter he had received from the general manager of the *New York Herald*, reporting that "a series of important experiments in wireless teleg-raphy is now being conducted on the coasts of Virginia and North Carolina by a Professor Fessenden, under the auspices of the [U.S.] Government." Hall commented: "Everything points to the desirability of taking speedy action in America." Marconi—in Crookhaven, totally preoccupied with preparations for his transatlantic leap noted cryptically in the margin of Hall's letter: "?=". (Fessenden already held a number of patents, but Fleming didn't think they were any cause for concern.)[15]

It soon became clear, however, that Fessenden was not just a flash in the pan. On April 27, 1902, the *New York Times* reported the "extraordinary results" being obtained in tests of the new "Fessenden system" at Roanoke Island, North Carolina.[16] The US Navy (which, since its testy interactions with Marconi during his first American visit in 1899, already seemed to be taking the view of "anyone but Marconi"[17]) pronounced it superior to all other systems, and was recommending it for adoption. So did the army. The article quoted Fessenden as stating that his system and Marconi's were not in any way alike, and presenting details of their differences.

Fessenden was, however, temperamentally ill-equipped to remain part of a large bureaucratic organization like the Weather Bureau and was soon trying to build his own independent corporate structure. In November 1902, with two wealthy backers from Pittsburgh,[18] he formed the National Electric Signaling Company (NESCO). But he wasn't too good at business either.

On August 12, 1902, the US Patent and Trademark Office awarded Fessenden thirteen patents, for various improvements in wireless telegraphy, on applications going back as far as 1899.[19] Marconi engineer Charles S. Franklin reported to the company a month later that several of the Fessenden patents appeared to be "theories and ideas which are not workable at present." After reading Franklin's report, Marconi wrote to Hall that he didn't think any action at all should be taken, "for the present at any rate." He was far more concerned about possible infringements of his patents by another young American inventor by the name of Lee de Forest.[20]

Lee de Forest was a very different piece of work. He was nearly ten years younger than Fessenden, but still a year older than Marconi.[21] Historian Susan Douglas has described him as a "cultural outcast"; his father was president of Talladega College, a school for black students in Alabama, and the de Forests lived on the campus in a black part of town and were shunned by whites, "making him the wireless inventor most attuned to the aspirations and frustrations of masses of Americans," according to Douglas. As a youth, his generic idol was the inventor-hero exemplified by Thomas Edison. De Forest was hungry for fame and celebrity.

De Forest went to Yale, graduating with a BS degree in 1896 and a PhD in engineering with what is considered to be the first US doctoral thesis on the topic of Hertzian waves, in 1899.[22] Like Fessenden, de Forest also tried to contact Marconi during the latter's first US visit in 1899. But while Fessenden was already established at that time, De Forest's contact was for reasons far more mundane. He was looking for a job.

In 1897 or 1898, de Forest had sought out and met Nikola Tesla, and was awestruck at this brush with greatness. Tesla was "quite cordial," de Forest told his diary, and on April 11, 1899, he wrote Tesla "the long planned letter for job."[23] It is not known exactly when de Forest first heard about Marconi, but on September 22, 1899, he wrote to him from Chicago. Marconi was in Washington. De Forest introduced himself humbly, presenting his credentials and desires. "My Dear Sir, I am a graduate of Yale University.... For a year and a half I have made a specialty of the study of Hertz waves, the coherer etc.... Knowing that you are about to conduct experiments for the U.S. Government in the Wireless Telegraphy, I write you begging to be allowed to work at that under you."[24] As in Fessenden's case, there is no record of a reply.

Between 1899 and 1901, de Forest had a number of jobs with companies involved in electricity and telegraphy, but he was either unwilling or unable to keep them because he always insisted (wisely) on keeping control of his inventions. In 1901, with two colleagues, he founded his own firm and began taking out patents for various aspects of wireless telegraphy. In 1903, backed by a stock promoter by the name of Abraham White, he launched the American De Forest Wireless Telegraph Company.[25]

De Forest reacted with scorn to the news of Marconi's transatlantic feat. He confided to his diary on January 13, 1902: "People are all credulity and are sanguine on account of Marconi's alleged feat. That is a help to us, but I do not expect to see him repeat his famous 'S' within a [illegible]. This will give us, some at the eleventh hour, our last chance to show what we can do. Whether he actually received those signals or not, he has certainly offered no real proof which scientists can accept; and all this great haloo and adulation with his wild talk of transatlantic messages at 1 cent a word smacks decidedly of chicanery and the methods of the professional newspaper [illegible]." However, Marconi was the only inventor de Forest considered a worthy rival, and his sole aim was "to make my name at least rank with that of Marconi."[26]

Like Fessenden, de Forest was being boosted by the American press. The *New York World* described his system as completely different from Marconi's, as faster and free of interruptions.[27] By the fall of 1902, he too claimed he had orders from the US Army.[28] An important part of his promotional strategy was to emphasize patriotism and reliance on "American brains and American capital." There was undoubtedly an undercurrent of bigotry to his strategy; Susan Douglas has written that de Forest privately referred to Marconi as "the Dago."[29] He was also abrasive, and the Marconi company soon viewed him as deliberately seeking to use the legal system to get around Marconi's patents.

De Forest was unscrupulous, Fessenden unenterprising. Neither was an immediate major threat to Marconi—they had poor or no business skills—but the company felt that it was time to seek validation of its monopolistic position by a court decision. In 1902 the Marconi Wireless Telegraph Company of America launched a suit under Marconi's original US patent claiming infringement by de Forest. London, and Marconi in particular, seem to have been only vaguely aware of the action until it was well under way. Although he was vice-president of the American company, Marconi only learned about the de Forest litigation in a letter from Hall.[30]

De Forest, meanwhile, was marshalling support from other potential Marconi rivals. On January 27, 1903, he wrote to Fessenden, asking if he would appear as an expert witness in Marconi's infringement suit; perhaps I can do the same for you some day, he wrote. "I believe our motto should be 'American wireless telegraphy for Americans'!" Fessenden wrote back the very next day, stating that he would be glad to do anything he could to help. He then elaborated in a second letter a few days later: "Marconi has never had or constructed any stations capable of commercial operation without infringing the American patents." But, he added presciently: be careful "not to prejudice yourself in case of any possible complication between your company and mine which might arise later." Fessenden reminded de Forest that Marconi's advances were based on his, Fessenden's, own inventions. He pointed out that "even with the inventions which he has appropriated from others," Marconi had yet to make his system a commercial success.[31] In this letter to de Forest, Fessenden made an important point that could have had sweeping implications if validated by the courts: Marconi's contribution was "not sufficient to give Marconi what may be called vested rights in the art"; it would be unjust to consider his patent as "fundamental" and hence bar other inventors from building on it. This would be the very crux of the de Forest patent case.

Fessenden continued to advise de Forest—for example, drawing his attention to a new book by Augusto Righi in which, he said, Marconi's former mentor referred to Marconi's apparatus as being identical to his (Righi's) own. Fessenden thought this could have an important bearing on de Forest's defence, as Righi "points out that vertical wires were used previously by Lodge, Loomis and others."[32] Fessenden showed himself here to be a shrewd interpreter of patent law; however, unfortunately for de Forest's case, Fessenden's reading of Righi was based on an erroneous interpretation drawn from a review of Righi's book that had just been published in England in the journal the *Electrician*.

Righi's book *Telegraphy without Wires*, published in 1903, was making waves in Europe, and on April 1, 1903, Righi and his co-author, Bernardo Dessau, wrote a letter to the editor of the *Electrician*, correcting the very remarks that Fessenden had seized upon. "Nowhere in our book did we attribute to Prof. Lodge priority in utilizing the vertical wire known as antenna.... as far as we knew, nobody prior to Mr. Marconi had utilized the antenna at the same time both in the generating and the receiving apparatus."[33]

This was a critical validation of the originality of Marconi's contribution, especially coming as it did from one who arguably had a claim to being a precursor. Interestingly, the letter was forwarded to the editors of the *Electrician* by none other than Marconi himself, after the authors sent it to him with a request that he forward it. It was possibly Marconi who drew Righi's attention to the review in the first place.[34]

Righi had become an important ally for Marconi. On May 2, 1903, he replied to a letter from Marconi thanking him for his letter to the *Electrician*. It wasn't necessary, Righi said, "It was a question of bringing the truth to light, and naturally you will always find me ready to do that." Marconi apparently took issue with some other passages in the book and Righi reassured him; he was prepared to modify any passages "susceptible of malicious interpretations" insofar as that could be done truthfully. He ended with high praise "for your genius and for the ardour with which, from your youth, you have dedicated yourself to scientific things, and that henceforward you represent our Italy in the struggle between the inventors of telegraphic systems."[35]

As the issues came to be framed increasingly along national lines, Marconi had the undoubted advantage of being able to mobilize more than one set of nationalistic feelings and perceived national interest—British and Italian—as opposed to his more rooted rivals.

By May 1903 Marconi had a second American litigation going, after Fessenden and a collaborator, Harry Shoemaker, sued *him* for patent interference.[36] As we've seen, Marconi arrived in New York on August 29, 1903, to give evidence in the Fessenden/Shoemaker action. The American inventors were claiming priority over Marconi's "magnetic detector" patent. Although Fessenden and Shoemaker's own comparable patent was of a later date than Marconi's, US patent law was based on priority being established *in the United States*. They claimed they had invented something equivalent to Marconi's "maggie" in 1901, but Marconi had only patented his detector in 1902. As Marconi habitually insisted on absolute secrecy until a patent was awarded, he faced a difficult time establishing when his invention had been made. His case would be based on two pillars of evidence: his own word and George Kemp's diaries. Marconi's deposition attested that he conceived the invention in England in 1898.[37]

Fessenden and de Forest continued to share information. De Forest wrote to Fessenden on September 9, 1903, asking to see the results of his attorneys' patent searches: "You will appreciate that if we lose this case, it would be a severe blow for all systems of Wireless Telegraph competing with the Marconi system." Fessenden, meanwhile, was beginning to fear that his lawyers were not up to the task; he wrote to a business associate on October 9 stating that the case was being badly handled and suggesting that they pay and get rid of them.[38]

Hall was arguing increasingly that it was time to take patent action in Britain. "We must get a decision in the English courts.... It is universally recognized that we are ahead of everyone else, but it is an almost equally universal impression that the patent position is a weak one." A patent action would be expensive, time-consuming, and troublesome, but "it would be an exceedingly good investment.... If we get a decision now we should probably knock a good many people out of the business altogether."[39]

The United States Patent Office ruled on December 22, 1903, that Marconi's patent number 141,398 for "Wireless Telegraphy," filed on February 2, 1903, "is adjudged to interfere with others."[40] Shoemaker immediately filed another interference case and the company's New York lawyer, Frederic H. Betts, wrote to London frantically on New Year's Eve. Marconi had testified that he had told his assistant, Kemp, and chief American engineer William Bradfield on January 13, 1902 (when they were all in New York after Newfoundland), that he had an idea for a varying field magnetic detector; however, "We have carefully looked through the diaries and memorandum books which Messrs. Kemp and Marconi introduced in evidence in these prior interferences, and which were left by them in our custody, and failed to find in them any sketch or reference to this specific form of varying field magnetic detector." Please search for them, Betts urged.[41]

Fessenden's response to the interference ruling was edifying. He had been frustrated when Edison sold his wireless patents to Marconi in May 1903;[42] now, early in 1904, he wrote to the American Marconi company suggesting they jointly try to buy out de Forest. In London, the British Marconi board decided "it would not be judicious at this juncture to entertain the proposition."[43] Fessenden was trying to join forces with Marconi but, publicly at least, Marconi continued as though it was business as usual with respect to Fessenden.

In June 1904, Marconi was back in New York, testifying again. De Forest had sued American Marconi for circulating false reports about him, and was seeking $1 million in damages for libel. His lawyer told the *New York Times*, "The De Forest Company is the only wireless telegraph company to achieve practical results of any importance in this country—it is distinctly American, the product of American brains, and backed by American capital."[44] Marconi was furious and told Hall he should tell the press the company would take legal action in England if de Forest tried to launch a station there. Hall thought that in England it would be easier and preferable to quietly squeeze de Forest out of getting any contracts.[45]

A few months later, de Forest and Marconi squared off at the 1904 St. Louis World's Fair. The Fair authorities had initially offered the Marconi company an exclusive wireless display, but de Forest, hammering on his nativist chord, convinced them that an "American" company should also be represented. Marconi withdrew, leaving de Forest to put up a tower with his name spelled out in lights. Susan Douglas describes him at this time as "a self-deluded, crass, and entrepreneurially shortsighted inventor who knew and learned little about legitimate long-term corporate strategies."[46]

Marconi was, however, truly concerned about de Forest. He kept a running notebook while he was in the United States in September 1904.[47] The notebook, he said, "was not kept as a regular diary, but only as a convenient means of noting and ascertaining certain facts as they occurred."[48] There are twelve pages of notations, beginning with the entry "Sept 1904 D F case." The writing is thick and quick, in heavy pencil, often illegible. The notations are mostly references to documents (reports, lectures, and articles by reputed authors such as Slaby, Fleming, and Braun), as well as enigmatic thoughts, sometimes in Italian, the items separated by heavy black lines. There are four references to de Forest, including one to his doctoral thesis. And then it ends. There is nothing further in the notebook. (De Forest, meanwhile, was taking the high road. He visited England in late October 1904, at the invitation of the Post Office, lecturing in London with Sir William Preece in the chair.[49])

Marconi's testimony in both 1904 cases followed a pattern he had developed, with the encouragement of his American legal team, and would use in testimony of various sorts for the rest of his life.[50] He basically recited his life story and the entire history of his work, recounting how he had developed what he invariably called "my system," and addressing the specifics of the case in question at a selectively appropriate place in the narrative. In the Fessenden/Shoemaker case, it went like this: I first conceived the idea (of the magnetic detector) in 1898 in Poole. I was extremely busy at the time, equipping stations for the British Royal Navy, and with the station for long-distance experiments at Poldhu, Cornwall. The subject of the alleged interference, "that is, the idea of utilizing the property of a varying magnetic field for constructing a wireless telegraph receiver," was conceived in Newfoundland in December 1901. At this point, Fessenden's lawyer objected to Marconi referring to matters occurring outside the United States. Marconi then referred to his pocket diary to establish that signals were received on December 12, and that he had conceived the magnetic detector two days before. On January 13, 1902, in New York, Marconi communicated his new idea to William Bradfield. He also entered in the diary "a very rough sketch of the receiver which I intended to construct."[51]

Marconi's parallel use of patents and secrecy occasionally backfired, especially in the United States, as he learned painfully in the Fessenden/Shoemaker case. He was occasionally at odds with his advisors on this point. Fleming particularly believed that a method or instrument had to be made public in order to establish priority, whereas Marconi scrupulously kept his discoveries secret until final patent approval had been achieved; and because of his multiple commitments and his own personal attachment to research over paperwork, he often procrastinated over patent filing. Hall, meanwhile, believed strongly in both, keeping knowledge proprietary and protecting it with patents.[52]

With the Townsend decision, the De Forest Wireless Telegraph Company was restrained from further use of Marconi's invention; more important, the decision sent a message to the industry and to would-be trespassers on

Marconi's turf. Townsend established Marconi's patent as "fundamental" when he wrote:

> It would seem, therefore, to be a sufficient answer to the attempts to belittle Marconi's great invention that, with the whole scientific world awakened by the disclosures of Hertz in 1887 to the new and undeveloped possibilities of electric waves, nine years elapsed without a single practical or commercially successful result, and that Marconi was the first to describe and the first to achieve the transmission of definite intelligible signals by means of the Hertzian waves.[53]

This was the clearest legal, if not scientific, statement ever of Marconi's accomplishment, and the Townsend decision became the cornerstone of Marconi's patent protection strategy. Townsend recognized that Marconi had brought something new to the practice of communication. The decision meant that Marconi's basic patent 12309 of 1896 had established "an entirely new art."[54]

De Forest managed to also claim victory, as Townsend found that he had not infringed Marconi's patent in every one of the suit's twenty-four claims.[55] But it was enough to get de Forest to drop his million-dollar libel action and strengthen Marconi's confidence that the company could move forward with assurance. If nothing else, they understood that a court ruling was only as strong as the means that could be marshalled to enforce it. Even as Townsend was preparing to announce his decision, Bradfield was on the phone to Reginald Fessenden, disingenuously asking if he would provide a sample of his "barretter," the piece of equipment that Fessenden claimed was an improvement on Marconi's coherer.[56] Fessenden said he would be happy to oblige.

Although the Townsend decision did not give Marconi total victory, the company decided not to appeal regarding the disallowed claims, as Hall felt there could be "serious tactical disadvantage in appealing after the public statements that had been made that the decision gave us control of all practical wireless telegraphy."[57] In other words, the public perception of the judgment was more important than the judicial reality, which would always remain vague and subject to practical execution, enforcement, and defence. Fleming expected that Marconi's rivals would "whittle away" the effect of the decision "by making such trifling alterations as will enable them to contend that they are no longer using the combination." Marconi appreciated the importance of Fleming's observation; he wrote to Hall that "we shall have to use our wits in order to secure the full benefit of Judge Townsend's decision."[58]

This meant returning to the United States to give more testimony in the Shoemaker interference. "I fear you will have to go," Hall told Marconi; if they should lose the case, "the injury will be irreparable."[59] Once again, Marconi crossed the Atlantic.

On October 23, 1905, the Marconi team called a sensational witness: none other than Lee de Forest, who gave his occupation as "wireless telegraph expert." De Forest's company was the assignee of the Shoemaker patent applications and de Forest may well have been a hostile witness; he certainly had no favourable disposition toward Marconi, and the feeling of displeasure was mutual. The following day, October 24, 1905, Marconi's testimony began. Once again, he outlined the trajectory of his work since 1896, all the designing, erecting, testing, and operating of experimental stations he had done; all the governments he had advised as chief technical expert to the Marconi company—the US government was conspicuously absent from the list.

On October 30, de Forest published an open letter to Marconi challenging him to demonstrate that vessels equipped with his apparatus could transmit messages as successfully as those using de Forest equipment. Marconi replied, through the press, with a skewering putdown: "My Companies are engaged in the strictly commercial enterprise of the transmission of telegrams by Wireless Telegraphy, and not in conducting so-called competitive tests with persons who claim to be rivals of my Company."[60]

Marconi's deposition continued until November 6, when he was finally able to return to England. Just before leaving he wrote a final letter to Hall saying how bored he was at having to spend another week in New York alone "for this beastly magnetic interference case." It took more than another year before the case was decided in Marconi's favour, in what Hall described as "a very close and hard fought litigation." In a memo to department heads, Hall wrote: "Marconi has been adjudged the prior inventor and is entitled to patents and claims in all four of the so-called magnetic detector interferences.... By this decision we secure patents which prevent the use of magnetic receivers of the varying magnetic field type by the owners or users of any system of Wireless Telegraphy other than the Marconi."[61]

The battles were over; now the war began. Hall realized that the company had to act aggressively on every available front to nail down the benefits of the court decisions. As he wrote to Marconi, "The de Forest decision and the magnetic detector decision are worth more than all the expressions of opinion of the whole Royal Society and of the technical press, but unfortunately far more attention is drawn to adverse criticism than to favourable decisions in the Law Courts." And act aggressively they did. A typical example of the company's approach is seen in a letter to a Johannesburg agency, providing arguments for why the Government of Natal should adopt the Marconi system: "We have a fundamental patent…a sweeping judgement in our favour has been delivered…anyone working any other system is pirating our invention."[62] This would be the company's basic position after the Townsend decision; it was a position that prefigured a lot of the talk that we hear about piracy and intellectual property rights in the media industries today.

Reginald Fessenden and Lee de Forest continued to make some remarkable contributions to the development of communication, particularly in "wireless telephony"—the early term for what later became radio broadcasting. On Christmas Eve 1906, Fessenden made the world's first public radio broadcast, playing a recording of a largo by Handel, and himself performing "O Holy Night" on the violin (and singing the last verse), to a number of ships at sea equipped with his apparatus.[63] De Forest made the next breakthrough innovation in radio with an invention he called the audion vacuum tube—which was based on a design that Fessenden successfully claimed was copied from his "barretter."* Fessenden won a patent infringement suit, but de Forest's finances were so chaotic that he was unable to collect anything.

Fessenden and de Forest were the first innovators to seriously shake Marconi's technical and conceptual foundations, and they drove the development of wireless communication onto a new course, which Marconi would soon be following as well. But they never again represented a threat to Marconi's dominance of the field, not in the United States or anywhere else. They both spent much of their time during the next several decades in draining court cases, several of which pitted their respective companies against one another. Fessenden continued to have a soft spot for Marconi. In May 1906 he proposed an amalgamation of the Fessenden and Marconi systems, a proposal that the American Marconi company considered interesting and recommended for London's consideration.[64] He also wrote a letter to the *Times* supportive of Marconi's position following the second

---

* The audion, or triode, was also based on Fleming's most important invention, the thermionic, or diode, valve. Under Fleming's consultancy contract with Marconi, the thermionic valve belonged to the company and Fleming saw very little money from it as a result (Hong 2001).

Berlin international radio conference in 1906.[65] As for de Forest, his attitude toward Marconi mellowed over time, and so did Marconi's toward him; in his later papers—once de Forest was no longer a threat—Marconi often paid tribute to de Forest's contribution. The two never met, although Marconi apparently tried to contact his erstwhile rival during a US visit in 1933.[66] On the day Marconi died, de Forest issued a warm statement, recalling their "many a keen competitive battle in and out of the patent courts," hailing Marconi's "pioneer work and daring vision," and crediting him as "unquestionably" the first to transmit signals through space without wires, well entitling him to be called "the father of the wireless telegraph."[67]

14

The Marconi Aura

The fallout from the 1903 Berlin conference plagued Marconi and poi-soned the relationships with the world's most powerful governments that he had been developing so strategically; there was no returning to business as usual. In Britain, the company redoubled its efforts to obtain commitments from government agencies that would override the regulatory constraints that might follow if the Berlin recommendations were implemented. It had a firm contract with the Admiralty, but the Post Office was already using Berlin as an excuse to further stall the long-delayed arrangement that Cuthbert Hall had been trying to establish. Worse, Hall was now accusing the Post Office of meddling in the company's relations with the Admiralty. The situation was aggravated by reports in the German press that the only reason the British were disinclined to endorse the Berlin proposals was because they were trying to protect a Marconi monopoly. Marconi's affairs, and the stakes of international communications, were now at the centre of the most important geopolitical rivalries and it was all becoming increasingly indigestible.[1]

In New York, where he had his own sources of information, Marconi was far more sanguine than Hall. Italy and Britain had not agreed to the German proposals and the American delegates were saying that the US Constitution would not allow their government to go along either, as only Congress could approve an international treaty. Marconi said that he had heard from Italy (likely via Luigi Solari) that Admiral Grillo's "rather luke-warm attitude" was intended more as an act of diplomatic politeness and *realpolitik* than an indication of any Italian intent to accept the proposals.[2]

This was somewhat disingenuous on Marconi's part, as the Americans were squarely on board with Berlin and the Italians were already putting pressure on Marconi to release them from their contractual obligations so that they could go ahead and join the other powers.[3] Marconi was likely trying to calm Hall, who was rapidly burning bridges with the Foreign Office, the Colonial Office, and the Post Office. The situation was aggravated further at the end of September 1903 when Prime Minister Arthur Balfour shuffled his Cabinet, undoing Hall's lobbying efforts.[4] Marconi was not well disposed to spend his time or energy bickering with the British government. Yet that seemed to be where the efforts of the company's London office, as directed by Hall, were being concentrated.

The company's lobbying strategy revealed the state of British politics in the early years of the twentieth century, as the role of permanent officials took on increasing importance. Marconi preferred to deal directly with ministers; Hall believed there was no point unless one had the senior permanent staff on side rather than ephemeral Cabinet appointees to present their case.[5] Marconi was also inclined to be more conciliatory while Hall continued to take a hard line. "At the present moment I consider we hold the cards in our hands," he wrote to Marconi, in a letter briefing him for an upcoming lunch with the new postmaster general, Lord Stanley, arguing that "it is no mere bluff to say that if we do not get security here for the fruits of our enterprise the business will be established elsewhere.... They cannot afford to drive the business out of the country." There was a fear, however, that His Majesty's Government might decide "to take the whole thing over" and work it themselves.[6]

The company was shifting its position in light of the Berlin proposals. For one thing, it no longer insisted on a technical argument against intercommunication between equipment based on different systems, but rather on the right to protect its patents. Marconi wrote to the Italian naval minister, explaining why he could not entertain the suggestion to cancel the exclusivity clause in his contract: if stations equipped with his apparatus were allowed to communicate with ships carrying other systems, it would in effect amount to recognition of these systems and could seriously affect his rights.[7] Marconi was showing that he could play hardball, even with his beloved homeland.

Marconi was taking an increasingly global view of his affairs. He was now looking beyond individual agreements with particular governments, exploiting national differences and rivalries in order to push forward his own research and corporate agenda. He had established an unassailable position in Italy; he was admired in the United States; he was receiving invitations, accolades, and business offers from across the world. And, once again, his technology was a central factor in a major imperial war.

❀   ❀   ❀

While the western European powers continued jostling for position, imperial ambitions in the Far East boiled over into war between Russia and Japan. By the turn of the century, the Russian Empire extended across Central and East Asia as far as the Kamchatka Peninsula. Japan, recently emerged from isolation and transformed into a modern industrialized state, humiliated China in a war over control of Korea. France and Germany, seeking to contain Japan, supported a Russian occupation of the Liaodong Peninsula, in southern Manchuria. In February 1904, Japan attacked the Russian naval base at Port Arthur, in Liaodong.

Marconi's system was used by both sides in the Russo-Japanese War of 1904–05. The company supplied the Russian army with field wireless sets operated by Marconi engineers, but these were never used in combat.[8] The Russian navy, on the other hand, equipped its vessels with the Telefunken apparatus, and this proved to be a miserable failure.[9] The Japanese navy, meanwhile, was using a cloned version of Marconi's system. It had been seeking since 1902 to purchase equipment from Marconi in preparation for the hostilities, but found his price too steep and decided to develop its own technology. The Japanese fleet was equipped with Marconi-like apparatus by the end of 1903, two months before the outbreak of war.

The Russo-Japanese War was the first major conflict where journalists used wireless telegraphy to file their reports, and this sharpened the determination of governments to seek control of the new means of communication. The Russians said they would treat anyone found using wireless as a spy, and the United States, although not involved in the war, said it was considering measures to control wireless telegraphy in wartime. The London *Times* commented: "[Wireless] escapes the control of the censorship and renders illusory the surveillance of communications by the belligerents, thus putting an end to their security."[10]

Early in the morning of May 27, 1905, the Japanese cruiser *Shinano Maru*, patrolling the East China Sea near the mouth of the Tsushima Strait between Korea and Japan, spotted the passing Russian fleet. The *Shinano Maru* wireless operator signalled the news to Admiral Togo Heihachiro's flagship *Mikasa*, and the Japanese fleet was able to launch a surprise attack and sink or capture almost the entire Russian force. The Japanese navy "utterly annihilated" the Russian fleet in the Battle of Tsushima, which is considered one of the most conclusive victories in naval history. More important for this story, the wireless message that the Russian fleet had been sighted played a crucial role in the Japanese victory.[11]

Tsushima effectively ended the war, marking not only a decisive victory of Japan over Russia, but of Marconi over his German competitors.[12] Geopolitically, the war jolted the balance of power in Europe, as British intelligence had collaborated with Japan while Germany supported Russia. It also marked the first defeat of a European power by an Asian nation and hastened the collapse of the Russian Empire. Wireless was now a key strategic, tactical, and military factor in what soon came to be known as electronic warfare.

One sad by-product of the Russian experience was its impact on Marconi's admiring rival Alexander Popov, who had been manufacturing wireless equipment for Russian warships in Kronstadt since 1900. He was unable to provide equipment quickly enough when the war broke out, and that led the Russian fleet to order the ill-fated Telefunken equipment. Popov was reportedly devastated by the outcome of the Battle of Tsushima, where a number of his students were killed, and died a few months later, of a brain hemorrhage.[13] Meanwhile, convinced by the war experience of the system's dependability, the tsar had Marconi apparatus installed in three palaces in Moscow and St. Petersburg in 1905, to ensure a system of secure communication that could not be disrupted by revolutionaries.[14]

Although it had essentially pirated his system, Japan "felt profound gratitude and respect" for Marconi, whom it recognized as the inventor of wireless telegraphy. As soon as the war was over, the Japanese navy proposed that a formal gesture of appreciation be made toward him. But the idea was only implemented in 1933, when Marconi visited Japan while on a world tour and was awarded the Order of the Rising Sun, the highest honour that could be conferred by the Japanese emperor.[15]

All told, and despite Marconi's claim that his invention was primarily an instrument of peace, the war was again (as it had been in South Africa a few years earlier) good for business. "The War between Russia and Japan has brought a good deal of money to the Company and things are looking very well," he wrote his mother.[16]

❀   ❀   ❀

Marconi continued to work his main strengths—making connections and networking—while building his global concern post-Berlin. His aura was now also increasingly global. Early in 1904, the company was negotiating a contract with the Argentine government along the lines of Marconi's Italian agreement of the year before, with the idea of establishing a connection between Buenos Aires and Coltano.[17] As the company began to develop business in Argentina, an independent publishing company produced an eighty-page *Album Marconi* "celebrating Marconi and the development of wireless telegraphy" and referring to him as "the king of space." Intended to introduce him to a South American audience, it is a revealing historical document, featuring set pieces of the Marconi mythology as well as iconic photographs of the dramatis personae in his story, such as Augusto Righi, Solari, and members of his family.[18] Argentina was soon the focal point of Marconi's activities in Latin America.

On August 3, 1904, with great pomp and witnessed by royalty, Marconi inaugurated another "first," regular international commercial service across the Adriatic between Bari, Italy, and Antivari, Montenegro. Italy and the small Balkan kingdom were allied by family ties (Nicholas, the Crown Prince of Montenegro, was the father of Queen Elena of Italy), and Marconi cemented the alliance by establishing the telegraphic link. More important for prestige than for commercial benefit, the Bari–Antivari link was critical to Marconi's international agenda; it was so important that Marconi's proxy at the Berlin conference, Solari, had actually left the conference early, in August 1903, to go to Montenegro for discussions with the king about establishing the service.[19]

Marconi left London on July 20, 1904 (after his initial rebuff by Beatrice), reaching Bari on the twenty-fifth. On the twenty-ninth, along with Solari, Marconi left Bari for Antivari, across the Adriatic, on a small cruiser provided by the Italian navy. They arrived in Antivari on the thirty-first, and the inauguration was set for August 3. (Although the link had been operating for nearly three months, it had not been officially inaugurated.[20]) While they waited they were entertained at the royal palace at Cetinje, which they reached by travelling on some of the worst roads Marconi had ever seen. The royal party soon arrived, having crossed Lake Scutari in row-boats, in a procession led by "the royal larder" of sheep, goats, pigs, and cattle, destined to feed the court. At the opening on August 3, Crown Prince Nicholas noted that just as Montenegro had been the first Balkan state to adopt the printing press, it was now the first to inaugurate "the newest and most marvelous method of electrical communication." An interesting bit of Balkan protocol—symbolic of the strategic importance of the region, and ominous in some ways—then followed, with the prince sending telegrams to the King of Italy, the Emperor of Austria, the Tsar of Russia, and the King of Serbia.[21]

In typical fashion, Marconi reported to his mother that he had enjoyed himself and that the prince had been "very nice" to him.[22] He then returned to Bari, where he was received with great enthusiasm and a gala performance at the local opera house. He shipped out again with Solari, this time for Venice, doing experiments en route, stopping at Ancona (near Solari's ancestral home at Loreto) where he was feted once more, and at Cattolica (near Rimini), where his sister-in-law Letizia and her children were vacationing (the Marconis often vacationed there, going back to Guglielmo's childhood). Coming only a few months after Giuseppe's death, it was an emotional family reunion.[23]

Some time later that year, the Italian government informed Marconi that the German kaiser had requested to visit the Bari station while cruising in the area on his yacht, the *Hohenzollern*. Under the terms of their contract, no one other than Italian government officials could visit the station without Marconi's permission. Pressed by the Italians as well as his own sense of diplomatic propriety—and although he was more than a bit irritated with the kaiser in the wake of the Berlin conference—he agreed to allow the visit, provided the emperor was not accompanied by any of his scientific or technical advisors. The visit never took place.[24]

Marconi's Balkan interlude was revealing in many respects. Business-wise it was a further assertion of his multi-pronged, multinational approach, adroitly networking high-level contacts, fitting into the cracks of interstate power shuffling, and leveraging his personal affairs with those of his company. Technically, he was once again going where no one had gone before. Personally, as he would for the rest of his life, he would choose to travel by sea rather than land, both for the comfort it afforded but also using the sea as a research base. Factoring in that he had just absorbed a rejected marriage proposal, it demonstrated his resilience, his capacity to compartmentalize, and his cavalier, adaptable approach to love and companionship.

❄   ❄   ❄

As the issues they were dealing with became increasingly transnational, so did the company's strategy. The starting point was still Britain. Any deal MWTC made with the British government would stand as a template for the company's relations with other governments; even more important now was British support for its international standing. While Marconi was in Rome in March 1904, Hall wrote that the British government was moving toward signing the Berlin protocol; would he see the king (of Italy) to discuss the matter as quickly as possible?[25] Hall said that if the company decided to fight the government, they would have the support of almost the entire press (and hence, he said, the public), as well as considerable backing from parliamentarians and colonial authorities, especially in Canada. The government knew this and surely didn't want a public battle that could also undermine empire solidarity; the company could turn this to its advantage.[26]

Late in the afternoon of March 29, 1904—while Marconi was in Italy mourning his father's death—the Royal British Post Office and Marconi's Wireless Telegraph Company reached an agreement. The company saw the Post Office deal as the start of a new epoch, at last putting some stability into its relationship with the government body that had been Marconi's original benefactor.[27] Marconi himself was upbeat as he had not been in years with respect to the British government, announcing to the press that it meant they would now be able to begin transatlantic wireless service.[28] He was being a bit precocious; the service was still a few years off. In theory at least—and this is what Marconi wanted the world to realize—he was in a position to compete in the overseas market with the cable companies, who were now operating fourteen transatlantic cables. The cable companies were not quaking at the prospect, however; Marconi had yet to show that he could provide a commercially competitive product.

Meanwhile, he continued to perfect his system. On his next ocean crossing in May 1904, he kept contact with Poldhu until he was seventeen hundred miles out at sea, and reached Cape Breton from fifteen hundred miles away. He told the *New York Times* on his arrival that his latest equipment had "the greatest range of anything of its kind ever made," and the paper headlined his declaration that "My next experiment will be that of girdling the globe."[29] This achievement, too, was still some years away, but he saw it coming. In this interview, Marconi articulated for the first time the idea of a linked "chain"—or network—of stations based in different parts of the world. This was another argument for the benefits of a single connected system—his.

The details of the Post Office agreement were presented to the MWTC Board on June 9, 1904.[30] The Post Office would give the company a three-year exclusive licence for its long-distance and ship-to-shore stations, renewable but with no guarantee of extended exclusivity. The British government would seek a two-year delay in implementing the Berlin proposals and would insist that non-Marconi ships pay double rates when communicating with Marconi stations. For all that the agreement was an achievement, it also marked the end of Marconi's dream of a wireless monopoly. He would have to settle for a head start—a pretty generous one that the company calculated would allow it to establish worldwide market dominance, with the support of the British government. In advancing the government's long-term policy aims, the agreement was even more crucial; coming on the heels of Berlin, it further enshrined the principle of national government regulation of wireless communication, a principle that would later be extended to cover radio and then television broadcasting.

On July 18, 1904, Lord Stanley introduced the bill that would become the Wireless Telegraphy Act.[31] The bill provided for the regulation of wireless, its objective explicitly "to give the Government control over wireless telegraphy." Any wireless operation, whether on British soil or British ships, would require a licence from the postmaster general. This meant two things: the government intended to maintain control, and it intended to oversee wireless operations. It remained to be seen whether this would ultimately mean exclusive operation by Marconi or by Marconi and competitors. But it was clear that wireless operation in British jurisdiction would be regulated and controlled by the state. For the Marconi company, the future lay in establishing and maintaining a close relationship with the government. The Wireless Telegraphy Act was adopted on August 15, 1904, and took effect on January 1, 1905.[32] The UK legislation was expected to have a snowball effect, and indeed it was followed shortly by mirroring legislation in Canada and Australia.[33] (The world's first telecom law had already been adopted by New Zealand, however, in 1903.[34])

The Post Office was anxious to complete its contract with Marconi before the act became law—especially as the company now had doubts about how good a deal it actually was. Hall caught up with Marconi in Bari as he was about to launch his Balkan connection. The Post Office was backtracking on exclusive licensing, "an oversight" since it would obviously prevent the cable companies from getting in to the business of wireless, but that was precisely what the Marconi company was looking for. Marconi's friend in Parliament, MP J. Henniker Heaton, a reformer interested in cheaper cable rates, wrote Lord Stanley, strongly objecting to the government's reluctance to protect "the rights of the great inventor who had suffered everything to bring the system to perfection." The bill was a scandal, in Heaton's view, and ought to be abandoned.[35]

On August 11, 1904, as the bill was heading for final consideration, Hall wrote to Marconi, who was now in Venice as the guest of the Italian navy on the RN *Sardegna*. Everything was settled. To read Hall, it looked as if the company held the fate of the legislation—and the future of wireless regulation—in its hands. Hall wrote that the best thing would be if they got signatures on the agreement before the bill went through Parliament. However, he added that it would have to be backdated so that the date was prior to the passing of the bill. Things having come this far, the company couldn't very well wreck the legislation.[36]

The Heads of Agreement between the postmaster general, Marconi's Wireless Telegraph Company and the Marconi International Marine Communication Company was indeed dated August 11, 1904. It granted the Marconi companies, for fifteen years, "facilities for the collection, transmission and delivery in the United Kingdom of messages to and from places in Newfoundland and North America" similar to those already granted to the submarine cable companies. The government had the right to take over stations, with compensation, in the event of any emergency. Only British subjects would be employed as operators at the companies' stations and messages of the British government should have priority over all other messages and be transmitted at half the public rate.

The agreement had a critical paragraph on the eventuality of an international convention based on the Berlin protocol. In the event of the United Kingdom's signing such a convention, the companies undertook to observe its provisions and to accept—importantly, "without prejudice to their patent rights"—the obligation to communicate with ships and stations equipped with other apparatus. However, "The Companies shall not be bound under any provision of the Convention to give information as to details of apparatus."[37] The proprietary rights embedded in Marconi's patents remained sacred. His system would remain closed.

Hall's confidential report to the board of directors shows how close the deal—both the agreement and the government bill—came to collapsing, and also reveals a stupendous concession Lord Stanley agreed to make at the eleventh hour: as compensation for the withdrawal of the originally promised exclusive licence, Stanley agreed to forego the company's undertaking not to erect any long-distance stations in foreign countries. Instead, the company agreed to simply give the postmaster general notice of intention before carrying out such a step. This infuriated the Admiralty—the ministry most affected by the change. Finally, Prime Minister Balfour had to intervene.[38]

While all this parliamentary intrigue was going on in London, Marconi was making headlines again in Italy. The British agreement was fundamental to his plan to "girdle the earth," but in Italy he was establishing another pole. Marconi's station at Coltano, near Pisa, would be, when completed, the largest and most powerful in the world. It would communicate not only with the rest of Europe and the Americas, but also with east Africa and ships sailing in the Mediterranean, the Red Sea, and the Indian Ocean. Despite the pomp and circumstance, however, the Italian government was hardening its position toward Marconi. It informed him that it would not erect the high-power long-distance station at Coltano unless a corresponding station was built in Argentina, as stipulated in their agreement. They also wanted a company to be formed in Italy to handle Marconi's Italian rights.[39] A few months later, a flurry of negative reports about Marconi's contract in the opposition press prompted him to offer to cancel it. However, the Italian government was quick to publicly confirm that the contract would remain in effect.[40]

These were annoying and financially perilous contingencies for the company. It was still short of cash and a new stock issue was proposed in October 1904. Marconi himself agreed to take up a third of the unsubscribed shares, ten thousand, as it turned out. Hall told him this would be "a considerable inducement to others," but Marconi was now personally overextended—a situation he would often find himself in over the next decade. He enjoyed a relatively comfortable income of around £4,000 a year at this time, as a director and technical advisor to the Marconi companies, but he habitually lived beyond his means.

The company continued to lobby the British government through 1905, but it was on shaky ground because of Berlin and starting to lose traction. Marconi's latest position was articulated in a letter from Hall to Post Office official H. Babington Smith on August 20, 1905. The company remained opposed to intercommunication, which it viewed as an infringement on its patents and an "enforced partnership, in which no account is taken of the value of the contribution made by each partner." At the very least, the company should be compensated for the surrender of its rights. Hall underscored the political motivations behind the international policy to which Britain was now about to adhere. Britain's control of cable was "no artificial State-fostered monopoly, but the natural outcome of British foresight and enterprise." Trying to show the same qualities with respect to wireless telegraphy, the Marconi company was "indisputably ahead of all competitors in technical achievement and commercial application." The bottom line, Hall argued (not for the first time), was that it was in the British national interest to support Marconi.[41]

The argument had begun as one of technical capacity (intercommunication was technically impossible because the transmitting and receiving equipment of different systems were supposedly incompatible), then became one of property rights (the fundamental patent of wireless telegraphy belonged to Marconi). It was now a patriotic one. Hall followed up with a brilliant plea to Prime Minister Balfour. The crux of the matter was this: the Post Office had achieved its goal of bringing wireless under its wing; the Admiralty was eager to establish the system throughout the British Empire; but the government, as a matter of high policy, wanted to sign the Berlin protocols and could not do so without the company's acquiescence. The Berlin proposals were objectionable from the standpoint of British imperial and commercial interests ("represented in this matter by our Company"). Wireless communication was important for Britain to keep in touch with its scattered empire, and to maintain its naval predominance. Marconi's, a British company, had established a virtual monopoly in this new enterprise, and it was in the British imperial interest to support it. Anything else was both unjust to the company and contrary to British interests. In other words, Marconi's interests and British interests were the same.[42]

15

A New World Order

This was a turbulent period in Marconi's life. In addition to the political and diplomatic intensity of his corporate activities, he was still driven by the powerful existential motivation to continue extending long-distance communication. He was also beginning to adapt, sometimes painfully, to married life. His North American working honeymoon with Beatrice in April 1905 provided a taste of what lay ahead. One sign was their visit to Cape Breton, where the company had relocated its Glace Bay station to a new site in the woods a mile south of the city, inland, and Richard Vyvyan had built an immense twenty-four-mast antenna as well as a comfortable new residence.[1]

At Glace Bay, Marconi was immediately engrossed in experimental work, leaving his new bride to entertain herself in utterly foreign surroundings. The experience left her with some sharp observations about her husband; one of the moments she recalled was when some bad news arrived about one of the company's overseas stations and he reacted by playing on the piano with one finger until a solution was found.[2]

When Marconi and Beatrice returned from America, they stayed first at Poldhu, with Annie and Alfonso as well. Beatrice found the hotel gloomy and bare, the food atrocious. But something about the surroundings must have agreed with them and she was soon pregnant (or as she put it sweetly in her memoir: "started first baby"). Marconi was working day and night, running back and forth between London and Poldhu, and Beatrice was suffering from morning sickness, so they decided she would stay in London, where she could be near her mother.

In September 1905 they moved into a rented townhouse at 34 Charles Street, in Mayfair. The house was near Berkeley Square, just around the block from Marconi's former bachelor digs at 90 Piccadilly, but light years from the modest flats he had shared with his mother in Bayswater and Westbourne Park. It was a chic area, soon to be favoured by the likes of William Somerset Maugham (who lived around the corner in 1911). Beatrice's mother, who had found the house, thought it suited the couple's station, but Marconi seemed to be following the advice often given by bank loan officers: always live in the biggest home that you cannot afford. Today the house is part of a swanky boutique hotel.

On February 4, 1906, Beatrice gave birth to a daughter. Less than four weeks later, on March 2, the infant was stricken with a fit of convulsions and died. Marconi was at Poldhu and rushed back to London, too late. He was devastated. He wrote to his mother in Bologna: "I am so sorry, darling Mamma, that you never saw our Baby. She was such a pretty and bright child, and even after death she looked so sweet and peaceful. Bea had wanted to call her Lucia."[3] No cemetery wanted to bury Lucia, as she had died without being baptized, but Marconi managed to find one in the City of Westminster at Hanwell.[4]

Marconi took to his bed and remained out of commission for nearly three months. He had been suffering recurring bouts of malaria since contracting it in Montenegro two years earlier and was frequently laid up with high temperatures, but now he collapsed completely, incapacitated by what Beatrice described as a nervous breakdown.[5] Marconi had clearly suffered a burnout. Annie and Beatrice were in constant touch about his condition and the doctors monitored him almost daily, ruling on whether and when he could be allowed up for a few moments. He rallied briefly in late April and tried to do a bit of work but didn't get very far. It was impossible to keep the illness under wraps; the Buckingham Palace Court Circular for May 4, 1906, reported that "Mr. Marconi has been very ill in the past month but is now recovering," and the news was reported in Italy as well. As late as May 24, however, Hall told a caller that Marconi was "unfit to do any work at all," and that it would be "madness" for him to try. Hall kept Marconi informed of business doings with the occasional long letter to Charles Street, signing one with the postscript: "Don't trouble to answer letters of this kind." But a week later Marconi was back at work at the Haven.[6]

More than anyone else, Beatrice noticed her husband's mental state and perceived its lasting effect. "This also I think tended afterwards to alter his character and render him irascible and difficult," Beatrice wrote in her 1950s memoir. "He was a tremendously difficult patient." She was concerned about little things that she had previously dismissed as eccentricities. He railed about the shortcomings of the British medical profession and, to amuse himself, he collected funeral ads, leaving Beatrice to conclude that he was not only difficult but could be morbid as well. One day when she came home from having a prescription filled, she found him standing on his head and thought he had completely lost it until he explained that he had broken his thermometer while taking his temperature and was trying to cough up the ball of mercury he had swallowed.[7]

❁   ❁   ❁

Hall kept up the company's vigorous negotiations with the government while Marconi was off sick after Lucia's death. When Marconi returned to work at the end of May 1906, he was not yet at full strength but, anxious to get back to his research, he was soon spending most of his time at the Haven and Poldhu. He wasn't happy with Hall's negotiation style, which now started to drive a wedge into the significant cleavage that had for some time been separating him from his energetic managing director. In mid-June, Marconi suggested the company might do better if Hall weren't always using a "whip hand," provoking this piqued response from Hall: "It is impossible to make a strong case without arousing antagonism in certain quarters. Anyone can sit on his chair and twiddle his thumbs and preserve an attitude of philosophical impartiality, but no big cause was ever made in this fashion."[8]

Marconi tried to calm Hall down, coddling him a bit as well in the process. The company should try to get whatever it can out of the government, "without if possible irritating the officials, of course making them understand that the only thing they can do is to give us what we want—without telling them in so many words." The contracts Hall had got were "excellent and to your everlasting credit." Hall did calm down, somewhat: "If one can get one's own way without stirring up antagonism it is of course always preferable.... There have been times, however, when we have failed to obtain our ends by conciliatory methods, and when it has been necessary to carry the position by assault."[9]

Parallel to the political negotiations, Marconi was also thinking about organizational questions. He developed a detailed proposal that he considered "absolutely necessary for the Company to adopt if my work is to be of any real efficiency." Nearly fully recovered from almost four months' absence due to illness, he told his manager that he would not take up active work again "unless I have a completely free hand in dealing with the members and organization of the technical staff." He was open to advice as to ways and means, but this was a non-negotiable demand. Marconi was insisting on absolute control over everything that fell directly within the scope of his activities as "technical advisor" to the company. He wanted the company's experimental staff, including a new chief engineer, to report directly to him and to him alone.[10]

Nothing was done. It was not quite clear at this point who was in charge. Marconi was nominally the boss, but Hall was actually running the company, especially as Marconi was away from London and at his research stations as much as possible. In fact, Hall had written to him just days before, deploring the situation: "It is very unfortunate that at this juncture you should hardly be available even for conversation. Of course you must get fit, but I have a feeling that you have withdrawn yourself mentally as well as physically, which is very depressing and discouraging." In August 1906, Marconi wrote from Poldhu, where he was following "the arduous fight you have been carrying out in the true interests of the Marconi companies and I might add of the British Empire." He touched various business and research issues, and again enclosed a copy of his proposals for the technical staff. "It would greatly ease my mind if you would send me a note stating that you will agree to it."[11]

This was more than a personnel decision. Marconi was also questioning his role and his future. Just as he was engaging with Hall on his organizational plans, it was reported that Marconi had joined the experimental staff of the Columbia Phonograph Company, flagship of the US recording industry, as "consulting physicist." What was he doing? The Columbia announcement made clear that Marconi's role would be "to further develop the science of recording and reproducing sounds," so presumably this moonlighting would not be in violation of his contract with his own company. Maybe he was just testing the transatlantic waters. At any rate, the appointment was real; Marconi visited the Columbia plant during his US visit in September 1906.[12]

❀   ❀   ❀

The International Radiotelegraph Conference was set to begin in Berlin on October 3, 1906. Six months earlier, in February, Hall had nailed down the arguments for the company's position and against the proposals, from the British perspective at least. There were eight points in all: 1. The proposed regulation would retard technical development; 2. The owner of a valid patent would be prejudiced; 3. It was not practicable; 4. It imposed an unfair enforced partnership; 5. Many governments and companies with practical experience were opposed to it; 6. It was contrary to the interests of British Imperial Defence; 7. It was contrary to the interests of British business; 8. It was contrary to the interests of the Marconi company. Looked at across a century of communication regulation, it's clear that these arguments provided a blueprint for corporate strategy on public policy, combining common sense with appeals to high national interest and leaving the most obvious, private interest, to the very end. Hall recalled, with some accuracy, the problem that faced the international community with regard to regulation of a highly technical and politically fraught technology—from the perspective of a private company.[13]

The conference opened on October 3, 1906; its object was now "radiotelegraphy," updating the wireless lexicon.[14] Thirty-one governments assembled at the invitation of Germany; in 1903 there had been nine. The press recalled that there had been little progress in the interim with regard to the raison d'être of the conference: to open wireless to "intercommunication." "Little has come of these recommendations," reported the *New York Times*. "The Marconi Company has continued to set up stations and to refuse the messages of rival systems," leaving no doubt as to why the conference was called. The United States was now strongly on board, supporting the 1903 Berlin proposals and, in fact, enforcing them on the ground, having asked the Marconi company to remove its instruments from the lightship *Nantucket* off the coast of Massachusetts after receiving an official German complaint.[15]

The conference was chaired by the same German officials who had presided three years earlier, Reinhold Kraetke and Reinhold von Sydow. Italy, the United States, and the United Kingdom all had new delegations, the UK's headed by Post Office official H. Babington Smith. One oddity of the conference was the presence of the delegate from Montenegro—a certain H. Cuthbert Hall. Hall was seated at the fourth session on October 6, 1906. He was replacing Marconi, who had been slated to represent Montenegro; it was as though Marconi now had his own miniature client-state.[16] From the first preliminary exchange of views, the British delegates seemed more favourable than expected toward the idea of international regulation, while upholding the agreements they had already made with Marconi. The British course would probably be followed by Italy, the *New York Times* speculated, adding, as an afterthought, "Mr. Marconi has not yet arrived."[17]

At the second session, Babington Smith reported that the British delegation was authorized to provisionally accept the principle of exchange between ship and coast stations regardless of the system being used. But the British emphasized that this acceptance was conditional to the adoption of appropriate regulations, as well as to the reservation for each government to designate stations that were excepted from the obligation to exchange. This could have been referred to as the "Marconi loophole," and pertained to the British separate agreement with Marconi. Italy's position had not changed. Senator Joseph Colombo, the chief Italian delegate, told the conference that Italy recognized the importance of intercommunication for international relations, but that it was bound by the conditions of its contract with Marconi and could not change them without his agreement.

Again—or still—even if absent, Marconi loomed over and above the diplomatic conference. The participation of two powers in an international treaty depended on him. On October 13, Marconi, back in England, received a telegram from the Italian naval minister, his old friend Carlo Mirabello, alerting him to the risk of Italy's diplomatic isolation. "Telegraph without delay what concession you would be prepared to grant to Italian Government in order that it may be able to adhere to an international accord." Marconi received a second telegram the same day, from the Italian minister of posts and telegraphs, Carlo Schanzer, asking whether he would accept to reopen his contract. "We trust your patriotism will cause you take favourable decision."[18]

Marconi blinked. He decided to wire to Rome asking for time, but he was wavering. "I do not think it would serve our best interests to let Italy get isolated, or, as the Government considers, put them in a hole," he wrote to Hall in Berlin. As his letter to Hall contained the decoded text of his Italian telegrams, he sent it to Berlin by special messenger, anxious that the Germans might get wind of its contents if he used the regular post.[19]

The record of the 1906 conference shows much less heated and more homogeneous debates than the conference in 1903. Hampered by Marconi's ambivalence, Hall proved to be a less forceful advocate for the Marconi interests than Solari had been; he also had a far less prestigious tribune from which to speak. The most he was able to put forward was that the Marconi company was not in violation of the public service (or common carrier) principle, as had been suggested, because it accepted to transmit all messages, from whomever, over its links (as did the cable companies). Just as the cable companies did not allow anyone to link to their networks, Hall argued,

Marconi's did not allow anyone to communicate by wireless with its stations, and it was unjust to oblige a long-established enterprise to endanger its business by making such a compromise. But Hall's position didn't fly.

By agreement, the conference proceedings were supposed to be secret, but leaks to the press—both real and fabricated—were endemic. "Judging from the information which was brought to me before I was accepted as a delegate . . . this obligation is not being very closely observed," Hall reported to Marconi.[20] The transcripts show a more or less harmonious, even dull, train of events, but sensational reports in the British press had the deliberations fraught with the tensions of power relations between England and Germany.[21]

The fact was that no one, except for the delegate from Montenegro, was opposing the German proposals any longer. "The chances of smashing the Convention altogether have been lost," Hall wrote to Marconi on October 17. He added, more out of temperamental consistency than to make any effective point: "The English Delegation is not very communicative, and is taking rather a mysterious attitude. . . . Probably, however, this is attributable to the delegates not knowing what they are doing."[22] Hall's determination, pushed vigorously in his relations with the British government, was not realistic—as Marconi himself now realized. Marconi was able to promote and cross-promote his corporate interests through compromise; where he was uncompromising was with anything to do with new technical advances. The outcome of Berlin on these counts would be a blow to Marconi's claim that his patents were fundamental to all wireless telegraphy (despite validation of that claim by the US court's Townsend decision of 1905). The system of international communication regulation spawned at Berlin thus attacked not only a particular company but also the very corporate model of property rights based on private ownership and control. This was its historic importance.

However, the company was not ready to give up. As the conference was winding down, MWTC chairman Charles Euan-Smith met with Prime Minister Henry Campbell-Bannerman (whose Liberal Party had succeeded Arthur Balfour's Conservatives in December 1905) at 10 Downing Street and told him that the deal being struck in Berlin was "one of the maddest things that any Government had ever been called upon to recognize or sanction," in that it sacrificed British imperial interests without getting anything at all in return. "Personally I don't understand the damn thing at all," he said.[23] Campbell-Bannerman casually blamed the confusion of the

British position on the change of government; Euan-Smith was pleased with the meeting but nothing changed.

The conference ended on November 3, 1906, its proposals endorsed by twenty-six countries, including Great Britain and Italy (subject to the restrictions of its Marconi contract), not to mention Germany, France, and the United States (Montenegro was not among the signatories).[24] If ratified, the proposed convention was to take effect on July 1, 1908. The treaty's most important provision governed wireless communication between the shore stations of contracting nations and vessels of any state, regardless of the wireless telegraphy system these ships employed. The United States insisted on including ship-to-ship transmission. One clause that attracted far less attention provided for the establishment of a bureau to oversee international wireless telegraphy, including future changes to the convention. This bureau, which was eventually integrated to the International Telegraph Union, would be the first international regulatory body for broadcasting and telecommunication. This was the lasting legacy of the process that had started because Marconi refused to allow his licensees to communicate with competing systems. It is possibly the only example in history of an international regulatory apparatus being launched to frustrate the commercial objectives of a single company.

There was now a new world order in wireless communication. The main users of wireless were still military forces and shipping companies, and Marconi had used contracts and patents to establish a monopolistic position in Italy and Britain (and the British Dominions like Canada and Australia), but the rest of the world was wide open. The US Navy was using Telefunken and De Forest apparatus. The Japanese had developed their own system. So had the French. All of these systems were in some way derived from Marconi's, but he had not managed to establish the firm control that he was heading toward before Berlin.[25] Telefunken was a manufacturer of instruments, not a service provider, Hall wrote in rebuttal to a journal article, but that could not camouflage the tectonic shift that was taking place.[26]

The international debate confirmed the commercial and strategic importance of the new technology of wireless communication, still barely ten years old. Marconi was at the centre of the international jostling of states, empires, and finance capital at Berlin. Buttonholed by reporters in Milan, he commented cryptically: "The conflict between my company and the Slaby-Arco Company [sic—he meant Telefunken] cannot be practicably settled by a conference, but by the courts. It is not a question of theory but one of

Top left: Giuseppe Marconi, c. 1869
(Fondazione Guglielmo Marconi)
Top right: Guglielmo Marconi, c. 1889
(Fondazione Guglielmo Marconi)
Bottom: *(left to right)* Guglielmo, Annie,
and Alfonso Marconi, Bologna, c. 1877
(© Hulton-Deutsch Collection/CORBIS)

Left to right: Giuseppe, Guglielmo, Alfonso (standing), and Annie Marconi, Villa Griffone, 1897 *(Fondazione Guglielmo Marconi)*

Top left: Marconi's cousin Henry Jameson Davis, the first managing director of the Wireless Telegraph & Signal Co., London, 1898 *(Bodleian)* ***Top right:*** H. Cuthbert Hall, managing director of Marconi's Wireless Telegraph Company, 1906 *(Bodleian)* ***Bottom:*** Poldhu station, Cornwall, England, c. 1901 *(Bodleian)*

THE LATEST MODERN MIRACLE.
Signalling through Space without Wire.
MARCONI'S GREAT DISCOVERY.

LECTURE & PRACTICAL DEMONSTRATION

BY

MR. WILLIAM LYND,

Late Principal of the West London College of Electrical Engineering, Author of "The Practical Telegraphist," "Ancient Musical Instruments," Editor of "The Telegraphist," "The Phonogram," "The Family Circle of Science," &c., &c.

Who has delivered 1,100 Popular Science Lectures, and visited over 600 towns in Great Britain and Ireland since March, 1889.

(By Permission of the Proprietors of the "Strand Magazine.")

MARCONI AND HIS WONDERFUL APPARATUS.

Publicity for event featuring "Marconi's Great Discovery," London, c. 1900. Photo is from Marconi's first interview in *Strand Magazine*, March 1897 *(Bodleian)*

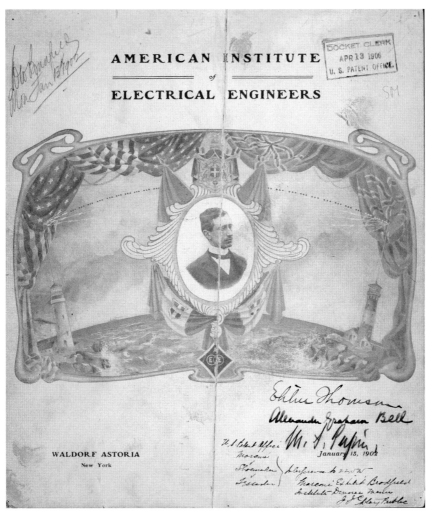

Program and menu for banquet honoring Marconi, Waldorf Astoria,
New York, January 13, 1902, autographed by Alexander Graham Bell *(Bodleian)*

Top left: Marconi, c.1900, around the time of his engagement to Josephine Holman *(Photo by ullstein bild/ullstein bild via Getty Images)* **Top right:** Marconi's fiancée Josephine Holman, January 1902 (The Courier *[Lincoln, Neb.], January 11, 1902*. Chronicling America: Historic American Newspapers, *Library of Congress*) **Bottom:** Marconi (left) and assistants hoisting kite-aerial during re-enactment of the first transatlantic wireless signal transmission, Signal Hill, St. John's, Newfoundland, December 17, 1901 *(© Hulton-Deutsch Collection/CORBIS)*

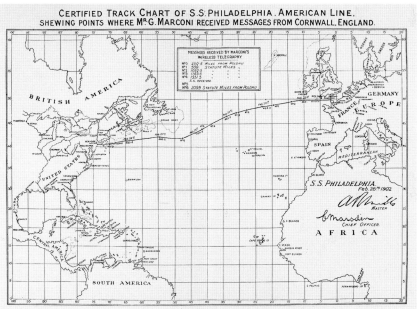

Top left: Marconi and Luigi Solari, The Haven, Poole, England, 1902 *(Bodleian)*
Top right: Marconi (third from left) and staff at Glace Bay, Nova Scotia, December 1902. Station manager Richard N. Vyvyan is to Marconi's right. *(Bodleian)* **Bottom:** Track chart of signals received by Marconi from England while crossing the Atlantic on the *SS Philadelphia*, February 1902 *(Bodleian)*

Top: Marconi and managing director Godfrey C. Isaacs, leaving a meeting of the company's directors, London, after the *Titanic* disaster, May 1912 *(Bodleian)* **Bottom:** *(left to right)* Degna, Giulio (back, standing), Beatrice, Gioia, and Guglielmo Marconi, Eaglehurst, c. 1918 *(Degna Marconi Paresce Fund)*

facts."[27] Put another way, if politics couldn't settle the issue in his favour, patent litigation would.

The company's efforts now turned to preventing ratification of the convention by the various governments, starting with the British.[28] This meant demonstrating, publicly and in private lobbying of MPs, that the convention was not in the British national interest. According to Hall's intelligence, the Cabinet was divided and Marconi had strong supporters, but it was "a very large order to refuse to ratify the acts of plenipotentiaries." Hall continued pressing Marconi to use his influence in Italy, to open a second front of diplomatic pressure. "Italy can help us by saying she considers the whole scheme impracticable. . . . You can render a great service at this juncture if you can induce the Italian Government to say: 'We do not want to be alone.'"[29]

The exact terms of the Berlin convention were still secret, but Marconi managed to get his hands on a copy in Rome. In coded telegrams to Hall, he said he was trying to do as Hall suggested but was not overconfident of success. Was he being candid or cagey? Marconi was travelling around Italy, consolidating his own affairs. Hall suggested that it would be a good idea to put word out through the Italian press that the future development of the Marconi system would be tied to Italian interests, and to emphasize the point that Italy would gain nothing by signing the convention. "I wish you would rub this in to people," he wrote.[30]

National interests aside, Hall's strategy was to show that the whole scheme was bad, period. "We must, if possible, destroy the chances of the Convention on the ground that the whole proposal is bad in essence—apart from our interests. We must show that it is against the public interest."[31]

On December 18, 1906, after a vigorous lobbying campaign by the Marconi company and with hardly any MPs in the House of Commons, Sir Edward Sassoon moved the creation of a parliamentary Select Committee on the 1906 Radiotelegraphic Convention. To everyone's surprise, the government disarmed its critics by immediately acceding to the demand.

Responding to public ambivalence, to the desultory view of the government (which still saw the convention largely as a politically fraught legacy of its predecessor), and to the persistent pressure of the Marconi company, the committee's brief was to report "what, from the point of view of national and public interests, would in their opinion, be the effect of the adhesion or non-adhesion of this country to the Convention." The committee held

thirteen meetings and examined eighteen witnesses, including representatives of the Post Office, War Office, Admiralty and Colonial Office, Lloyd's,
experts who could explain "various systems of radiotelegraphy other than
that known as the Marconi system; representatives of the Marconi Companies,
and Mr Marconi himself." Once again, "Marconi himself" incarnated a
unique status.

The Select Committee reported on July 8, 1907.[32] Its report included
a sort of disclaimer that actually wasn't one, to the effect that the committee
did not wish to express any opinion on controversial questions of priority
and patent rights, but it appeared to be generally admitted that "Hertzian
waves were first experimentally applied for practical telegraphic purposes in
1895–6 by Mr. Marconi." After many further inventions and improvements
by Marconi and others, there were now several "systems" in operation: two
British, two French, one German, one Danish, and four American. The report also included some stark statistics: Marconi stations were preponderant
in the United Kingdom, Italy, and Canada, "but in all other parts of the
world, with few exceptions, the stations are on other systems."

The report then made an unequivocal statement that was damning to
Marconi: "Approaching the question from the point of view of British
commercial interests, there can be no doubt that the freest utilization of
wireless telegraphy is to the general interest of commerce and of the mercantile marine." The Marconi company had no "claim" to a wireless monopoly, even though it effectively had one in Britain, Italy, and Canada. That
said, the committee was reluctant to make any recommendation that would
harm "the fair working" of Marconi's business. The committee noted that it
had heard only four witnesses argue for rejection: Cuthbert Hall, J.A.
Fleming, Henniker Heaton, and "*with much less intensity* [emphasis added],"
Marconi. The committee concluded that Britain's adhesion to the convention "would be advantageous to national and public interests, and that its
non-adhesion would be detrimental to those interests." There was no reason
to believe the convention would be prejudicial to Marconi's interests, but if
it turned out to be the case, the committee proposed compensation for
three years' lost revenue.

The committee hearings put the spotlight on the principal players of
the wireless telegraphy industry. Its witnesses included many of the characters we have already encountered in this tale: Cuthbert Hall; Ambrose
Fleming; Henry Muirhead of the Lodge-Muirhead syndicate; Nevil
Maskelyne, who was advising a company created to take over the de Forest

and Poulsen interests in the United Kingdom; Reginald Fessenden by proxy, via the letter he sent to Marconi asserting his opposition to the convention; Henniker Heaton, Marconi's man in Parliament, who described the relations between the Post Office and the Marconi company as "strained"; and, of course, Marconi himself. Testifying on April 30, 1907, and presenting himself as "chief technical adviser of the Marconi Company," Marconi said the convention proposals were not practicable, unnecessary, and "would practically put an end to the development of wireless telegraphy."

Handwritten notes in Marconi's script in the Oxford Archives indicate his thinking as he prepared his arguments to the committee. "General remarks. If the Convention is ratified it will, so to say, wrap up a new science in swaddling clothes and, in my opinion, very seriously hamper its progress . . . it will be a tortuous and most unsatisfactory way of keeping wireless telegraphy on a steady and efficient basis. . . . I venture to say that in the future intercommunication will tend to become more and more impracticable." Furthermore, "the Art has not by any means reached that stage where International control is either desirable or necessary."[33] In other words, regulation stifled innovation. This was a discourse one would get used to hearing with regard to new communication technologies.

The archives also contain a note, evidently from Hall to Marconi to prepare him for testifying, summing up what had been said by witnesses so far, and formulating counter-arguments. Some members of the committee had alluded to a "wireless war" that would ensue if England didn't ratify the convention. The note refuted this as "utter bunkum." The author added, "By-the-way, you are not supposed to have read the verbatim report of the evidence."[34]

A week later, a figure from Marconi's past appeared before the Select Committee, the venerable Sir William Preece, who distinguished between Marconi the inventor and the Marconi company. Marconi's credit was well deserved, but "he could not, and did not, patent wireless telegraphy, for wireless telegraphy existed then"; Marconi's patents were for "improvements." Preece said it would be "a very serious evil to this country" if the convention were not ratified. The convention would "break down the monopoly that the Marconi Company are trying to secure" and simply extend to wireless what already existed with regard to postal services and wired telegraphy—that is to say, the worldwide regulation of communication. Preece's was the voice of a believer in public enterprise; he would have preferred the government to buy Marconi's system, in which case wireless would have been a government monopoly. But now that it had been developed by a private company, he said, "let there be free trade in wireless telegraphy." To an MP who asked him what he meant, Preece replied: "I mean by free trade that under the regulation of the Post Office, those who apply for licences should have licences." The convention would lead to a uniform world system worked by various companies; its object was to establish the rules and regulations.[35]

❊   ❊   ❊

The aftermath of the 1906 Berlin Convention was not only an apogee for the company's business affairs, it marked a profound shift in Marconi's relations with his closest business collaborator, Henry Cuthbert Hall.

As we have seen, Marconi could be tough and single-minded. But when he was not convinced about a project's viability, or when he was otherwise preoccupied, he would frequently procrastinate. In these situations, Hall's tenacity kept the wheels of business turning. In fact, Hall was the only one in Marconi's entourage who would dare shake him. Marconi was always busy, and his distraction became a source of power for Hall. A modus vivendi was established soon after Hall's appointment in 1901: all important matters flowed through Hall.[36]

This led to many small disagreements and bickering; when Marconi sent a directive with which Hall disagreed or thought unimportant, Hall would simply ignore it. But Marconi never forgot and would eventually follow up, sometimes testily. Marconi could be obstinate on questions that Hall considered to be within his managerial prerogative—for example, with respect to salaries; Marconi intervened frequently on behalf of particular employees he thought should be better paid. Marconi was also obsessive about being kept informed of every aspect of the company's business.[37]

Along with Marconi himself, Hall was crucial to crafting the company's image. He would advise him on the posture to take at an upcoming shareholders' meeting: aim for "cheerful optimism...avoid all bitterness.... The general tone of a speech is often more important than the actual statements."[38] Hall not only spoke frankly to Marconi, he also didn't fall short of calling him to task. Marconi was notorious for insisting on secrecy but he enjoyed sharing confidences with people he was trying to impress. To Hall, secrecy was a fundamental tenet of corporate policy and should not be breached under any conditions. He concluded a memorable seven-page letter about the need for secrecy at the Poldhu experimental station with this admonition: "I shall be much obliged if you will take what I am now saying into your most serious consideration...for certainly you have been the principal sinner in admitting people to Poldhu!"[39]

But Hall was also fiercely loyal to Marconi, and to the cultivation of his mystique. There was a touching side to this definitely complex relationship. Marconi didn't have male "buddies," only colleagues, and (along with Solari) Hall was probably the closest he had to a friend. (All of Marconi's relationships were compartmentalized: Davis was family, Kemp an assistant.) Hall fretted about Marconi's health almost as much as did his mother. Before Marconi married, they would often dine together at one or another central London club. They shared an implicit trust. As we've seen, Hall was one of the trustees of Marconi's marriage settlement; he was also a witness to Marconi's 1906 will.

The clash over corporate organization that was already brewing when Marconi returned from sick leave in June 1906 drove a wedge between them. But the last straw, the issue that ultimately alienated Marconi, was Hall's aggressive attitude toward the British government. Hall was an ideological free enterpriser, to whom government interference of any kind was anathema. If dealing with the government could bring benefits to the company, then fine. But there was nothing intrinsically beneficial to the relationship. Marconi, though not at all ideological, felt intuitively close to political power of every stripe. In his mind, nothing could be more powerful than a partnership with government—*any* government. He had formed his company because governments were not forthcoming in wanting to deal with him; now, governments were his principal and preferred clients.

Marconi was at first only mildly annoyed by this difference in approach with his managing director, but as negotiations with the British Post Office tightened in 1907, he began to see it as a fundamental and mutually exclusive difference that threatened his long-term fortunes. Hall's abrasiveness was not limited to his relations with the British Post Office. The Canadians, too, were familiar with his sting.[40] It became a titanic battle of personalities, styles, ideas, and vision, and by early 1908 the situation was untenable. On February 12, 1908, after reviewing Hall's financial entitlements with the company secretary,[41] Marconi wrote a thoroughly detailed letter to board chairman Sir Charles Euan-Smith. It was nothing less than an ultimatum. One of them, Hall or Marconi, had to go.

The company was in "a very serious state of financial embarrassment" because of its management policies, characterized by inefficiencies in operations and a destructively combative attitude toward governments, Marconi wrote.[42] On the one hand, Hall's ignoring of Marconi's recommendations for reform of the company's technical operations forced him to conclude that, unless things changed, "any further connection of mine with the technical business would inevitably result both in loss of time on my part and in little or no benefit whatever to the Companies." At the same time, the company's "defiant tactics" had so alienated the British government, as well as other governments, that its business dealings were now impaired—the development of the company's business *depended* on its relation to governments. A radical and immediate change of policy was necessary, and Marconi tied it all to the character of Hall, who was "both in temperament and according to his own convictions utterly unable to cope with the task." Marconi asked the board to request Hall's resignation. Otherwise, he said, "I shall be compelled... to sever, before the end of this week, all my connections with the Marconi Companies, and to make public my reasons for so doing."

The stakes were very high, and Marconi placed all his chips on the table. He would have taken this step some time ago, he said, if it would not have hampered the company's fight against the Berlin Convention, which Hall had led, and were it not for his own preoccupation with establishing commercial operations across the Atlantic. Marconi proposed to carry on management of the company himself until a competent person could be engaged; to establish a sound financial footing, to increase efficiency, and to rebuild amicable relations between the company and governments. This was not ego but conviction and determination speaking. It was an irrevocable decision.

What followed next was a classic corporate bloodletting. A special MWTC board meeting held at the home of director Albert Ochs, one of the company's key financial backers, unanimously decided "that the continuance of Mr H. Cuthbert Hall as Managing Director was unfortunately no longer conducive to the best interests of the Company" and resolved that Hall's engagement be terminated as of March 2, 1908.[43] He was requested to hand over his office and papers to Marconi, who would assume the duties of managing director the very next day. An MIMCC board minute spelled out Marconi's arguments elaborately, emphasizing "the immediate advisability of re-establishing friendly relations with the British and other Governments, and of modifying the policy hitherto pursued in opposing the ratification of the Berlin Convention." The secretary was instructed to prepare a notice of "the appointment of Mr G. Marconi as Managing

Director of the Company in succession to Mr H. Cuthbert Hall who has vacated the appointment."[44] A laconic report in the *Times* the following day formalized the announcement in the precise words of the company's press release.[45]

Hall took the removal stoically. He managed to negotiate thirteen months' severance pay (£1,625), ten thousand MWTC shares (with a par value equivalent to nearly seven times his annual salary of £1,500), and a 5 percent commission (£600) on the company's recent sale of its Argentine rights. The company agreed that Marconi would withdraw his letter to the chairman "containing certain representations and charges," and that it would be announced that Hall had resigned over a question of policy. In addition, oddly, Hall could remain a company director, should he wish to do so.[46] The company's solicitors opined that "the pecuniary arrangements are exceptionally favourable to Mr Hall, but if the Directors consider it is in the interests of the shareholders of the Company to carry out the proposed terms, we think it is within their power to do so."[47]

Marconi's position had the full support of the board, and only he had the stature to force the removal of Hall. But Hall remained true to character. Five weeks after leaving his job he resigned his company directorship as well, stating that he had lost confidence in the administration of the company. In a note to the chairman expressing regret at resigning his directorship, he wrote: "Marconi has however made his bed and must lie in it, and I cannot stultify myself by sharing it."[48] Hall announced that he would attend the company's annual meeting in two weeks' time, and, as a shareholder, attack the board for its new conciliatory policy of "loyal cooperation with His Majesty's Government." The company went into damage control mode and Hall never carried out the threat—or at least there is no record that he did, either in the report of the meeting or anywhere else.[49] Instead, he left London the day after resigning from the board, leaving instructions to have the balance owing on his severance paid into his account at the London & Country Bank.[50]

Hall then completely disappeared from view, sinking like a meteor that has dropped into the sea. There is no further mention of him in the company archives—even his personnel file has disappeared. He surfaced occasionally over the years, promoting a new and ill-destined mode of cable telegraphy known as the "Knudsen system" in 1910; testifying for the defendant in the 1913 libel trial of journalist Cecil Chesterton (who was charged with libelling his successor as managing director of the Marconi companies);[51] and, most notably, publishing an article on "The Flaw in British Government" in the *National Review*, a right-wing journal edited by Leopold Maxse, in 1915.[52] Summing up the corporate management philosophy he had tried, unsuccessfully, to implement at Marconi's, he wrote: "The most successful companies are those in which one man—a managing director, chairman, manager—acts as a dictator."[53] Marconi, in a much more subtle and unassuming way, evidently believed the same thing.

16

On the Way to Somewhere

The road to Derrigimlagh runs south of Clifden through soft hills covered in lush vegetation, typical of this, the prettiest part of Connemara, in Galway, on the west coast of Ireland. One leaves the main road on the way to Ballyconneely and only the smallest sign points to the site of the one-time Marconi station;* before you know it you are on a two-track road (the former light-rail track) barely wide enough for a single vehicle and shared with several dozen sheep, heading toward boggy Slyne Head. At the end of the road, the only obvious traces of what was once the world's most powerful wireless station are a few stone foundation pieces, remains of some of the buildings and masts, and a plaque put up in 2007, on the one-hundredth anniversary of the start of service. The bog is equally famous as the landing spot of British aviators John Alcock and Arthur Brown on the first non-stop transatlantic airplane flight in 1919. Alcock and Brown crash-landed with a squelch in the middle of the spongy bog, in what Alcock later said "looked like a lovely field," within sight of the Marconi station some five hundred yards away. The first people to greet them were staff from the station.[1]

Marconi began looking for a place like this soon after deciding that he was ready to establish a regular, commercial transatlantic wireless service with Glace Bay as the North American pole. In May 1904, a decision was made to relocate the Glace Bay station to a larger site with increased capacity, and after construction began in the fall, Marconi realized that he needed a more powerful station on the European side as well. In November,

---

* At least that was the case at the time of writing; in 2014, a plan was approved to develop the site for tourism.

after returning from a trip to Canada (with a brief pit stop to receive an honorary degree from the University of Glasgow), he travelled to the west coast of Ireland to look for possibilities. It was difficult physical and socio-economic terrain; the political situation could be volatile, there was little infrastructure, and skilled labour would have to be imported and therefore expensive and difficult to obtain.[2] But these disadvantages were outweighed by the fact that the west of Ireland was considerably closer to Glace Bay than Poldhu. It was also the sort of challenge Marconi liked.

On July 1, 1905, shortly after Marconi's return from the North American leg of his honeymoon (and from introducing Beatrice to the new manager's residence at Glace Bay), Hall informed him that the company had found an "apparently suitable" site near Cashel in Connemara. A few weeks later Marconi and two associates visited and carried out tests, and a week after that Henry Jameson Davis went out to have a look at the various sites under consideration. After walking the site at Derrigimlagh, a bog three miles south of the coastal town of Clifden, he placed an offer to buy 310 acres.[3]

Derrigimlagh had a clear line of sight to the Atlantic, a freshwater lake, and, most important, enough peat to provide fuel for the station's massive boiler requirements for at least twenty years (peat was less than half as expensive as coal). It was also only about eighty miles from the ancestral home of Marconi's in-laws, at Dromoland. A railway line from Galway to Clifden had opened in 1895 and it was relatively easy for the company to build a light two-foot gauge rail line from the public road to the site. (Significantly, a telegraph line to Ballyconneely went past the entrance to the station.[4]) Construction began in October 1905 and by the time it was completed, it was said to be the largest and most powerful wireless station in the world. The aerial was comprised of eight wooden masts each more than two hundred feet high; there was a 350-by-75-by-33-foot building clad in galvanized iron to house the giant condenser; and a set of generators driven by six steam engines provided five hundred kilowatts of electrical power at twenty thousands volts to supply the most powerful electric battery ever built. When the station was operating the noise was deafening.[5]

Organizing the Clifden station was Marconi's greatest logistical challenge to date. Clifden itself was little more than a small west-of-Ireland town, and Derrigimlagh even more remote. The bog was damp and mossy. Material arrived in town by train or by steamer and then had to be brought in either by light rail, by donkey, or by a type of ancient sailing vessel known locally as a hooker, then dragged up the rocky cliff to a track leading to the site. There is little trace of Marconi's personal visits while the station was being built between 1905 and 1907, but Connemara artist Alanna Heather has written that Marconi stayed with her grandmother, Henrietta Wall Heather, at Errislannan, a comfortable manor house a mile or so from Derrigimlagh, "as she and his mother, an Irish girl, had been friends when they were young."[6] Later Marconi would stay at the Railway Hotel (now Foyle's) in Clifden.

The station had a strong impact on the local economy and social structure of Clifden, making it the most prosperous region in Connemara, one of the poorest parts of Ireland at the turn of the twentieth century. But most of the station's senior staff of engineers and operators were brought in from England and Wales; skilled staff came from all over Ireland, and apart from a local engineer, only the unskilled labour (mainly peat-cutters) was local to Clifden. The station employed but a handful of women, mostly as housemaids.[7]

In February 1906 the Clifden Board of Guardians adopted a resolution intended to bring to the government's attention the fact that "for us here in Connemara the Land Act of 1903 might as well have never been passed"; owing to the landlords' exorbitant demands, the act remained practically a dead letter. Clifden was in need of a proper pier. Steamers laden with material for Marconi's works were obliged to anchor in the bay and unload cargo into smaller vessels, causing much damage, delay, and expense. A copy of the resolution was sent to Marconi but there is no sign that he ever did anything about it.[8] (He would have received it during the period when he was off sick following the death of his infant daughter, Lucia.)

The station was scheduled to open on October 15, 1907, but work was not completed in time. However, a "limited" transatlantic press service was inaugurated on October 17, charging half the rate of the cable companies. The first messages sent from Derrigimlagh ranged from one by David Lloyd George, future prime minister and then president of the Board of Trade, lauding improvement in communication as helping to consolidate and strengthen the British Empire, to another from Henry Murphy, a member of the Galway County Council, appealing for help in obtaining self-government for Ireland. Marconi sent messages to supporters like Lord Kelvin and MP Henniker Heaton, and celebrities were eager to use the new service. The French actress Sarah Bernhardt was among the first. Arriving in London for a stage performance, she sent a message to the United States:

"This fraternal kiss of Europe and America across space is the most poetic manifestation of science."[9]

The opening of the service was lauded by the press as a new and valuable means of distributing news across the Atlantic; for the *New York Times*, the Marconi company was now "an active and potent competitor with the cable companies." Once again, the paper hailed Marconi as "the greatest inventor of this hour."[10] The launch also brought a new wave of techno-euphoria. Thomas Edison predicted that within ten years Marconi would be sending and receiving a thousand words a minute across the Atlantic. Edison raised the question of proprietary benefit. "There is no franchise-giving power for the use of the air, but there is a common law. . . . The common law will see that the founder of a service is not hurt by a follower." In other words, the air would be regulated in Marconi's favour.[11]

Locally, the Clifden station had an enormous impact on the area. At its peak it employed between sixty and eighty permanent and perhaps as many as 140 temporary staff (the latter mostly engaged in harvesting peat).[12] It staunched the drain of population to North American emigration for the next fifteen years, until its technical plant was bypassed by new technology.

❀   ❀   ❀

Marconi himself was overseeing operations in Glace Bay when the transatlantic service was launched. He had been virtually commuting from Europe for over five years, but the Cape Breton station was one of the places at which he was most at ease—despite the fact that the London *Times* considered it "a long way from everywhere and on the way to nowhere."[13] He had his piano, a tennis court, the camaraderie of colleagues, the adulation of the local population, and the kind of social life he enjoyed.

A reminiscence by a station employee, William Appleton, gives the flavour of the Glace Bay station. Appleton emigrated from the United Kingdom to Nova Scotia with his parents in July 1906, at the age of sixteen. He quickly got a job in the coal mine, but took advantage of the Labour Day holiday on September 3 to walk over to the new Marconi station and see if there were any vacancies. He was advised to proceed down the drive to the manager's residence. "About halfway I met a young man in tennis flannels and asked him if I was going in the right direction. . . . He was most polite and helpful and then went on his way. In about fifteen minutes I met [the manager] Mr. Vyvyan . . . [who] surprised me by saying 'I see you have already met our Mr. Marconi.'" Appleton had vivid memories of Marconi's presence: "His personality kept everyone in a state of enthusiasm."[14] Another employee, Alexander Dooley, remembered Marconi as shy and unassuming. Dooley started working at the station as a schoolboy, delivering lunches to the staff; he eventually got a job in the machine shop and stayed with the company into the 1950s. "To see him around at Glace Bay on his visits in 1905 and 1908, you'd think he was a regular labourer; he wore a slouch hat, a fisherman's sweater, and boots like a lumberjack's with the trousers tucked in...[he] shared in the rough life and work of the wireless station."[15] These portraits of Marconi in tennis flannels, open and casual, generous and unassuming, contrast with the sense one often gets from descriptions and formal photographs depicting him as constricted and aloof.

Vyvyan himself described life at Glace Bay as "on the whole quite pleasant." The fishing was great, and "no more delightful holiday could be imagined than to camp out for two or three days" on the banks of the nearby river.[16] It sounds almost like a rustic early version of the Google campus. It is still a pleasant place, although all that remains of the twenty-two-building complex is the eight-bedroom house that Vyvyan built in 1904. It was bought for a song by a former employee, Russell Cunningham, when the Canadian Marconi Company closed the station in 1945, and is still in the possession of the Cunningham family. They say the house is just like it was in Marconi's time, with its oak panelling and steam radiators, asbestos-lined fireplaces, the manager's dining-room table, and a portrait of Pope Leo XIII (who died in 1903) hanging on a wall.[17]

While the site was active, it was served by a railway spin built for Marconi by the Dominion Coal Company, connecting its main line to the station. Today it lies on the aptly named Marconi Towers Road, just off the scenic Marconi Trail that runs between Glace Bay and Port Morien. Like Table Head, Marconi Towers has also been designated a Canadian national historic site, but one couldn't possibly know it; there is not even a plaque to commemorate the fact. The Cunningham house is the only original structure still standing at any of Marconi's transatlantic stations.

Beatrice's sister Eileen O'Brien visited Glace Bay with Marconi and Beatrice in July 1907 and kept a diary.[18] They left Liverpool on the *Empress of Ireland* on June 28, 1907, arriving in Rimouski, Quebec, in the early hours of Thursday, July 4.[†] They stayed in Rimouski one night and, after a visit to the

---

[†] Launched in 1906, the *Empress of Ireland* sank off Rimouski on May 29, 1914, after colliding with a Norwegian ship. More than a thousand died in this largest Canadian maritime peacetime disaster.

wireless station at Father Point (Pointe-au-Père), they left by train for Truro, Nova Scotia. (Degna Marconi describes her father in a family photo taken at Pointe-au-Père as "smiling self-consciously from a tattered carriage behind an ancient discouraged horse."[19])

After a twelve-hour train ride from Truro on July 9, the party arrived in Glace Bay where they were met by Vyvyan and another O'Brien sibling, Barney (Marconi had got him a job at the station). They drove to the station in a cape cart,[†] a two-wheeled, four-seat, horse-drawn carriage, during a severe electrical storm so great that Vyvyan feared the towers would come down. Beatrice had invited the Duke of Abruzzi to visit as well and he had wired to say he would let them know in a day or two if he could come.

Eileen reported: "Guglielmo's work going on well only material not arrived from England which delays work considerably. Very hot here mosquitos make it impossible to sit out...." A compilation of photographs from this trip in the Bodleian Library shows a happy, relaxed crowd.[20] But after only three days, Marconi was called back to England. On Saturday July 13, 1907, Eileen wrote: "It was decided that I shall leave Cape Breton with Guglielmo on Tuesday or Wednesday on S.S. *Victorian* from Montreal, Bea remains here." They left on Wednesday, July 17, accompanied by Barney as far as Montreal, where they dined with Marconi's Canadian company manager, John Oppe, at his home before boarding the *Victorian* around 11:00 p.m.

Eileen O'Brien's diary also illustrates a typical transatlantic crossing in Edwardian times:

> We slept in ship in dock and moved out on our journey at 6.30 am July 19th...Lonely passage up St Lawrence liner stopping at Quebec weather boiling hot! Had tea up at the Hotel [likely the Chateau Frontenac] a climb up with Dickie G.M.[§]...got back about 6 and dined at 7 o'clock on board.

> July 20th left Rimouski. 8.30 am wet no fog, sighted Newfoundland 7 pm... Very very cold all in fur coats again! Sent Marconigram to Barney...[off Labrador Coast] small iceberg passed quite close...Left Belle Isle Straits well on our way to England...

> July 22nd Monday. Passed 20 icebergs night of 21st. Dined with Captain of ship, amusing dinner. Wednesday 24th Guglielmo told us at dinner story of how he and the German Emperor did not get on...

[‡] So called because it originated in South Africa, where the cape cart was used to carry passengers and mail before the arrival of the railways.

[§] One of the O'Briens' nicknames for Marconi was "Dickie" or "Dick."

July 25th 100 miles from Ireland went over Wireless Cabin with G. who
explained works. Went to a concert on board. Guglielmo made a speech proposing a vote of thanks etc to all....[21]

The party arrived in Liverpool at 7:00 a.m. on July 27 and took the train to
London. Beatrice stayed on with Barney and the Vyvyans in Glace Bay while
Marconi moved between England and Clifden, putting the final touches to
the Irish station. He didn't return to Canada until mid-September, and
arrived in Glace Bay exhausted. Asked by the *New York Times* whether he
had chosen a name for his commercial wireless messages, he shrugged:
"No, not yet. 'Marconigram' seems to stick on the tongue, but so far we
have no other."[22] The term had been in use since 1903, soon after Marconi's
breakthrough in transatlantic messaging. And marconigram it was.[23]

Marconi continued travelling back and forth between Cape Breton,
England, and Ireland while Beatrice remained in Glace Bay. He was back in
the United Kingdom in November, and then, in early December, he was in
Glace Bay again, until he and Beatrice finally sailed for England together
after New Year's, arriving on January 12, 1908. In all, she had spent six
months in Glace Bay, three of them without her husband. After the rough
period that followed the death of Lucia, Beatrice's long and quiet stay in
Cape Breton revitalized her. (Among other things, Beatrice spent a week
salmon fishing in Newfoundland with her brother Barney.[24]) When she and
Marconi returned to England, her mother marvelled at how very well and
happy she looked.[25] Beatrice was also pregnant.

Early in 1907, the Marconis rented an eighteenth-century house called
Somborne Park, in the Hampshire parish of Kings Somborne, off the old
Roman road from Winchester to Salisbury, in an area of gently rolling chalk
hills typical of southern England. The house was owned by their brother-in-
law, the baronet Frederick Hervey-Bathurst, then married to Beatrice's sister Moira. The property goes back to the eleventh century, and has been
owned by the Hervey-Bathursts since 1832. The house itself dates from at
least 1759 and has been tenanted for most of that time. In Marconi's day it
had a walled kitchen garden and a notable sundial; today, the garden is gone
but the sundial survives. The house remains surrounded by two hundred
acres of parkland.[26] Beatrice found it a "lovely simple country house in perfect taste...spacious, cosy and comfortable....We had on the whole a very
happy peaceful time in this delightful completely peaceful district." She

loved Hampshire, but it was a bit far for a daily commute to London and Marconi had to be in the city every day when he wasn't travelling.[27]

Marconi was now busier than ever. The full commercial service between Glace Bay and Clifden was launched on February 3, 1908, and he was soon thrust into the dramatic power struggle leading to Hall's removal. On September 2, he sailed again for New York, with Henry Jameson Davis, having installed an eight-months pregnant Beatrice in a rented house in Upper George Street, London, for the "event." It was at least his eighteenth transatlantic trip but his comings and goings were always news and he was still seen as an exotic creature; reporting his arrival, the *New York Times* said Marconi "now speaks English like a native," oblivious to the fact that he had done so all his life.[28]

One of the last things Marconi did before leaving for America was to send a registered letter to his sister-in-law Lilah in Ireland, asking her to be available to go to Beatrice on a moment's notice should there be any difficulty with the pregnancy. He enclosed ten pounds in cash for the journey, asking her to be sure not to lose or squander it—a note of typical Marconi condescension but, at the same time, familiarity.[29]

The day after Marconi arrived in New York, September 11, 1908, his daughter was born in London. She later wrote that her father did an immediate about-face and sailed for home at once, but archival records and newspaper reports place him in the United States, Glace Bay, and Montreal through the end of September and early October. He was in South Wellfleet, on Cape Cod, with Davis and Bottomley on September 13, then back in New York, then in Glace Bay and Montreal at the end of the month, meeting financiers and announcing plans for an overland system across Canada.[30] He even visited Niagara Falls, and sent Beatrice a postcard: "Wish so much you were here with me. All *il mio amore* to you both. Guglielmo."[31] Eventually he did return. Reading Pompeo Molmenti's history of Venice on the ship home, he chose an old Venetian name, Degna. She was christened at St. George's, Hanover Square, where her parents had married three and a half years earlier.[32]

Back in London Marconi began staying at the Ritz, and going back and forth to Clifden, while Beatrice returned to Somborne Park with Degna. Marconi's letters to Beatrice during this period illustrate the emerging complexity of their relationship, as he tries to balance his new family life with his established patterns of work and socializing, as well as his unique epistolary style. His earliest surviving letter to her, from London dated

November 13, 1908, meanders across a characteristic mélange of concerns and insecurities, attempts to reassure both her and himself, and contains both breeziness and signs of stress:

> Things are going rather well and shares have gone up to 13 sh. I've been having lots of work and a bit of fun too. The Sassoons' dinner (for men), was rather nice and I afterwards had supper with the Whitlaw Reids (the American Ambassador). I had lunch before with Alfred Lyttleton the uncle of Maud. You know he was Colonial Secretary. I found him very interesting and clever and we talked a lot of your friends which are his relations, and of heaps of things about Italy and he was clever enough never to put his foot into it when talking about Italian matters. . . . Some of the papers have said that I have got the Nobel prize of £8000. It rather makes one's mouth water to think about it just now, but I suppose it's not true. . . . Don't stop loving me for I could not possibly do without your love. Your own hubby. Guglielmo.[33]

In London briefly after a trip to Clifden, he writes on December 10, 1908:

> My own sweet darlingest Buzzel, . . . I'm quite well and only want you to be perfectly happy. We had great trouble with the aerials at Clifden, it was so difficult to get them up in the rain and snow, and the machinery I expected was several days late so I shall have to go there again tomorrow for a few days. Isn't it awful my having to go back to Ireland without seeing you again darling— please do forgive me going [emphasis in original] but I think it's for the best. . . . Don't think anything wrong about me if you heard I was lunching with ladies at the Ritz. It was the day of the christening of Mrs Coghlin's baby.[34]

The following day, he sent her a copy of *Corriere della sera,* with an article about him that he thought might interest her. Finally, from Clifden, December 14, 1908, while she was still in Somborne Park: "My own darling Buzzel, I'm here am very well and working hard and thinking of you all the time. You are such a darling sweet to write me such lovely letters, it makes me so very happy to hear that my chicks are well, and that big chick loves me and wants me all the time. . . . I do miss you so awfully [emphasis in original]."[35]

There is a building tension in these letters that reaches a peak just before Christmas. Beatrice was now visiting with her eldest sister, Clare, and her husband Noel Armar Lowry-Corry, in Llangattock (south Wales) and Marconi was supposed to join them. On December 23, 1908, he wrote:

> Dearest,
> I was so very glad to get your wire this morning and to know that perhaps someone cares for me after all.

You see, in your letter dated last Saturday you said that I could not be put
up before tomorrow the 24th, as there would be a man coming who would
have my room—but that you "<u>would let me know for certain later</u>" [emphasis
in original]. Nearly five days have passed, in which I've been longing to come
to you, and rather unsettled, without your letting me have a line—sooner or
later—as to whether I might come today or not. This made me think you did
not care when I came. . . .

. . . I've been thinking and wanting you all the time. . . .[36]

Marconi's discomfort was palpable. The letter was an exercise in ambivalence and manipulation: he was piqued and at the same time trying to
knock her off balance.

Unfortunately, there is no record of Beatrice's replies to any of these
letters. Either he did not keep them or they have been lost.[¶] But she did
reminisce at length about the strains of the relationship in a memoir she
wrote in the 1950s when their daughter Degna was preparing her book on
Marconi. In one story, Beatrice, pregnant (with Giulio in October 1909)
and wanting to surprise Guglielmo with the happy news, took a boat out
from Cork Harbour to meet his ship returning from America, only to be
surprised herself. "My reception was not as I had imagined and was most
ill-timed," she wrote. Marconi was furious at her inopportune arrival—
interrupting a cheerful party with him and fellow international superstar,
opera tenor Enrico Caruso, at its centre, surrounded by adoring young women.
In his fury, Marconi charged that the baby was not his and Beatrice retreated
in tears to her husband's cabin. "After a while he calmed down. Rather
ashamed of his behaviour he tried to be nicer and get me to join his party."[37]

In addition to highlighting the emotional turmoil he was feeling at
this time, Marconi's letters illuminate the social aura he was constructing
around himself. Despite his financial complaints, he was having an evening
suit and smoking jacket made by Scholte, a tailor in Savile Row credited
with creating the early twentieth-century look known as the "London cut."
As we saw in his earliest letter about dinners with the Sassoons and the
Reids and his lunch with Lyttelton, Marconi was cultivating political contacts, an activity he found fun, as opposed to work, which he was beginning
to find dutiful and exhausting.[38]

¶ There is only one letter from Beatrice to Guglielmo in the entire Marconi archival stock, written
on their son Giulio's first birthday, May 10, 1911. The letter is in the Royal Institution, London
(BOM to GM [May 10, 1911], RI 93).

One of the most curious of his new relationships was with the Duke and Duchess of Manchester, a couple surrounded by scandal and controversy. In September 1903, William Montagu, the ninth Duke of Manchester (known to his friends as "Kim"), and his American duchess, formerly Helena Zimmerman of Cincinnati, moved in to Kylemore, a nineteenth-century castle around twelve miles from Clifden. At twenty-six years of age, the duke had a reputation as a rogue, a gambler, and a spendthrift, and was famous for having squandered his inheritance even before coming of age. The purchase of Kylemore was a complicated transaction involving a mortgage of more than £50,000 that was guaranteed by Helena's father, oil and railway tycoon Eugene Zimmerman.

There was not much in the way of high society in Connemara in the early 1900s, and it wasn't long before Marconi and the Manchesters found each other. On December 11, 1908, Marconi wrote to Beatrice that the Manchesters would be calling on him in Clifden, "so I expect to see someone civilized this time"—an interesting comment about his attitude toward the locals (or his presumptions about Beatrice's attitude; this is classic Marconi, writing what he thinks his correspondent wants to hear).[39] Marconi not only considered the Manchesters among the more "civilized" inhabitants of the region, he also did business with the duke, leasing land from him for a station at Letterfrack, which stood on the Kylemore estate, in 1913.[40] By then, Manchester was again on the verge of bankruptcy. Local legend (according to a story-board at Kylemore) has it that he lost the deed to the estate on a single hand of cards; more likely it was by an accumulation of debt. At any rate, the Manchesters had to give up Kylemore in 1914. It was eventually sold to an order of Benedictine nuns from Ypres in 1920 and converted into an abbey; today it is one of the principal tourist attractions of west Galway.

Manchester spent the rest of his life evading creditors and hitting up wealthy acquaintances for money—Marconi evidently among them. An undated letter signed "Kim," on stationery of London's Savoy Hotel and addressed to "My dear Guglielmo . . . a pal," says that he is "in a terrible fix" and asks for a few days' loan of £120 in order to pay his hotel bill. He says he can't ask the bank to increase his overdraft, which has been running for a year.[41] We don't know Marconi's response to this particular appeal but he does seem to have kept up friendly relations with Manchester. They were occasional London theatre companions and Marconi kept Manchester's name in his diary/address book.[42] The Manchesters divorced in 1931, and in a memoir published the following year, the duke candidly stated that he was "unrepentantly addicted" to gambling and unable to look after his finances.[43] In 1934 he wrote to Marconi in Rome trying to promote an invention, a missile-launcher, that he thought might interest Mussolini. Marconi, ever helpful, replied with a warm note saying he would be delighted to speak with Mussolini about it.[44] Whether a sign of poor judgment, loyalty, or naïveté, Marconi's eagerness to help Manchester was typical of his relationship with the secondary characters who passed through his orbit.

A sensational event in 1909 became another important part of the Marconi mythology. In the early morning of Saturday, January 23, 1909, the RMS *Republic*, a luxury ocean liner headed for a Mediterranean cruise with twelve hundred passengers on board, was rammed by an Italian ship, the SS *Florida*, in heavy fog off Nantucket. As the ship was going down, John R. (Jack) Binns, its twenty-four-year-old wireless operator, managed to send off a CQD (the signal for *sécurité* distress, introduced and still used by the Marconi company despite the international adoption of "SOS" by the Berlin Convention of 1906). The signal was picked up by the Marconi station at Siasconnett (now Siasconset), Massachusetts, where the operator on duty, Jack Irwin, relayed the message to ships in the vicinity, which then entered communication with the *Republic*. Operators H.J. Tattersall and Gilbert Balfour, on the RMS *Baltic*, manned their wireless station indefatigably throughout the rescue operation. Only six lives were lost, a remarkable outcome given the gravity of the accident.

By the time it was over, a new class of hero—the "Marconi man," or wireless operator—had captured the popular imagination. The dashing Jack Binns, in particular, became an overnight sensation, after publishing a first-person account in the *New York Times*. Testifying at the US Senate inquiry into the sinking of the *Titanic* three years later, Binns told how the article came about. His testimony provides a window into how the Marconi company operated—as well as the relations between news media, newsmakers, and news: "Coming up the coast I received wireless messages from various newspapers asking me for my own personal story. . . . I also received a message from the Marconi Co. asking me to reserve the story, if possible, for the *New York Times*, owing to their friendly connection with the Marconi Co., by whom I was employed at that time."[45]

Marconi himself was besieged by congratulatory marconigrams and requests for interviews. He would later refer to the *Republic* incident as one of the first concrete demonstrations of the value of wireless for saving human lives. Jack Binns continued to work for the company until 1912, when he became a popular columnist for William Randolph Hearst's *New York American*.[46]

17

The Perfect Laureate

On December 10, 1896, as Marconi had been prepping for the London lecture that would introduce him to the world, another European inventor-entrepreneur died, from a cerebral hemorrhage, at his Italian villa. When Alfred Nobel's will was opened a few days later, his family was astonished to learn that he had left nearly his entire fortune of more than 30 million Swedish crowns (some US $180 million today) to establish a series of prizes in his name.[1]

Born in Stockholm in 1833, Nobel had made a fortune commercializing more than 350 inventions, the most famous of which was dynamite. He was also a major arms manufacturer, leading a French newspaper to label him, in 1888, "the merchant of death." (Two of his brothers, Robert and Ludwig Nobel, were even richer, from developing the Baku oil fields in Azerbaijan.) Childless and concerned about his legacy, Nobel rewrote his will a year before he died, leaving relatively small amounts to his nephews, nieces, and various people close to him, and the bulk, more than 90 percent, to award "those who, during the preceding year, shall have conferred the greatest benefit to mankind."[2]

It took the executors nearly five years to overcome the objections of Nobel's heirs and set up a foundation to administer the five annual prizes specified in his will (for contributions to physics, chemistry, physiology or medicine, literature, and peace—economics was added in 1968). In the process—and coherent with the experience of other large philanthropic undertakings in Europe and America in this period—the prizes came to be seen in Sweden and internationally as an important public enterprise that superseded private interests and concerns. A complex machinery was put in place, specifying an international nomination procedure with six categories of nominators, and establishing a series of committees for recommending the laureates. Four institutions would make the final decisions: the Royal Swedish Academy of Sciences (for physics and chemistry), Stockholm's Karolinska Institute (for medicine), the Swedish Academy (for literature), and a committee appointed by the Norwegian parliament (for peace). In 1900, the foundation's statutes were finally promulgated by the Swedish king, Oscar II, and the first prizes were set to be awarded in 1901.

One of the first people recruited to nominate candidates was Marconi's scientific advisor, James Ambrose Fleming. Fleming didn't have to think long to come up with a candidate. On October 4, 1901 (while in the midst of preparing the transatlantic station at Poldhu), he wrote to inform Marconi—despite the conflict of interest—that he would be "very pleased" to nominate him for the honour.[3] It was the beginning of an eight-year lobbying effort on Marconi's behalf. Marconi's name was put forward in 1901 but the Swedish Academy awarded the first Nobel Prize in Physics to the German physicist Wilhelm Conrad Roentgen for his 1895 discovery of X-rays.*

Marconi was nominated again in 1902 and 1903,[4] and it was even reported in the press that he was to be awarded the prize that year, but the academy gave it instead to Marie and Pierre Curie (who shared it with the then much more famous Professor Henri Becquerel of the Sorbonne) for their research on radioactivity, a term coined by Marie Curie in 1898. The Curies were completely unknown to the public when their discovery of the element radium thrust them into the limelight. (They had previously been nominated in 1902 but Marie Curie's name was not on the initial 1903 nomination; although not strictly within the rules, her name was added at the express wish of her husband and collaborator, Pierre Curie.)† According to one Nobel scholar, "It was also recent enough news to compete with Marconi's new 'wireless telegraphy' for popular attention."[5] Marconi was at the height of his fame in 1903; a single newspaper, Bologna's *Il Resto del Carlino,* published more than two hundred articles on him that year.[6]

---

* Marconi was nominated in 1901 by Italian mathematician, physicist, and senator Pietro Blaserna, founder of the University of Rome's Institute of Physics, where Enrico Fermi and his colleagues would do their famous work on nuclear fission in the 1930s. Fleming was too late for the 1901 competition but nominated Marconi in 1902 (Crawford 1987).

† Marie Curie was also awarded the prize for chemistry in 1911, for the discovery of radium and polonium; she is the only woman to have been awarded two Nobel prizes, and is, of course, one of the iconic symbols of twentieth-century scientific achievement as well as the professional accomplishments of women.

The Nobel prizes quickly became a ritual of anointment surrounded by mystery and secrecy.[7] The process also generated a small lobbying and public relations industry. In February 1904, Marconi's managing director Cuthbert Hall was contacted by one Thor Lutken, a well-connected lawyer "who has been trying to do business with us in Norway." For the modest sum of £100 to cover expenses, Lutken offered to pave the way for nominating Marconi for the Nobel Prize. Hall, alert, clear, and direct as usual, wrote to Marconi: "I should be disposed to thank him for his offer; point out that your name has already been proposed for the prize by various scientific societies of high standing of their own initiative, and that we think it better that we should not move personally in the matter." Hall added, however, that Marconi's getting the prize would be of "considerable indirect advantage" to the company's interests.[8]

The Nobel prizes were already a popular subject for the international rumour mill, especially where a high-profile media figure like Marconi was concerned. On November 12, 1908, the *New York Times* reported "unofficially" that Marconi was to receive the next Nobel Prize in Physics. A few days later, the paper recanted, correcting itself to state that the prize would in fact be awarded to the German theoretical physicist Max Planck. That, too, was wrong. Both Marconi and Planck were indeed nominated in 1908, and the committee recommended Planck for the prize, but it was overruled by the full academy, which gave the award to a French scientist, Gabriel Lippmann, for developments in colour photography.

Aside from the honour and prestige, the Nobel prizes were also lucrative—awarding the winners approximately £8,000 or US$40,000 in the early years. Nobel scholar Elisabeth Crawford has estimated the value of the prize in 1901 to be thirty times the annual salary of a university professor, or two hundred times that of a skilled worker.[9]‡ In the case of the Curies, for example, the award was the equivalent of twice their combined salaries for the next ten years. For Marconi, starting a family and plagued by corporate financial issues in 1908, the money was more important than the honour.[10]

On November 10, 1909, finally, the long-awaited letter arrived, announcing that the Royal Swedish Academy of Sciences had decided to

‡ One might say that academics have done relatively better than Nobel laureates over the years. The prizes are currently worth US$1.5 million, around twenty times the average annual salary of a full-time American professor (U.S. Department of Labor, *Teachers-Postsecondary: Earnings*, December 18, 2007). In terms of purchasing power, the prize was worth about twice as much in 1901 as in 1983.

award Marconi the Nobel Prize in Physics. The results would be made known in Stockholm on December 10—the anniversary of Alfred Nobel's death—and the laureate was sworn to secrecy until that time.[11§] To quite some astonishment, however, it was to be a shared prize, awarded jointly to Marconi and one of his German rivals, Karl Ferdinand Braun, "in recognition of their contributions to the development of wireless telegraphy."[12] Marconi and Braun had never collaborated and by almost any standard their "contributions" were certainly unequal. Marconi was tempted to refuse this "half-prize,"[13] but his pragmatism came to the fore and he accepted graciously. In later years he would tell an anecdote about finding himself on the same train as Braun on the way to Stockholm; Braun suggested they play cards for the whole prize but "I was such a poor card player that I declined."[14]

What were the politics behind this curious joint award? Braun was a legitimate candidate, the holder of several patents building on Marconi's, and one of the founders of Telefunken. But he was certainly not a contender at the level of, say, Lodge, Fleming, Tesla, Popov, Righi (who was nominated every year between 1905 and 1920 but never won),[15] or even his colleague and countryman Adolf Slaby. Other nominees for physics in 1909, in addition to Max Planck, included Edouard Branly (who, as inventor of the "coherer" used by Marconi, would have been a far more likely co-laureate), Oliver Heaviside, Henri Poincaré, Valdemar Poulsen, Righi, and both of the Wright brothers.

The decision was unquestionably tied up with the intense politics of the Nobel awards. When the Nobel Committees opened their records in the 1970s, it was discovered just how arcane some of these politics could be. In 1909, Gösta Mittag-Leffler, Sweden's most famous mathematician and an influential figure in Scandinavian academic politics, although not a member of the committee, had decided to start a campaign in favour of Poincaré, the French mathematical physicist. As the prize had just been awarded to a French scientist, Gabriel Lippmann, in 1908, Mittag-Leffler decided to aim for 1910. An interim laureate had to be found. Mittag-Lefler wrote to his friend and colleague Paul Painlevé in Paris: "Do you think one could give the Nobel prize for the invention of the airplane?" Painlevé wrote back enthusiastically, suggesting the prize be divided between Orville and Wilbur

§ The Nobel prizes are still awarded annually on December 10; however, the names of the winners are now publicly announced in October, two months before the ceremony. Secrecy was abandoned as "impractical" in 1910 (Crawford 1984, 7).

Wright, on the one hand, and French aviation pioneers Gabriel and Charles Voisin and their pilot Henry Farman on the other.[16] It was an inspired idea. Aviation had captured the popular imagination (Marconi himself attended a flight demonstration by Wilbur Wright while in the United States in September 1909[17]) and awarding the prize to the Wright brothers and the Voisins would attract and endear a mass audience. Poincaré himself was even recruited to sign the nomination. The aviation proposal was submitted in the names of Mittag-Leffler, Poincaré, Painlevé, and four mathematicians from the Swedish technical college. It was received with skepticism, however; in view of the loss of life that could result if anything went wrong (and there had been a rise in air accidents in 1908), the committee concluded that "in its present state, this invention can hardly be considered to be of benefit to mankind."[18]

The Swedish Academy was ready to recognize a practical application in 1909, but it was just too soon for the airplane. Wireless telegraphy, however, was exactly the kind of invention that met the Nobel criteria; contrary to the risky business of aviation, wireless was now seen as an enhancement of safety in travel. It had also concretely benefited a great number of people. The public had been captivated by the role of Marconi's wireless in saving twelve hundred lives after two ocean liners, the *Republic* and the *Florida*, collided in a fog off Nantucket earlier that year—demonstrating to the world "the benefit to mankind" represented by wireless. Whether or not this explicitly had a hand in convincing the Swedish Academy is not clear, but when the awards committee's report was placed before it in early November the result was unequivocal. The aviation proposal garnered only four votes against some fifty cast for Marconi and Braun.[19]

Marconi was an example of a candidate whose accomplishments had already been established when the prize was created in 1901, and who had accumulated nominations over several years. The 1909 award was also one of several examples from these early years where the committee and the academy disregarded the majority of the nominators. The politics surrounding the prizes disfavoured obviously deserving candidates like Max Planck and Henri Poincaré. A Marconi candidacy, on the other hand, held a certain attraction to the Nobel awarders. It would be popular, notorious, and not a problem for any faction.

Most important, Marconi represented a new type of laureate. By 1909, the prize-givers had become sensitive to the perception that the awards were esoteric, and with rare exceptions (such as Roentgen or the Curies), overly academic.[20] Now, for the first time, the prize was recognizing a type of activity not foreseen by Alfred Nobel—although implicit in his categories of "discoveries, inventions and improvements": Marconi was the first Nobel prizewinner who represented innovation in *technology*. Marconi was arguably a "discoverer," but in Nobel terms he was primarily an "inventor." He was also something else, something Nobel himself would have appreciated but which the academicians awarding the prizes had so far eschewed. He was an entrepreneur. At one of its very first meetings in February 1901, the Nobel physics committee had agreed that "patented inventions should not be taken into consideration," thus possibly settling Marconi's case before it even got on the docket.[21¶] But this was not a statutory rule and proposals to reward patented work were not rejected out of hand; in the early years, however, they were considered only where the candidate had not benefited financially from the invention, a paradoxical position to be sure, since one of the measures of a successful invention, almost by definition, was commercial success. Marconi was surely one of the first laureates to be selected despite the perception that his invention had brought him great wealth. For neither the first time nor the last time, Marconi found himself in the right place at the right time; it was pure good fortune, and nothing he did or could do had any bearing on the result. As so often in his life, Marconi's outsider status made him a useful favourite of the insiders.

Travelling with Beatrice and her sister Lilah, Marconi left London for Stockholm on December 5, 1909. In the Swedish capital they were put up, along with the other Nobel laureates, at the Grand Hotel and taken in charge by a young military officer by the name of Thor Thörnblad, who had previously been in touch with Marconi about a book he was writing on wireless telegraphy. In what was certainly a breach of protocol, Thörnblad invited the Marconis to dinner at his home along with the famous explorer and orientalist travel writer Sven Hedin.[22] They also met Marconi's fellow Nobel laureate Selma Lagerlöf, the first woman to win the prize for literature,[23] and were received by the royal family, but that was nothing new.

¶ Marconi hadn't stood a chance in 1901; Roentgen received sixteen nominations in the first prize competition and the committee decided to propose him for the award in January even before all of the nominations, including Marconi's, had been received.

The awards ceremony took place on December 10.** The laureates were presented by Hans Hildebrand, an archaeologist and president of the Royal Swedish Academy of Sciences. After the traditional nod to Faraday, Maxwell, and Hertz, he recited what was now the standard description of Marconi's accomplishments—taking care to laud their entrepreneurial as well as scientific sides. Marconi was the man of the moment, able to shape wireless into "a practical, usable system." However, he went on: "The development of a great invention seldom occurs through one individual man, and many forces have contributed to the remarkable results now achieved. Marconi's original system had its weak points. . . . It is due above all to the inspired work of Professor Ferdinand Braun that this unsatisfactory state of affairs was overcome."[24]

The next day, when it was the turn of the laureates to recount their achievements, Braun paid tribute to Marconi's pioneering work before focusing on his own efforts. He had taken an interest in wireless telegraphy in 1898, when he began trying to increase the range of transmission and struck on the idea of a sparkless transmitter. After experiments run by his assistants at Cuxhaven, in 1899 and 1900, he described the advantages of sparkless telegraphy at a public lecture in Strasbourg in November 1900 as "advantages which Marconi had by then also recognized." In other words, Braun acknowledged that his work was built on Marconi's.[25]

Marconi, who had of course been involved in an acrimonious patent dispute with Braun, must have been squirming in his seat. In his own Nobel lecture, delivered immediately before Braun's, he hardly mentioned his corecipient. The lecture, now part of the Marconi canon, was hardly riveting stuff, even for specialists. Anyone even the least bit familiar with Marconi from reading the popular press would have learned very little from it. It is historically significant because of the context and, like many of his lectures, articles, testimony, and interviews, it added a bit of incremental knowledge to the biographical and scientific narrative that Marconi was constantly revising throughout his life, right up until his death.

Marconi rehearsed the story of his early experiments: his 1895 Pontecchio breakthrough, the discovery of the use of antenna and ground, the 1896–97

** The ceremony is always the same. Two thousand dignitaries gather in the Stockholm Concert Hall at 4:00 p.m. The new laureates sit on the stage in a fixed order, according to how the prizes were listed in Nobel's will. Physics is first, followed by chemistry, medicine, literature, and peace; economics is last. After a brief ceremony, the king hosts a banquet; the meal is traditionally venison, provided by his own hunters. The next day the laureates each give a major address, the scientists explaining (or trying to) the work for which they were awarded (Feldman 2000).

English trials. He noted that the practical value of his innovation was not understood by most physicists until its recognition by Professor Moisè Ascoli of Rome, who in *L'Elettricista* of August 1897 "correctly attributed the results obtained to the use of elevated wires or antennae." He recalled that Adolf Slaby had come to similar conclusions after witnessing Marconi's tests in 1897, and that his work on the use of condenser circuits "was carried out simultaneously to that of Professor Braun, without, however, either of us knowing at the time anything of the contemporary work of the other." This was his only wink at Braun. At the very end of the lecture Marconi tried to articulate his vision of what lay ahead: one day, it might be possible to send messages "right round the world... by means of a very small amount of electrical energy, and therefore at a correspondingly small expense." Thus having foreseen inexpensive long-distance communication, he checked himself and apologized for leaving the realm of facts and entering that of speculation.[26]

Marconi delivered his lecture in English, noting that he didn't, alas, speak Swedish, and English would be the next language best understood, moreso than Italian. The Nobel committee actually considered Marconi British, since he had conducted most of his career in the United Kingdom (just as Marie Curie was considered French, despite her Polish birth and the fact she had immigrated to France as an adult). Interviewed in Berlin a few days later by an Italian journalist, Marconi said: "Please write that I am glad to have won the Nobel Prize, not for me, but for Italy."[27] It was another step in the zig-zagging trajectory of his national identity.

The Nobel Prize was important in providing Marconi with the academic legitimacy that had thus far eluded him (despite the honorary degrees and other accolades), but he didn't need it for the notoriety. By 1909 he was already one of a handful of world figures who were beginning to define modernity in the twentieth century. The prize acknowledged that wireless communication was beginning to define the period, as television later defined the 1950s, and as the Internet defines our own. Rarely has a technology been so strongly associated with a single individual; in fact, while ushering in a new era of technological modernism, Marconi was at the same time the last of the individualist hero-inventors (after Samuel Morse, Thomas Edison, and Alexander Graham Bell). The only comparable figures in our own time are probably Steve Jobs and Bill Gates, who, like Marconi, each invented a relationship with technology that continues to define an age.

No single person can create a technology. That depends on social forces greater than any individual, and on the unfathomable desires of the millions of people who are eventually labelled the "users," the "audience," the "public," or the "market." While contributing to the production of modernity, wireless—especially its incarnation as "radio"—simultaneously became a *product* of modernity. At the time of the Nobel Prize, the idea of wireless had eclipsed Marconi's own vision. The technology would not be fully realized for another ten years—until it was encapsulated in the black box that began speaking to millions of ordinary people in their homes the world over—but by 1909 it was on the minds of poets and social visionaries. Up until this point, for most people, wireless was not so much a product as an *idea* for a product.[28] Marconi's "system" was not yet commercially viable (and he was obsessed with that, for obvious reasons), but he never ceased to believe in, and trumpet—as he did in his Nobel lecture—its tremendous potential.

Although he barely realized it himself, even in his most visionary moments, Marconi's wireless was about to become the technological foundation of the greatest vehicle for mass communication the world had ever known. The telegraph had extended communication on a practical basis, but it would always be controlled by a small number of state or corporate owners. The telephone had democratized one-to-one communication and removed the obstacle of distance from the equation of everyday life. Radio, however, was untrammelled and potentially uncontrollable, and so was mobile communication. Through radio one could potentially speak to everyone, and mobile communication allowed one to do that anywhere, at any place, at any time—instantaneously and simultaneously.

The Italian futurists were among those who adopted Marconi as a symbol of the new modernism. Their movement was launched with a 1909 manifesto by poet Filippo Tommaso Marinetti, published in Paris just a few months before Marconi again captured global attention with the Nobel Prize.[29] Italian futurism was the first avant-garde art movement to embrace technology, and, to the futurists, Marconi's invention evoked a world shrunk by the dissolution of time and space constraints, a world marked by "the new aesthetic of speed."[30] Marinetti lauded Marconi, promoting a new literary form he called "telegraphic language." Wireless communication was mythicized and fetishized in pre-war European poetry and literature from Marinetti to Guillaume Apollinaire, becoming "the trademark of a crucial modernist sensibility."[31]

Marconi's influence on Marinetti (as well as contemporaries in the new field of "science fiction" such as H.G. Wells) popularized the notion that wireless telegraphy represented not only a revolution in communication but in how future generations were going to live.[32] To modernists, wireless was a symbol of progress, not only a tool but a marker of human advancement.[33] Today we might call this the birth of a "wireless imagination." The impact of wireless on the popular imagination in the first decade of the twentieth century underscores that it should be seen as a distinct historical media and cultural form, not just as a transitory bridge between telegraphy and radio.[34] In fact, its resurgence one hundred years later indicates that in the long sweep of history, the telegraph and broadcasting may yet be seen as ephemeral and transitory where technologies that combine speed, instantaneity, mobility, long-distance, and ubiquity endure. In our own time, the radical economy of wireless is behind such communicative notions as the sound byte, the pitch, the snapshot, or the tweet—a communicative economy where concision takes precedence over depth and reflection.[††]

[††] Arguably, the ideas of the sound byte—"give it to me in two words"—and the 144-character tweet derive from the economy of telegraphy, where every letter is costly.

PART
III

The Patriot

18

The Godsend

By the time he was awarded the Nobel Prize, Marconi had been running the business side of his company for a year and a half, filling the position of managing director since he had, himself, orchestrated the forced departure of Cuthbert Hall in March 1908. It was never intended to be more than a stop-gap measure, as he fully expected to return to working exclusively on his research, but finding an appropriate replacement for the flamboyant and volatile Hall was not easy.

During this period, Marconi fell back on the support of old company hands like Samuel Flood-Page and, especially, his cousin, Henry Jameson Davis, who knew the company better than just about anyone. After Hall's departure, Davis was given full power to discharge the duties of managing director during Marconi's frequent absences from London.[1] But in October 1909, shortly after travelling to the United States and Canada together, Marconi fell out irrevocably with Davis. On October 29, 1909, following a tumultuous board meeting during which Marconi cast doubt on Davis's good faith, his oldest collaborator resigned as director of the Marconi companies. In a handwritten letter on black-bordered personal note paper— almost like an obituary notice—Davis wrote:

> Why you acted towards me as you did today I can't imagine....It is a new experience for me to hear a statement seriously made by me in a business matter dismissed, or to have doubt thrown on my ability to meet my business engagements. Altogether it appears to me to be a very poor return for all I have done for you and for the company.[2]

It was a rather sad affair. Nonetheless, Davis still addressed Marconi as "My dear Guglielmo," and signed, as he always had, "Your affectionate cousin." No longer simply "Henry," he was now "H. Jameson Davis." The argument involved money, and Marconi responded immediately:

My dear Henry,

I regret to learn from your letter just received that you take exception to something I said at the meeting today.

On my part I was sorry that you should have made remarks which as I understood them, seemed to belittle the financial sacrifices which I have had to make for the Company....

I had no wish or intention to offend you or to doubt your word and I have said this to many of the directors today after you left. I hope that under the circumstances you will not insist on resigning for which course there does not appear to be any reason.[3]

Mirroring Davis's signature style, as he always did, he signed: "Your affectionate cousin G. Marconi."

Marconi didn't submit the correspondence to the board for three months. On January 25, 1910, he reported Davis's resignation, which the board accepted with thanks and regrets. In the very next item, the board elected a new director, Godfrey Charles Isaacs, appointed him joint managing director (to share the position with Marconi), and approved his contract.[4] Marconi lauded Isaacs's qualities in a letter to Luigi Solari: "This man is able and honest and belongs to a financial group of the first order. I would be grateful, Solari, if you would support this manager as much as possible ... the Company now has two managing directors, him and me."[5] Marconi was never without a close business collaborator for very long.

Isaacs was a godsend. Marconi was itching to return full-time to his research. He had managed, heroically, to keep the company afloat during the past two years, but it was stagnating financially and there was still not the glimmer of a dividend in sight. As Davis and Hall before him, Godfrey Isaacs became Marconi's corporate alter ego. He would turn around the company's financial fortunes and served as managing director longer than anyone else, more than fourteen years, until he was obliged to retire in November 1924 because of failing health.

Isaacs was born in 1866, the eighth of ten children of Joseph Isaacs, a prosperous East London fruit merchant, and Sarah Davis. The Isaacses were one of London's oldest Jewish families, tracing their ancestry back to central Europe in 1659 and their presence in England to the early eighteenth century.[6*] Godfrey Isaacs's uncle, Sir Henry Isaacs, had been Lord Mayor of

---

* The family patriarch, Isaac Isaacs, was born in 1659; his grandson, Israel Michael Isaacs, the first English-born member of the family, was born in Chelmsford (Essex) in 1735.

London in 1889–90 and was known as a reformer. One of Godfrey Isaacs's older brothers, Harry, carried on the family business; another, Rufus, went into a stellar career in politics, first as attorney-general, then Lord Reading and Viceroy of India. While Godfrey married a Catholic (French-born opera singer Lea Constance Perelli) and embraced the traditions of her faith, the family's Jewishness played an important role in their—and Marconi's—destinies.

Isaacs had a lean, patrician air about him and the resemblance with his famous older brother Rufus was striking. One commentator wrote: "He had the same dark eyes, beautiful, large, expressive, very vigilant, with a suggestion of ardent fire, and yet soft and pleasant. He had a very sweet smile, though the expression of the face was often somber.... His features were regular, with nose well shaped but not large, the mouth thin-lipped and very firm; the cheek-bones were rather protuberant and the cheeks somewhat hollow."[7] A full-page colour drawing of Godfrey in a 1915 newspaper supplement portrayed him looking young, studious, and businesslike in morning coat, pinstriped woollen trousers, black leather shoes with grey spats, pearl-grey vest and watch chain, collar, charcoal tie, and wavy hair.[8] Energetic and enterprising, a man of great charm and a tireless worker, he had travelled widely, spoke several European languages, and had a generally cosmopolitan worldview—like Marconi. He had grown up working in the various departments of his father's substantial business and had accumulated an interesting corporate curriculum vitae.

Marconi personally recruited Isaacs after meeting him in 1909 through his brother-in-law, Donough O'Brien, a London City stockbroker with whom Isaacs had been associated in business.[9] Marconi took an immediate liking to Isaacs and, from a business standpoint, his hiring turned out to be the most astute and important move Marconi made since launching the company. Isaacs's handling of the business side of things freed Marconi to concentrate on research and technical experimentation and, eventually, to develop the other (political and diplomatic) interests that came to take more and more time in the later stages of his career.

Isaacs took over as sole managing director in August 1910, seven months after joining the company. On August 3, 1910, Marconi returned to his preferred role as technical director and scientific advisor. In this capacity he was to enjoy sole and full authority over the company's technical staff, with power to engage, dismiss, fix duties and remuneration, and move them about from station to station as necessary. The terms of his contract also specified that he was free to devote as much time as he liked to other activities and hold other appointments.[10]

Marconi's salary was now £2,500 a year ($200,000 in today's money). Isaacs would be paid double that plus a 5 percent share of the company's profits. Only days before the change in management, the company declared a 7 percent dividend, its very first—a sign to the business and financial community that things were going to be different.[11] As Marconi withdrew from day-to-day business involvement, Isaacs imposed his presence in a very short time, presenting new initiatives, launching patent suits, recommending new installations, seeking new capital...but Marconi was never far from the centre of the company's affairs. He remained chairman of the board, a position he assumed in February 1909 and would keep until 1927.[12]

Having freed himself from the responsibility for the day-to-day management of the company, Marconi might have taken the opportunity to spend some time with his family—he had again missed the birth of a child, his son Giulio, on May 21, 1910, at Villa Griffone, where he had installed Beatrice and Degna while he made another trip to America. But he was too eager to return to his research and experimental work, which he had not been able to focus on for the past two years. He now had his own special staff, and he was beginning to develop new collaborators who would leave their own marks on the history of wireless, men like Charles S. Franklin and H.J. Round.[13] With the expansion of the company's ambitions under the new leadership of Isaacs, a more conventional research department was put in place as well.[14]

One of Marconi's first efforts now that he was free to focus on his research was to renew his experiments with long-distance communication. In September 1910, he and Round sailed from Genoa to Buenos Aires on the *Principessa Mafalda*, Italy's largest and most luxurious passenger ship. The long-planned connection between Coltano and Buenos Aires was about to be established, and Marconi now envisaged a direct link between Argentina and Ireland as well.[15] En route, they tested reception from Clifden and Glace Bay, flying kites as high as six thousand feet and receiving signals from Clifden at four thousand miles during day and 6,735 miles at night. The experiments showed that the difficulties of long-distance communication in sunlight could be overcome, and confirmed Marconi's belief in the feasibility of a global network. Before leaving Buenos Aires in October, they arranged for a high-power station to be built at Punta del Este, in neighbouring Uruguay. Marconi's visit to Argentina in 1910, when he managed to communicate with Ireland, is considered the foundational event of Argentine radio broadcasting; once again, he achieved the longest distance communication to that time, using a rudimentary kite-mounted antenna and earphones as he had nine years earlier in Newfoundland, only this time the distance covered was some sixty-eight hundred miles.[16]

❀   ❀   ❀

Marconi now brought his family back from Italy and rented an eighteenth-century Georgian brick mansion in Richmond Park, Surrey, on the out-skirts of London. Officially known as Trumpeters' House (for the statues of two trumpeting figures that once adorned the property gates), everyone called it the "Old Palace." When Prince Klemens von Metternich was living there during his 1848–49 exile, Disraeli visited him and wrote to his sister: "I have been to see Metternich. He lives on Richmond Green in the most charming house in the world."[17]

The Marconis were soon fixtures of the London society pages, enter-taining guests such as the King of Portugal for lunch or lawn tennis.[18] Beatrice's two youngest sisters were just coming out and Beatrice was happy to be their hostess. At first Marconi tried to keep up with the soci-ety life, but he soon tired of it and withdrew. "This meant that I rarely saw him," Beatrice remembered, "as I returned late from dances etc etc and he left early for his office work. This divided us, he became nervous, irri-table, irrationally jealous...."[19] He was also still frequently away, spending more time in Italy now that he had shipped his family back to England, or at one or another of his stations in Ireland or Canada. Yet in his letters, Marconi scolded Beatrice, gently but firmly, for not writing to him more often.[20]

Degna later wrote that "the tensions between my mother and father [now] became truly serious"; they were in love but had grave troubles. A rare surviving letter from Beatrice, however, depicts a tender relationship; they were much more lovers than the story usually suggests, if also possessed of two strong wills, wanting both much from life and from each other. Returning from a trip to Clifden with her husband, she addresses him as "my darling" and describes arriving to a domestic scene of mild turmoil about guest lists and children's maladies. It was Giulio's first birthday (hence, May 21, 1911), "bless his little heart and we forgot all about it. Degna asked at once, *e dove è Babbo* [and where is Daddy], and kept on asking, could she go and have a look in your room to see if you weren't really there after all. My darling I miss you and long for you to be with me."[21]

An interesting sidebar to Marconi's story from this period illustrates both the complexity of his personal life and the enigmatic air that he tended to cultivate around it. In the summer of 1911, Marconi became entangled in a heated emotional situation involving a couple, Charles and Florence Clover, who had invited him to be the godfather of their six-month-old daughter, Betty. The Marconi Archives in Oxford house five letters written by Marconi to the Clovers in July and August 1911 about some unspecified but "very great trouble" they were having, and offering his assistance.

Charles Matthew Clover, born in 1877, was from Wrexham, North Wales. Florence Eliza Bell was from Canada; she was born in 1887. Florence's father was George Bell, a Toronto lawyer (and relation of Alexander Graham Bell); her mother, Marion Sproat, was the daughter of a Conservative member of the Dominion of Canada's first Parliament. The Clovers were married in York, Ontario (now part of Toronto), on October 30, 1908.[22] A son, Charles George, was born in 1909, and a daughter, Betty, was born in January 1911.

The story was a typical Edwardian saga. On July 27, 1911, as Marconi was preparing to leave London for Clifden, Florence Clover contacted him with some disturbing news. He wrote to her a few days later from Connemara. "Dear Mrs Clover, I am extremely distressed at receiving your letter and so awfully sorry to know of your very great trouble. I shall certainly help you in the manner you ask me as soon as I get back to London in a day or two.... I am sure that if Charley has really forgiven you so much, you will not find him very hard after all. Especially after some time... I must say I consider Charley has so far been most good to you. Don't you really think that the person who has led you astray and has been responsible with you for all the trouble should do something to help matters? But in any case both you and Charley can rely on my doing my best to help you both.... Yours truly, Guglielmo Marconi."[23]

At the same time, he also wrote to Charles: "My dear Clover... I think that you are wonderfully good and kind, but at the same time I quite agree with you that she hardly realizes things. I shall have of course to consider my position as godfather to Betty, but after all it's not the poor girl's fault. She is quite innocent and pure. I am afraid I spoke rather strongly to Florence (I mention her Christian name as the knowledge of one's pain and misfortunes tends to make one less formal) but think she is quite repentant. I am really glad you have decided to give her another chance.... Yours very truly, Marconi."[24]

Back in London, he wrote again to Florence: "My dear Mrs. Clover, I am glad I can help you in the way you have asked me.... I hope to see Charley before I sail.... He is awfully upset poor fellow but I think inclined to be kind to you. Don't oppose his wishes too strongly for after all it's a very great thing that he should be ready to forgive you.... You are both going to make a new start and I wish you well from the bottom of my heart. I shall write you again before I sail. Yours very truly, G Marconi." Finally, he wrote one more time: "Dear Mrs Clover, This is just to say *au revoir* and to wish you every good and kind wish.... I do believe that Charley will forgive you, as I have forgiven you for Betty's sake. If I can be of any use to you ask me.... Yours very sincerely, G Marconi."[25]

This is clearly novelist's fodder and, indeed, Vita Sackville-West had just such a subplot in her 1930 novel *The Edwardians*.[26] After Lord Roehampton finds out about his wife Sylvia's affair with the young Sebastian, and she objects to his proposal that they move to the country, he tells her she has no appreciation of his generosity. "No other man," he says, "would have given you a second chance. Another man would have turned you straight out of the house." She in turn accuses him of being a Victorian husband: "You are making such an absurd fuss.... Don't you know how people live?...why behave as though we were living in eighteen-fifty?" But in Roehampton's world, manners, pretense, and facade were everything.

Marconi was advising Florence not to behave like Sylvia, to be grateful for her husband's magnanimity and submit to his conditions. As happened to him so often, Marconi found himself caught in a crisis of passage between three eras, the Victorian, the Edwardian, and the modern. And as was often the case, his own role in the drama is rather obscure. As far as the archival record is concerned (and there we have only his side of the story), the matter lasted barely a week and the story is mundane; a friend helping a couple get past a difficult patch in their marriage. But one can imagine many alternative narratives. For example: What if Marconi himself were "the person who has led you astray" and he was now suggesting that Florence allow him to take his responsibility and "do something to help matters"? The formality of Marconi's tone of address to Florence Clover recalls the masked intimacy of his early letters to Josephine Holman a decade earlier. His eagerness to help is in character (he was often helpful to relatives, friends, and people he liked[27]), and so is his discretion—there is no sign of the Clovers anywhere else in the Marconi record, except for a handful of cryptic entries in his diaries.[28]

Every life has its store of secrets and mysteries, which is perhaps why people get so exercised at the thought of some government agency having access to their phone records or hard drives. Marconi's diaries hold clues to questions unasked and unanswerable, often hinting at relations and interests that then vanish without a trace. There is a backstory to Marconi's elusive life that we can only begin to glimpse. He loved recounting and reinventing it, but he also took great care to keep parts of it shrouded in obscurity. He would record meetings and make notes to himself in small leather-bound diaries, but he often used a series of indecipherable codes meaningful only to himself; a word, a name, a single letter, a number or an X. The diaries are inscrutable, strewn with references to people who turn up nowhere else in any of the accounts of his life. One of these ephemeral figures was Betty—whose full name was Marion Elizabeth Jessie Marconi Clover. Marconi would occasionally make a note regarding his children—Degna, Giulio, Gioia... and, once in the same breath, Betty Clover.[29] On the surface, what could be more reasonable than taking one's fourteen-year-old goddaughter to Cartier, London's finest jewellery dealer (as he recorded doing on January 24, 1925[30]), or so it would appear.[31]

Marconi was now travelling constantly between London, Clifden, and Glace Bay, trying to incorporate his latest research into the company's crucial transatlantic commercial service. The Clifden–Glace Bay service had used state-of-the-art long-distance wireless technology when it was inaugurated in 1907. Technically, it was a "simplex" system, that is, transmission and reception could not be done simultaneously. In 1911, Marconi and Round perfected a "duplex" system whereby messages could travel bi-directionally, in opposite directions, along the same wavelength (a feature already developed in wired telegraphy). If Marconi could integrate "duplex" technology into his transatlantic messaging, it would be that much more attractive to customers and make Marconi's more competitive with the cable companies. Rather than try to convert Clifden and Glace Bay, with their huge sunk investments, Marconi started looking for sites close by where smaller receiving stations could be set up.

On the Irish side, Marconi began scouting locations close to Clifden in April 1911. After visiting various sites in the area, he decided that the likeliest spot was at Letterfrack, eight miles northeast of Clifden on the Kylemore property of his dissolute buddy the Duke of Manchester. On April 11, 1911, Marconi climbed Diamond Mountain along with engineers Charles

Franklin and William Entwistle, and George Kemp, lunched at the Letterfrack Hotel, and decided that this was the spot.[32] The new experimental station was located at a school run by the Christian Brothers, "very near Kylemore Castle on the estate of the Duke of Manchester," he wrote to Beatrice. "The Duke's agent has been most kind and given us all possible help such as carts and poles and donkeys to carry the stuff up the mountain."[33] The site had many advantages, beside the fact that he was friends with the landlord; its location allowed it to receive signals from the transmitting station at Clifden, the risk of interference was minimal, and it had an unobstructed view of the sea. It could also be easily connected to Derrigimlagh by land line. Once again, Marconi was contributing to the Irish local economy; about eighty men and twenty-five donkeys would be employed in the construction of Letterfrack.[34]

Marconi set Kemp to work at Letterfrack immediately, using tempo-rary aerials to do tests with Clifden and Glace Bay, while he remained in Clifden, checking on progress every few days. Kemp proved adept as usual at cobbling together a working system on difficult terrain. But there was also a delicate social situation to manage: in order to complete the installa-tion Kemp needed to erect poles and a wire across land belonging to the former owners of Kylemore, now occupied by tenant farmers. Kemp proved to be a congenial and able negotiator; he had no trouble getting the permis-sion—or so he thought.

Marconi and Beatrice arrived at Letterfrack on May 17, 1911, then vis-ited Kylemore the next day. The installations were proceeding well but they kept running into surprises with the complicated Irish land arrangements. Marconi was interested in finding a possible alternative site and Kemp went to investigate at the nearby village of Leenane; on May 21, he told his diary: "I went by motor car to the Quay at the mouth of Killary Bay and discov-ered that the mountain Milray belongs to Lord Sligo but is rented by Mr. Houston for sheep grazing etc."[35] All these subtleties would have to be dealt with, delicately.

On May 22, 1911, Marconi did an important demonstration for Royal Navy Captain C.G. Crawley, of the naval torpedo school. The duplex sys-tem worked; Marconi was able to send full power from Clifden at the same time as receiving from Glace Bay at Letterfrack. The scope of the two-way signalling was mind-boggling: Clifden was using a 15,750-foot wave; Glace Bay's was twenty-one thousand feet. Marconi said these were only preliminary experiments and that "very soon simultaneous transmission and reception on waves much closer together than those shown would be practicable," Crawley reported. "It was apparent that the aerials and temporary station at Letterfrack had only very recently been rigged up and the results of these preliminary tests would appear to fully justify Mr Marconi's optimism on the success of the system," he continued, a strong endorsement.[36]

The system was now working but Marconi was looking for perfection. Before locking himself into a permanent installation at Letterfrack he wanted to have a look at the site Kemp had scoped out at Leenane. He brought in his most seasoned research engineers, Entwistle, Franklin, and Round, and on May 26, they motored to the head of Killary Harbour, where they climbed a mountain known as the Devil's Mother. They took a bearing and found that the line of signalling from Glace Bay would just touch the north side of the harbour. This might be more reliable. "I returned to Letterfrack and discharged the men for a time until something happened to the lines or I received definite instructions from Dr. G. Marconi," Kemp wrote.[37]

Marconi, meanwhile, returned to London, where he spent most of the following month. On June 2, he gave a lecture at the Royal Institution and then took part, with Beatrice, in a frenetic social schedule tied to the coronation of King George V on June 22, 1911. While he was away, work continued at Letterfrack. Kemp and Franklin were satisfied with the progress; on July 4, 1911, Kemp proclaimed it a "great day," having accomplished the first practical duplexing. On July 15, Marconi returned and once again visited Leenane along with Franklin and Kemp. This time, they also met with the local tenants whose agreement they would need in order to set up a station there. When they returned to Letterfrack, "ideas were discussed regarding another station for duplexing." The following day, they visited the Manchesters at Kylemore before Marconi returned to Clifden. The duke seemed thrilled to have this important work going on on his estate; Kemp was basically given the run of the place and told he could use any of the outbuildings to store the station's gear.[38] Pleased with how things were going, Marconi returned to London for his company's annual general meeting.[39]

His biggest concern was still the rights to use the property. On July 21, the Congested Districts Board for Ireland wrote to Marconi from Dublin, asking him to provide an ordnance map showing the location of his proposed station, and offering him a meeting to discuss the erection of poles on the property. Marconi supplied the map, and the board wrote again two weeks later informing him what he already knew: the land was partly occupied by tenants and he needed to make arrangements with them. The board also confirmed Kemp's discovery that part of the Leenane land was on the estate of the Marquis of Sligo.[40]†

Marconi left London to return to Clifden on July 27 (carrying with him the problem dropped in his lap by Florence Clover). On July 29 Marconi and Kemp climbed the small mountain of Currywongaun, where the Letterfrack station would be placed, to take bearings to Clifden. They were receiving strong signals from Glace Bay. On July 31, Kemp brought Manchester to the spot that Marconi had marked for his benefit. The following day Marconi and Kemp lunched at Kylemore with the duke and his American father-in-law, Eugene Zimmerman, who was visiting. Marconi then left for Dublin while Kemp stayed on to continue making arrangements.

Marconi and Beatrice sailed for Canada on August 5, 1911. While he was away, the land issues concerning the Irish stations came to a head. Manchester had several meetings in London with company officials, including Kemp, Isaacs, and head of research Andrew Gray. By mid-September they had given up the idea of Leenane and settled definitely on Letterfrack. Kemp kept up a stream of letters to Marconi, detailing his negotiations with Manchester and the tenants—only two of them were giving him trouble, he reported on September 25, but it was serious. The duke had been advised to settle the matter by purchasing the bit of tenanted land, but the tenants had told Kemp that "they did not want us there and that it would be much better for me to stop work." The other tenants, too, expected compensation. Kemp tried to keep things quiet by staying clear of cultivated land, but under the Irish land laws, tenants could not be turned out as long as they paid their rent. On September 26, 1911, Kemp wired Isaacs to say that the tenants wanted them to stop working on their land.[41]

In Canada, meanwhile, Marconi was making arrangements for the new duplex receiving station equivalent to Letterfrack, which was to be built at Louisbourg, site of an eighteenth-century French fortress about twenty-five miles from Glace Bay. The station was located in the old town, inside the fort. Louisbourg was much less problematic than Letterfrack. There were challenges, to be sure; in February 1913, the lines connecting Glace Bay and Louisbourg collapsed under the weight of a "silver thaw" (ice

---

† The line dividing counties Galway and Mayo runs across the ridge of Devil's Mother; the Marquis of Sligo had a house at Westport, bordering Connemara but in county Mayo (Shane Joyce, "Marconi in Connemara," correspondence with author, 2013).

storm).[42] One curious visitor to Louisbourg was Alexander Graham Bell, whose summer estate Beinn Bhreagh was at Baddeck, about two hours away.[43] Marconi himself only made one further visit to Cape Breton, but Louisbourg retains a place in company lore for having received the first east-west transatlantic voice transmission, from a Marconi station at Ballybunion, Ireland, in March 1919.

Amid the mundane challenges of organizing his new Irish and Canadian receiving stations, and the ongoing drama in his personal life, Marconi suddenly found himself with a new preoccupation. On September 29, 1911, the Kingdom of Italy declared war on Ottoman Turkey over claims regarding their conflicting interests in North Africa. Marconi dropped everything he was doing to offer his services to the Italian cause.

19

Signals of War

By 1911, Italy was looking for overseas colonies with which to nurture its growing imperialist ambitions. Colonial possessions were seen as offering the thousands of Italians with no foreseeable economic future at home a potential alternative to foreign emigration. Italy was part of the Triple Alliance (Italy, Austria-Hungary, and Germany) that had come together in 1882, but its relations with Austria in particular were often tense and it was moving closer to the rival Triple Entente (Britain, France, and Russia) through a series of both secret and more transparent deals. Given this ambivalent position in the murky environment of European politics, the likeliest place for Italy to advance its colonial efforts was in parts of the Ottoman Empire.

On September 28, 1911, using the pretext of perceived threats to Italian nationals by alleged Muslim extremists in the Ottoman North African provinces of Tripoli and Cyrenaica—known internationally as Libya—Italy issued an ultimatum demanding that it be allowed to occupy the provinces in order to protect its citizens. The ultimatum was rebuffed and the Kingdom of Italy declared war on the Ottoman Empire the following day. The conflict upset the international balance of power, revealed Ottoman Turkey's growing weakness (already evident in a series of conflicts over the previous fifty or sixty years), and unleashed nationalist sentiment in Italy that had been simmering since unification in 1871, when Rome became the country's capital. A small cohort of Italian industrialists saw the war as an opportunity to extend Italian economic influence. In many regards, the Italo-Turkish War was a precursor not only to the First World War, but to the second as well.

The Italian press had campaigned for months in favour of a military invasion, and Prime Minister Giovanni Giolitti, although temperamentally and politically more disposed to diplomacy than the use of arms, had received favourable signs from other European powers regarding their eventual reaction to an invasion.* Italy would be a welcome proxy for France and England in the growing challenge to the power of the declining Ottoman Empire and its principal supporter, Germany. Italy's opposition Socialist Party was itself divided and an ineffective opponent of intervention—although one of its leading orators, Benito Mussolini, took a prominent anti-war position. Most patriotic Italians, like Guglielmo Marconi, were in favour.

Italy sent its fleet to Libya and quickly occupied the coastal cities of Tripoli, Derna, and Benghazi, meeting, however, with unexpected resistance in the hinterland. The Ottomans had few troops in North Africa, but they were enough to mobilize and support an Arab resistance movement, and Italian forces were soon confined to a series of coastal urban enclaves and the sea, which Italy controlled. Turkish military officers—including Mustafa Kemal, one of the leaders of the Young Turk modernization movement and soon to be known to the world as Atatürk, "father of the Turks"—snuck into Libya to help organize guerrilla activity by groups like the Sanusi brotherhood, a militant Sufi order. Kemal left Constantinople (now Istanbul) on a Russian ship bound for Alexandria, disguised as a journalist, on October 15, 1911, arriving in Libya around the same time as Marconi; he later said that defending Libya was a hopeless task but a patriotic duty and a political necessity. The Turkish–Arab resistance in Libya also led some Western observers to mistake a coincidence of self-interest with religious solidarity.[1]

The Italians immediately started building wireless stations along the North African coast (the first, establishing a link between Tripoli and Sicily, was launched within days of the start of the occupation). The Italian minister of the marine asked Marconi to go to Libya to inspect the new stations, test communications with Coltano, and carry out experiments in the desert.[2] On November 5, 1911, Italy declared its suzerainty over Libya, and on November 20, the *Times* reported—curiously, in its "Court Circular" column—that Marconi was preparing to leave for North Africa.[3] He was already in Coltano, where on November 19 he demonstrated to the King of Italy communication with Clifden, Glace Bay, and Massawa (in Italian Eritrea, East Africa, more than twenty-two hundred miles away).[4] He then sailed from the Italian naval base at Taranto (Puglia) on November 30,

---

* Giolitti was prime minister five times between 1892 and 1921 (almost uninterruptedly between 1901 and 1914) thanks to a talent for building centrist coalitions out of the left and right factions in the mainstream of Italian politics. A liberal (in the American sense), his governments were responsible for various progressive social reforms and public intervention. Less concerned about territorial ambition, he was virulently contested by Italian nationalists, and opposed Italy's entry into the First World War.

leaving at such short notice that he barely had time to notify his company colleagues.

It was the third time wireless was being used in war, after the Anglo-Boer War in South Africa and the Russo-Japanese War in Far East Asia. Communication technology was becoming a crucial element in warfare, and the Italian navy placed an eleven-thousand-ton first-class armoured cruiser, the R.N. *Pisa*, at Marconi's disposal. He would also focus upon the elimination of interference and jamming problems that were still an issue with short-distance wireless.

It was meant to be a short trip. Marconi arrived at Tobruk, near the Egyptian border, on December 2, 1911, and he immediately set to work testing and trying to improve conditions for short-distance overland field communications. The Italians had about a dozen field stations in operation, as well as one or two "knapsack" stations that gave good results up to about twelve miles.[5] Marconi was able to communicate between field stations at short distances without using masts or poles, with all the instruments and "aerial" wires placed directly on the surface of the sand. This development overcame the disadvantage of revealing the stations' locations through use of kites or balloons and excited Marconi because of its significance for one of his major research interests: mobility. At Tobruk they could receive good signals at all times from Coltano (twelve hundred miles) as well as from Clifden (two thousand miles). This had important implications: it convinced Marconi that the insulating qualities of dry ground allowed for great distances to be covered.

The newly opened station at Coltano was the most powerful in the world, consistently communicating with the Italian installation at Massawa, a distance of 2,238 miles.[6] But the North African experiments also had a humbler, less exciting, and more functional outcome for the critical military purpose of short-distance communication between units in the field. In later years, Marconi frequently referred to his three-week North African stint (the *Times* called it "a tour of inspection"[7]) as a high honour, one during which he was able to answer the call of the Italian war and naval ministries.[8] Documents show, however, that as late as May 1911 the Marconi company was in active competition with Telefunken for a contract to supply portable field stations to the Ottoman army. In three weeks of tests in a mountainous area near Constantinople, the Marconi system was found to be quicker to set up, its signals stronger and clearer, and more reliable in every respect.[9]

Marconi had been trying to do business with Turkey since 1902, when the Ottoman sultan first demonstrated interest in developing wireless technology.[10] Luigi Solari—now troubleshooting for the company in sales negotiations in all of continental Europe—had approached the Ottoman government as early as 1908 with no success; a detailed proposal to the minister of posts was received dismissively, and while the technical advisors of the Ministry of the Marine were in favour of the Marconi system, the Ottoman Ministry of War felt the German Telefunken system was superior.[11]

According to Marconi's ghostwritten 1919 autobiography, the 1911 test came about after Italian diplomats in Constantinople informed Solari that the Germans were pressuring Turkey to adopt the Telefunken system. Solari, then in Lisbon, sped to the Ottoman capital and suggested a practical test of three systems: Marconi's, Telefunken, and the Danish "Poulsen arc" system (there is no mention of the Poulsen system in the official report of the Turkish test). The result, according to Marconi, was "a lively discussion" between the Ottomans' Germanophile army officers and naval officers sympathetic to Marconi, in which the press also joined in on the side of the Germans. However, the Turkish government prevailed on Berlin to accept the challenge.[12]

Marconi gave his version of what happened next in testimony to a 1913 British parliamentary committee. Earlier that month, he said, "a competition" had been conducted by the Turkish army over a ten-day period during which the Telefunken, Poulsen, and Marconi systems were tested to see which one the army should adopt. "The Telefunken system failed to get a complete message through until the last two or three days of the trial, and then only at night time; the Marconi system maintained good communication throughout the whole period; the Poulsen apparatus failed entirely; their arc blew out on the average once a minute." As a result, Marconi reported, his system was adopted. Marconi's 1919 ghostwriter states that the Turkish navy placed an order for ten ship stations and a high-power station in the capital. But the Ottomans were strongly beholden to their German advisors, and as Marconi's own letters from the war zone attest, the Ottomans were using the Telefunken system in North Africa.[13]

It was typical of Marconi to have two versions of a story, to use for different purposes, undoubtedly both strictly accurate but each glossing over a crucial nuance or interpretation. The Turkish navy "decided to" adopt Marconi's system because it was superior to the German, but did they? And then, of course, there was the moral question: If Marconi's Italians were the "right" side in this war, why were he and his company ready to supply the enemy with war technology? The answer is that Marconi was playing on both sides.

Marconi completed his tests on December 6, 1911, but the ship couldn't leave Tobruk because of Turkish activity in the area—ten thousand Turkish soldiers were encamped just five miles away. When Italian reinforcements of cavalry and artillery arrived a few days later, the *Pisa* could finally sail. They arrived in Derna on the twelfth, Benghazi on the thirteenth, and Tripoli on the morning of the fifteenth. All of these stations were in wireless communication with each other,[14] and as a result Marconi seemed to be always just a step ahead of the fighting. The day the *Pisa* left Benghazi, Turkish forces launched a massive attack on the city; Tobruk was recaptured on December 22 by troops led by Kemal.

Marconi's letters from the Libyan front—to Isaacs, Beatrice, and Annie—are among the most vivid and descriptive he ever wrote. Unlike his usually brief notes, telegraphic in style, noting personalities he had met, offering encapsulated renditions of his work in progress along with weather and health reports, the letters from North Africa express the thrill of sheltered danger and contact with the exotic other. They also provide a memorable portrait of a desert war—not quite T.E. Lawrence's *Seven Pillars of Wisdom* but engrossing nonetheless. "I am given a horse and escort when I land and have witnessed one or two small scraps at the outposts. . . . The arrangements for the troops are excellent and the health and spirit of the men could not be better. There are now over 100,000 men landed at or in the vicinity of Tripoli, Benghazi and Derna. They complain if anything of been sent too many provisions."[15]

Marconi's enthusiastic orientalism emerges. "The Turks and Arabs seem to have done little to cultivate the place and if the system of artesian wells works out successfully on a large scale large tracks of land will be turned to profitable account." He was particularly fascinated at the accuracy of modern guns. "The most interesting sight I have seen during the war was the bombardment by our and other warships of a Turkish posting during the night. I never saw anything more like what could be described." Again, as in the South African and Russo-Japanese wars, technology pitted Marconi against the Germans. "I visited yesterday the remains of the Telefunken station at Derna which was destroyed by this ship. . . . A few days later . . . the admiral called . . . by wireless and informed the operators, in <u>French</u> that if they wished to save their skins they had better get out of the way as the station would be shelled in 10 minutes. They were seen to close up without a moment's hesitation." In a letter to Beatrice, Marconi wrote:

> I shall be a bit more busy at Tripoli with the new arrangement which I'm making them use for the field stations. As you know I tried the arrangement with the aerial wire on the ground in the desert and it seems to work awfully well. It's such a great advantage to be able to do without the poles and masts as these are very difficult to carry about, and show one's position to the enemy.
>
> It seems that about ten thousand Turks and Arabs are near here with 18 guns. A few of them come occasionally and fire at our outposts, at night, but there has not yet been any real battle. Yesterday a few Arabs appeared on horseback on the hill close to which we are anchored, we immediately fired at them with the ships guns. Before they bolted away they fired a few shots at us with Mauser rifles but were much too far off to hurt us.
>
> This morning about sixty more came near, and managed to get in between us and the outposts, but were soon chased off.
>
> I'm in the best of health and spirits and see quite a lot of Pippo [Filippo] Camperio [older brother of his childhood buddy Giulio]. There is only one woman in the whole place—she is an old Arab. We have a splendid hospital ship here beautifully equipped, but with no nurses or *signore della croce rossa* on board. They say they had to send them away for as they had no wounded to attend to and nothing to do, they flirted outrageously with the officers.
>
> We have six aeroplanes, and I'm learning heaps about them. I can go anywhere and do anything I like as if I were a General or an Admiral. It would be such fun if you were here too. The officers are such good fellows and the soldiers and sailors as keen as mustard. They all work together in a wonderful way. They seem never to tire. They work all day loading guns and supplies, and fish all night. As I told you we get splendid fresh soles here fished at the sides of the ship.
>
> Nothing much is said of this place in the papers, as no reporters have been allowed to come. The torpedo boats are continually stopping ships going in and out of Alexandria in Egypt which is quite close to here. The Egyptians seem to be doing all they can to help the Turks. I wonder what the English Admiralty think at having Alexandria blocked by Italy.
>
> It is a fine sight to see the soldiers and sailors making their preparations, and the enthusiasm and eagerness of all to get to the front every time an alarm is sounded.[16]

It was the most newsy letter he had yet written to his wife—less whiny and self-preoccupied with personal complaints—and a rare chronicle of the Italo-Turkish War. For Marconi, the war was short and exhilarating. He got to do patriotic duty, reposition himself with respect to his German rivals, and engage a new hobby, photography. He also got to make a number of trips in army airplanes, accompanying scouting parties.[17] He even had time to write a long, reassuring letter to his mother and send a postcard to his loyal assistant, George Kemp, in Ireland.[18]

The Italian invasion of Libya was the first step in the dismantling of the Ottoman Empire (which would be at war almost uninterruptedly for the next twelve years). The Turkish command was advised by German mentors; the Italians took the pulse of the neutral British and the French. The boundaries of modern warfare were also being set. Military applications of twentieth-century technologies were tested under war conditions for the first time. This was, for example, the first war in which airplanes and dirigibles were used; Italian pilots made the first ever wartime reconnaissance missions behind enemy lines, and dropped the first aerial bombs. Aeronautics and wireless were the two areas that observers had their eyes on, the *Times* cautioning that the Italians "are much further advanced than we are in wireless work."[19] The success of the Italian wireless campaign convinced both Britain and Germany to invest heavily in developing the new technologies.

By mid-1912 the Libyan war had come to a stalemate. Italy controlled the sea but its ground forces were besieged in seven coastal enclaves, the largest at Tripoli. The war spread to other parts of the Mediterranean, to Lebanon and the twelve Turkish Aegean islands that came to be known as the Dodecanese. In October 1912 the four nations of the Balkan League (Greece, Serbia, Bulgaria, and Montenegro) opened another front in Ottoman Europe, and Turkey sued for peace with Italy.[20]

This was all good news for the Marconi company. Shortly after the start of the Balkan war, they received a telegram in Rome from the Serbian prime minister asking Solari to come at once to Belgrade; when he arrived four days later he was taken immediately to the front at Kumanovo on horseback to inspect the telegraph station. "Beside the instrument there stood a cup of coffee still filled and an open French novel showing that the formidable battle of Kumanovo which had been raging for a couple of days at a short distance had failed to disturb the placidity of the Turkish telegraphist," according to Marconi's ghostwritten memoir.[21] Solari helped the Serbs build a network of mobile stations mounted on carts; it was so successful that when the war ended Belgrade ordered a powerful Marconi station to communicate with Sofia, Rome, Salonica, and Athens. Before it could be built, however, the First World War intervened. As it had done in previous conflicts, the Marconi company provided equipment to both sides in the Balkan war of 1912.[22]

On October 18, 1912, the Treaty of Ouchy (also known as the First Treaty of Lausanne) ended the Italo-Turkish conflict. The immediate outcome was the withdrawal of Turkish military forces from Libya, with the maintenance of civil authorities loyal to the Sultan. As the only regular military force, Italy was then able to extend its occupation while Arab resistance continued, with the ongoing encouragement of the Ottomans.

As wars go, the Italo-Turkish War was a minor conflict. It seemed to be an easy victory for Italy, highlighting the weakened state of the Ottomans and sparking the Balkan nationalism that became even more significant in the run-up to the First World War. But the war was financially ruinous for Italy, bringing disillusionment with Giolitti that would create the instability of the next ten years and foster the radicalization of Italian politics and the rise of Mussolini. With the outbreak of the First World War, Libya became a relatively marginal arena. Italian colonization continued, but by the 1930s the population had been cut in half by emigration, famine, and war. In 1950 the population of Libya was the same as it had been in 1911, 1.5 million people.

Swept up in his enthusiasm for the war, Marconi was upset by press reports of "alleged cruelties of Italian troops" that he believed were "entirely and absolutely false.... I absolutely believe that the charges of systematic cruelty brought up against the Italian soldiers are baseless," he wrote, with righteous indignation.[23] Still, it was a short but a nasty war. The extent of the brutality has never been fully established, but at least one commentator had a different opinion. "Italy has 'won' the war... [a] typical colonial war.... A perfected, civilized blood bath, the massacre of Arabs with the help of the 'latest' weapons," wrote V.I. Lenin in *Pravda*.[24]

❀   ❀   ❀

Back in England at the start of 1912, Marconi moved his family again. On February 26, 1912, a tiny item in the London *Times* reported that Marconi had leased and moved into a new home, Eaglehurst, "one of the most beautiful places on the Hampshire coast."[25] The Marconis rented Eaglehurst after his friend Pippo Camperio suggested their marital tensions might be eased if they had a place in the country. Beatrice found the "romantic, if impractical, house" (according to Degna, who spent part of her childhood there)[26] at Fawley, just south of Southampton on the shore of the Solent, not far from Bournemouth. It was an hour's commute from the centre of London, a comfortable ride in Marconi's new car, a chauffeur-driven Rolls-Royce. These were his old stomping grounds, a short drive away from the site of his very first experimental station and almost in sight of the Isle of Wight.

As Degna described it, the house was long and wide, on one level for the most part but with two-storey octagonal wings at either end, crowned "with the crenelated tops that so obsessed a Gothic-minded generation of builders."[27] A long drive bordered with oleanders and rhododendrons was flanked on either side by the type of lawn it is impossible to imagine anywhere but the English countryside. Peacocks paraded on the grounds until Beatrice banned them out of superstition. According to the Hampshire County Council, Queen Victoria had considered making Eaglehurst her country home before choosing Osborne House, on the Isle of Wight, instead.

The eighteenth-century property's quirkiest and most historic feature was Luttrell's Tower, a six-storey turreted structure by the water's edge, an architectural anomaly so odd that it is known to this day locally as "the Folly."[28] The tower was built around 1780 for Temple Simon Luttrell, a member of Parliament said to have smuggling interests on the south coast of England. From the top of the tower one could watch the ships setting out from Southampton to cross the Atlantic and here, indeed, Degna and her mother would wave unsuspectingly to a steamer called the *Titanic* as it passed by on April 10, 1912.[29]

Although he was rarely there himself, Marconi was serious enough about Eaglehurst to obtain a Post Office licence for experimental wireless telegraphy and install a lab and small transmitter in the tower.[30] But the family never knew when to expect him; he would drive down from London when it suited him (and to keep Beatrice off-balance), often bringing along George Kemp or another staff member to continue working on whatever problem was concerning him at the moment. According to Degna, when they saw him at all, he was tired, intensely irritable, and increasingly alienated from Beatrice; in a word, miserable. To gloss over their unhappiness, the Marconis kept up a glittering social calendar with regular guests at the house, including mostly his cronies like Pippo Camperio, Italian ambassador Guglielmo Imperiali, and the Isaacs brothers, Godfrey and Rufus.

Degna described her parents' relationship at this time as stormy and riven with jealousies, especially on Marconi's side. Yet he was the one who carried on a series of apparently joyless and barely camouflaged affairs. In London, Marconi shared a pied-a-terre on Sloane Square with Bea's sister Moira Hervey-Bathurst, a stage designer who introduced him to the theatre people in her circle. As he was in London most of the time, when he wasn't out of the country altogether, he ended up spending more time with Moira's theatre set than with Beatrice. The arrangement got him mixed up in some indiscreet liaisons. Beatrice seemed more upset by the indiscretion than the infidelity.[31]

He left traces of his love life here and there; draft fragments of a letter to an unidentified lover in London, for example, handwritten in ink with pencilled corrections and additions, undated on stationery of his Clifden station: "*Ma très chère amie*—There was no post today so I couldn't write. I am sending this by the first possible one as promised. . . . " The affair was hardly secret; one time, when he turned up hours late, Beatrice "imagined I had been run over or had been with you—which in her eyes is worse— I have never seen her so upset—instinct perhaps—she could not [have] known [. . .] I did not deny it but told her nothing—but she knows." His feelings wavered between guilt and ambivalence. "She makes me feel very ashamed of myself," he wrote.[32]

Yet despite his philandering, Marconi's letters to Beatrice are also loving, and Beatrice's 1950s memoir portrays the nuances of the relationship. Marconi was not aloof; he was certainly insecure about losing her, trying to make her feel she would enjoy being with him, presenting a certain skewed image of his life and what he was going through. In short, he wanted her to love him a certain way. It is a pity that we don't have more of Beatrice's correspondence.

❦   ❦   ❦

With Isaacs now directing the company, Marconi still had some unfinished business with his oldest collaborator, his cousin Henry Jameson Davis, who wrote to him in December 1911, while Marconi was in North Africa. Davis started out congratulating Marconi on the company's recent financial success but quickly got to his point: he wanted to share in it. Davis had stepped in to help fill the breach left by the forced departure of Cuthbert Hall, resulting in the neglect of his own affairs. "I feel sure you will agree that my efforts on behalf of the Company during a very anxious time should meet with some recognition now that the corner has been turned." He left it there, wished Marconi and his family a happy Christmas and the company continued prosperity, then signed off with his traditional "Your affectionate cousin."[33]

Marconi took some time before he eventually brought the matter before the board—the board that Davis himself had founded. On August 6, 1912, Marconi wrote Davis to say he couldn't get the board to agree to compensate him further but would send him a personal cheque. He added that it was a pity they never met any more. Davis wrote back, stating that he didn't understand why the matter couldn't be settled by the company directly but not objecting to receiving a personal cheque. As to getting together, he said, "I should be pleased to meet you any evening we are both in town."[34]

Marconi promptly sent Davis £500—an absurdly small sum in consideration of Davis's contribution to Marconi's success over the previous fifteen years—which Davis replied that he could not accept "in settlement of any claim, moral or legal." He suggested five thousand shares and £2,000 to £3,000 cash, recalling that Flood-Page had received £3,000 in shares when he joined the company plus £2,000 a year as managing director, and Hall had been paid £1,500 a year and received ten thousand shares when he left. "I think my claim a very modest one, more particularly as I helped the Company with funds when it was required and did not ask for the promised remuneration until the Company was well able."[35]

Marconi replied on September 20, 1912, apparently rejecting Davis's arguments (Marconi's letter is lost). On September 27, Davis wrote one last time, recapitulating the history of the company and his role in six tight pages: "I am sorry to see the line you take and that your memory is so short. . . . I had a great deal of hard work to bring anyone to believe in it or take any interest in your discovery. You know how long it was in my Flat, which was turned into a show-room for the purpose of Wireless Telegraphy. . . . I became responsible, in the first instance, for everything, and in the event of nothing coming of it all my time and money would have been thrown away." He recalled how close Marconi came to being "landed by the Post Office," and how his investors didn't want Marconi to get £15,000 plus half the share capital of the company. "After the Company was formed, you will remember, the then Board was anxious to have the Invention turned over quickly at a profit, and resented any expense . . . and I had all the up-hill work to convince the Board that there was something in Wireless Telegraphy beyond the idea that it might be exploited in a small way for quick profit." He had also personally underwritten the company's initial £25,000 working capital, and paid from his own pocket for the foreign patents, he recalled.[36]

Company shares under Davis's management rose sevenfold hitting seven and a quarter. When he left the managing directorship he wanted to sell his shares and leave the company altogether but accepted the pressure of Marconi and the board to stay on and hold his shares. Marconi then promised a substantial present later on. He was in a position to do so, Davis wrote, with the money he had received on the formation, having no other expenses, and it was "practically your own business, you holding half the share capital." Now, despite his strong feelings, "I should be very sorry to let any claim of mine, however well founded, be a subject of law proceedings...I consequently accept the £500 in settlement." No longer Marconi's "affectionate cousin," he signed in a more formal business fashion, "Yours faithfully...."[37]

It was the final word on the matter. Marconi and Davis had nothing further to do with each other, and rarely met again, except for a social occasion or two.[38] Davis remained close to Marconi's family, however, always present in times of crisis when Marconi himself was absent.

❋   ❋   ❋

Marconi would have probably responded to Davis's letter under normal circumstances, but he had a good excuse for his silence. Two days before Davis wrote the letter Marconi suffered a major life event. Marconi and Beatrice travelled to Italy in September 1912, ostensibly for the inauguration of the high-power long-distance station at Coltano, but also on the recommendation of Pippo Camperio, who thought a holiday would help their marriage. They decided to drive. They met up with Solari in Pisa on September 23.[39] The following day the party visited Coltano, and Marconi demonstrated long-distance transmission to Glace Bay and Massawa for the king and queen, before joining them for lunch at their palace at San Rossore.

On Tuesday morning— September 25, 1912—Marconi and Beatrice, a driver, and a secretary set off on an eighty-mile drive to Genoa, where they were to meet Solari, who had gone ahead by train. It was a beautiful day, and despite the notoriously narrow mountain roads in the region, Marconi loved driving. He insisted on taking the wheel of his new fifty-horsepower built-for-speed Fiat, as he loved to do whenever he could, with Beatrice beside him and the others in the back.[40] As they passed the village of Borghetto di Vara, near La Spezia, in an area known as the Bracco ascent, the road curved sharply and the car slammed into an oncoming vehicle driven by an expatriate businessman, Antenore Beltrame, who was visiting with his family from Argentina, where he had lived for the past thirty years. It was 12:30 p.m.[41]

A Royal Navy ambulance arrived almost immediately. The secretary had suffered a dislocated shoulder. Beatrice and the idle driver were fine. But Marconi was badly hurt; his right eyeball was pierced by a splinter of broken glass and he was in insufferable pain. Beltrame, himself injured and his car totalled, told the press, "I would have preferred to die or to have both my legs cut off, rather than to be even the unblameable cause of the accident whereof the victim is Signor Marconi, whom we all fervently admire."[42] Marconi was taken to the military hospital in La Spezia, where he remained in great pain and at times unable to see at all, as doctors debated what to do. Finally, a Viennese specialist, Ernst Fuchs, was summoned to remove the eye after it was determined that the sight of Marconi's other, good eye was threatened.[43]

Beatrice stayed at Marconi's side, nursing him, throughout his ordeal. When the bandages came off ten days after the removal of the damaged eye, Marconi could see nothing. Practical as ever, he began planning a sightless life, at Villa Griffone, supervising research done by his assistants. But his optic nerve soon recovered and the sight in his left eye returned. He began wearing a rakish black eye patch while a master glazier from Murano worked on a glass eye for him. By the end of November, after picking up the new prosthesis in Venice, Marconi was back in London. Soon he was driving again.[44]

Solari wrote that after the accident, Marconi's youthful character changed. He seemed "physically and morally aged," sadder than ever, and work became his only escape.[45] However, although the glass eye became a nuisance that caused him problems for the rest of his life, he took to it remarkably well, using a stylish monocle instead of reading glasses. After trying doctors in London and Wiesbaden, he eventually found one he liked in Paris, and would visit him at every opportunity.[46] Degna wrote that it was years before she realized her father had an artificial eye. In photographs, it is often undetectable, but it sometimes gives Marconi a haunting, zombie-like look, as he appears to be staring blankly into space. Strangely, this was a look that had followed him from earlier times. One of the first published sketches of Marconi, in a December 1897 edition of the *Glasgow Evening Times*, shows him, eerily, missing his right eye.[47]

20

Wireless and Disaster

With the arrival of Godfrey Isaacs, Marconi's company embarked on a
new approach to establishing global supremacy, tying itself more
closely to British foreign and colonial policy, and promoting the idea of an
"Imperial wireless chain" of stations. One of Isaacs's first acts, just two
months after joining the company, was a bold proposal to the Colonial
Office, in March 1910, that they give Marconi's licences to link the British
Empire with a network of wireless stations, which the company would
erect, maintain, and operate entirely at its own expense. It is not entirely
clear whose idea this was initially, but the formulation of the plan in terms
the government could embrace was certainly Isaacs's.

The company had been thinking of such a plan for some time but had
never expressed it so succinctly.[1] "We do not ask for any subsidy," the company wrote in its proposal, making a case that pressed all the right patriotic
buttons. The Imperial wireless chain would comprise eighteen stations,
one thousand to two thousand miles apart, that would circle the world.
Marconi had already shown that his system could easily cover the distance.
The beauty was that each station would be situated on British territory. The
stations would be worked commercially in peace time and could be handed
over to the government in times of war (as was already the case with stations
owned by the cable companies). The proposal aimed to undermine German
efforts to challenge British communication supremacy.[2] In fact, as soon as
word of the proposal was announced, the German government, too, decided to build a chain connecting its colonies.[3]

In Marconi's business plan, the scheme was the key part of an even
grander ambition. As far as the company was concerned, its goal was to
create a global communications network. The biggest obstacle was what it
called the "Telefunken wall"[4] of high-level German political support for
the rival firm, which Marconi's was running into increasingly as it tried to

establish itself in foreign countries, like Spain and Portugal. Telefunken was under-quoting Marconi (thanks to a German government subsidy, it was whispered) and getting the contracts, while Marconi's was stagnating. Even some British ministers were considering giving Telefunken contracts if they quoted lower prices than Marconi, it was said.[5] On the other hand, the government as a whole felt this to be an unthinkable alternative for Britain in view of the political context, especially given Germany's ongoing naval rearmament.[6]

For the Marconi company, the easiest, most efficient, and most profitable way to move forward would be in partnership with the British imperial government. All that said, however, the company was going to establish a global network one way or another. "It is a very big undertaking, and although the British Government is only concerned with erecting stations in its own territory, the Marconi company will erect them in any country that wireless telegraphy can benefit," Marconi said.[7] This was always the company's official position—that it was open for business with anyone—and it was echoed by Marconi's managing director. "It was my intention to erect those stations and to create this service all around the world. . . . I wanted to do it on British soil if I could, but I was going to do it, and if I could not do it on British soil I should do it on foreign," Isaacs later said.[8] In this regard, the company's corporate strategy and Marconi's personal philosophy were the same: be noble where possible, place self-interest first. This was also Marconi's cosmopolitan, internationalist vision: wireless should serve the whole world, not a particular national interest. (The only time he diverged from that position was where Italy was concerned.)

He still had to figure out how to deal with the Germans. After all, it had been the successful efforts of the German government to marshal international support for the idea of an interconnected global wireless system that had undermined Marconi's emerging monopoly in 1906. As the new regime under the leadership of Isaacs began thinking about this, a new strategy emerged: Marconi's would use its privileged position in Britain and Italy *and* international corporate conciliation to maintain and extend the company's dominant position.

The opening of a German high-power long-distance station at Nauen, near Berlin, in September 1906 was a concrete sign of German determination and Telefunken power. Nauen was intended to be a state-of-the-art transmitting as well as experimental station. During the Berlin conference, representatives of all twenty-seven nations in attendance were invited to witness the station establish a wireless telegraphy connection to Saint Petersburg (1,085 miles away).[9] On April 3, 1907, Nauen transmitted messages to a German cruiser travelling on the Atlantic Ocean near Lisbon, at 1,600 miles.[10] This was still way below Marconi's standard; he was then communicating regularly between Clifden and Glace Bay, a distance of 2,250 miles. But on May 27, 1910, Nauen successfully transmitted a telegram to the vessel *Bosnia*, which was at sea on its way to New York, over a distance of 3,125 miles.[11] Now Germany was really in the game.

Marconi began thinking about making a deal with Telefunken. Scattered throughout the company archives between 1908 and 1910 are various schemes and strategic discussions for dealing with the company's German rival. One of Marconi's distant Jameson cousins proposed a project for amalgamation.[12] Marconi himself met in Berlin with Georg von Arco, on his way home from receiving his shared Nobel Prize with another Telefunken principal, Ferdinand Braun. (Although Marconi and Braun were clearly not of equivalent importance in the development of wireless, the shared prize in some way reflected the new corporate context, a global market shared by the two companies.) The company's scientific advisor, James Ambrose Fleming, was called on to study Braun's patents and advise on infringement suits and countersuits.[13]

Then, suddenly, in January 1911, the business and financial world was excited by the stunning announcement of the formation of a new company jointly owned by Marconi's Brussels-based subsidiary, la Compagnie de télégraphie sans fil, licensee for its German patents, and Telefunken. Marconi and von Arco (who was later, notably, a pacifist during the First World War) were among the directors of the new entity, known as DEBEG (Deutsche Betriebsgesellschaft für drahtlose Telegraphie).[14] Initially limited to the German mercantile shipping industry, this deal was the first of a series between Marconi and Telefunken aimed at eliminating direct competition between them. By November 1912 the two companies were able to announce the withdrawal of all pending patent proceedings, effectively burying their disputes. In countries where there were still other competitors, like the United States and France, they agreed to work together to respect their patents.[15]

The agreement was the result of intense negotiations between Isaacs and Hans Bredow, the managing director of Telefunken. Bredow was a mandarin and an organizer, not a scientist or technical expert (he was, in a sense, Telefunken's equivalent of Isaacs). It was followed, on March 6, 1913, by a more elaborate, global agreement. Written in French, the original copy bound in a lush red leather folder, the agreement basically divided the world between the two (today we would say telecom) giants, giving each other rights to each other's patents.[16] Marconi reserved the United Kingdom and Italy, Telefunken kept Germany and Austria-Hungary; there would be no competition in those countries. The companies agreed to disagree in the United States and France; in those major markets, they would continue to compete, primarily with the strong US and French national champions.[17] In the rest of the world, Marconi and Telefunken would work together according to local conditions, forming jointly owned subsidiaries where necessary, sometimes competing politely where political conditions seemed to call for it (like in Ottoman Turkey), or at least staying out of each other's way.[18] Marconi soon had new contracts in Turkey, Rumania, Greece, and Portugal. Competition between Marconi's and Telefunken thus effectively ceased, but as Isaacs later said, "there was apparent competition . . . nobody knew of this agreement . . . nobody knew of any definite agreement except the directors of the two companies."[19] The ten-year agreement was intended to last until 1922, and it did. The Great War of 1914 to 1918 mirrored the wireless war that had preceded it, but when it ended, Marconi's and Telefunken were able to resume their collaboration as though it had never taken place.

❈   ❈   ❈

The idea for a British imperial wireless scheme lingered for a year until, in the immediate wake of the Marconi-Telefunken alliance, a parliamentary committee suggested, in March 1911, that an empire-wide wireless system would be desirable. It was then discussed at an Imperial Conference in London in May and June 1911, and a resolution was passed on June 1, 1911, in favour of establishing a chain, without specifying who would own and operate it. On June 20, Isaacs reported to the board that he had "in consultation with Mr Marconi" (who was absent) submitted to the government "a scheme to put all the British Possessions into wireless communication with each other and applied for a concession for the erection and working of the necessary stations."[20]

Negotiations started between the Post Office and the company in autumn 1911. While the government kept a proper silence on the talks, Marconi's publicized them as much as it could, starting a wave of public speculation that soon spilled over into the stock market.[21] Share prices began rising steadily, and by early 1912 were as much as five times what they had been only a few months before. From London on February 1, 1912, the *New York Times* reported on "the best authority" (in a dispatch transmitted to the newspaper by wireless, under a recent transatlantic agreement between the company and the newspaper that took effect early in 1912) that the negotiations were now complete and a new "All Red" wireless system (as the Americans called it) would soon be inaugurated. "Mr. Marconi, whom I saw today, modestly deprecated being a spokesman in regard to the developments of the scheme," the anonymous correspondent reported.[22]

A key figure in the discussions was Post Office secretary Sir Matthew Nathan.[23] Nathan told Isaacs the government wanted the chain to be government-owned, and proposed that Marconi's build the stations for a fixed amount plus 10 percent of the gross profits. While this seemed like a much surer financial deal for Marconi's, the critical distinction was that the stations would belong to the government. The company finally agreed and submitted a tender that was accepted by the Post Office on March 7, 1912. Marconi's would build six stations for £60,000 each and a royalty of 10 percent of the gross receipts for twenty-eight years. The company immediately sent a letter to its shareholders informing them that the proposal had been accepted (but mentioning that it still needed parliamentary ratification), and share prices in both British and American Marconi began to rise again.[24] The next day, Isaacs told the *New York Times*'s London correspondent: "The Postmaster General's notification to us marks the first step toward the establishment of a girdle of wireless stations round the British Empire." The terms were "very satisfactory" to the company, he said.[25]

Marconi was only indirectly involved in the negotiations, which were largely conducted by Isaacs. Isaacs kept him thoroughly informed, but Marconi later stated: "I did not personally take any active part in the negotiations or correspondence leading up to the Contract with the Post Office but whenever I was in England during the period of the negotiations I from time to time had discussions with Mr Godfrey Isaacs and other members of the English Company's Board upon the subject of the negotiations."[26] Marconi spent most of the week leading up to the agreement at his Clifden station.[27]

Despite all that was going on in London and at his various stations around the British Isles, Marconi still had urgent matters to attend to in Canada and the United States. He had crossed the Atlantic more than forty times by now, and was a much-appreciated client of the highly competitive transatlantic passenger liner business, so much so that early in 1912 the

White Star Line invited him and Beatrice to be its guests on the maiden voyage of its new flagship, the *Titanic*. Marconi didn't want to wait for the *Titanic*, which was only due to sail in April. Two days after the company's tender to the Post Office was accepted, on March 9, 1912, Marconi and Isaacs sailed for New York on the *Lusitania*. As usual, Marconi took advantage of the crossing to experiment with his stations as well as with passing ships, and learned a new thing or two that would be of use. He was feeling fit and rested and joked in a shipboard letter to Beatrice that "I've had a much quieter time than usual on board this time. There are practically no 'flirtable' girls."[28] (Beatrice had intended to follow on the *Titanic* but cancelled at the last minute when two-year-old Giulio came down with a high fever.[29])

The main purpose of the trip was to tie up a lingering court case with the United Wireless Telegraph Company (UWTC). Marconi prepared for these court appearances meticulously, using diaries, notebooks, and other documents to trace the origins of his inventions. In this case, he arranged for George Kemp's diaries to be sent to New York and brought with him the exercise book containing details of their first experiments at the Post Office in 1896.[30]

UWTC was the successor to the American De Forest Wireless Telegraph Company, which went out of business in 1906, shortly after losing the landmark Townsend case to Marconi.[31] De Forest's assets were taken over by stock promoter Abraham White, who flamboyantly announced a bogus merger with Marconi's before giving the new company up to an even more outlandish promoter by the name of Christopher Columbus Wilson, in February 1907. By 1910, UWTC was the largest US wireless operation, with more than seventy land and four hundred shipboard stations, but its managers seemed more interested in fraudulent stock schemes than operations, and in June 1910 the US government indicted seven UWTC officials for mail fraud; five of them, including Wilson, were convicted and sent to prison.* UWTC declared bankruptcy and went into receivership in July 1911; Marconi then sued them for patent infringement, a brilliant move intended to once again assert his patent supremacy while laying claim to UWTC's assets. The case was due in court in March 1912.[32]

Marconi's arrival on March 15, 1912, was duly reported by the *New York Times*, which sent a reporter out to interview him as the *Lusitania* approached

---

* Wilson, then sixty-five, famously married his eighteen-year-old secretary the day he was indicted; he died in prison on August 26, 1912.

New York Harbor. He commented somewhat contemptuously about the pending court case. "Years ago, we had to fight the De Forest Co for infringement of our patents, and we won the case. Later the United Wireless springs phoenix-like from its ashes, and now we have to go through the same business all over again." The newspaper also briefly interviewed Isaacs, noting that he was the brother of the British attorney-general. It was Isaacs's first visit to America and he was accompanied by his son, Marcel, a student at Cambridge University.[33]

Marconi was out to dinner with his New York lawyers the evening he arrived when he ran into his one-time fiancée Inez Milholland, dining at the next table with another woman and two men. Marconi had remained friends with Inez and her family and he was evidently comfortable enough to mention his encounters with her enthusiastically in his letters to Beatrice. "As soon as I saw her I went over to say how you do. She was looking charming. Next day she came to lunch with me but at once went for me for having been so stiff and stand offish when I met her the evening before. She said she had told her friends I was one of her greatest friends etc, and that they said it did not look like it from my manner to her."[34] He had been told before that he had difficulty displaying warmth to people he liked whom he hadn't seen in a while.

Now twenty-six, Inez was one of New York's leading feminist activists. A familiar figure in the bohemian circles of Greenwich Village, she had been arrested twice in 1909 and 1910 during demonstrations in support of the Triangle Shirtwaist Factory strikers and was romantically linked for a time with the radical journalist Max Eastman, soon to become the editor of *The Masses*. One of the most vocal and visible advocates of women's suffrage in the United States, she had led a march of three thousand women up Fifth Avenue in demand of the vote in May 1911 (and would do the same in 1912 and 1913). She was also studying law. As her biographer later put it, Inez easily crossed the line between socialite and socialist.[35] Her friendship with Marconi was cherished by both of them; each seemed to see an unfinished bit of themselves in the other; they enjoyed each other's celebrity and enjoyed spending time together whenever they could. If there was any sexual tension between them, it was never explicit or clear. She was also a progressive influence on his family, introducing them to the avant-garde Montessori method, which so impressed them that they enrolled Degna and Giulio in the Montessori school near the Villa Borghese in Rome in 1913.[36]

The next day, Marconi was feted at a banquet given by Adolph S. Ochs, publisher of the *New York Times*. For the previous three months, the newspaper had been receiving nearly all its European news, roughly twenty thousand words a week, via Marconi wireless. The event was the scene of one of those sensational publicity stunts for which Marconi had become famous. The company recruited a handful of British political personalities (including Sir Rufus Isaacs, Godfrey's distinguished brother) to send wireless messages of congratulations during the dinner, and arranged with the British Post Office to prioritize the land link from London to Clifden, establishing a new record for London–New York communication: ten minutes, as compared to the previous best record of fifty-five minutes. The stunt was also calculated to pressure the Post Office to provide a better landline service at its end; the transatlantic link from Clifden to Glace Bay was almost instantaneous, as was the Western Union land connection from Glace Bay to New York.[37] Marconi always maintained an almost humble posture in these situations, disarming potential critics while endearing himself to supporters with his genuine modesty; asked before the dinner how quickly a message might get through, he optimistically predicted forty minutes. The newspaper waxed: "We are privileged to live in a wonderful age, and none of its developments is more wonderful than the adaptation of wireless telegraphy to commercial use by the ingenious and purposeful Italian inventor Guglielmo Marconi."[38]

It is hard to imagine, in this day of tweets and hashtags, instant messaging and social media, the impact of such a stunt. While in the United Kingdom it took a government licence to operate a wireless system, in the United States the air was open to anyone with a bit of mechanical skill and thirty dollars or so to lay out for an "app" (short for "apparatus," in those days). Amateur (or "ham") operators were able to build cheap and functional "crystal" sets, so called because they used bits of crystal substance from minerals such as iron pyrite to forge fine cat's-whisker detectors of wireless signals. One of the most effective crystal systems, known as the Perikon, unwittingly signalled our own information age: it incorporated a piece of silicon.[39]

Thousands of amateur operators were on the air each evening, listening in to ship-to-shore and ship-to-ship transmissions and sharing them with their online community. The *New York Times* dinner "was the talk of all the wireless operators on the Atlantic Ocean and along the coast while the dinner was in progress," according to one amateur working out of his apartment on Riverside Drive. This anonymous hacker showed an enthusiastic reporter how he could send messages over seventy miles and receive from ships as far as a thousand miles away at sea, a rough lever on his homemade apparatus allowing him to switch frequencies.[40]

By 1912, US amateurs were linked through a series of formal and informal networks not unlike those that characterized the late twentieth-century Internet. No one knew just how many there were, but clearly they numbered several hundred thousand, and had their own gathering points. One leading figure in the amateur virtual community was Hugo Gernsback, who had emigrated from Luxembourg to the United States in 1904, at the age of eighteen, and set up shop selling wireless components on Fulton Street in New York City.[41] By 1905, Gernsback was providing complete wireless systems, including both transmitter and receiver, to amateurs across the United States. In 1908 he started the first of three magazines he would launch during the following decade, *Modern Electrics* (the others were *Electrical Experimenter,* 1913, and *Radio Amateur* News, 1919), and in 1910 he founded the Wireless Association of America (WAOA).[42] A proliferation of local amateur clubs followed; one of them adopted the motto: "Marconi was once an amateur."[43] Most of the amateurs were evidently male, but there were stories of bold women who entered this man's world, able, in the days before Morse code gave way to voice, to conceal their gender identity if they wished. One woman amateur was quoted as stating: "Just because a man, Signor Guglielmo Marconi by name, invented commercial wireless telegraphy does not mean for a moment that the fair sex cannot master its mysteries."[44] By March 1912, there were said to be as many as four hundred thousand wireless amateur operators in the United States,[45] and they were beginning to be seen as a problem by both corporate and government interests.[46]

Marconi and Isaacs were well prepared for their case against United Wireless, and the liquidators offered to make a deal, recognizing that a loss in court would wipe them out as it had De Forest. They didn't have many bargaining chips; UWTC was in receivership, its former head and most of the directors were in federal prison, and they had no defence. The UWTC trustees conferred with Isaacs, and on March 21, 1912, they gave in entirely to his terms, which had Marconi's taking over all of UWTC's assets in exchange for Marconi shares.[47] John Bottomley, who had the job of incorporating the UWTC assets into Marconi's US operations, said UWTC had been characterized by "a spirit of carelessness...almost unparalleled in business history."[48]

Marconi wrote to Beatrice, "They have admitted to having copied or stolen my patents and all their stations are to be called Marconi stations. This will do our Company heaps of good." Marconi still had a case pending with Reginald Fessenden, "but there are signs which seem to show that he too is going to cave in. If all goes well we shall control all the wireless in America."[49] That was indeed what happened, and despite his grace and modesty in public Marconi was happy to gloat in private. His letters home reveal that he was having a jolly good time ("been very gay," he wrote, in the vernacular of the day), going to dances, dinners, and the opera. He was meeting "heaps of new people" and, typically, his letters were littered with impressive names. The opera season was in full swing and he'd had lunch with Caruso.[50] "I've only got to go to a dinner and about six people ask me to others—and it's nothing to get two or three invitations for the same evening."

The press, especially the *New York Times,* glowed over him. On March 24, 1912, the paper devoted a full page of its Sunday edition to a feature on Marconi's future plans, trying to draw him out but without much success. Wireless telephony as well as telegraphy would soon be universal, Marconi said, but asked to reveal specifically what he was working on he was elusive: "I have my eyes and my ears open...the inventions that I am now looking forward to making realities are still in the theoretical stage, and so I do not care to speak of them."[51] (The process of voice transmission that Marconi was still referring to as "wireless telephony" would soon be reconceptualized as "radio broadcasting.")

Marconi did, however, describe a new invention in an interview published by the *New York Times* a week and a half later: a "wireless compass," a kind of primitive GPS intended to help ships at sea establish their location by a process he called "triangulation" (if a ship sent a signal to two separate receiving stations, its location would necessarily be at the point where the signals crossed). Typically, he was sparse on details pending patent approval, but he took the opportunity to go on record in support of limited international regulation of wireless. "The wireless shouldn't be regulated to death, as it easily could be. But it simply must be regulated in some manner, and the one body fit to do the regulating would be an international board. It's a bigger job than any one nation could handle. All must be considered and must join in the proceedings."[52] His main target: the amateurs, who were now persistently, if usually benignly, hacking his business.

Paradoxically, Marconi, a powerful opponent of government and international regulation when it constrained his own commercial operations, was one of the most vocal high-profile advocates for regulation of amateur wireless activity.

News of the pending United Wireless settlement sparked rumours that a deal with the US government was about to be made, and another spike in Marconi shares in London.[53] Isaacs, meanwhile, was working on a reorganization of the American company and a new share issue that MWTC of America planned to issue in April. He also made a deal with Western Union to use their land lines for onward transmission of wireless messages—the details to remain secret until after the new US offering was approved. In England, news of the Post Office contract had boosted Marconi share prices, which had been trading vigorously in the preceding months. Even Marconi's usually placid brother Alfonso took the opportunity to sell a block of four thousand shares he had purchased—at Marconi's insistence—in 1909; now, three years later, he made a twelvefold return.[54] The fact that Marconi's was about to absorb its most important US rival meant the same could be expected for the company's US shares. Suddenly, it looked like there was finally money to be made. It was either capitalism run amok or the day before the initial public offering of a twenty-first-century dot.com wonder—or both.

Shares in the American company were undervalued and trading at a 20 percent discount,[55] and Isaacs proposed to increase the subsidiary's capital by $6 million. The American company's directors were nervous, however, and demanded that the parent company guarantee that the offering would be fully subscribed. Marconi insisted that Isaacs take personal responsibility for unloading five hundred thousand of the 1.4 million shares and the company would guarantee the balance. Isaacs agreed with Marconi's condition and began placing the shares with dealers on the London and US markets, keeping one hundred thousand for his own disposal when he got back to England. It is unlikely he could have done this without the prestige of the Post Office agreement. A year later, when Marconi was asked by a parliamentary committee how Isaacs got to personally dispose of five hundred thousand shares without their being offered to the company, he replied: "He got them by having the pluck to take them when no one else was likely to take them."[56]

Isaacs had made a great success of his maiden voyage to the United States.[57] ("They all think Mr Isaacs is quite a genius over here," Marconi wrote to Beatrice, with perhaps a bit of envy.[58]) On April 3, 1912, Isaacs sailed for London, while Marconi travelled to Montreal and Glace Bay.[59] This enabled Marconi to later claim he knew nothing of what happened next—which could very well be the case, although, as we have seen, he was indeed involved in, even the origin, of Isaacs's share holdings. Marconi himself eventually bought ten thousand of Isaacs's US shares (he also owned approximately twenty-eight thousand shares in MWTC at this time).[60]

Marconi wrote to Beatrice from Montreal, where he was staying at the Saint James Club. The city and countryside were frozen and white. Far from the cheery social life of New York, he missed her and was given to romantic thoughts. "When I travel over here everything seems to remind me of you, and to make me long oh long so very much to have you with me."[61] It was a trying journey to Glace Bay. His train from Montreal was late and he had to lay over in "lovely Truro" (Nova Scotia) after missing his connection to Sydney. When he finally arrived in Cape Breton, the harbour was frozen; he said he had never seen anything like it and was able to walk across the ice to see a new site in which he was interested.

Meanwhile, in London, Postmaster General Herbert Samuel finally informed the House of Commons, at its last sitting before Easter adjournment, of his arrangement with the Marconi company. The proposed chain of wireless stations would be of both military and commercial importance and would place the British Empire ahead of any country in the world in global wireless communication. Some parliamentarians had been pushing for a state-owned transatlantic cable, but the government dismissed that idea as too costly, too vulnerable in times of war, and too reliant on a dated technology. In New York, it was also reported that American Marconi would be expanded as a result of its absorption of United Wireless.[62]

Now back in London, Godfrey Isaacs invited his brothers Rufus and Harry to lunch at the Savoy on April 9, 1912, the day after his arrival, and offered them batches of the forthcoming American Marconi shares, which were already trading informally at a premium to the future issue price. Rufus later said he sought assurance that the American company was in no way benefiting from the British Marconi's fresh deal with the Post Office. Godfrey reassured him but Rufus declined nonetheless. He then left the meeting. Harry stayed on and agreed to take fifty thousand of Godfrey's shares (an amount he later increased by six thousand). Godfrey had no difficulty placing the remaining shares, keeping twenty-five hundred for himself.

During the following week, while Marconi was travelling in Canada, there was "an unofficial boom" in American Marconi shares, though they were still not trading on the stock exchange. On April 11, 1912, shares in American Marconi were worth double what they had been on March 7. Both the British and American companies now had the wind in their sails, and it was apparent that the US issue was going to be very successful. Harry Isaacs now persuaded his brother Rufus to take ten thousand of his shares; Rufus insisted on paying the appreciated price, and later insisted that the fact he bought them from Harry, not Godfrey, made a significant ethical difference. That same evening, Rufus visited two of his Liberal Party colleagues, Chancellor of the Exchequer David Lloyd George and the chief government whip Lord Murray, and offered them each a thousand shares at the price he had paid; both accepted.

Marconi, meanwhile, returned to New York from Glace Bay and was tying up his US business when news reached the city that the *Titanic* had struck an iceberg in the north Atlantic.

❉   ❉   ❉

There were few luxuries greater than a transatlantic ocean voyage in 1912, and the *Titanic* was designed to be the new top of the line. Among its other innovative features, the *Titanic* was equipped with the most advanced and powerful apparatus then available for seaboard wireless communication. Two transmitter masts rose two hundred feet above the deck and were guaranteed to work over a range of 250 miles under any conditions, reaching up to two thousand miles at times. The White Star Line's contract with the Marconi International Marine Communication Company was standard: the company provided the equipment, and hired, trained, and paid the operators, while the shipping company provided space and electricity and maintained the masts. The service for passengers was operated by MIMCC as a commercial concession, but messages regarding ship business took priority. Although they were not ship employees, Marconi operators were under the authority of the ship's captain in the same way as regular crew members. The *Titanic* was one of about fifty transatlantic ships that carried two operators.[63]

The *Titanic* left its construction dock in Belfast on April 2, 1912, and slowly made its way to Southampton, where the majority of its passengers embarked on Wednesday, April 10. The following day, after picking up more passengers at Cherbourg, France, and Queenstown (now Cobh), Ireland, it headed out into open sea, carrying 2,208 passengers and crew. At 11:40 p.m.

(ship's time) on Sunday, April 14, the *Titanic* struck an iceberg, ripping a three-hundred-foot hole in its side.[†] Minutes later, as the ship began to sink, wireless operator Jack Phillips was instructed to send out distress signals CQD (the Marconi standard) and the more recently introduced SOS. The message was received by the Marconi station at Cape Race, Newfoundland (as well as some other North American coast stations), but, more important, by several ships in the area.[64] The *Titanic's* sister ship, the *Olympic*, heard it but it was five hundred miles away, far too distant to be helpful. The closest to the *Titanic*, the *Californian,* never got the message, as its sole operator had retired for the night. The Cunard liner *Carpathia* also had only one operator on board, Harold Cottam. He was preparing for bed but had not yet shut down the system and heard the call: "Struck iceberg, come to our assistance at once," followed by the *Titanic's* position.[65] The *Titanic* continued calling for help for more than an hour until its signals ended abruptly. At 2:20 a.m., the *Titanic* went down. The *Carpathia* rushed to the scene, arriving around daybreak. It was the only ship to pick up survivors. Seven hundred and five passengers and crew members in lifeboats, including the *Titanic's* second operator, Harold Bride, were rescued from the sea and the *Carpathia* began heading for New York. Jack Phillips was one of more than fifteen hundred who perished.[66]

The lack of news from the *Carpathia* became an element of controversy in the days that followed, but it was attributable to two things: the relatively short range of the ship's wireless equipment and a decision by the captain, Arthur Henry Rostron, to prioritize messages from survivors to their families. After transmitting a list of survivors, Cottam and Bride (whose feet were seriously injured from his ordeal) worked round the clock sending survivors' messages via relay by neighbouring ships. News of the *Titanic's* sinking and its aftermath was thus second-hand while the *Carpathia* sailed to New York.

The first published news report, in the *New York Evening Sun*—based on misinformation allegedly spread by amateur wireless enthusiasts—said that everyone had been saved and the ship was being towed to Halifax, but by the evening of April 15 it was clear this was no more than wishful thinking.[67] Marconi was informed by his US manager, John Bottomley, that the ship had gone down with very heavy loss of life. On the positive side,

---

[†] Ships crossing the Atlantic operated on their own time clock. According to the *New York Times* (April 15, 1912), the *Titanic* struck the iceberg at 10:25 p.m. New York time. Thus, the *Titanic's* clock was 1:15 ahead of Eastern Standard Time at the time it struck the iceberg.

the press began reporting lists of survivors who had been picked up by the *Carpathia*.[68] Later that evening, Marconi's office informed him that the *Carpathia* was returning to New York with the survivors. He wrote to Beatrice the following day:

> I've witnessed the most harrowing scenes of frantic people coming here to me, and to the offices of the Company to implore and beg us to find out if there might not be some hope for their relations.
>
> Lots of people I know, Captain Smith and other officers, the two wireless operators have gone down [he was, at this early stage, misinformed; one of the operators, Harold Bride, survived]—but although only a few were saved everyone seems so very grateful to 'wireless.' I can't go about New York without being mobbed and cheered. Worse than Italy.[69]

The aura surrounding the role of wireless in saving the *Titanic's* survivors further fuelled the speculation in American Marconi shares, and the shares reached another record high even as the *Carpathia* was sailing toward New York. Marconi shore stations off the Canadian mainland were besieged with requests for information from friends and families of *Titanic* passengers and from the press. London newspapers were offering as much as £1,000 for survivors' stories. Having transmitted the names of survivors on board, the *Carpathia* itself was no longer reachable.[70]

Marconi had a long-scheduled address to the New York Electrical Society on April 17, 1912. When he arrived at the Engineering Societies Building the hall was jam-packed, and, unusually for a business speech, the *New York Times* reported that "almost half of the audience were women." When Marconi appeared, "the crowd in the balcony saw him first, and the cheering began. It spread to the main floor, and for at least two minutes there was continuous applause. Marconi bowed in acknowledgment, but had to rise again and again before the crowd was satisfied." He gave a relatively uninspired, technical, boilerplate speech but ended it with a passage that was clearly added after the main text had been composed, referring to the role that wireless had just played in the saving of hundreds of lives.[71] Physicist Michael Pupin, thanking him, suggested that air waves be henceforth no longer called Hertzian waves but Marconi waves.[72]

The following day, Marconi headed the guest list at the annual luncheon of the League for Political Education at the Hotel Astor. Speaking extemporaneously, he criticized the US and UK regulations that had allowed the *Titanic* to sail without an adequate supply of lifeboats, and pointed out that many ocean steamers were inadequately fitted with wireless equipment as well. The *Carpathia*'s equipment, for example, had been strong enough to receive the *Titanic*'s call for help but too weak to transmit to full capacity. Had the *Carpathia* been the one to hit the iceberg, he said, there would have probably been no survivors.

The American Marconi shareholders met that day as planned, and approved the proposed share restructuring, with part of the proceeds to be spent on acquiring the United Wireless assets and the rest on expanding the company's US activities. The stock's par value was reduced from $25 to $5 while the company's capital was increased to $10 million. The stock continued to rise on New York's informal "curb" market.‡ The new $5 shares were already at $13-1/8 and they hadn't even been issued.[73] "Old" shares, which had been lingering sleepily around $40 a few weeks earlier, now hit a sensational $245. Fortunes were being made in arbitrage trading.[74]

Marconi was dining with a group of friends and colleagues at Bottomley's house that evening, April 18, when the *Carpathia* docked at 9:25 p.m., sooner than expected. He proceeded by elevated railway to the Cunard pier at the foot of West Fourteenth Street, where he was joined by Inez Milholland. Although he professed to being shaken by the news and personal loss, he was alert enough to be filmed by the newsreel cameras "in characteristically smiling mien," with Inez on his arm.[75] It was, after all, a complex situation for him. Despite the tragedy, he was being hailed as the saviour of more than seven hundred lives. He described the scene at the pier as "a weird and uncanny sight."[76]

If there were still any skeptics about the value of wireless, the *New York Times* was not among them. The paper editorialized: "There is hardly a chance that one member of the *Titanic*'s great company would be alive today had there been no wireless telegraphy ... the personal achievement of Mr. Marconi in a sense and to a degree that few other such achievements are the work of a single man ... he stands alone as the originator of ethereal communication."[77] Such elegies were echoed by calls for a public monument to be erected in his honour, and resolutions of appreciation.[78]

The day after the shareholders meeting, April 19, 1912, a "working agreement" to tie Western Union's US land stations to Marconi's transatlantic connection was announced. Plans revealed simultaneously in London, to link Pacific islands, the US west coast, and Asia meant that, for the first time, it would be possible to send a wireless dispatch entirely around the

‡ Outdoor "curbstone" brokers then operating in the financial district of Lower Manhattan.

world. American Marconi stock skyrocketed to $350, prompting the *New York Times* to describe the rise as "the wildest boom which the curb market has ever witnessed."[79] But conservative investors were keeping a close watch on their purse strings; some even dared to whisper that American Marconi had never paid a dividend. Twenty-four hours later the boom was over and the stock sunk to $150. Wall Street had no explanation; Marconi himself just shrugged.[80]

In London, meanwhile, American Marconi shares traded at twice the price paid by Rufus Isaacs and his friends days earlier (and four times the price paid by Harry Isaacs); Isaacs, Lloyd George, and Murray decided to sell part of their holdings. None of this, of course, was known at the time to anyone but the interested parties, who were also running the country while trading Marconi shares. There was what author Frances Donaldson later called "a benighted innocence in all this."[81] None of the ministers tried to hide their activity; in fact, none of them seemed to give it much thought.

Britain had just ended a crippling national coal strike,§ but between the Post Office contract and the *Titanic* drama, Marconi's name was on everyone's lips in London. Another great proof of the humanitarian value of wireless had been furnished, said the *Times*; but the recent speculation in Marconi shares had now "degenerated into a dangerous gamble." Wireless had revolutionized world communication, but its developments had not been matched by dividends. That said, the paper continued, "It may be worth while recalling that an Empire scheme of wireless has been agreed to between the British Government and the English company and is to be commenced forthwith." The list of station locations alone invoked glamour and wanderlust: London, Egypt, Aden, Pretoria, Bangalore, and Singapore. The government and the company were now partners in an exciting new adventure, but it remained to be seen how well this would translate into benefits for shareholders.[82]

❧   ❧   ❧

The US Senate acted with astounding speed to launch an investigation that began the morning after the *Carpathia* docked, on April 19, 1912, with hearings at the Waldorf-Astoria in New York, in order to facilitate the participation of survivor-witnesses in no shape to travel to Washington.[83] Marconi was one of the star witnesses on the first day, appearing immediately after J. Bruce

§ The thirty-seven-day national coal miners' strike, Britain's first, ended on April 6, 1912, with the adoption of a minimum-wage law.

Ismay, head of the White Star Line, and Captain Arthur Henry Rostron of the *Carpathia*. He was questioned by the committee chair, Michigan senator William Alden Smith.

Smith asked about the "providentiality" of *Carpathia* wireless operator Harold Cottam hearing the call as he was retiring: "Ought it not be incumbent upon ships at sea who have the wireless apparatus to have an operator always at the key?" Marconi replied: "I think it certainly should be. Of course it might come rather hard on small ships. The ship owners will not like the expense of two men. . . . they do not like to carry two operators when they can get along with one." (Only 10 percent of transatlantic liners had two operators working in shifts.)

Smith then asked a loaded question. "Do you regard the Berlin Convention [which the United States had only approved less than two months earlier] as a step in the right direction of the international utility of wireless telegraphy?" "I think in regard to shipping and shore stations," replied Marconi, "it is a good regulation. It is a means for regulating the working and preventing interference; provided, however, that it is administered in a fair manner by the governments concerned." Here Marconi chose his words carefully; this may have been the first time non-specialized politicians and the public were hearing about the convention, as well as his first public admission that it might be a good thing. When Smith recalled the Berlin Convention provision that distress calls be given precedence over all other messages, Marconi replied that his company had been following such a policy since "before there was any Berlin Convention."

Marconi was anticipating one of the sharpest arguments made by critics in the wake of the disaster—that his company's operators on the *Titanic*, the *Carpathia*, and other vessels were there primarily to produce revenue by sending commercial messages on behalf of passengers. In fact, there were two functions to seaboard wireless—ship safety and passenger convenience. Marconi's contracts with the shipping companies provided clearly that ship business took precedence over commercial activity, and the wireless operators, while Marconi employees, took their orders from the ship's captain so there could be no question whose priorities were in charge.

Marconi continued his testimony on the sixth day of hearings, April 25, 1912, after the committee reconvened in Washington. (As a result, he had to miss a press dinner in New York where he was to be honoured along with Alexander Graham Bell and Thomas Edison.[84]) Here he provided a rich account of the workings of wireless, his company, and his own role in the selection and equipment of wireless operators on ocean vessels. Smith seemed to think it was important for the committee to hear Marconi's life story, or at least the story of how he had invented wireless. This was Marconi's performative strong suit. Time and again he had had courtrooms and parliamentarians eating out of his hand. Now he worked his magic on the United States Senate.

> SMITH: ...who was the first practical operator of wireless telegraphy covering long distances?
> MARCONI: ...I think it was myself in England...1896 and 1897. I carried on tests for the Army and for the Navy.
> SMITH: Had you been an operator before?
> MARCONI: No; I had not. I had not been an operator. I took an interest in electrical subjects generally. I had studied a great deal. I was what I might rightly describe as an amateur.
> SMITH: If you can state briefly, I would like the record to disclose it.

And Marconi repeated his well-rehearsed story, how he carried out his first tests with electrical waves in Italy in 1895; how "by a modification of the apparatus" he was able to radically increase the distance over which these waves could be sent and received, from twenty or thirty yards to two or three miles. Then, how he came to England and demonstrated his idea to the British Post Office, the army, the navy, and to Lloyd's. How further improvements were then perfected "by myself, and some by others," gradually increasing the range and making it apparent that it would eventually be possible to communicate over thousands of miles.

> SMITH: What do you ultimately expect of it?
> MARCONI: I expect it will be one of the principal means or methods for communicating between distant parts of the world....I think that with the increase of speed and the understanding of electricity it will some day become the chief means of communication.

This was Marconi at his best: articulating a global vision for communication that placed his technology at its centre. Whatever one what might think of his self-interest or hubris, he was correct. The Senate committee found that "this catastrophe makes glaringly apparent the necessity for regulation of radiotelegraphy." It recommended that ships have "an operator on duty at all times, day and night, to insure the immediate receipt of all distress, warning, or other important calls" (meaning more business for Marconi's companies), and a range of specific security measures to ensure the smooth operation of wireless on shipboard, as well as legislation to ensure the secrecy of wireless messages and prevent interference by amateurs (who hampered communi-cation, in Marconi's view). When, a few months later, on August 13, 1912, the United States adopted its first comprehensive Radio Act, the *New York Herald* trumpeted: "President [Taft] moves to stop mob rule of wireless."[85] The "chaos" that had followed the *Titanic* disaster was the most frequently cited justification for the legislation which had, unsurprisingly, been strongly opposed by Gernsback and the WAOA.[86]

❊　❊　❊

Marconi wireless operators Cottam and Bride were the heroes of the day, and everyone wanted their stories, which appeared on April 19, 1912, in the *New York Times*.[87] Marconi's role in how they got there became one of the minor controversies in the aftermath of the *Titanic* tragedy. In his testimony before the Senate committee, Marconi described how he had gone to the *Carpathia* and spoken with Bride.[88] Later, he said, Cottam called him and asked if it was all right if he told his story to a reporter who had offered to pay for it. Marconi told him he could go ahead. But it turned out that this was part of a more elaborate staging, designed to control the shaping of the story and ensure that it appeared first in Marconi's preferred newspaper, the *New York Times*. (The paper was one of the most important clients of Marconi's overseas press service, its news pages full of the now familiar by-line "By Marconi Transatlantic Wireless Telegraph to the *New York Times*.")

When Cottam made his second appearance before the committee on April 25, 1912, Smith read into the record four messages sent from shore to the *Carpathia*, and intercepted by a US Navy vessel, between 8:12 and 9:33 p.m. on April 18, as it was entering New York Harbor:

> 8:12: Say, old man. Marconi Co. taking good care of you. Keep your mouth shut and hold your story; It is fixed for you so you will get big money. Now, please do your best to clear.
>
> 8:30: Arranged for your exclusive story for dollars in four figures. Mr. Marconi agreeing. Say nothing until you see me. [Signed] J.M. [sic] SAMMIS [American Marconi's chief engineer] Where are you now? O.P.R. "C"
>
> 9:00: Go to Strand Hotel. 502 West Fourteenth Street. To meet Mr. Marconi. [Signed] C
>
> 9:33: Meet Mr. Marconi and Sammis at Strand Hotel, 502 West Fourteenth Street. Keep your mouth shut. [Signed] MARCONI.

Cottam couldn't recall the precise details but told the committee, "I had a message from the company asking me to meet Mr. Marconi in the Strand Hotel and I was preparing to get ashore as she touched." Asked, "Did you keep your mouth shut," as instructed, Cottam replied: "Certainly." He told the committee he had gone to the Strand Hotel (just across the way from the Cunard docks) but there was no one there to meet him. He then telephoned Marconi and received his personal consent to give out his story.

Marconi was in the room during Cottam's testimony and was briefly recalled later that day. Smith asked him one question: whether any officer or employee of the White Star Line had requested him or anyone associated with him to delay any message or enjoin silence on the part of Bride or Cottam. Marconi replied, "No, I am absolutely certain that I have received no such request." The *New York Times* was furious about the senator's line of questioning. "The preposterous Smith," as they labelled him, had made a spectacle of himself with a ludicrous performance. His questioning of Marconi seemed motivated "by a sort of stupid malevolence...harrying and hectoring as if he had before him some criminal instead of the great benefactor of the human race whose praises are upon the lips of all men." The *Times* speculated that Senator Smith was put up to it by the paper's own jealous competitors.[89]

Indeed, the controversy had originated with the *Times* rival—and now Marconi nemesis—*New York Herald*, and Smith had picked it up for his own purposes. To counter the senator's claims, the Marconi company circulated (and the *Times* dutifully published) an affidavit from the operator who had actually transmitted the four messages to the *Carpathia*. He had telephoned the company from his post at the Marconi Sea Gate station on Coney Island, where he had been trying unsuccessfully to make contact with the *Carpathia*. Contrary to Smith's claims, the company had instructed operators to try to obtain any possible information about the actual accident. As late as 7:45 p.m., an hour and a half before the *Carpathia* docked, the operator had addressed a message to Bruce Ismay from the White Star Line asking for a concise account, but there was still no response. The operator then phoned the Marconi company to say he had been unable to obtain any news or anything like a news dispatch, and was requested by chief engineer Frederick M. Sammis "to advise Messrs Bride and Cottam not to talk of their personal experiences to reporters...as an amount running into four figures would be paid by some newspaper for such a narrative." The wording of the messages he sent was "of my own construction."[90] In the Senate transcript the operator's name is given as "Mr. Davidson," but as published in the *Times* it was David Sarnoff—a young station operator for American Marconi who would eventually rise to the very top of the US broadcasting industry under Marconi's personal mentorship. (Sarnoff was based at the Marconi office in Wanamaker's department store at Broadway and Eighth Street and was seconded to the Sea Gate station in order to keep in contact with the *Carpathia* as it entered range.)[91]

Marconi appeared again on the tenth day of hearings, April 29, 1912, to put on the record that he had sent a message to the *Carpathia* in the early hours of April 18, the day it would arrive, asking why there had been no news sent from the ship; he had received no reply. He said this should suffice to prove he had no intention of holding back information; it was the only message he had sent to the *Carpathia* and at no time did he send or authorize a message instructing the operators to keep their mouths shut. Asked whether the news was withheld in the hope of profiting from the story (as Jack Binns had in 1909), he replied: "I should say not.... It seems to me that the public interest or the newspaper interest, becomes so great when an individual finds himself placed in the position of these men, that whatever they say that has a public interest is paid for by these enterprising American journals."

Smith then asked one further question: "I ask you whether from the developments of this inquiry you do not feel that it is incumbent upon you to discourage that practice; indeed, to prevent it altogether, so far as you are able?" Marconi replied: "Certainly; I am entirely in favor of discouraging the practice [of selling stories to newspapers], and I naturally give very great weight to any opinion expressed by the chairman of this committee." Sammis was then called. He insisted that he had only communicated to Cottam an arrangement made by Marconi, Bottomley, and the company at large. "I believe an error was made," he said. "I believe it would have been better to have sent this news to the Associated Press and let them settle with the boys, if they liked. The news then would have had more general distribution, and there would not have been any sore toes."

In further testimony by Bottomley (on day twelve), it was also clear that Marconi, "while he did not altogether care for the business," saw no objection to the operator giving his story to the *Times*. So he, Bottomley, called the Marconi company office and told them to go ahead. Smith then asked Bottomley: "Would you favor an international agreement for the control of information of disasters at sea?" To which he replied: "Yes." Bride, the surviving *Titanic* operator, was recalled on the fourteenth day, having recovered somewhat from his injuries. He confirmed that on the *Carpathia's* arrival in New York, Marconi, Sammis, and a reporter for the *New York Times* had come on board, "and I received $500 for my story, which both Mr. Marconi and Mr. Sammis authorized me to tell." This was probably equivalent to about a year's salary for Bride. (The paper had offered a total of $1,000 for the two operators' stories.)

The Senate committee's report judged the operators' behaviour severely, concluding that they had withheld information that should have been made public for their own advantage. The disposition of Marconi company officials to allow for the sale of the story and even make the arrangements "subjects the participants to criticism, and the practice should be prohibited. The committee are pleased to note that Mr. Marconi approves of such prohibition." The committee seemed to be oblivious to what it had heard about Marconi being directly involved in the incident. He, as usual, had been juggling a million things, making judgment calls on the fly in what seemed to be his best interests. He had managed to detach himself from the decision and he got away with it, coming off looking heroic and respectable even as his company had its knuckles rapped.

Having been at the heart of what the *New York Times* called "the most important 'news event,' probably, in the history of modern journalism,"[92] Marconi finally sailed for Europe on April 30, 1912, on the North German shipping line's *Kaiser Wilhelm II*, a ship that was assuredly equipped with enough lifeboats for every one of its passengers and crew. His friend and countryman Enrico Caruso was also on board, but the passenger at the centre of everyone's attention was Marconi.[93]

On April 23, 1912, Britain's Lord Chancellor appointed a commissioner, Lord Mersey, to look into the *Titanic* disaster, and on April 30 the Board of Trade requested a formal investigation. The inquiry began on May 2, while Marconi was crossing the Atlantic. Curiously, the Board of Trade appointed Attorney-General Rufus Isaacs as counsel, a task for which he was handsomely paid.[94] Once again, the Marconi company's evidence was paramount to the investigation. "The wireless installation on board the SS *Titanic* was our property," the company wrote to the Board of Trade on May 1, 1912. It had a transmission range of around 370 nautical miles, and a receiving range of as much as fifteen hundred miles from both Poldhu and Cape Cod. In other words, wireless allowed the *Titanic* to be in touch with neighbouring vessels as well as one shore at all times. The operators were engaged, paid, and instructed by the company. "The shipowners recognise in us an expert international concern," able to fulfil their obligations better than themselves, the company wrote. A few days later the company named its delegate to the inquiry: Guglielmo Marconi, just back from America on May 6, 1912.[95]

Although Marconi was perfectly capable of testifying unrehearsed, both the company and the Board of Trade prepared for his appearance carefully. Deputy Manager George Turnbull prepared a detailed briefing note and a tight, six-page memo recounting the *Titanic*'s wireless activity from the time it hit the iceberg till the time it went down.[96] Marconi now had a very well-oiled machine at his disposal. His message, which seemed so spontaneous when he delivered it, was in fact well crafted. But he still didn't know precisely what the inquiry was looking for from him. In the end, Marconi's appearance was anti-climactic—especially when compared to the Venetian drama of the US Senate investigation.

Marconi was examined by Sir Rufus Isaacs on June 18, 1912, the twenty-sixth day of hearings. Isaacs's first question: "Mr. Marconi, you are the inventor of your system of wireless telegraphy?" And the answer: "Yes." Most of the examination was equally mundane. Isaacs asked a string of routine questions about the general functioning of wireless and very little about the *Titanic* specifically. After the US hearing, it was hard to say that anything new or remarkable emerged. The members of the inquiry board seemed mostly in awe, one of them stating simply, "I only want to say we are very glad to have had the honour of seeing Mr. Marconi."[97] The British inquiry reported on July 30, 1912. It recommended that all ships over a certain size be equipped with wireless "and that such installation should be worked with a sufficient number of trained operators to secure a continuous service by night and day."[98] This was precisely what Marconi had suggested.

❀   ❀   ❀

While the British *Titanic* inquiry was still sitting, the long-planned International Radiotelegraph Conference—the third, after Berlin in 1903 and 1906—opened in London on June 4, 1912, under the patronage of British Postmaster General Herbert Samuel and chaired by former GPO permanent secretary H. Babington Smith. This conference was far less controversial than its predecessors; on the other hand, the *Titanic* disaster hung over it like a cloud waiting to burst.

The main item on the agenda was the updating of the 1906 Convention, to formalize the practical realities of "intercommunication." Article 3 of the Convention was updated to read that: "Coast stations and ship stations are bound to exchange radiotelegrams reciprocally without distinction as to the radiotelegraph system adopted by such stations."[99] (The conference also formally adapted the Saint Petersburg International Telegraph Convention of 1875 to explicitly include wireless, or *radio*telegraphy; and it established a system of unique international call signs for transmitting stations that is still in place today: each country is designated a prefix by the ITU and then assigns two or three additional letters to identify every station under its jurisdiction.)

Marconi had long ago accepted the established fact of intercommunication. Now, on his own turf, he put himself at the centre of the conference's social calendar. While Isaacs and company manager William Bradfield were official delegates, Marconi personally attended the conference sessions on June 19 and 26, and hosted two gala events for delegates, one at the Savoy on June 22, and another at Eaglehurst on June 30. The four hundred delegates also paid official visits to the Marconi plant at Chelmsford and the transatlantic station at Poldhu. The conference called on individual governments to oblige certain classes of ships to carry wireless installations. This would be formalized by an International Conference on Safety of Life at Sea, also in London, in January 1914.[100]

In the midst of all this activity (and despite recurring health issues that had him taking to his bed between engagements), Marconi also continued to extend his international reach. Almost as soon as he had returned to England in May he was off again, this time to Spain and Portugal, with Isaacs and Solari. A Spanish company, La Compañia Nacional de Telegrafía sin Hilos, had been formed in Spain in December 1910 to secure Marconi rights for Spain and the Spanish colonies, and to erect a network of high-power and ship stations under concession from the Spanish government. Stations had been built at Tenerife, Majorca, Cadiz, Barcelona, Madrid, and other sites and were ready for international service.[101]

Marconi arrived in the Spanish capital on May 13, 1912. It was his first visit and he stayed long enough to be received by King Alfonso on May 20. Then his party left for Lisbon, where a contract was being debated in the Portuguese parliament. Portugal, too, was looking to establish a system similar to the British imperial scheme, linking the country and its African colonies as an alternative to undersea cables. Marconi had made a proposal, in 1910, to link the mainland, the Azores, Madeira, and Cape Verde. This was the first of three trips Marconi would make to Portugal, but it took another fifteen years before commercial service would begin.[102]

Marconi's short Iberian trip provided a snapshot of how the Marconi company was now integrating its global operations. The day Marconi returned from Portugal—Friday, May 24, 1912—the company published a giant ad in the London *Times*, intended to convince investors and parliamentarians of its worldwide reach in the context of the British imperial wireless contract. A text entitled "The Pulse of Empire" was embedded in a frame featuring a sharp line drawing of Marconi House, the company's spanking new headquarters in the Strand, with two transmitters capped by Marconi banners on the rooftop, and inset scenes depicting the scope of Marconi's far-flung activities: an ocean liner, an oriental gateway, and two stations with huge masts (probably intended to represent Clifden and Glace Bay).[103]

A few days earlier, the company had moved into its new headquarters, located just west of the Aldwych Theatre on the north side of the Strand in the heart of London's "Theatreland" district. The nine-storey fifty-five-thousand-square-foot property was directly opposite Somerset House and next door to one of the city's most popular entertainment spots, the Gaiety Theatre, which opened in 1903 on the site once occupied by the rollicking Strand Musick Hall. The building Marconi's took over had most recently housed the Gaiety Restaurant and a block of unoccupied residential flats on its upper floors. On March 25, 1912, the London County Council granted Marconi's a ninety-nine-year lease for £6,500 a year, and, after a complete overhaul, the company moved in on May 21. Marconi House retained the building's Italian renaissance features, and commentators underscored the striking "artistic atmosphere" one noted upon entering. The entrance hall was panelled in Honduras mahogany and had a counter where customers could hand in marconigrams for transmission to anywhere in the world. A ten-foot-wide staircase dominated the central space, and there was, of course, a modern twelve-passenger lift. Under the window on the first landing was Puck's line from *A Midsummer Night's Dream*: "I'll put a girdle round the earth in forty minutes." There was stained glass on every floor, two dozen fire hydrants, more than a hundred internal telephones, and two aerial masts on the roof. Marconi's private space on the fourth floor facing the Strand included a 240-square-foot lab that by his own admission he rarely used. The company's new home was only a few blocks from its most recent one at Watergate House, a sprawling office complex in Adelphi, but light years away from Henry Jameson Davis's rooms in Mark Lane, next to the Corn Exchange. Marconi House was soon "one of the most conspicuous landmarks of the metropolis."[104] It still is, now as an apartment complex. In 2016, a two-bedroom unit was going for £2.5 million ($3.6 million). The aptly named "Radio Rooftop Bar," opened in 2013 in the adjoining ME Hotel, regularly features on lists of the world's coolest night spots.

21

"The Marconi Scandal"

Even as the British *Titanic* inquiry was in deliberation, skepticism was being voiced in Parliament and the press about the Marconi contract.[1] There was public confusion, with most people and even some parliamentarians believing the March 1912 tender was the actual contract, which it was not. Details remained obscure, and public concern was fuelled by the government's failure to reveal the substance of the agreement. Postmaster General Herbert Samuel asked the government whip, Lord Murray, why there was so much opposition. He was told it must be due to the rumours that the contract was excessively generous to the Marconi company— rumours that had begun in March when shares in both the British and American companies suddenly shot up then dropped again. It was also being rumoured that government ministers had used privileged knowledge to speculate in Marconi shares.

The contract was finally tabled in the House on July 19, 1912, and the very next day the rumours hit print. The most damaging article appeared in the weekly *Outlook*, over the signature of journalist Wilfred Ramage Lawson, who wrote: "The Marconi Company has from its birth been a child of darkness. Its finance has been of a most chequered and erratic sort." The article made much of the family ties linking the government to the company and ominously alluded to the fact that the two Isaacses and the postmaster general were all "of the same nationality"—barely code for the fact that the three of them were Jewish. In the coming weeks Lawson published further articles in the *Outlook* that were rife with smugness, imperialist sentiment, and antisemitism. On August 31, 1912, he questioned whether the Marconi shareholders were "the right sort of people to be entrusted with an all-British scheme of wireless telegraphy." Half of them were Irish nationalists; a fourth were "foreigners, and perhaps not always friendly foreigners"; and of the remaining fourth, "comparatively few are Gentiles." All together they were "a decidedly polyglot multitude" in which to stake the fate of the British Empire.[2]

An even more scurrilous series appeared in the *Eye-Witness*, a weekly founded in 1911 by Hilaire Belloc, the high-profile writer whose views on Jews were infamous.[3] The paper was now edited by journalist Cecil Chesterton, brother of writer G.K. Chesterton, and, like his brother, a frequent collaborator of Belloc's. The first of Chesterton's articles, on August 8, 1912, was entitled "The Marconi Scandal." The label stuck. *Outlook* and *Eye-Witness* were at first the only journals to adopt this tone, but others followed. The *Spectator*, the most respected of the British weeklies, suggested that Isaacs and Samuel would not have consciously lent themselves to secret financial manoeuvres but may have both been "outwitted by shrewd business men."[4] At the other end of the political spectrum, the Marxist newspaper, *Justice*, described the Marconi contract as "one of the most disgraceful jobs to which a great public department like the Post Office has ever been made a party,"[5] showing that the ruling Liberal Party was not concerned with social welfare at all, but solely with the benefit of the capitalist class.

Lawson scrutinized every clause of the contract and summed up the arguments against it in the October 1912 issue of Leopold James Maxse's respectably conservative *National Review*,[6] comparing "the Marconi scandal of 1912" to the eighteenth-century South Sea Bubble. A third minister, David Lloyd George, was now drawn in—as chancellor of the exchequer he was responsible for Treasury expenses (and, as was yet unknown, he was also a purchaser of American Marconi shares). When it was openly suggested that the opposition to the Marconi contract was motivated by antisemitism, Hilaire Belloc commented "anybody less Jewish than the Chancellor of the Exchequer I cannot conceive."[7] The criticisms raised by Lawson and others included questions about the cost and duration of the contract, why there had been no competing tenders, why the claims of other inventors were being ignored, and why the Marconi company was being granted a monopoly. The criticism obscured something that should have been obvious: the Marconi company and the government, more specifically the Post Office, had a long history of hostile and difficult relations and yet were now entering into a close partnership.

There were three possible explanations. The one that attracted the most sensational media attention was the least plausible and could be summed up in a single word: corruption. The one put forward by the government, the most rational, would soon be upheld by a committee of experts: Marconi's was the only company equipped to deliver this type of scheme. The third, and closest to the right answer: once it had been decided to green-light the project, the company's established position in public opinion made it the only logical and politically acceptable choice. According to one author, in England in 1912, the words *Marconi* and *wireless* would have been synonymous to a reasonably informed child of twelve.[8] Marconi's had recently acquired the patents of the only other British wireless company, the Lodge-Muirhead Syndicate. There remained two possible foreign competitors: the Danish Poulsen group, which was completely untried, and, of course, Telefunken. Samuel had considered looking abroad but he would have had a difficult time getting the idea past his Cabinet colleagues. Home Secretary Winston Churchill, for one, had a strong view, having stated, when Samuel consulted him: "I think it is very undesirable to give Government patronage to foreigners."[9]

The affair also has to be considered in the context of British politics. The Liberal Party, headed by Sir Henry Campbell-Bannerman, had been invited to form a minority government in December 1905 and won a decisive election victory a few months later. Over the next few years, the Liberals would begin to lay the foundations of the British welfare state. Campbell-Bannerman was succeeded by Herbert Asquith in 1908, but after two general elections in 1910, the Liberals were able to govern only with the aid of the Irish Nationalist Party. The Marconi contract thus arrived in a tough political context, as well as amid growing unease in England over the threat of rising German military strength.

In June 1912, taken up as he was with the *Titanic* inquiry, amid the swirl of the brewing controversy and possibly in connection with his nomination to Cabinet that month, Rufus Isaacs told Samuel of the ministers' transactions in American Marconi shares.* Asquith's biographers have claimed he was not told until six months later; Samuel said he informed him immediately. They both thought it was a foolish thing to do, but no more than that. Most important, Samuel knew that Marconi's was the only possible contractor for the Imperial wireless project, and recognized the urgency of getting it going.

After the *Eye-Witness* appeared with its allegations of corruption on August 8, 1912, Isaacs and Samuel discussed whether or not to take legal

---

* Isaacs was the first attorney-general to be elevated to Cabinet status; he joined the Cabinet in June 1912 after being named to the position in October 1910.

action, asking Asquith's opinion on this point as well. All agreed it was best to do nothing. They may have underestimated the influence of the paper; its contributors at the time included authors like Arthur Ransome, G.K. Chesterton, and H.G. Wells, and it was widely read by a small but influential circle of elite intellectuals. The fact that the politicians did not counter its attacks showed poor judgment, but there is no indication that they considered the matter further even as both the *Eye-Witness* and *Outlook* persisted in their assault. In October 1912, *Eye-Witness* went bankrupt and Cecil Chesterton, without missing a stroke, borrowed money from his father and started another paper, *New Witness*, in which he continued his anti-Marconi crusade.

Despite the Opposition rumblings, Herbert Samuel thought that ratification of the contract would be a mere formality. However, the press revelations blew that hope away and he couldn't get it approved before the House rose on August 7, 1912. When the House returned in October, Samuel proposed to refer the contract to a parliamentary select committee.[10] Asked what the Post Office was buying from Marconi, Samuel replied: "We are buying the use of all their patents. . . . We are buying . . . the unpatented inventions, the secret inventions, which I am told are really of considerable importance. We are buying what is more valuable still, their experience of long-range wireless telegraphy, which they alone have. . . . Not only are we buying the existing patents, but all future patents."[11] Furthermore, the Post Office, not Marconi's, would work the patents. Rufus Isaacs told the House he only learned a contract was in the offing when his brother Godfrey informed him about it at a family function a few days before the tender was accepted in March (he was already attorney-general but not yet in the Cabinet at that time). Isaacs stated unequivocally that neither he, Lloyd George, nor Samuel had ever transacted in Marconi shares—making a distinction, in his mind, between the parent company and its American subsidiary, but not considering that distinction worth sharing with his parliamentary colleagues.[12]

The select committee began sitting on October 29, 1912.[13] The hearings were long and tortuous and it soon became impossible for anyone to follow them without devoting full time to the enterprise, and even then it was a challenge. The committee reported on January 13, 1913, concluding that an Imperial wireless chain should be established and recommending that a technical advisory committee be struck to report on the merits of the existing systems of long-distance telegraphy. The Post Office accepted the recommendation, and a committee chaired by Judge Robert Parker was appointed on January 23, 1913. (Lord Parker had ruled in Marconi's favour in a 1911 patent case.)

On January 15, 1913, Godfrey Isaacs wrote to Samuel stating that the company was incurring huge costs due to the delays in executing the contract and asking for it to be cancelled. Marconi himself was called before the select committee on January 27, 1913, to state why the company wished to withdraw from the contract. But he refused to limit himself to this topic. When the chair asked him to state his reasons for a withdrawal, Marconi replied that answering that question meant replying in full to "all that has been said in this room,"[14] meaning the attacks against his honour. After withdrawing to consider Marconi's request (which was really a demand) in camera, the committee decided that Marconi would be heard in full at a later date.

On February 12, 1913, the question of the ministers' share dealings resurfaced, when Maxse, in the *National Review*, made the point that ministers should be quick to deny their dealings in shares of *any* Marconi company—emphasizing the word *any* in the written text accompanying his testimony. The following day, a Parisian newspaper, *Le Matin*, implied that Maxse had presented proof of collusion between the Isaacses and Samuel. Despite the paper's prompt retraction, Rufus Isaacs and Herbert Samuel sued for libel. The case came quickly to court and on March 19, 1913, the purchase of the American shares was almost casually introduced by Isaacs and Samuel's own lawyer, Sir Edward Carson—mentioning that Isaacs actually lost money on the deal after the share prices crashed. The strategic decision to reveal the stock transactions was backed up by political efforts to control how the story would come out in the press. Winston Churchill was dispatched to visit Lord Northcliffe, owner of the *Times*, at home in his bedroom to ensure sympathetic coverage in the paper. Informed that this was happening by Conservative Opposition leader Andrew Bonar Law, MP Max Aitken (the future Lord Beaverbrook) hurried to see Northcliffe as well, only to learn that Churchill had already been. On March 20, 1913, the *Times* lead editorial said the ministers might have shown more delicacy in the selection of their investments, but mere lack of judgment did not justify the monstrous offences that had been imputed to them. The ministers won their libel case, and on March 21, 1913, Lloyd George wrote thanking Northcliffe for the "chivalrous manner" in which he had treated them. While it now looked good, he added ominously: "The atmosphere is now a morbid one owing to the controversy that gathers round Marconi enter-prises." Rufus Isaacs also wrote to thank Northcliffe.[15]

On March 25, 1913, Rufus Isaacs appeared before the parliamentary committee. At that date he stood with a net loss of £1,300 on his Marconi shares. This was the first time the committee, and the public, heard the details of the events—the brothers' meeting at the Savoy, Harry's purchase of Godfrey's stock on April 9, and Rufus's purchase from Harry on the seventeenth. After the American company's restructuring was announced on the eighteenth, Rufus said, Harry and he, as well as Lloyd George, then sold some of their shares. Lloyd George testified that he had later bought back more shares, making the series of transactions look more like speculation than investment. Lloyd George had a record of moralizing about the ethical standards of politicians; it now came back to bite him. The other minister who had bought shares, government whip Lord Murray, had left politics and was now representing a British company in South America. Murray, who also went by the odd noble title the Master of Elibank, was a well-known figure in London of his day, described by a contemporary as "always cheerful and considerate.... His ample figure and full-moon face, with its fringe of curls, were always a pleasant vision and he had a persuasive manner that was hard to resist."[16] However, it emerged in testimony from other witnesses that Murray's purchases of American Marconi shares were made using funds from a trust he controlled on behalf of the ruling Liberal Party (he had topped up his initial lot with some additional purchases of his own). After the ministers' performances before the committee, Northcliffe felt he had been duped and wrote to Churchill: "Your Marconi friends stage-manage their affairs most damnably."[17] The *Times* began taking a harsher editorial line.

Meanwhile, at Marconi's Irish installations, the land issue hadn't gone away. On Sunday, March 2, 1913, Kemp arranged for Marconi to meet at the Railway Hotel in Clifden with a group representing tenants from the villages of Currywongaun and Mullaghglass. There was much discussion about the placement of masts. Two local clergy were appointed to act as arbitrators, but before they could give their verdict Marconi was called back to London by the death of his mother-in-law, Lady Inchiquin, on March 3, 1913.[18] On March 4, the arbitrators recommended an award to the tenants of £162 (around one pound per square foot) for the company's encroachment on what Kemp considered to be very inferior bog land. It was a huge sum to Kemp, who was then earning less than six pounds per week.[19] He was indignant and urged Marconi to lobby Irish members of parliament "to write to these two arbitrators in the strongest possible manner and get them to go and drive fear into these pirates." Kemp's reaction places this small conflict in the wider context of developing Irish home rule. Effectively, Kemp was proposing to bully the tenants into accepting a lower settlement by convincing them that they were injuring the Irish case. If Marconi decided to pay, Kemp suggested that he funnel the money through one of the Irish nationalist organizations "and so make them think they have robbed their own cause; I am sure this will make them respect us more in future and prevent them ever trying another piratical outrage when they want some more money."[20]

Marconi had no appetite for being distracted by Ireland's volatile and highly contentious politics. Despite his deep and ongoing ties to Ireland—and his penchant for courting whoever was in power in any situation—he managed to remain uncompromised, maintaining excellent relations with all sides of the Irish question (a bit in the tradition of his O'Brien in-laws). Marconi wisely let the tenant issue blow over; to him it was just another part of the cost of doing business, and he certainly wasn't going to let it involve him in the quagmire of Irish politics. Land issues continued to harass the Letterfrack operations, but Marconi was never personally involved after March 1913.[21]

Marconi returned to Clifden from mourning his mother-in-law almost as quickly as decently possible, on March 11, 1913, bringing with him his sister-in-law Lilah. On March 12, Marconi, Lilah, and Charles Franklin observed the progress Kemp had made with duplexing. The system was working beautifully and Marconi was confident enough to invite the Parker Committee to come out to witness a demonstration. The committee visited on March 26, "a great duplexing day" according to Kemp, and again on April 26, when they witnessed the transmission of a sixty-word press dispatch to Glace Bay over so-called continuous waves by Marconi himself.[22] Continuous wave communication, developed in the United States by Reginald Fessenden and perfected by Henry Round's experiments and James Ambrose Fleming's thermionic valve, became the basis first for wireless telephony and eventually broadcasting. Always taking the high road, Marconi told the select committee on May 7, 1913: "I have now at Clifden a system utilizing continuous waves and employing no spark whatever in the transmission of messages, and I am right in saying that it is still a Marconi system."[23]

Marconi rarely (if at all) returned to Clifden, or indeed Ireland, after these demonstrations. His last recorded traces are a note on Clifden stationery addressed to an admirer by the name of Monica Brodgen, dated August 20, 1913 (auctioned at Bonhams, in London, in 2013), and a signature in the Dromoland guest book on September 1, 1913, when he visited his brother-in-law Lucius, Lord Inchiquin, to discuss arrangements for looking after the recently orphaned O'Brien daughters.

❀   ❀   ❀

April 1913 was a difficult month for the government. Asquith, during one of his weekly meetings with the king, told his sovereign the ministers had offered their resignations early on and though he felt their behaviour "lamentable," he had refused to accept them.[24] Marconi told his friend John Milholland that it had been Asquith who had advised the ministers to maintain silence when he learned that their dealings had been with the American and not the British Marconi company. "It was bad advice and all realize it now," Marconi told Milholland.[25] During all of this, it took Winston Churchill, now First Lord of the Admiralty, to make the main point: the delay in ratifying the Marconi agreement had deprived Britain of the advantages it had hoped to gain by developing an imperial chain of wireless stations.[26] Churchill also created one of the liveliest scenes at the select committee hearings when he appeared after the editor of the *Financial News* told the committee that Churchill, too, had bought Marconi shares. Churchill, having stormed in fuming, vigorously—and truthfully—denied the claim: "I have never at any time, in any circumstances, had any investment or any interest of any kind, however vaguely it may be described, in Marconi telegraphic shares or any other shares of that description in this or any other country of the inhabited globe—never," he said, as only Churchill could. After a few brief exchanges with committee members who tried to appease him, the First Lord of the Admiralty asked, rhetorically, "May I assume that your examination of me is finished?", then rose and left.[27]

The Parker Committee visited installations in Germany, Denmark, and Clifden, took technical evidence from parties including Marconi, and reported on April 30, 1913, that the Marconi system was the only one capable of fulfilling the requirements of the imperial chain; only Marconi's had the requisite practical experience in long-distance communication.[28] The committee recommended reopening negotiations with the Marconi company. Marconi finally had his day in court, or rather before the parliamentary committee, on May 7, 1913. Here he stated unequivocally that he had never speculated in his company's shares. In response to questioning from Lord Robert Cecil, Marconi qualified that statement with some fancy semantic footwork.

CECIL: I see you say you have never bought or sold a share during the boom of any of these companies.

MARCONI: No. What I mean there is this. . . . I never bought or sold shares from the time they went well over £5 to when they went to £9, and back to £5 or £5½. I mean to say that when they were £7 or £8 or £9, I did not have anything to do with them.

CECIL: The only reason I thought it right to put the question is that evidence has been given that you did buy 10,000 of those American shares on or about the 9ᵗʰ April.

MARCONI: That is perfectly correct, but I really bought those shares before because I agreed to take them. . . . It is one of the things I always do in these companies—I take a part of the issue as a rule. . . . Just to show I have confidence, and because I have confidence . . . I told Mr Isaacs that I wanted a certain number of these shares before that time.[29]

His appearance before the committee cleared Marconi of any suggested or apparent wrongdoing in his own purchase of the new American share offering, and he easily side-stepped any implication in what had happened next. Isaacs had returned to England while Marconi remained in the United States, where he became publicly embroiled in the wake of the *Titanic* disaster. Isaacs alone handled the contract negotiations with the Post Office, informing "nobody at that time but Mr. Marconi personally."[30] Strictly speaking, this meant that if Isaacs himself was not the source of leaked information that could fuel speculation in company shares, it could have only been Marconi. But Marconi denied telling anyone about the shares. "I would not have dared to do such a thing. . . . No, it is rather against my principles."[31]

There was also a juicy sideshow to the affair. After Cecil Chesterton published an article in the *New Witness* in October 1912, repeating the corruption allegations and recounting Godfrey Isaacs's "ghastly record" in business before joining Marconi's (a record, according to the paper, distinguished by failed mining speculations and taxi-cab companies), Chesterton was charged with criminal libel. The trial began at the Old Bailey on May 27, 1913, and, as a result, Chesterton never appeared before the parliamentary committee. Chesterton was considered a good journalist, a practitioner of "lively and pungent" invective, and an orator who could hold his own in London debating societies with the likes of George Bernard Shaw and Beatrice and Sidney Webb.[32] But his writing was openly, viciously anti-Jewish; he was also considered conceited and arrogant, and in court, he seemed to ignore the law. The Marconi ministers, and Marconi himself, gave evidence on behalf of Isaacs, and a jury found Chesterton guilty of criminal defamatory libel, a conviction that could have landed him five years in jail. When the judge sentenced him to a fine of £100 plus costs, Chesterton's supporters considered it a victory, despite the fact that he had withdrawn his allegations.[33]

With the Parker Committee endorsing Marconi as the only possible contractor for the imperial system, negotiations between the Post Office and the company resumed, and on July 30, 1913, the details of the contract were published. It was between three parties: Postmaster General Herbert Samuel, Marconi's Wireless Telegraph Company Limited, and Commenda-tore Guglielmo Marconi. This unusual tripartite agreement, committing not only the Marconi company but its founder and chief scientific advisor per-sonally (and, inexplicably, using his formal Italian title),[†] underscored both the unique relationship of Marconi to his company and the fact that when contractors dealt with the company they expected to be dealing with *him*.[34] The contract was finally ratified by Parliament a few days later. However, the lost year had curbed the enthusiasm of all sides.

After nine months of hearings, the parliamentary select committee looking into the ministers' share issue completed its work and withdrew to produce a report. Both the Liberal majority and the minority Conservatives agreed that the ministers' purchase of American Marconi shares had been indiscreet, but they could not agree on the tone and terminology to adopt, nor on measures to propose. After several drafts, there were two reports, one adopted by the majority of the committee, and a second signed by inde-pendent Conservative Lord Robert Cecil (who was later one of the archi-tects of the League of Nations and Nobel Peace Prize winner in 1937).[35] The chief claim of Cecil's report was that the Post Office contract would be of great benefit to all Marconi interests because of its global reach and impli-cations for future business. It gave them a great advantage over competitors present and future and would induce inventors to sell their patents to them. Cecil mentioned that during the boom, shares of other Marconi subsidi-aries, such as the Spanish and Canadian, also rose. He said Marconi's recollec-tion on the subject of how Isaacs got to be in charge of selling the shares did not appear to be very clear, while the ministers' impropriety was making an advantageous purchase on the basis of information received from a govern-ment contractor. Cecil seemed to have characterized the ethical issues well.

[†] Marconi was big on titles, but in England at this time he was usually addressed in the conven-tional manner as "Guglielmo Marconi, Esquire."

When the reports appeared, Marconi's name again dominated the news and commentary on current affairs. The *Times* published six editorials on the subject in June 1913. The majority report was termed a "whitewash," and Rufus Isaacs and Lloyd George, while cleared, were tainted. Isaacs, though maintaining an outward calm, was privately "sadly changed," while Lloyd George, according to one of his biographers, "lost weight, lost vitality, fell ill...his black hair grew grey, the lines began to mark his face."[36] On June 18, 1913, the Conservative Opposition moved in the House that it "regrets" the transactions of certain ministers in the American Marconi company and "the want of frankness" they had displayed in their statements on the subject to the House. The motion was debated for two full days; Godfrey Isaacs was in the visitors' gallery the whole time. Three new ethical rules were suggested: a minister should not use confidential information received as minister; he should not invest in companies benefiting from government contracts; he should not receive any favour from someone contracting with the Crown.

Rufus Isaacs finally agreed that he and his colleagues had been mistaken in not making a full disclosure when the issue of share purchases was raised, but he managed to put the best possible spin on it: "It never occurred to me that any human being could suspect me of corruption because I purchased American Marconi shares some six weeks after the announcement was made of the acceptance of the tender of the British Marconi Company by the British Government."[37] Had he known that he would be so strongly and unfairly accused, he said, "I should not have gone into the transaction." Lloyd George, more vehemently, made a similar statement: "I acted thoughtlessly, I acted carelessly, I acted mistakenly, but I acted innocently, I acted openly and I acted honestly."[38]

Asquith spoke on the second day of the debate, denouncing "the most disgraceful appeals [that] were made from the beginning to racial and religious animosity." The rules of conduct the Opposition was suggesting for ministers and other officials, he said, would create "a most extravagant and hysterical standard" that would make it impossible for business men to take part in the government of the country. He then laid out the rules that *he* thought should be observed, in terms similar to the ones the Opposition had outlined, and these "Asquith rules" became the unofficial ethical guidelines for British parliamentarians for the next fifty years.[39] The House adopted a majority amendment to the original resolution, accepting the expression of regret of the two ministers.

The "Marconi scandal" continued to echo in British politics.[40] Its immediate impact was the defeat of Asquith's candidates in two by-elections in which the Marconi case was treated as a major issue. But most of the principals did remarkably well. Rufus Isaacs was named Lord Chief Justice in 1913, Baron Reading in 1914, British ambassador to the United States in 1918, viceroy of India in 1921, and finally foreign secretary in 1931. Lloyd George went on to a historic role as Britain's last Liberal prime minister (succeeding Asquith), from 1916 to 1922, although his premiership was compromised by the memory of the Marconi affair and a sense that he owed his political career to powerful defenders who had forced closure to scrutiny of his affairs.[41] Herbert Samuel became home secretary and was the first elected British politician to put forward the idea of establishing a British Protectorate over Palestine, in November 1914; in 1920 he was appointed first high commissioner of the new protectorate. Lord Murray, the Master of Elibank, returned to England from South America, was tried in the House of Lords in 1914, and was acquitted of dishonourable conduct.

Marconi's contract with the Post Office did not survive the outbreak of war. The Post Office cancelled the agreement at the end of 1914 and the company sued for breach of contract and to recover its costs (Isaacs said in 1918 that the company had spent £140,000 before the contract was cancelled[42]). It was awarded £600,000 in compensation when the suit was settled in 1919, at which time the company approached the Post Office with a new offer. Fraught with political obstacles, as any dealing with the Marconi company now was, another seemingly endless round of committees and inquiries saw to it that nothing happened for seven more years. The imperial wireless chain was not built until 1926, and then only after Marconi intervened directly, informing the British government that he had discovered a new method for long-distance communication, using shortwaves.

The Marconi scandal left legacies on two fronts. It was the first case of dubious financial dealings by government members to get a full airing in the British Parliament, creating a new baseline in ethical standards for politicians. The tone of the Opposition and the press also marked the introduction of a new style of politics, where personality attacks and veiled innuendo began to replace the rational discussion of political differences.[43] And the affair also established a new boundary for respectable expressions of superiority and concern over the growing place of Jews in the upper levels of British society. A half-dozen Jewish MPs won seats in the general election of December 1910 (re-elected among them was the cable-rate reformer

Sir Edward Sassoon[44]). Of these, Herbert Samuel and Rufus Isaacs were the first self-acknowledged Jews to serve in a British Cabinet (the nineteenth-century prime minister Benjamin Disraeli left Judaism at the age of twelve, long before he ever entered politics). The slow but steady inclusion of Jews in British political life was even more marked in the upper levels of the civil service, where a Jewish businessman's son like Matthew Nathan could rise to become the governor of Hong Kong (Kowloon's main street, Nathan Road, is named after him) and then, in 1909, permanent secretary of the Post Office—where he was the government's chief negotiator for the Marconi contract, a fact that somehow eluded the scandal-mongers.[45]

So the Marconi scandal set the stage for a new type of antisemitism in Britain.[46] Where anti-Jewish prejudice was previously aimed at excluding Jews from the preserves of established society, like politics and government, it now aimed to set a higher standard for those who had been admitted, implying that they were there on sufferance, their good fortune a result of poor judgment on the part of the Gentile establishment. As just one example, outraged at the appointment of Rufus Isaacs as Lord Chief Justice in the wake of the scandal, Marconi's one-time admirer Rudyard Kipling published a poem rife with virulent antisemitic stereotypes, in which the title biblical character, Gehazi, Elisha's avaricious servant, gloats: "My zeal hath made me a Judge in Israel. . . ."[47] In short, a Jew—any Jew—in Edwardian England was still a fair target for racial humiliation.[48] This sentiment was further coloured by a feeling among the pre–First World War British elite that their Jewish fellow citizens could not be fully trusted in view of their roots in central Europe (although many of Britain's most prominent Jewish families, like the Sassoons, were of Sephardic or Middle Eastern origin).

G.K. Chesterton, whose brother Cecil had been one of the leading journalistic voices to turn the Marconi affair into an ethno-religious crusade, signed another exemplary sample of this thinking in an open letter to Rufus Isaacs (now Lord Reading) in *New Witness* on December 13, 1918, after it was rumoured that Isaacs was about to be appointed a member of the British delegation to the Paris Peace Conference. The spectre of a Jew negotiating the fate of the defeated Central Powers was too much for this blue-blooded Englishman to bear. Characterizing Isaacs as "the chief Marconi Minister," a "stray stockbroker who has somehow turned into a Lord Chief Justice," Chesterton wrote: "Are we to lose the War which we have already won? . . . Is there any man who doubts that you will be sympathetic with the Jewish International? . . . Do you seriously imagine that those who know, that those who care, are so idolatrously infatuated with Rufus Daniel Isaacs as to tolerate such risk, let alone such ruin?...Daniel son of Isaac, Go in peace; but go."[49] In short, Jews will be Jews but stay out of British politics.

The Marconi scandal was a reference point in discussions about political corruption in Britain well into the 1930s and beyond. G.K. Chesterton himself declared it a watershed in British politics: "It is the fashion to divide recent history into Pre-War and Post-War conditions. I believe it is almost as essential to divide them into the Pre-Marconi and Post-Marconi days," he wrote in his 1936 autobiography.[50] This indelible stain was much to Marconi's chagrin, but painful as it was, it became just another irritation he got used to and dealt with, like his lost eye or his recurring bouts of malaria. Like almost everyone touched by the affair, Marconi regretted the short-sighted dabbling of government ministers in his company's shares. To his credit, however, he never bought in to the antisemitic hectoring that was too often associated with the disgrace brought on to his good name—a name that had now actually become a *target* of antisemitic polemic. To the contrary, both publicly and privately, he always remained fiercely loyal to Godfrey Isaacs, proud of having brought him in and gratefully crediting him at every opportunity with having finally put the company on a path of financial success. According to Frances Donaldson, author of the definitive account of the Marconi scandal, "the one person whose reputation was never assailed or lowered was Signor Marconi himself."[51] Marconi emerged personally untainted, but his name, already interchangeable with wireless telegraphy, now became synonymous (in Britain, at least) with share speculation and dirty politics. The experience was an important milestone in the drawn-out process of his alienation from the British.

22

The Invisible Weapon

Marconi returned to the United States in June 1913 (to give evidence in the company's suit against Reginald Fessenden's company, NESCO[1]), and this time it was with Beatrice.[2] On June 12, 1913, while dining at the Holland House in New York, the Marconis introduced Inez Milholland to Eugen Boissevain, "a strapping Dutchman with a bacchanalian laugh and a poet's soul." When Boissevain learned that Inez, who had recently finished law school, was sailing to England with the Marconis on July 2, 1913, he booked himself passage. According to Inez's biographer, Linda Lumsden, "Boissevain matched Milholland in his unconventional views and lust for life," and the chemistry between them was explosive. After a chaste crossing, during which they could barely contain themselves (Inez was staying in the Marconis' suite), the new couple travelled with their hosts to Eaglehurst and after a night of love-making in the tower (which both of them recalled frequently, and graphically, in letters to each other for the rest of their life together), they married as soon as they could in the Kensington registry office. There were only two witnesses; Marconi, one of them, was best man.[3]

Back in New York, John Milholland was appalled. He had met Boissevain while seeing Inez off on the *Mauretania* and found him to be "an excitable young man of French appearance." Informed of the marriage by cable from London, he wired his daughter for confirmation. "I fear the worst," he told his diary. The only thing Boissevain had going for him, in Milholland's eyes, was a character endorsement from Marconi (who was mentioned in press reports as a close friend of the bridegroom). When Milholland finally met his son-in-law, in London at the end of July, he described him as "one of life's bitterest disappointments."[4] Milholland was exercising conventional paternal concern. Although a progressive figure in American politics, he was also a teetotaller and an evangelical Protestant. He respected his daughter's progressive politics but could not comprehend her lifestyle.[5] But he admired Marconi, whose attraction to Inez was just the opposite. Bemused by Inez's politics, which he regarded as dalliance, Marconi was strongly attracted to her as a woman.

Indeed, on the surface Boissevain was as far from Marconi as one could imagine—a friend described him as "handsome, reckless, mettlesome as a stallion."[6] Temperamentally Marconi's opposite, Boissevain brought something out in him. Like Marconi, Boissevain was Irish Protestant on his mother's side (his father's family was Dutch), and Eugen and Guglielmo were said to enjoy going on an Irish pub-crawl when they found themselves together in New York. Marconi purposefully set up the meeting between Inez and Boissevain, expecting them to hit it off. He had been fascinated by her sexuality since meeting her when she was an adolescent, but despite his fame and celebrity, he felt insecure about her. Marconi told Boissevain he thought she needed someone more masculine, or "stronger" than himself, whatever her feminist protestations; and this was evidently the case.[7] Eugen and Inez's relationship was driven by a powerful erotic connection that she, particularly, lived out through sexual fantasies, masturbation, flirtations, and fleeting extramarital adventures when Eugen was unavailable.[8] Not only were Marconi and Boissevain opposites, their relationships with their respective wives were about as different as marriages could be.

❧   ❧   ❧

Marconi was increasingly involved with his in-laws, the O'Briens. On September 1, 1913, he stopped at Dromoland on his way to Clifden, in order to discuss with Lucius and Ethel arrangements for the financial support of Beatrice's sisters following the death of Lady Inchiquin.[9] Then he moved Beatrice and the children to Rome and settled them in to the Hotel Regina. Degna and Giulio were enrolled in the Montessori school (on Inez Milholland's recommendation, as we've seen), and quickly became little celebrities. When Maria Montessori herself arrived from the United States on January 2, 1914, the newspapers reported that she was greeted by teachers and pupils from the school, among them the children of Guglielmo Marconi.[10]

Marconi, meanwhile, was mainly in England until early March, when he travelled to Rome to deliver a prestigious lecture at the Augusteum on March 3, 1914, to which he invited the entire royal court. Augustus's tomb had terrible acoustics, but Marconi worked the audience of three thousand to "a pitch of popular enthusiasm." After the evening lecture was over, he left the hall in secret, to avoid a repeat of the near-riot his presence had created at his last major speech in Rome, in 1903.[11] He then joined the Italian naval warship *Regina Elena* for a research cruise along the Italian coast and around the shores of Sicily.[12] Accompanied by his associate H.J. Round, Marconi was to do tests on voice transmission (or, as they called it, wireless telephony), which was beginning to attract a great deal of attention among both military authorities and the public. Round had been experimenting with wireless telephony since doing tests between Battery Place and Times Square in New York, and "eavesdropping" on Reginald Fessenden's initial broadcasts in the United States in 1906. Round's reports in the Marconi Archives also make clear the driving role he played in this research. He is justly considered a pioneer of signals intelligence, direction finding systems, and broadcasting technology.[13]

Marconi saw new commercial as well as military possibilities for transmitting voice over wireless, and especially mobile wireless communication. "The popular anticipation of pocket wireless telephones by means of which a passenger flying in an aeroplane over France or Italy might 'ring up' a friend walking about the streets of London with a receiver in his pocket cannot be said to have been as yet practically realized but there is nothing inconceivable or even impracticable about such an achievement and the progress of wireless telephony seems to be pointing in that direction," he wrote after completing these experiments.[14] The parallels with what we do today are obvious; that Marconi foresaw it a century ago is at least uncanny, if not downright eerie.

Working with Round, Marconi kept even his longest-standing associates at arm's length. Fleming, for one, learned about the new experiments from the press, even though he was himself working on a similar problem.[15] Fleming read that Marconi had succeeded in transmitting speech over one hundred miles. This was a great exaggeration; the actual results were much less sanguine. While on the *Regina Elena*, Marconi and Round received "extremely loud and clear" voice signals at distances ranging from a quarter of a mile to twenty miles, but there was a big problem: voice transmission seemed to jam the conventional wireless communication (by Morse code) with long-distance stations at Clifden, Glace Bay, Tobruk, and Massawa—all of which were in normal range of the *Regina Elena*. Press reports, though, were triumphant. Interviewed on his return to Rome, Marconi was quoted as saying, "The problem of wireless telephony or radio-telephony is practically solved. . . . My conviction is that the day is not far off when the human voice will cross the Atlantic." His new apparatus had been adopted by the

Italian navy, and "will shortly replace entirely all other kinds of signaling." The Marconi company claimed this was the first practical use of wireless telephony.[16] Marconi was being disingenuous in these prophecies. The first acknowledged radio transmission of human voice across the Atlantic was indeed made in 1915, but it was done by the Western Electric Company, a subsidiary of AT&T, from the US Navy station at Arlington, Virginia, to the Eiffel Tower in Paris.[17]

❋   ❋   ❋

Coincidentally (or not), Marconi's movements kept overlapping those of his family; wherever they were, he always seemed to be somewhere else. In the spring of 1914, Beatrice and the children finally returned to England and Eaglehurst. Throughout, whether Beatrice was with him or not, Marconi filled his evenings in London with socialite friends, dinners, and theatre outings. If he wanted to see a headline show, say George Bernard Shaw's new hit *Pygmalion* at His Majesty's Theatre, he went directly to the top for tickets, in this case requesting them from Herbert Beerbohm Tree, who not only owned and ran the theatre but was also starring in the role of Henry Higgins.[18] (Tree was, like Marconi, a bridging figure between Victorian and Edwardian London.)

In July 1914 Marconi finally received the British knighthood that had long eluded him. Marconi's friend, the Conservative MP John Henniker Heaton, had been lobbying for this for years, informing him in June 1912 that "my request will be granted." But there was a big problem. A knighthood had to be initiated by the prime minister—whose government, in 1912, was already under scrutiny for perceived favouring of Marconi, as we've seen. On June 18, 1912, Heaton received a note from Lord Stamfordham, private secretary to the King, advising him to take up the matter with Asquith. Heaton updated Marconi on June 24: not to worry, he would work it out with the PM. "I asked him finally to trust me and if there was anything alleged to give me the opportunity of explaining and denouncing the slanderers." Then he added, "Destroy this letter." No more was heard of the matter for two years, when Stamfordham again wrote to Heaton: "I spoke to the Prime Minister, who I know took the case into consideration, but I suppose it was found impossible to submit [Marconi's] name for an honour." Heaton wrote immediately to Marconi: "Everything that could be done was done.... I told him [Stamfordham] that you would feel slighted and would leave this ungrateful country if we did not do honour to ourselves." Heaton then wrote directly to Asquith, literally harassing him with what must have seemed quite a trivial detail when he should have been totally preoccupied with the international situation (not to mention growing domestic unrest and the festering Irish question). The reply from Downing Street came quickly: Marconi was not a naturalized British subject and was hence ineligible. Heaton pointed out that neither were hundreds of others who had received the honour. Finally, on July 24, 1914, Marconi was received by the King at Buckingham Palace and awarded the Grand Cross of the Victorian Order.[19] Coming so shortly after the Marconi scandal, this was at least some recognition that as far as the King's Court was concerned business was business and everything was all right.

It was too early to tell, but whether it was still avoidable or now inevitable, Europe was headed for war. The inter-imperial rivalries that had been contained for the previous forty years—Marconi's entire lifetime, in fact—were about to explode. The latest crisis was sparked on June 28, with the assassination of the heir to the Austrian throne, Archduke Franz Ferdinand, in Sarajevo, ancient capital of Austria's Balkan province of Bosnia. There had been previous crises, many of them, and most sophisticated Europeans were sure this one would be weathered like the others.[20] On July 28, 1914, the Austro-Hungarian Empire declared war on Serbia; but despite the intensifying international situation, for Marconi it was still business as usual.

In late July, with parts of Europe already at war, a delegation of Marconi senior engineers visited Berlin, where they were hosted with courtesy and hospitality by Telefunken. The grand finale of the visit was an inspection of the German high-power station at Nauen, with its massive, newly installed antennas. Nauen was the centrepiece of Telefunken's increasingly global commercial network, but as soon the Marconi delegation left, the station was closed to normal operations and taken over by the German military, who had been standing by, waiting for the foreign visitors to leave.[21] On July 30, a wireless message from the British Admiralty to the Grand Fleet cancelled all naval leave, an ominous precaution; on August 1, the use of wireless was suspended for merchant vessels in British territorial waters, and on August 2, the government took control of all wireless communication. On August 3, experimental stations in the United Kingdom were closed.[22] A few weeks earlier, Marconi had said in a public speech, "The value of wireless telegraphy may one day be put to a great practical and critical test; then perhaps there will be a true appreciation of the magnitude of our work." That test had now come; wireless became what writer Harold Begbie called "the invisible weapon of war."[23]

For England, the war began at 11:00 p.m. Greenwich Mean Time on Tuesday, August 4, the deadline it had given Germany for a guarantee of Belgium's neutrality. As the deadline passed, First Lord of the Admiralty Winston Churchill sent a wireless message to the fleet: "Commence hostilities against Germany."[24] The German wireless station at Nauen immediately sent a message to all German merchant ships, calling them to make for the nearest neutral port.[25] Listening in at a Marconi station, engineer H.J. Round intercepted a message in German that read: "State telegram from Berlin, war is declared against France, Russia and England." The Marconi company thus claimed it was among the first outside of official circles to learn of the outbreak of war.[26] Within hours, British ships cut German cables in the Atlantic Ocean and the North Sea, leaving the four-thousand-mile wireless connection between Nauen and Sayville, New York, as the only German communication link to the neutral United States (France and the United States had proposed that cables be considered neutral in wartime; Britain refused).[27] A few weeks later, German technicians destroyed the Telefunken wireless station in the West African protectorate of Togoland, as it was about to be captured by French and British troops; soon the only wireless station Germany had left outside Europe was at Windhoek in German South West Africa (now Namibia). Effectively, Nauen was Germany's only direct link to the rest of the world.

Italy formally proclaimed its neutrality on August 3, the only major European power to do so.* Italy had joined Germany and the Austro-Hungarian Empire in the Triple Alliance, formed in 1882 in the wake of the French conquest of Tunisia the year before. But Italy's relations with Austria were tense, going back to the struggle for Italian unification in the 1860s and '70s, and made worse by the Austrian annexation of Bosnia-Herzegovina in 1908 and the Italo-Turkish War of 1911–12. In leaving the Triple Alliance, Italy leaned strongly toward a victory for the Anglo-French-Russian Triple Entente, or at least a standoff. The truth of the matter was that it had interests close to both sides and little to gain unless the other powers would agree to giving it territory (Tunis, for example). Secret negotiations began immediately, aimed at drawing Italy in to the Entente.[28] Italian opinion was divided, with the government and political establishment, personified by Giovanni Giolitti, favouring neutrality, while a new, largely extraparliamentary

---

* The Ottoman Empire made a secret alliance with Germany on August 2 and entered the war on October 28.

opposition fomented for war. Marconi, anglophile and viscerally anti-"teutonic," cabled the Italian naval ministry in Rome, offering his unconditional services to the Italian government while not hesitating to show his colours despite Italy's official neutrality. He obtained from the British government permission to put Clifden in contact with Coltano, and the Italian ministry replied asking if a similar concession could be made to connect Poldhu and the Italian high-power station at Centocelle, outside Rome.[29] The Italians were hedging their bets, and Marconi, as always, kept his own counsel.

In England, Italy's neutrality brought suspicion on Marconi and his family, especially in light of his wireless outpost at Eaglehurst. As rumours swirled about spies circulating off England's southern shore, local distrust among the Hampshire country folk focused on the little family behind the gates of the estate. Beatrice happened to give her last big party the weekend before war was declared. Marconi was at Eaglehurst the first few days of the "general panic" that followed, then he went off to London before leaving for Rome. For several months, Beatrice barely left the grounds.[30]

On August 14, 1914, Marconi left England for Italy. Travelling through France with an official note from the French embassy, his only complaint was the slowness of the journey. "We had a special carriage attached to troop trains and it was most interesting going slowly through the country and villages of the French Riviera with all the people cheering the soldiers. Here in Italy except for many more soldiers about everything is quite normal. All the trains run as usual," he reported to Beatrice on August 22, four days after arriving in Rome and settling in at the Grand Hotel. Marconi showed himself to be an appreciated observer of the military and political situation. Upon his arrival in Rome he was debriefed by foreign minister Antonino di San Giuliano, confirming Italy's suspicions that England and France, as well as Russia, would all be concentrating their forces on Germany rather than Austria.[31] This was extremely important information for the Italians, as Italy was much more concerned—and threatened—about Austrian designs and had its eye on the potential spoils that might come its way in the event of an Austrian defeat.

The question of Italian loyalties was crucial to the unfolding of the war, and Marconi's special position, constant travelling, and transnational connections made him an influential figure in diplomatic circles. One of the most interesting pieces in the Marconi Archives is an undated, anonymous typescript memorandum discussing conditions under which Italy might enter the war on the side of the Entente, evidently written in the early months of the war. The memorandum, inscribed "Very private," in Marconi's hand, was found among his personal papers without further explanation. It outlined the pros and cons, from a British viewpoint, of issues concerning Italy and called for a concerted English-French approach to bring Italy into the Entente as a step toward building an enduring postwar peace. The memo suggested that England should "tactfully" pressure France to accept the transfer of Tunis to Italy: "If England could bring about the cession of Tunis, it would lead to a rapid conclusion of the War and a permanent Peace and it would be a glory of British Diplomacy."[32] We have no inkling of the authorship of this document, its precise purpose, or when, how, and why it came into Marconi's possession. But the document speaks to a new dimension of Marconi's career. For the rest of his life he would be more or less actively engaged in diplomacy, often acting as a bridge, a go-between, or an informed advisor to governments and embassies on military and political questions.

❀   ❀   ❀

Marconi arrived in Rome on August 18, 1914. "Every preparation for war has been made," he wrote to Beatrice. "They think that probably Italy will fight, and if so, of course together with France and England, but that the time has not yet come." One of the first things Marconi did was inspect the large wireless station at nearby Centocelle, then the equipment on Italy's warships at Ancona. "I shall not be in any danger so don't worry," he wrote to Beatrice. "The country is very quiet and calm. Life goes on in its normal way, and if it were not for the newspapers one could not realize the awful storm which is raging around us on every side." His friend the Duke of Abruzzi had been made commander in chief of the fleet and Marconi expected to see him in a few days. Meanwhile, he had lots of time to socialize in the summer heat, going to the movies with Prospero Colonna, the mayor of Rome, and flirting with princesses—as he didn't hesitate to report to Beatrice. He ended a letter to her on an ominously prophetic note: "I suppose you know a good deal about the war but from what we learn here it seems that the French and English armies are in for an awful time."[33]

Marconi wasn't sure of his future movements, which would be tied to the war developments. "If Italy should really remain neutral I will of course come back to you and my darling chicks," he wrote home on September 2. Italy was still biding its time and relatively relaxed. Between station inspections, Marconi seemed to have lots of time for visiting friends and family on the Adriatic coast, including Gino Potenziani (at Rimini), his sister-in-law

Letizia (at Cattolica), and the Duke of Abruzzi (at Taranto). At Rimini he even had time to bathe in the sea. Among the news he received from home was a letter from H.J. Round, including instructions for the use of two sets of equipment sent to Italy in early September. Round also reported on the early stage of the war as seen from England: "Volunteering proceeds at a quite extraordinary pace.... There is no noticeable change in life here except that everybody is a little more docile and good natured."[34]

As the world watched the conflict unfold, Marconi had an increasing sense that Italy's entry into the war was inevitable. He received news from London indicating that some feared it might be on the German side, but from his own correspondence it is clear that he felt it would—and should—be on the side of the Entente. Italian feeling against Germany and Austria was growing very strong; there were demonstrations and troop movements every day, and Italy had close to a million men under arms. "The Government intends to maintain neutrality for the present, but this is made always more difficult by the daily discovery of German ships with papers and plans showing they are plotting against Italy.... Numerous secret wireless stations have also been discovered (I have been helping the Government in this) and dismantled," he wrote to Beatrice.[35]

While the cauldron simmered, he took some time off to go to Bologna and look in on the Villa Griffone. Everything there seemed to be all right. The roof was leaking and needed some repairs, but "the house was quite clean and well aired. I would love to have you and the chicks there for a bit." Once again, he scolded Beatrice for not writing and for not keeping on top of his movements. "You address some telegrams to the Hotel Regina and some to the Grand Hotel in Rome. Why? You know it's always the Grand Hotel."[36] Apparently she was miffed, because in his next letter he apologized for the scolding. "You must remember, darling, that although I'm very busy here, I'm <u>very very</u> lonely." By early October he was planning his return. "I've got heaps of things to arrange and heaps of things to do in England for the Italian Government. Mostly things not connected with wireless."[37] Before leaving Rome, he dropped a line to Winston Churchill, expressing appreciation for remarks made to an Italian newspaper and conveying "the depth and sincerity of friendly feeling which exists all over Italy towards England and France... we are all hoping for the day in which we shall be doing our part."[38]

Meanwhile, Marconi was approached about accepting the presidency of a new Italian bank, the Banca Italiana di Sconto, which was about to be set up in an effort to strengthen Italy's national presence in the financial sector. One of the main drivers of the project was Francesco Saverio Nitti, a politician, economist, and, until recently, minister of agriculture, industry, and trade in the Giolitti government. Marconi was skeptical about his qualifications to assume this role, but Nitti convinced him that he would be performing a great service to the nation because of his contacts in US and British financial circles. In return, Marconi received a little less than 10 percent of the initial share package. The founding of the Sconto, which was launched at a raucous meeting in Rome on December 31, 1914, was seen as throwing down the gauntlet against the country's French- and German-controlled private investment banks.[39]

❀   ❀   ❀

Marconi arrived back in London on October 19, 1914,[40] to discover that his movements around England were subject to the provisions of the Aliens Restriction Act. These were not terribly onerous, only obliging him to register when residing within a prohibited area, but he must have felt the sting when the Home Office refused the company's request for an exemption to enable him to travel about the country without constraint.[41] He remained in England for the rest of the year; had he been in Italy he might have taken note of a new newspaper that published its first issue on November 14, 1914, *Il Popolo d'Italia*, edited by the journalist and former socialist activist Benito Mussolini.

At the end of 1914, Marconi received one of the most important honours of his career. On December 30, the King of Italy appointed him Senator of the Realm, a lifetime position for which only people over forty years of age were eligible, as recognition for eminent service to the country. Marconi turned forty in April 1914 and was named at the earliest possible date. (Nitti had promoted his appointment.[42]) As well as providing him with the title *Senatore*, which he loved using, the appointment established him as a player in Italian politics. On the same date, the British government cancelled the company's imperial chain contract, as we've seen; Italy was now becoming decidedly more interesting to Marconi than the United Kingdom. By early January 1915 Marconi was back in Italy, helping with relief efforts after an earthquake devastated the Avezzano district, burying twenty thousand people. He was soon called back to London on company business.[43]

As the war settled into a grim stalemate, Italy made up its mind to leave the Triple Alliance and join the Entente. On February 16, Italy's ambassador in London, Guglielmo Imperiali, received the list of Italian demands from Rome; these were communicated to British Foreign Minister Sir Edward Grey on March 4. The Allies thought they were excessive but agreed to consider them and make counter-proposals.[44] Continuing to criss-cross wartime Europe, Marconi arrived back in Rome on March 3, after travelling via Paris with two women he identified in a letter to Beatrice only as "Hermione" and "Iris." His report home was full of gossip worthy of an Italian opera: "Both Gino and Arrivabene marriages are going to be annulled for reasons which are rather difficult to explain in a letter. Madda is at St Moritz. They say she is expecting a baby by Scordia. Giuseppe Scordia is very depressed, because if Gino's marriage is annulled, he will have to marry Madda which he doesn't want to do...." He ends with a profession of fidelity. "Darling, I only wish you were here too. I'm very good and not flirting with anyone."[45]

It was looking more and more like Italy was about to enter the war, but Marconi was concentrating on business. He had a new wireless agreement with the Italian government, which was coming under a parliamentary scrutiny not unlike what had happened in England and it was giving him grief.[46] He was also having a difficult time with his new duties as a bank president.[47] But the big news of this trip was his swearing-in at the Senate. Marconi was welcomed to the Italian upper house on March 15, 1915, where his life's accomplishments were recited. Rome's *Il Messaggero* reported that the new senator appeared happy and youthful as he was greeted by the unusually great applause of his colleagues.[48] Marconi wrote proudly to Beatrice, on Senate letterhead: "As you see I am now in the Senate all right. They were all most kind and cheered the day I came to take my seat. The papers too have said some very kind things about me." Then he was back to London, to receive the Albert Medal from the Royal Society of Arts. A few days before the ceremony, his old business colleague Samuel Flood-Page passed away at the age of eighty-two.[49]

❧   ❧   ❧

Marconi's US lawyers were now urging him to come to New York to give evidence in an infringement suit against the Atlantic Communication Company, a wholly owned subsidiary of Telefunken. Atlantic operated the long-distance station at Sayville, Long Island, that communicated with Nauen, and regular commercial communication between Nauen and Sayville had begun in 1914. (The German postal administration was still reluctant to get involved in wireless telegraphy, regarding cable connection as more secure—a false sense of security in that it assumed the cables would not be cut in time of war, which they were, eventually.)[50] Marconi's contended that Atlantic was infringing on two of its patents (one of which was in Marconi's own name while the other one had been bought from Oliver Lodge). One of the experts testifying for Atlantic was Ferdinand Braun, sent over by Telefunken expressly for that purpose. (Braun arrived in the United States in December 1914. He was detained after the United States entered the war in 1917 and seized the Sayville station, and died in Brooklyn in April 1918.[51])

As early as August 7, 1914, only days after war broke out in Europe, US Assistant Secretary of the Navy Franklin Delano Roosevelt received confidential information that German agents were operating a clandestine wireless station at 90 West Street in Lower Manhattan, which although low-power was capable of communicating with German cruisers in the western Atlantic. In November, Roosevelt informed the British embassy that the navy was investigating complaints of violations of neutrality by illegal wireless stations. Sayville was the crucial US link in Germany's global communications strategy.[52] The United States insisted that Sayville respect American neutrality by strictly limiting itself to commercial operations, but although no illegal acts on its part could be proven, the navy believed the station was operating "without doubt... under the control and direction of the Imperial German Government." After Atlantic Communication sought to renew Sayville's licence when it expired in December 1914, the navy suggested that the licence be refused and that it take over the station.[53] Meanwhile, Marconi took on Atlantic Communication for patent infringement in US federal court.

Marconi sailed from Liverpool on the *Lusitania* on April 17, 1915, bound for New York. On boarding the ship, he was troubled to notice that its wireless aerial was not in place, and observed that it was not fitted until the following morning, more than twelve hours after leaving the landing dock. He complained to the captain and began drafting a letter to Alfred A. Booth, chairman of the Cunard Steamship line. "Had the ship been struck by a torpedo or by a mine, or met with any ordinary mishap during this time, the great probability is that no life would have been saved," Marconi wrote. The *Lusitania* arrived safely on April 24, 1915, Marconi's forty-first birthday; the letter was never sent.[54]

While the *Lusitania* recharged in New York Harbor, the German Foreign Office sent an encrypted wireless message from Nauen to Sayville with the instruction: "Warn *Lusitania* passengers through Press not voyage across the Atlantic."[55] Berlin's embassy in Washington placed advertisements in US newspapers warning that, in view of the state of war between Germany and Great Britain, US citizens were at risk if sailing on ships flying the British flag or entering British waters, which was considered part of the war zone. (The German government had issued a proclamation to this effect on February 4, 1915.[56]) On the *Lusitania's* return trip, a German U-boat torpedoed and sank the ocean liner off the coast of Ireland on May 7, 1915, taking twelve hundred lives, most of them civilians. The *Lusitania's* operator got off an immediate SOS, which was received by a coastal station, but it was in vain. There were enough lifeboats for all, just not enough time to get everyone into them. It took only eighteen minutes for the ship to sink. The sinking of the *Lusitania* had a crucial impact on turning world and particularly US public opinion against Germany.[57]

After sailing on the *Lusitania's* 201st and final complete voyage, Marconi had again escaped disaster at sea. Beatrice cabled him at the Holland House: "Deeply thankful your escape *Lusitania* my most loving thoughts please be careful about returning wire me in detail your plans when you mean to return and how stop very worried. Bea Marconi. 809pm." He wrote to her: "I'm afraid several letters to you must have gone to the bottom on the *Lusitania*. Wasn't it too awful. I've lost so many good friends on her. I saw her off the morning she sailed. I'm sure you remember poor old McCubbin the purser, and the doctor, and the pretty typewriter girl, and so many stewards and stewardesses that had known us for years."[58]

Two days after Marconi arrived in New York, on April 26, 1915, British, French, Russian, and Italian diplomats in London signed a secret treaty by which Italy definitively abandoned the Triple Alliance and pledged to enter the war on the side of the Entente. Italy saw the Treaty of London as its path to gaining or regaining territories it perceived to be under foreign rule, such as Trentino, the South Tyrol, and the Adriatic regions of the Austro-Hungarian Empire. The secret treaty promised Italy the Tyrol, Trieste (but, significantly, not neighbouring Fiume), northern Dalmatia, the Dodecanese islands, Libya (which Italy had held since 1912), parts of Albania, German colonies in Asia and Africa, and a share of Turkey. Serbia and Montenegro, while not signatories (although Serbia was informed), would also get parts of Dalmatia and Albania. The pact officially remained secret until its details were published in the Bolshevik newspaper *Izvestia* after the October Revolution of November 1917. However, just as in today's world of porous

diplomacy, its existence was known and even discussed in public fora—
Marconi himself referred to the treaty, approvingly, in his maiden speech in
the Italian Senate on December 16, 1915.[59]

Despite its quarrel with France over Tunisia, Italy had always kept a
flirtatious eye on the Triple Entente. Already in the early days of the war,
neutral Italy had acceded, in a first secret agreement on September 5, 1914,
not to make a separate peace. The Treaty of London was thus what scholar
René Albrecht-Carrié called "an application of the doctrine of the balance
of power."[60] The Entente wanted Italy, but was it prepared to pay the price?
As a prominent and well-connected anglophile, Marconi found his political
stock on the rise. Within a few days of the treaty's signing, he was briefed by
Italy's ambassador to the United States and quietly began to prepare for his
imminent return to Europe. On May 3, 1915, Italy formally denounced its
former allies, but its entry into the war was still not inevitable. Italy remained
split between neutralists and interventionists (Mussolini among the latter;
Giolitti the leader of the former).

In New York, meanwhile, Marconi continued to plod away at his court
case. As he described it to Beatrice: "The court is in Brooklyn and we have
to be there at 10 o'clock in the morning. We have a rotten lunch and stop
at the court till 5 pm."[61] Still, he managed to find the time for at least one
piece of important business. On May 18, 1915, he travelled to Schenectady
in upstate New York to visit the General Electric plant and see a machine
that had been developed by GE engineer Ernst Alexanderson. Originally
from Sweden, Alexanderson had graduated from the Royal Technical
University in Stockholm and had also studied under Adolf Slaby in Berlin.
He joined GE in 1904 and was put to work on an order placed by Reginald
Fessenden for a powerful high-frequency alternator capable of generating
continuous radio waves. Fessenden used the first version of Alexanderson's
alternator in his pioneer voice broadcast on Christmas Eve 1906; Fessenden
then went bankrupt, but GE decided to continue developing the alternator,
and Alexanderson patented it (in 1911) while they looked for another customer. (The American Marconi company also purchased an early prototype
for its shore station at New Brunswick, New Jersey.) By 1915, the work of
Alexanderson and others in GE's research lab had resulted in a complete
continuous-wave (that is, broadcasting) transmission and reception system.
GE was essentially an equipment manufacturer and not interested in becoming a communications company like Marconi's (or AT&T, the US
leader in telephony). There is some disagreement as to whether Marconi

was invited to inspect the alternator, or whether he invited himself, but after spending the better part of the day with Alexanderson and his alternator he decided he wanted to buy it. Before a deal could be made, however, he was called back to Italy.[62]

Marconi was giving evidence in the US District Court in Brooklyn when he learned that Italy was about to enter the war. He told Judge Van Vechten Veeder it was only a matter of hours and the Italian government had asked him to come to Rome immediately to oversee its wireless arrangements.[63] The news was not unexpected and, at Marconi's request, the trial was suspended so that he could return to Europe.[64] He made arrangements to sail as soon as possible, and at the last minute Inez Milholland decided impulsively to leave her job with a New York law firm and sail with him, hoping Marconi would be able to open doors for her to report on the war as a freelance correspondent for the *New York Tribune* and other outlets. The *Tribune* pictured the two of them on its front page as they boarded the *St. Paul* on May 22, 1915. He looked weary, she radiant.[65]

Italy declared war on Austria and severed diplomatic relations with Germany on May 23, 1915. The *Tribune* reported that day that "Marconi has been called to Rome by the Italian government." The paper recalled that he was a reserve lieutenant in the Italian navy, and quoted him saying his work would be "the supervision of the wireless, with headquarters in Rome." His travelling companion was described as a "lawyer, writer and suffragist."[66] The *St. Paul* arrived in Liverpool on May 30, and on June 2 the *Tribune* carried Inez's first signed report. Under a huge photo of her and the byline "Inez Milholland Boissevain," the paper's glamorous new foreign correspondent described how a German U-boat had chased the *St. Paul* as far as the mouth of the Mersey in an effort to kidnap Marconi. "Mr Marconi received a warning from the Italian consul before leaving New York and for this reason his name was kept off the passenger list and every endeavor made to prevent his presence on board becoming known," she reported.

> As we approached the war zone rather elaborate precautions were taken to safeguard Mr. Marconi...there was a general tacit agreement among the passengers that if the *St Paul* was stopped by the submarine we all would "lie like gentlemen." Meanwhile Mr. Marconi removed all the labels from his luggage, gave his private papers into my care and got into clothes suitable for slipping into a hiding place somewhere down in the bowels of the ship next to the keel, where the chief engineer said the captain himself would be unable to find him.[67]

The curious thing about this sensational article was that the *Tribune* had published a front-page photo and article announcing that Marconi would be sailing to Liverpool on the *St. Paul*. Under these circumstances, it is not likely that German intelligence would have been led off the track by a bogus passenger list. Be that as it may, the German submarine was driven off by British torpedo boats before catching up to the *St. Paul*. When the ship landed, Marconi was quoted as saying the voyage had been uneventful.[68]

After only a day or two in London, Marconi and Inez left for Rome—seen off by Beatrice at Victoria Station. Inez's first-person account of their trip across war-torn France was a classic piece of anti-war journalism. After crossing the Channel from Folkestone to Boulogne, with their own car in tow, they made their way to Paris. A hare crushed by their vehicle became a metaphor for the conflict. The young women they saw by the roadside as they passed through the villages of northern France appeared "cheated of throbbing life," left behind, their men gone. "I cannot begin to describe to you the sadness of Paris. . . . The hotel is very still. Perhaps there are no people here; perhaps there are just sad people here. We'll leave for Rome in half an hour by train. It takes too long and is too doubtful by motor."[69] The article was datelined Paris, June 23, 1915. That was certainly an error, as records show that Marconi and Inez left London on the first or second and were already in Rome together on June 6. On June 11 he wrote to the Ministry of War from Rome, confirming the offer of service he had made in person a few days before, and the ministry replied on June 17.[70] Inez's article does not name Marconi as her travelling companion, but to anyone in the know, there could be little doubt that it was him. As they were leaving Victoria Station, "Bea waved us a last goodby and we were off."[71]

In Rome, Marconi installed himself and Inez at the Grand Hotel. She wrote her husband, Eugen Boissevain, that she was in a luxurious room in a luxurious hotel, making friends and contacts, but she was lonely and depressed, tortured and unhappy. "I am being taken care of by Guglielmo. Shall I always be parasite do you suppose? Shall I never earn? Have I no value?" Her articles had yet to appear. Whatever doubts she had, her relationship with Marconi was not part of them; it appeared to be chaste and platonic, something like that of an uncle and a favourite niece. In her letters to her husband she always wrote candidly about flirtations and lovers, but with Marconi in Rome she seemed bored and anxious. "Love of my life I miss you so," she wrote Eugen.[72] Boissevain, for his part, recognized Marconi's usefulness to the couple. He was trying to set himself up as an import–export agent in the United States, and maybe the Italian government would be interested in what he had to offer. "Maybe when the right moment presents itself you might mention it to Billy,"[†] he wrote to her.[73]

Inez's career was soon well on track. The *Tribune* published a second major piece under her byline, datelined Rome, June 27, 1915. It was even more moving than the first, and this time she interpellated Marconi and his circle directly. "They tell me—everyone tells me—that Italy was mad for war. Maybe. But let me tell you what I have seen and also what I have heard." In Rome, she said, the humble folk have broken lives and nobody wants war but the extreme nationalists. "These gentlemen view it from an imaginative and story-telling standpoint, and not from a standpoint of reality. They live in the past and the traditions of the past. They contemplate the greatness of Rome that is gone, and dream of living that greatness over again." Ordinary Italians love to demonstrate and display their emotions but that doesn't mean they want war. "If they see Marconi or [the nationalist poet Gabriele] D'Annunzio or somebody who they are able to hero worship in the street they give him an ovation, be the time peace or war." Italy's entry to the war exacerbated an already fertile ground for disaffection and social unrest,[74] but when Inez reported this aspect of Italian reality, when she told her "tale of no enthusiasm," the authorities in Rome protested that she was being untrue and unjust.[75]

Marconi, on the other hand, was welcomed as an unequivocal war booster. The Ministry of War heralded his arrival in Rome, commissioned him as a lieutenant in the army's Corps of Dirigible Engineers, and accepted his offer to make Villa Griffone available as a military hospital. On June 19, 1915, the date of his appointment, he received a note informing him that the king was convinced he would render important services to Italy. A short time later, he gave an interview to the Central News Agency, in which he told the reporter, in a tone reminiscent of his earlier reaction to warfare in North Africa: "The enthusiasm of the Italian people at the present time to prosecute the war to a successful and victorious conclusion is intense." Meanwhile, Inez's vocal opposition to the war antagonized the Italian government and annoyed Marconi, and she soon found herself shut off from high-level sources.[76]

Shortly after Marconi and Inez arrived in Rome, they met D'Annunzio, whom she was hoping to interview for the *Tribune*. D'Annunzio was an

[†] The name the Milhollands often called Marconi.

emblematic figure in Italian modernist culture, as well as an adamant high-profile advocate for Italy's participation in the war. He and Marconi were well acquainted with each other by reputation.[77] They shared a presence in the small pantheon of internationally visible Italian icons in the early decades of the twentieth century. (Marconi's friend Enrico Caruso was another, as was the futurist poet Filippo Tommaso Marinetti.) Among other things, they were both savvy at using the media to promote themselves and their projects. Like Marconi's, D'Annunzio's doings and pronouncements made headlines as easily in London, Paris, or New York as they did in Rome or Milan. Marconi had been interested in D'Annunzio as early as December 1901: his Newfoundland files contain a news clipping reporting a commotion at the first performance in Rome of D'Annunzio's new play, *Francesca da Rimini*.[78]

Ten years older than Marconi and of humbler origins, Gabriele D'Annunzio, whose real name was Gaetano Rapagnetta, grew up as the son of a small merchant in the Adriatic town of Pescara and came to Rome as a young man, ditching his wife in favour of a series of liaisons. Most of these were far too fleeting and casual to really be termed affairs, except for two that made him famous among the international paparazzi, with stage sirens Sarah Bernhardt and Eleonora Duse. D'Annunzio was addicted to casual sex, aesthetic posturing, outrageous behaviour, and the public attention that came with them.[79] It is hard to imagine someone superficially more different than Marconi, yet the libertine poet—aesthete, sybarite, sensualist—and the proper, geeky inventor became close friends, with a powerful chemistry between them.[80] D'Annunzio became a critical ideological influence on Marconi and his burgeoning Italian nationalism.

D'Annunzio was a vocal promoter of irredentism, a current of opinion that aimed to unify all the Italian-speaking people and territories of Europe. Irredentist claims could go as far as Corsica, Nice, or Malta, and certainly the Italophone portions of Slovenia and Croatia. The movement had its roots in Giuseppe Garibaldi's campaigns of the 1860s that led to the unification of Italy, and went further than the Italian demands that were covered by the Treaty of London. The Adriatic city of Fiume, or Rijeka, as it was known in Croatian, was a crucial part of irredentist Italy. The liberation of territories that irredentists viewed as occupied was perhaps Italy's strongest motive for entering the war, and would become the critical issue for Italy in the peace talks afterwards, when irredentists as well as less orthodox nationalists felt betrayed by England and France and infuriated by Italy's exclusion from colonial gains. D'Annunzio had been a vigorous agitator for Italy's entry into the war, much more politically activist than Marconi in this regard. He was also more eager to put his body on the line. He learned to fly and gained a reputation as a swashbuckling daredevil—he would lose an eye in a flying accident in 1916, gaining another attribute, a physical one, that he would share with Marconi. As D'Annunzio later put it, his poetry and Marconi's radio were both employed as instruments of war.

Now here they were, both in Rome and in uniform. "After many days and many years of waiting, I met Gabriel D'Annunzio," Inez wrote in a text that was never published. "The meeting was a source of keenest pleasure to me," but also a disappointment. She had expected more, a critical perspective on patriotism, "a vision of a nobler destiny for Italy than any that may be brought to her by war." Instead she met a man fanning the flames of zealotry on behalf of a political system on the verge of collapse.[81] Inez had considered D'Annunzio "a sort of expert in human souls"; what she found was a master of populist mobilization. (His most successful disciple, Benito Mussolini, was soon to appear on the scene.)

Milholland, Marconi, and D'Annunzio made a powerfully intoxicating triangle in wartime Rome. For one thing, radically different as they were, they were all attracted to publicity as though it were a drug. Marconi and Inez shared a comfortable platonic attraction and loyal devotion to each other; but D'Annunzio represented something that each of them was striving to reach. Inez and D'Annunzio oozed sensuality, passion, and the possibility of sexual abandon, where Marconi was reserved, rational, even puritanical. D'Annunzio and Marconi shared a vision of a glorious future for Italy; while Inez was skeptical and even outraged at the cost. Inez's politics came from the heart, D'Annunzio's from the spirit. Marconi, on the other hand, allowed himself to be drawn into political schemes. He held few strong views but some fundamental convictions: love of country was one of them. He was drawn to D'Annunzio's patriotic fervour, as Inez was to his charisma. At the end of the day, Inez and D'Annunzio each symbolized aspects of Marconi's unrealized self. He longed to have such passion in his life; except where his research was concerned, it always eluded him.[82]

Attracted to Inez's vivacious personality, her uncommon culture, and her rare beauty,[83] D'Annunzio, unsurprisingly, tried to seduce her, while she used him as a foil to explore the Italian soul. "She is having a flirt (quite harmless) with D'Annunzio," Marconi wrote to Beatrice, perhaps a bit naively.[84] Meanwhile, he cultivated his own relationship with the charismatic poet. As they waited for their respective assignments, Marconi and D'Annunzio hung out together in Rome, nurturing a precious friendship that would serve both of them. Inez was the catalyst.[85]

One sultry day, D'Annunzio accompanied Marconi on an inspection visit to the wireless station at Centocelle, two dashing soldiers in uniform, with sabres on their knees. As they drove through the outlying Roman landscape and its relics of homage to antiquity, they talked about the future. Arriving at the airfield they could hear the dots and dashes of the wireless receiver. Marconi, as he always did, examined the transmitter with his delicate, agile hands, moving D'Annunzio, who saw sensuality in the gesture. "It was a blustery day," he later recalled, and "the whirling wind, lifting the ash from the sepulchers, transformed it into the seeds of the future."[86] D'Annunzio was so moved that he wrote an epic prose paean to Marconi that he called *L'eroe magico* ("The Magic Hero"):[87]

> We were alone in a fast-moving vehicle, seated side by side. To my former admiration of the magician of space there was added the fraternity of the field of arms.... We were both in military uniform, we both had our hands on the hilts of our swords, we were equals in rank and in the desire for service, in the dedication of ourselves entirely to our country. We were two Italian soldiers. His science and my poetry had become instruments of war....

> [At Centocelle,] beneath the clouds one could see the high iron towers, the antenna, the wires, the new sublime acropolis; the ultimate height of the mind; the lofty temple of mystery ... constructed with a new art, raised toward the sky with a naked and dainty architecture like that of the antenna which was radiating that miraculous energy and which, with the disposition of vertical and horizontal wires, gives the impression of being a gigantic musical instrument, of a never before seen Aeolian harp resonating with breaths of human spirit....

> The magic hero was in one of the buildings, beneath the steel towers, between the strands of wire, in the midst of the continual commotion of electric vibrations, shocks from the terrible scintillating sparks of lightning.... The immense cosmic energy, constrained by exact instruments, measured, and observed, talked with this tranquil and self-possessed man in a language which he understood like the speech of his own child.... We listened attentively. Signals were being sent.... Very well, then ... this Magician has his secrets....

Marconi always spoke of his invention as an instrument of peace, but now he was enveloped in the discourse of D'Annunzio, which was unequivocal in its glorification of war. Back in Rome they parted, each off to his own assignment at the front.

                              ❀   ❀   ❀

Marconi was upbeat about the war. "Here the war is really (bar the loss of men) going very well and I think we have every reason to be very proud indeed of the Army.... We expect important developments soon," he wrote to Beatrice. He described the army's dirigible unit, where he was working, as "a kind of cavalry" with smart, practical uniforms. He hardly had time to see anyone; all the men he knew were at the front, the women in the *Croce rossa*, the Italian Red Cross. His letter ended with the usual plea: "and do please love me."[88]

Marconi's assignment brought him closer to the fighting than he had been in North Africa in 1911—still at a safe distance, although on one occasion, he told a reporter, he had "narrowly escaped being blown to pieces by an enemy shell." He also had a soldier seconded to look after him.[89] Early in July 1915, he watched the Italian bombardment of Gorizia at the first battle of the Isonzo, on the eastern Austrian (Slovenian) front, through field glasses from a neighbouring mountain, describing it as "awe-inspiring ... crash upon crash, echoing and re-echoing throughout the surrounding valleys."[90]‡ It was the fiercest Italian-Austrian engagement to date. The mountain slopes were strewn with dead, and Italian troops bayoneted Austrians by the hundreds; nonetheless, Italian losses alone numbered more than one hundred thousand.[91] Marconi was a bit closer to the action at Monte Grappa, in the Veneto, where he had to crouch in the trenches while heavy Austrian shelling whistled around him.[92]

A few weeks later, Marconi was back in England, remarkably agile at transitioning from the war front to the corporate boardroom. After years of frustration, he finally had a new and reliable private secretary, Leon de Sousa.[93] On July 27 and August 18, 1915, Marconi chaired the annual general meeting and a special general meeting of MWTC, wearing his livid (blue-grey) Italian army uniform, with two rows of medal ribbons on his tunic, and wrapped in a voluminous cloak. The war was good for the company

---

‡ There were twelve battles of Isonzo between June 1915 and November 1917, part of the Italian effort to occupy Austrian areas inhabited by ethnic Slovenes and promised to Italy by the Treaty of London. Their ultimate goal was Trieste. The Isonzo campaign cost more than five hundred thousand casualties, three hundred thousand Italian and two hundred thousand Austro-Hungarian. Ernest Hemingway's *A Farewell to Arms* is partly set there.

and Marconi was able to announce a 10 percent dividend on common shares and the introduction of an overseas wireless connection between the United States and Japan.[94]

In September, he was off on a dizzying tour of the French front, meeting all the great generals and lunching with Marshall Ferdinand Foch. (Foch reminisced about his experiences in the Franco-Prussian War of 1870.[95]) After that, he was back to Italy, in Turin and Genoa in November, and finally in Rome, where he made his maiden speech to the Senate on December 16, 1915.[96] It was a mildly nationalistic address, but mainly focused on economic questions: the lira's poor rate of exchange against the pound; the high cost Italians were being forced to pay for resource materials like coal; the financial sacrifices they were making. The speech attracted press attention in both Italy and the United Kingdom, where the *Times* hinted that Marconi was acting as something of a go-between on this sensitive topic. A few days later, he received a letter (addressed to him as "Sir William Marconi") from Lord Kitchener, the British secretary of state for war, thanking him for his efforts to smooth relations with the Italian government.[97]

Alongside all these duties, Marconi also made a new experimental breakthrough. At the end of 1915, working with Charles S. Franklin, one of his long-standing research associates, Marconi began experimenting with short wavelengths, returning to the part of the spectrum he had used in his earliest work. Using short wavelengths was important insofar as it freed other parts of the radio band for other traffic, but especially because short-wave transmission required substantially less power than the longwaves Marconi had been using since 1901. Marconi took an active part in the experiments, but the work was driven by Franklin, who was in charge of an independent research unit within the Marconi company, based at Poldhu. Franklin spent part of 1916 working at the Italian naval installation in Livorno, with Marconi coming and going amidst his myriad activities; here Marconi and Franklin discovered that by using a certain type of reflector, radio waves of two metres or less in length could be concentrated in a "beam." As it developed during the next few years, this discovery would become the kernel of the global shortwave radio system that emerged in the 1920s, removing the barriers of cost and excessive power from long-distance wireless. Franklin is considered by some to have been the true inventor of what came to be known as the Marconi beam system.[98]

Only briefly in London early in 1916, Marconi attended a social evening at the Belgravia home of a wealthy philanthropic couple (and friends of his in-laws), the Howard de Waldens. One of the other guests was a young Irish opera singer, Margaret Burke Sheridan. Observing her perform, Marconi could barely contain himself; as soon as she finished, he rushed up to her, kissed her hand, and declared with his irresistible charm, "This is the voice I have been waiting to hear all my life."[99] Once again he was smitten with a young and beautiful, talented and ambitious woman.

Margaret Sheridan was twenty-six years old. She was born in Castlebar, County Mayo, in October 1889, a daughter of the local postmaster and his wife. Her family background was stable and pleasant but her parents died young, leaving her orphaned though not penniless at eleven years of age. Her father's will provided for her to be sent to board at a Dominican Convent in Dublin, where she studied music and voice, and then went on to the Royal Academy of Music in London in 1909. Remembering the path his mother had taken, and the chance she had missed, Marconi urged Sheridan to pursue her operatic studies in Italy, and offered her an opportunity she could hardly refuse: come with him, immediately, under his patronage and protection. She had a sweetheart, a thirty-five-year-old Irish Nationalist MP from Galway by the name of Richard Hazelton, but with just a bit of hesitation she left Hazelton behind for adventure and career.

Marconi was now adept at travelling across war-torn Europe, and on January 21, 1916, he and Sheridan crossed the Channel between Folkestone and Dieppe, risking German submarines. Passing through Paris, Turin, and Genoa, they eventually reached Milan, where Marconi's arrival at the central train station was hailed by a cheering crowd of admirers. By now, this was all routine for him. He turned to Sheridan and said, "It's nice to be famous, Margherita, some day you must be famous too."[100] Then it was on to Rome, where he installed his newest protégée in the posh Hotel Excelsior on the Via Veneto (where he also installed his family in 1916[101]) and introduced her to his influential society and artistic circle, where she became known as "Peggy from Ireland," a friend of Marconi. He helped her financially and arranged for her to study under opera teacher and conductor Alfredo Martino, who elevated her from mezzo to full soprano. One of the superstars she met through Marconi was the composer Giacomo Puccini, who soon became her main benefactor. She eventually played two of Puccini's operatic heroines, Mimi in *La Bohème* (debuting in Rome on

February 3, 1918, with Marconi in the audience) and Cio Cio San in *Madame Butterfly* at Covent Garden after her return to England in 1919.

Sheridan perfectly fit the profile of a Marconi woman. As with Inez, there is no evidence that they were physically intimate, at least not in a sustained relationship, and he clearly relished his role in contributing to her professional success. He not only helped her establish her career, he continued to follow it with interest and admiration. His diary shows that she was a frequent dinner and theatre companion of his in London in the 1920s, and she was one of the last old friends to be in touch with him in Rome only months before he died in 1937.[102] Remarkably, she is not mentioned, even in passing, in any of the Marconi biographies.

❀   ❀   ❀

Marconi's new shortwave experiments soon became entangled with his business and political activity. On March 3, 1916, he laid out his thoughts in a long letter to Godfrey Isaacs, written while he was recovering from a bout of tonsillitis at the Hotel Miramare in Genoa. (He actually considered the letter so exceptional that he made a rarely effusive note in his diary that day: "Wrote Mr. Isaacs important private letter."[103]) He explained that his ideas on shortwave had been percolating for some months, if not years, but he had been unable to try them out in England because of the obstacles placed in his way by the government since the beginning of the war. British wartime restrictions required him to make "complete and prior disclosure of all I was doing or intended to do," and he was not prepared to submit to this for his own very practical reasons: wartime or not, he was simply not going to tell the British government—a potential competitor—what he was up to until he had achieved definitive results and protected them fully by patents.

His latest findings, he said, "are far reaching and concern the future practice of the entire success of wireless telegraphy and telephony, whether over long or short distances or conducted by means of ordinary spark system, grounded sparks or continuous waves." The Italian government was giving him a completely free hand, so long as nothing he did interfered with naval or military signalling, and even that he could do with prior consent. The Senate was about to resume sitting and he was going to Rome; although he had not been "troubling myself with politics to any extent," he had been asked whether he would be prepared to enter the government. "This I have refused to do at present as I believe my best chance of being of use is mainly confining my efforts to those scientific pursuits in which I am best qualified," he said.[104]

Isaacs replied, impressed and elated. He wanted to issue an immediate statement to the company's shareholders and the press, announcing that Marconi was applying for new patents concerning "the future practice of the entire science of wireless telegraphy and telephony, over both long or short distances," and specifying that the new innovations would "probably be applied at once in Italy to military purposes." He expected such an announcement to have a far-reaching effect, "both 'at home' and elsewhere and particularly desirable at this moment." There had been rumours that Marconi was being solicited to go into the Italian government and Isaacs was pleased to learn that Marconi had declined: "There are perhaps others who might be able to do equally well or well enough, but there is nobody to take your place in work which is of such value not only to your country but to the world at-large."[105]

Marconi was still in Italy when, on Easter Monday (April 24) 1916, Irish nationalists launched an armed insurrection, the most serious threat to British rule in more than a century. Militants in Marconi's ancestral home town, Enniscorthy, played a small but significant role, taking over the town and holding it for four days while they flew the republican tricolour flag over their headquarters at the Atheneum theatre. The heart of the uprising was, of course, Dublin, where a force led by Patrick Pearse of the Irish Republican Brotherhood (IRB) and socialist trade unionist James Connolly's Irish Citizen Army occupied the General Post Office in Sackville (now O'Connell) Street and proclaimed the Irish Republic.

The question of self-government, if not full independence, for Ireland had been on the British political agenda since a first home rule bill failed to achieve a parliamentary majority in 1886; a second attempt in 1893 passed in the House of Commons but was defeated in the House of Lords. The Third Home Rule Bill, introduced by Prime Minister Asquith in 1912, was adopted on May 25, 1914. The question had pretty much dominated political discourse on Ireland throughout Marconi's lifetime, and it was quite a challenge for him to remain on the sidelines.

Home rule was opposed by Irish Unionists, whose main political spokesman was Sir Edward Carson, a talented orator, Conservative politician, and litigator who had occasionally acted for the Marconi company or its friends.[106] Unionists formed the Ulster Volunteer Force, the first armed group in the twentieth-century Irish struggle, in January 1913; this sparked the formation and arming of the pro–home rule Irish Volunteers under the impetus of the IRB, among others. Marconi's cousin and recently estranged business associate, Henry Jameson Davis, was a leading member of the Irish Volunteers in County Wexford. Davis was also a supporter of the Irish Nationalist Party, and was a close friend of its leader, Wexford MP John Redmond.[107]§ By the start of the war, the presence of two opposing armed camps in Ireland was threatening, especially in light of equally opposing views on Irish participation in the conflict. Only a few months after its adoption, the Home Rule Act was suspended for the duration of the war. The only party in Parliament that did not join Asquith's wartime coalition government of 1915–16 was Redmond's Nationalists (which, nonetheless, supported the government and did not see itself as the opposition).

Marconi's wireless stations in Ireland were taken over by the British government and used for billeting troops as well as for communication.[108] When the Easter Rising broke out, great care was taken to keep the news from being sent out from Clifden—in Montreal, the Canadian Marconi company sought clearance from the wartime press censor before publishing anything related to the events in Ireland. (The press censor saw no problem, "from a military standpoint.")[109] But the insurgents in the Dublin GPO managed to get their hands on a transmitter from the nearby Irish School of Wireless Telegraphy and diffused the news that the Irish Republic had been declared, not aiming it at any particular receiver but "broadcasting" it at large (albeit in Morse code) in the hope that it would be picked up by a passing ship and relayed to the press.[110] The Irish republican insurrectionists thus claimed to have made the world's first "broadcast"—a claim acknowledged by Marshall McLuhan in his book *Understanding Media*, among others.[111] There is, however, no record of any reception of the twenty-five-word message,[112] and controversy about what actually happened. (Some say the news reached America when it was leaked by an operator working the transatlantic cable at Valentia.[113]) But as the Irish journalist and broadcast executive Maurice Gorham wrote: "Whether or not the broadcasts reached their destination it showed great imagination for the men who planned the Rising to think of using wireless for such a purpose as early as 1916. . . . They were, of course, ahead of their time."[114] In 1977, the writer Conor Cruise O'Brien, then Irish minister for posts and telegraphs, said of the event: "The

§ Lest one be tempted to see Irish politics too closely through a binary Catholic-Protestant grid, it is worth noting that Davis's parents were Quakers, Redmond was Catholic, although his mother was a converted Protestant from a unionist family, and Marconi himself, though baptized Catholic, was raised in the traditions of the High Church of Scotland, with Presbyterian and Waldensian influences.

painful conclusion is, I think, inescapable. . . . Broadcasting was conceived in sin. It is a child of wrath. There is no knowing what it may not yet get up to."[115] Undoubtedly, this was the first use of wireless for broadcasting international propaganda.

The Easter Rising was put down within six days. Wireless transmissions continued for twenty-four hours, until British artillery fire forced the rebels to stop.[116] The force occupying the GPO surrendered on April 29, 1916, and most of the insurrectionists outside Dublin, including those at Enniscorthy, gave up the following day. Fifteen leaders, including all seven signatories of the republican proclamation, were executed by firing squad.[117] During the subsequent British military occupation, nationalist opinion moved toward support for independence, as former home rule supporters became convinced that parliamentary methods would not suffice to remove the British presence. After coalescing republican support and winning 70 percent of the Irish seats in the British general elections on December 14, 1918, the Sinn Féin party formed the Irish Parliament (Dail Éirann) and declared Ireland's independence on January 21, 1919. The Irish War of Independence that followed culminated in the Anglo-Irish Treaty of December 6, 1921, and creation, in 1922, of the Irish Free State as a self-governing entity with the same constitutional status as Canada or Australia; the conflict didn't end there, however, but only after the subsequent civil war between pro- and anti-treaty advocates.[118]

On May 17, 1922, the Marconi company applied to the postmaster general of the new Free State for a broadcasting licence,[119] but the civil war broke out in June and soon spread to Galway. On July 24, 1922, "irregular" (that is, anti-treaty) republican forces occupied the Marconi station at Clifden (which the company had recently announced it would be phasing out). First reports were that the station was totally destroyed and burned to the ground, but the damage turned out to be not quite that bad. Marconi was at sea returning from the United States and expecting to receive test messages as well as news from Clifden as he approached the Irish coast, but there was nothing; he told the *Times* he had the curious experience of being "cut off" from communication—a throwback to twenty-five years earlier when ships at sea had no contact with the shore. "I was surprised to find that nothing was reaching me. I could not make it out. On arrival [at Southampton] I found that Clifden Station had been seized by Irish revolutionaries, and some of the buildings burned. The staff were expelled. I understand that the main buildings have not been damaged, but minor ones have been destroyed. It is hoped that control of the station will soon be regained."[120]

The damage—and the IRA occupation itself—was serious enough to immobilize the station. Several buildings were set on fire, the condenser and receiving houses were effectively destroyed, and the staff were sent home, although the engines and generators as well as the various bungalows around the station were undamaged.[121] The pro-government *Irish Independent* headlined: "Terrible havoc," and the *Manchester Guardian* called it "reckless destruction" and an outrage, part of an IRA scorched-earth policy to destroy as much as possible as their campaign in Galway was routed. Transatlantic communication was quickly re-established from a new Marconi station in Essex. The consequences were most serious for the hundreds of Connemara families whose breadwinners lost their employment as a result of the shutdown. Monsignor McAlpine of Clifden's St. Joseph's Church condemned the IRA action and paid tribute to the Marconi officials who, he said, had always been on friendly terms with the local people.[122]

Within a few days it became evident that the attack had been politically motivated. The *Manchester Guardian* reported that "the ostensible reason for the partial destruction of the station was that it was regarded by the irregulars as a British concern, the rights of which, they said, were withheld from Ireland under the treaty."[123] The Anglo-Irish Treaty did indeed include an annex in which control of existing, as well as future, cable and wireless communication with places outside of Ireland remained with the British government.[124] Aside from its implications for the limits to Irish sovereignty, this annex underscored Marconi's ultimate dependence on the British state for this key part of his operations; it also foreshadowed a new development—the assimilation of cable and wireless issues under the same corporate and policy umbrella that would take place a few years later.

On August 10, the Marconi company was informed that no broadcasting licences would be awarded "owing to the present disturbed conditions in Ireland." (Lord Beaverbrook's *Daily Express* made a similar application and received the same response.)[125] The IRA irregulars held Clifden until August 15, 1922, when the town and the station were retaken by the Irish Free State's regular army. The last marconigram sent from Clifden, dated August 17, 1922, announced the capture of Clifden town—"the last stronghold of the irregular forces in the west. . . . All barracks as well as the wireless station are now held by national troops."[126] The Clifden station, which became a garrison for government troops, remained a pawn in the war between the Free State and the IRA.[127] On November 29, 1922, a Republican force of 350 men recaptured the town, and the station garrison withdrew to Galway. Barely ten days later, on December 9, the town and the Marconi station changed hands yet again, this time taken over by government forces.

At the end of the civil war, the Free State government invited broadcasting applications, but the Marconi company found itself excluded because of the reservations of postmaster general James J. Walsh, who saw the company as primarily commercial, as well as foreign, an outside body (outside to the Irish state, as well as to the public service). Walsh told a parliamentary committee: "We are not able to penetrate the political or financial wall of the Marconi mind, except we feel we are in the grip the whole time of a very dangerous institution, and nothing will shake us free of that feeling."[128] With characteristic panache, the company tried to create a *fait accompli* by installing a low-power transmitter in the Royal Marine Hotel in Dún Laoghaire on August 14, 1923—the first serious attempt to establish a broadcasting station in Ireland—but the Post Office ordered it to stop after two days.[129]

The Irish government and the Marconi company made some desultory efforts to re-open Clifden but without real enthusiasm on either side. In retrospect, it was clear that the civil war and its aftermath only hastened the downscaling of Marconi's Irish stations; it was not the cause. The implications of Irish independence and the spirit of national self-determination that accompanied it was an important impetus. Ireland was now less accommodating to Marconi, whose company was—correctly—perceived to be an offshoot of British domination. But technology and the development of important new applications for wireless played the crucial role. By 1922, Marconi could reach America just as easily from England as from Ireland. The company's commercial imperatives were shifting toward broadcasting, and that focus was certainly going to be based in the United Kingdom (although the company's conventional smaller markets like Ireland and Canada would still be important). Marconi himself was increasingly interested in the research that would soon culminate in shortwave "beam" communication, and the Irish stations were not equipped to do that work.

All that said, negotiations between the Irish and British governments regarding the station's future continued up until 1925. The talks were fruitless, and the Marconi company complicated the issue by seeking compensation from the Free State for the damage caused by the rebels, sparking what the press termed "a constitutional struggle of some interest." The Irish courts were sympathetic to such cases, where the owners of the damaged properties were prepared to rebuild; Marconi showed no sign of such intention, however, and the Irish postal minister, notoriously suspicious of what he called "the Marconi Trust," indicated that he would be just as happy to see the Company leave Ireland for good.[130] To top it off, Marconi's engineers convinced him that Clifden had become obsolete and, reluctantly, he sold most of the equipment for scrap; what remained—lead from the batteries, galvanized iron sheeting, copper cables—was eventually removed by people helping themselves.[131] The Irish Post Office began public service radio broadcasting on January 1, 1926, ironically using a 1.5 kilowatt Marconi transmitter. Marconi cabled congratulations from Rome, but he never returned to Ireland.[132]

23

"L'eroe magico"

Marconi was back in England in early May 1916, to spend some time looking after company business and visit his family. He had missed the birth of his daughter Gioia, his fourth child, in London on April 10, 1916; as usual, he was away, this time in Italy. Meanwhile, the war continued apace. On May 16, 1916, France and Britain concluded the secret Sykes-Picot Agreement,* dividing their future spheres of influence in the Middle East in anticipation of the defeat of the Ottoman Empire. In June, the "Arab Revolt" against the Ottoman sultan broke out, aided by the covert tutelage of British strategist T.E. Lawrence (soon to be a bigger celebrity even than Marconi). As far as Marconi was concerned, the most sensational event of this period was the Battle of Jutland on May 31 and June 1, where wireless played a crucial role that was not fully revealed until after the war.

Marconi engineer H.J. Round had developed a wireless direction-finding system shortly before the outbreak of hostilities; the War Office was aware of his work and Round was attached to military intelligence at the start of the war. A network of wireless direction-finding stations was developed along the entire Western Front as well as along the British coastlines. These stations were able to obtain bearings from German submarines and Zeppelins, providing vital information regarding their movements. On May 30, Round noticed an unusual amount of activity with the German fleet at Wilhelmshaven, on the North Sea. The tracking system was able to follow the ships as they moved north, toward the Danish peninsula of Jutland. The signals allowed the British to ascertain the position of the German ships with a remarkable degree of accuracy. Anticipating the movements of the German fleet, the British attacked in what was the only major naval engagement of the war. The Battle of Jutland was the first direct confrontation

---

* After diplomats François Georges-Picot and Sir Mark Sykes, who negotiated the accord for France and Britain, respectively.

of battleships since Tsushima in the Russo-Japanese War of 1905, and the last of its kind. Both sides suffered heavy losses and both claimed victory. Round's role was publicly revealed in 1920 by Admiral Sir Henry Jackson, the wireless pioneer and early backer of Marconi, who was First Sea Lord (professional head of the Royal Navy) at the time of the battle.[1]

Wireless was also being used in conjunction with the war's other new technology, aircraft, providing an unprecedented concentration of military power; airborne wireless communication was simply the most effective tool yet for both reconnaissance and fire-power precision. One important development for aerial navigation was a gizmo known as "the Marconi-Bellini-Tosi radiogoniometer"—basically a wireless compass that enabled the user to take the bearings of oncoming aircraft, a kind of early reverse GPS system. Developed by two Italian officers, Ettore Bellini and Alessandro Tosi, the apparatus got its name after the inventors sold their patents to the Marconi company in 1912. During the war, Marconi wrote, "it was quite possible from the receiving station on top of Marconi House, in London, to listen to the wireless signals of approaching Zeppelins while they were crossing the North Sea, preparative for a raid on England, and to determine accurately the direction of their approach."[2]

In London, Marconi continued to pursue his new vocation of diplomat. On May 19, 1916, he attended a dinner for a visiting Russian delegation offered by the Lord Mayor of London, along with other political dignitaries, including outgoing foreign secretary Sir Edward Grey and Field Marshall Kitchener. (Kitchener was killed less than three weeks later, on June 5, when the ship carrying him to Russia for negotiations struck a German mine and sank, taking six hundred lives.) Then he was back in Rome, where he was now being "persistently mentioned" as a probable member of the new government in formation. On July 4, 1916, he made an important speech in the Senate on Italian-British relations, suggesting that Italy adopt a more astute approach toward publicizing its war effort in the United Kingdom in order to convince the British to provide Italy with badly needed economic support.[3]

A few weeks later (on July 27, 1916) Senator Marconi (a.k.a. Lieutenant Marconi) was promoted to the rank of captain in the Italian Navy and asked to begin working on installing wireless communication on dirigibles. He now travelled regularly between Rome and La Spezia and managed to squeeze in a holiday at Viareggio, which would become his favourite vacation spot in later years. The new assignment also allowed him to collaborate with Charles Franklin, who was in Livorno from August 1 to October 22, 1916. Marconi was as busy as ever, but, he wrote to Beatrice from Livorno's swank Palace Hotel, he was enjoying the new work very much indeed. Beatrice, meanwhile, remained in England, installed with the children in a suite at Claridge's.[4]

Despite the Marconi company's involvement in the British war effort and Marconi's personal enthusiasm on the Italian front, unwelcome questions were being raised in London about the company's loyalty—specifically regarding its 1913 agreement with Germany's Telefunken. In November 1916, a series of articles in the *Financial News* recalled the complicated arrangement that gave Marconi's and some of its directors an interest in a Telefunken subsidiary. Godfrey Isaacs personally came under attack, as did his brother Rufus (now Lord Reading).[5] Although the company insisted that its association with the Germans had gone into abeyance at the start of the war, it remained invested in regulating competition with Telefunken in neutral markets, particularly the United States. Once again, unease regarding the affairs of the Marconi company surfaced in Parliament.

Godfrey Isaacs felt compelled to publicly explain the company's arrangement with Telefunken, which he did this way: "Whilst it is true that the Marconi Company were interested in the German Company and had directors on the German Company's board, the English Marconi Company has never had any German director upon its board, nor have the German Company at any time held any interest in the English Company," he stated with just a touch of circumlocution in an open letter published on November 17, 1916. Writing privately to Marconi (who was in Rome) the following day, he was more direct. Questions had been asked in Parliament about "our old agreement" with Telefunken and "answered in a way which I consider anything but satisfactory, with the usual want of courage on the part of the Government." The flurry of bad feeling contributed to further souring Marconi's taste for the British. The government's attitude toward the company during the war "was such as to disgust me," he later wrote, and he several times expressed a desire to resign. It was only the urging of Isaacs and some of the other directors that prevailed upon him to remain.[6]

❀   ❀   ❀

As the year drew toward a close Marconi received an unexpected blow. On October 23, 1916, Inez Milholland collapsed on stage while making an impassioned plea for women's suffrage in a Los Angeles auditorium, on the last leg of a long speaking tour undertaken against the advice of her doctors.

Suffering from a complicated cocktail of ailments (inflamed tonsils, bleeding gums, aplastic anemia, and pleurisy, among others), she remained in hospital for over a month, her condition deteriorating, while her father and husband argued over her treatment. Inez died in Los Angeles's Good Samaritan Hospital on November 25, 1916. She was barely thirty years of age.[7][†]

We don't know precisely when or how Marconi received the news, but on his next visit to the United States, eight months later, he visited John Milholland in New York and revealed "his deep devotion to [Inez], his undying appreciation of her gifts and above all of what she had been to him as a profound inspiring influence." She was, he told Milholland, "the great factor in my life. We should have been married. I think this might have been possible when she was a girl of 17 but it was evidently not to be. Yet what it might have meant to both of us. There was no such woman as Inez in all the world." When Milholland showed Marconi a plaster cast he had made of his daughter's hand, "he kissed it fervently and the tears ran down both his cheeks." Moved, Milholland told his diary he had been willing that Inez should marry him.[8]

There is no doubt that Marconi was sincere in the sentiments he expressed about Inez Milholland. In some respects, she was the great unrequited love of his life. At the same time, as we have seen, he needed to feel passion in the moment, however ephemeral or elusive it might be. Not even two months after Inez's early death, Marconi was again infatuated, this time with a woman half his age whom he barely knew. Nene Tornaghi was young and well bred, a daughter of the Roman aristocracy. We don't know how they met, although we do know where she lived—in a pleasant cottage at 6 via delle Tre Madonne, near the Villa Borghese in Rome's Parioli district. Marconi's brief fling with Tornaghi only came to light in 2008 when nineteen love letters deliriously chronicling his side of what the Italian press called a "rocambolesque and tormented liaison" came up for auction in Rome.[9] (Three more letters surfaced at auctions in Rome, Turin, and Munich between 2010 and 2014.) Absolutely nothing else is known about her—even the Internet has nothing to add. We can only speculate as to what Nene Tornaghi was like, how many others like her must have passed through Marconi's messy life, and who they were.

The death of Inez, followed almost immediately by the uptake of an affair with Nene Tornaghi, illustrated a pattern, where Marconi embarks on a new but fleeting romantic interest to make up for the abrupt end and

---

[†] Inez's widowed husband, Eugen Boissevain, later married the poet Edna St. Vincent Millay.

great loss of an important one. Marconi's intimate relationships also reflected the templates he had absorbed from his parents: the women he admired, like Josephine and Inez, resemble the idealized memory of his mother in her youth, before he knew her; his relationship with Beatrice and his children was reminiscent of those of his father to his own wives and children—authoritarian and protective.

Marconi now again moved his family to Italy. He was already spending most of his time in Rome, living at the Excelsior Hotel in the Via Veneto, where he now took over an entire floor. The Excelsior was right across the road from an apartment where a great friend of his, Lilia Patamia, held a glittering salon. Marconi and, eventually, Beatrice as well spent many afternoons there.[10] After a few months, they gave up the Excelsior and Beatrice and the children moved to Casalecchio, near Bologna, where they stayed with the Gregorinis, old family friends, while Marconi remained in Rome. Marconi was frequently at the front or on various diplomatic missions, but Beatrice recalled that he would visit whenever he could get leave, taking the children on long walks in the country and showing them his childhood haunts. It was perhaps the period when he was closest to them.

※   ※   ※

The United States of America entered the war on April 6, 1917. Count Vincenzo Macchi di Cellere, Italy's ambassador to the United States, described the American contribution to the Allies as "a usurer's mortgage." President Woodrow Wilson's purpose in bringing the United States into the war was to be able to dictate the peace; his professed disinterest only served to conceal his real purpose, di Cellere wrote.[11] Nonetheless, Britain, France, and Italy all promptly organized missions to Washington, seeking financial and trade agreements as well as influence on US war priorities. On April 30, 1917, Italy's foreign minister, Sidney Sonnino, instructed Ambassador di Cellere to inform the US government that the Italian mission would be headed by Ferdinando di Savoia, the Prince of Udine and a cousin of the king, and would include a Cabinet minister, two former ministers, the undersecretary of state for foreign affairs and Senator Marconi.[12] The mission became a focal point of Italy's volatile internal politics; one of Marconi's fellow delegates, his friend Francesco Saverio Nitti (now finance minister), successfully used his time in Washington to lay the foundation of his forthcoming return to the very centre of power.[13]

The Italian mission sailed to Halifax and continued by train to Washington, where it arrived on May 23 amid reports that Marconi had perfected and brought with him a new anti-submarine device (he would not confirm or deny the reports). In the American media's eyes, Marconi was clearly the star of the mission, and he was in the press nearly every day. On May 24, the mission was entertained at the White House by the Wilsons (it was not, of course, Marconi's first visit there). On May 27, they visited George Washington's tomb at Mount Vernon, where Marconi spoke, extolling "those democratic principles which are the life and the hope of all progressive communities." On June 2, he was the only member of the mission, along with Savoia, to address the House of Representatives (making a brief, impromptu, and largely innocuous speech praising America, after being spontaneously called to take the floor by the Speaker). The following day, June 3, the *New York Times* devoted a full page to a lengthy interview with Marconi, in which he confessed that the submarine problem was still unsolved. On June 5 he received an honorary degree from Columbia University. His message on all of these occasions was clear and simple: We want America to help solve our industrial problems, as so many of our people have come to America and helped you solve your labour problems (through immigration). "We need your coal and wheat and shipping now, but beyond that Italy will need your enterprise and capital after the war."[14]

The Italians considered the US mission a great success, well received, and the occasion for a "warm affirmation of italianitude."[15] Marconi, always attentive to such matters, noted that in New York toward the end of their visit, an estimated five thousand to six thousand members of New York's Italian colony lined the route the delegates took to their lodgings.[16] He was the star at a glittering dinner hosted by the city of New York at the Waldorf, at which he made real news by revealing for the first time how Italy had informed France of its intention to remain neutral at the outset of the war. On July 30, 1914, two days before Germany declared war on France, the Italian foreign minister (Antonino di San Giuliano) informed the French ambassador in Rome that Italy would not side with the Central Powers. This information allowed the French to begin deploying hundreds of thousands of troops against the German army's advance through Belgium, rather than station them in defence of its Alpine frontier. Marconi argued that this ultimately changed the course of the Battle of the Marne, and cited it as an example of Italy's decisive role in the war.[17]

Probably his most interesting and least likely intervention during this US trip was a long interview with the weekly *American Jewish Chronicle*, in an article on "The Italian Commission and Zionism" (for which Nitti was also interviewed). In view of his recent experience in England with British antisemitism, it is worth quoting at some length.[18] He began: "I have found the Jews eminently satisfactory in their square dealing. I can say from personal relations with innumerable Jewish business men, that the prevailing notion in the comic journals and on the vaudeville stage of their 'sharpness' is very unjustified...." Because of his frequent visits to England and his continual contact with Jews, he was familiar with Zionism, Marconi said. Was he sympathetic toward it, the journalist asked:

> Senator Marconi paused for a moment and looked out of the window near which he was sitting. "Do I sympathize with the movement?" he repeated after me in his pleasing, quiet, manner. "Put me down in the affirmative as many times as you like.... My heart is with the Zionists in their enterprise," he said with deep feeling. "And so far as Italy is concerned," he concluded slowly, "if it's ever up to her, you need not worry."

This was Marconi's only known public statement on the question of Zionism and, more broadly, Jewish matters. As was often the case with Marconi's political interventions, it is difficult to say whether he was being remarkably astute or pathetically naive. But unlike those who protest too much that "some of their best friends are Jews," in Marconi's case, some of his best friends actually *were* Jews. And Marconi himself suffered personally as a result of antisemitism, having his name drawn through the mud during the Marconi scandal. Where the spectrum of socially acceptable attitudes toward Jews in England ranged from benign tolerance to virulent hostility, Marconi was off the spectrum altogether, rarely for once, on the left side. And unlike many Englishmen of his generation and class position, he did not say one thing about Jews in private and maintain an opposite stance in public.

In this regard, Marconi was much more like his Italian compatriots. In Italy, it was not considered at all unusual that a politician of Jewish origin, Sidney Sonnino, was wartime foreign minister, unlike the United Kingdom where, as we saw earlier, Lord Reading's political rise raised the spectre of an international Jewish conspiracy among a segment of the respectable intelligentsia.[19] The general secretary of the Italian Commission to the United States, Cavaliere de Parente, was Jewish, descended from one of Italy's oldest Sephardic families. Also interviewed for the *Chronicle* article, Parente said that if there were any opposition to Zionism in Italy, it would come from Jews, not Gentiles: "Italian Jews would consider Zionism as something marring their comfort." Apart from the Jews themselves, the Italian government and public opinion were in frank approval of Zionism, Parente said. This was also evident in the interview with Nitti. "Nitti and Marconi, two of Italy's leading political figures, have committed themselves, and the sentiment of Italy, as being in favor of Zionism," the *Chronicle* reporter concluded.[20] On July 2, 1917, the mission, minus Marconi, left America secretly on a ship bound for Bordeaux; Marconi stayed behind to look after personal business. There was no public announcement of the mission's departure.[21]

Part of Marconi's personal business, as we've seen, was a visit to his old friend John Milholland, Inez's father, on Saturday, July 14. Marconi phoned Milholland at the family's New York flat from the Ritz-Carlton Hotel, where he was staying, and went over for tea. Milholland found him with "the same smiling, boyish face, frank, delicate manner, sensitive... but comprehensive grasp of things, revealed in language so simple as to make one wonder if he had really come to realize the import of his contribution to the discussion... he is great, really great and 'Nan' [Inez] did recognize the fact, even as a slip of a girl." Marconi stayed with Milholland for two hours. After sharing their grief over the loss of Inez, they talked about the war. Marconi spoke frankly. Italy faced the possibility, if not probability, of a revolution. The country needed 550,000 tons of coal a month. The people were restless, failing to see how the war was going to benefit them. Marconi's view of the war, as observed by Milholland, now more closely resembled Inez's firm opposition than the jingoism she had attributed to Marconi in her reports from Italy two years earlier. Whether he had evolved or was changing his stripes to suit the person he was talking to is hard to know, but he certainly expressed himself in terms that endeared him to his once prospective father-in-law. "Nan was right," he told Milholland.[22]

His business done, Marconi left New York on the *St. Paul*, his favourite of the transatlantic liners—the same one on which he had sailed with Inez two years earlier. He arrived in Liverpool on July 28, 1917. The Italian government was considering sending him back to the United States, in a new position of "High Commissioner"; Marconi wrote to Prime Minister Paolo Boselli that if it were deemed "in the national interest" he would accept, but under strict conditions that he laid out in a letter on August 18. He would serve for four to six months and be accountable only to the prime minister. The government decided not to appoint a high commissioner to the United States, but this became a template for Marconi's diplomatic activity: he would be available for short assignments provided he could remain free to speak his mind.[23]

Marconi's US pronouncement on Zionism was timely. On October 31, 1917, British foreign secretary Arthur Balfour suggested to the War Cabinet that a pro-Zionist declaration by the British government would be good for propaganda in Russia and the United States, and would also appeal to Jews in Germany. Control of Palestine would be strategically important after the war as well (as an outlet for Iraqi oil and a gateway to Suez). On November 2, 1917, Balfour wrote a letter to Baron Walter Rothschild, a prominent British Jewish community leader, intended for transmission to the Zionist Federation of Great Britain and Ireland. The so-called Balfour Declaration was published in the *Times* on November 9 and eventually incorporated in the peace treaty with the Ottoman Empire and the League of Nations Mandate for Palestine. It was brief and succinct: "His Majesty's government view with favour the establishment in Palestine of a national home for the Jewish people, and will use their best endeavours to facilitate the achievement of this object, it being clearly understood that nothing shall be done which may prejudice the civil and religious rights of existing non-Jewish communities in Palestine, or the rights and political status enjoyed by Jews in any other country."[24] The declaration both fell short of Zionist expectations and angered the Arabs.

By now, wireless was playing a role in all the theatres of war, particularly Palestine, where it was critical in the advance of British troops across the Sinai in 1917. Sir Archibald Murray, commander of the British troops in Egypt, characterized wireless as the modern substitute for the biblical "pillars of smoke and flame" that had guided the children of Israel on their way to the Promised Land: "Through the medium of signal stations and wireless installations, the desert was subdued and made habitable whilst adequate lines of communication were established between the advancing troops and their ever-receding base."[25]

After the battle of Caporetto (or the twelfth battle of the Isonzo) in October and November 1917, the Marconis returned from Bologna to Rome, into a three-storey house belonging to the Duke and Duchess Sforza Cesarini on the Gianicolo Hill, just beside the Piazza Garibaldi with its famous fountain and a magnificent view overlooking the city. Marconi outfitted a laboratory and studio on the top floor and spent a lot of time at home, one of only a few dignitaries allowed to circulate in wartime Rome by car. Beatrice was happy there and they had frequent, interesting visitors: Nitti, who was always trying to involve Marconi in government activities; Gabriele D'Annunzio, who "spent hours with us declaiming his poems"; Cardinal Pietro Gasparri, the Pope's secretary of state, who came to tea to talk about his idea for a Vatican wireless station.[26] At noon, they would sometimes watch the firing of the cannon marking the exact time (which one can still do today), and they often spent summer evenings in one of the small trattorias nearby in Trastevere.

By the beginning of 1918, Marconi's diplomatic career was in full swing and taking up most of his time. He was still chairman of the board and technical advisor of the London-based Marconi company, president of the Banca Italiana di Sconto, and overseeing a talented bevy of research engineers, but by far most of his energy was now going into helping to position Italy favourably in anticipation of the end of the war. In January he was in Rome, speaking at various political functions, including a dinner clearly intended to curry favour with the Americans.[27]

Beatrice felt that Marconi seemed to become more attached to the children during this period, perhaps because he often had time on his hands between wartime assignments; yet he was still restless and dissatisfied, for no apparent reason, or so she thought. In fact, in his mind, his family situation had reached a permanent state of paralysis. He would later state, in sworn testimony, that it was at this time that he realized his marriage was over. He was increasingly consorting with the most fashionable women in show business and high society.[28] On February 3, 1918, he attended the premiere of Margaret Sheridan's performance in *La Bohème*.[29] But the relationship that was taking his attention was with silent-screen actress Francesca Bertini, reportedly the highest-paid movie star in the world. (Bertini was said to have earned $175,000 in 1915, more than Mary Pickford.) Photographed with a cigarette dangling from her lips, she had a languid, modern look—a ringer for Inez Milholland and reminiscent of Greta Garbo.

Marconi and "Nuccia," as he called her, met at the 1918 premiere of the Italian film version of *Tosca*, in which she played the lead role.[30] With the hyperbole he usually employed when he was newly infatuated, he told her it was the most beautiful evening of his life. A typical Marconi romantic courtship followed. One warm autumn day (later in 1918) they went for a walk on the *Trinità dei Monti* (at the top of Rome's Spanish Steps) and he proposed to her, she wrote in an autobiographical memoir.[31] "Nuccia, I love you. . . . I can't live without you. I want to marry you," she said he told her. Bertini was in consternation. She saw him as a dear friend and admired him, but he was already married and much older (she was twenty-six, he was forty-four); it would be a scandal and, besides, she was a Catholic and marrying a divorced man was out of the question.

That same evening, Bertini wrote, she ran into Beatrice Marconi—
"tall, blonde and beautiful"—at a British embassy reception.[32] Indeed, notwithstanding his own philandering or, at the very least, attempts to stray,
Marconi continued to be wracked with jealous speculation about what
Beatrice may or may not have been doing. "B went to Civ with De and
Giu—arranged by phone with D.T. to go to Circus moderno," he wrote on
May 26, 1918, in a rare, unusually detailed entry spread across two pages of
his diary. "I saw them all Thur. Being questioned B denied having met any
one she knew." And a few months later (September 2, 1918): "B Tel D arr
met Grand Hotel told me nothing attempted deny all—met—D left hotel
1.1/4 hour after B arrived there."[33] Yet, at the same time, he continued vigorously pursuing Francesca Bertini and who knows who else.

He phoned Bertini every day, persisting. Finally, she decided to write
to him frankly: "Dear Guglielmo . . . I am too young, I am married to my art,"
and it was over. "He was not a man like the others . . . (but) the affection I felt
for him was not love," she wrote. Some years later they ran into each other
in the dining room of the Savoy in London and recalled that day on the
Spanish Steps. Marconi told her he would soon be a bachelor again, adding
sadly: "Be happy, Nuccia." Bertini married a Swiss banker, Paul Alfred
Cartier (also a friend of Marconi's), and continued to have a stellar screen
career. In her 1969 kiss-and-tell memoir she referred to him as "my great
unforgettable friend Guglielmo Marconi," but insisted that nothing had
happened between them. In 2004, the Milan daily *Corriere della sera* remembered her as "The actress who said No to Marconi."[34]

These relationships were important to Marconi, but we don't really
know how far they went. Bertini wrote that while he was courting her, he
would tell her how lonely he was, and reminisce about Inez (who, of course,
had recently passed away). "Nothing could happen between us. But I fell in
love. . . . "[35] Much later, long after his death, Marconi's daughter Gioia liked
to tell an anecdote about her father: Marconi, attending a glittering dinner
function, found himself seated next to a stunning young woman who appeared
to be in awe simply to be in his presence. At one point the woman turned to
Marconi and said, "Tell me, Mr Marconi, what does a brilliant man like you
think about while sitting around a table like this?" To which Marconi apparently replied, "My dear, I think about how I can get into bed with a beautiful woman like you."[36] This was certainly his reputation, and the aura he
projected. But think about it as he might, these flirtations did not necessarily
lead to sex, nor did they make him happy.

In Italian politics, however, his star was rising. On February 5, 1918, Macchi di Cellere, in Washington, wrote to Sonnino in Rome after the Associated Press reported that he was about to be replaced by Marconi as ambassador to the United States. Arguing that such reports could harm the prestige and authority of the office and its incumbent—himself—di Cellere asked Rome whether the news was well founded, and if not to issue an official denial. A few weeks later, in Rome, Marconi made an important speech on a long and convoluted corruption scandal involving munitions purchases, eliciting an irritable reaction from di Cellere because of the bad press the affair had attracted in the United States. The ambassador's frustrated implication was that Marconi's political interventions in Italy, automatically deemed newsworthy by the American press, made it difficult for him to represent the country properly.[37] In fact, they probably led to Italy being taken more seriously in US public opinion.

Marconi also made a series of important political speeches in the United Kingdom during several visits in 1918. On June 30 he arrived in London at the head of the Italian delegation to a conference of parliamentarians from the allied countries (the Inter-Allied Parliamentary Commercial Conference) called to begin exploring the new economic order that would emerge from the war. A clear sign of his shifting priorities was his absence from the annual meeting of MIMCC on July 2, although he was in the country. Instead, he was busy preparing a speech to the parliamentarians' conference.[38] On July 5, 1918, the conference delegates were received at Buckingham Palace, and then entertained at a dinner hosted by the British government in the Royal Gallery of the House of Lords. After a long speech by Chancellor of the Exchequer Andrew Bonar Law forecasting that the war would be over in three months (he was almost exactly right) and outlining the economic situation, Marconi spoke briefly, underscoring the "bond of sympathy and solidarity" linking England and Italy. He might have been speaking about himself, a physical manifestation of that bond.[39]

Marconi had strong words for Germany and the role of German science in the ongoing war. His antagonism toward Germany—and German science in particular—was deep-rooted in his own early-career experience, as we've seen. But now it took on a political tone as well. "It is now thirty seven months since Italy entered the war that was being waged to save the world from becoming a Prussian possession, and I can only say that we shall go on fighting as long as is necessary, that is until Prussian militarism is utterly destroyed." On the typed copy of his remarks, he added by hand: "One word about science—It was generally believed that the progress of science inevitably meant peace. That notion has failed—Germany has utilized science to the utmost, but I say with sorrow and regret that Germany in this war has prostituted her scientific achievements."[40]

The Marconi Archives in Oxford contain several versions of this speech, which Marconi revised, recycled, and repurposed for different occasions between April and September 1918 (when he was back in London for a series of public celebrations of Italo-Anglo relations), emphasizing aspects of economic co-operation that would be particularly important for Italy after the war, like the exchange rate between the lira and the pound, or shipping freight charges. He bought in completely to the new ideology of international trade that was beginning to replace the idea of colonialism: "Commerce will always become more and more an essential element of the life and prosperity of nations, as it has been in the past one of the most powerful agents in disseminating civilization," he wrote in a draft for one of these speeches.[41]

In his new diplomatic role, Marconi returned regularly to his binational status: "I am here as a representative of the Italian Parliament, but I am also proud of my blood relationship with the British nation," he said in the same London speech in 1918. He increasingly saw himself as a bridge between the two political cultures, both of which he claimed to understand, but ironically, and painfully, he would never be fully accepted by either—used but not fully trusted, considered a valuable spokesman, source of information, and useful messenger, but ultimately neither fish nor fowl. As a global figure, however, he was also from this time on frequently solicited to lend his name to causes related to peace and internationalism.[42]

The company's name was still tainted in England; it was "a name of ill omen in politics" according to historian A.J.P. Taylor, biographer of Lord Beaverbrook, Britain's wartime minister of information. In August 1918, when the ministry was planning to send someone who had been connected to the Marconi company on a foreign mission, Beaverbrook was advised by deputy prime minister Stanley Baldwin that "it would be a fatal error to employ anyone connected however remotely with the Company," and Beaverbrook "gratefully accepted the advice."[43] But Marconi's name was still magical, an opener of doors and magnet for all manner of glittering invitations and opportunities. Marconi himself was never tainted with the tar brushes that regularly covered his company's misfortunes.

❈   ❈   ❈

On November 11, 1918, at 5:40 a.m., Marshall Foch sent a message announcing the end of hostilities and the taking effect of the armistice as of 11:00
a.m. that day. The Marconi company proudly lauded the fact that "a wireless
message, the first open act of war, was also the last."[44] An urban legend promoted by the company had it that Marconi (who would have had to be up
somewhat earlier than usual if this were the case) heard the announcement
at his home on the Gianicolo Hill in Rome, and was the first to inform the
Italian Cabinet. Actually, Marconi told British writer Harold Begbie he had
been the first in Italy to learn of the abdication of the kaiser, two days before
the armistice, on November 9. "I was sitting in my room in Rome, and the
box [Marconi's personal wireless receiver] was on the table at my side, when
suddenly a message began to arrive.... It was a message from Germany and
the news was...that the Kaiser had abdicated. I took up the telephone and
rang up the Prime Minister of Italy...."[45]

Company historian W.J. Baker wrote that until his death, Marconi
"never ceased to look upon (wireless) as a potential means of promoting
peace and understanding between the nations."[46] But we shouldn't forget
that he also saw it as an instrument of war. This ambivalence had been clear
from his very earliest interventions in politics and diplomacy, ever since he
began using his fame as a bully pulpit. In December 1906 he had been in
Venice to receive a gold medal from the city. He generally used such occasions to speak of the benefits of communication for enhancing peaceful
relations among nations, but this time he went a step further and in a different direction. "I hope that wireless telegraphy will always be a herald of
peace and civilisation between nations; but if, some day, cries of war should
once more re-echo on this sea on which Venice has always triumphed, I
trust that wireless telegraphy will serve to transmit to Venice the news of the
victory of the Italian arms." What Marconi considered a "merely academic
remark"[47] embedded in a boilerplate patriotic speech was received as a bellicose statement by neighbouring Austria, further embittering that country's
already strained relations with Italy and leading to a minor diplomatic incident. Vienna rankled with resentment at the suggestion that war might one
day rage on the Adriatic. There was speculation that the speech was a programmed move in an Italian propaganda campaign.[48] It was not, which
begged the question that would perplex observers of his adopted role of
statesman for the rest of his career: Was Marconi a subtle and sophisticated
political operator or just self-interested and naively patriotic?

24

The Statesman

The global scene, and Marconi's own agenda, now shifted radically to shaping the peace. It was well understood that the agenda for the upcoming conference in Paris would be determined by Woodrow Wilson's "Fourteen Points," a blueprint for a new postwar world order published by the president early in 1918. But Italy intended to insist at least on the conditions of the 1915 Treaty of London, as well as respect for the moral commitments it had made at the so-called Congress of Oppressed Nationalities in Rome in April 1918.

Organized under the patronage of the Italian and French governments, the Congress of Oppressed Nationalities gathered representatives of various national groups under Austro-Hungarian rule—including Italians, Poles, Rumanians, Czechs, and Yugoslavs—mutually recognized their right to national unity as independent states, and agreed to settle territorial issues amicably on the basis of national self-determination. The resulting Pact of Rome was not a treaty and not signed by governments, but a moral obligation in the eyes of Italian Prime Minister Vittorio Emanuele Orlando, who was also an authority on constitutional law; there were, however, some inconsistencies between the Treaty of London and the Pact of Rome. The Yugoslavs opposed the secret treaty, while for Italy it was fundamental.

As early as November 1, 1918, Orlando had expressed Italian reservations regarding point nine of Wilson's Fourteen Points. The United States recognized that point nine proposed "less than the Italian claim, less of course than the territory allotted by the Treaty of London, less than the arrangement made between the Italian Government and the Jugo-Slav State [that is, the Pact of Rome]."[1] On the eve of the peace conference, the Italians were guarded and suspicious, their reserve and indecision intensified by the unstable political situation in the country. The press was divided, the nationalist *Giornale d'Italia* (a paper regarded as foreign minister Sidney Sonnino's mouthpiece, as he was its largest shareholder) describing the Pact of Rome as a weak compromise dictated by "paroxysms of cowardice," while the more liberal *Corriere della sera*, for example, defended the Pact of Rome, albeit half-heartedly and apologetically.[2]

The Treaty of London, considered favourably by Italian opinion, faced a different set of obstacles. Since it was concluded before the United States entered the war, Washington had not been a party to the treaty, and Wilson did not feel in any way bound by it. Furthermore, it had been a secret agreement, an instrument of diplomacy now widely discredited. That said, the British and French (who certainly wanted their own secret treaties to stand) did feel some obligation to respect it. "Whatever may be said of secret treaties," British Foreign Secretary Arthur Balfour wrote to Wilson in January 1918, "it must be remembered that they were engagements representing a promise of payment for value." In particular, "Italy had joined the war, not on a plea of self-defense, but deliberately and with a clear agreement as to what the price of her assistance was to be."[3]

Wilson and his entourage sailed for Paris on the *George Washington* on December 4, 1918. Macchi di Cellere and the French ambassador were on the same ship. For Wilson, the peace conference was an opportunity to set up a new global order in which international relations would be governed by democratic ideals and the rule of law. His allies in the recent war had far more mundane considerations, such as land, oil, and recrimination. Di Cellere took advantage of the opportunities for informal exchanges with the Americans afforded by the ocean crossing to push the validity of Italian claims vis-à-vis the Yugoslavs.[4] But he managed to mislead himself, and, consequently, his superiors, about the success of his efforts, assuring Rome, prematurely, that Wilson was sympathetic to Italy's aims. (According to historian Margaret MacMillan, di Cellere was "a man with an extraordinary capacity to ignore the facts."[5])

On December 14, 1918, Wilson arrived in Paris, and the same evening asked for the text of the Treaty of London. The question of whether Wilson knew about the Allies' secret treaties before arriving in Paris has been the object of controversy for nearly a century.[6] During the US congressional debate over ratification of the Treaty of Versailles in August 1919, he denied having had any such knowledge. This is possibly true. However, the existence of the secret treaties was a matter of public knowledge. Parts of them had been published by the Bolsheviks in 1917, and, as we've seen, Marconi mentioned the Treaty of London in a speech in the Italian Senate as early as December 1915. Certainly, Wilson's deputy and chief negotiator in Paris, Edward M. House, was aware of them, as Balfour had discussed them with him in 1917; the American commentary on the Fourteen Points, drafted under House's order, also mentioned the terms of the Treaty of London.[7]

Italy's claims were among the first issues Wilson had to deal with when he arrived in Paris. On December 29, 1918, Wilson, House, Orlando, and Sonnino met for two hours. House later wrote: "The President talked well but he did not convince the Italians that they should lessen their hold on the Pact of London."[8] Wilson then visited Rome from January 2 to 6, 1919, but was unsuccessful in his attempt to reach an understanding; however, he reported back to House on January 7 that he was pleased with the trip.[9]

The Italian situation was complicated by the resignation from government of socialist Leonida Bissolati in December 1918, and of Francesco Nitti as finance minister the following month. Orlando continued to oversee the negotiations, although Sonnino remained involved as well. The other three members of the delegation, Antonio Salandra, Salvatore Barzilai, and Giuseppe Salvago-Raggi, were largely insignificant to the talks. The delegation was neither strong nor effective; its members were divided among themselves, and they didn't trust their allies.[10] The Italians were also following developments in London: Who would be the British delegates to the conference? Ambassador Guglielmo Imperiali wrote to Sonnino while he was still in Rome that it seemed the candidacy of Lord Reading had been eliminated in favour of Balfour; the conservative newspapers kept recalling the Marconi scandal at every opportunity. Imperiali wasn't doing much more than providing Sonnino with a British press review, but his selection of items worth mentioning was interesting.[11]

The year ended sourly for Marconi, as well; he spent New Year's Eve alone and ill in his room at the Savoy Court in London.

❀   ❀   ❀

The Paris Peace Conference started on January 18, 1919. However one looks at it, the conference was a transformational moment in world politics and governance arrangements. As Margaret MacMillan details in *Paris 1919*, it marked the coming out of nearly all the national entities that are still struggling for recognition a hundred years later. It also at least nominally recognized the opening of international diplomacy to non-governmental actors, although substantively it was still an affair between states. At the bottom line, the ultimate decisions were made by at most four men, sometimes three, sometimes one.[12]

On February 7, 1919, the Italians presented their official memorandum to the conference, reiterating the planks of the Treaty of London without mentioning it explicitly.[13] They now demanded even more, integrating the irredentist claim to the Adriatic city of Fiume, which until the end of the war was a part of the Austro-Hungarian Empire. Orlando's stump speech slogan now became "The Treaty of London plus Fiume." Italy's problem was that the British and French felt its contribution to the war had been limited, while Wilson did not recognize the Treaty of London and supported Yugoslav claims over Italian requests for territory in the name of self-determination. Fiume was doubly problematic in that the city had an Italian majority but did not figure in the Treaty of London. It was another example of the postwar conundrum whereby two parties could each make a legitimate claim and what the Allies decided would make one side happy and the other side furious.

On April 3, 1919, Balfour addressed a long and reasoned letter to Italy's London ambassador, Imperiali, trying to reassure him, in the face of Italian skepticism, about British intentions. Although Italian issues had not yet been discussed, Britain (and presumably France as well) would remain bound by the Treaty of London. "The frontiers may be good or bad for Italy, fair or unfair to Italy's neighbours. But if Italy, in spite of all that has happened since 1915, still wishes to obtain them, we are bound by Treaty to support her." If Italy now wished to modify these, for example, to gain Fiume, did that not change British and French obligations? Britain was not hostile or indifferent to Italian interests, despite the prevailing view in Rome. It didn't matter to Britain whether Fiume was Italian or Slavic. Britain's only motive was the well-being of Europe and its friends.[14]

According to René Albrecht-Carrié, author of a 1938 monograph *Italy at the Paris Peace Conference*: "The Italian question was now the outstanding issue of the Conference."[15] On April 13, 1919, Orlando asked for an interview with Wilson. The meeting was set for 11:00 a.m. the following day. Georges Clemenceau, meanwhile, alerted the Americans that Orlando was threatening to withdraw Italy from the conference and go home.[16] At his meeting with Wilson on April 14, Orlando brought along Andrea Ossoinack, former member for Fiume in the Hungarian parliament (and Fiume representative to the conference). Orlando was still asking for the Treaty of London plus Fiume. Wilson now made the American position clear to the Italians for the first time.

Ossoinack had proclaimed the annexation of Fiume to Italy in a statement in the Hungarian parliament on October 30, 1918. On March 15, 1919,
he had presented to the conference a document signed by himself, Antonio
Vio (mayor of Fiume), and Antonio Grossich (president of the Italian
National Council of Fiume), arguing that Fiume had always been and was
at present an Italian city, and asserting the validity of the annexation proclamation on the basis of a people's right to self-determination.[17] As Ossoinack
made his case, he had a series of rough exchanges with Wilson. Ossoinack
insisted that all of Istria, Dalmatia, and the offshore islands were Italian:
"Jugoslavia begins on the mountains and not on the coast. I am proud of
being born in Fiume and not on the mountains." When Ossoinack said it
made sense for Italy's contiguous territory to extend to Fiume, Wilson interrupted him: "The Peace Conference has not yet decided that this territory [bordering Fiume] will be Italian, and therefore we cannot yet speak of
territorial continuity." The arguments were economic as well as nationalistic. When Wilson interrupted to make a point about trade and railways
lines, Ossoinack countered: "My dear sir, you are wrong." Orlando inter-
jected at one point to say he was prepared to guarantee the rights of all the
nationalities living in Fiume, but Ossoinack raised the spectre of civil unrest
if the city were not annexed to Italy—this evidently able politician closed
the discussion with the remark that "Fiume will not accept any other solution and I for my part decline all responsibility for the results which may
ensue on any decision other than that of uniting Fiume to Italy." Marconi
had a typewritten copy of a report of this meeting in his personal papers.[18]

The day after the meeting, April 15, 1919, House wrote, "I talked with
Orlando and found him exceedingly bitter."[19] On the evening of the nineteenth the Italians had a stormy meeting and decided to stick to their position on the Treaty of London.[20] They were now also insisting on Fiume. On
April 22, House wrote: "The whole world is speculating as to whether the
Italians are 'bluffing' or whether they really intend going home and not sign
the Peace unless they have Fiume. It is not unlike a game of poker."[21] On
April 24, Orlando left Paris and returned to Rome, leaving Italy unrepre-
sented at the conference until the Italians returned to witness the presentation of the peace treaty to the Germans on May 7, 1919. Along with issues
related to the organization of the newly created League of Nations, the
conference's attention now focused on finding a compromise solution to
the Adriatic question. Dealing with these two problems was House's main
activity during the next three weeks.[22]

At this stage, quietly and without fanfare, Marconi entered the picture. On May 23, 1919, Orlando reported to the King of Italy that a deal proposed by House had been blocked due to Wilson's intransigence on the Adriatic question. "A conversation today between Senator Marconi and President Wilson confirms this disturbing finding," Orlando informed the king.[23] Marconi's meeting with Wilson was highly discreet; in fact, it left no direct trace in any of the official or unofficial records of the conference. Those were heady days, described by House as a "treadmill," with all sorts of important callers coming and going: General John J. Pershing, Alexander Kerensky, Felix Frankfurter. . . . House's diary for the twenty-third mentions visits from Serbian, Russian, and Montenegrin diplomats, but no Marconi.[24]

It is not clear in what capacity Marconi was received by Wilson or how the meeting was set up. Marconi had several meetings with Wilson and US officials in Paris regarding the future of radio regulation, which may be why this one eluded House's record. But there is no doubt that it took place. "We discussed the question of Fiume and the Dalmatian coast fully," Marconi is quoted by his first British biographers, Jacot and Collier. He told fellow parliamentarian Silvio Crespi, a member of the Italian delegation, that he was left with a bad impression; Wilson was intransigent, bitter toward Orlando and toward Italy. Solari later wrote: "The interview was long, calm, but inspired with great force. After it was over, I went to see him [Marconi] at the Hotel Ritz in Paris. He said to me brusquely, 'unless someone takes matters in their own hands there will be nothing done at all.'"[25]

The date of Marconi's meeting with Wilson, May 23, 1919, was also the anniversary of Italy's entry to the war, and there were demonstrations all over the country. Orlando told Clemenceau that Italy was totally exasperated, adding that this explained the explosion of public sentiment in a form that he certainly deplored.[26] A rally planned for the Augusteo in Rome, with D'Annunzio as orator, was banned, apparently by order of Orlando himself. According to Crespi, US Ambassador Nelson Page threatened that the United States would break diplomatic relations if D'Annunzio were allowed to speak, after the poet had virulently insulted Wilson and his wife.[27]

Orlando reported to the king that British Prime Minister David Lloyd George was urging him to hang on for two or three more days. "However, from hard experience, I do not rely on such assurances. . . . Your majesty sees well that I am doing what is possible to find a solution; but it is impossible for me to descend below these really insurpassable limits."[28] House wrote that Orlando told Clemenceau he was sticking to the Treaty of London, and

Wilson was "disturbed but not 'panicky.'" On May 28, 1919, Lloyd George and House met with Orlando, then with Wilson and Clemenceau. "We have the Adriatic question whittled down to the vanishing point," House wrote in his diary, indicating that it was all but resolved. Marconi, meanwhile, returned to Rome.[29]

The Italian National Council of Fiume decided to appeal directly to the United States Senate, bypassing the president who, it said, was obstinately denying Fiume's right to self-determination. The only thing Fiume was asking, the council said, was that the conference recognize the October 30, 1918, proclamation of Fiume's annexation to Italy. The council's statement was couched in just the sort of menacing ethnocentric rhetoric that so exasperated Wilson in his efforts to find a just way out of Europe's inter-ethnic quagmire. It was strange, the council wrote, that an Italian city "should be subjected to influence of people inferior to it in civilization." It was a "monstrous iniquity" and "Fiume will offer resistance to this attempt on its liberties." Ossoinack presented the statement to the conference on June 2, 1919, in the name of the national council, and cabled it to Senator Henry Cabot Lodge in Washington on June 3.[30]

Back in Rome, Marconi was embroiled in a flurry of personal activity. His diary for the week of June 8 to 14 was exceptionally loquacious and characteristically enigmatic. It contained complicated and lengthy (for him) entries every day, mostly in Italian and much of it incomprehensible—full of initials, numbers devoid of any context, bits of Morse code, and comments indicating he was, jealously, practically stalking Beatrice.[31]

On June 19, 1919, the Italian government fell, and on June 23 Orlando resigned as prime minister, acknowledging his failure to obtain Fiume as one of the reasons. He was replaced by Francesco Saverio Nitti. Sonnino remained in Paris to sign the German treaty but there was a new foreign minister, Tommaso Tittoni, a career politician who had served briefly as prime minister in 1905, and more recently as ambassador to Paris.[32] One of the new government's first tasks was to name a fresh delegation to return to Paris and resume negotiations. The delegation was named immediately: it would be headed by Tittoni and composed of Vittorio Scialoja, Maggiorino Ferraris, Silvio Crespi, and Marconi. (All were senators except for Crespi, who was a deputy of the lower house.)[33]

The only member of the new delegation actually present for the signing of the German treaty (which history now knows as the Treaty of Versailles) on June 28, 1919, was Crespi (the other Italian signatories were

Sonnino and Imperiali[34]). The rest of the delegation only arrived in Paris the following day, June 29. It was based at the Hotel Edouard VII, except for Marconi; always his own man, he insisted on staying at the Ritz. Hardly any of his friends were left in Paris, he wrote to Beatrice on June 30, and nearly all the Americans had gone.[35]

The treaty had created a commission on colonial mandates to deal with the colonial territories of the defeated Central Powers, and Marconi was delegated to sit, replacing an ailing Crespi. He travelled to London on July 6 and had a day to himself before the commission met on July 8. The task was daunting, the players at the top of their game. House was there for the Americans, Lords Alfred Milner and Robert Cecil for the British, François Simon for the French, Viscount Sutemi Chinda for Japan, and Marconi for Italy. In three days the commission settled the future of Germany's former colonies in Africa, deferring the fate of the former Ottoman colonies (Syria, Palestine, and Mesopotamia) to a later discussion.[36]

On July 11, Marconi returned to Paris, where Tittoni had been cooling his heels waiting for talks to recommence regarding the remaining issues. On July 15, the Council of Four (Britain, France, the United States, and Italy) met, but Tittoni considered the discussions to be on issues the Italians felt were of secondary importance. Italy seemed confined to a passive role, mainly of observation, its key questions not even on the table. The conference had entered a new phase following the departures of Wilson and Lloyd George and was now in the hands of the ministers. Tittoni was everywhere, talking to everyone, accomplishing little.[37]

The stagnation in Paris was evident from the nature of the documents accumulated in the Italian Ministry of Foreign Affairs archives: instead of grand politics, the delegation was now inundated with decrees about internal organization, details about per diem allowances, an evacuation plan for their quarters at the Edouard VII, and a request from the French government that they try to use less electricity.[38] But even as late as the end of July, Marconi remained hopeful. "I am going to London with Tittoni in a day or two," he wrote Beatrice on July 29. "I expect to be in Italy about the 15th of August by which time most of the important things here should be decided. Have been working very hard to get coal and corn of which we are in the greatest need."[39] In London a few days later he attended the final meeting of the Supreme Allied Council, then the company's annual general meeting—"in his personal capacity," while Isaacs chaired.[40] Then he was back to Paris.

With the departure of Wilson, House, and Secretary of State Robert Lansing, the Americans' new chief negotiator in Paris was Frank L. Polk, an old friend of Marconi's. On August 15, 1919, Tittoni cabled this news to Nitti in Rome, adding hopefully that Marconi had asked Polk, and Polk had agreed, to raise the Italians' concerns once again directly with Wilson. Marconi reported to Beatrice with uncharacteristic political frankness: "We are in an awful fix. If we don't give in on the Adriatic question, America won't send us grain and coal, and we would simply starve in a very short time as England can't send us anything like enough just now. So we are in a very very difficult position."[41]

Marconi had another useful acquaintance with the US delegation: Ray Stannard Baker, who had travelled with him in Newfoundland and Cape Breton in 1901 and written an important feature on him for *McClure's*. Baker had spent 1918 in Europe for the US State Department, meeting political figures and reporting on potentially destabilizing radical movements. As director of the American Press Bureau at the peace conference, he was effectively Wilson's press secretary. He became a strong advocate of the League of Nations and, eventually, Wilson's authorized biographer. But there is no evidence that Baker and Marconi were actually in contact.

On August 20, 1919, Marconi made the most important intervention of his young diplomatic career, writing directly to Nitti, by hand: "*Carissimo Presidente*...Our affairs are not going well, principally because of the attitude of W̲ [emphasis in original], who has all the cards in his hands and also because of lack of support from England and France." The American position was that Italy would get the credit and coal it urgently needs "only when we will have made an agreement on the Adriatic question." Polk had told Marconi privately that he believed it possible to obtain more from Wilson, "but not everything that Tittoni is requesting." Polk said he did not understand why England and France were not prepared to be a bit more generous with territorial concessions for Italy in Africa after all Italy had sacrificed and suffered in the common cause. "Pardon me for such a pessimistic letter. If we have to cede it would be best to decide to speak clearly and explain everything frankly to the country." And the letter was signed with a strong underlined flourish, Guglielmo Marconi.[42]

This was a forceful statement of Marconi's nationalist sentiments, as well as a demonstration of his diplomatic skill, suggesting Italy should consider making a political compromise in order to obtain crucial economic support. Polk sent Tittoni's proposal—that Fiume become a neutral city under the League of Nations and that Dalmatia should go to Yugoslavia—to Wilson.[43] While the Italians waited for Wilson to reply, Marconi headed to Rome for a very brief visit with his family; he was back in Paris again before the end of the month. The government was now pressuring him to head an economic mission to the United States.[44] He didn't want to go but—characteristically compartmentalizing on the one hand, multi-tasking on the other—looked at the challenge as an opportunity to repair his marriage. "I've done my best to get out of it but it's very difficult as both Nitti and Tittoni keep on begging me to go, and saying that they can't possible find anyone else who could take my place. It's not decided yet but I fear I may have to go. Would you come too? I would love you to come," he wrote to Beatrice.[45] They would probably spend a month in the United States.

Beatrice didn't reply. Five days later he wrote again, in a confused and anxious state. He had tried to bow out of the mission for health reasons but Nitti was "begging me to go for the sake of Italy and of patriotism! I really don't know what to do as it breaks my heart to know I won't see you and the chicks until perhaps the middle of November and the awful feeling of leaving you in Italy all that time (You understand)."[46] The mission had now blown up into a larger conference, involving the English and the French as well, and wives were not expected to attend; he would have insisted to bring her, he said, but as she had not answered he didn't know what to do. He signed the letter, oddly, "Guglielmo Marconi," not the usual "Guglielmo." This episode contradicts the view that he no longer cared about her and was happy to be away as much as possible; in fact, it seems more the opposite. She was beginning to set the boundaries of her own life. Meanwhile, Marconi said no to Nitti; he would not go to the United States.

Nitti was now thrust into an impossible situation. On September 11, 1919, the Italian delegation in Paris signed the Austrian treaty. The following day, September 12, Gabriele D'Annunzio led an army of "legionnaires" into Fiume, occupying the city and creating an international political sensation that Marconi (along with a large part of the Italian population) applauded. The Italian government was embarrassed, annoyed, and secretly pleased that someone had finally pricked the abscess. Tittoni continued to negotiate in Paris while Nitti focused on the upcoming US mission.

Although it was announced that Marconi would be heading the Italian mission, he publicly declared that he would not go, pleading illness and also hinting to the New York Times that he didn't want to embarrass the United States and Italy over his strong position on Fiume. On October 20, 1919, his old nemesis Vincenzo Macchi di Cellere died unexpectedly on a Washington operating table after a sudden illness, and Marconi's name was raised again as a possible ambassador.[47] There was no question, however. He wasn't going anywhere but Paris. On October 21, 1919, Beatrice Marconi received a brief telegram at her home on the Gianicolo Hill. It was from Paris, and read: "*Arrivato benissimo abbracci* [Arrived well hugs]. Guglielmo Marconi."[48]

25

The Spark

The war had had a profound effect on Marconi. His exposure to it was just authentic enough to provoke some deep reflection. Shortly after the end of the hostilities, Marconi gave a long interview to British writer Harold Begbie, a social reform advocate and religious moralist who had publicly defended the rights of pacifists and conscientious objectors to oppose the war. The interview was done for a book Begbie was preparing on the topic of wireless in war, which was to be published—cozily enough—by the Wireless Press, a subsidiary of the Marconi company. Begbie wrote the book quickly, beginning in January 1919 and completing it a few months later. The Begbie manuscript provides a thoroughly detailed and rather glorified account of the role of wireless in the war, and as an instrument of war in general. The book was never published, however, for several reasons. The company considered it too praiseworthy of its own role in the war and too critical of the company's treatment by the British government. Most important, perhaps, parts of it were considered "undesirable from the point of view of the national interest" by the military intelligence authorities to whom the manuscript was submitted for approval. The manuscript was considered so problematic that several chapters were removed from the draft copy and the proofs were destroyed after Begbie's death in 1929.[1]

Nevertheless, the interview, done around April 1919, provides a unique window into Marconi's thinking coming out of the conflict. It also hints at a spiritual attitude that we have not yet seen—though how much of it is Begbie's and how much is actually Marconi's is hard to say. Marconi speaks of "the mystery of wireless telegraphy and its power to assist the mind of man in threading his way through the complexities of the universe." The future of wireless, he told Begbie, was "profoundly unknowable. We are simply at the beginning of our fumblings with an infinite and eternal force." Marconi told Begbie he was receiving vibrations that could only be coming from other planets in outer space; it was not inconceivable that intelligent people living on other stars should be striving to communicate with us. He said the fresh knowledge we might gain from these older, wiser, extraterrestrial beings would soon solve "the mystery of existence."

While it was apparent to Marconi that "behind all phenomena there is a great spiritual reality," he also expressed to Begbie that science was the means toward understanding the universe. This curious mix of spirituality and rationality was typical of the ambivalence with which Marconi went through life. But despite what he might say to an interlocutor, Marconi was rooted in a material reality: he loved his research better than anything else, and the world was only a distraction from "the intense strain of serving science," he told Begbie. In fact, Marconi's experience in the war and its aftermath not only convinced him that he had a role to play in politics, it reconfirmed his sense that his unique contribution—the perfection and extension of mobile, wireless communication across greater and greater distances—had a decisive role to play in the new world order.

Marconi also revealed to Begbie that he had just purchased a yacht. He was ready to move on to another stage of his life and as usual, he did it with éclat. On April 14, 1919, from London, he wrote excitedly to Beatrice in Rome:

Darlingest,

I've bought the yacht *Rovenska*! The last one of which I showed you photographs and plans. I went to see her at Liverpool and she is really very fine and comfortable.

The price is 30 thousand.* I'm a bit uncertain as to whether I shall bring her to Italy, as soon as possible, or have my new invention put into her right away. If I could do this it would of course be better, as it would be cheaper to run and I would have no trouble over coal.

Darling, I wanted to tell you at once. Quite well. All my love, Yours ever, Guglielmo.[2]

Marconi had been shopping for a yacht for some time. Ever since he was a youth in Livorno, he was never happier than when at sea alone. His interests oscillated between England and Italy, two maritime powers. He could do his

---

* Pounds sterling, twice what he got for his UK patents, equivalent to US$150,000, or around $2.4 million today.

research anywhere, so long as he had the proper equipment. Now that the war was over he was aching to get back to experimental work. Most of all, he needed to free himself from the chains of domesticity. "I want to live at sea," he once wrote in a letter to a friend. "I shall sell my house in Rome and get in its place a yacht, which will in future be my home. There I shall be able to study and experiment without the fear of unwelcome interruption." Beatrice later wrote: "He imagined perhaps quite honestly that we and a growing family could live all the year round afloat."[3]

The *Rovenska* was built in Scotland in 1904, as a pleasure-cruising craft for the Archduchess Maria Theresa of Austria. Maria Theresa's husband, the Archduke Karl Stephan, was a grand admiral in the Austrian-Hungarian navy and had been mentioned as a possible future king of Poland under the Hapsburg Empire. He already owned several British yachts. "Rovenska" was the name of the bay on the island of Lošinj in Croatia, where the Archduke had a castle.[4] The yacht was 220 feet long, and displaced more than seven hundred tons. With its twelve-hundred horsepower coal-powered steam engine, it had a cruising speed of ten knots and devoured twelve tons of coal a day. The ship had been taken over by the British Admiralty during the war and used as a mine-sweeper in the North Sea. It was not difficult to reconvert for private luxury use. Marconi engaged a Neapolitan naval officer, Raffaele Lauro, to be the captain, hired a crew of thirty, and renamed the ship *Elettra*—the Latinized form of the Greek word for amber, the substance that, when rubbed, creates a spark.

With its two huge masts and steam funnel, the *Elettra* looked like a giant swordfish. It was soon one of the most famous private yachts in the world. Marconi had it outfitted with his and hers master cabins—his furnished in dark wood, the height of masculinity; hers in flowery ceramics—plus four cabins for guests and a regal, Renaissance-style dining room. Most important, he had it equipped with a wireless cabin and laboratory in which he could work on his research.[5] Shortly after buying the yacht, Marconi told Begbie that he intended to use it "as Darwin used the *Beagle*. I am going to sail away from cities and in quiet bays and inlets conduct certain experiments which I permit myself to dream may carry the world a little farther forward on the road of discovery. . . . I don't have to worry about the commercial side of things, and the days when I had to race up and down the world trying to demonstrate what could be done with wireless are over."[6]

The London press was soon making a great fuss over the *Elettra*, confirming Marconi's conviction that this was what he needed.[7] He later wrote: "I suppose the majority of inventors are able to do most of their work in one fixed laboratory, but in my case the very nature of the work has entailed constant movement." Until now, Marconi had done much of his experimental work at installations purpose-built to be stations, like the Haven or Poldhu, where he was never more than an ephemeral presence; on commercial or military vessels, where he was not at home; or places he was literally passing through, like the Libyan desert or Signal Hill, Newfoundland. Paradoxically, he now combined mobility and stability. The *Elettra* was Marconi's first personal and most permanent experimental laboratory, the first one that was truly and exclusively devoted to research and truly and exclusively his. By 1923, he could say that the *Elettra* was "my principal laboratory, and it is here that my real work is done."[8]

The *Elettra* was equipped with the most modern instruments and apparatus, including direction-finding and shortwave aerials. It had a three-kilowatt transmitter operating on continuous waves, not spark, and the latest system of wireless telephony, soon to be used in broadcasting. It also carried a "conventional" spark transmitter for ordinary ship-to-ship and ship-to-shore communication. But the main research problem Marconi worked on there in the early years was shortwave, or "beam," reception. "For this work nothing is so suitable as a yacht," he wrote.[9] The *Elettra* was not only Marconi's private floating laboratory, it was also his retreat and the closest thing he ever had to a real home. Most of all, it was quintessentially his own. It was the perfect reflection of his character, temperament, and taste—a paean to mobile communication. In this space, at one time or another, he would entertain both his wives, several mistresses, all of his children, colleagues and assistants, business associates, film stars, journalists, politicians, and world leaders. D'Annunzio called it the "snow-white miracle ship."[10]

While Marconi was negotiating the purchase of his yacht, a crucial upheaval in his corporate affairs was playing out on the other side of the Atlantic. The American Marconi company had been formed on November 22, 1899, and taken over by its British parent company on April 18, 1902. Through a combination of research innovation, aggressive patent protection, and integration into Marconi's global network, American Marconi was by 1919 the dominant force in US wireless communication. Its only possible competitor was the United States Navy. In February 1914, after the American company applied to the US government for a concession to build a high-power station in the Philippines, Captain William H. Bullard, superintendent of the

Naval Radio Service, wrote a memo articulating the view "that radio communication is a natural monopoly and that all stations should be government owned and controlled." If a Philippine concession were granted to
Marconi's, Bullard argued, it would surely be followed by other requests; he
advised against it, and with the outbreak of war in Europe, the company did
not press the issue.[11]

Marconi was also thinking ahead to the next phase in the development
of radio. After his May 1915 visit to the General Electric plant in Schenectady,
New York, he approached GE's chief counsel, Owen D. Young, about acquiring the rights to its new generator of continuous radio waves, the so-called
Alexanderson alternator.[12] They were on the verge of an agreement by
which GE would manufacture the alternator, developed by its engineer
Ernst Alexanderson, while Marconi would have exclusive rights to use it,
but before the deal could be finalized, Marconi had left urgently for Europe
as Italy entered the war.[13]

Marconi's offer to buy the Alexanderson alternator was a turning
point. It was more than a routine business deal: Marconi's organization was
now prepared, for the first time in its history, to commit itself to an outside
supplier.[14] Telefunken was working continuous-wave alternators across the
Atlantic from its Sayville station, and while Marconi was confident the
Sayville operation would be shut down, either through his patent infringement suit or US government action, it was time to catch up. In the past few
years, he had become increasingly interested in continuous wave as opposed
to spark technology, a change that was about to enable the paradigm shift
from telegraphy to voice as the dominant mode of wireless transmission.
If Marconi could acquire the Alexanderson alternator, he would be able to
convert his existing and future plants to continuous-wave transmission—
soon to be known as "broadcasting." Through his ironclad, court-upheld
patents, Marconi controlled the wireless industry in the United States. But
the deal he was now proposing would make GE the sole supplier of the key
piece of infrastructure hardware in the next stage of the technological
development of wireless.

The proposal was intensely political and essential to Marconi's global
strategy. Marconi's UK base was constricted by British wartime restrictions,
but the war also presented opportunities for technological development and
the company was still determined to build a global network anchored by a
British imperial wireless chain. At the same time, the huge US domestic
market and likely emergence of the United States as the dominant world

power after the war foreshadowed an increased role for Marconi's American operations. A deal with GE would palliate American nationalist concerns and reduce Marconi's exposure to the British public sector, with which he had such a fraught relationship. At the same time, the company was anxious to position itself once and for all against the anticipated postwar resurgence of Telefunken.

The United States' entry to the war in April 1917 put a major crimp in Marconi's plans. The US Navy took over all wireless operations on April 7, and as the war proceeded it was not entirely clear what would happen to them once the conflict ended.[15] It was clear, however, that now was the time for all parties—advocates for an expanded postwar role for government, commercial companies, the new legions of "amateur" users—to position themselves.[†] At the war's end, all sides were lobbying Congress and the White House regarding the future of communication (or *tele*-communication, as it was starting to be called). The three main US electrical companies— GE, AT&T, and Westinghouse—all had a stake in wireless; but in international communication, a foreign company, Marconi's, dominated the field.[16] When Woodrow Wilson went to Paris, one of the fundamental points on his agenda was "to safeguard American pre-eminence in radio."[17]

At the height of Wilson's negotiations with the other Allies in February 1919, he received a succinct brief from his communications advisor, Walter S. Rogers, urging that the United States assert itself in global communication especially with regard to Britain's global control of the international cable system. The key to this was the development of wireless (now increasingly referred to as "radio communication"). "High-power radio, with its unlimited possibilities for broadcasting messages to the ends of the earth, presents a startling opportunity for disseminating intelligence," Rogers wrote, in what was surely one of the first policy-oriented documents to explicitly recognize the power of the new technology for the spread of *ideas* as well as unprocessed information.[18] In Rogers's vision, "national ownership" of radio and "world-wide service" were crucial to the success of the embryonic League of Nations, the formation of which was still at the very top of Wilson's agenda.

[†] The American Marconi company and US amateur wireless advocates found themselves allied when Congress introduced a bill to make government control of radio stations permanent after the war. The company and the Wireless Association of America led opposition to the bill, which was tabled indefinitely on January 16, 1919, signalling the end of the navy's ambitions. Wilson approved the return of radio stations to their owners in July 1919 (Howeth 1963, chapter XXVII).

The president left Paris three days after receiving Rogers's memo, on February 15, 1919, and by the time he returned a month later, on March 14, he had been convinced that national ownership of wireless was also, and primarily, in the American national interest. Wilson was accompanied on his return voyage to Europe by Josephus Daniels, the leading political advocate for government control; Daniels, secretary of the navy throughout Wilson's presidency, was determined that the war serve to consolidate the navy's dominance in US broadcasting.[19] His view was echoed by Postmaster General Albert S. Burleson, who summed up the problem in a cable to Wilson on March 15: the United States was dependent on "the courtesy of foreign-controlled means of communication.... The world system of international electric communications has been built up in order to connect the Old World commercial centers with that World business. The United States is connected on one side only."[20]

Marconi's, meanwhile, had reopened its negotiations with GE. Their international position was stronger than ever as a result of the war. They now offered to purchase twenty-four alternators at $127,000 apiece, a contract worth more than $3 million.[21] On March 29, 1919, GE's Owen Young wrote to Acting Secretary of the Navy Franklin D. Roosevelt, informing him of the proposed sale; Roosevelt wrote back on April 4, requesting that GE confer with navy representatives before making a commitment to Marconi.[22] On April 8, the US director of naval communications, now Rear Admiral Bullard, and Captain Stanford C. Hooper, the navy's chief expert on wireless operations, met with GE officials in New York and asked that they not sell the Alexanderson alternator to Marconi.[23][‡] According to the standard account of the meeting, Bullard took Owen Young aside during a break and confided that he had just seen Wilson in Paris and Wilson had personally asked him to dissuade GE from going through with the deal. The president had become convinced, said Bullard, that world pre-eminence would be determined by three factors: oil, transportation, and communication. The United States was predominant in oil, but Britain could not be challenged in transportation or cable communication. Wireless, however, was still up for grabs, and if the United States could achieve dominance there, the result would be a standoff between the two powers.[24] Bullard said the US government was seeking a

[‡] Hooper, himself a former wireless operator, has been called "the father of naval radio." Bullard, superintendent of the Naval Radio Service from 1912 to 1916, was named director of naval communications a few weeks before the events described here; he was later the first chair of the Federal Radio Commission, but died within eight months of taking office, in November 1927.

policy for wireless that would be "similar to the Monroe doctrine," whereby control of radio in the Western Hemisphere would remain in US hands.[25]

The request left GE in a quandary. They were, after all, in business. The war was over and they had something to sell for which there was only one possible buyer—Marconi's company. The navy had failed in its attempt to establish pure government ownership, but the government clearly did not want to see the sector remain in foreign hands. What was the solution? Young came away with the idea that the government would support a move by GE to buy out Marconi and create an American wireless company. He also believed that the parent British Marconi company could be persuaded to sell its American interests if they were convinced that the US government intended to intensify its opposition toward British domination of wireless. Young informed American Marconi president Edward J. Nally that GE would not provide Marconi's with alternators, and a few weeks later suggested to Nally that GE and American Marconi combine to form a new company.[26] This was instrumental to breaking the impasse. By proposing a consolidation that would leave American Marconi essentially intact but under GE ownership, Young was able to persuade Nally, and later British Marconi, that going in with GE would be the best course in view of the US government's determination.[27]

Daniels returned from Paris at the end of May and, apprised of the developments, expressed skepticism about creating a private monopoly and began voicing his old views about government ownership. The navy's official communications historian, L.S. Howeth, writes: "Young must have been completely surprised by this turn of events. He had every reason to consider that Bullard and Hooper had spoken authoritatively and that the Secretary would follow the advice of Roosevelt." GE's interest was in creating a privately owned company with a commercial monopoly—something like Marconi was trying to establish, but American-owned (Rogers's proposal, "national ownership," was ambiguous enough to be interpreted either as government ownership or ownership by a nationally based private company). There were, after all, limits to patriotism in business. Howeth writes: "What transpired at the next meeting of the board of directors of the General Electric Co. has never been completely divulged but they did decide to proceed with the organization of the Radio Corporation without further governmental blessing." GE notified Daniels that it would be negotiating directly with the Marconi company and "there is no record of further official correspondence requesting Navy support."[28]

Nally prepared a first draft of a possible agreement between GE and American Marconi and met with Young to discuss it on June 13, 1919. Nally estimated that the American company was worth $12 million, with about half its stock held in the United Kingdom; the British Marconi company probably owned as much as 25 percent of the total. GE was prepared to pay handsomely to buy Marconi out. If it refused to sell, however, the American company would have few prospects. The government still controlled Marconi's key US assets and was not likely to give them back to a foreign-owned company. The sale was not likely to be too painful to the British parent company; on the contrary, it could be helpful—Marconi's American subsidiary had never been terribly profitable.[29]

It was more important for Marconi to nail down a good agreement with the Americans on postwar spheres of influence. This was the focus of negotiations that continued in London in August 1919, with Godfrey Isaacs representing Marconi. The talks went on for two months, with the Marconi officials being constantly reminded that the US Congress could at any time enact legislation prohibiting the foreign control of wireless facilities located on US territory. Finally, the Marconi directors decided to make the best of the situation and take what they could get. A preliminary agreement was signed on September 5, 1919, by which British Marconi sold GE control—365,000 shares—of its American subsidiary. The Radio Corporation of America (RCA) was formed on October 17, 1919, incorporating the assets of the Marconi Wireless Telegraph Company of America (including the stations appropriated by the US Navy during the war) into a new publicly held company in which General Electric owned a controlling interest. RCA replaced Marconi's as the major US domestic wireless company and gave the United States a solid foothold in the sphere of global communication. At the company's invitation, the government nominated Admiral Bullard to represent its interests at RCA board meetings.[30]

Why did Marconi agree to this? Simply, he was persuaded that it would be fruitless to persist in the face of US postwar intentions and preferred to make a deal rather than an enemy. He believed in powerful alliances, and one of his great strengths was knowing when and with whom to make them. He abhorred fighting with governments on questions of principle (which was the reason he had broken with Cuthbert Hall in 1908); when it came to dealing with governments he was comfortable working in partnership and was prepared to give up part of his stake in exchange for the protection that came with staying on their good side. In this respect he was one of the first global entrepreneurs to understand—and to master—the benefits of a regulated monopoly.

Owen Young was made chairman of the RCA board, while Nally became president; David Sarnoff, Marconi's one-time office assistant and wireless operator, was appointed managing director. Some eyebrows were raised in Washington over Nally's position—the perception was that Marconi was taking over GE rather than vice versa—but they settled down when Young, a consummate politician, made it clear that as chairman he would represent "the composite views of the old Marconi Company, of the General Electric Company, of the Government, and of American business interests broadly."[31] These were, after all, the interests that expected to be looked after.

The deal was consecrated by an agreement signed on November 21, 1919, under which RCA took over British Marconi's interest in its American subsidiary, including the rights to use Marconi's patents in the United States. The agreement specified that each party would continue to improve its respective patents as well as develop new inventions "in order to establish economical world-wide public commercial wireless telegraphic and tele-phonic services by means of radio devices."[32] Overnight the new company became a giant. RCA took over not only the American Marconi patents, but also Marconi's high-power stations, a contract with the US Shipping Board for the maintenance of radio equipment on four hundred vessels, the Wireless Press, and Marconi's 37.5 percent interest in the Pan-American Wireless Telegraph and Telephone Company. Foreign domination of American radio was no longer an issue.[33]

From the perspective of the US military, the creation of RCA was a compromise. The navy was thwarted in its effort to keep its wartime mo-nopoly, but the alternative the deal brokered was a new kind of creature, a privately owned company that was uniquely identified with the US national interest.[34] RCA was soon the largest radio company in the world and the vehicle for US national telecommunications policy.[35] In the view of business historian Robert Sobel, it was nothing less than the first ele-ment of what President Dwight Eisenhower later ominously called "the military-industrial complex."[36] Ten years after the arrangement was made, a US Senate committee was told that there had been considerable discus-sion in Paris between Bullard, Wilson, and Marconi with regard to radio regulation and control. Whether this was during Marconi and Wilson's meeting of May 23, 1919, or at another time is not clear, and precisely what was said between them on the subject of radio has never been revealed.[37]

* * *

While the RCA deal was being sealed in London, New York, and Washington, Marconi himself was back in Paris. The Adriatic question was still unresolved and high on the Italians' agenda; it was also a sticking point in settling the boundaries and spheres of interest in postwar Europe. And it was a factor in the Italian national elections of November 16, 1919, in which Nitti's governing centrist coalition just managed to remain in power in the face of the steep rise of the Italian Socialist Party.[38]

On November 20, Marconi wrote to Nitti, recommending that if the US Senate refused to ratify the Versailles Treaty, which it had been debating since July, Italy demand the application of the Treaty of London by England and France.[§] This would improve Italy's position in the event of new negotiations, he wrote. Nitti replied promptly, acknowledging the merit of Marconi's suggestion.[39] The American press, meanwhile, was describing the Italian domestic situation in alarming terms, forecasting imminent revolution, the abdication of the king, and war with Yugoslavia in the wake of the US's ultimate refusal of Italy's claims on Fiume.[40] Marconi was situating himself on Nitti's nationalist flank as he took on the role of strategic advisor in the volatile Italian political situation. Just as Marconi was advising Nitti, Tittoni wrote urging him to return to Rome to intervene at the upcoming opening of the Italian Senate.[41] Instead, Marconi remained in Paris, representing Italy at the signing of the Bulgarian Treaty.[42] He then decamped for London to join a mission that Nitti was planning there.

As usual, Marconi travelled to London with an entourage, including Beatrice, Leon de Sousa, and a valet. The Italian delegation in Paris made all the arrangements, from first class Pullman seats on the Folkestone–London train to rooms at Claridge's. They travelled on December 4, 1919, with high-level appointments with Lloyd George and foreign secretary Lord Curzon (who had succeeded Arthur Balfour in that position a little over a month earlier) set to take place as soon as possible after they arrived. Nitti didn't make it to London, but the others had a series of meetings with the British over the next five days. The Italians were trying to save face over D'Annunzio's "pirate action" (the term used by Frank Polk, who was still heading the US legation in Paris, to describe the occupation of Fiume). Officially, they could

---

[§] The United States never ratified the treaty.

neither denounce nor condone it. On December 10, Nitti, in Rome, was on the verge of announcing that he was going to send troops to the port city. On December 12, Vittorio Scialoja cabled him from London: "Don't do anything definite about D'Annunzio until you hear from me again later. This is most essential." Lloyd George, Clemenceau, and Wilson were now all involved in the discussion, directly or through their respective intermediaries.[43] A cable from the Italian embassy in Paris to London and Rome summed up the situation this way: "Lloyd George despite his usual uncertain thinking says he needs to distance himself from the US and build stronger ties with European allies. Wilson following his stroke [of October 2, 1919] can't concentrate for more than 10 minutes at a time. Lloyd George said . . . that it was necessary to help Italy resolve the Adriatic difficulties and showed himself well disposed in our regard."[44]

Nitti finally went to London to meet with Lloyd George in January 1920.[45] The peace conference was now effectively over but Italian issues remained unresolved. While the rest of Italy's delegation remained in Paris, Marconi spent most of early 1920 in London, or on the *Elettra*.[46] The only further reference to him in the Italian archives on the conference has to do with communications—in January 1920 he arranged for a circuitous route for wireless messages from Paris to Rome, by way of London and the Marconi station at Carnarvon, Wales. Why the bother? The archives are silent on this point, but they do urge the delegates to limit their communication to the most important messages because the service was already overloaded.[47]

On April 14, 1920, Marconi wrote to Nitti, resigning as Italian plenipotentiary delegate to Paris; his resignation was accepted by the king the next day.[48] Although his year of intense diplomatic activity had been exhilarating, it was also extremely frustrating and, especially, disappointing. Unlike his research or business activity, there was not much to show for it at the end of the day. One biographer, Giancarlo Masini, wrote that "Marconi left Paris nauseated"; another, Giuseppe Pession, said he was "disillusioned and disgusted."[49] Marconi told Harold Begbie that his experience as a delegate in Paris had left him with two conclusions: "One, that diplomacy is not an exact science; and, two, that it is easier to handle nature than human nature."[50] But Marconi's incursion into the world of high-power diplomacy gave him a taste for more. His experience in Paris had a lasting effect.

Although his participation was constantly highlighted in the press and political dispatches, in a broad historical sense Marconi was a relatively minor figure in Paris. The Italians used his stature to try to make a few points with the British and the Americans, but he was not a decider or a player, and as a delegate he was somewhat undisciplined (coming and going according to his own business and personal agendas; insisting on staying at the Ritz). Yet his role was influential, for example, regarding the commercial mission to the United States, or in securing a direct channel to Wilson in May 1919, at a crucial moment in the talks. He had contacts in the United States and England and could open doors for Italy like few others could, at a time when Italy needed cash and aid. At the end of the day, despite his protestations, he found a new vocation that would stay with him for the rest of his life. He also became a prototype for another generic twentieth-century figure: the international media mogul who uses his position to try to influence global politics.

Marconi was certainly not the only political figure to come out of the war with a cynical and angry view of British-French greed and American haughtiness; his experience during these years went a long way to fixing his politics on the nationalist right, where they would remain. Certainly, this was the trajectory that drove him away from sympathy with his British side and toward the Italian. And he was not shy about expressing his disappointment. A 1919 speech about Italy's postwar relations with the United States, for example, begins: "Among all the unexpected and surprising results of the great war which ended a year ago there is none more unexpected or more surprising than the disagreement between Italy and the United States on the subject of the Adriatic."[51]

He was also soon dealing with a loss. His beloved mother, Annie, died of heart problems on June 3, 1920. She had been suffering from an array of ailments for a number of years, looked after at the rented home she shared with her son Alfonso at 75 Harley House, St. James (where Marconi and Beatrice occasionally stayed when in London during Annie's last years[52]). An obituary in the trade journal *Wireless World* said, "It is largely due to the encouragement given to the illustrious pioneer of wireless by his mother, the deceased lady, that he was able to prevail, in the days of his early discoveries, against the incredulity of the scientific world." But the illustrious pioneer wasn't present when his mother died. Marconi was on a research cruise on his yacht in the Mediterranean when Annie passed away, and did not make it back to London for her burial, which took place on June 9, 1920, at London's Highgate Cemetery. Among those in attendance were Alfonso, one of her sisters, and two of Beatrice Marconi's O'Brien siblings, Donough and Lilah.[53] Annie Fenwick Jameson Marconi was laid to rest in a gravesite purchased by her nephew, Henry Jameson Davis, some years earlier. The records of the London Cemetery Company state that she was in her eighty-first year.[54]

❀   ❀   ❀

Although Nitti tried to govern for half a year after the November 1919 elections, Italy's aggravated political situation finally got the better of him. On June 17, 1920, he resigned and was replaced by Giovanni Giolitti, a four-time prime minister and Italy's most powerful politician of the previous thirty years.[55] Giolitti had favoured neutrality in the war, and once Italy joined, he had receded into the background. He had played no part in the war or the peace conference,[56] and had argued, during the 1919 electoral campaign, that Italy had been dragged into the war against the will of the majority. The change in government set the stage for the dramatic political shifts that would soon follow.

Nitti's destiny had been sealed on September 12, 1919, when Gabriele D'Annunzio and his "legionnaires" occupied the contested city of Fiume. Under the terms of a report by an Inter-Allied Commission of Inquiry, dated August 25, 1919, Fiume was to be kept under international control, with a small Italian military role. The British were to take charge on September 12, but D'Annunzio pre-empted this when he organized his force and marched into the city just before the British were due to arrive. Following "some theatrical gestures on both sides," Italian General Vittorio Emanuele Pittaluga, who was in charge, surrendered his command to D'Annunzio.[57]

Fiume now became the hotbed of Italian nationalism, an incubator of the nationalist wing of a new movement, fascism, that would soon grip Italy. D'Annunzio put into practice a number of features that Benito Mussolini would adopt a few years later: cult of personality, dictatorial rule, Roman salutes, balcony speeches, programmed mass demonstrations, the black-shirt uniform, and even, it was said, punishing opponents by making them drink castor oil.[58] D'Annunzio, in fact, became a role model for Mussolini, but he himself was not a fascist (at least not in the formal sense; he was a leader, not a joiner). He also promised women the vote, artists free speech, and anarcho-syndicalists social reforms.[59] D'Annunzio's Fiume was what one of his biographers called a microcosm of the "madness and magic of the twentieth century, an early laboratory in which the germs of mass politics—of both right and left—were tested on human subjects."[60]

Fiume was "a sort of experimental counterculture with ideas and values not really in line with then current moral codes."[61] Margaret MacMillan describes the scene as "a mad carnival of ceremonies, spectacles, balls and parties. The town's buildings were covered with flags and banners, its gardens ransacked for flowers to throw at the parades. In a fever of nationalism and revolution, fueled by drink and drugs, priests demanded the right to marry and young women stayed out all night. The city reverberated, said observers, with the sounds of love-making."[62] Despite an Allied blockade, D'Annunzio rallied to his cause an assortment of celebrants that included Filippo Tommaso Marinetti, the Futurist artist; Arturo Toscanini and his orchestra; Mussolini; a gaggle of opposition politicians, gangsters, and prostitutes; and Guglielmo Marconi.

Following Marconi's resignation from the Italian delegation to Paris, Nitti had asked him to go to Fiume and try to persuade D'Annunzio to give up his adventure in the greater interests of Italy. Marconi's heart, however, was not in it; he was sympathetic to D'Annunzio's occupation and did not want to embarrass Nitti. But he had no particular ties to or admiration for Giolitti, and now, Marconi agreed to go. On September 22, 1920, on what he described as a "delicate mission" for the government,[63] which feared both further international repercussions and the possibility of a local civil war, Marconi sailed the *Elettra* into Fiume harbour, where he received a heroic welcome unlike any he had ever seen.[64] If there was any doubt as to D'Annunzio's influence on Marconi, it was put to rest in Fiume. When Marconi arrived he was ostensibly on a mission to convince D'Annunzio to give up his cause; by the time he left he was irrevocably converted to it.

Marconi's arrival in Fiume was a well-staged media event. D'Annunzio had been holding the city for a year, maintaining a state of excitement among the people with a series of proclamations that were seen as more literary than programmatic. He also had to minimize the impact of an Allied blockade. The arrival of an international celebrity in a white ship was a master coup and D'Annunzio played it perfectly, while Marconi went along. D'Annunzio decorated Marconi with the colours of Fiume, then went aboard the *Elettra* to make his first radio broadcast "to the world"; Marconi, no longer mindful of his "mission," promised to install a powerful wireless station in Fiume at his own expense.[65]

Luigi Solari, himself no slouch when it came to rhetorical hyperbole, described the scene in his 1928 book: "Loud cries of 'Long live Italy! Long live Marconi! Long live D'Annunzio!' came from the bare throats of thousands of young legionnaires, and resounded many times in the port of

Fiume." D'Annunzio spoke from the balcony of his palace, with Marconi by his side, to the admiring throng:

> Citizens, Legionnaires—we salute and honour in Guglielmo Marconi the genius of Italy, diffused throughout the whole world with the speed of light.... We have suffered, O Guglielmo Marconi, magic hero that you are, because we did not have on the tower, or on the lighthouse, one of your metallic spires, which are in themselves the pulsing summit of the spirit of propagation. But today, having disembarked at Fiume...does it not seem...that he bears with him all the vibrations of those most mysterious messages? He has come to amplify indefinitely the harmonious waves of the voice of Fiume.[66]

D'Annunzio's delirium has been read as an example of "wireless writing," or "the product of a media ecology altered by the introduction of wireless," and that is an interesting appraisal in itself; but it becomes utterly fascinating when one considers it alongside a rare photograph of the "Comandante of Fiume" with the "magic hero" of his text standing coolly at his side, looking over his shoulder as he reads his speech.[67] Marconi remained in Fiume for only forty-eight hours, but the relationship born in Rome and Centocelle in 1915, now cemented, became one of the most important of his life. Marconi and D'Annunzio carried on a busy correspondence in the 1920s and '30s, and the relationship was critical to both of them, as their letters demonstrate (Marconi's are in an utterly different voice from anything else he ever wrote, while D'Annunzio's are perfectly true to character). D'Annunzio was a poetic alter ego to Marconi, and Marconi the scientist gave the poet his unconditional support.[68] Marconi also made a speech at Fiume, a short and vibrant one that mirrored D'Annunzio's lofty rhetoric (Marconi described him as "the true Hero of action").[69] Then he was gone.

Marconi's attraction to the trappings of D'Annunzio's statelet was odd and, on the surface, anomalous—Marconi typically supported whomever was in power, not the marginal opponents. On the other hand, his loyalty to friends always trumped ideology, especially when the loyalty was reciprocated and the friends were adoring. But his enchantment with D'Annunzio also helps explain his later adherence to fascism. No one ever showered Marconi with adulation the way D'Annunzio did. The libertarian atmosphere of Fiume was an oddly contradictory foreboding of the authoritarian culture of fascism, and Marconi—certainly more than D'Annunzio himself—embodied these contradictions.

Less than two months after Marconi's visit, the Fiume adventure was over. The US presidential elections of November 2, 1920, repudiated Wilson's

European policies, installing Republican Warren G. Harding in the White House. The way was now clear for a new agreement. On November 8, Giolitti and his foreign minister, Count Carlo Sforza, a career diplomat who favoured co-operation on the Adriatic question, met with their Yugoslav counterparts at the seaside town of Santa Margherita Ligure and laid out a treaty recognizing, among other things, an independent Free State of Fiume. The Treaty of Rapallo was signed on November 12. The only remaining obstacle was D'Annunzio's continued occupation. He had already proclaimed his own independent state, which he called the Italian Regency of the Quarnero (Carnaro), that looked a lot like the one the diplomats had now created, with one important difference: the Quarnero was his to rule.

On December 1, 1920, D'Annunzio declared war on Italy, but four weeks later, facing Italian determination to end its international humiliation by enforcing the Treaty of Rapallo with arms, he left the city, declaring that "Italy is not worth fighting for."[70] The Italians entered Fiume on January 1, 1921, and the city council recognized the Treaty of Rapallo. But the eviction of D'Annunzio was a thankless and unpopular task for which Giolitti paid dearly; in the national elections on May 15, 1921, his governing coalition missed getting a majority. The revelation of the elections was the new right-wing National Bloc, which received 19.1 percent of the vote and 105 seats. In June 1921 the Giolitti-Sforza government resigned.[71] The "treachery" of the Giolitti regime became part of the foundational narrative of what lay ahead, as nationalists, irredentists, and soon fascists refused to accept the independence of Fiume.[72] Fiume was annexed to Italy in 1924 but its status was settled only after the Second World War, when it became part of Yugoslavia and was renamed Rijeka. Today the city is part of Croatia, a remnant of faded elegance replete with drafty villas and an urban beachfront where D'Annunzio used to famously go for nude midnight swims while his lover of the moment held his robe.

PART
IV

The Outsider

26

The Master of the House

One of the new members of the Italian parliament elected in 1921 was a journalist and political organizer by the name of Benito Mussolini. He was born in 1883 in the village of Predappio, seventy-five miles south-east of Bologna, in the revolutionary heartland of Romagna. Mussolini's parents were as different as parents could be, and yet they each left a mark on their eldest son; his father was a rough-and-tumble anarcho-socialist black-smith and his mother a Catholic schoolteacher. Benito (who was named after the nationalist president of Mexico, Benito Juárez) was influenced by his father's politics and tough-guy ways, read a bit of philosophy, sociology, and Marxism, and became a socialist activist; like his mother, he also worked briefly as a schoolteacher, and learned enough about the Catholic religion to be able to use the knowledge to his advantage in later years.[1]

Mussolini and Marconi had something important in common: they were both profoundly affected by Italy's experience in the First World War. Two years after becoming the editor of the socialist newspaper *Avanti!* in 1912, Mussolini was expelled from the party for advocating Italian entry to the war, a position shared by the unambiguously capitalist president of the Banca Italiana di Sconto. In November 1914, Mussolini started his own paper, *Il Popolo d'Italia*; around the same time he organized the Fasci d'Azione Rivoluzionaria, an extraparliamentary association supporting Italy's entry to the war. Mussolini went to war, returning with shrapnel wounds in 1917. Swept up in the postwar social and labour strife of northern Italy, Mussolini's natural constituency was the thousands of returning veterans; he played on their nationalism, their sense that the country owed them something, and their love of camaraderie under arms.

Marconi had a similar view of the country's state of affairs after the war. But while Marconi's postwar attention was focused on international diplomacy and the Paris peace conference, Mussolini began building a political movement, convening the first formal meeting of his group (now called the Fasci di Combattimento) on March 23, 1919. The Fascists ran in the national elections of November 1919, electing only one deputy. Undismayed, Mussolini began the process of turning the movement into a political party. Unclear about his goals at this point, Mussolini took advantage of Italy's unstable political situation; unlike any other leader, he would champion both law and order and national pride. These, again, were qualities attractive to Marconi. In the chaotic political landscape, marked among other things by the fear of socialism and the humiliation many Italians of Marconi's social class felt about Italy's treatment in Paris, the dark side of Mussolini's movement, with its black-shirted bands of marauders, was overlooked. After the 1921 elections, thirty-five Fascists entered parliament, which was dominated by the Giolitti government's anti-socialist coalition. In November 1921, the National Fascist Party was formed.

Marconi had given some clues as to his politics in an October 1920 interview with the Hearst syndicate. "The suggestion of Bolchevism in Italy is thorough nonsense. There is no more and no less Bolchevism in Italy than there is in America," he told reporter Charles Bertelli. The recent social unrest that had culminated in workers' occupation of their plants was a manifestation of the same unrest now pervading the entire world, and which was a direct consequence of the war. But Italians had fought for the abolition of despotism, and consequently, "the idea of any dictatorship is repellent to the Italian worker... the doctrine of a dictatorship of the proletariat is against his nature just as it is absolutely against his nature to attempt a violation of the right to private property."[2]

Part of Marconi's attraction to Mussolini involved his own precarious circumstances. Marconi's impatience with what he saw as the sloth and moral corruption of the Italian political class went back to his earliest experience of trying to find support for his invention and the discouragement he had received from the highest levels at the time. The political choice he would shortly make started there, but it was driven most of all by a perception of self-interest. On December 28, 1921, the Banca Italiana di Sconto, presided over by Guglielmo Marconi, senator of the realm, stopped honouring its commitments. Depositors had withdrawn 900 million lire from the bank during the previous month, in response to a crisis at the firm Ansaldo, in which the Sconto was heavily invested. (Some even said that one of the reasons the bank had been created was to bolster Ansaldo, a munitions manufacturer and important arms supplier during the Great War.)

The suspension of payments was framed as a government-backed temporary moratorium, and the *Times* reported from Milan that it had caused nervousness among small deposit–holders but no alarm in financial and industrial circles; yet it was soon clear that the bank's seven-year adventure had utterly failed to live up to even the mildest expectations.[3] Worse, there was now a corruption scandal. On January 3, 1922, Italy's *Official Gazette* announced that the bank's directors had been relieved of their posts and their property sequestered. The Italian press welcomed the move, although in some reports it was "regretted that it may cause inconvenience to men who, like Senator Marconi, are above suspicion."[4]

The failure of a government-backed bank—especially one chaired by such a high-profile figure—sent shock waves into international financial circles. An official statement by the Italian Commercial Delegation in London tried to gloss over the issue, stating that the difficulties were only temporary and the bank's assets were more than enough to cover its liabilities. Marconi issued a statement that was far less reassuring, and seemed more designed to cover his own backside than to clarify the matter: "In consequence of my technical work requiring that I should be absent from Rome for very considerable periods of the year," he said, "it was agreed . . . that I should not be expected to take any part in the direction of the bank's affairs."[5]

From London, where he was holed up at the Savoy Hotel, suffering from the flu, Marconi wrote to Beatrice in Florence. While the trouble of the Banca di Sconto was a "serious matter" in that his company had invested heavily in the bank, he said, his own finances were stable. (He was about to sign a new agreement with the Marconi companies, increasing his annual salary as technical advisor to £4,500, or $360,000 in today's money; he was also provided an additional £3,000 for expenses, which he used mainly to keep the *Elettra* afloat.[6]) "Fortunately my affairs here are going well and I hope to make up for all losses."[7] He didn't expect any move to be made against him regarding sequestration, but nonetheless cautioned Beatrice: "I think however that until the matter is cleared up it would be better for you not to go to Rome for your duties as Dama di Palazzo." (Beatrice had been named a *Dama di Palazzo*, lady-in-waiting to the queen, in 1911.)[8]

Marconi had actually tried to bail out of the sinking bank a few months earlier, tendering his resignation on August 30, 1921, due to his increasing duties in England and other countries. However, the bank's executive board prevailed upon him to reverse his decision. When the crunch came, the company poured millions of lire in deposits into the Sconto to help shore up its accounts.[9] Marconi was now called on to prove his usefulness by organizing a consortium of English banks to get the Sconto out of its mess. Although his doctors warned him he was not well enough to travel, he orchestrated a round of meetings with British bankers to try to set up a bailout. In early January he met with Sir Herbert Hambling, head of Barclays Bank and "a good friend of mine and of Italy." Barclays was a shareholder in the Sconto, "and the figure I mentioned, £1 million sterling, did not frighten him at all."[10]

The affair soon took on the air of a scandal. The Italian government wanted Marconi in Rome, but after consulting Luigi Solari and his lawyers he decided that going to Italy would not serve any useful purpose and, if anything, would actually be harmful. He would go to Rome if and when the Sconto inquiry began looking into his affairs, but for the time being it wasn't. "Considering their behavior towards me, I do not intend to have anything else to do with them," he told Beatrice.[11] He was approaching the end of his tether as far as relations with the Italian government were concerned. On April 8, 1922, the Senate's high court of justice took up the case. Forty-two administrators and directors, including some of the country's leading industrialists as well as four senators, were called to testify on allegations of misappropriation of funds. Marconi, Solari, and a former minister of the marine with whom they had had many dealings, Pasquale Leonardi Cattolica, were among those summoned.

Marconi's frequent absence from Rome was at the heart of his claim that he knew nothing at all about whatever may have been going on. In support of this argument, he had the Italian embassy in London submit an affidavit stating that he had been a British resident since 1896.[12] It took the better part of the year for Marconi to find a time when he could come to Rome to testify, and the Senate committee did everything it could to accommodate him.[13] Finally, on December 11, 1922, he appeared before the court, armed with a lengthy prepared statement that he insisted on reading into the record before answering any questions. This was Marconi's forte, and he didn't disappoint on this occasion. His statement reiterated the well-rehearsed details of his biography with a particularly patriotic Italian spin. He cited the long list of honours he had received, highlighting those such as the Nobel Prize that he felt brought special pride to Italy. He emphasized that he had never sought money from Italy for military use of his patents and recalled his war efforts in Libya and the Austrian campaign, listing every one of his military promotions. Finally, he recalled that he had accepted the

Sconto position as a service to the nation, despite the fact that he had no technical knowledge of banking and would not be able to attend board meetings regularly, and that after trying to resist the call he had accepted because of the feeling that his international contacts would be useful for Italy.[14]

A few days later, on December 20, Solari, too, testified that he had entered the bank's service not for personal gain but out of patriotic duty, adding astutely that he had joined the syndicate only after Marconi had accepted the presidency (implying that he was just a pawn following instructions from Marconi). He had acquired some shares, he said, according to "my modest financial condition," and had lost 250,000 lira (about $200,000 in today's money), not an indifferent sum. He had developed relations with foreign banks, and had carried out the support operations of the Marconi companies; everything he had done, he said, was out of loyalty to Marconi. He presented himself as a victim of the bank's crisis, the only breadwinner for four families from Loreto whose menfolk were decimated by the war. "I am just an honest Italian dedicated to implanting the invention of he who has so honoured the name of Italy." Solari's maudlin testimony, closing with this brief but pointed elegy to Marconi, augured the hagiographic style he would adopt in his biographies.[15]

Marconi's and Solari's lawyers presented another version of their clients' claims to innocence in late May, and on June 1, 1923, the Senate court summoned Marconi and others to appear regarding the misappropriation of 360,000 lire ($300,000 today). The court then made an even more damning charge: that the directors had declared a false dividend in 1920. Arrest warrants were issued on the second charge but Marconi and Solari were not included because they were not present at the board meeting where the false dividend was approved.[16] On November 27, 1923, the Senate court completed its work. Thirteen directors were indicted to go to trial on charges of fraudulent bankruptcy, distribution of fictitious dividends, and stock manipulation. Marconi was completely exonerated, the key piece of evidence in his favour being that he had not been present at meetings where the criminal activities were discussed. One more time, reminiscent of the Marconi scandal, he managed to keep his nose clean and his reputation intact.[17]

The extent of Mussolini's involvement in the denouement of the Sconto affair is murky; one of his government's first decrees was to issue an amnesty for Fascists accused of crimes under the old regime, and many of the Sconto principals were Fascists or sympathizers (and more of them—including Marconi—became so as the hearings and trials proceeded). On

March 2, 1926, the case ended with the acquittal of all the accused. One of the most compelling conclusions of the court was that the bankruptcy of the Banca Italiana di Sconto could have been averted had the government of the day intervened with bail-out measures. This was seen by supporters of fascism as a condemnation of the *disfattismo* (defeatism) of the liberal system, and the acquittal was hailed as a triumph for the party.[18]

Marconi learned several important lessons from this experience. One was that he had best tie his Italian future as closely as possible to that of the new regime. Another was that the fascist interpretation of Italian politics was close to his own. If Mussolini's Fascists were what they seemed to be, his own days of frustration with *disfattismo* could be over. He may have found a political home.[19] And the master of the house was Benito Mussolini.

❖   ❖   ❖

Nothing in Mussolini's background and early life indicated that he was destined to head the Italian state. But by July 1922, Mussolini's Fascist Party boasted seven hundred thousand members, and the following month it fought off an anti-fascist general strike, paving the way for the historic March on Rome in October 1922. On July 29, 1922, a socialist call for a national strike backfired as Fascist bands burned down union halls and forced strikers back to work. On October 16, Mussolini convened leading members of his party to finalize plans for an insurrection, built around a "march on Rome" that would take place on October 28. As eventual head of a new government, Mussolini was to remain in a safe place, while a Fascist "quadrumvirate" led the march. An elaborate fascist mythology later grew up around the march, and Mussolini's role has been the subject of controversy; some believe he may have had doubts in the days leading up to the march, and that his long-time lover and closest political advisor, Margherita Sarfatti (also, incidentally, a childhood friend of Marconi), was the one who convinced him to go ahead.

On October 24, 1922, Mussolini declared, "Our program is simple: we want to rule Italy."[20] As Fascist "blackshirts" occupied strategic government offices and approached Rome, Prime Minister Luigi Facta believed he could face down the ragtag group of marchers (less than thirty thousand versus an equivalent number of government troops) and decided to declare a state of emergency. However, King Vittorio Emanuele III refused to sign the order, apparently worried about the messiness of an armed confrontation and preferring to see Mussolini in government than outside. This view was shared by much of the army and business elite, who were tired of years of social unrest and anxious to put an end to socialist threats of subversion.

Facta resigned. The king tried to bring back Giovanni Giolitti to replace him and give the Fascists a few Cabinet positions, but Mussolini refused to accept this option. Although the numbers taking part in the March of Rome were totally inflated after the fact, the king still feared it could be the first move toward a civil war, while the Fascists were not seen as a threat to the Italian elite; to the contrary, they held out a promise of restored order, and this was part of their attraction to establishment figures like Marconi.[21] On October 29, the king invited Mussolini to form a government, sensing, correctly, that he would have the support of the military, the business class, and the nationalist right wing (although there were still only thirty-five Fascist deputies in the 535-member parliament). Mussolini arrived by train from Milan on October 30 and began putting together a Cabinet, reserving for himself two areas that would particularly interest him to the end of his career: internal security (the police) and foreign affairs. In mid-November, he asked parliament for a vote of confidence; 316 members voted in favour. No one in Italy, or in most of the world, could imagine where he would lead them, but in Germany a now forgotten Nazi leader told a rally a few days later: "Germany's Mussolini is called Adolf Hitler."[22]

❈   ❈   ❈

One of the first prominent Italians abroad to send congratulations to Mussolini was Guglielmo Marconi, who was in London looking after company business (among other things, they were preparing for the launch of regular BBC broadcasts).[23] The gesture was noticed by MI5—Britain's domestic counter-intelligence organization—which was already keeping an eye on Mussolini's followers in England. Marconi's motivations were, typically, many, varied, and complex; they involved a combination of business, convenience, patriotism, and conviction—in that order. The extent to which Marconi actually supported Mussolini has been played down, certainly by his British and American biographers, but Marconi became Mussolini's critical link to the United Kingdom, and, later, to the United States, a role he relished. When he was named honorary president of the fifteen-hundred-member British branch of the National Fascist Party, in 1927, Marconi celebrated by organizing a "blackshirt gala" at the Savoy—attended by some of the leading figures of London society—and arranging a midnight wireless link with Rome so that his guests could be personally greeted by Il Duce.[24] According to the historian of the Savoy, Stanley Jackson, "Boys in black shirts and pretty girls wearing the national colours kissed and hugged with patriotic hysteria under a huge portrait of Mussolini."[25]

Relations with Britain were important to Mussolini and the party immediately began organizing support activities in London.[26] Barely a month after assuming power Mussolini travelled to London for an important conference of the Allied leaders to discuss lingering issues regarding German reparations. He spent three days there in December 1922, meeting with the prime ministers of the United Kingdom, France, and Belgium (and being received by King George V); it was, said the *Times*, "an inspiring reminder of the part Italy played in the war and a fresh assurance that, with new resolution to heal her wounds, she is determined to play an active part in the consummation of peace."[27]

Mussolini quickly recognized that the support of well-known Italians like Marconi could become an important international buttress for his unknown party.[28] This was all the more so as there were serious questions abroad about the stability of Mussolini's government. The Italian embassy in Washington, for example, was anxiously compiling US press reports on Italy's internal situation for the Ministry of Foreign Affairs in Rome; an article in the Philadelphia *Evening Public Ledger* reporting that Gabriele D'Annunzio was now Mussolini's chief opponent and preparing to oust him especially caused concern.[29] Marconi in particular provided an "aura of respectability" to the London activities of the fascist movement among Britain's Italophiles, conferring "luster and even a certain intellectual glamour to the Italian blackshirts," in the words of journalist Alfio Bernabei.[30] It helped that relations with the Italian government, whoever was in power, were always crucial to Marconi, although he had never actually joined a political party. While Mussolini was in London, Marconi was on a quick trip to Rome, to give his deposition in the Banca di Sconto case. It must have occurred to him that the new government might be able to help resolve the smouldering financial scandal in his favour. It might also unblock his critical business negotiations that had stalled amid the chaos and social unrest in the country.

On January 21, 1923, Marconi received a telegram in London: "Am glad of assurance that you will be in Rome about the 31st instant accompanied by Director with full powers. Am certain that through direct negotiations we will arrive at satisfactory results.... Mussolini." Excitedly, Marconi cabled Beatrice in Rome: "Darling, I hope to be in Rome again for a few days from the 11th of February for discussing certain wireless matters with Mussolini. He has wired me begging me to come." Marconi and Godfrey Isaacs met, twice, with Mussolini in Rome on February 11 and 12, 1923, to discuss the Italian government's plans for wireless. Marconi was apprehensive about the meetings, which were taken at Mussolini's initiative, but once they were over he was upbeat. "Have seen Mussolini yesterday and today, and I really think all difficulties between the Company and the Italian Government are going to be settled," he wrote to Beatrice.[31]

But that's not quite how things played out. Mussolini proved to be a far more adept manipulator than Marconi had anticipated; he was susceptible to his own circle of influences and was always suspicious of Marconi's dual national affinities, however useful they were to him. Marconi became an important supporter in that he was a rare individual who could be seen both as British and Italian: in May, Mussolini announced the formation of an Italian "Friends of Great Britain" group, as a counterpart of the existing UK "Friends of Italy Movement." Marconi was listed as one of the supporters of the new friendship society, whose offices were to be housed in the Palazzo Venezia in the centre of Rome, the site that Mussolini would soon select as the seat of his government.[32]

Mussolini's minister of posts and telegraphs, Giovanni Antonio Colonna di Cesarò, was the politician in charge of consolidating Italy's wireless activities. Cesarò was a throwback to the old regime. He had been involved in various parties of the non-Marxist left before and during the war and was elected to parliament in 1909, 1913, 1919, and 1921. Named minister in the Giolitti government after the elections of May 15, 1921, he was one of the few who carried over after Mussolini took power in October 1922. He remained in office until February 1924, long enough to be a serious pain in the side for Marconi.[33] Under Cesarò's ministry, Marconi's initiatives to consolidate his position in Italy were consistently rebuffed. Marconi wrote to the government in 1921 and again on August 17, 1922, proposing to build a national wireless company not reliant on foreign capital; his efforts met with no success.[34]

Tainted by its role in the still-unresolved Sconto scandal, the weak liberal government had been reluctant to back another adventure led by Marconi. In this regard, it seemed that the regime change in Rome couldn't have been better for him. On November 18, 1922, barely two weeks after Mussolini's appointment as prime minister, a Rome lawyer, Filippo Bonacci, wrote Mussolini's office a long letter on behalf of a private group of investors who had recently formed a new company, known by the acronym SISERT (Società Italiana Servizi Radiotelegrafici e Radiotelefonici). Bonacci outlined the history of botched efforts to create an Italian national communications company, and proposed that a concession be awarded to SISERT, whose president was Guglielmo Marconi. Bonacci assessed and summed up the history of Marconi's activities in Italy; his Italian company (Società Italiana Marconi, or SIM) was worth 1 billion lire (about £10 million sterling, or US$50 million—around $800 million in today's currency[35]) and had fifteen hundred employees. Despite Marconi's concession of his patents, previous governments had been unable to organize a national communication system; there was now an opportunity to correct the situation.[36]

As in so many areas of modernization, Italy was behind European countries such as Britain, France, and even war-ravaged Germany in the development of wireless, notwithstanding—and perhaps because of?—its free use of Marconi's patents. Marconi had granted Italy this right in 1897, but under his 1903 contract with the Italian government, modified in 1916, Italy was bound to use his system to the exclusion of all others. This was a mixed blessing, setting Italy on a separate development path than the other European powers. (We have already seen how this agreement prevented Italy from signing the Berlin Convention of 1906.) SIM had a de facto monopoly on Italian commercial maritime wireless communication, but, to Marconi's surprise, Mussolini's government didn't want that monopoly extended to a formal national system, or repeated in the coming field of broadcast radio.[37]

The government did not respond to the Bonacci proposal. In the wake of its silence, Marconi suggested that his Italian representative Luigi Solari— not only a board member of SISERT but also, as he himself put it, "a fascist and not of the most recent moment"—write to Mussolini directly, arguing that a SISERT monopoly was not only technically, economically, and financially sound, but would also allow for the creation of a strong propaganda instrument. Solari wrote to Mussolini on January 19, 1923, after which Marconi (not yet a party member) wrote to Mussolini as well, asking him to take a position on the proposal and arguing, only slightly more subtly than Solari, that the Fascist regime should take clear responsibility for the exploitation of radio.[38]

Mussolini's industrial policy was to reclaim the Italian economy, sector by sector, by overseeing the creation of strong national companies with close ties to the state, free enterprise monopolies formed by coalitions of major groups with international ties. On February 8, 1923, Cesarò issued a decree announcing the government's intention to create a single state-controlled entity for radio-telegraphy. To Marconi's great annoyance, Cesarò invited two rival companies—the German Telefunken-controlled Radio

Elettrica, and Radio Italia, which was tied to the French Société générale de télégraphie sans fil—to discuss forming a consortium with the Marconi-led SISERT.[39] Far from giving Marconi the monopoly he was hoping for, Mussolini's government proposed to create a consortium in which the Marconi interests would be partnered with their French and German rivals. Marconi's meetings with Mussolini in Rome a few days later were not so reassuring after all.

Marconi now needed Mussolini more than Mussolini needed him, and his next move seemed inevitable. On June 15, 1923, the day after returning to London from a two-month research trip on the *Elettra* in the Atlantic, he joined the National Fascist Party (as Solari had been urging him to do) and promptly asked Solari to update Mussolini about his latest research.[40] He then headed for Italy. On July 4, 1923, in San Remo en route to Rome, Marconi wrote to Mussolini himself, informing him that he had offered his forthcoming new patents on shortwave communication to the government. He was unequivocal as to what he expected in return. "If my work is considered useful to the country, I can put myself at the disposition of the Government again as soon as the Courts will have clarified my position in Italy," he wrote, referring blatantly to his recent summons to appear in the Sconto trials. That wasn't all: Cesarò's conditions were unacceptable. To emphasize the urgency of the situation, Marconi mentioned that he would be leaving imminently for London.[41]

Solari's version of what happened next is that Mussolini called a meeting with Marconi, Cesarò, Minister of Marine Paolo Thaon di Revel, and himself. The meeting was brief, Mussolini conducting it standing up, as was his habit when he wanted to keep things short. According to Solari, Mussolini declared: "Italy must create the conditions for the best radio service. Marconi has proven that he deserves the trust of his country. The Società Italiana Marconi has presented a proposal to the Minister of Posts. A decision will now be taken that is worthy of this proposal."[42] Recounting this nearly twenty years later, in a book intended to rehabilitate the myth of Marconi's close relations to Mussolini, Solari interpreted this comment as a directive to Cesarò. Italian scholars cast some doubt over the accuracy of the quote and certainly of Solari's interpretation.[43] At any rate, Marconi was not pleased with the outcome. Cesarò continued to try to unite SISERT, Radio Elettrica, and Radio Italia, but SISERT (that is, Marconi) absolutely refused to be part of such an alliance. The French and German groups got the message, however, and coalesced to form a new company, Italo Radio. A contract with Italo Radio was signed on August 29, 1923 (Mussolini, Cesarò, Thaon di Revel, and three other senior officials signing on behalf of the government).[44] The government wanted Marconi to serve as president, but on September 19, Solari wrote to Mussolini stating that Marconi would never accept being president of a company he did not control.[45] The high-stakes game continued, and on September 23, 1923, a decree granted Italo Radio a twenty-three-year concession to operate Italy's principal wireless stations, including the prized high-power station that Marconi had built at Coltano.[46]

Marconi arrived back in Rome in early October, and on October 12 Cesarò personally handed him a copy of the contract.[47] On October 17, 1923, after studying the terms of the proposed concession, Marconi wrote a frank letter to Mussolini, declining the presidency of Italo Radio. A spirit of patriotism might have allowed him to overcome the repugnance he felt at working under an agreement that included foreign companies, he wrote, but he had no confidence in the technical program of the proposed company. "I cannot lend my name, and thereby encourage the investment of Italian capital, to an undertaking which, in my opinion, has no probability of success." Unlike his disclaimer for responsibility in the Sconto failure, in this case he "would not even have the moral alibi of incompetence." Invoking his earlier experience of having to go abroad as a young man, he now prepared once again for exile if necessary:

> In the same way as at the very outset of my invention, after having unsuccessfully offered it to the Italian Government, I had to develop it and apply it in the first instance in Great Britain and in America, I shall now with the deepest sorrow be compelled to relinquish every initiative in Italy after repeatedly deluding myself that I would be able to create in my country the basis of the technical development of work and the centre of long distance communications in southern Europe.[48*]

A few days after writing this letter, Marconi gave a long interview to an Italian journalist, detailing his complaints about the Italo Radio contract. "Is it true that you are about to leave Italy for good?" he was asked, and replied, "It is not me who is leaving Italy... it is Italy that is leaving me."[49] Nearly thirty years after moving to London to develop his invention, and four months after joining a political party for the first time in his life, Marconi was openly contemplating leaving Italy for good. He was at this point one

---

* As mentioned in chapter 1, this is the only time that Marconi explicitly claimed he had offered his invention to the Italian government before leaving for England in 1896.

of the most famous and recognizable Italians in the world, arguably better
known and more admired than the upstart politician who had just administered such a crisp slap in the face.[50] Marconi had wanted, expected, and
thought he deserved to be the linchpin of Italian radio, on the basis of
national consideration and in exchange for the privilege he had offered
Italy in the free use of his patents. But his company was seen as British—and
eventually, he would look at it that way himself. Marconi's interests had
embroiled him in the thickening soup of Italian politics. Mussolini, still only
the head of a coalition government, was increasingly insisting on a warranty
of political loyalty in exchange for the favour of his regime—and as Marconi
was discovering, even that was not enough. Marconi's relations with Mussolini
were as low as they would ever be. He decamped for London, to take part
in the Imperial Conference and celebrations of the first anniversary of the
British Broadcasting Company.[51]

Broadcasting, in fact, was excluded from the Italo Radio concession, which
was concerned only with the "old" technology of wireless telegraphy, but
not the "new" medium of radio. So while a monopoly had been created for
wireless, the way was open for an alternative in broadcast radio.[52] Here
again, Marconi and his interests should have been in the front seat.

    The field was being developed by an Italian engineer/entrepreneur by
the name of Luigi Ranieri, who had adapted a novel broadcasting-like system he called the "circular telephone."[53] The circular telephone was technically and legally a wired telephone system, but it was an uncanny precursor
of radio. Based on a prototype introduced in Budapest in 1893 (and adopted
in other countries as well), Ranieri's system offered a package of news and
entertainment, transmitted over telephone wires, to subscribers. Seen as a
novelty for the rich, the service peaked at around thirteen hundred subscribers in December 1914, after a celebrated live broadcast of Gabriele
D'Annunzio arriving in Rome to convince politicians to intervene in the
war. A 1917 court case ruled that Ranieri's technology was not telephony,
because it transmitted one-way only, and that it was hence a transmission
medium, not a communication medium (a view with which Marconi would
have surely concurred). This groundbreaking ruling put Ranieri out of business; after a hiatus, however, Ranieri reintroduced his system as a more user-
friendly service, organized like a broadcaster and financed partly by
advertising, in Rome at the end of 1922. He then began lobbying Mussolini—
who happened to be a subscriber—for a broadcasting licence.

Encouraged by the evidently poor relations between Marconi and Cesarò, Ranieri persisted, writing to Mussolini directly on December 12, 1923, about setting up a radio station in Rome to complement his circular telephone system. He wasn't the only one who had this idea. A month later, on January 7, 1924, a new entity backed by the US company RCA was formed for the same purpose. The RCA initiative was known by the acronym SIRAC (for Società Italiana Radioaudizioni Circolari.) On January 16, 1924, Solari too wrote to Mussolini again, this time wearing his hat as head of Marconi's radio news agency, urging the importance of establishing a radio news service for international propaganda as soon as possible.[54]

While all this was sorting itself out, in February 1924, Cesarò suddenly resigned. He was replaced by a Mussolini loyalist—and Marconi supporter— Costanzo Ciano. Ciano was a native of Livorno, and had entered the Naval Academy there at the age of fifteen, in 1891; one of his classmates was Luigi Solari and, like Solari, Ciano, too, may have heard the urban legend about the boy who made sparks fly from the rooftops. He was a naval officer by the age of twenty, then a ship's lieutenant, and soon a commander. Like Marconi, he took part in the Italo-Turkish War of 1911–12, but he likely saw a lot more direct action. Ciano served in Libya again after Italy entered the Great War in 1915. Immediately after the war, Ciano created a political party to promote his ardent nationalism, but was soon one of the first adherents to fascism. He became a Fascist MP in 1921, heading the party's Livorno section and taking part in the March on Rome. He was one of Mussolini's first political appointees, named undersecretary of the marine on October 31, 1922. When Cesarò resigned on February 5, 1924, Ciano was named minister of posts and telegraphs, a portfolio he soon converted into a Ministry of Communications and held for ten years.[55]

One of Ciano's first acts as minister was to write to Marconi in London, reassuring him about the Italo Radio agreement and his eventual role in the development of Italian communications. "I do not consider the difficulties insurmountable," the newly minted minister wrote.[56] Keenly aware of the propaganda value of radio, Ciano was determined to create a national service as soon as possible, and he wanted to do it on the BBC model; that is, by creating a single state-controlled national company coalescing broadcasting and manufacturing interests.[57] Ciano clearly intended that the new medium of broadcast radio should become an instrument of fascism.[58] Getting the competing radio interests to work together was critical to his plan.

As national elections approached, Mussolini was due to give a major speech in Rome on March 23, 1924. Ciano arranged for the Marconi group to broadcast the speech and, a week before the date, Ranieri offered to cover it for his subscribers as well. When the speech, at the Teatro Costanzi, was transmitted by both the Marconi station at Centocelle and Ranieri's Radio Araldo, a mysterious "induction phenomenon" disturbed the signal and made it inaudible, resulting in what Solari called a "total fiasco."[59] The Marconi company blamed the fiasco on Araldo, saying its amplifiers had created interference with Marconi's. Mussolini's oratory (fashioned after D'Annunzio's) was central to building the fascist identity, but Mussolini himself was not yet convinced that radio was a reliable medium, and the incident did not help. Solari was summoned to Ciano's office, where he was greeted by a glum minister; after a few minutes of silence, Ciano instructed him to repair the damage with a perfect demonstration, and Solari said it would be done immediately. Marconi and Ciano had close friendly ties, and Ciano likely endorsed Marconi's view placing the blame on Ranieri. Nonetheless, the regime wanted a radio system organized.[60]

Mussolini's National Bloc (an alliance of his National Fascist Party with Catholics, Liberals, and Conservatives) won an absolute majority in the national elections on April 6, 1924, after a campaign and election day marked by violent intimidation of voters and opposition candidates. The Fascist-driven coalition received 64.5 percent of the vote and 374 seats. The Catholic Popular Party was second with 9.0 percent and 39 seats, and Giacomo Matteotti's social-democratic United Socialist Party was third with 5.9 percent and 24 seats after breaking away from the old-guard Italian Socialist Party, which followed closely with 5.0 percent and 22 seats. The Communist Party of Italy, headed by Antonio Gramsci, received 3.7 percent of the vote and 19 seats.

Ranieri, meanwhile, formed a new company with backing from Angelo Pogliani—Marconi's one-time managing director at the Banca di Sconto—and joined forces with SIRAC. Together, Araldo and RCA presented a credible force. On May 31, 1924, they submitted a proposal to Ciano. As it had successfully stipulated in wireless telegraphy, the government wanted to create a single entity for broadcasting, and asked for the inclusion of Radiofono, a new company recently launched by Marconi. In the ensuing negotiations, Radio Araldo was dropped from the project. On June 17, 1924, the new Fascist majority government granted a concession creating Italy's first broadcasting entity, the URI (Unione Radiofonica

Italiana). Nine of the broadcaster's twelve directors came from Radiofono; Solari was the first vice-president. Had Cesarò still been the minister, it may have been Marconi who was excluded. Instead, Marconi benefited from the rapidly growing consolidation of Fascist Party power in Italian politics. As for Luigi Ranieri, he joined the long list of innovators Marconi left in his wake, another interesting figure who had played his cards poorly.[61]

❀   ❀   ❀

Not a week before the URI concession was announced, one of the few remaining active voices of opposition, Socialist MP and party leader Giacomo Matteotti, was kidnapped and brutally murdered in circumstances that have never been fully clarified. Matteotti had called for annulment of the April 1924 elections because of the numerous incidents of voter and candidate intimidation. More worrisome to the government, it was said that he had uncovered and was about to reveal incriminating documents showing that the regime had sold exclusive rights to Italy's oil reserves to a US company—a devastating indictment in view of the government's professed nationalist ideology. On June 10, 1924, Matteotti had lunch with his wife at their home near Rome's Piazza del Popolo. He was due to make a speech in parliament that afternoon, and after lunch he set out on foot for the Chamber of Deputies a few blocks away. As he walked through the busy streets in broad daylight three men forced him into a sedan and sped off, the driver leaning on his horn to cover up Matteotti's cries for help. Matteotti's battered body was found twenty miles outside the city two months later. Mussolini tried to distance himself from the assassination, firing his chief of police and undersecretary for internal affairs, but the blame pointed to known Fascist goons and at least one henchman of the prime minister: Amerigo Dumini, a prominent member of Mussolini's secret police, the Ceka. Dumini and four others were charged with the murder; three were convicted but amnestied by the king.[62]

After the Matteotti affair, many middle-class supporters who had embraced fascism as a conservative nationalist movement began to desert the party. Others, like Marconi, held fast, buying in to Mussolini's argument that only fascism could provide Italy with the stability it needed to move forward and flourish. During the months of confusion and unrest that followed, Mussolini set the country on the path to dictatorship. Less than a year after making threatening noises about leaving Italy for good, Marconi was more important than ever to Il Duce, and on his way to becoming one of the most influential figures in fascist Italy.

27

The Beam Indenture

On May 27, 1922, Marconi left Southampton on his most ambitious research journey to date.[1] Accompanied by a team of engineers and assistants, he intended to cross the Atlantic on the *Elettra*, doing experiments with direction finders and short- and longwaves along the way. The departure was announced in the British and US press, and Marconi's arrival in New York was awaited. A lecture was scheduled there for June 20, 1922, at which he would report on his test results. The trip was also meant to be a moment of personal reconciliation for Marconi, at least in part. Beatrice would join him in the United States; she sailed on the White Star *Olympic*, arriving in New York on May 30, 1922.[2]

Marconi's crossing was troubled by bad weather. The *Elettra* could make only twelve or thirteen knots in the best of conditions, and after refuelling in the Azores, it had to pull in to Bermuda to weather a storm. The yacht arrived in New York on June 16, 1922, ten days late. As it approached the harbour, Marconi received a wireless from the Associated Press, inquiring whether he had picked up any messages from Mars during his trip: "No sensational announcement to make," he messaged back.[3] The *New York Times* had gone out on a limb before Marconi's departure from England, reporting that he was hoping to receive signals from Mars.[4] After landing, he quickly skewered the story, though in characteristic enigmatic fashion. "He said that any suggestions of interplanetary communication were 'absurd' and 'ridiculous' but added that the 150,000-meter wave which he had picked up on the *Elettra*'s radio apparatus in the Mediterranean several months earlier certainly did not originate on the earth." Well then, how absurd or ridiculous was it? "He is unable to say whether that immense wave, five times the length of any wave used on earth, originated on the sun or some other planet. But he received it."[5]

The *Elettra*'s arrival was greeted with the type of commotion that had marked so many of Marconi's US visits. "The inventor was boyish and hilarious in mood as he came up the harbor on this, his eighty-fifth visit to the US," the *New York Times* reported, exaggerating the number, which was closer to thirty.* "He posed for photographs and moving pictures, chatted gaily with radio engineers who had come to meet him, and at intervals listened in on his apparatus to broadcasting programs. His holiday mood continued throughout the afternoon...."The yacht arrived off Sandy Hook early on June 16 and dropped anchor off the Columbia Yacht Club in the Hudson River shortly before noon. As it entered the range of the city's estimated eight hundred thousand amateur radio enthusiasts, "receiving sets were cluttered with messages of greetings.... It seems that every amateur within sending distance was eager to pay his wireless homage to the man who has done so much for the science."[6]

As soon as the *Elettra* docked, the real significance of Marconi's latest research began to emerge. He had set up on the yacht "a secret and as yet unpatented contrivance which virtually cuts out all static," a major impediment to wireless competition with cable-based telegraphy. It was reported that radio engineers who visited the *Elettra* upon landing came away convinced that this was now Marconi's most important work; if he succeeded, "Marconi will have achieved as great a triumph in radio engineering as when he first used the Hertzian wave for communication purposes," they said. Marconi also told the press that his latest experiments at sea had convinced him that it was possible "to send a radio message around the globe." The secret key to this new success was wavelength; Marconi was now working with uncommonly short waves.[7]

On June 20 Marconi presented his heralded lecture to a joint meeting of the New York Section of the American Institute of Electrical Engineers and the Institute of Radio Engineers (IRE). It was a stellar performance, in both form and substance. Using a transmitter described as a "baby wireless set," he awed his audience by demonstrating "how a flying shaft [beam] of radio waves may be hurled in a desired direction, straight at a receiving station intended to receive it." This was the new directional "beam" system he had been developing primarily with his associate Charles Franklin since 1916. Marconi recalled that his very first experiments in Italy had been with shortwaves, and announced that he was now returning to them for what the press declared might be his crowning achievements. The "baby wireless"

---

* One possible explanation is that Marconi might have said it was his eighty-fifth *crossing*, which would naturally be double the number of visits, but even that figure would have been inflated.

demonstration amazed observers, who had never seen anything like it before: using a one-metre wave, "the directional sender shot its shaft of rays
perfectly across the twenty feet of space, sounding loud, clear notes." With
3.5-metre waves, Marconi said, he had achieved nearly one hundred miles.[8]

Near the end of his talk, he made an astounding announcement: "It
seems to me that it should be possible to design apparatus by means of
which a ship could radiate or project a divergent beam of these rays in any
desired direction, which rays, if coming across a metallic object, such as another steamer or ship, would be reflected back to a receiver screened from
the local transmitter on the sending ship, and thereby immediately reveal
the presence and bearing of the other ship in fog or thick weather." Although
it was not widely noticed or recognized at the time, Marconi was describing
a process that would become known as "radar."[9] This was one of the most
significant speeches of Marconi's career, and it has been published, republished, and abundantly cited. It is an example of the uncanny foresight that
some have called "prophetic" and that more pragmatic analysts have noted
as an ability to think ahead.

The origin stories of groundbreaking technologies are often rife with
embellishments and fanciful details, and radar is no exception. Hertz had
shown that radio waves were reflected by solid objects; Tesla, as early as 1900,
had suggested that they might be used to detect a moving object; and a
German engineer by the name of Christian Hülsmeyer patented a system
for doing this in 1904.[10] The amateur radio enthusiast Hugo Gernsback,
who was also one of the pioneers of the modern genre of science fiction,
accurately described what came to be known as radar in a 1911 novel (the
bizarrely titled *Ralph 124C 41+*).[11] Radar has typically been defined as a
method of locating an object by determining the time interval between
transmission of a radio pulse and the receipt of the echo of that pulse by the
same equipment after reflection by the object whose position is desired.[12]
According to radio historian L.S. Howeth, the term was coined in 1913 by
two US naval officers, E.F. Furth and S.P. Tucker, to describe what the British
were then calling "radiolocation" and the Americans "radio position finding";
it was meant to be a contraction of "radio detection and ranging." But—once
again—no one had developed a working system before Marconi's 1922 suggestion of using shortwaves.

Marconi's associate H.J. Round had developed a primitive position-
finder during the First World War, as we saw earlier, but the breakthrough
now—already noticed by Marconi and Franklin—was that shortwaves were

reflected back to a transmitter by obstacles lying in their path. Marconi did not pursue this property of shortwaves at the time as he was focused on the development of the shortwave "beam" for long-distance communication. Howeth—an authority notoriously skeptical of extravagant claims and no fan of Marconi—says the idea remained dormant until Marconi revived it in his speech of June 20, 1922. In 1945, the US Office of Scientific Research and Development recognized three seminal moments in the pre-history of radar: Hertz's 1886 discovery, Hülsmeyer's 1904 patent, and Marconi's 1922 suggestion of the "use of short waves for radio detection."[13]

Marconi returned to his research on what he called "blind navigation" in Italy in the 1930s, but by then military researchers were on their way to perfecting the process. Only a year after Marconi's New York lecture, British engineer Robert Watson-Watt developed a device for telling the direction, distance, and velocity of an object. Watson-Watt (a descendant of the inventor of the steam engine, James Watt) went on to develop the first practical system for aircraft detection by radar while director of radio research at the British National Physical Laboratory in 1935. Tests of his system were so secret that even the prime minister was unaware of them. Successful use of radar was considered one of the key pieces of Allied military superiority in the Second World War; today, of course, it is essential to air traffic control and navigation, among many other things.[14]

After New York, Marconi went on an inspection tour of US wireless installations, hosted by his friends and former colleagues at RCA. Having finally met up with Beatrice, they travelled together, stopping over briefly in Atlantic City (where they rode the roller coaster—an apt symbol for the state of their marriage at this time). Then Marconi visited RCA's transatlantic station at New Brunswick, New Jersey—a site he had selected himself, years earlier, for transmission of the first commercial wireless messages from the United States to Europe. He also visited the old Telefunken station at Tuckerton, taken over during the war and now also operated by RCA.

The tour reminded him how much had changed since his early visits to the United States. After New Jersey, he returned to New York City to visit the RCA "nerve center" in Broad Street. He then sailed up the Hudson on the yacht, to visit the GE plant at Schenectady.[15] On the way upriver, the *Elettra* broadcast a kind of tourist program on points of interest. Less than three years after its creation on the foundation of Marconi's American company, RCA was now the largest commercial radio enterprise in the world, with five transatlantic and one trans-Pacific circuits. It was in contact with Marconi's long-distance station at Carnarvon, Wales, as well as stations at Stavanger (Norway), Bordeaux, and Berlin, and would soon connect with Rome, Warsaw, and Ghent. The company had just made a new break-through, a small tube developed by engineer Irving Langmuir, that was about to replace the Alexanderson alternator. Marconi was tremendously impressed with Langmuir's invention, as he had been a decade earlier with Alexanderson's.[16] Although he was still revered, and referred to systemati-cally as the "inventor of the first system of wireless communication" (as well as, occasionally, "the wizard of wireless"), he seemed to be on a path toward elder statesman.[17]

On June 24, 1922, the New York papers noted the presence of two "foreign notables" on the passenger lists of the ocean liners departing for Europe: Beatrice Marconi, wife of the noted Italian inventor, and Sir Arthur Conan Doyle, creator of Sherlock Holmes. Doyle had written to the Marconi company in August 1921, asking them for "an address which will find Mr Marconi. I had a matter of some importance to communicate."[18] Either Marconi never received the message or he did not reply (which is not likely, or at least odd, as he nearly always responded to correspondence, especially from famous people). Now they were in New York at the same time. In April 1922, Doyle had arrived on a two-month speaking tour. His topic was spiritualism. Doyle was one of the leading figures of the spiritualism move-ment, which claimed as many as ten million adherents in England and North America in the years following the First World War, as bereaved families sought solace by seeking to contact their lost loved ones. Doyle had been interested in psychical research since the 1880s, but after losing a son and a brother-in-law in the war he became almost fanatical about the most ex-alted claims of the spiritualists.[19]

Doyle gave at least seven lectures on his US trip, speaking three times to overflow crowds at Carnegie Hall alone. An inveterate headline-grabber, he happily gave his views on all and sundry topics, and they were duly re-ported in the press: Doyle thought it would be wonderful to introduce the game of baseball to England; he was attracted to prohibition (although he admitted to enjoying a drink) and intended to advocate for its adoption in Britain when he got home. He also had views on the possibilities of radio for contact with the dead. Doyle was already on record stating that wireless allowed for communication with the spirit world. Broadcast radio provided even more possibilities.[20]

In mid-June, in Atlantic City (just a few days before Marconi's arrival in the United States), Doyle had a crash course in the technical marvels behind the new medium, and enthusiastically announced his intention to apply it to his psychic investigations. After listening to some broadcasts on a mammoth receiving set in his hotel suite, he arranged to have a complete outfit installed in his English home immediately on his return.[21] Doyle and his family were still at the Ambassador Hotel in Atlantic City on June 17 and 18, 1922, when they were joined for the weekend by Harry Houdini and his wife, Bess. One Houdini biographer has said about Doyle and Houdini: "No two people could have been more different. They were opposites in every way. But they were fascinated by each other."[22] On June 18, a séance was held at which Doyle later claimed that Houdini—who famously did *not* believe in spiritualism, only man-made illusions—had communicated with his long-dead mother. The two families then returned to New York.[23]

Marconi, meanwhile, was also in New York. On June 20 he delivered that widely reported talk about radar, and it is almost unthinkable that Doyle was not aware that Marconi was in town. The more likely scenario is that he did learn Marconi was there, and wrote to him again, from New York:

> Dear Mr Marconi
>   I wrote to you once before in London but my letter probably never reached you.
>   I wished to suggest—what has no doubt been often suggested—that these waves which have been so often noted—are from human minds & human skill discarnate, rather than from Mars [as the press had suggested Marconi believed].... Having to construct their own means to impress our machines, they have not yet had full success. I was told—I don't know with what truth—that one operator got clearly SIG on the morse code—an attempt at Signal. If it is so it will develop rapidly.
>   Yours faithfully,
>   Arthur Conan Doyle[24]

Marconi received many palpitating letters of this sort, and usually replied politely, keeping the writer at bay. There is no sign that he ever replied to Doyle, however, nor any mention of a meeting in either of the men's extensive writings. This may mean any number of things, but there is no question that Marconi received the letters—they are both in their original pristine condition in the Marconi Archives in Oxford. What we do know is that Doyle and Marconi were in New York at the same time and just missed each other in Atlantic City. Marconi and Beatrice were in the New Jersey resort on June 22,[25] but by then the Doyles were back in New York. So it is un-likely that "the wizard of wireless" and the world's leading spiritualist ever met, although their paths could have crossed at the departure docks in Hoboken on June 24. At 9:00 a.m. that morning, Houdini escorted the Doyles and their three children to the RMS *Adriatic* (in the presence of some two hundred newspeople and well-wishers).[26] At around the same time, Marconi was seeing his wife Beatrice off, somewhat more discreetly, on the RMS *Olympic*.[27]

To this day, many proponents of spiritualism claim that Marconi believed it was possible to communicate with the dead and attended séances,[28] yet there is no evidence that he did. This view is at best a wishful misinterpre-tation of his scientific curiosity about unexplained radio signals apparently emanating from outside the earth's atmosphere. Marconi's evident coolness toward Doyle indicates, if anything, that Marconi, like Houdini, was not a believer in spiritualism, otherwise he would certainly have met up with Doyle at some point. But that didn't prevent Doyle's son Denis (who had been on the trip to New York and Atlantic City as a lad of thirteen) from describing his father years later as a "spiritual Marconi." "Just as our temporal existence is regulated and controlled by certain physical and material laws," Denis Conan Doyle wrote in 1937, after both men were dead, "so there are basic and uni-versal spiritual laws which affect existence in the higher realms."[29]

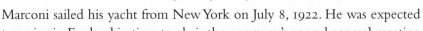

Marconi sailed his yacht from New York on July 8, 1922. He was expected to arrive in England in time to chair the company's annual general meeting on July 25, but the *Elettra* was delayed again by weather and he missed the occasion. He was happy to be back. "It will be lovely to get a good Irish meal again!" he wrote to Beatrice, who had by then gone on to Italy.[30] He was elated with the results of his transatlantic experiments and his reception in New York. "No research work, so far as I could ascertain, appears to have been carried out in America with my directive system employing short waves, or anything similar to it," he reported to the company board the day after his return. In fact, he said, the findings that he described in New York "seemed to come as a surprise to all." Marconi himself seemed surprised—and pleased—that he was far ahead of the Americans.[31] He recognized that the introduction of his shortwave "beam" system would revitalize his career. Now that he was back in England, he threw himself full force into devel-oping the system and getting it adopted by the British government. No longer just a respected pioneer, he was becoming a player again.

The work had started in Italy in 1916, when Marconi and Franklin developed a system for short-range directional communication between naval vessels, using wavelengths of two to three metres. At the time, they managed to attain a distance of six miles; later, in 1917, working from Marconi's new high-power station at Carnarvon, Wales, Franklin achieved twenty miles and, by 1919, seventy. In 1920, Franklin began using a rotating parabolic reflector to achieve more precise accuracy and greater distance. Meanwhile, Marconi himself was doing similar tests on the *Elettra*.[32]

On the business side, the company was having difficulty resuscitating the imperial chain project; halted by the war, stalled by a legal action for damages the company took against the British government as a result of the contract's cancellation, and called into question because of rising projected costs and competition from rival technologies like the Alexanderson alternator and the Poulsen arc systems, the issue was like a festering sore. In 1919— after receiving a £600,000 settlement to compensate for the cancellation of the 1914 contract—the company submitted a new proposal and another government committee was set up to study it.[33] The committee considered the Marconi proposals "too comprehensive and ambitious to be viable," and expressed a scathing view of private enterprise in general, recommending that construction and operation of the imperial chain stations be mandated to the Post Office. The GPO, meanwhile, built a new high-power station at Leafield, in West Oxfordshire, and was ordering equipment from Marconi rivals. Faced with government inaction, Marconi took another tack.[34] Britain controlled the UK end of long-distance communication, but each of the colonial Dominions had its own arrangement. The company began negotiating contracts through subsidiaries in Australia, India, South Africa and Canada, creating a *fait accompli* for London. But the key was the new system Marconi was developing.[35]

A debate was taking shape in the UK regarding the relative merits of public versus private enterprise. In December 1922, Britain had taken a big step in creating the British Broadcasting Company. Now, just three months later, on March 5, 1923, the new Conservative prime minister, Andrew Bonar Law, announced that the government would issue licences to private companies to erect stations for wireless communication with the Dominions. This begged the question whether a similar model of public-private partnership would emerge in telecommunication as it had in broadcasting. The policy of private ownership but national control also suggested a role for Marconi similar to that of RCA in the United States. British opinion was skeptical and the *Times* wrote: "Private enterprise...means the Marconi and associated companies."[36]

A few weeks later, Marconi set off on a two-month cruise on the *Elettra* in the western Mediterranean and Atlantic. He kept up a stream of correspondence with Franklin at Poldhu, Isaacs in London, and company chief engineer Andrew Gray. The energy and enthusiasm evident in the exchanges was impressive. By the time he returned and reported to his board in the middle of June 1923, he had made up his mind that the beam system "was destined to carry a considerable part, perhaps some day all of the high-speed long-distance traffic of the world."[37] The research on the *Elettra* had been intended to test the reliability of shortwave signals transmitted over considerable distances; to investigate the conditions affecting their propagation; to ascertain the maximum ranges obtainable by day and by night; and to determine the technical characteristics of the radiation "beam"—all in regard to the possibility of establishing "long distance directional wireless services." The results of Marconi's tests convinced him that it was now possible to carry out "continuous commercial services at high speed by day and by night" between stations situated twelve hundred miles apart. The practical limit of the night range had not yet been approached; it could be four thousand miles, six thousand miles, or even more.

The tests had demonstrated that the prevailing views of most technical experts regarding the behaviour of radio waves was erroneous. Marconi was using the ocean as his research lab, moving the yacht as necessary in order to test different types of conditions. At St. Vincent, in the Caribbean, he found the signals from the GPO station at Leafield "weak and often unreadable," while there was "no difficulty whatever" receiving messages transmitted to him expressly at his request from his own experimental station at Poldhu. He was convinced that with a strong receiver at the other end, Poldhu could communicate with Brazil and Argentina, and regretted that he could not carry out tests in South America now because he had to return to London.

Marconi was functioning at full steam, using his proven methodological approach, which was to test a simple hypothesis and then develop its practical application. He was simultaneously thinking about a new technology, the design of a mechanical apparatus, the operational and organizational steps necessary to establish a new system, and the political hurdles that needed to be crossed. He now had a vision for "wirelessing the world." He was still thinking of radio communication as "wireless," despite the recent development of "broadcasting," a mode of communication with a completely different logic, different end-purposes, a different relationship to users, and, above all, different commercial as well as political possibilities. But he was ready to state that "there exists no theoretical reason" why, with shortwaves, the speed of working and the distance covered should not be several times greater than the limits previously thought possible using the longwaves now in general use for long-distance communication.[38]

Marconi decamped for Italy to try to impress Mussolini with his latest discoveries, while the UK debate shifted to Parliament. On July 24, 1923, the postmaster general told the House that since the announcement of the government's new policy in March, two applications for imperial broadcasting had been received: one from the Marconi company for a general licence covering the whole of the British Empire, and the other from the Eastern Telegraph Company for communicating between India and the United Kingdom. The government was negotiating with Marconi regarding its proposal. A deal was taking shape by which the wireless services of the British Empire would be conducted through stations involving both the company and the government.[39]

In September, Marconi was back in London; on the twelfth, the *Times* reported that he had joined the Milan section of the Italian National Fascist Party. The news was three months old, but it was news. On September 17 Marconi and Isaacs met with the postmaster general to pitch the latest version of their imperial scheme. Marconi was insisting that its licence cover worldwide communication, irrespective of any government station, and postmaster general Sir Laming Worthington-Evans said the government would never agree. "That, you see, is not private enterprise, that is private enterprise excluding Government enterprise altogether which we have never agreed to," the postmaster general said.[40]

This was to be the British model, one based on public-private collaboration. It was a model that Marconi was always comfortable with; he preferred the solid backing of government rather than the expectations of fickle shareholders. Isaacs was in charge of the company, however, and, like Hall before him (although far more congenially), he took a harder line. On October 19, 1923, the company released a statement to the press, over Isaacs's signature, stating that the postmaster general had misrepresented the company's proposal; Marconi's was prepared to enter into "a pooling arrangement" with the Post Office or, failing that, "a general non-exclusive licence" under which it would provide facilities for the Post Office station to communicate with the Dominions. Marconi's was offering as much and indeed more than the Post Office could reasonably expect from a commercial company, Isaacs said. The arrangements proposed by the government, on the other hand, "would mean a loss to both parties."[41]

An Imperial Economic Conference, involving the British Dominions and colonial satellites, was deliberating in London while this debate was taking place. (British imperial policy played out through a series of such conferences on a range of topics, including communications, during the early 1920s.) On November 9, the conference passed a resolution affirming the importance of promptly establishing "an efficient Imperial service of wireless communication" and suggesting that the empire's governments should act to remove any obstacles to its immediate accomplishment, "while providing adequate safeguards against the subordination of public to private enterprise."[42] It seemed to be an appeal to both parties to water their wine.

Marconi, meanwhile, spoke on the air on the British Broadcasting Company's first anniversary broadcast on November 14, 1923, and two days later the company issued a letter to the press over his signature as chairman. Tying the issue of empire communication to the general election campaign, he asserted that the parties' proposals for strengthening the British economy would be "intimately affected" by the decision regarding imperial wireless; if this problem were not satisfactorily solved, none of the others could be. Marconi was publicly tying his interests to the country's and argued that there seemed to be a "conflict of memory" regarding what took place at Marconi's and Isaacs's meeting with the postmaster general on September 17—contrary to what was being suggested, his company was not aiming for a monopoly of wireless services. Whether this was Marconi himself or the company using his name to put weight behind its position is unclear; either way, though, it was astute.[43] They were now positioning themselves in a larger, emerging debate between the developing British welfare state and corporate finance.[44]

Suffering from throat cancer, Bonar Law resigned in May, before his wireless policy could be put into effect. He was succeeded by Stanley Baldwin, and elections on December 6, 1923, produced a hung Parliament, in which the first British Labour government was formed, headed by Ramsay MacDonald (with the support of Herbert Asquith's refurbished Liberal Party). The new government appointed yet another committee to examine the issue of "imperial wireless telegraphy," chaired by newspaper editor Robert Donald, head of the Empire Press Union—an ardent advocate for

an imperial wireless chain. Reporting in February 1924, the Donald Committee proposed that the British government should own and operate all
wireless stations for communication with the Dominions, while communication with "foreign countries" should be developed by the private sector.[45]
This was a reversal of the previous government's policy yet the generally
conservative *Times* found the report "admirable."[46] It still remained to be
seen how and on what basis those stations would be built.

Meanwhile, Marconi was working feverishly to get his new system into
workable shape. He had already reported receiving strong signals at twenty-
two hundred miles from Poldhu, sent using a single kilowatt of power—a
stunningly small bit of energy. He was about to overcome the main obstacle
to creating a global network: the extraordinary power needed to sustain signalling across long distances.

As soon as the Donald report appeared, the company immediately issued a statement to the press. The general public is not interested in the
controversy between the Post Office and the Marconi company, but "that
the Empire shall have the benefit of the latest, most efficient, most rapid, and
cheapest means of wireless communication." The statement added that the
Donald Committee was evidently unaware of Marconi's latest work, which
was about to be put into operation and "promises to revolutionize the entire
practice of wireless telegraphy."[47]

Behind the scenes, Marconi and Isaacs wrote to the prime minister on
March 4, 1924, seeking a meeting to state their case directly to the government (had the Donald Committee heard them, this wouldn't have been
necessary, they seemed to imply). "The Company is anxious to do all in its
power to assist the Government in its policy of providing the Empire with
the most efficient system of wireless telegraphic communications," they said
reassuringly, adding a modest reminder: "It hopes it is not presumptuous in
pointing out that it has done the whole of the pioneer work in wireless
telegraphy, and that its advice and assistance should have some value."[48]
On Monday April 14, 1924, at 11:30 a.m., Marconi and Lieutenant-Colonel
Adrian Simpson, the company's assistant managing director (replacing Isaacs,
who was too ill to attend) met with the Cabinet committee studying the
Donald report, in Whitehall. A record of the "strictly private" meeting,
marked "Secret," was found among Marconi's papers.[49]

The new postmaster general, Vernon Hartshorn, a Welsh trade unionist
only a couple of years older than Marconi, opened the meeting by reiterating the government's position on the imperial scheme: it wanted to erect,

own, and control all the domestic stations for communication with the Dominions and throughout the empire. The Marconi company was already, in nearly every case, either contractor or partner with the respective Dominion governments, "and we are only anxious that such arrangements should be entered into between yourselves and us as will make your arrangements and ours effective throughout the Empire." Marconi had trumped the government by getting inside its system via the Dominions, but that wasn't the core of his strength.

Marconi informed the committee that "the technical question" had reached an important new level in the last month or two. He then read a lengthy statement (appended to the official secret transcript), detailing his development of "an entirely new system of long distance communication" in the previous year. "The distances recently covered are practically the greatest separating any two places on earth. . . . I have no hesitation in stating that the new system is destined to bring about a complete revolution in the methods hitherto employed for communicating by wireless with distant countries." From a commercial standpoint, it would be a mistake to continue the imperial scheme according to the existing plan. The new system was remarkably and unexpectedly reliable; its principal characteristic was that "the electric waves are projected in a beam [by "short waves," Simpson added] in any desired direction only, instead of being allowed to spread around in all directions." Another advantage was the "secrecy of communication unobtainable with any other wireless system. . . . The importance of this fact to the Empire, particularly in War time, is obvious." It was signed G. Marconi, April 14, 1924.

This was high-stakes poker at its best. Marconi's bluff was to convince the government it didn't know what was going on, it don't know what he knew, it didn't know what he would know tomorrow, and therefore the British Empire needed him. When he was finished, the chancellor of the exchequer, Philip Snowden, described Marconi's statement as "startling." Marconi was proposing a 180-degree turn in strategy—that the government should abandon its plan to build a chain of conventional longwave long-distance stations and replace them with his new, untried shortwave beam system. The colonial secretary, J.H. (Jimmy) Thomas, asked if Marconi was saying that "these experiments that we have just heard about have so affected the original plans that even notwithstanding the expenditure already incurred you would consider it wise to scrap it. Is that so?" "That is so," Simpson replied.

With the new system, stations could be built at one-tenth the cost, and could transmit a greater volume of traffic at greater speed. Marconi argued that only his company had the necessary practical experience and technical knowledge to build them, and he tendered an olive branch to Labour's political position: after the stations were built, the government could take them over if it wished. The experts in the room were skeptical. When Snowden suggested that the news was probably as much a surprise to them as to the politicians, F.J. Brown, assistant secretary of the Post Office, remarked drily: "I do not think so." W.H. Eccles, a theoretical physicist who had worked for Marconi at the turn of the century, before going on to a distinguished academic career and most recently serving on the Donald Committee, was more elegant and less grumpy but equally skeptical: "This is a very good illustration of how private enterprise does assist in the progress of an art...an inventor's hopes of his new invention are expressed very forcibly, and I think Senator Marconi will not resent it if I say that although I do wish him the utmost success, I think it will take him longer than at present he has indicated to make the trial prove successful." To which Marconi rejoindered: "I wish you knew what I know!"

Eccles's comments were aligned with a theory of incremental technological change, but Marconi was proposing a "revolution" (and, importantly, claimed that only he could orchestrate it). But that was not quite how things worked, Eccles argued. Continuity and change represented two approaches, the incremental and the revolutionary, and raised not only practical but also ideological questions, as well as a fundamental one: Do new technologies replace old ones, or build on them, like sediment, like geological layers? Marconi seemed to become impatient with his professorial one-time assistant as he hammered his point home. "The future will show who is right," he said. But if *he* was right, it was urgent that England go ahead with this new system "because nothing can stop the progress of science." If Marconi was right, then presumably other countries such as France, Germany, and the United States would soon get on to it and use the new technology first. Marconi added that he was willing to publicly stake his reputation on the success of his new claim.[50]

The meeting adjourned, with an agreement that Marconi would provide evidence of his results to Eccles and Brown. The Labour government presented its first budget a few days later. The question of imperial wireless was still in abeyance, though high enough on the agenda to merit mention in news reports; a decision was delayed while the Marconi company did certain experiments for the government's technical advisers, the evidently well-informed *Times* reported. Meanwhile, Marconi started putting out word of the success of his beam system; in June, it was reported that he had communicated from Poldhu to Buenos Aires. At the same time, the government also received confidential reports that the system was working.[51]

On July 24, the government announced that it had decided to adopt the main recommendations of the Donald Committee regarding government ownership and was looking at the Marconi company's proposals for "short wave directive stations—so-called beam stations" for communicating between England and the Dominions and India. On July 28, the government and the company signed an agreement—which came to be known as "the beam indenture"—for construction of a wireless station on the beam system.[52] The agreement was ratified by the House of Commons on August 1, 1924, and its terms were publicized around the world. The agreement called for the company to erect, as contractors for the Post Office, a bidirectional beam station for communication with a corresponding station in Canada, with provision for extending the contract to provide similar communication with South Africa, India, and Australia. The contract was for cost plus 15 percent (estimated at between £45,000 and £50,000), as well as a royalty of 6.25 percent of gross receipts "so long as any Marconi patents essential for the working of such station are employed therein."[53] The company was bound to have the link to Canada operational within twenty-six weeks.

Financially, it was a windfall for the perennially cash-strapped company. Chairing the MWTC annual general meeting on August 15, 1924, Marconi was able to forecast a "cheerful outlook," saying the company's prospects were "brighter than they had ever been" thanks to the new agreement. Ever since it had first proposed the imperial chain in 1910, the company's main objective had been "the creation of a world wireless telegraph service." Now it was on the way to realizing that goal. The new agreement was "only the beginning of a great scheme of wireless communication," thanks to its small capital requirements and operating costs. It was also the beginning of a new era of friendly relations with the Post Office. Almost by way of celebration, Isaacs moved that the company pay a 10 percent dividend.[54]

The short-lived Labour government was defeated on October 9, 1924. Its signature of the beam indenture is not widely remembered, overshadowed by its sweeping if tentative welfare measures and controversies surrounding its "socialist" overtones. The agreement, nonetheless, was of major significance. Within two years, when the imperial beam system began operations, it could be said that the sun never set on British wireless communications.

28

Radio

Marconi did not see what the fuss of broadcasting was all about; to him it was just another application of wireless, not a new technology. Radio was about communication, not the one-way delivery of light entertainment—which is basically how he saw what broadcasting was doing. In Marconi's practical vision, the first clients (for the technology) were governments, which had a range of possible uses for sophisticated means of communication, including military; the next clients (for the infrastructure) were commercial—the suppliers and operators of the heavy equipment that would make the technology work; the end-users were the people and companies who would pay a small fee for the service of sending messages for business and personal purposes. The idea of providing content didn't enter into it; content would be provided by the users. Marconi thus leap-frogged over the twentieth-century broadcasting model; his nineteenth-century idea would eventually become the dominant model for communication in the twenty-first century—although it must be said that he did not foresee the commercial potential of creating a base of consumers who would wish to acquire personal communication appliances, let alone the sophisticated spinoff industry of advertising. He was also skeptical about the benefits of a medium that he saw as "not always instructive."[1]

But when broadcasting appeared Marconi embraced it, used it, and integrated it into his vision for global communication. By 1924, noting the popular appeal and rapid progress of "wireless telephony," as he still called it, he was writing that "broadcasting, properly handled, will make a material contribution towards greater understanding and amity between Nations, the cementing of home life and the happiness of the individual."[2] Just how he saw that happening was ambiguous, however. In one of his many speeches in the 1920s, he unwittingly foresaw an ominously instrumental purpose to which radio would soon be put: "For the first time in the history of the world man is now able to appeal by means of direct speech to a million of his followers, and there is nothing to prevent an appeal being made to fifty millions of men and women at the same time."[3]

Broadcasting—"the reproduction at a distance of speech and music," in the definition of Marconi associate R.N.Vyvyan[4]—was technically made possible by the development of continuous wave transmission in the United States between 1905 and 1910, notably by Marconi's archrivals Reginald Fessenden and Lee de Forest. It quickly became popular with US "amateurs." For the first ten years of its existence, however, broadcasting went largely unnoticed by the state, and its commercial possibilities were not recognized until 1919 or 1920. Marconi himself later said that he had never taken a personal interest in the business side of broadcasting, or "the entertainment field" as he put it.[5]

While radio broadcasting was an offshoot of wireless technology, ironically, the telephone figures larger than the telegraph in its prehistory. The world's first broadcasting system, Telefon Hírmondó, was created by Theodore and Ferenc Puskás in Budapest in 1893; conceived as a "telephone newspaper," the system provided program content to subscribers via telephone lines.[6] This method was reproduced, among others, by Luigi Ranieri's Radio Araldo in Italy a generation later, as we've seen. Unlike the telegraph, which was always seen as a means of point-to-point communication of basic information, the telephone was imagined as both an instrument for two-way conversation and, by some, for entertainment as well. In this way, it was, conceptually, much like the modern Internet.

Aside from personal communication, the full potential of both the telephone and the telegraph as instruments of mass diffusion had to await the development of wireless voice transmission. The key moment in this evolution was the discovery of continuous waves—as opposed to Marconi's spark, or "damped" waves—which first came into play around 1905 as part of the efforts to address some of the technical inefficiencies of wireless telegraphy; for example, they allowed for precise tuning.[7] The fact that wireless signals could be picked up by anyone—that they scattered, or "broadcast," like certain seeding techniques in agriculture—had long been considered a disadvantage; now it came to be seen as an asset (and soon, as the rationale for a new industry).[8] While the capacity of continuous waves to carry the human voice and other forms of sound was at first seen as an additional benefit of wireless, this aspect was soon recognized as the most important. Now telephony, too, could be wireless.

Three figures—all connected in one way or another to Marconi—were critical to the technical development of sound broadcasting. In November 1904, Marconi's scientific advisor J.A. Fleming patented something he called the "thermionic valve" (soon to be known, in the United States, as the vacuum tube). Fleming's valve, which looked like an electric light bulb, was based on work he had done decades earlier when consulting to the British Edison company; by isolating two electrodes, connected by a carbon filament, in a sealed glass tube, the valve was able to rectify (Fleming's term) the electrical oscillations that took place between transmitter and receiver during wireless communication. Characteristically, Fleming revealed his discovery to Marconi before telling anyone else. "It may become very useful," he wrote from his laboratory at London's University College in early November 1904. A few weeks later, on November 16, a joint patent was filed in the names of Fleming and the Marconi company.[9] Under the terms of Fleming's contract, the patent belonged to the company. In a sense, the vacuum tube is to broadcasting as the microchip is to computing—the crucial piece of technology that makes the practice possible. Fleming's thermionic valve is considered the "stem cell" prototype of the various improved vacuum tubes that followed.[10]

The most important of these improvements was made by Lee de Forest in 1906. De Forest added a third, grid-like element between the anode and cathode of Fleming's valve; hence, the de Forest vacuum tube—which he called the "audion"—came to be known as a "triode," as opposed to Fleming's "diode" valve. De Forest's audion allowed for the amplification of wireless sound signals at the receiving end, and is hence considered by many scholars to be the most important innovation in the development of radio broadcasting.[11] De Forest would call himself "the father of radio," but, as we saw earlier, he lacked the entrepreneurial talent to capitalize on his invention.

The inventor who actually pioneered the practice of radio broadcasting was Reginald Fessenden, who performed the first wireless sound broadcast on Christmas Eve 1906, as we've seen. Having accomplished that, Fessenden returned to experimentation, and it was de Forest who recognized that broadcasting was a new medium and that, once the technology was perfected, it could become an important business.[12] Both de Forest and Fessenden were, however, constantly in the courts and made various deals that resulted in their US patents ending up in the hands of large corporate entities like AT&T and RCA.

Following the adoption of the US Radio Act in August 1912, it was required for anyone operating a wireless system to apply for a licence from the Department of Commerce. The number of licensed amateurs and amateur stations in the United States rose sharply from 322 in 1913 to more than 10,000 in 1916; nearly 8,500 amateur stations were licensed in 1915 and 1916 alone. In 1917 the number reached 13,581, with estimates that there were as many as 150,000 unlicensed receivers. The purpose of the 1912 Radio Act had been to control unregulated entry to the field of wireless communication, but despite the restrictions amateurs continued to dominate the airwaves.[13]

Initially limited by early wireless technology to Morse code, by 1914, amateurs were using their apparatus to broadcast voice and music. As amateur numbers mushroomed, radio clubs proliferated in the United States, and inventive individuals found creative ways to tap into existing resources. In San Jose, California, to cite one example, Charles David "Doc" Herrold hooked into the streetcar lines of the Santa Fe Railway to power a transmitter and broadcast news and music from his engineering school.[14] In 1915, the American Marconi company's promotional magazine, *Wireless Age*, sponsored formation of a National Amateur Wireless Association; the same year, Hugo Gernsback organized the Radio League of America.[15] Also in 1915, Hiram Percy Maxim, a man with considerable business experience, decided to create a national amateur radio system in the United States. It was a disruptive notion, and today we could call Maxim's idea an alternative media and communication network. On Washington's birthday 1916, Maxim used relays to transmit what is considered the first nationwide US broadcast, a very Jeffersonian message, which read: "A democracy requires that a people who govern and educate themselves should be so armed and disciplined that they can protect themselves."[16] A more sustainable alternative stream was launched by the University of Wisconsin, which began experimental broadcasts in 1917, a feat that allows it to call its station WHA, to this day, "the nation's oldest broadcast station."[17]

It was Lee de Forest who first anticipated, and practised, the form of radio broadcasting that would soon be delivering regular "programs" into people's homes. In 1910, de Forest broadcast a performance by Enrico Caruso direct from the New York Metropolitan Opera; in 1915, he erected a 125-foot tower on the roof of his factory in the Bronx and began airing nightly concerts; and in 1916, he broadcast the Yale-Harvard football game. He also provided six hours of live presidential election coverage that year (calling the outcome wrong, however; despite the close race and de Forest's pronouncement, Woodrow Wilson, not Charles Evan Hughes, was elected).[18] Although the art form had been pioneered by Fessenden, de Forest was the first of the inventors to anticipate how radio would be used to broadcast music, entertainment, sports events, news—and advertising—into people's homes. But de Forest was incapable of organizing a system to commercialize the new technology.[19]

Meanwhile, over at the American Marconi company, David Sarnoff was closely following the developments. Born in Uzlian (Minsk), Belarus, in 1891, Sarnoff came to the United States with his family as a rough, Yiddish-speaking lad in 1900. In 1906, at the age of fifteen, he went to work as a messenger with the Commercial Cable Company. After a few months he moved to American Marconi as an office boy, and quickly endeared himself to Marconi.[20] According to his biographer, Eugene Lyons, Sarnoff tagged after Marconi, hanging around the workshop on Front Street and doing odd personal chores like carrying his briefcase and delivering candy or flowers to Marconi's interest of the moment. Sarnoff also acquired Marconi's view of science and progress, where the creative impulses of the dreamy inventor coexist with inexplicable mysteries. According to Lyons, Marconi told Sarnoff: we know *how* things happen, we don't know *why*. An "extraordinary friendship" flowered from what was at first a "lop-sided association"— like many of Marconi's relationships. They were seventeen years apart, there were unquantifiable layers of social class between them, and Marconi had few intimates, but Sarnoff became one of them.[21]

Congenial and industrious, Sarnoff rose steadily with the company, from operator to radio inspector (the position he held at the time of the *Titanic* disaster) and, by 1915, assistant traffic manager. Sarnoff later claimed that in 1915, he wrote a memo to American Marconi vice-president and general manager Edward J. Nally, suggesting a plan "which would make radio a household utility in the same sense as a piano or phonograph. The idea is to bring music into the home by wireless."[22] There is no actual record of the memo or of a response, though early broadcast historians suggested that Nally read it as an interesting but hare-brained scheme by an ambitious and energetic young employee, and put it aside.[23] In 1920, after American Marconi had morphed into RCA and radio broadcasting was beginning to attract some interest, Sarnoff reiterated his suggestion, this time to RCA chairman Owen Young, predicting the sale of one million "radio music boxes" at seventy-five dollars each within three years of a product launch.[24] It was a lucrative insight, that people would buy a gadget that delivered entertainment content live into their homes. Whether it occurred in 1915 or 1920, there is no denying that Sarnoff had outlined some of the foundational concepts as well as the logical structure of the commercial broadcasting industry.[25]

During 1916 and 1917, stories in the popular press framed the skilled broadcast amateurs as a potential wartime resource (amateurs listening in along the eastern seaboard could potentially pick up communications between enemy vessels out at sea). When the United States declared war on Germany in April 1917, the amateurs were ordered to shut down and dismantle their stations. Within a few weeks, police in New York City alone had shut down more than eight hundred broadcasters, and the amateurs would never regain their prewar position. By the time the United States Navy lifted the ban on amateur broadcasting in April 1919, the corporate infrastructure of the broadcasting industry was being put in place. In May 1920 a newspaper in Pittsburgh reported that a Westinghouse engineer, Frank Conrad, was regularly broadcasting live concerts, and Horne's department store began offering its customers sets capable of picking up the broadcasts at a price of ten dollars (about $150 today). A Westinghouse vice-president, Harry P. Davis, saw Horne's ads and realized that this new means of instantaneous communication, properly marketed, represented a "limitless opportunity" for the sale of radio receivers.[26] Westinghouse encouraged and supported Conrad to perfect his broadcasts and branch out into other genres, and on November 2, 1920, its newly licensed Pittsburgh station KDKA broadcast the US presidential election results. RCA and KDKA broadcast a fight between Georges Carpentier and heavyweight champion Jack Dempsey on July 2, 1921; the Jersey City fight, before more than ninety thousand fans, was reputedly the richest gate to that time in boxing history, as well as the first national radio broadcast in the United States.

Word spread at first in the amateur community, as well as by word of mouth, and by the spring of 1922, the US "radio boom" had taken hold. Smartly designed receiving sets, stylishly reflecting modern tastes in fashion and home furnishings, became highly desirable consumer products. (Think of the Apple line of iProducts from the early 2000s for a sense of what this meant.) In 1922, the first year for which there are figures, one hundred thousand radio sets were sold in the United States at a unit price of fifty dollars each. In 1923 there were five times as many sets and the price had dropped by half.[27] Broadcast radio, no longer an amateur affair, now became part of the mass entertainment industry. Department stores, newspapers, and the large electronics companies—not only Westinghouse but GE, AT&T, and RCA—began setting up stations and selling receiving sets.[28] By 1941, there were thirteen million sets in the United States. During the first twenty years of US broadcasting, the value of radio sets in circulation went from $5 million to $460 million.[29]*

In the United Kingdom, meanwhile, the foundation of broadcasting also began to be laid during the heady period of the First World War. At the war's end, when the technical innovations perfected under military conditions were translated to civilian use, as in the United States, the idea of integrating set manufacturing and program production emerged, although within a very different ownership and control model. In the United States, the former American Marconi company, now RCA, took its place as a government-sanctioned but strictly private corporate commercial concern; in the United Kingdom, it was those two ubiquitous partners, Marconi's and the Post Office, that would develop what became the public service broadcasting model.[30] Unlike in the United States, where broadcasting was initially local, then "networked," in the British model, under the influence of Marconi, broadcasting immediately aimed to extend its reach nationally and even internationally, as far as technically possible.

The Marconi company established a long-distance wireless telephony connection from Ballybunion, Ireland, to Louisbourg, Nova Scotia, on March 19, 1919. A few months later, the company received a licence for a low-power (six kilowatt) station to begin experimental broadcasting from its factory in Chelmsford. Between February 23 and March 6, 1920, musical broadcasts from Chelmsford were received up to fifteen hundred miles away. In April, Italian journalists in the United Kingdom broadcast wireless messages from Chelmsford to the Marconi station at Centocelle, Italy, and the president of the Spanish Chamber of Commerce in London sent a wireless message to the king of Spain, then visiting Marconi's yacht at Seville. By June 1920, a more powerful (fifteen kilowatt) transmitter was ready to operate, and on June 15, 1920, the Marconi company made what is considered to be "the first pre-announced broadcast of public entertainment" (as a plaque at Marconi House in London's Strand still proclaims) in Britain, if

---

* The unit cost of sets evolved like a sine curve: from $50 in 1922, it climbed steadily, peaking at $136 in 1929, the year of the Great Crash; after that it dropped year by year until 1932, then rose again to $56 in 1937, the year of Marconi's death. It dropped to $35 in 1941, the lowest price since 1923 (Maclaurin 1949, 139).

not the world—a recital by the Australian opera soprano Dame Nellie Melba. Her voice was reportedly heard clearly as far away as Sultanabad, in northern Iran.[31]

The company continued its experimental broadcasts, but unlike in the United States, the object of the new technology was not yet clearly established. Marconi himself still saw "broadcast telephony" as yet another extension of wireless, rather than as a new technology with a new purpose. To Marconi, the use of broadcasting for mass entertainment was simply superficial (the company, however, was not at all opposed to any strategy that would result in the sale of huge numbers of receiving sets). However, he recognized the power of voice as a vehicle for discussion and persuasion. In December 1920, just as Westinghouse's KDKA was beginning regular consumer broadcasting in the United States, the Marconi company was using radio to report the proceedings of the first meeting of the League of Nations in Geneva. Lord Burnham, owner of the *Daily Telegraph* and successor to Robert Donald as president of the Empire Press Union, and Alexander Graham Bell sent wireless voice messages to the meeting from Marconi House in London, and a message was read from Chelmsford on behalf of Marconi himself, who was in Rome at the time. The company saw the event as foreshadowing a time when wireless and landlines would be permanently linked, and the range of telephony extended across the entire earth.[32†]

The company continued its experimental broadcasts through 1921 and, on February 14, 1922, began the first regular broadcasts in the United Kingdom with a series of weekly half-hour musical concerts from a new station known by the call letters 2MT, at Writtle, a village near Chelmsford.[33] On May 11, 1922, it launched a London station, 2LO, from the roof of Marconi House, borrowing a page from KDKA with a blow-by-blow description of a boxing match at the Olympia between Georges Carpentier and Ted "Kid" Lewis. 2LO operated from Marconi House from May 11 to November 15, 1922, when it became the first station of the British Broadcasting Company.[34] Five days later, the Metropolitan-Vickers Company was licensed to begin test broadcasting from a station in Manchester (2ZY). But the Marconi broadcasts were far and away in the lead of the nascent industry.[35]

---

† Actually, the messages travelled over a combination of land lines and wireless to get from London to Geneva: by Post Office wire from London to Chelmsford, by wireless from Chelmsford to the Marconi receiving station near Geneva, and from that station to the League of Nations Conference Hall several miles away by land line.

On May 4, 1922, Postmaster General Frederick George Kellaway announced in the House of Commons his intention to allow the establishment of a limited number of broadcasting stations. Two weeks later, on May 18, Kellaway called the principal British companies engaged in wireless manufacturing to a private conference at the GPO "so that a scheme might be evolved" for broadcasting. In addition to satisfying the growing popular interest in listening, the press reported, radio could be employed "for a much more serious purpose." Right from these early days, the British government saw a strategic purpose for broadcasting; at any moment, the government would have the power to use broadcasting to transmit "information that is deemed to be of national importance."[36] This didn't preclude private sector involvement, and at this stage the Post Office (which attached jurisdiction over radio to its existing responsibilities for telegraphy) entered into complex negotiations with the six most important British electrical manufacturers (including some that were US-controlled). Marconi's was by far the dominant of these, and a compromise was brokered by which Marconi's patents could be used for national broadcasting without creating an effective Marconi monopoly.[37]

The issue of foreign involvement in the embryonic British radio industry arose almost immediately, and Marconi's managing director, Godfrey Isaacs, became a leading advocate for restricting the market to British-made products: "Either the production of sets for wireless broadcasting will be protected from foreign competition or the manufacturers will not be able to establish broadcasting stations," Isaacs told London's *Daily News* on July 14, 1922, warning that there was a "flood of German and American sets which are waiting to come in." Isaacs denied rumours that Marconi's was planning to import cheap sets from Hungary. On July 18, 1922, Kellaway outlined in the House of Commons the main features of a proposed scheme that had been approved by Cabinet; that same day, the government placed a two-year embargo on foreign-made sets in order to protect British manufacturers.[38]

By August, Kellaway was able to report that while no licences had yet been issued to establish broadcasting stations, he understood that the principal wireless manufacturers would be combining to form a company to provide broadcasting services.[39] At least one London newspaper reported—erroneously—that Marconi would be the chairman.[40] Finally, on October 18, 1922, the British Broadcasting Company was incorporated for the purpose of seeking a licence to create a national broadcasting service. Its shares were held equally by the six companies—Marconi's, Metropolitan-Vickers, the Radio Communication Company, British Thomson-Houston, General Electric, and Western Electric. Its chairman was Lord Gainford, a former postmaster general in Asquith's Liberal Cabinet.

A month later, it was the end of the line for Lloyd George's National Liberal Party, which went from government to fourth place in general elections held on November 15, 1922 (the day after the British Broadcasting Company's inaugural broadcast). Kellaway lost his seat—and, within days, was named to the board of the Marconi company. Once again, the Marconi company was in the ethical crosshairs of the British press. "Is the public service, then, to be regarded as a training-ground for the managers of competing or feeding industries?" thundered one newspaper.[41] Another coined a new term, "Marconism," to describe "a tendency that seems to be in danger of becoming a concomitant of political life, namely, that of making the entrance to parliament the short cut to commercial success."[42] The legacy of the Marconi scandal was still vividly in the public memory.

Kellaway would be remembered as the "Father of British Broadcasting,"[43] as the new Conservative government of Andrew Bonar Law (with Neville Chamberlain as postmaster general) went ahead with the plan to create the British Broadcasting Company. With state involvement as a proxy for corporate competition, this first version of the eventual British Broadcasting Corporation was thus formed as a manufacturer's consortium, on the basis of pooled patents and noncompeting stations. The Post Office—once again—had feared a Marconi monopoly and took the initiative to create a single authority under its own tutelage. A state monopoly was seen as the only possible alternative to a Marconi monopoly, which would have been the outcome of a purely commercial system. The new company was nevertheless granted exclusive rights, with broadcasting to be financed by a ten shilling ($2.50) licence fee paid by everyone who owned a receiver.

The British Broadcasting Company's mission was to build and operate "a public utility service for the 'broadcasting' supply to the public by means of wireless telephony and/or wireless telegraphy of news, information, concerts, lectures, educational matter, speeches, weather reports, theatrical entertainments and any other matter."[44] This broad public service mandate would be carried over when the "utility" was transformed into a public broadcasting corporation with a royal charter—the British Broadcasting Corporation, or BBC—on December 31, 1926. In both the United States and the United Kingdom, broadcasting was now a cultural industry, sharing that space with recording, cinema, the performing arts, professional sports, publishing, and the press. There was now, and forever more would be, a blurry line between entertainment and communication.

❀   ❀   ❀

The sudden mania for broadcasting seemed to come out of nowhere. Once the press took notice, its fascination with radio echoed its earlier fascination with wireless; this time, however, it was an entire industry and legions of users who were its heroes, not a single individual. Or almost, but not quite. Pioneers like Marconi and Tesla experienced a surge of interest, their most supposedly outrageous views, or even hints of them, trumpeted and encouraged by a sensation-seeking press. Marconi, meanwhile, tried to remain grounded. Asked to reflect on his "dreams" in a *New York Times Magazine* article provocatively entitled "Marconi Takes a Look into the Future," he stated to the contrary, "I live and work in the present."[45]

The proliferation of broadcasting as a viable and popular technology attracted the attention of the political class as well. In the postwar world order, radio seemed like a powerful new weapon, with ideological and propagandistic, as well as military and diplomatic, possibilities. One of the first coherent exposures of this potential was laid out in the memo from Walter Rogers of the US State Department to President Wilson on February 2, 1919 (referred to in chapter 25). It is worth recalling. Rogers had underscored the "opportunity for disseminating intelligence" presented by the "unlimited possibilities" of radio "for broadcasting messages to the ends of the earth," and suggested that the nations of the world should each nationalize their radio facilities and act together to develop "a truly world-wide radio service." Alongside "the steady extension of democratic forms of government and the increasing closeness of contact between all parts of the world," the ultimate basis of world peace, Rogers wrote, "is common knowledge and understanding. . . . Hence the distribution of intelligence in the form of news becomes of the utmost importance."[46]

Wilson had already formed his own impression of the power of radio after learning that his Fourteen Points were being received all over Europe in direct wireless transmission from the United States. His chief radio advisor, Admiral William H. Bullard, told broadcast historian Gleason Archer he had seen school children in the Balkans studying the Fourteen Points "as they would learn their catechism—made possible by the Alexanderson alternator at New Brunswick, New Jersey, which, defying all censorship, was stimulating in everybody everywhere a deep anxiety that the war should end."[47]

Rogers, a former editor of the *Washington Herald* and official of the Committee of Public Information (the US wartime propaganda agency) as well as technical advisor on communications at the conference in Versailles, was one of a small group of staff within the State Department who were developing a global vision of the scope of radio. Other members of the group included Ernest Power and Breckinridge Long (later ambassador to Italy in the 1930s[48]), both third assistant secretaries at State. The approach they developed to global communication policy became the kernel for the doctrine of "free flow of information," which later became the centrepiece of US foreign policy in communication.[49] Rogers and his colleagues believed that communication would be a decisive factor in the emerging global system, where it would play several key roles: as the informational infrastructure of the new global order; as the foundation for an important market in commodities; as the basis for dialogue within multilateral institutions; as a facilitator for expanding the postwar wave of democratization; and as a tool for the administration and modernization of empires. They also understood and articulated a comprehensive view of the barriers to communication that had built up over the fifty-year recent history of media—the monopolies, high rates, hidden subsidies, and propaganda purposes to which media had been put.[50] (Rogers is also considered to be an originator of the idea of international press freedom.)

Without seeing broadcasting as a separate sphere of activity per se, the State Department now became concerned that the major corporate players were trying to create a global wireless cartel: at this point, monopolies, not foreign ownership—as in England—was their target of concern. The growing power of RCA was a pivotal factor here. The company had a different view of globalization. It did not want open markets, multilateralism, free media, and democracy, but a version of globalization where private firms would organize cartels and manage markets much as the cable industry had done in the nineteenth century.[51] The view propounded by Rogers and taken up by Wilson was that stable markets required regulation and state oversight, but the idea of a global regulatory framework was anathema to the new communication giants like RCA (much as it had been to Marconi's at the turn of the century). This now became a global issue.

Paradoxically, a completely different model emerged in the United Kingdom, initially in the absence of social and political concerns. There, wireless telegraphy—and hence, radio—had required a licence since 1904, and the phenomenon of an "amateur" broadcasting frontier simply did not exist. Marconi closely followed developments in the United States, but his own interest in wireless voice transmission went in the direction of long-distance wireless telephony rather than "broadcasting," and his own research at this time was focused on the development of shortwave.[52] In 1919 the company realized that entertainment broadcasting augured a potentially important new market for equipment.

By 1922, millions of people in the United States, the United Kingdom, and, increasingly, elsewhere as well, were "listening in." In the United States, the cultural and economic importance of the new phenomenon was driven by Herbert Hoover, the secretary of commerce in Warren G. Harding's and then Calvin Coolidge's Cabinet, and the hero of European postwar recon-struction; it was validated commercially by the aggressive promotion of RCA. The public demand for radio receivers seemed insatiable—there had been nothing like it before or since (until the iThings of the early 2000s). US sales in 1922 totalled $60 million; they doubled the following year, to $136 mil-lion; then tripled, reaching $358 million in 1924.[53] In the UK, anticipating the demand, the Marconi company created first a department and then, in 1923, a subsidiary to design and produce its own line of "marconiphone" receiving sets; the company estimated that there were already between three hundred thousand and four hundred thousand receivers in use at that time. (The Marconiphone Company was sold to RCA in 1929; the sale included not only the trademark but also the right to use the signature "G. Marconi" on each radio set.[54])

The emergence of two opposing and contradictory models in England and the United States marked the history of broadcasting—first, radio, and later, television—during the following century. In the United States, com-mercial radio became part of the entertainment industry, while its educa-tional possibilities were lauded by social reformers.[55] Some hoped that radio might improve politics, make politicians less remote and more accountable, and even lead to political empowerment. Radio had a culturally unifying effect as a new collective entity took shape: the radio listeners' audience. The audience was huge; it was invisible; it was unknown. Radio became a symbol of progress and modernity as the corporate ambitions of the broadcast com-panies and the dreams and desires of the set-buying public merged seam-lessly.[56] In the United States, with the support of Congress, commercial broadcasting emerged triumphant over the intentions of the United States Navy, the Department of Commerce, and the State Department that radio should be used to further national interests. In the United Kingdom, however, a powerful counter-idea emerged: the idea of "public service broadcasting," referring not only to the sending and receiving of broadcasting messages under the auspices of government oversight, but to a mission of information and education as well as entertainment. The placing of broadcasting on a non-commercial footing was the key to the distinction.

In April 1923, barely four months after the launch of the British Broadcasting Company, the government struck a committee chaired by Major-General Sir Frederick Sykes "to consider broadcast in all its aspects." The committee members included W.H. Eccles, the former Marconi engineer and president of the Radio Society of Great Britain, MP Sir Henry Norman, and J.C.W. (John) Reith, general manager of the British Broadcasting Company. The broadcasting committee reported four months later, on August 23, 1923—a remarkably short time for such an extensive study. Its main conclusion was that "the control of such a potential power over public opinion and the life of the nation ought to remain with the State, and that the operation of so important a national service ought not to be allowed to become an unrestricted commercial monopoly."[57] The report included an important statement justifying the consideration of "wavebands" as public property: the fact that these were *limited*, a scarce resource. Their use should therefore be assigned according to the public interest.[58]

The British Broadcasting Company licence was due to expire on December 31, 1926, and in anticipation, another advisory committee, chaired by the Earl of Crawford, was established early that year. It took the logic of the Sykes Committee a step further. "Broadcasting has become so widespread, concerns so many people, and is fraught with such far-reaching possibilities," wrote Crawford, "that the organisation laid down for the British Broadcasting Company no longer corresponds to national requirements or responsibility . . . we are impelled to the conclusion that no company or body constituted on trade lines for the profit, direct or indirect, of those composing it can be regarded as adequate in view of the broader considerations now beginning to emerge."[59]

In short, although broadcasting remained non-commercial, the fact that it was carried out by a consortium of commercial companies was inappropriate. The US system of "free and uncontrolled transmission and reception" was seen as unsuitable for Great Britain, where a monopoly controlled by a single authority was deemed to be the best organizational structure.[60] After considering several alternatives, the Crawford Committee recommended that broadcasting should be conducted by "a public corporation acting as Trustee for the national interest, and that its status and duties should correspond with those of a public service."[61] It therefore recommended reconstituting the BBC from a privately held company to a public corporation, governed by a royal charter. The government accepted the recommendation and the British Broadcasting Corporation was launched on January 1, 1927. John Reith, general manager of the outgoing company, was reincarnated as first director-general of the BBC. The wireless manufacturers ceased to be directly responsible for broadcasting in the United Kingdom, although they enjoyed the phenomenal growth in demand for equipment. There were now more than two million British radio listeners. But broadcasting in Britain would be the purview of a public corporation, with not only a new legal status but also, under Reith, a new public service ethos.

In most of the rest of the world, the systems that emerged were situated somewhere between the US and British models. In many places, broadcasting was largely governmental, or modelled on the BBC notion of arms-length public service. But the most developed national system was in the United States, and there it was a commercial model that prevailed. Even in the United States, however, a general model for the regulation of privately owned commercial broadcasting had emerged; in fact, this model would soon be adopted almost everywhere in the capitalist world and its colonial dependencies: privately owned stations would be required to obtain licences to use the frequencies allocated for broadcasting by international agreement but managed at the national level in each country. Broadcasting frequencies soon came to be considered as scarce public resources, to be regulated by national authorities in the public or national interest. The allocation of broadcast frequencies *between* countries thus became a hotly contested international issue as well. One of the most intense sites of friction on this question was between the United States and Canada—two countries where present and former Marconi interests were competing for space in the new market for broadcasting.

Canada, where Marconi had established deep ties since 1901, provided an instructive example. The Canadian Marconi company began experimental radio broadcasts from its plant on William Street in the St. Henri district of Montreal in the fall of 1918,[62] but these could not be considered "programs" in the British or American sense. Canada had a licensing system similar to the United Kingdom's, and Marconi's station XWA was indeed authorized to experiment with voice and sound in December 1919, but the company's later claim that it began "broadcasting" at that time is based on the date of the licence, not on what the company actually did.[63] According to an internal company document, Canadian Marconi started producing "test programmes" that year, playing records and reading weather reports.[64] XWA was indeed the first station in Canada to experiment with continuous wave technology, in 1919 and 1920,[65] but as we have seen this was already being done by then in the United States. Furthermore, the station's early news bulletins were still transmitted in Morse code, and there was almost no audience for broadcasting in Canada beyond a small number of amateurs who were able to receive and understand the Marconi signal.[66] In 1922, now known by its modern call letters CFCF, the station moved into its first purpose-equipped broadcasting studio, in the Canada Cement Building on Montreal's Phillips Square.[67] There was certainly no fanfare comparable to that surrounding the November 1920 launch of Pittsburgh's KDKA, and Canadian Marconi's annual reports did not even mention the station until 1925. All that said, however, former Marconi news editor Arthur Burrows, first director of programmes for the BBC and then secretary-general of the International Broadcasting Union, credited Canada with having the world's first regular uninterrupted broadcasting service.[68]

Canadian commercial radio was quickly drawn into the US orbit, with RCA's David Sarnoff playing a catalytic role. Sarnoff worked closely with a technical whiz and former Canadian Marconi employee, Jacques-Narcisse Cartier, who was known colloquially as "Marconi Jack." According to Quebec radio historian Pierre Pagé, Cartier collaborated with Sarnoff on the development of the *idea* of broadcasting, leading to RCA's launch of stations in Philadelphia and New York in 1920 and 1921. In Montreal, the daily newspaper *La Presse* announced the creation of radio station CKAC, the first French-language station in North America, on May 3, 1922. Its director was Jacques-Narcisse Cartier. The *Montreal Herald* reported Cartier's vision for radio: "Radio is here to stay. Contrary to other luxury items, radio has become a necessity, a permanent fixture of household furnishing."[69] CKAC began broadcasting on September 27. More than sixty broadcasting licences were issued by the Canadian Ministry of Marine and Fisheries in 1922.[70]

With the emergence of commercial network broadcasting in the United States and public broadcasting in the United Kingdom in 1926, Canadian radio reached a crossroads. Canada had political, social, and cultural affinities with both systems. In 1928, a royal commission was mandated to propose a framework for the future organization of broadcasting in Canada. The three commissioners visited RCA's National Broadcasting Company (NBC) in New York, where they were shocked to learn that the network's plans included extending its coverage to encompass Canada; they also visited the BBC in London, and heard first-hand John Reith's concept of broadcasting as public service. They held hearings across Canada, consulted widely, and finally reported in 1929 that "Canadian radio listeners want Canadian broadcasting," and that the best way to ensure they got it would be to organize radio on the basis of national ownership and public service. It took another three years before the government followed up on the proposal, and then only under the prodding of a national lobby group, the Canadian Radio League, which convinced it that its choice was "between the State and the United States." In 1932, it created the Canadian Radio Broadcasting Commission, which was transformed four years later into the Canadian Broadcasting Corporation (CBC). But unlike in the United Kingdom—and despite enabling legislation which would have made it possible—the CBC never established a monopoly and Canadian broadcasting evolved as a hybrid system, incorporating parts of the British and American models.[71] This system is still in place to this day.

29

The Merger

The beam contract revitalized Marconi's standing as a visionary technical innovator. His stubborn persistence was already legendary, but in 1924, with the greatest innovation in communication since 1897, Marconi became one of the rare people to twice reimagine the way the world would communicate.[1] Within two years, the beam system became the foundation for the long-delayed, long-stagnating imperial wireless chain—and the prototype for global long-distance wireless communication. Extending out from Britain to Canada, Australia, India, and South Africa, it became one of the anchors of the emerging British Commonwealth. Once again, Britain seemed to be at the apex of global connectivity, and once again Marconi, however much he was slouching toward Italy, was at its centre, and the company was on its soundest financial footing in decades.

Marconi's colleague Richard Vyvyan observed that the long delay in establishing the imperial wireless chain had been a blessing in disguise, not only for Britain but for Marconi himself. Had the chain stations been established with longwave technology, their performance would have been far more costly and far inferior to what could be done with shortwave, regardless of whether the stations were built by the government or the Marconi company, Vyvyan wrote in a 1933 memoir. The long delays in building the imperial chain had the unintended consequence of a beneficial effect on both research and commerce.[2] Marconi's contract with the Post Office may have placed him at the centre of British imperial communication, but the company itself went further. Not only was it directly involved in every one of the links in the British imperial chain, it also built its own stations for communication with the United States, South America, and the Far East, extending the transoceanic network it had started in 1907 with the connection between Clifden and Glace Bay.

The sensational success of the beam system elevated Marconi to an unprecedented height in the public eye. It no longer mattered who had done what, how, or when in the messy history of wireless. Marconi was now ensconced in the pantheon of technical innovation with a handful of peers such as Bell and Edison, and he would remain there for the rest of his life. The giant corporations built on his inventions, and their rivals, the governments who sought to control the powerful forces they unleashed, would continue to wrestle in the commercial and political, national and transnational arenas of the 1920s and '30s, but Marconi's stature—if nothing else—was assured. He had provided the technical, and overseen the political and commercial basis, of the worldwide networked system of wireless communication he had first imagined twenty-five years earlier. The development of shortwave was his last defining moment.

The arrival of the beam system also disrupted the sleepy coexistence that had settled in between the cable companies and wireless during the previous two decades. Thanks to shortwave, wireless was now incontestably quicker, cheaper, and just as efficient as cable telegraphy; it was also the basis for the international extension of broadcast radio, which thanks to shortwave could now be transmitted worldwide. With the arrival of beam communication—the precursor of what we now call "telecommunication"—every country in the world also had to decide between private and public enterprise, and choose a company to either undertake or partner in its construction and operation. This was invariably one of the four giant companies that had already divided the world into spheres of interest.

The international consortium that carved up the world's wireless activities was created in 1921, with the sanction of, among others, the United States Navy. It consisted of the Radio Corporation of America, Marconi's Wireless Telegraph Company, France's Compagnie générale de télégraphie sans fil, and Germany's Telefunken. (The swift rehabilitation of Germany's role in global communication so soon after the war was particularly impressive.[3]) Under an agreement of October 14, 1921, a new entity was created, with these four companies as equal partners. Each one obtained the exclusive right to use the other companies' patents within its own territories, and "mutual traffic arrangements" would be made wherever possible. This highly pragmatic agreement transcended politics and commercial rivalry and was a reflection of the ascendancy of transnational business and diplomacy over the inter-imperial struggles that characterized the 1920s. It was nothing less than the first global radio cartel.[4] Its initialism, AEFG, reflected the

members' awareness of their place in the geopolitical order: the initials stood
for America, England, France, and Germany.

Each partner named two members of the governing board, which
would be presided by a ninth member, a nominally neutral chair who had
veto power if in his judgment an unfair measure was contemplated. The first
managing director of the consortium was E.J. Nally, late of both American
Marconi and RCA. The Americans were especially delighted with this arrangement. RCA had soared to prominence within two years of its creation
on the foundation of Marconi's US operations. Not only was it the peer of
its more venerable partners, it could actually aspire to control by its affinity
with the chair.[5] As one British opposition MP noted in a 1923 House of
Commons debate, the cartel was "one world-wide wireless combine with
increasing monopoly.... Competition has been eliminated."[6] The Marconi
company was the linchpin; it was the only one of the four companies with
deep ties to all the others.

In the United Kingdom, Marconi's had to counter charges that it
was "not altogether a wholehearted British concern," an allegation that it
contested vigorously. "That is absolutely untrue," Colonel Simpson told
the Cabinet committee during its historic April 1924 meeting with Marconi.
The British company had entered into arrangements with its French,
German, and American counterparts "under, as one might call it, a convention, to cover traffic working and to cover an exchange of inventions
in our respective territories." This was not at all a "trust," as had been suggested, and there was certainly no foreign control of Marconi's, he said.
Marconi's held a large interest in many foreign companies and was represented on their boards, "but not one of those companies has one single
share in the Marconi company and they have no representation in our
management."[7]

The world was becoming smaller, thanks in part to Marconi's technology, and also thanks to the corporate allegiances that technology made
both necessary and possible. Simpson could play an unequivocally patriotic
card; he had a distinguished war record in British intelligence, having been
involved in the creation of MI8, the section responsible for signals intelligence. Marconi himself was in a more ambiguous position. Considered
"foreign" in Britain, "British" in Italy, and "not American" in the United
States, he was indeed the first truly cosmopolitan communications entrepreneur. His personal interests lay in developing his system; for that, he
would work with whichever government offered support.

On November 13, 1924, after nearly fifteen years at the helm of MWTC and the associated Marconi companies, an ailing Godfrey Isaacs followed his doctor's orders and retired as managing director and deputy chairman.[8] He was replaced by Frederick George Kellaway, the former postmaster general who had joined the Marconi board after brokering the amalgamation of corporate interests that led to formation of the British Broadcasting Company. A long shadow was cast over the company's new success, when on April 17, 1925, Isaacs succumbed to his accumulated ailments. "It is hardly too much to say that the company owes as great a debt to Mr Isaacs on the business side as it does to Mr Marconi on the scientific side," the *Times* wrote in a long obituary.[9]* While aggressively promoting the company's corporate commercial interests at every turn, Isaacs had also loyally supported Marconi's core idea that the company was—first of all and foremost—an enterprise of innovation, and if it could turn a profit for its shareholders, so much the better. Some felt he had pushed the company too fast and too far, but Marconi would never again have an associate with such affinity and common purpose. While he had a cordial and constructive relationship with Kellaway, he soon, at first almost imperceptibly, began distancing himself from the company and its destiny, tied as that was to British domestic and imperial politics.

With the death of Isaacs, Marconi lost not only the pillar of his corporate empire, but his major remaining point of connection with England. The company still provided him with his main source of income and crucial support for his floating research lab (and the closest thing he had to a home), but he now began to slide closer to establishing himself permanently in Italy. Previously, with Isaacs looking after his affairs in London and Solari doing the same in Rome, he could literally sail between the two capitals; now the British end of his empire was rudderless. Intending to continue spending most of his time at sea developing his inventions, he became more dependent than ever on Solari, and, hence inevitably, on Mussolini.

Marconi's public enthusiasm for Mussolini and his fascist regime was evident in a remarkable, widely reported speech he made in Bologna on June 13, 1926, commemorating the thirtieth anniversary of his initial breakthrough patent. After telling his usual story and acknowledging not only Hertz and Maxwell but also "our great Bolognese Physicist, Augusto Righi," he described the work that had led to the recent success of his beam system.

---

* Isaacs left an estate of £200,000, some $15 million in today's money (*LT*, June 23, 1925).

By a fortuitous linguistic quirk, the Italian translation for *beam*, as Marconi used it, is *fascio*. "Yes!" Marconi announced, "*Sistema a Fascio*...I always claim for myself the honour of having been the first Fascist in radio telegraphy."[10] Marconi was rarely ironic or playful in his speeches. He was now fully identifying with the fascist regime.

The beam system proved so successful that all the contracts for high-power communication between Britain and the Dominions were rewritten.[11] The Dominions were inclined to develop their own communications capacities and this played in Marconi's favour. The building of new stations began in April 1925, and the first links in the new imperial chain were opened in Canada eighteen months later, in October 1926, with two stations at Drummondville and Yamachiche, near Montreal; the GPO was responsible for managing the British end.[12] Stations were under construction in India, South Africa, and Australia as well. Marconi's personal announcement of the opening of the beam system to Canada was widely reported around the world, and Mussolini was among those who cabled congratulations. Feeling vindicated, Marconi left for Rome on November 13, 1926. On November 17, Mussolini made his first appearance in the Italian Senate following an attempt on his life, and paid warm tribute to Marconi, who was present, for his recent discoveries. Marconi was reportedly "visibly moved." On November 21, wearing full fascist insignia, Marconi made another important speech on beam communication, in Rome. He then returned to London.[13]

In his *"fascio"* speech in Bologna in June 1926, Marconi mentioned some of the various uses to which broadcasting was being put, including keeping the public calm "during any popular disturbance which interferes with the peace-making work of the Press." He gave the example of the recent general strike in England.

A general strike had been called by the British Trades Union Congress (TUC) on May 3, 1926, in support of a million coal miners who had been locked out as they were protesting a planned wage reduction and worsening working conditions. Earlier in the year, the government had created a royal commission chaired by Sir Herbert Samuel (the former postmaster general who had awarded Marconi the original imperial chain contract in 1913) to look into problems of the mining industry. After the Samuel Commission reported in March, mine owners announced a reduction of wages and, unable to reach agreement with the unions, locked out the miners on May 1. The general strike was then called. The Labour Party tried to broker an

agreement but failed after printers of the *Daily Mail* refused to print an editorial suggesting that a general strike was a revolutionary move aimed at
destroying the government.[14] Conservative Prime Minister Stanley Baldwin
(who had replaced Ramsay MacDonald in November 1924) called the strike
"the road to anarchy and ruin."[15] The "general" strike was limited to certain
key sectors: railway and transport workers, printers, dockers, ironworkers,
and steelworkers; 1.7 million workers in all went out. On May 7, the TUC
met Samuel and worked out a set of proposals for ending the dispute. The
miners' union rejected the proposals, but on May 12, the TUC decided to end
the strike, without an agreement, after concluding "that a satisfactory basis
of settlement in the mining industry can now be formulated." Not surprisingly, the official *British Gazette* called it a "surrender" and an "unconditional
withdrawal."[16]

The British government, police, and intelligence services played a
tense and complicated cat-and-mouse game with the unions and the public
during the ten days of the strike. A confidential letter in the Marconi Archives
in Oxford reveals that the company played a key role in detecting unau-
thorized radio transmitters for the police. (The Archives also contain an
organizational chart of the company's communications personnel, prepared
to support its services to the authorities during the strike.) "You will be glad
to know that the D.F. [directional finding] was entirely successful in locating
the block of buildings in London in which an unlicensed transmitter was
installed," the author wrote to the head of the company's field and air division on May 16, 1926, three days after the end of the strike. "The transmitting station was then raided whilst transmission was in actual progress."[17]

The letter was written from a private address in Croydon by an intelligence communications specialist by the name of Harold C. Kenworthy, a
somewhat cloak-and-dagger figure who seems to have been simultaneously working for both the company and the London Metropolitan Police
(Scotland Yard). (According to the *Historical Dictionary of British Intelligence*,
Kenworthy headed a secret station established in 1923 by the London Metropolitan Police in Grove Park, to intercept broadcast signals being transmitted on behalf of the Comintern from an unlicensed transmitter operated
by the British Communist Party in Wimbledon.) Kenworthy's letter continued:
"There are several things which make it imperative that nothing should leak
out as to where it was found and that it was located by D.F." The apparatus
had been built by a company employee named Wright (in all likelihood
George Maurice Wright, one of the founders of British signals intelligence

during the First World War, and a future head of the company's research department) and would probably be "taken over." "The most important incident is the special message that Sir Wyndham Childs [assistant commissioner of the London Metropolitan Police] received from HM the King thanking the wireless staff for the successful conclusion of the work they had carried out and that His Majesty had been following our efforts with great interest. It is understood that this piece of work was actually the most important thing in the whole strike."[18]

The company's ties to British intelligence have not been fully explored. The First World War contributions of key employees like H.J. Round and Adrian Simpson have been accounted for earlier, but the relationship continued and, indeed, flourished in the peace that followed. In addition to the Grove Park intercept station operated by Kenworthy, which was still in service during the early part of the Cold War, the British Secret Intelligence Service (SIS) in the 1920s used a Marconi factory in Barnes High Street as a control centre for transmissions received from overseas.[19] But the company's most prominent known service to British intelligence came in 1952, when it received a classified contract (code-named "Satyr") to uncover the secrets of a microphone and transmitter that had been planted in US Ambassador Averell Harriman's residence in Moscow.[20] The problem was cracked by a Marconi radio technician by the name of Peter Wright, who was recruited by the SIS in 1955 and went on to great notoriety as the author of the 1987 bestselling memoir of his career in counter-intelligence, *Spycatcher*. Wright was the son of George Maurice Wright, who had developed the Marconi directional-finding system used by Scotland Yard during the 1926 general strike. Peter Wright joined the Marconi company in 1946, and wrote that he began working for UK intelligence soon after that.[21]

Marconi was sailing on a research trip in the Mediterranean during most of this time (May 3 to 17, 1926), but he arrived in England in time to chair a company board meeting on May 20, at which the board was informed, in general terms, of the "special services" the company had provided to New Scotland Yard during the general strike. Kellaway read the letter from Childs "expressing his appreciation of the services rendered to his Department during the emergency and stating that as a result of the experience gained Scotland Yard would be equipping themselves with the Company's latest apparatus."[22] The news couldn't have been better. Not only had the company provided much-appreciated service to the forces of order, but it would also enjoy an interesting commercial benefit as a result. This was Marconi's favourite formula—having his technology yield commercial benefits while serving the institutions of power.

The 1926 British general strike was an important moment for the Marconi company. It provided an occasion to test the application of some of its latest technology on directional finding (or what we would today call GPS) to political surveillance and police work (and in this regard was a precursor to the collaboration of US information companies in providing metadata to the National Security Agency in the 2000s). The watershed role of broadcasting in channelling the government's viewpoint to public opinion would be mentioned for years in tributes to Marconi, the company, and the medium itself.[23] The secret contribution to Scotland Yard and British intelligence that has just been described was unknown until now.

❧   ❧   ❧

Since ceding the day-to-day management of the company to Godfrey Isaacs in 1910, Marconi had occupied the position of chief engineer and technical advisor, at a salary of £2,500 per annum (about $200,000 a year in today's money, or about half the salary of the managing director). In this role, he enjoyed "the entire charge of the technical staff of the Company," with power to hire and fire, set remuneration, and move people about from station to station as needed. He also continued to occupy the position of chairman. Most important, he was free to set his own agenda, to devote whatever time he felt necessary to whatever activity, and to hold outside appointments as well. In addition to his salary, he also received a research budget, an allowance for the expenses of the *Elettra*, and a personal secretary.[24]

Marconi later rewrote the story of his involvement with the company during these years; in a 1934 deposition aimed at establishing that he was not a British resident, for tax purposes, he declared: "After the war my association with my native country became considerably closer than it had been for many years before. I began to fill various official positions in Italy and later became interested in the Fascist regime and took part in political work in which I became engrossed as well as in my scientific work and as regards the latter I formed the impression that I received more encouragement for my work at the hands of the Italian Government than I seemed to receive in this country."[25] According to his own records, he attended fewer than half the company's board meetings between 1909 and 1917, a proportion that continued to drop as time went by.[26] He also declared that he gave up his residence in the United Kingdom before 1915 (which was not quite true, as he continued to maintain a home at Eaglehurst as late as 1918, although he was indeed rarely there), and that he had not owned or rented any property in the United Kingdom since 1923. His visits to England were now brief, sporadic, and "for purely social or diplomatic purposes."[27]

Marconi's personal records show that between 1905 and 1921 he regularly made loans to the company, and bought, sold, loaned, and transferred shares to family members in England and Italy; he often transferred shares from his name to those of nominees, so that when he sold shares other shareholders would not think there was a lack of confidence in the company. Between 1908 and 1909 alone, "for the sole purpose of assisting" the company, he sold or pledged more than twenty-nine thousand shares at a price of ten shillings (half their par value). At one time he held only forty-eight shares. During the three-year period encompassing 1907 to 1910, he paid the company £34,525 for shares, and loaned it £48,200, for a total of more than £82,000 ($6.5 million today)—a fortune far greater than his original stake, which he believed "saved the Company." The result was, for him, "on the whole a considerable loss." Marconi's relations with the company were enigmatic to say the least; in addition to his financial dealings, Marconi's notes show that he offered to resign as a director at least twice, in 1914 and 1926.[28]

As of January 1922, Marconi received a total of £4,500 a year as technical advisor to the Marconi companies, plus £3,000 a year for expenses.[29] The company was also paying for his personal staff. (In MWTC's organizational chart of April 1923, Marconi is connected directly to the board, and holds no specific position.[30]) According to a note in the Marconi Archives, Marconi had almost no shares in his own name at this time (903 preference shares and no ordinary shares); however, Barclays Bank held some fifteen thousand ordinary shares as his nominee, and more than ten thousand preferred shares were still being held in trust by his brother-in-law Charlie White (presumably under the terms of Marconi's 1905 marriage settlement on Beatrice).[31]

The success of the beam system created an industrial paradigm shift that had a profound and paradoxical effect on Marconi's personal and corporate future in Britain. By the mid-1920s, Marconi's personal position was already shifting. The transition from Isaacs to Kellaway had marked a subtle restructuring of the company as a modern corporation. Where Isaacs had been an imaginative managing director, Kellaway was a more conservative, conventional executive; one of the first things he did was set up a committee to reorganize the company's financial practices and accounting procedures, and consolidate its myriad activities that were now scattered around some sixty companies in twenty countries. In the words of one old hand and long-time observer, "the time had arrived for Marconi wireless to serve the needs of finance, instead of the other way about as it had done so far."[32] For Marconi himself, this was a much less interesting prospect. In July 1927, the board reconstituted itself with the addition of new members from the financial community, and Marconi stepped down as chairman. He was replaced by a prominent and successful self-made entrepreneur, Andrew Weir (Lord Inverforth), founder of an eponymous firm of shipowners and merchants, who had been brought in to the War Office as minister of munitions in 1917, and then chaired the commission charged with disposing of surplus war property after the war.[33] Never one to care much for the duties and intrigues of the boardroom, Marconi was now freed to devote himself fully to the technical side of the company's work (as well as to his increasing activities in Italy).[34]

On July 26, 1927, at the same board meeting where he resigned as chair and proposed Lord Inverforth to succeed him, Marconi informed the board that he had been invited by RCA to attend the forthcoming (fourth) International Radiotelegraph Conference in Washington, DC. The board agreed that it would be desirable that both Marconi and Kellaway attend on its behalf.[35] After the two Berlin conferences of 1903 and 1906, and the London conference of 1912, it was clear that some form of international regulation was required to traffic the increasingly congested airwaves, especially with the vast multiplication of stations expected with the advent of shortwave. After the arrival of radio broadcasting and then shortwave, the environment had changed so dramatically that a whole new way of thinking about communication was required. The Washington conference would lay down certain regulations, principally dividing the spectrum into sections allocated for specific uses.[36] The main purpose of the conference was to establish international regulations for the use of the radio spectrum.

Aside from the need for technical regulation, in the world political context of 1927, a number of important principles came into play. A League of Nations study a few years later noted that the international nature of wireless technology meant that "any scheme for its organization on international lines must arise from the development of national broadcasting undertakings," a major affirmation of the sovereign rights of nations to determine the use of the spectrum on the territories they controlled.[37] The fourth international radio conference had been due to convene in 1917 with the US government as host, but the plan was disrupted by the world war. After the war, the Allied powers established a committee to look at updates to the 1912 London convention, but their work was continuously eclipsed by ongoing technological innovations. The 1927 international conference in Washington was the largest intergovernmental conference ever to take place outside Europe. Nearly three hundred delegates from eighty countries and representatives of more than forty communication companies would take part.

The conference opened on October 4, 1927, with an address by US President Calvin Coolidge (Americans were listening to the New York Yankees defeat the Pittsburgh Pirates in game one of the 1927 World Series that day; led by Babe Ruth, who had just hit a record sixty home runs in the regular season, the Yankees would defeat the Pirates in four, to the delight of a live radio audience reached by two national networks, NBC and the Columbia Broadcasting System, or CBS). Scarcely thirty years after the transmission of information by radio began, "Communication is one of the important supports of civilization," Coolidge told the delegates. Radio held a particular promise of "reaching into [the] dark places of the earth" because of its low cost. "An instrument of such far-reaching magnitude, fraught with so great a power for good to humanity, naturally requires national and international regulation and control," Coolidge said. "In many fields our country claims the right to be the master of its own independent development.... But in the radio field the most complete development, both at home and abroad, lies in mutual concession and co-operation.... Your main object will be to raise this great industry into the realm of beneficent public service." Coolidge's thrust was echoed by Secretary of Commerce Herbert Hoover, the conference's official chair as representative of the host country. Recalling that in 1912, at the time of the previous conference, radio was only a means of communication with ships at sea, Hoover remarked that it had now reached the point where it was necessary to regulate the air in order to "avoid chaos." The number of channels was limited and they had to be organized. The conference's main task was to decide the allocation of uses to different classes of frequencies.[38]

The ambivalence—between control and freedom—in the US position was noticed, among other places, in Canada, where the *Ottawa Citizen* editorialized that "there are far more dangerous rivalries in the air, such as the competition between private interests" creating monopolies for the wireless business of the less developed economies of the colonial territories and the Far East. "It is doubtful whether the control of radio broadcasting can be safely left in the hands of private corporation interests," wrote the paper's editorialist, signalling a global communication divide that continues to hang over international negotiations in communication governance to this day.[39]

Marconi arrived in New York on October 6; the conference was already under way (his duties in Italy had kept him busy there until the end of September). He gave an interview to the *Evening World* in which he affirmed the perennial nature of newspapers and the written press. Radio could never take the place of the newspaper, he said. People might listen to news on the radio, but they will always go to the newspaper to find out more. The newspaper had the distinct advantage of constituting a record, he said. "You cannot paste radio announcements in a scrapbook; the newspaper puts the news down in black and white."[40]

On October 15, 1927, Marconi was the keynote speaker at a banquet given by RCA for the conference delegates at New York's Plaza Hotel. It was a fairly succinct speech. Marconi recalled that when he had started out there were no laws, no regulations, no restrictions governing radio waves, but it soon became imperative to "regulate and discipline" their use. "Space is becoming very congested over a wide range of wave-lengths, and as we have only one ether for us all, we find ourselves compelled to come to an agreed understanding in order to prevent what would otherwise result in a state of chaos." That said, he continued, we still know far too little of the mechanism governing the propagation and behaviour of the waves, and it would be dangerous to decide on international legislation and rules based on the limited knowledge we possess. Latitude and liberty is required, he suggested, and urged the delegates to allow the fullest possible scope for research and experimentation from both a scientific and a commercial point of view. In short, while recognizing the need for regulation, he was advocating against it. He added, almost as an afterthought, that radio would surely be a weapon of peace.[41]

Two days later, Marconi gave a more substantial lecture to the American Institute of Electrical Engineers and the Institute of Radio Engineers.[42] Here he traced the history of radio, but emphasized his most recent experiments since about 1922 (beam system, multiplex system, et cetera). "We are yet, however, in my opinion a very long way from being able to utilize electric waves to anything like their full extent, but we are learning gradually how to use electric waves and how to utilize ether and, thereby, humanity has attained a new force, a new weapon which knows no frontiers, a new method for which space is no obstacle, a force destined to promote peace by enabling us to better fulfill what has always been essentially a human need: that of communicating with one another." He didn't talk about his companies, their lobbying efforts, the international regulations, or propaganda. His focus was on the ideal and the promise of radio.

Marconi returned to England at the end of October, while the conference continued for seven weeks before producing a final convention signed by eighty governments. It made one fundamental decision that has had lasting consequences. Rather than divide the spectrum among countries, which would have inevitably given rise to a power grab, to rivalries, and to injustice, it was decided to divide the channels into groups, each group for a particular use, and to leave each nation free to use them accordingly, for these specific services. The only remaining issue—and it was a big one—involved the allocation of frequencies within classes among neighbouring countries where use of a particular frequency by more than one country would create interference. This issue would have to be resolved by regional conventions.[43] Hoover considered the conference outcome "a safe middle ground between avoidance of restriction and the maintenance of orderly traffic." And indeed it was—a *Pax Americana* in international communication that has structured the agenda for everything from transborder telecommunication rates to Internet domain names down to today. With some modification, the international radio regulations of 1927 are still essentially in place.[44]

Marconi arrived in London in time to attend a board meeting on November 1, 1927. He then took to his bed for the rest of the year. He had experienced violent chest pains on the journey over from the United States and his doctors told him he had suffered a heart attack. As a result, he now missed a series of meetings that dramatically changed the destiny of his company. Over Christmas, with Marconi still ailing,[45] Kellaway and Inverforth had a serendipitous meeting with Sir John Denison-Pender of the Eastern Telegraph Company at which the idea of a merger between Marconi's and the cable companies was broached.

By 1927, cable had lost half its international business to shortwave wireless, and the cable companies were claiming they might be forced into liquidation. Cable was a far more reliable ally to the British government than Marconi was; the spectre of the industry floundering, even going under, created "a crisis of national, even imperial, proportions" for Britain.[46] The cable companies had been floating the idea of a merger for some time, but there was no obvious advantage to Marconi's. The company's new leadership, however, had no ties to the old ways of doing things; from a shareholder perspective, in fact, there were a number of potential benefits. The company was still in a precarious financial position; it could not count automatically on government support as new uses and markets for wireless developed; there was a lingering antipathy toward it in some political quarters; and the Eastern's deep coffers made the outcome of a protracted inter-corporate struggle fraught and unpredictable.[47]

Previously, such a venture would have been unthinkable. The Marconi company's most important goal since 1897 had been its own survival as an independent entity; this common purpose united its principal actors despite their varying personal goals—to have an independent instrument for research and technical development, in Marconi's case; to run a company with full autonomy, in the case of men like Henry Jameson Davis, Cuthbert Hall, and Isaacs; to maximize the chance of huge profits, for the company's main financial backers. The company had been entertained with merger suggestions in the past, and had never considered them seriously; its principals, Marconi first among them, were always ready to put their personal resources on the line rather than give up any of the company's independence. This now changed. The company leadership was in the hands of a far more cautious and pragmatic group than ever before; with two former Cabinet ministers in its key executive positions, it was for the first time led by political insiders, not corporate frontiersmen like Hall and Isaacs. Marconi himself, spending ever more of his time in Italy and his energy on his Italian affairs, was no longer in a position to devote himself exclusively to the company; indeed, he no longer had the interest to do so.

While the Eastern and Marconi's pursued their negotiations in the back channels, the cable companies lobbied openly, through the press, for the government to rationalize the communication industries. The Dominions were also agitating. In January 1928, the British government called an Imperial Wireless and Cable Conference to examine the situation that had arisen as a result of the introduction of wireless beam communication. On March 14, 1928, while the conference was still meeting, Denison-Pender, chairman of Eastern, and Lord Inverforth, chairman of MWTC, announced a provisional agreement, creating a new company merging all the international communications of the British Empire (based on a scheme drawn up by accountants, as some old wireless hands noted with disdain).[48] The notion of "national champions"—industry powerhouses large enough to do battle with other countries' firms—that had long prevailed in the United States and in France now triumphed in Britain as well.

In July, the conference submitted its report, and on August 2, 1928, the House of Commons debated and approved its recommendation to support the formation of a new company to acquire the capital of Marconi's Wireless Telegraph Company and the Eastern and Associated Telegraph Companies.[49] The merger capitalization was valued at £53,700,000, a colossal sum at the time—more than $4 billion in today's money—but dwarfed by today's information industry mergers. A company called Imperial and International Communications Limited (IIC) was formed to acquire "the communications assets" of the two groups—an interesting formulation, for it meant the exclusion of Marconi's increasingly important manufacturing arm. The formation of IIC was designed to ensure that Britain continued to be the leading power in global communications. IIC would be owned by a holding company to be called Cables and Wireless Limited (C&W).[50]

Wireless technology was now institutionally subordinate to cable. The merger radically modified the character of the Marconi company, which would now be limited to research and manufacturing, while IIC would handle actual long-distance wireless communication (sharing control with the Post Office—another example of private-public partnership). Marconi's would henceforth have little to do with the application and exploitation of its own innovations.[51] Its role would be much smaller than before, despite the fact that the bulk of the traffic would be carried by the Marconi beam system (now owned by IIC). The deal negated everything Marconi had worked for on the corporate level since 1897. Competition between cable and wireless would cease. Marconi's was a minority shareholder in the new company—"a junior partner in an organization where the cable interests predominated," in the words of his associate H.M. Dowsett[52]—and Marconi himself had little left to do.

Personally, however, the merger was a financial windfall for him. On August 1, 1928, Kellaway wrote to Denison-Pender, chairman of the merged board: "Marconi's continued association with the Merger is vital if we are to keep level with world developments in telegraphic and telephonic communication." Marconi's contract bound him until December 31, 1931, "but if he feels that the terms he is proposing are not acceptable to the Merger and they are either rejected or consent grudgingly given, the Merger will certainly lose the best he is capable of for the unexpired term of his agreement and at the end of the period he will join up with a competing organization." Denison-Pender replied immediately: "I expect we shall have to accept your strong recommendation, but the figure is very big."[53] His contract was

completely rewritten once again; his salary as technical advisor soared to
£20,000 per annum ($1.5 million in today's money) and his research allow-
ance was increased to £5,000. He was also named chief technical advisor to
all companies of the C&W group; on the surface, this appeared to be a sinecure—the expectations were minimal, the deliverables nil—but it was
crucial to the public face of the new company, and especially to the shareholders of MWTC.[54] The company did insist, however, on a clause to cover
the contingency of Marconi's being unable to carry out his duties for a
prolonged period because of illness, an insistence that annoyed him terribly;
but for the company it was a real concern, in view of his heart condition.[55]
Under the new agreement, C&W also received all the patent rights and
benefits Marconi had previously assigned to his own company in 1897 and
confirmed in 1922.

The Marconi board was reconstituted again in the wake of the
merger. Despite the company's shrunken role, the number of directors
swelled to twenty-three; of these, only one—Marconi—could be said to
have any technical expertise in the field.[56] Marconi continued as technical
advisor, not only to the Marconi company but now to the whole C&W
group. On June 26, 1929, he assumed the new, largely ceremonial position
of president of MWTC as well; Inverforth and Kellaway continued in the
operational positions of chairman and managing director, respectively.
Marconi now had to swallow a similar fate for his UK operations to what
he had gone through a decade earlier with the effectively forced liquidation of his assets in the United States. Once again, he was the victim of his
own success.

The C&W merger marked the definitive transition in Marconi's corporate presence in England, but the shift was already perceptible to those
around him. He still made a point of attending the highlights of the English
social season, such as the royal summer regatta at Cowes (where he could live
and entertain on his yacht), but even relatively casual observers now noticed
a change. According to one visitor to the *Elettra*, Lord Louis Mountbatten,
"The impression I had was that...he had lost close contact with all the
developments although they were being carried out under his aegis."[57]
Publicly, Marconi's aura was as powerful as ever, his opinions still eagerly
sought and willingly given,[58] but after 1928, with the merger firmly in place,
he began disengaging even more from the company's projects and interests.

The merger was nonetheless a development that Marconi accepted, if
with resignation. In a sense, it made things easy for him. His lack of affinity

for the cable interests was well known. There were fewer specific demands on his time and energy, less need for his presence in England, and no more expectation that he would step up to support the company financially. To the contrary, ironically, he was now for the first time a highly paid corporate executive.

30

The Anchor

With the cable and wireless merger, Marconi's new corporate status allowed him, for the first time, to begin thinking seriously of resettling in Italy. This was a welcome opportunity as his personal life (one couldn't really call it "private," as every step he took made news) had been a tempest for the previous decade. He really hadn't stopped running since the end of the war. Despite his mastery of compartmentalizing, his heart attack was a clear sign that life's realities were catching up to him.

To pick up the thread, in 1919, after years of forcing his family to live a nomadic if gilded existence, Marconi had finally decided to buy a house in Rome, on Via Pietro Raimondi, bordering the Villa Borghese gardens. Beatrice was beside herself with joy. "In those days Villa Borghese was much more private than it is now and the gardens were beautiful and secluded," she wrote in her memoir. "I was in the 7th heaven to have at last a home of my own." However, they kept it—the only house Marconi ever owned—only two years. Beatrice did everything to try to dissuade him from giving it up but it was no use. "The selling of it was a terrific blow to me and one that proved disastrous to our marriage as it broke up our having an established home."[1] Beatrice wrote of "a strong influence at work behind him trying to make him take this step."

The "influence" was Marconi's twenty-five-year-old mistress Paola Sanfelice di Viggiano, a Neapolitan princess and one of Marconi's most complicated extramarital love interests. The relationship was more serious than Marconi's usual inconclusive flirtations. Paola was married to the aristocratic Marquis Luigi Medici del Vascello, a rising nationalist member of parliament (as well as, like Marconi, an unsuccessful suitor of actress Francesca Bertini),[2] and the Medicis were frequent guests on the *Elettra*, where Marconi and Paola carried on their affair in full view. Beatrice, who usually tolerated Marconi's diversions as long as they remained nominally innocent or at least discreet, had a particularly dark spot in her heart for the stylishly acerbic Paola, whom she found a repugnant rival. Marconi seemed to think nothing of having his wife, his mistress, and her husband all in close quarters on the yacht. Beatrice quite understandably was not interested in this arrangement.[3]

After the "tragic" sale of the house on Pietro Raimondi to the Dutch consulate in the summer of 1920, "our vagabond life started again...we practically lived separate lives," Beatrice wrote.[4] For Marconi, it was a relief. "Now I will return to living *a zonzo* [*en flâneur*, or like a boulevardier]," he told Solari happily, adding that he had made a windfall gain of 1.5 million lire on the sale (the equivalent of what he had received for his original patent twenty-three years earlier).[5] After Beatrice refused Marconi's suggestion that they take an apartment in the nearby Medici property, she moved to the Hotel de Russia while Marconi stayed at the Grand Hotel, living there with Paola and, as Beatrice discovered later, "spending vast sums on her." Beatrice "made various efforts to get him to keep the family together at least in name," but this was effectively the end of Marconi's married life. After a few weeks, Beatrice took the children and moved to Florence, where they lived in a small pension for three miserable years. Marconi continued seeing Paola at least until 1923 (his diary mentions her five times that year), when she fell in love with the classical pianist Arthur Rubinstein, with whom she had her only child. The Medicis' marriage was eventually annulled by the Vatican and Paola continued to live what Rubinstein's biographer Harvey Sachs termed "a life with no center of gravity."[6]

Beatrice had a good grasp of Marconi's state of mind as their marriage was winding down: "He wanted at this particular period to be quite free, unfettered by even the simplest exigencies of married life. There was no reason at that time for us to separate. He was as free as air. I never made scenes or tried to force his hand. I made no objections and was ready to overlook his curious way of living a completely bachelor life. I only begged and implored him to keep a home for the eyes of the world and for his growing children." It was no use. Between bouts with the apparently mercurial Paola, there were others as well. He would turn up from time to time, be particularly charming, bearing an expensive gift, usually at the end of some adventure which he would recount in great detail, asking Beatrice's advice; she felt she was playing "the role of a mother more than a wife."[7]

On February 8, 1922, Marconi wrote to Beatrice from his sick bed at the Savoy Hotel. The letter was typical, in tone and in substance, of their relationship at this time—courteous and business-like, he was basically dictating settlement terms: "Dearest B, I am now able to state what I can do in the way of placing your money matters on a more satisfactory and stable basis than they have been in the past. I feel that what I propose is generous and fair and absolutely the utmost I can afford to do, in proportion to my means at present." He offered to allocate her a not-ungenerous sum of £4,400 per annum, almost equivalent to his company salary at that time, to cover her rent, car, and expenses for herself and the children. "I think this arrangement should enable you to be much better off than you have been in the past. . . . Of course, should anything happen or go wrong in my affairs I might have to reduce the amount, but at the same time I don't mind telling you that should things go well, I might even increase the amount if necessary. . . . I hope you will be pleased, as I've done the utmost possible."[8]

It was a dutiful arrangement. After she replied that she was pleased, he followed: "It will be a strain to keep it up but I think I can do it, and believe it's right that I should do all that's possible for you and the children . . . [I] hope the arrangement will make you feel happier and more independent, and enable you to save something now and then for a rainy day."[9] However, having ensconced Beatrice in Rome, he was, in view of his problems with Italy's liberal government, now thinking about leaving. Paradoxically, she was becoming more entrenched there just as his alienation was reaching its height. They spent a bit of time together during Marconi's visit to the United States in June 1922, as we saw earlier, but there was no real hope that they would ever be a couple again. They were still entangled, however. She wanted a stable home and couldn't possibly build one without his agreement and financial support. On August 17, 1922, he wrote to her: "Have just wired you that I'm favourable about taking the house in Rome, but it's not probable that I shall ever stay for long in Rome in the future. As you know I'm getting out of all my Italian interests. But it may nevertheless be wise to have the house for the furniture and for us if we should be there. . . . if you like the house in Rome, take it."[10]

He became more sanguine about Italy once Mussolini came to power at the end of October, and on December 13, 1922—two days after his deposition in the Banca di Sconto corruption case—Marconi and Beatrice filed a document with a notary in Rome in which they declared that in view of his long absences for reasons of work, they agreed to live separately, that the children would spend most of their time living with her, that he would pay their expenses, that she didn't care what he did when he was away, and that she expected the same treatment from him.[11] The agreement meant two things as far as Beatrice was concerned: Marconi would never take the children from her and she could now freely frequent her new suitor, Liborio Marignoli, the Marchese di Montecorona, who had for some time been showering attention on her that she was now ready to reciprocate.[12]

Occupied as he was in the coming months with his activities in England, his research, and negotiations with the new Mussolini government, Marconi continued managing Beatrice's housing and financial arrangements by long-distance correspondence. Then, on February 28, 1923:

> Dearest B, have just received here forwarded from London a rather excited letter from you, without either date or address.
>
> What you have been told is <u>entirely and absolutely false</u> as I have never told Paola anything of the kind in regard to any intention of mine of as you say catching you in '*flagrante*.'
>
> The only conversation I had with her, in which you were mentioned, was that she told me she had heard you were, or had been at Perugia, and that a certain person was there too. As I expected that that would be the case, '*non ci feci caso*' ['let's not make an issue of it'].
>
> ... we both suffer from the vile and putrid atmosphere of gossip and *maldi-cenza* [slander] which ferments everything and everybody in Italy just now. ...
>
> What I told you of myself during our conversations in Rome is true and is that I have not at present any flirt or attachment for anyone outside ordinary friendship. But perhaps you think you may know better.

And he signed, "Aff^tly yrs, Guglielmo."[13]

A few days later, Beatrice's lawyer made Marconi a proposition. She wanted a divorce, but he was being typically rational about it. "The decision is so awfully important that it must be very carefully thought over," he wrote to her on March 6, 1923. He was trying to talk her out of it. "What I am trying to find out here [in London] is as to whether a divorce obtained at Fiume and approved by the Italian courts would be effective in England," he wrote in a second letter on March 18. "It is an important point because if you should think yourself free and then ever marry again you might be arrested for bigamy if you came to England." One of the curious new twists of the Italian situation was that civil divorce was now possible for citizens of Fiume, and Italian nationals were rushing to take advantage of it. On March 29, he wrote again, clearly ambivalent about divorcing: "You will, I am sure quite understand how serious, and at the same time ridiculous, the position would be, especially from your point of view [since she is the one who wants to remarry] if the contemplated divorce should be when granted only valid in Italy and in a few other countries. Such a very important step in our lives should only be taken if we are reasonably certain of a complete success."[14]

Beatrice was pressing him, to his annoyance. On April 11, 1923: "According to what you say I hardly can believe you quite understand me, besides showing an attitude which I can only consider as unfair." He chastised her for suggesting that her lawyer's knowledge extended to English law. "If we obtained a divorce in Italy which was not valid in other countries we would indeed be the laughing stock of the world! Aside from this matter you can not truthfully say I have not recently done all you have asked for." On June 6: "I am still quite at a loss to understand what is the great trouble and danger you write so persistently about, unless it is caused by some abnormal *mentalità* of some one under whose influence you may happen to be."[15] He keeps referring, cryptically but unambiguously, to Marignoli.

The correspondence went on in this vein for a few months, while Beatrice moved to Spoleto, the ancestral seat of the Marignolis, in Umbria. He wanted to see her and the children but he was reluctant to visit her there: "Spoleto and indeed Italy is such a small place socially and the object of my visit would be misunderstood." Despite the extreme intensity of his business in Rome (he had just refused Mussolini's invitation to the presidency of Italo Radio), he had seen his own lawyer, who told him there was no doubt they could get a divorce in Fiume and have it approved by the Italian courts; however, it would then be almost impossible for him to ever regain his Italian nationality, he was told. "I'm now trying to get an opinion from the Government in regard to what they might do [about his regaining Italian citizenship if he became *Fiumano* for the purposes of divorce], but it's rather a difficult moment for as you know I am not on the best of terms with Mussolini."[16]

The ever-astute Mussolini himself assured Marconi that there would be no problem retaining his citizenship[17]; there might have been a price to pay for that. The status of Fiume had been temporarily settled by the Treaty of Rapallo, as we've seen, signed November 12, 1920, between Italy and the new state of Yugoslavia. One of the many scattershot agreements that complicated the politics and geography of Europe in the wake of the Paris Peace Conference, the Treaty of Rapallo partially recognized Italy's postwar claims. Rapallo established the Free State of Fiume, which was recognized by the United States, France, and the United Kingdom, ending D'Annunzio's occupation and creating, among other things, a haven for Italians seeking to obtain a civil divorce. The only thing necessary was for them to become citizens of Fiume.

On December 18, 1923, Marconi and Beatrice filed a document declaring the terms of the divorce they were seeking, jointly by mutual accord, in Fiume. Marconi agreed to pay Beatrice three million Italian lire (around US$150,000 at the time——or roughly what he had paid for his yacht in 1919), in two instalments. Because his working conditions obliged him to travel and to move continually, and for that reason he could not oversee the education and upbringing of the children, it was agreed that they would remain entrusted to their mother; Marconi agreed to provide, in monthly instalments, £1,600 ($8,000) per annum, for the support, education, and instruction of the children (until each reached the age of majority or married), and to pay the costs of the judgment.[18] (This was considerably less than he had proposed to Beatrice the year before.)

They had to move quickly. Mussolini was determined to reclaim Fiume for Italy and was negotiating a new agreement to that effect with the Yugoslavs. The Treaty of Rome was signed on January 27, 1924. That same day, Marconi arrived in Trieste from London; on January 29, accompanied by Solari and his lawyer, he travelled the short distance to Fiume where he took out "temporary citizenship" and the divorce was settled in a few hours. He then headed back to London by way of Trieste.[19] The divorce decree was issued on February 12—one of the last acts of the Free State of Fiume before the Treaty of Rome was ratified on February 22, and the city was annexed to Italy.

Marconi didn't have any problem retaining his Italian citizenship—with the annexation of Fiume, the question was moot, anyway. Less than three weeks later, on March 3, 1924, Beatrice married Liborio Marignoli and became the Marchesa di Montecorona. She informed Marconi immediately. In London, he annulled the holograph will he had made in 1917 and then wrote to her on March 19, somewhat passive aggressively: "Please forgive me not replying sooner to your letter but to tell you the truth I was somewhat uncertain as to how to address you on the envelope, as I do not know whether or not you are already generally known in Italy by your new name...my most serious wish is that the step you have taken will be conducive (in so far as that is possible in this world) to your permanent happiness...."[20] He seemed to be having difficulty accepting the day-to-day implications of the divorce.

Beatrice, meanwhile, had tied up her funds in a seaside villa owned by Marignoli at Capo di Posillipo, in Naples, and was pressing Marconi for early remittance of the second instalment of the divorce settlement. He wasn't pleased. "Am much perturbed over what you say in regard to money matters," he wrote on June 14. The company was doing badly; he'd had to sell shares at a low price in order to pay her the first two million lire, "and if I paid the further million at once I would be left without a penny and no prospect of leaving anything to the children, even if I had to sell the yacht!" But by the end of the year he was happy to tell her he hoped to pay the third million "in accordance with what was stipulated when certain proceedings were decided upon."[21] He still couldn't bring himself to write the word *divorce*.

Divorce (and in her case a new husband) notwithstanding, Marconi and Beatrice continued a kind of co-dependency that had marked the last, dysfunctional years of their marriage. If anything, their relationship became more amicable. After the divorce Beatrice found "he was never nicer, more considerate and kind. We seemed to come closer than ever before." Marconi continued to write to her regularly, even suggesting he visit her and her new husband, an erstwhile acquaintance, at Spoleto. "We dissuaded him from this although he thought it was the most natural thing to do," Beatrice remembered. However, she did continue to care about him. In January 1925, when Marconi was ill and alone in London, Beatrice offered to come and look after him. He was touched and wrote to reassure her that it wasn't necessary. "He seemed much brighter, freer and happier than I had seen him for years," she wrote in her memoir, but he was lonelier than ever, and before long, his newly eligible bachelor status was making headlines once again.[22]

At the age of fifty, Marconi was officially a bachelor. Despite the attention he continued to receive from women, he was not a natural bon vivant and could easily withdraw into his thoughts, distant from wherever he was at the moment. He was still accumulating artistic protegées. Solari described dining with him quietly at the Savoy on a rainy evening in December 1924 when he spotted a beautiful Argentinian actress across the room and approached her. Solari wrote that Marconi never spoke of her again, but that he, Solari, ran into her in Paris some months later and realized that she and Marconi had met several times since then. As he often did, Marconi had taken the actress under his wing to help her in her film career.

Soon, always susceptible to the flirtatious flattery and distractions offered by very young women, he found himself enmeshed in a bewildering array of entanglements. On April 8, 1925, the *Daily Express* reported that Marconi was about to be engaged to an eighteen-year-old girl from West

Cornwall, Elizabeth (Betty) Paynter. In fact, it was the day after her eighteenth birthday, as Marconi noted in his diary. The paper said Marconi was an old friend of her parents, and had known Betty since she was fourteen. "Nothing definite regarding an engagement can be said yet," her mother was quoted.[23]

The Paynters were an old family from Saint Buryan, a village five miles west of Penzance on the way to Land's End. Their ancestral home, Boskenna, sat high on a cliff frequented by Cornish owls, rooks, and horses, and an array of interesting human fauna as well. The novelist Mary Wesley lived at Boskenna for a time and recalled that Betty's father, Colonel Camborne Paynter, was known to be something of a swordsman, the seigneur of the area whose distinctive features could be seen on many of the children of Saint Buryan. Wesley also described him as "an eccentric, hospitable and incredibly tolerant father."[24] Wesley's biographer, Patrick Marnham, wrote that "Betty had been brought up to hunt for a wealthy husband ... [she] was not beautiful but she had a fine slim figure, huge dark eyes, thick black hair and a large nose that she had broken while out fox hunting." It was the look of a typical Marconi girl—in fact, like his mother when she was young; many of Marconi's girlfriends as well as his two wives bore an uncanny resemblance to the young Annie Jameson. "She also possessed an inexhaustible sexual energy that made her very attractive to men," Marnham continued, suggesting that a minor character in Wesley's *A Sensible Life* seems to have been modelled on Betty Paynter: Joyce was "'a very agile girl' who flitted from flower to flower and brightened people's lives, or, to put it another way, fucked with 'all and sundry.'"[25]

Marconi knew the Paynters from his many trips and long stays at Poldhu. Lately, he often docked the *Elettra* at Falmouth, in southwest Cornwall. On March 30, 1925, on his way back to England after a long cruise that took him to Sicily and Cannes, he wrote to his friend Lady Falmouth asking if she would look after Degna for a week or so after Easter "as I will not be able to have her with me all the time on account of my work."[26] He didn't want his daughter in the way while he entertained his lover; they were both the same age. On Easter Monday, April 13, 1925, the *Evening Standard* ran a photo of Marconi and Paynter, with the headline "Marconi's romantic Easter visit." Compartmentalizing as he did so well, he wrote to Kellaway that he had been slightly ill but was on his way to Poldhu to consult with Franklin about their ongoing beam research. Kellaway wrote back with characteristic directness: "Are you a budding bridegroom or are you sick?"[27] (Marconi left Penzance on April 16, 1925.) Colonel Paynter angrily denied the engagement reports (according to Mary Wesley he was so furious he horsewhipped a group of reporters who had nothing to do with the story when they turned up at Boskenna for a fox hunt), but not before the news was published in Rome. Marconi, annoyed, wrote to his lawyer in Italy, asking whether he should take any action against the press.[28]

Surprising as it may seem, the affair with Betty Paynter was not the only one Marconi carried on at this time.[29] Early in June 1925, while he was in London, Marconi received a letter from a lover with child-like handwriting who signed herself "Tuff," and was staying with family or friends in Branksome Park, a suburb of Poole: "My own Darling," it begins. "Darling I adore you more than any one on earth and I always will for ever & ever and I just know we will be happy too I miss you so & am longing to have you back by jove it will be a relief when we've married." Marconi wrote at least three letters to "Tuff" between December 1925 and September 1926. The notation style he uses in his diary suggests that "Tuff" and Betty Paynter were not the same person.[30]

We know by now that Marconi was susceptible to infatuation with the attractive young daughters of people he found interesting. In fact, his diary for 1925 is full of names and birthdates, some of them easy to identify, some of them more cryptic—the easy ones are his two Betty's (Paynter and his goddaughter Betty Clover), but there is also a Myriam, an Ann, a Kathleen, and a recurring Lois (an item going back to 1923, when, at the age of twenty-two, she suddenly replaced Paola as his favourite, or at least most frequent diary entry). One of the more enticing was "Cri Cri"—possibly Cristina Casati, whose £250 account with a London decorator Marconi had recently guaranteed.[31*] Cristina Casati, a twenty-four-year-old beauty with violet eyes, was the daughter of Luisa Casati, an outrageous figure who collected exotic animals, wore live snakes as jewellery, lived in a *palazzo* on the Grand Canal in Venice (the one later made famous as the home of Peggy Guggenheim), hosted the Ballets Russes, was openly bisexual, and was a long-term lover of Marconi's soulmate D'Annunzio. Marconi's diary has entries for "Cristina" in March 1921, December 1923, and March 1925 (before he met the Cristina he would later marry), and ten entries for "Cri Cri" between October 25, 1925, and July 23, 1926.[32]

---

* As cheap as he was with his family, Marconi was notoriously generous, to the point of extravagance, with his love interests. During a two-month period in 1923 his bill for jewellery from Cartier was close to £1,000—around $75,000 in today's currency (OX 22/179).

Marconi's courtship of Betty Paynter, and his reaction to the attendant publicity, highlighted the anguish he often suffered with these relationships – or at least, his ambivalence. One might wonder about his true motivations but they were compelling enough that he felt he had to clarify them to Beatrice in August 1925: "Don't be surprised or upset if you hear I have become engaged to Betty Paynter. I care for the girl, an awful lot, more than I ever thought I could, and she for me. I've been fighting myself over this for a very long time, but am afraid it's no use. Of course, it may be said she is too young. But you too know how one can feel. She is an only child and by no means poor even now. There would of course be no question of settlements from me, or anything of that kind if I married her. After all, even you perhaps know how lonely I am." A month earlier he had written to Degna: "P.S. It's not true, what the papers said about my engagement to Betty Paynter, but I'll tell you more about the whole thing when we meet."[33]

According to Beatrice, Marconi was always asking her for romantic advice, but this was a bit too much. In early October, when he visited her in Florence and asked her what people were saying about this affair, she asked back: "Why Guglielmo do you come and ask me about such a delicate question and he replied Because I know you are the only person who will tell me the truth."[34] Beatrice was stunned; she received the news with dismay and replied with unaccustomed sharpness:

> I am surprised to hear you have decided to take the step you write me of in your last letter and create new ties and most probably a new family. I would like to wish you every happiness but this news distresses me for I wonder after all the years we were together when your own desire expressed continually was for freedom to concentrate on your work as your family impeded and oppressed you, why you should suddenly feel this great loneliness and need of a home—this craving for fresh ties! These ties were eventually what broke up your home and ended in our divorce. I fail to understand.[35]

She didn't need to be concerned—at least not about Betty. Marconi was distressed when he received Beatrice's letter.[36] She knew him better than anyone and she was the only person whose judgment he trusted, or indeed, cared about. He also desperately wanted to settle down. His diary notations indicate that he was in a frenetic state; on some days he would write to as many as five different women, including Beatrice; on March 18, 1926, for example, he records: "Scritti Bea-Tuff-Lois-Cristina-Cri Cri." This "Cristina" was someone new. Marconi met Maria Cristina Bezzi-Scali in March 1925, shortly before her twenty-fifth birthday, and first mentioned her in a letter

to Beatrice on April 5, 1925, in passing, as part of a list of visitors he had recently entertained on the *Elettra*.[37] She obviously made a powerful impression. Even as he was steaming toward Penzance for his assignation with
Betty Paynter, he recorded in his diary on April 1, 1925, underlining the
entry in heavy bold ink: "25th Birthday Cristina." She was a bit older than
his usual infatuations, so unusually so that when news of this romance came
out newspapers invariably reported her age as much younger, an impression
that stuck and in some cases took decades to correct.

Degna Marconi described Cristina as "a quiet and serious girl...the
antithesis of the gay and sophisticated women Marconi cultivated at that
time." He had to tread delicately; "she was not a girl who appeared in public
unchaperoned." To get close to her, he made friends with her parents, inviting them to visit the yacht and sometimes, when they were all in Viareggio,
seeking them out at a small local restaurant, the Savoia, that the Bezzi-Scalis
frequented. "Father was like a schoolboy, hoping to catch glimpses of her,
his romantic attachment feeding on her inaccessibility.... Cristina was decidedly unworldly—she had seldom traveled far from Rome —and to her,
Father was a knight in shining armor." Marconi asked Degna what she
thought, and after reflection she solemnly approved, hoping that a nice girl
like Cristina would make him happier than "the fly-by-night ladies with
whom he was solacing himself." Except for her beauty, Cristina, "thoroughly nice...fresh and dewy," was a new type for Marconi, "completely
naïve about the ways of the world he frequented, more so, even, than his first
wife had been." Marconi desperately wanted to be married and the only
thing standing in his way with Cristina was the fact that he was divorced.[38]

The Bezzi-Scalis were a traditional Roman family, part of the "black
nobility" (*nobiltà nera*), the part of the Roman aristocracy that remained
faithful to the pope after 1870, did not recognize the new Italian state, and
occupied most of the key lay positions in the papal administration.[†] Members
of this group included the Orsini, Colonna, Ruspoli, Tedesche, Guglielmi,
Pacelli, Lepri...all names one encounters in the thin-air stratosphere of
Roman society to this day. Cristina's maternal line, the Sacchettis, an old
Florentine family, were one of the most influential of the group, one of four
families of even higher nobility known as the Marchesi di baldacchino
(Marquesses of the Canopy), who had the privilege of serving in the pope's

---

[†] Until 1968. The term refers simply to the colour of priestly dress, thus distinguishing the group
from the "white nobility," which chose to be loyal to the royal House of Savoy.

antechambers. The black nobility's base of influence was the Vatican, where Cristina's family was represented on both sides. The Bezzi-Scalis didn't accept civil marriage, let alone divorce.

Cristina's parents were basically of Marconi's generation, although their forebears had been firmly anchored on one side of the turmoil that led to the dissolution of the pope's temporal power and the establishment of the Italian state in 1870, while Marconi's had managed to hedge their bets and remain in the middle. Cristina's mother, Anna Sacchetti, was five years younger than her future son-in-law; she was born in 1879, the daughter of Marchese Urbano Sacchetti (namesake of a famous seventeenth-century cardinal) and Princess Beatrice Orsini (whose family occupied the ancient Teatro di Marcello in the centre of classical Rome). The Sacchettis formed a branch of the descendants of the papal Prince Barberini, one of the most powerful and fabled families of old Rome. Cristina's father, born in 1869, was Count Francesco Bezzi-Scali, a brigadier general in the Papal Noble Guard (originally a cavalry unit, then a mainly ceremonial unpaid volunteer corps). Bezzi-Scali had been a member of the Camerieri Segreti Partecipanti, the inner circle that waited on Pius XI and accompanied him in his movements. In 1911 he had been part of the papal delegation to the coronation of George V, along with the young Monsignor Eugenio Pacelli, the future Pope Pius XII. Among other things, as a lay member of the "papal family," he was exempted from the worldly activities demanded of ordinary citizens of the Italian state, such as jury duty.[39]

The Bezzi-Scalis and their four children lived in a seventeenth-century late baroque *palazzo* at number 11 Via dei Condotti, a stone's throw from the Spanish Steps and just across the street from one of Rome's most famous watering spots, the Antico Caffè Greco, where Cristina's mother was an afternoon *habituée*. Built in phases between 1597 and 1708, the mansion comprises a square block with façades on three streets. In the late nineteenth century it was the site of a famous literary salon called "Il Casino degli Inglesi"—a name that is difficult to appreciate in English as one of the common meanings of *casino* in Italian is "brothel." The *palazzo* was acquired by the Bezzi-Scalis in the early twentieth century. Today its main entrance is on one of Rome's most fashionable shopping blocks, through a discreet portico between storefronts for Bulgari and Angeletti.[40]

Things between Marconi and Cristina developed slowly at first. Cristina described their first meeting: "I was introduced to Marconi who looked at me with interest and gave me a particularly charming smile ... I was immediately

fascinated by his great personality and charm and his rather English elegance." Soon they met again in Rome at the home of her friend Maria
Cristina del Drago, whose mother used to give afternoon parties at her
mansion in Via Quattro Fontane, which Marconi often attended when he
was in Rome. "He possessed a mysterious force of attraction which drew
people to him and a marvelous sense of humour," wrote Cristina.[41]

Marconi was still trying to sort out his complicated love life. On
October 24, 1925, he told his diary *"scritto Cristina e Cri Cri"*—perhaps in
an effort to keep them straight. A week later it was "Cristina B," to distinguish her from the other one, and continued that way until the end of 1925.
But by January 1, 1926, the romance was alive. "New year entered with C,"
he wrote on New Year's Day, and after that there is at least one mention a
week of Cristina Bezzi-Scali, more than fifty in all through 1926.[42] It is a
pity that we don't have access to Marconi's 1926 letters to Cristina, because
his letters to Beatrice during that same year are among the most interesting
he ever wrote.[‡] Once he had made up his mind that he wanted to marry
Cristina, he applied himself to making it happen the way he always approached
whatever he wanted, sparing no effort, leaving no string unpulled. On June
27, 1926, he wrote to her from London: "Dear Cristina . . . as you know, I am
determined (as soon as I am completely free), to ask you something very
serious and very important. The greatest, most important and serious question that a man can ask and a woman can answer in all her life."[43] Getting
himself "completely free" was easier said than done. Marconi and Beatrice
had managed to obtain a civil divorce that was recognized both in England
and Italy, already no mean achievement; but the Catholic Church would
not tolerate a marriage with a *divorziato*, and Cristina Bezzi-Scali could not
be married anywhere but in a Catholic Church.

Somehow, the original documents regarding Marconi's divorce from
Beatrice ended up in the possession of their youngest daughter Gioia; there
is a fascinating document there that seems to have been copied from a published text, three pages written by hand, in pencil, on school notebook
paper. Here one reads: "In Church law there are just as many back doors as

---

‡ Marconi's letters to Cristina are in the possession of their daughter, Princess Elettra Marconi
  Giovanelli, who has never let anyone see them. Elettra includes only one of the 1926 letters in
  her edition of her mother's book (M.C. Marconi 1999, 26). Cristina wrote that she had over two
  hundred letters from Marconi, most of them written from London before and during their engagement. Elettra has published brief excerpts from thirty-three of these letters (one from 1926,
  twenty-four from 1927, and eight from 1934–37; M.C. Marconi 1999, 26–43).

in other law. One just has to have the key to open them. This key is delivered to highly placed persons either by reason of their traditionally good relations to the papal chair or by reason of certain reciprocal favors or it is put at their disposal against deposit of a large sum. For the rich and for members of ruling houses the rule of the indissolubility of a Catholic marriage was never valid."[44] This text, which reads like a religious equivalent to Machiavelli's *The Prince*, describes precisely what Marconi set out to do. He began putting together documents and a plan for convincing the Vatican Sacra Rota to annul his marriage.[45] (*Annul* is not quite the word here; the goal was to get the Rota to declare that there had never been a marriage at all.)

On May 30, 1926, Marconi met in London with Francis Cardinal Bourne, the Archbishop of Westminster and Britain's highest Catholic cleric.[46] Bourne was a conservative, even by the standards of the Catholic Church, and was opposed to anything modern, be it the consumption of alcohol, interfaith dialogue, ecumenism, birth control, and especially divorce. This was going to be a very hard sell; about the only thing Marconi had going for him was that Bourne, like him, was the grandson of a Dublin merchant. Marconi then discussed the matter with Beatrice in Rome; she had nothing to gain from his scheme (which would bastardize their children in the eyes of the church) but was willing to facilitate his happiness, as she told herself. They would also have to involve others, close friends and family members who would be able to testify to the nature of their marriage. He followed up their meeting in Rome with a list: "I think the witnesses whom it might be well to have called might be: Lucius [O'Brien, the current Lord Inchiquin], [Beatrice's uncle] Charlie White, Lady Leicester, [Beatrice's siblings] Barney and Clare...." He asks her to write to each of them saying he would be coming to see them to explain what it is all about.[47] He is recruiting her to assist with his plan while, typically, keeping full and absolute control.

Marconi was confident his plan would work, although he hadn't even proposed to Cristina, as his June 27 letter to her makes clear. He went about assembling his witnesses, in view of a hearing set for July 7 before the Roman Catholic Ecclesiastical Commission at the archbishop's residence. He also began coaching Beatrice for her *interrogatorio*, which would take place in Spoleto. "The most important thing for you to say is that <u>before</u> we married...we were quite agreed that in the event of our marriage not turning out to be a happy one, we would have availed ourselves of the possibility of divorce, and that our marriage was not entered into with the idea that it was indissoluble in the sense understood by the Roman Catholic Church."[48] This was one of the longest, most detailed, and thoroughly energetic letters he ever wrote to Beatrice. He thought of everything, covered all bases, and programmed her to do exactly what he believed necessary to achieve *his* goal. Divorced, remarried, and with problems of her own, she agreed to go along. His final instruction to her was: "Don't show this letter to anyone [emphasis in original]."

The day after the hearing at Archbishop House, Marconi wrote again to Beatrice. He had been asked what he thought of the words "till death do us part" in the Church of England marriage service. "I said that in my opinion those words although retained in the Church of England services, are not considered to be binding under all circumstances, even by the Church itself, because the Church of England is a Church established by the law of England, which admits divorce, which the same Church recognizes and in many cases approves, and that knowing this I did not consider them binding in the literal sense any more than the wife's promise to obey."[49] This remarkably modern and legally sound argument, if entirely self-serving (with echoes of Henry VIII), had only one flaw: it did not address the precepts of the Catholic Church, which had to be convinced that no marriage had in fact taken place. The case then turned around the church's definition of marriage.

The ecclesiastical court heard a host of O'Briens testify that Beatrice's mother had been dubious about the marriage; although they too had nothing at all to gain, Marconi had excellent relations with his former in-laws and found them most willing to help. Donough O'Brien told Marconi he was prepared to say that their mother "would not have given her approval if we had got married in a Church or country in which divorce would have been impossible, and that this possibility of divorce was in their minds all the time." If this were the case, what kind of marriage *was* it? Was there ever really love or any hope or intention of permanence? There was never any mention of the substantial settlement Marconi made for Beatrice before the marriage, nor of his will, which left everything to her and their children. Be that as it may, Marconi again cautioned Beatrice to "do try to confirm all I have said" when she was called (emphasis in original).[50]

The documents from the Westminster hearings were sent to Spoleto but then there was a lull. Marconi wrote Beatrice on August 10 to say he had been told she might be called any day now; obsessing, he reminded her once again: "Please remember all I have already written you in regard to the matter and particularly in regard to our understanding or agreement to divorce in the event of our marriage not turning out to be a happy one." He was optimistic. "From all I can hear, the case, so far, has gone very well for us, and they now only need your testimony before coming to a favourable decision." But now that the case had moved into the hands of Italian authorities there were inevitable delays; the papers needed to be translated, Beatrice had to be summoned. Finally, in mid-September, she was called and gave her evidence.[51]

On November 10, 1926, Marconi wrote again to Beatrice. He seemed relaxed and upbeat. His work was going splendidly, he had "lots of new inventions," and relations with the company were less tense. He also had news: "I'm awfully glad to let you know I've just heard from Msgr Surmont that the Court at Westminster Cathedral has granted the religious annulment of our marriage, in so far as it concerns the Roman Catholic Church." The decision still had to be confirmed by the Vatican and this could take another two or three months. "In the meantime of course you could not marry in Church," he wrote—not that she had any interest in doing so; this was typical Marconi, projecting his desires on to her. But it was already more than half the battle won. "All this too has taken up a good deal of my time."[52] Relaxed and upbeat as he was, Marconi was also keenly aware that the biggest hurdle still lay ahead. He faced it in typical fashion, taking the high road. He asked for an audience with the pope.

The publicly accessible file on the Marconi marriage case in the Vatican Secret Archives illustrates the bureaucratic and political dimensions of the case from the perspective of the Holy See. On November 9, 1926, Marconi's letter requesting an audience with Pius XI was transmitted by an anonymous Vatican functionary to Monsignor Camillo Caccia Dominioni, the pope's majordomo. On November 23, 1926, the head of the Vatican administration, Secretary of State Pietro Gasparri, cabled Cardinal Bourne in London: "Please wire me whether Marconi has introduced his matrimonial case and whether the Court has granted a decree of nullity."[53] (According to what Marconi later told Beatrice, Gasparri had set the whole thing up, suggesting that he ask for the audience.[54])

Marconi met with the pope at 6:00 p.m., on Wednesday, November 24, 1926. Somehow, the press got wind of it. In his diary, he recorded: "6 pm audience Pope—(Pubblicity [sic]. Regrettable)." Two days later he wrote to Beatrice: "As you may have seen by the papers on Wednesday evening I was received by the Pope. He was quite charming and I found him quite broad minded." Incredible as it may seem (what, after all, was the purpose of the meeting?), he added: "Of course I never mentioned anything about our case....Don't tell this to anyone." The case was now in the hands of the Vatican. Marconi informed Beatrice that she would soon receive a summons "which Avvocato Ferrata is bound by their rules to send you." But, he said, Ferrata's instructions were that Beatrice should reply that she would not appear but "rely on the justice of the court." However, "if the invitation or *citazione* should be by a judge of the court, then you will have to attend. I hope I have made myself quite clear." It was comprehensible, but anything but clear, and he added, again, in a postscript: "Please don't mention to anyone that Avvocato Ferrata does not want you to appear on his *citazione*."[55]

There was now a new wrinkle. On November 24, the same day Marconi met with the pope, the Sacra Rota announced the annulment of an equally high-profile marriage, between Charles Spencer-Churchill, the ninth Duke of Marlborough, and American railroad heiress Consuelo Vanderbilt. Some of the characteristics—notably, the class origins of the supplicants—were similar, but there was a twist; in the Marlborough case, it was claimed that the wife had been forced into a marriage she did not want by her mother and that she had stayed with her husband only because she did not realize the marriage was invalid. The Marlboroughs separated in 1906, were divorced in 1920, and both parties had since remarried; now they both wanted their new marriages recognized by the Catholic Church. The most striking similarity between the two cases was the string of witnesses who came forward in each case to affirm that the couple had agreed that the marriage should be annulled if it turned out to be unhappy.[56]

The Marlborough case set off a hue and cry. In the new democratic spirit of the 1920s, the idea that the Catholic Church seemed to be supporting two sets of values, one for the rich and one for everyone else, was received with cynicism. "It is an ancient commonplace among those who are not members of the Roman Catholic Church that she makes a practice of circumventing her doctrine of the absolute indissolubility of marriage," wrote the London *Spectator*. "The annulment of the Duke of Marlborough's marriage to his first wife, however, has created more astonishment and bewilderment among non-Roman Catholics than any previous annulment."[57] Marconi couldn't see it. "I can't understand all this fuss over the Marlborough case," he wrote Beatrice, "as the decision only concerns the R.C. Church, and no other, and does not affect their children in any way."[58] He was being just a bit disingenuous; if his first marriage were to be annulled, his own children would be illegitimate in the eyes of the Church of Rome.

Beatrice still had not been called before the Rota and her new husband, Liborio Marignoli, was becoming increasingly exasperated by what he (rightfully) saw as the harassment of his wife. Beatrice showed him a letter in which Marconi apparently stated that if their marriage were not annulled, "he would take reprisals against the children" (his precise words, according to Marignoli).[59] Unable to remain silent in the face of such threats, Marignoli wrote to Marconi stating that he would not convince Beatrice to give evidence by resorting to illegal and inhuman means. "You are a powerful man and perhaps it is easy for you to commit such an injustice. But do not forget that your skin can be pricked the same as mine." Two days later Marignoli received a call saying that Senator Marconi wanted to speak with him; they met the following day at the Grand Hotel in Rome. Marconi reproached Marignoli for writing what he called such an aggressive letter, and Marignoli suggested Marconi think about the gravity of his own letter, adding with a smile: "God forbid that I had lost the letter and it had ended up in the hands of journalists." At the end of their conversation Marignoli returned Marconi's letter to him, with the suggestion that he think of what the world would say if it got out, and Marconi returned Marignoli's to him stating that he would like to be able to say that he had never received it nor read it.[§] The only thing missing from this baroque exchange was a challenge to a duel. But Marconi was comfortable enough with the outcome to write to Beatrice afterwards: "Today I had a very nice talk with Marignoli."[60]

Marconi testified to the Rota that from the very beginning he had had grave doubts that the marriage would be happy, as he didn't think that Beatrice liked him well enough. When he expressed his doubts to her, she confirmed that she was not quite sure of her feelings. Beatrice said Marconi had always had her believe he accepted the Anglican rites, which accepted the possibility of divorce. This was the agreement with her family, of which her siblings were aware, she said. They both testified that this was not a vague but a precise agreement. The court also heard testimony from Marconi's in-laws Donough O'Brien and Moira Hervey-Bathurst (herself divorced), as well as Luigi Solari, the only non-family member called as a witness. They all repeated the same mantra. Donough confirmed the view that there was a verbal agreement to

---

[§] This explains why the offending letter from Marconi to Beatrice is not in the FP archive with all the others; if Marconi kept it, it has long since disappeared.

the effect that the couple agreed they could divorce if they were unhappy. Moira said her sister loved Marconi, but not as he loved her. He was smitten, not she. She reiterated the party line: the couple had agreed that they could be divorced, their church allowed divorce, and their mother would never have allowed the marriage if she thought it was irrevocable.[61]

On April 11, 1927, the three-member Rota handed down its decision, accepting the proof of the nullity of the marriage. According to the ecclesiastical court of the Catholic Church, there had never been a marriage between Guglielmo Marconi and Beatrice O'Brien. The judgment was published on April 30, 1927, and the Rota's file on the case is still secret, but an official summary of the decision is accessible online—a wonderful example of the contradictions of privacy and transparency in the world of modern communication made possible in part by Marconi.[62] The published summary is in Latin, but extracts from the verbatim witness testimony are, oddly, in French and Italian, including ones that were almost certainly made in English. The annulment proceeding showed Marconi at his most determined, crafty, and ruthless, and of course he got his way. To many it reflected poorly on the church. To this day, the Marconi case is cited as precedent in examples of church favour of the privileged.[63]¶

From the moment the decision was announced, and to this day, there has been a feeling in Rome that Marconi literally bought his annulment. There is no concrete evidence to back this up, but a document in the state archives in Rome demonstrates how deeply the conviction runs. In 1927, Mussolini's political police opened a file on Marconi that they continued to expand for the rest of his life and even beyond, until the end of the regime. The very first item in the file is dated May 6, 1927, and records the unfavourable reaction *at the Vatican* to the announcement of Marconi's engagement. "In fact—it is said—a 'black' family should be able to do better than intermarry with a *divorziato*.... Marconi greased the Rota, which dissolved the marriage of the great inventor.... It is a very bad precedent.... But with money you can get everything; and when they see money at the Vatican they swoon." It remains to be seen, the police report continued, which church personality will sanction the annulment by performing the wedding and whether the newlyweds will be received by the pope.[64]

❋   ❋   ❋

¶ Strictly speaking, Marconi's marriage was annulled by the Court of the Roman Catholic Archdiocese of Westminster on the ground of defective intention, and the decision upheld by the tribunal of the Sacra Rota in Rome.

Italy's public security police force, the Pubblica Sicurezza (PS), was formed in the 1880s and charged with the repression of subversive activity—a euphemism for containing the anarchist and syndicalist groups at the heart of Italy's social conflict that reached a peak after the end of the First World War, when "bolshevism" joined the list of subversive threats. Soon after Mussolini came to power in 1922, he formed an extra-legal secret police force, which he called the Ceka (borrowing the title from Vladimir Lenin). The Ceka operated independently of the police—officially, it did not exist—but as the Fascist regime expanded its reach, the distinction blurred and the opponents of fascism became the enemies of the state. Totalitarianism meant that no form of dissent, however small or insignificant, could be tolerated. Police activity grew beyond the containment of presumably subversive elements; it extended to everyone and everything "of interest."[65]

By November 1926, the transformation of fascist Italy into a totalitarian state was complete. This was marked by the creation of two new, parallel structures in 1926 and 1927: a new division within the PS, the Polizia Politica (or "political police," known colloquially as "Polpol"), and the secret Organization for Vigilance and Repression of Anti-Fascism (OVRA), a more sophisticated successor to the thuggish Ceka. OVRA's role was operational and intended to anticipate and prevent anti-fascist activity through infiltration and reporting on every aspect of Italian life. Polpol was set up to be the intelligence arm of the PS; its main task was "to fight and repress political deviancy" by gathering every imaginable bit of useful information about prominent individuals (including party members), especially those of uncertain allegiance.[66] Both OVRA and Polpol were placed under the direction of Arturo Bocchini, a capable—and politically indifferent—career officer who already headed the regular public security police. Between 1927 and his death in 1940, Bocchini was one of the most powerful figures of the regime.[67]

While OVRA focused on controlling and preventing political dissent, Polpol became the instrument for feeding Mussolini's obsession with knowing as much as he could about everyone who might in any way be important—or harmful—to the regime. Mussolini and Bocchini met every morning to review the most recent informants' reports, and before long, Bocchini's archive was a fundamental source of Mussolini's power.[68] Mauro Canali, one of Italy's leading historians of fascism and author of the magisterial *Le spie del regime* ("The Spies of the Regime"), has uncovered 815 "direct" agents of Polpol, each identifiable by a unique number; some of them short-term, some longer, and some who served for the entire duration of the regime. Many more "sub-agents" worked under the direction of these principal agents, possibly as many as five thousand in all, according to Canali's estimate. Eventually, Polpol had files on around a hundred thousand Italian nationals.

In 1927, a few months after going into action, Polpol opened a file on Marconi.[69] He wasn't exactly "followed"; it was more that his activities were monitored. More than one hundred reports, ranging in length from a single sentence to several pages, and in importance from the précis of a newspaper article to thoughtful political analysis of something Marconi had said or done, are filed alongside reports on high-level discussions in the councils of the Vatican. The police file also contains a typewritten copy of Marconi's last will—as well as several posthumous reports, expense chits for the agent assigned to him as a bodyguard during his travels, and wildly fabulous rumours that he had invented a "death ray," or committed suicide.

More than thirty agents reported on Marconi, including one of the regime's most alluring and legendary super-spies Bice Pupeschi, Bocchini's mistress, who kept a large stable of sub-agents as well as an apartment in Rome that served both as love nest and safe house for discreet meetings with her informants.[70] The most important agent on Marconi's trail was Virginio Troiani di Nerfa—the author of no less than eight reports, among the lengthiest and most important ones in Marconi's file. Troiani—a journalist and cousin by marriage of Gabriele D'Annunzio—was one of the regime's most effective and reliable spies. He was recruited to the PS in 1919 and was active with Polpol from 1927 to 1944, operating under the cover of a fake commercial agency based in Rome. Troiani's network focused on two sets of subjects: anti-fascist dissidents, and personalities with political ties to the regime. Marconi clearly fell into the second category. Troiani also demonstrated a flair for recruiting effective sub-agents, and an important arm of his network was in the Vatican, where his sub-agents had entries at the highest level.[71]

Marconi's police file, one of the most important sources for the last ten years of his life, and especially his relations with the Vatican, also illustrates the extent to which he was not trusted by the regime. (In interesting contrast, a file on Marconi's colleague Luigi Solari contains a single item, from 1936, stating that the subject is serious, well versed in the techniques of radio, and fully devoted to fascism.) Some of the most detailed and revealing reports contain exclusive information and insights about Marconi's thinking, his activities, and how he was viewed; these are simply unavailable anywhere else.

❋   ❋   ❋

In order to marry Cristina, Marconi not only had his previous marriage annulled, he also formally affirmed his faith in Catholicism, taking confirmation at a ceremony performed at the Bezzi-Scali palace. (He had been baptized a Catholic at birth, even if he was raised Protestant.) Cristina's spiritual counsellor was Eugenio Pacelli, soon to be a cardinal and the future Pope Pius XII, but then the papal nuncio in Berlin; another family friend was Cardinal Pietro Gasparri, the current Vatican secretary of state. Marconi wisely made allies of both men.

As soon as the annulment was approved and the engagement announced, Marconi wrote to Pacelli in Berlin asking if he would come to Rome to perform the blessing. Pacelli was away when the letter arrived and didn't read it for several days. Upon his return he wrote to Gasparri saying he was well disposed to go but that it was evidently up to the Superiors to decide if he could leave Berlin. If this were a good time for an exchange of views on the German question, then he would be happy to travel but "I don't want to come to Rome simply for a wedding, albeit that of the famous Marconi." Pacelli asked Gasparri what answer to give and was told to reply that urgent affairs prevented him from coming to Rome.[72] His letter was fortuitous. The Vatican and Mussolini were holding secret, highly sensitive negotiations intended to normalize relations between the church and the Italian state, and the Vatican had learned that Mussolini would be a witness at Marconi's wedding. In view of the delicacy of their ongoing political negotiations, it was not opportune that Pacelli perform the blessing.[73] But, in fact, Mussolini declined the invitation to attend the wedding as well; he also said he was too busy, and likely for the same reasons. Despite a persistent urban legend spread by Marconi biographers and amateur history buffs, Mussolini was neither a witness nor the best man at Marconi's second wedding.** He did however, send a telegram of congratulations.[74]

Marconi and Cristina Bezzi-Scali were married, first in a civil ceremony, in Rome on June 12, 1927. The only member of Marconi's immediate family to attend the wedding and reception at the bride's residence afterwards was his brother Alfonso, who came from London. Then, on June 15, the religious ceremony took place at the Basilica of Santa Maria degli Angeli—only yards away from the Grand Hotel, Marconi's favourite secular worshipping place in Rome. Cristina proudly wrote in her memoir that she

---

** The earliest reference I have seen to this erroneous fact was in a *New York Times* obituary published the day after Marconi's death.

wore a diamond tiara designed by Marconi himself and an antique Irish lace veil that had belonged to his mother. The day before the wedding, "in very great haste," Marconi found a moment to write to the mother of his children.[75]

Divorce is never easy, and nothing can complicate relations with children like a parent's new marriage. Marconi's relations with his children were already complicated enough; now they became even more fraught. Despite his long and frequent absences, and his distraction from them and their needs when he was around, Marconi dealt with his children as he dealt with everything else—popping in and out of their lives, directing things long-distance, delegating whatever he could, and always insisting on remaining in charge. His letters to Beatrice are a manual of parental micromanagement. In different ways, the struggles between the parents took a toll on each of the children. By the time Marconi and Beatrice were effectively separated in 1920, the children were twelve, ten, and four years old. His letters to Beatrice from that time on have frequent references to their schooling, health issues, financial arrangements (as we've seen), and arrangements for them to spend time with him on the yacht.

As Marconi and Beatrice went about their separate lives (he mostly in London, she in Italy), the children shuttled around in the spaces between them. A major concern was Giulio's schooling. Marconi wanted his son to enter the naval academy in Livorno and take up the career that he himself had refused against his own father's wishes a generation earlier. The entrance exam was stiff, but Giulio had several years to prepare, Marconi wrote to Beatrice in October 1922, when Giulio was twelve years old. Giulio's schooling would cause tension for years to come, and trying to please his father eventually took a toll on him. He at first tried to assert himself; shortly after he turned fourteen, Giulio announced that he was not interested in a navy life. "This is quite a surprise and a great disappointment to me as I had always thought he loved the sea," Marconi wrote a month later. The navy provided certainty of employment and the education at the naval academy was better than at the universities. "Had I known he did not want to go to the *Accademia* I would have much preferred sending him to a good English school." Then Giulio changed his mind and decided to try to get in to the naval academy after all. Marconi disingenuously denied he had anything to do with the decision: "He came to this decision quite on his own, without the slightest pressure from me, after I had confirmed to him the pros and cons of both courses." Giulio failed to pass the exams, however, provoking Marconi to remark that "the system is rotten"; he was at a loss as to what to suggest for his now fifteen-year-old son. He bounced back, however. Giulio would reapply and "I will also write to some Minister or General who may help to get him accepted...." Giulio's academic future continued to see-saw. Marconi set him up in Livorno, where he continued to prepare for the naval exams, although at age seventeen he was still unsure about going.[76] Marconi continued to try to direct him, surely with the best of fatherly intentions. Giulio was by all accounts a sensitive young man who did not find it easy living in the shadow of his demanding, directive father and his expectations. In family letters and conversations, he was often referred to as "poor Giulio."[77] He was soon experiencing mental and emotional difficulties. He was not happy preparing for a naval career but could not manage to set a clear alternative course for himself. Father and son clashed frequently, often through the intermediation of other family members.[78]

A few months later Giulio had to spend time in a clinic. Marconi had two concerns: that Giulio be able to go on with his naval career, and the cost of the clinic. He asked Beatrice's sister Clare for help finding less expensive care, adding: "As to payment, I will of course have to be responsible, but as I too am now hard up, I strongly feel that Bea should at least share part of the expense."[79] Giulio also had to suffer Marconi's transference of issues left over from his relationship with his own father. Giulio's aunt, Beatrice's sister Lilah, described how she once found Giulio living in a boarding house on the London Embankment, run by an Irish landlady, which made it easier for him to eke out a subsistence, but "This affected the boy's health and morale without his father ever being aware of it. The same thing with the other children, having been brought up to every luxury imaginable, were reduced in unbelievable and unnecessary ways, G.M. always measuring their needs according to his own very simple ones of by-gone years."[80]

Marconi's worries about his eldest, Degna, were of another order. "I think she studies too much and does not get quite enough of fun for a girl her age. She so often looks sad...and is so awfully shy of people...." A year or so in a good English school and lots of outdoor activity would do her a world of good and greatly improve her chances in life, he wrote Beatrice. In September 1925, Marconi seemed to be thinking seriously (however fleetingly) about becoming a hands-on single dad; he wrote to Beatrice, from his room at the Savoy: "In regard to the children's future, I am quite willing and desirous of taking them over myself." To start with, he proposed to find a suitable house in London, "independently of whether I remarry or not."

The nesting instinct didn't last long. A month later, while insisting that Degna come to London (to have her teeth fixed), he wrote: "I am perplexed as to what I ought to do. Is there any of your relations with whom she could live in London? I'm ready to do anything possible for the child even should it cost a great deal."[81] That said, he would reduce the children's allowance when there were extra expenses like these.

Marconi kept up a lively correspondence with Degna, the only one of his children he wrote to regularly. He empathized with her trials at school ("I used to hate exams but now I've got lectures and speeches and meetings instead, which are almost as bad"[82]). Along with Giulio, she frequently spent holidays on the yacht, suffering bravely from sea sickness. The oldest of the children, she was often called on to mediate difficult situations with Beatrice; by the time she was fifteen, her father was sometimes writing to her as though she were an adult. On March 1, 1927, she received a letter from him informing her that he was engaged to marry Cristina Bezzi-Scali. "You— that have seen her and know something of how sweet she is, will I feel sure understand, and take heart in my joy, because I also feel certain she will be very fond of you."[83]

Degna was given opportunities to study in Paris and London and, if anything, it was her social shyness that was a cause of concern. That soon shifted, as she developed a problem living within her means. On August 22, 1929, Marconi wrote that while his responsibilities would end when she turned twenty-one, he would continue to give her fifty pounds per month. "These payments will depend of course on my financial conditions remaining about the same as they are at present." Obsessively precise as always, he asked her to pay Beatrice for eleven days of allowance because she would be turning twenty-one on the eleventh of September. His preoccupation with money where the children were concerned, echoing his own father's years earlier, was all the more remarkable given how free he was about money in general. Degna replied: "My dearest daddy,... I am most grateful to you for what you are willing to give me and thank you very much. As you know there are always difficulties with my mother but I hope to arrange that she gets the usual amount for my keep as long as I live with her. If I see that I can't come to an agreement with her I will probably leave my mother's house, as above all, I don't want to be a pest to anybody and in any way."[84] Degna was uncomfortable, caught between her parents, feeling not fully welcome in her mother's new home and unwanted at her stepmother's.[85] Her father was still her confidant, the one she turned to for consolation after a broken heart,[86] but she began seeing less of him after the birth of her half-sister, Elettra, in 1930. She also developed a problem with money.

The youngest, Gioia, was still just a little girl when her parents separated. When she turned ten, in 1926, Marconi wanted to send her a pony, but he still couldn't resist needling Beatrice about the cost: "I'm still trying to arrange for sending out the Pony for Gioia but its not at all easy and frightfully expensive.... The pony and journey would cost about £150, which is a lot of money for me just now." The pony was a proxy for other issues. Even as he was organizing his papal annulment and engagement to Cristina, Marconi was still talking about getting a flat, or a small house, in London, "just sufficient for me and two of the children." But his thoughts were scattered, his concerns jumbled. Giulio was in Livorno but not fully settled. Beatrice wanted to send Gioia to join her father on the yacht. He was soon having second thoughts about renting a house. "I'm worked almost to death just now and never have a moment...I can't possibly do everything now without the grave risk of getting really ill through over work...the Co's affairs are, as I told you, in a bad way and things can only be got right again by my sticking to my work." In December 1926, he told Beatrice things were so bad that a committee of investigation was going through the company's accounts. Beatrice was nevertheless insistent on coming to London with Degna, who was now eighteen; Marconi said he would pay for the journey, "although I can ill afford it."[87]

Within a few months, such questions became moot. After Marconi's second marriage, his life was focused on Rome.

PART V
The Conformist

31

A Servant of the Regime

The Via dell'Arco della Ciambella, a stretch of Roman street barely a hundred yards long, lies five minutes by foot from the Pantheon, in the direction of the Tiber.* The "arch" in question is a fragment of a vast complex of baths built in the first century BCE by the general Agrippa, lieutenant to the Emperor Octavian, for the relaxation of his soldiers between engagements. (Agrippa's most important victory was at the Battle of Actium in the Ionian Sea, where Octavian's fleet under his command defeated the combined forces of Antony and Cleopatra.) The *ciambella* (sometimes translated as "teething ring," sometimes as "doughnut") refers to the ring-like shape of the long-buried baths, believed to be the oldest in Rome.

The "modern" (that is, medieval Renaissance) street was designed in the sixteenth century, around the remaining bits and pieces of Agrippa's baths. Today it bustles with tourist apartments, a chic hotel, a mélange of old-fashioned and hip bars and restaurants, local services like a beauty parlour, laundry, and shoemaker, and a small dress shop run by a Russian expat who fabricates original silk concoctions based on vintage designs at 800 euros a pop. A tangle of motorbikes are parked higgledy-piggledy on every available inch of cobblestone. This is just a small corner of the city of layers where, as Goethe wrote in 1786, one epoch follows another and even the most casual visitor "becomes, as it were, a contemporary of the great decrees of destiny."[1]

It's a ten-minute walk to the Piazza Venezia, a grander example of Roman layering. It is one of the city's busiest areas, well known to visitors as the site of one of Rome's most recognizable and vilified monuments: the decadent modernist "wedding cake" eyesore built by King Vittorio Emanuele III at the turn of the twentieth century in memory of his grandfather,

---

* I lived in the Via dell'Arco della Ciambella in May and June 2014 while I was completing the research for this book.

Vittorio Emanuele II, the first monarch of united Italy. On another side of the square is Rome's most perfect Venetian palace and one of its Renaissance jewels, the fifteenth-century Palazzo Venezia. One of Mussolini's most inspired decisions was to locate his headquarters here in 1929; the palace has a balcony that was just right for framing a speaker as he harangues a crowd, and the square was just the right size for the fifty thousand or so faithful he knew he could mobilize at a moment's notice and whip into a froth. The only problem was the traffic; once, in 1934, when Marconi was trying to make a case to him for more funding for scientific research, Mussolini replied that he would welcome some research that could reduce the traffic in Piazza Venezia.

Mussolini's project was to restore and extend the glory of Rome, and while doing this he was responsible for both preserving the city's ancient archaeological heritage and building some of the twentieth century's most stunning architectural monstrosities. Mussolini was a better archaeologist than architect. He ordered the unearthing of the Trajan Market and nearby Roman Forum, building one of Europe's widest avenues to connect Piazza Venezia and the Colosseum. To do this, however, he had to destroy dozens of heritage churches and houses of Renaissance and baroque Rome. Much of the administrative functions of the Italian capital today are based in the EUR district in the city's southern precincts, an area of ersatz reminders of grandeur somewhat similar to contemporary Las Vegas.[2] The district was designated to host the 1942 World's Fair, which never took place; one of its hubs is the Piazza Guglielmo Marconi, a roundabout centred with a 150-foot-tall faux-Egyptian obelisk covered in panels depicting mythical scenes from the life of the inventor. (Commissioned by Mussolini upon Marconi's death in 1937, the structure was completed only in 1960, just in time for the Rome Olympics.) One of the main buildings in EUR houses the Central Archives of the State, which contain Mussolini's personal papers and the remarkably accessible files of the fascist regime's political police.

Rome's layered nature extends, of course, to the political sphere. Between the Republic that ended with the rise of the Emperor Augustus and the Republic recently caricatured by Silvio Berlusconi, Rome played host to a succession of more or less imperial figures. The last of these was Mussolini, whose political style can be described as a blend of methods borrowed from sources ranging from Gabriele D'Annunzio to Napoleon Bonaparte to Josef Stalin. In the 1920s, Mussolini progressively eliminated all meaningful opposition, managing to remain aloof enough to maintain a statesman-like stance and the fascination of the developed world. His tough stance on "bolshe-vism" and labour agitation, his imprisonment of the communist philoso-pher and politician Antonio Gramsci, and his crushing of the Italian labour movement quietly endeared him to the leaders of the world's great powers.

After gaining a parliamentary majority in the 1924 national elections, Mussolini, backed by the king and supported by the country's conservative and business elites, restructured Italy's political, administrative, and public institu-tions in subordination of a totalitarian dictatorship, eradicating their former roles and placing all power in the hands of the executive branch of govern-ment, under his own direction.[3] By 1926 Mussolini was referring to himself as Il Duce (the Leader).[†] The editorship of the main newspapers was secured by fascists, and hostile papers were persecuted or driven underground; eventually, only party members would be allowed to work as journalists. In December 1926, the fascist insignia (a bound bundle, or *fascio*, of wooden rods) became the official symbol of the Italian state.[4] The March on Rome was reimagined as the founding moment of the fascist revolution, and 1922, the year Mussolini came to power, was designated year one of the *era fascista*, and this is reflected in the dates one finds on official documents, public buildings, and even private cor-respondence from the period (including much of Marconi's).[5]

Mussolini also began to build a cult of personality. He appeared more often in the fascist militia uniform of black shirt and cavalry boots, and es-tablished the personal routines that would mark his lifestyle as leader. When he first came to power, while his wife and children remained in northern Italy, he lived at the Grand Hotel (also Marconi's favourite) sneaking out at night to visit Margherita Sarfatti, to the dismay of his security people.[6] Mussolini may have had a voracious sexual appetite, but he had little use for bourgeois trappings. He soon moved to a dingy apartment, looked after by a tough housekeeper Sarfatti found for him; she became known as *la ruffiana* (the procuress), because a key part of her job was to arrange her boss's many assignations. Mussolini collected violins and would play occasionally for re-laxation; in those days the only people he trusted were his brother, Arnaldo, now editor of *Il Popolo d'Italia*, and Sarfatti, who was his chief political advi-sor as well as his favourite lover. The diplomatic community in Rome con-sidered Mussolini intriguingly enigmatic.[7]

As he learned the mechanics of government, Mussolini distanced him-self from the rabble that had been so critical to his rise to power, making new

---

[†] He was also routinely known as *Il Capo del governo*, the head of government.

alliances with business and industrialists and, crucially (and at first secretly), with the Catholic Church. He became adept at feigning ignorance of incidents of fascist thuggery, like the sacking of the home of former prime minister Francesco Saverio Nitti in the centre of Rome in November 1923, and clucking disapprovingly at reports of priests being forced to drink castor oil by black-shirted bullies who accosted them in the street. As he became more respectable and brought in radical new measures, his government's activities were followed with fascination in the democracies of Europe and the Americas, but also among the darker political movements, which began to see Mussolini and his party as a role model. In Germany, the methods and symbols of Italian fascism were emulated by Adolf Hitler's Nazis, from the uniforms and the straight-armed salute to the attempt to replicate the march on Rome with a "beer-hall putsch" (on November 8, 1923). The Nazis began to style Hitler as "Germany's Mussolini"; as Mussolini began calling himself Il Duce, Hitler would be der Führer.

Already a key figure in Italian business and finance, and one of the most visible and popular of his countrymen to the world outside Italy, Marconi was now called to play a crucial role in the cultural, scientific, and intellectual reshaping of Italy to fit the mould of fascism. During the last ten years of his life, Marconi became one of the most influential figures in Mussolini's Italy. One of his strengths was that until now, he had managed to remain aloof from the more contentious aspects of Italian politics, including the jousting between pro- and anti-fascist intellectuals.

On April 21, 1925, under the initiative of Giovanni Gentile, the self-styled "philosopher of fascism," the "Manifesto of the Fascist Intellectuals" appeared. Gentile was a towering figure in Italian academic life and, until recently, Mussolini's minister of public education. Gentile's "Manifesto"—and a later article he ghostwrote for Mussolini, entitled "A Doctrine of Fascism"—became the foundations of fascist ideology as well as for the totalitarian state. The manifesto was drafted following the Conference of Fascist Culture in Bologna on March 19, 1925, and was the first attempt to formally define the cultural aspirations of fascism. Chaired by Gentile, the conference was attended by some four hundred intellectuals; 250 signed the manifesto, including Filippo Tommaso Marinetti, Gabriele D'Annunzio, and the playwright Luigi Pirandello. It was published in the national press on April 21.

Ten days later, on May 1, 1925, a "Manifesto of the Anti-Fascist Intellectuals," written by Italy's leading liberal philosopher, Benedetto

Croce, appeared. Croce, an intellectual mentor of both Gentile and Antonio Gramsci, came from a wealthy and influential family. A critic of Italian participation in the First World War, he was minister of public education in the last Giolitti government (ironically, the same position held by Gentile in Mussolini's first ministry). Croce supported Mussolini's government in 1922, and Gentile's extensive educational reforms were based in part on Croce's suggestions. However, like many other early supporters (but not Marconi), Croce's support was shaken by the Matteotti affair. There was an interesting symbolic counterpoint to the two publication dates: April 21 was the anniversary of the founding of Rome; May 1 was International Workers' Day. Marconi did not sign either manifesto.

One of Mussolini's early initiatives was to create a national research council, an idea first promoted by mathematician Vito Volterra shortly after the end of the First World War. Volterra was one of Italy's most brilliant scholars, rising from humble origins in a poor Jewish family in Ancona to become professor of mathematical physics at the University of Rome, senator of the realm, and a member of the Accademia dei Lincei, the royal academy of sciences. He was one of Italy's most internationally prominent scientists and, after doing patriotic duty in the Italian army in the Great War, was elected vice-president of the Lincei in 1916 and then president in 1923.[8]

Volterra suggested the idea of a national research council in 1919 but it found no support within the government of Giovanni Giolitti. The idea was taken up by Mussolini, however, and the Consiglio Nazionale delle Ricerche (CNR) was instituted by decree on November 18, 1923, to coordinate Italy's activity in "the study of all matters pertaining to the sciences and their practical applications."[9] Mussolini promptly named Volterra CNR's first president, with headquarters at the Lincei offices in the Palazzo Corsini. Never a supporter of fascism, Volterra had been worried about the developing political situation since Mussolini's arrival in power. But like many of Italy's elite intellectuals, he was prepared to work with the Fascist government in order to achieve greater ends. At the same time, he was openly critical of some of Mussolini's policies; for example, joining with other intellectuals in publicly objecting to new rules brought in for the education sector by Gentile.[10] Volterra became actively involved in anti-fascist activity after the murder of Matteotti in 1924, however, and, in 1925, signed the "Manifesto of the Anti-Fascist Intellectuals." This made him "an official enemy of Mussolini," and when Volterra's term as president of the CNR came to an end, it was not renewed.[11] The council was "restructured" by decree on March 31, 1927, and on September 1, 1927, a new president was named: Guglielmo Marconi.[12]

Well known as he was, Marconi was nevertheless a complete stranger to Italian academic circles. Volterra had tried to involve him in the council's international activities, but Marconi was always too busy to do much.[13] The presidency was something else, for it came at a time when he was trying to position himself and his business interests in the new emerging "mixed system" in Italian communications ("mixed" in both the sense of public/ private and broadcasting/telecom applications of wireless; today we would use the term *convergence* to describe what was going on). He was scaling down his relationship to the British Marconi company in view of the coming cable and wireless merger; he had just married Cristina Bezzi-Scali and was reintegrating himself to Italian high society; he was conducting a full, cutting-edge research program on the *Elettra*; his health, shaky in the best of times, was weak. Some historical assessments, and most Marconi biographies, have seen his role as chief figurehead of Italian fascist science policy as largely ceremonial, but his importance in this role should not be underestimated.[14] Not only was it one of his principal activities during the last ten years of his life, it signified his attempt to influence the course of scientific and technological development in Italy and worldwide.[15] To Marconi, the appointment was not an honorific or a sinecure but an important platform for promoting his agenda; it was thus one of the linchpins that bolted him to the regime. For Mussolini, it was clearly strategic—part of a propaganda effort to improve his international image; he saw it as a move that would meet his political interests and, indeed, under Marconi's presidency, according to Italian historians Giovanni Paoloni and Raffaela Simili, the CNR became "a body at the service of the Fascist state."[16] Yet, the choice of Marconi also made sense in terms of industrial policy. Marconi's international experience in the use of patents and intellectual property rights was a unique asset for a government intent on developing synergies between science, technology, and industry.[17]

Marconi's appointment as head of the CNR came as Mussolini was beginning to outline a policy regarding the political conduct of public figures. Loyalty to the regime, which could range from active, vocal support to quiet acquiescence, was paramount. Marconi was not expected to make public pronouncements or proclamations; it was enough that he accept to collaborate in the defining projects of fascism. It is worth looking at Marconi's trajectory in this regard alongside that of his contemporary, Volterra. After the non-renewal of his presidency of the CNR, Volterra returned to his academic career. In 1931 he was one of only twelve Italian university professors, among more than twelve hundred, to refuse to sign a loyalty oath to fascism required by the regime; all were dismissed from their posts and essentially banned from public life.[18] The Fascist government also campaigned to have Volterra removed from the positions he held in international academic organizations.[19] He continued to live in Rome and remained a senator until his death in 1940, two years after the promulgation of the antisemitic race laws. A few years after being banished from the university, he famously wrote, on a postcard, what some consider to be a fitting epitaph for the Fascist regime: "Empires die, but Euclid's theorems keep their youth forever." Marconi could have taken a similar path. He was powerful enough to resist Mussolini's advances. He didn't have to accept the presidency of the CNR, though the offer provided him with an opportunity to promote his own business and research interests. He also thought, naively, that he would be able to bend Mussolini to support an independent research council that would be more than a political instrument or mouthpiece.

Marconi made his choice and embraced it. A few days after his appointment as CNR president, he served as honorary chair of a major event at Lake Como to commemorate the one hundredth anniversary of the death of Alessandro Volta. The International Conference on Physics—personally authorized by Mussolini—was a virtual summit of the world's leading physicists; its sixty participants, from fourteen countries, included such international luminaries as Niels Bohr, Robert Millikan, Wolfgang Pauli, Max Planck, Ernest Rutherford, and Werner Heisenberg (who first publicly presented his "uncertainty principle" here), as well as rising Italian stars like Emilio Segrè and Enrico Fermi—all past or future Nobel Prize winners. The conference was crucial to Mussolini's quest for credibility, but it was even more important for Marconi. It placed him at the centre of another paradigm shift in theoretical physics, legitimized him in the eyes of the international scientific community, and was good for his business; at the same time, his international visibility made him the official poster boy for fascist scientific theory. He was now the regime's leading scientific figure, a role he took to with public enthusiasm if with some private ambivalence, as we'll see.[20]

As the conference was about to get under way Marconi gave a lecture at Perugia's University for Foreigners on the state of technical progress of beam communication. Once again, the title of the talk—*"Le radiocomunicazioni a fascio"*—reflected the play on words that he had joked about not a year earlier, but this time (although the talk was not part of the conference) he kept a sober tone appropriate to the serious roster of visitors assembling in Como.[21] He also made the closing speech at Rome's Campidoglio, after the conference venue shifted to Rome, on September 19, 1927. Here his focus was history (he knew better than to address a conference of scientists about current science). He recalled Volta's association with the great scientists and philosophers of his time, like Voltaire, and Volta's meeting with Napoleon, noting that "a singular friendship was immediately formed between their two similarly dissimilar personalities. Napoleon who envisaged the World as a field of domination through physical power... certainly appreciated in Volta his control of a mystic force and a reconstructive agent." The parallel with how Marconi saw himself and perhaps his relationship with Mussolini was too obvious to miss. Marconi also recalled that when Volta, who was best known for the invention of the battery, turned sixty and expressed a desire to retire, Napoleon deterred him with the comment: "A valorous soldier dies on the field." Marconi was entering Mussolini's service at only fifty-three.[22]‡

After Marconi's speech Mussolini received the conference participants at his official residence in the Villa Torlonia.[23] The following day, September 20, 1927, Marconi chaired the first board meeting of the revitalized CNR, declaring the council to be an organ "of highest importance for the future of our country."[24] He then decamped for the United States, with Cristina, to attend the fourth International Radiotelegraph Conference in Washington and give a major speech in New York. The trip was supposed to be part honeymoon but it was mostly work, and after suffering a heart attack on the return sailing he was out of circulation for the rest of the year.

On January 1, 1928, Marconi's appointment to head the CNR figured prominently in Mussolini's New Year's message, where it was framed as an economic initiative. Mussolini publicly wrote to Marconi, defining and outlining the scope of the CNR, declaring the importance of scientific research while clearly subordinating it to his own authority. The role of the CNR would be "to coordinate and discipline scientific research," and make evident the progress of Italian science, technique, and industry. The council was also meant to play a role of political gatekeeper: no official scientific delegation could go abroad, no scientific congress could be held in Italy, no Italian scholar could propose an international meeting unless formally authorized by Mussolini himself, on recommendation of the CNR.[25]

‡ Volta continued to work until he was seventy-four.

This directive put Marconi in the powerful but ambiguous position of making recommendations that Mussolini could choose to follow or not; pressure would be put on Marconi not to make too many "wrong" recommendations, and sometimes his, Marconi's, word could be the deciding factor in a major decision. In Mussolini's grand design, Marconi was meant to be an instrument for marshalling and channelling the support of science, culture, and intellectual life in general toward the construction of a fascist totalitarian society.[26] Mussolini's message clearly went against any notion of independent, or even policy-targeted, research. The CNR now assumed the role of permanent consultative body to the Fascist government on technical and scientific questions; put another way, the restructuring of the CNR and, crucially, the nomination of Marconi as president was designed to be politically functional.[27] Marconi nonetheless took the message as an assurance of government support for research, but Mussolini was clearly more interested in fealty to the regime. Marconi and Mussolini had different ideas about research coordination and oversight (or "discipline," as the council's mandate put it). The chronic lack of funds not only obstructed the realization of the regime's stated science policy, it quickly became a source of tension between the council's new president and the head of government. (The CNR's budget under Vito Volterra was around 175,000 lire per year (about $120,000 today); when Marconi took over it was increased significantly, to around 575,000 lire (around $400,000), still a small sum in view of the expectations.[28])

Marconi took his mandate seriously, arguing for the establishment of and support for scientific research laboratories outside the conventional academic setting—intuitively thinking of his own experience as a practical model. He recalled that he himself had done and continued to do "research with some useful results . . . in non-university laboratories."[29] Years later, Marconi clashed with a powerful minister of education, Giuseppe Bottai, who argued that research should remain within the universities, but Mussolini himself seemed to agree that science should develop without the concerns of teaching—so long as it clearly served national imperatives.[30] Like so many political bankrollers before and since, he failed to understand the relationship between research and development, as well as the need for untrammelled government support.

However naïve he may have been in thinking he could impose a rational, progressive agenda on Mussolini, Marconi was now at least nominally the principal gatekeeper of Italian scientific research. Still basically in favour of Mussolini's broad approach—to use the corporate state as a lever of progress and development—he bought in to the rising cult of personality and almost mystical invocation of the idea of fascism itself. He had a clear vision for the role of scientific research in Italy's internal development as well as its place on the world stage, and professing loyalty to Il Duce and fascism was simply a means to an end. Marconi was keenly aware that the success of the council would depend on the level of government funding he could muster, though he didn't realize that this would depend almost entirely on the state of his personal relations with Mussolini (the experience of his business affairs should have alerted him to the way things were now working in Italy). He informed his colleagues on the CNR board that he intended to devote himself fully to the council if the government provided the necessary means.[31] Once the board was installed, however, Marconi left for "other scientific adventures" and didn't participate at any other sessions in 1927 or 1928, although he remained in touch.[32] Meanwhile, the council had to struggle for even the most basic amenities—as of mid-1928 it didn't even have a permanent office; after meeting with Marconi on April 13, 1928, Mussolini wrote to his minister of education, Pietro Fedele (who succeeded Gentile in that position), that he had to find quarters for the CNR, stating that "Today, its address is a post office box!"[33]

Marconi's frequent absence from the day-to-day activities of the CNR left a space others were eager to fill. One of the first targets of Mussolini's January 1, 1928, directive was the council's former head, Vito Volterra. On March 24, CNR vice-president Amadeo Giannini wrote confidentially to the Ministry of Foreign Affairs, invoking Mussolini's directive to propose removing Volterra from the Italian delegation to an international commission, given that his membership had become "problematic" in view of his "lack of attachment" to the government.[34]§ (The letter was ambiguously signed "*Il Presidente. VP Giannini.*" Marconi was actually the president at that time.)

Within the council's administration, an even more sinister agenda was starting to take shape. In an undated, handwritten sheaf of notes from the same period, CNR secretary-general Giovanni Magrini wrote about the noted biologist Paolo Enriques: "He is a Jew, and has all the qualities and faults of his race. He never says what he thinks. . . . He is one of the most

§ Giannini was a delegate of the Ministry of Foreign Affairs to the very first meeting of the CNR on January 12, 1924, where Volterra was elected president and Giovanni Magrini secretary-general (Simili and Paoloni 2001, 48–49). He was vice-president of the CNR without interruption from 1924 to 1944.

notable jewish-masonic exponents of the international scientific world." Magrini framed his comments as evidence of a problem facing the strengthening of fascist culture in Italy and outside. He noted that while Volterra (who was also Jewish) had been replaced by Marconi as head of the CNR, Volterra was still on the executive committee of the International Research Council, where he continued to engage in anti-fascist activism. Magrini noted ominously: the Jewish-Masonic circles should be of concern for the future of the regime.[35]

Magrini was the main link between the Volterra and Marconi presidencies, serving as the CNR's secretary-general from its very first meeting in 1924 until his sudden, unexpected death in 1935. He was also an insider in the bureaucracy of the Ministry of Foreign Affairs, a great admirer of Mussolini, and the principal liaison between the CNR and Mussolini's office. Giannini and Magrini were two of a tiny group of apparatchiks who essentially ran the CNR under Volterra and Marconi.[36] The document cited here was not an isolated intervention. In another note, Magrini labelled Volterra "a leading figure of international freemasonry, along with Einstein."[37]

Marconi seems to have been oblivious to the thinking of these senior officials; if he was privy to it, he certainly left no trace. It is either another sign of his wishful thinking with regard to the way the regime was evolving, or another nasty element that he was prepared to ignore in the interest of some perceived or imagined greater good, or interest of his own. Giannini and Magrini were fascists of the earliest date, unshakably loyal to Mussolini. They provided, in a sense, an insurance policy for Il Duce, a reliable counterpoint to Marconi, who Mussolini knew could diverge from his program at any time. But they also had an ideological agenda on race and ethnicity that was well ahead of Mussolini's own.

The reconstituted CNR was officially inaugurated with a pompous ceremony at Rome's Campidoglio on February 2, 1929, where both Mussolini and Marconi spoke. Mussolini was loftier on this occasion, tying the mission of the CNR—the reprise of scientific research in service of the nation—to the presidency of Marconi, "the pride and glory of Italian science."[38] Marconi began his speech by thanking Mussolini and the Fascist government for placing a priority on scientific research. His speech was reported in the British press, which noted his emphasis on "the present decadence of research and to the paucity of research workers in Italy." Modern scientific research required adequate organization and vast means; researchers needed to know that they could live "by science and for science"; there

was a need for coordination and discipline (Mussolini's words). It was criti-
cally important that a discovery made or piece of research begun in Italy
should mature and develop in Italy.[39] The tacit allusion to Marconi's own
early experience was unmistakeable.

While carving a unique place for himself in Mussolini's Italy, Marconi
also developed a particular relationship to the Vatican, still technically at war
with the Italian state (as it had been since 1870). The Fascists' assumption of
power in 1922 provided an unprecedented opportunity to normalize Italian-
Vatican relations, and the process began almost immediately. On January 19,
1923, a highly secret meeting took place between Cardinal Pietro Gasparri,
the Vatican secretary of state, and Mussolini. The need for secrecy was dictated by the fact that the Holy See did not recognize the legitimacy of the
state of Italy, and so its chief official could hardly be seen in a meeting with
the head of the Italian government. To the surprise of many of his own supporters, the non-believer Mussolini had declared his intention to restore the
church's influence and prestige in Italy, and Gasparri's mission, approved by
his boss, Pope Pius XI, was to gauge whether Mussolini was serious, whether
he could be trusted, and whether, with church support, he could succeed.
That was quite a challenge, as Mussolini, in his early years, had been known
as a *mangiaprete*, a priest-eater. But restoring the harmony between church
and state was part of Mussolini's agenda for consolidating totalitarian
power—after all, one couldn't ask for a better group of subjects than a flock
of Catholic believers. Before long, the priest-eater was being referred to at
the Vatican as "the man sent by Providence."[40] (One of Mussolini's early concessions to the Vatican was public opposition to divorce, although he privately counselled Marconi on how to get around this and Marconi paid
dearly for the advice.[41])

In September 1923, an internal briefing paper circulated inside the
Vatican—a "Programme of Collaboration of the Catholics with the Mussolini
Government"—reported that Mussolini was looking for a new base of support to replace his undisciplined fascist core. This base could be best provided by Catholics, the paper said, for they were accustomed to submitting
to top-down rule. Mussolini's fascist revolution was about to become "a
clerico-fascist revolution," and a new partnership had begun.[42]

On the surface, it is difficult to imagine two more unlikely partners
than Mussolini and Achille Ratti, the pope. About the only thing they had
in common was that they both came to power the same year—and they
were both determined to resolve the endless and poisoned conflict between

the Vatican and the Italian state. Born in 1857, the son of a silk factory super-visor in Italy's industrial north, Ratti always knew he wanted to become a priest. Appointed prefect of the Vatican Library in 1914, then Benedict XV's envoy to Poland in 1918, archbishop of Milan and cardinal in 1921, he was surprisingly elected to the papacy on the fourteenth ballot a few months later, after Benedict's unanticipated death, in February 1922.[43] The relatively obscure church librarian—"the unexpected Pope," as historian David Kertzer calls him—the most junior member of the conclave, a compromise candi-date, was sixty-five years old. From the perspective of the Catholic Church in 1922, Ratti was a good choice. A headstrong traditionalist, he sought to restore the church's medieval authority. A key plank in this platform was the re-Christianization of Italian society.

Ratti was both bookish and athletic, an alpine climber who looked like a scholarly intellectual. In origin and temperament, Mussolini was diamet-rically different. "The Rattis' heroes were saints and popes; the Mussolinis' were rabble-rousers and revolutionaries," writes Kertzer in his Pulitzer Prize–winning book *The Pope and Mussolini*. Ratti rose every morning at 6:00 a.m.; Mussolini liked to sleep late and started getting serious about his day around noon. Ratti was the first pope to replace the Pontiff's horse-drawn carriage with an automobile, but he refused to speak on the telephone or be photographed with guests unless it was absolutely necessary. Every day he spent an hour walking in the Vatican gardens, "hands often clasped behind his back, a black fedora perched over the white skullcap on his head."[44]

The pope's decision to support Mussolini surprised his colleagues in the church. In fact, the two leaders, each absolute in his own domain, had many values in common: a lack of sympathy for parliamentary democracy, no interest in free speech or association, a visceral antagonism toward com-munism—which the pope saw as the greatest threat to civilization. Both of them thought Italy was caught in a quagmire and that the existing political system could not possibly get it out. (Marconi also shared this view.) Under the pope's explicit instruction, propaganda instruments like the Jesuit journal *La Civiltà Cattolica*, previously critical toward fascism, which was judged hos-tile to the Church, were transformed into cheerleaders of the Fascist govern-ment.[45] Mussolini, meanwhile, became the first head of government since Italian unification to mention God in an address to parliament.

Very soon after he came to power, Mussolini began restoring the privi-leges the church had enjoyed before unification, placing crucifixes in every classroom and courtroom, making it a crime to speak disparagingly about the Catholic religion. In return, the Vatican vocally supported the government measures of which it approved, and remained silent on those it didn't. By acquiescing in the fascist use of violence as a governing mechanism, the church contributed to legitimizing Mussolini's regime and encouraged others to ignore this fundamental characteristic of fascism as well.[46] Within the church, there was also a sinister antisemitic streak, quite recognizable to the right wing of the Fascist Party as well as to the branches of the bureaucracy that we encountered a bit earlier. In October 1922, the month Mussolini came to power, *Civiltà Cattolica* published a rant entitled "The World Revolution and the Jews," outlining, among other things, the purported extent of Jewish influence in Bolshevik Russia.[47] The journal continued in later issues to denounce "jewish-masonic socialism." Pius XI, the traditionalist, was opposed to anything to do with the Enlightenment. He shared the general worldview that forbade Catholics from engaging in interfaith dialogue. Italy's Jews, representing less than 1 percent of the population, were not much of a threat to what he was trying to do, and he actually liked the few he had met (the rabbi of Milan, for example, who had helped him decipher a particularly difficult piece of Hebrew liturgical literature). But he had noticed, as papal nuncio in Warsaw during the Polish-Soviet War of 1919 to 1921, that the Bolshevik leadership seemed indeed to be disproportionately Jewish, and this observation remained with him when he returned to Rome. So, while the church officially condemned antisemitism, it also blamed the Jews for spreading chaos and revolution as they plotted to dominate the world.[48] Mussolini could not have cared less about issues of race and religion, and he thought the idea of a worldwide Jewish-Masonic-Protestant conspiracy was hogwash. But such ideas seemed to be meaningful to at least a fringe element of the church, and so they became important to him.

In August 1926, the pope chose a layman, lawyer Francesco Pacelli (older brother of the future secretary of state and then Pope Pius XII, Eugenio Pacelli), to represent him in the negotiations with Mussolini. The Pacellis, like Marconi's future in-laws the Bezzi-Scalis, were "black nobility." The talks proceeded carefully but slowly, over several years. On February 7, 1929, Secretary of State Gasparri told the ambassadors to the Vatican that a historic agreement was about to be made. The next day, Mussolini announced the news in a telegram to all Italian ambassadors. The final details of the agreement, known as the Lateran Accords, were settled by Pacelli and Mussolini two days later. Catholicism was declared to be Italy's state religion, and Vatican City was established as a sovereign state. Relations between the

Holy See and the state of Italy were to be governed by a concordat. The accords also provided financial compensation (around US$1 billion in today's money) for the Vatican's claims stemming from the absorption by Italy of the papal states in 1870. The three accords were signed by Gasparri and Mussolini—on behalf of the official heads of state, Pope Pius XI and King Vittorio Emanuele III—on February 11, 1929, thus ending the sixty-year state of hostilities between the Vatican and the Quirinal. The signing took place in the Lateran Palace, ancient seat of the bishop of Rome—the pope. No pope had visited there since 1870.[49] Pius XI became the first to do so on December 20, 1929, when he crossed Rome to say a mass at St. John Lateran a few months after the signing of the historic accords. It was the first time he had left the Vatican since his election as pope in 1922.[50]

La conciliazione, as it came to be known, was a triumph for Mussolini, who was hailed around the world as a great statesman; no one else could have made it happen. But Marconi was quick to congratulate the pope. The Vatican Secret Archives hold a small card in Marconi's handwriting with a glowing message to Gasparri, hailing the momentous occasion.[51] He was certainly in a position to acknowledge both sides, but there doesn't seem to have been an equivalent note to Mussolini (or, if there was, there is no trace of it in the archives). The pope, it seems, was Marconi's true friend, and having the head of the world's newest sovereign state as a friend would be an interesting hedge against the unpredictable petulance of Il Duce.

Marconi's relations to Pius XI are flagged in, of all places, a satirical passage from Ezra Pound's Cantos. Pound, like so many modernist artists, was fascinated with Marconi, his aura, and his rock-star status; he was also, as is well known, an admirer of fascism. Pound's Canto XXXVIII imagines the following odd scene: "Marconi knelt in the ancient manner / like Jimmy Walker sayin' his prayers. / His Holiness expressed a polite curiosity / as to how His Excellency had chased those / electric shakes through the a'mosphere." According to Pound's editor, Ira B. Nadel, the poet also imagined wireless communication "as a method of conceiving the pagan gods appropriate to the limited bandwidth of the Modernist intellectual spectrum.... Potentially Marconi's radio-waves can stimulate divine excitement in those who possess sufficiently sensitive antennae."[52]

❋ ❋ ❋

The Lateran Accords were ratified by the Italian parliament on June 7, 1929. Four days later, Marconi made a visit to the Vatican, where he met privately with the pope.[53] It was not his first papal audience but this time, after the meeting, he made a special tour of the Vatican gardens, looking for an appropriate site for a radio transmitter. The papal entourage was impressed with his modesty as well as his thoroughness.[54] On June 14, Marconi received a letter from Francesco Pacelli, informing him that Pius XI had examined and approved his plans "for the installation of a radiotelegraphic and radiotelephonic station in the Vatican City." The letter asked Marconi to take charge of building the station, for which "payments of the sums required for the carrying out and completion of the work will be made to you, from time to time, at your request"—clearly indicating that the station was not a "gift" from Marconi, as has often been suggested, but that he was paid for building it.[55]

The idea of a Vatican radio station was first broached to Marconi by Cardinal Gasparri toward the end of the First World War. According to Beatrice Marconi, Gasparri came to tea one day when the Marconis were living on Rome's Gianicolo Hill in 1917 and 1918 and asked to be shown Marconi's experiments. "He was desirous of erecting a station at the Vatican at that time, but this did not mature till some years later," Beatrice remembered.[56] Marconi stayed in touch with the Vatican, however, and his company presented a state-of-the-art "marconiphone" receiver to the pope in January 1925, through Gasparri, by which radio broadcasts from all over Europe could be received. "It is understood that the set has been placed in one of the best drawing rooms of the Vatican and that it will be used during diplomatic receptions," a company press release fawningly declared.[57] The following year, as we've seen, when he was frenetically seeking to have his first marriage declared null, Marconi requested and received a private meeting with the pope. After his second marriage, he was a well-positioned friend of the Vatican.

Pius XI and his advisors were very conscious that the Holy See's new freedom under the Lateran treaties could be ephemeral. Successive popes had, after all, been virtual prisoners since Pius IX fled the Quirinal Palace and locked himself inside the Vatican after the troops of Vittorio Emanuele II took over the Papal States in 1870. Tensions were sure to flare between the new enclave, which was not much larger than a medium-sized urban neighbourhood, and Mussolini's Fascist government. Even before the ink on the treaties was dry, the pope's inner circle was looking for new means of support that they would be able to count on regardless of what happened in their relations with the Italian government. In a sense then, the concept and mission of a Vatican radio station was as old as that of the Vatican State itself.

Soon after the radio contract was settled, the pope decided it was time to replace Pietro Gasparri. The good-humoured *pecoraio* (shepherd) had, since 1914, served two popes well as secretary of state but he was seventy-seven years old and it was time to look ahead. The pope informed Gasparri in July 1929 and announced in December that his nuncio to Germany, Eugenio Pacelli, would be the new secretary of state. Pacelli was appointed cardinal and assumed the position on the first anniversary of the Lateran Accords, February 11, 1930. Eugenio Pacelli also happened to be, by now, well acquainted with Marconi, who had met him through his engagement to Cristina.[58] The Pacellis, like Marconi's new in-laws, had deep ties to the Vatican. Pacelli's grandfather had served in the papal government of Pius IX, fleeing into exile with his pope in 1848, and then returning to be one of the co-founders of the Vatican daily *L'Osservatore Romano*. His father, as well as his older brother Francesco, were important Vatican lawyers. The Pacellis were seen as "servants and slaves of fascism," according to Cardinal Bonaventura Cerretti, who had been Gasparri's choice to succeed him.[59] Mussolini's circle was said to be pleased with the appointment.**

Pacelli embraced the radio project and quickly became one of the leading proponents of the idea that the Holy See should look into ways of using the new medium, radio, for church propaganda. Pacelli himself was not averse to using technology—unlike the pope, who was suspicious of modern means of communication.[60] Radio had removed the meaning of geographic and political borders, and with radio, the church would no longer have to worry about such boundaries preventing the propagation of its messages. This was of particular significance to the leaders of the world's newest and smallest state, whose spiritual "subjects" nonetheless numbered more than 300 million people and were located around the world. Radio would mean that the pope could no longer be made subservient to a temporal power, at the same time giving him a global platform from which to propagate the interests of the church. The new secretary of state therefore enthusiastically supported Pius XI's proposal to build the world's first transnational broadcasting system, even more so since the job was to be done by his close acquaintance, Marconi.[61]

Marconi was of course delighted to take on the job, recognizing that it would enhance his worldwide notoriety as well as his position in Italy. The

---

** Pacelli was temperamentally the opposite of the often mercurial pontiff, and the two were also different in one respect that would become important much later: while they both understood the importance of maintaining Germany's loyalty to the church, Pacelli also had a strong personal and emotional bond with the conservative German Catholic community.

idea perfectly fit his global vision for communication. It was an excellent contract. It would cement his alliance with a major power centre, the Catholic Church; enhance the new status he was enjoying in both church and secular Italian circles as a result of his marriage to Cristina; and pay back the Vatican for accommodating the annulment of his first marriage. At the same time, it would provide an additional counterweight to his growing codependency on Mussolini and the fascist regime. In almost no time, an order was placed with Marconi's Wireless Telegraph Company in London.[62] A police informant noted a few months later that the Vatican was being "turned upside down" because of the scope of the work, for which "Marconi, naturally, is the contracting authority."[63]

Construction was swift and smooth. The station was manufactured at the Marconi plant in Chelmsford according to Marconi's specifications. The design followed the main features of the high-speed transmitters then in use by the British imperial beam stations. Marconi personally supervised the shortwave installation in the Vatican gardens. The site consisted of four towers, with seven broadcast transmitters, artfully arranged so as not to mar the beauty of the gardens.[64] There was also a separate building with a spacious transmitting room, a receiving room, machine room, general store, and general office.

On August 23, 1930, Marconi wrote to London from Civitavecchia, where he was summering with Cristina and their five-week-old daughter Elettra (born July 20, her baptism was performed in Civitavecchia by Pacelli), ordering thirty receivers suitable for receiving broadcasts from the future station. He added that this was a good opportunity for the company to design an up-to-date shortwave receiver. The order seemed to create some contractual problems for the company because of Marconi's agreements with the government of Italy, but Marconi informed the company's acting general manager H.A. White: "As you are, doubtless, aware the Vatican State is not part of the Kingdom of Italy and therefore according to my view our arrangements with associated companies and others in regard to Italy do not cover the Vatican State."[65] London evidently did not share that view and was slow to respond to the order. Eventually, a shipment of specially designed receivers was provided to the Vatican—a year late, but avoiding an embarrassing blow to the prestige of the company. For Marconi, the incident was an important step in his growing alienation from the head office.[66]

The pope decided that the Jesuits would be the best suited to run the new radio station, and named Father Giuseppe Gianfranceschi, president of the Pontifical Academy of Sciences, as its first director. Pacelli had the press called in to witness Marconi and the pope inspecting the finishing stages of work on the state-of-the-art station, pointing out that it was being done under Marconi's direct supervision.[67] On February 12, 1931, Vatican Radio was launched with an international broadcast by the pope. It was the world's first global broadcasting service.

The launch was the most vast and elaborate broadcast ever made, the first time—as Marconi pointed out in brief introductory remarks—that the voice of the pope would be heard live and simultaneously in all parts of the world.[68] It was extensively covered by the world press; had there been a competition for the most sensational headline, the prize might have gone to Toronto's *Globe*, which announced it was "without parallel in history." The *New York Times* called it the "greatest of all radio hook-ups . . . without doubt, the most gigantic broadcasting system ever assembled for one event."[69] More than 250 stations worldwide (some 150 in the United States alone) received the broadcast, sent via seven nondirectional shortwave transmitters, and translated in seven languages. Countless millions of listeners, from Melbourne to Manila to Madrid, heard the head of the Roman Catholic Church declare his pride to be "the first Pope to make use of this truly wonderful Marconian invention."[70] For some, who may never have heard a radio broadcast before let alone the voice of the pope, it must have seemed to be a miracle. The same untold millions also heard Marconi—exceptionally for him—invoke "the help of Almighty God, who places such mysterious forces of nature at mankind's disposal" that had enabled him to prepare this unique instrument. Correspondent Arnaldo Cortesi, reporting to the *New York Times* by wireless, described the scene: "Heralded by the blare of the Vatican's silver trumpets, the Pope arrived at the wireless station by automobile at 4.30 pm followed by his noble and chamber attendants. He was dressed in white except for a scarlet cloak over his shoulders . . . ."[71] The pope was greeted by Marconi, Pacelli, Gasparri, and other high Vatican officials as a platoon of Swiss Guards mingled with the papal chair-bearers outside the station. Cristina was present, as well as Marconi's associate Luigi Solari. While waiting for the pope to arrive, Marconi fidgeted with last-minute tests to places like New York, Melbourne, and Quebec, carrying out a two-way shortwave conversation with New York (the first time such an exploit was done under the difficult conditions of daylight).[72] Solari transmitted a brief broadcast account of what was happening. The pope's address lasted fourteen minutes, Cortesi reported, "considerably longer than expected."

When it was over, the pope was driven to the nearby Pontifical Academy of Sciences, where Marconi was admitted as a member.[73] The broadcast continued: Marconi spoke again, thanking the pope for the honour, and the pope replied that thanks should be addressed not to him but to Marconi and to God for placing "such a miraculous instrument as wireless at the service of humanity...man is small and in that sense, Marconi is small, but he is really a very great man."[74] The pope then conferred on Marconi the Grand Cross of the Order of Pius IX.[75] At sunset, the ceremony was over.

This was the beginning of the Vatican's long history of using the airwaves to propagate the Catholic faith.[76] It was also a key launching point for the modern understanding of "propaganda," a term originally applied to spreading the doctrine of the church. Coinciding with the rise of radio, the term now took on a sinister connotation, signifying the use of communication media to deceive and manipulate minds for political ends. Inspired by the example of the Vatican, totalitarian dictators—not least, Mussolini, and, soon after him, Adolf Hitler—as well as leaders such as Franklin Roosevelt and Winston Churchill were soon using radio to inspire, cajole, mobilize, or terrify. The great international shortwave broadcasters such as BBC World Service (the next to air, in 1932), Voice of America, Radio France, Deutsche Welle, and Radio Moscow followed in its wake.

In the short term, the launch of Vatican Radio strengthened Marconi's position with regard to Mussolini and his regime—even as it annoyed them. On March 5, 1931, one of the chief agents of Mussolini's political police, Virginio Troiani di Nerfa, reported that Marconi had decided to leave Rome suddenly, "without waiting for an *important meeting* of a political nature in which he was supposed to participate" (emphasis in original). It was said that before leaving town Marconi had an hour-long private meeting with the pope, of which he had requested that nothing be said. "As one can see, the skein is tangled," Troiani reported, somewhat cryptically.[77] Mussolini could not have been happy about Pacelli and Marconi collaborating to provide facilities for Vatican propaganda at a time when relations between the church and the regime were still tenuous, despite the Lateran treaties. Vatican politics were murky at the best of times, and within months of the launch of the Vatican radio station the publication of two papal encyclicals clouded the relationship of the Holy See to fascism even further.

Quadragesimo Anno, introduced by the pope in an hour-long Vatican Radio broadcast on May 15, 1931, called for the reconstruction of the existing social and economic order, describing the dangers arising from both

Top: Marconi, in uniform, with Italian Prime Minister Vittorio Emanuele Orlando, Villa Bianca Abano (Italian headquarters on the Austrian front), during Austrian armistice negotiations, October 27, 1918 *(Bodleian)* **Bottom:** Marconi and Inez Milholland, *New York Tribune* front page, May 23, 1915 (New York Tribune *[New York, N.Y.]*), *May 23, 1915.* Chronicling America: Historic American Newspapers, *Library of Congress)*

Top left: Gabriele D'Annunzio, poet and revolutionary who described Marconi as "the magic hero" *(Library of Congress, Prints & Photographs Division, LC-B2-1272-14 [P&P])* **Top right:** Francesca Bertini, "the actress who said no to Marconi" **Bottom:** Marconi protégée, journalist Lisa Sergio *(Library of Congress, Prints & Photographs Division, NYWT&S Collection, LC-USZ62-132346)*

Partito Nazionale Fascista

FASCIO DI MILANO

Il sottoscritto GUGLIELMO M A R C O N I

nato a Bologna *il* 1874

figlio di

indirizzo Regio Senato ROMA = (Marconi House Londra)

presa visione del programma fascista, dichiarando di accettarlo, **Senza Riserve** *chiede di essere iscritto al* Partito Nazionale Fascista, Fascio di Milano.

Si quota in £. .... *per spese di propaganda e si impegna di versare regolarmente i contributi stabiliti.*

In relazione a quanto fu disposto dai dirigenti il **P. N. F.** dichiaro io qui sottoscritto di non appartenere ad altro Partito politico.

Milano, ... *192*

(Firma leggebile dei soci proponenti)	(Firma del richiedente)

/11583)

Visto per l'ammissione

IL DIRETTORIO

Le domande non saranno prese in considerazione se non saranno corredate della firma di almeno **DUE** soci proponenti. — Il richiedente deve rispondere esattamente aile seguenti domande:

Se ha prestato servizio militare

Grado Capit. di Fregata *Corpo* Regia Marina.

Durata di servizio Complessivamente 6 anni.

Tempo trascorso al fronte 2 anni

Ferite .... *Decorazioni*

Su richiesta tutte queste risposte dovranno essere comprovate da documenti.

Siete elettore politico? SI *Amministrativo?* SI

Professione Scienze Elettriche

Impiegato presso

COMPETENZA SPECIFICA RADIOTELEGRAFIA.

Volete far parte delle squadre di principi? NO *o di triari?* Si

Siete ciclista? SI *Automobilista?* SI

Motociclista? NO *Aviatore?* NO

Possedete qualche veicolo di locomozione? SI

Appartenete ad Associazioni sindacali od altre? NO

Ad Associazioni sportive? NO

Siete mai stato in nessun partito? NO

Quale?

NB. - La domanda dovrà essere corredata dal certificato penale su richiesta.

Marconi's application for membership in the Italian National Fascist Party, June 15, 1923 *(Bodleian)*

Top: Marconi (in top hat) and Pope Pius XI, at inauguration of Vatican Radio, Vatican City, February 12, 1931 *(Photo by Vatican Pool/Getty Images)* ***Bottom:*** Marconi and Niels Bohr at the International Congress on Nuclear Physics, Rome, October 1931 *(Fot. Comm. A. Petitti, Lawrence Berkeley National Laboratory, courtesy AIP Emilio Segrè Visual Archives, Fermi Film Collection, Segrè Collection)*

Top: Marconi with Franklin and Eleanor Roosevelt at the Chicago World's Fair, October 1933 *(California Museum of Photography)* ***Bottom:*** *(left to right)* Marconi, Mussolini, and Cristina Marconi, on the *Elettra*, Fiumicino, June 7, 1930 *(ASSOCIATED PRESS)*

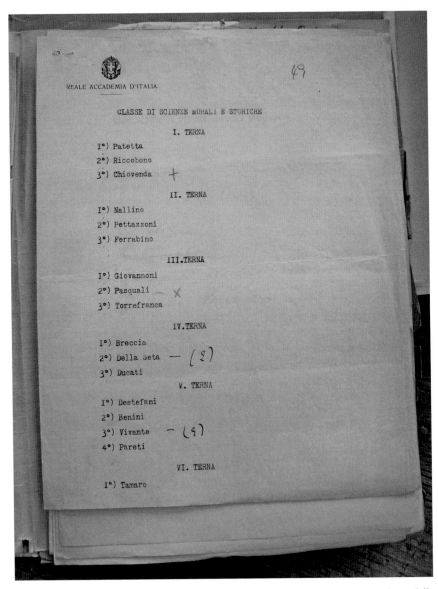

REALE ACCADEMIA D'ITALIA

49

CLASSE DI SCIENZE MORALI E STORICHE

I. TERNA

Iº) Patetta
2º) Riccobono
3º) Chiovenda

II. TERNA

Iº) Nallino
2º) Pettazzoni
3º) Ferrabino

III. TERNA

Iº) Giovannoni
2º) Pasquali
3º) Torrefranca

IV. TERNA

Iº) Breccia
2º) Della Seta
3º) Ducati

V. TERNA

Iº) Destefani
2º) Benini
3º) Vivante
4º) Pareti

VI. TERNA

Iº) Tamaro

Marconi's annotated list of candidates to Italy's Royal Academy, 1932 *(Archivio della Reale Accademia, Accademia Nazionale dei Lincei)*

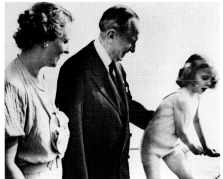

Top left: Marconi (second from left) hosting Nazi minister Hans Frank (fourth from left) at the Royal Academy, Villa Farnesina, Rome, April 4, 1936 *(Archivio storico Istituto Luce)* ***Top right:*** Cristina, Marconi, and daughter Elettra on the *Elettra*, Viareggio, March 1937 *(ASSOCIATED PRESS)* ***Bottom:*** The *Elettra*, moored off Viareggio, 1932 *(Photo by Mondadori Portfolio via Getty Images)*

Top: Mussolini saluting the arrival of Marconi's bier for funeral at the Basilica of Santa Maria degli Angeli, Rome, July 21, 1937 *(ASSOCIATED PRESS)*
Bottom: Villa Griffone and Marconi mausoleum, Pontecchio, 2009

unrestrained capitalism and totalitarian communism, as well as the ethical implications of reconstruction. It was one of the most important political interventions of the 1930s, approving the tripartite corporatism of government, industry, and labour (what we might call multi-stakeholderism today) favoured by Italian fascism and soon to be emulated by the proto-fascist regimes of Spain and Portugal. But it was also couched in a tone that could invite the praise of a liberal politician like FDR, who took to quoting it on the evils of excessive economic power. In many respects, the encyclical quite closely reflected Marconi's own sociopolitical stance.

Pius XI was quietly preparing a second encyclical, addressed to the regime's resistance to the church's efforts to increase its influence in Italian society. Afraid that Mussolini would try to block its publication in Italy, the Vatican had Monsignor (future Cardinal of Boston) Francis J. Spellman, then attached to the Vatican State Department, secretly smuggle copies into France, where the encyclical was published in June 1931. (It appeared first abroad on June 29, 1931, and later in the Vatican newspaper *L'Osservatore Romano*.) *Non Abbiamo Bisogno* (We Have No Need) denounced the regime's recent attacks on the church's proselytizing efforts—and, among other things, called attention to its use of radio to spread anti-Catholic messages.[78]

The line between sacred and secular issues became further blurred with the growing tension of the political situation in Europe. On July 20, 1933, the Vatican signed a concordat with Nazi Germany (negotiated by Secretary of State Pacelli), and in 1935 the church hierarchy (if not the Vatican per se) pronounced Mussolini's invasion of Ethiopia a "holy war." Vatican Radio played an important role throughout these events, striking "a strangely modern note in a world that seems dominated by the past," according to a 1936 article.[79] Officially, like the Vatican itself, the station maintained a policy of neutrality during the Second World War, but its personnel eventually found themselves involved in covert activities as the Vatican's position became more ambiguous. Some argue that Vatican Radio had an honourable and courageous record of war and Holocaust reporting, being the first radio station to reveal, for example, the confinement of Jews and Polish prisoners in ghettos in Poland in 1940. Later, during the Cold War, the shortwave station became an important piece of the West's global propaganda apparatus.[80]

❈   ❈   ❈

With the Lateran Accords behind him and what Kertzer calls his "peculiar partnership"[81] with the pope in place, Mussolini began concentrating in earnest on consolidating his dictatorship. On March 24, 1929, national elections

were held under an electoral reform by which only a single list of candidates, approved by Mussolini's Grand Council of Fascism, were presented to
the voters, who could vote "yes" or "no." Of those who voted, 98.4 percent
approved the list. The election was a plebiscite of adulation for Il Duce; his
cult of personality was now complete and he was commonly viewed as a
sacred idol (some of the few cynics who dared to voice their thoughts said
Italy now had *two* infallible leaders).

   Mussolini was also interested in creating a new fascist nobility, and one
of the beneficiaries of this policy was Marconi, who was named a Marchese,
or Marquess, by King Vittorio Emanuele III on June 17, 1929. It was a hereditary title, transferable down the male line, and Marconi's son, Giulio,
could expect to inherit it one day. Cristina, meanwhile, became the Marchesa
Marconi. The news attracted worldwide attention, pleasing Marconi, needless
to say, but he was annoyed that his friend Gabriele D'Annunzio was awarded
an even higher honour; D'Annunzio was named Prince of Montenovoso.[82]
On the international stage, Marconi, Mussolini, and D'Annunzio still stood
on equal footing as representatives of Italianness.

   In the weeks following the ceremony at the Campidoglio where the
priorities of the national research council were announced in February
1929, Marconi met with Mussolini and remained in close touch with his
office;[83] then he wrote to him (from London, where he spent most of 1929,
but on the letterhead of the Italian Senate): "Excellency," the letter began.
Marconi had just learned that the government would not give the CNR an
anticipated funding increase from 575,000 to 1.5 million lire. This was decidedly in contrast with Mussolini's speech at the Campidoglio, Marconi
reminded him. "In naming me I assumed the Government was aware of the
fundamental importance of scientific research for the life of the country," he
wrote. Now the CNR faced paralysis. "Let me tell you frankly that this is
creating disillusionment in the Italian scientific-technical world and abroad,
as well as providing evident pleasure to the critics of the regime." Marconi
ended with an urgent plea for Mussolini to consider increasing the CNR's
funding.[84]

   Marconi spoke crisply, without mincing words or kowtowing to
Mussolini, while necessarily taking care to use the regime's language in
framing his request. In this regard, he was behaving like a typical Italian intellectual of this period, using the language of fascism to promote his favoured
projects.[85] In fact, Marconi was one of the few figures who could clearly speak
his mind to Mussolini, and Mussolini appreciated his candour, although he

rarely let Marconi influence him. Marconi kept up his energetic research lobbying in the following months, and it was even reported in the fascist press.[86]

On May 27, 1930, after he received the CNR board for a meeting on the *Elettra* at Civitavecchia and demonstrated his latest methods for wireless transmission to London,[87] the reports pompously emphasized the "miracles" taking place on the floating lab. Marconi's magical qualities were still in the news. The press also recalled, somewhat exaggeratedly, that Mussolini himself often visited the yacht.[88] It was all excellent publicity for Marconi.

In September 1930, Mussolini elevated Marconi to a second major position of service to the fascist regime, naming him president of the Reale Accademia d'Italia, the new royal academy he had created to take the place of the venerable but testy Lincei, which had positioned itself as hostile to the regime. In 1926, inspired by what Napoleon Bonaparte had done in France more than a century earlier, Mussolini decided to set up a national academy, to showcase the role of Italian science and culture in Italy and the world. The academy would be headquartered in the Villa Farnesina in Trastevere, directly across the narrow Via della Lungara from the Palazzo Corsini, which housed the Lincei. The creation of the academy was announced in January 1926, but it was only launched on October 28, 1929 (the seventh anniversary of the March on Rome), when Mussolini named the former Liberal politician Tommaso Tittoni, now a fascist supporter, to be its first president.[††]

The academy was to have sixty members; the government would name the first thirty and they would nominate the others, with the head of government, Mussolini, having the final say. Like the research council headed by Marconi, the academy also had a strategic purpose. Positions were offered to figures who had not necessarily demonstrated open support for fascism; in order to accept they were required to swear an oath of loyalty. Almost no one refused an invitation to join.[89] The original thirty members included futurist superstar Filippo Tommaso Marinetti, playwright Luigi Pirandello, and a young physicist, Enrico Fermi. Marconi was not among the first group named to the academy (a rule intended to prevent the accumulation of public positions prohibited membership of certain categories of officials, including senators), but on September 19, 1930, Mussolini made him president.[90] Marconi had so far served him well. By 1930, Marconi's presidency of the CNR had become one

---

[††] The Royal Academy overshadowed and eventually absorbed the Lincei on June 8, 1939; the Lincei recovered its former role when the academy was disbanded after the war (Cagiano de Azevedo and Gerardi 2005, introduction).

of the principal instruments for developing a fascist view of culture, and Marconi had delivered, despite his private dissatisfactions, tacitly supporting Mussolini's increasingly restrictive constraints on academic freedom.

The presidency of the academy made Marconi an ex-officio member of the Grand Council of Fascism, which was more powerful than the cabinet and which drew Marconi even more deeply into Mussolini's circle. Set up in 1923, the Grand Council was, in fact, the main political institution of fascism; it was made up of a precise list of office-holders and party hierarchs and included among its powers the election of party candidates to public office, the nomination of party leaders, the approval of major policies as well as all decrees and laws, the ratification of international treaties, and essentially everything to do with foreign affairs. Its meetings were convened by Mussolini at his discretion. In theory, the Grand Council also had the power to recommend to the king the replacement of the prime minister. (Theory became practice on July 24, 1943, when, on a motion of Dino Grandi, the council voted 19–8 with one abstention to depose Mussolini.)

The presidency of the academy was thus far more explicitly political than the presidency of the CNR, which, at least nominally, was meant to play an objective role in promotion of research. Within the week, Mussolini wrote to his finance minister, Angelo Mosconi, about Marconi's financial laments, advising Mosconi that he was willing to grant the research council 570,000 lire "and not a penny more" from a slush fund he maintained for unforeseen expenses. Noting that Marconi was now also president of the academy, he added, pointedly, "It remains to be seen if this Council can continue to function, or if it can not better be absorbed by the Academy."[91]

It is not likely that Marconi ever saw or was even aware of this note, or was aware of Mussolini's thinking on the subject; Mussolini was clearly more inclined toward *la bella figura* of the prestigious but low-cost academy, rather than the expensive and politically risky expectations of an independent research council; it might have been enough for him to have Marconi in one, largely ceremonial, position, rather than lobbying for expensive projects from two perches. Marconi was officially installed as president of the academy, with great pomp, on November 29, 1930. When, around the same time, he submitted the annual report of the CNR, Mussolini wrote Marconi an elegiac note acknowledging the importance of the council's work—"in spite of the extremely modest financial means at its disposal"—adding that he believed the CNR should establish a close collaboration with the academy.[92] In coming years, Mussolini would frequently refer back to the idea of fusing the two; it never happened, although under Marconi's leadership the two institutions often functioned as though they were one, or at least were connected like communicating vessels. Not only did Marconi set thin boundaries between the activities of the academy and the council, he also used both for his own purposes or to promote pet projects.[93]

Marconi continued, of course, to maintain a high international profile, his global visibility being one of his strongest assets, both for himself and for Mussolini's regime. In June 1930, he took part, by radio from London, in the second World Power Conference (WPC), held in Berlin. (Marconi nearly always managed to avoid travelling to Germany.) On June 18, 1930, Marconi, Thomas Edison, Owen Young of RCA, Lord Derby (a former British postmaster general), and three other international personalities delivered broadcast talks to the conference plenary in Berlin by international radio hook-ups. Six years earlier, in a lecture to the Royal Society of Arts coinciding with the first WPC in London in 1924, Marconi had set the tone for the organization that saw its mission as peace-building and thought of itself as a "technical League of Nations." Radio, he had said in 1924, would bring about "the unity of world thought and opinion." Now he went further, in one of his most idealistic calls for international co-operation to date: "[Radio] is the greatest power we possess or control.…It is the power of the human mind, the power of love and of thought…the power which should for ever tend to promote and cement goodwill and co-operation between nations, and which I firmly believe is destined to prevent the horrors of war by bringing about a more and more perfect understanding between all peoples."[94]

In September 1931, Mussolini asked Marconi to represent the Italian government at ceremonies marking the one-hundreth anniversary of Michael Faraday's first experiments in England. Marconi loved nothing more than this type of ceremony, at which he could be assured of an opportunity to recall the latest version of his own experimental history. Marconi telegrammed police chief Arturo Bocchini to inform him that he was leaving for London on an official mission and inquiring whether his bodyguard, agent Annunziato (Nunzio) Lettieri, would be travelling with him as usual. The answer was positive.[95] On September 19, 1931, Marconi did a 10:00 p.m. BBC broadcast in which he characterized broadcasting as "the lusty offspring" of point-to-point wireless telephony. On September 21, he published an article in a special edition of the *Times* (quoted in chapter 2), appearing along with Lodge, Fleming, Kelvin, and others. As usual, he made the connection to his own work in both interventions.[96] On his return to Rome, he reported promptly to Mussolini, enclosing news cuttings of his performance and complaining that the event was totally ignored by the only Italian newspaper in London, the fascist journal *Italia Nostra*.[97]

❖   ❖   ❖

The way in which Marconi juggled his various roles in fascist Italy may be best illustrated by his relationship with the so-called Via Panisperna Boys, the group of young nuclear physicists working under the leadership of Enrico Fermi at the physics department of the University of Rome.

Via Panisperna is a pleasant, narrow street, a bit out of the way but still in the centre of Rome, just behind the Trajan Market and close enough to the major sights to support a host of small shops and restaurants catering to the tourist trade. At the far end of the street, in a quiet garden complete with shady almond trees and a goldfish pond, lies the university's Institute of Physics, headed in the 1920s by Mussolini's minister for the national economy, Orso Mario Corbino.[98] Himself a physicist interested in the commercial applications of electricity (among other things, he was on the boards of Italian General Electric and the Edison Company of Milan), Corbino gave up active research to pursue a career as a scientific manager and politician. A former liberal minister of education, he believed strongly in a role for scientific research and innovation as a pole of economic development and was an early supporter of Mussolini's economic policies.[99] Interestingly, he was not a member of the party,[100] an indication of his prestige and a significant counterpoint to Marconi in this regard. Marconi and Corbino became close allies in a joint effort to embed science policy in the basic framework of the new corporate state. Together, they promoted the role of scientific research and innovation as a key plank of economic policy and lobbied Mussolini to provide more support for young researchers.[101]

In 1926, Corbino established a chair in theoretical physics at the Institute, and recruited Enrico Fermi to fill it. With Corbino's active support, Fermi soon built one of the world's most advanced research centres in theoretical physics, assembling a team that included future Nobel Prize winner Emilio Segrè, Franco Rasetti, and Bruno Pontecorvo. The group came to be known, and is remembered in Italy today, as the Via Panisperna Boys. Fermi was one of the original members named to the newly launched Royal Academy in 1929 and, as was expected of him, he immediately took out Fascist Party membership; he was appointed secretary of the academy and still held that position as late as August 1936.[102] With the support of Corbino and Marconi, Fermi became one of the key players in Italy's scientific, industrial, and research establishment. He was also chairman of the CNR's physics committee, research consultant to several private sector firms, a member of the board of the national radio company EIAR, as well as head of the Via Panisperna research team.

In 1931, shortly after Marconi assumed the presidency, the academy was charged with organizing an annual conference with the Alessandro Volta Foundation, recently created and funded by the Italian Edison company. Edison enjoyed the favour of the regime and a near monopoly in Italian electrical industry, and Corbino was a member of the Edison board.[103] Together, Corbino, Marconi, and Fermi decided to build on the success of the 1927 conference at Lake Como (described earlier) and organize the first Volta conference on the topic of nuclear physics. Fermi was the chief organizer and conference secretary, Corbino was the effective chair, and Marconi, as honorary chair, was the public face of the conference. For Fermi, the conference was an opportunity to bring to Rome the world's most illustrious physicists, while Marconi's role was to convince Mussolini that such a world event fit with his view of the academy as a brain trust in service of the state. Fermi was constantly trying to limit Marconi's firm but diplomatic efforts to have the organization meet the regime's political demands. They were useful to each other but Marconi was definitely the representative of the power structure.[104]

The International Congress on Nuclear Physics met in Rome from October 11 to 18, 1931, with Marconi in the chair. No fewer than seven Nobel Prize winners, including the venerable Marie Curie, attended, and Mussolini put in an appearance at the opening session. The meeting, like many academic conferences of the period, was politically ecumenical; its participants included Bohr, the Danish later contributor to the Manhattan Project, and Heisenberg, who would head the German nuclear energy project under the Nazis. The meeting was also attended by two important figures from Marconi's early British days: James Ambrose Fleming (days before resigning as scientific advisor to the Marconi company) and Oliver Lodge, to whom Marconi had issued a warm personal invitation.[105] The conference was a stunning success and had enormous international resonance. What the world saw was the dynamic contribution of Marconi and Fermi to nuclear physics, not the bureaucratic back-room dealing and stifling constraints of Mussolini's science policy.‡‡

The relationship between Marconi and Fermi became a sort of proto-type for Marconi's vision of a merger between fascist doctorine and empir-ical science. In addition to doing fundamental research, Fermi was a research administrator, sitting on a series of interconnected committees within the academy, the CNR, and related bodies. Marconi, meanwhile, tried to repre-sent the high policy objectives of the state. In hindsight, Marconi's approach was a recipe for disaster. Despite repeated, even public, promises, Mussolini did not allocate more resources to science, and Marconi was left to rehash an unconvincing discourse on the fascist idea of "coordination and discipline" of scientific research, without giving any sign of understanding what that actually meant. Marconi had an honest belief that the state could be used for the public good, and his own experience tended to favour collaboration between the state, private enterprise, and individual creative initiative; but his inherent conservatism and his comfort in the orbit of existing power structures blinded him to the totalitarian nature of fascism. His acquies-cence with the regime would have onerous consequences.

‡‡ Marconi also played a key role in several of the later Volta conferences, which alternated be-tween topics in science and the humanities. The second one, in November 1932, focused on Europe; the third, in 1933, was on immunology. The 1934 edition (October 8–14, 1934), co-chaired by Marconi and Luigi Pirandello, was on the state of dramatic theatre in the world. D'Annunzio's play *The Daughter of Jorio* was presented, and the conference was attended by the likes of William Butler Yeats, Walter Gropius, Maurice Maeterlinck, and Marinetti. Marconi sent a personal invitation to George Bernard Shaw, but there is no record of a response. Paul Claudel and Jean Cocteau were too busy to attend but Cocteau sent a paper (GM to G. B. Shaw, May 14, 1934, LIN 14/154; *LT*, January 14, 1935; Andreoli 2004, 39).

32

Science and Fascism

After the successful introduction of the beam system, the cable and wireless merger, and his political appointments in Italy, Marconi considerably reduced his personal research activities, but in the early 1930s he found a renewed interest: very short, or *micro*, waves.[1] Most of his work in this area would take place either on the *Elettra* or at the Italian company installations at Santa Margherita Ligure near Genoa. The aura surrounding all things Marconi led to constant press speculation about what he was up to, especially as the political situation in Europe became more critical.[2] Mussolini's police informers kept up a flow of reports (mostly either cribbed from the newspapers or based on wild rumours) and Il Duce himself never missed an opportunity to see what was going on.

In the mid-1930s, the British Marconi company claimed it was spending £100,000 a year on research, much of it to support Marconi's "floating lab" on the *Elettra* as well as the experimental quarters at Marconi House that he rarely visited.[3] Early in 1929 Marconi's private research staff moved from London to join him in Italy. Although he was still chief technical advisor to the company (as well as to all the companies of the merged cable and wireless group), from that time on, Italy was the main location of Marconi's research.[4] Marconi was more the inspiration for than the executor of the work on microwaves. Once again he showed his intuition, his independent character, and his talent regarding strategic objectives, but also his reliance on the scientific competence of associates, and the support of his parent company. Marconi articulated this characteristically in a 1933 letter to H.M. Dowsett, then the company's research manager in London, about an issue concerning plans for a revolving directional aerial to be used at the new Italian government station being built at Civitavecchia: "The research work at this station is <u>entirely</u> under <u>my</u> direction," he wrote (his emphasis), "but the expense of its construction and working will be borne by the Italian

Government."[5] He was still keeping his finger in everything, juggling projects and trying to keep on top of the ball.

Most of Marconi's personal work was done with two key assistants,
Gaston Mathieu and Gerald Isted, and in frequent contact with old hands like
Dowsett and Andrew Gray in the company's research department in England.
Mathieu, a Belgian electrical engineer, became Marconi's closest research
associate during this period. While seconded to London to work in the British
War Office research labs during the First World War, he had designed wireless
apparatus for the British army and invented a wireless telephone for communication between airplanes. Mathieu's work attracted the attention of the
Marconi company and he came on board at the end of the war. He was soon
a member of Marconi's personal staff, becoming Marconi's research assistant
in 1921, and travelling with him during his 1922 visit to the United States,
where he produced a valuable report on the state of the art of US broadcasting.[6] By 1925 he was lecturing on the history of wireless, making sweeping
Marconi-like statements about the origins and development of the technology and its applications.[7] By 1927, Marconi was acknowledging Mathieu's
work in his own speeches and publications, placing him on a par with such
early key collaborators as Charles S. Franklin and Richard N. Vyvyan.[8]

Marconi's relationship with Mathieu reveals something about the master's relationships with his closest collaborators. In 1926, the company sent
Mathieu to Canada for last-minute tests at its new state-of-the-art stations
at Drummondville and Yamachiche, Quebec, before the launch of its first
transatlantic shortwave link. The Montreal daily *La Presse* described Mathieu
as "Marconi's favourite pupil...his disciple." The newspaper reported that
Mathieu spoke of Marconi "as a son of his father" and commented that
Marconi bred enthusiasm and zeal in those around him. It didn't use terms
like "worship" or "charisma," but that's what it was about.[9]

While Mathieu was in Canada, he and Marconi exchanged observations on the effect of the aurora borealis and "sun-spots" on the clarity of
the signals they were receiving.[10] As the beam service got going, "fading" of
signals was a persistent problem. Mathieu seemed to be working on the issue
on his own. On January 10, 1928, he wrote to Marconi expressing the view
that a dedicated long-distance channel for wireless telephony would not be
profitable so he had developed a "multiplex" system in which telegraphy
and telephony could be conducted simultaneously on the same circuit; the
telegraphy service would pay for itself and so the telephony would be net
profit. This came to be known as the "Marconi-Mathieu multiplex system"

and was first put into use by the imperial beam service. By 1929 the company was able to broadcast from London to Australia, by way of Canada,
using the system. The signal was reportedly received as clearly as a local
broadcast.[11]

In 1929, Mathieu joined Marconi permanently in Italy. He still travelled between London and Rome, however, occasionally bringing Marconi
small items of comfort that were unavailable in Italy between the wars: liver
extract, hair lotion, dental magnesia.[12] Much more important, he had a hand
in nearly all of Marconi's research activity, helping to edit lectures (such as a
paper Marconi presented in December 1932 at the Royal Institution in
London[13]) and accompanying Marconi on his travels. Mathieu had an able
young assistant, Gerald Isted. Isted was of a younger generation than the old
Marconi hands, and was himself a practising radio ham. He joined the company in April 1923, working first in production at its plant in Chelmsford.
In July 1926 he was transferred to Marconi's research lab in London, and
worked under Mathieu on developing the shortwave multiplex system. He
followed Mathieu to Italy in 1929; in a country where formalities were important, he was authorized to use the title "engineer."[14]

The microwave research got going in earnest in 1931, with transmissions from Santa Margherita Ligure to Sestri Levante, a distance of ten miles,
and then to Levanto, twenty-four miles away. The Marconi installation in
Genoa (essentially a factory), opened in 1908 under Marconi's agreement
with the Italian government, was the most advanced plant of its kind in Italy.
By 1927 it had four hundred employees and a diversified product line. Most
of its work was done on behalf of the British company. In the 1930s, however, Genoa became the base of Marconi's technical operations and, along
with Rome, of his Italian business operations as well. A report filed by
Mathieu on January 15, 1932, spoke of a new "quasi-optical" system Marconi
was working on, intended to compete with an anticipated IT&T wireless
telephone service that utilized eighteen-centimetre wavelengths to reach
across the English Channel. The old competition flame was still with
Marconi, but everything pointed to better conditions for his research in
Italy, where it was easier to obtain government permissions and testing facilities, thanks to the advantages Marconi could claim from his position as
head of the Italian national research council.[15]

Two years after the inauguration of Vatican Radio, Marconi installed
one of the first applications of microwave technology, a microwave telephone link between Vatican City and the pope's summer residence at Castel

Gandolfo, fifteen miles away. The pope himself inaugurated the link on February 11, 1933.[16] This was the first operational microwave duplex telephone and telegraph service in the world. Politically, it was also a subject of controversy; one of its chief purposes was to allow the pope to remain in contact with the Vatican from his summer palace without having to use the Italian state services for his communications. A usually well-informed Polpol agent, Arrigo Pozzi, reported discontent within the Vatican bureaucracy toward Marconi and his company. The new station was said to be very costly, with the company making a huge profit. The pope was protecting Marconi as a close friend and protégé, it was said. A scandal was brewing. Pozzi's report also returned to one of Polpol's preferred themes: Marconi's marriage. Theological experts were saying that Marconi's first marriage was valid, that the annulment ordered by the pope himself did not stand up, and that Marconi was in fact living with his second wife in a state of concubinage. The pope was quite annoyed with these rumblings, and that was one of the reasons he was giving Marconi a level of protection he had never accorded to anyone else, Pozzi reported.[17]

The company continued to spend large sums on Marconi's experimental work, but on May 23, 1933, company chairman Lord Inverforth was able to tell MWTC's thirty-fifth annual general meeting that profits were up despite the world economic crisis. Under Marconi's guidance the company's "unrivalled research organization" had been in the vanguard of progress in wireless telegraphy since 1897, he said. Marconi's work on the practical application of microwaves foretold another paradigm shift in global communication, like those Marconi had previously shepherded. Marconi had conclusively shown that microwaves could be used for radio and telegraphy over longer distances than anyone had ever expected or thought possible—as he had done previously with shortwave, and before that with his original work on long-distance wireless. Although Inverforth didn't say it in so many words, the implication was that there was another Marconi revolution in the air.[18]

In August, Marconi interrupted the sacrosanct Roman vacation period for a special meeting of the academy to divulge the results of his latest microwave experiments. He personally alerted his British company headquarters the day before so that the news could be announced simultaneously in London and in Rome, melding his global activities seamlessly.[19] Marconi read a paper entitled "On the Propagation of Microwaves at Long Distances," reporting reception of microwave transmissions on the *Elettra* more than

150 miles away from the transmitting station at Santa Margherita Ligure. He was anything but modest in his assessment of these latest results. "My tests are indeed extremely interesting, as new and peculiar facts, affecting the whole theory and practice of radio, are certainly emerging from them," he wrote from Sardinia to MWTC's new managing director H.A. White, who had taken on the position after Kellaway died in April 1933.[20] Marconi's police handlers seemed to agree, although they put a more serious spin on it. Marconi's latest secret studies on ultra-shortwaves were "the prelude to a great discovery of military nature and truly exceptional reach," reported Bice Pupeschi, one of the Polpol agents assigned to watch over him. Mussolini's interest was piqued; he kept a copy of Marconi's report to the academy in his personal files.[21]

Rumours of "secret" Marconi research had been appearing in Polpol reports as well as the British and Italian press for some time. From Polpol's perspective, the presence of French and English connections undoubtedly made it more interesting. On March 27, 1930, a Polpol agent reported from Genoa that Marconi was conducting secret research "of a bellicose nature" and that his most recent inventions could have a military purpose. The French consul general had been on the *Elettra* and was "most interested" in Marconi's experiments. It was said that Marconi also had an English engineer (presumably Gerald Isted) on board.[22] Mussolini himself was interested enough to show up at the yacht literally out of the blue, unannounced, while it was moored at Fiumicino on June 7, 1930; he was of course graciously received by Marconi, in full yachtsman's regalia, and an eight-months pregnant Cristina.[23] There didn't seem to be much research going on, though that didn't stop Marconi from later making political and social capital from the occasion. He had photos taken of himself with Mussolini on board the *Elettra* and had signed copies made for colleagues.[24] In the coming months Polpol continued to pursue the story of Marconi's secret research. On November 10, 1930, mainly picking up on press reports, Polpol reported that Marconi had "carbonized" a group of sheep with a burst of wireless energy.[25]

Speculation as to the military possibilities of Marconi's work spilled into the public sphere as well. In April 1931, Marconi was said to be on a "secret" mission to London (duly reported by Polpol[26]). It was so secret that the British tabloids couldn't get enough of it. The *Daily Herald* trumpeted: "Short waves that may put everyone on the phone to ends of earth." The *Daily Sketch* had a photo of him walking in the street, with the caption "The Ether Wizard. The Marchese Marconi, who is engaged on secret experiments with ultra short waves, photographed in London yesterday."[27] Marconi was by now moving his work on microwaves into high gear, as we've seen. By the time Polpol reported, on July 10, 1934, that Marconi had made "a sensational discovery" that could immobilize military apparatus from a distance, Marconi had Mussolini's full attention.[28]

In May 1935, newspapers around the world began carrying reports that Marconi had invented "a fantastic new weapon." Mussolini, himself, it was said, had attended experiments where Marconi caused internal combustion engines to cease functioning. There were further rumours of Marconi stopping aircraft in full flight, halting armies in their tracks, and having created a "death ray." Marconi denied the reports of his invention in letters to the press and elsewhere, but this only fanned the rumours.[29] Polpol was kept busy keeping up with it all. On August 13, 1935, an agent reported from Milan that Mussolini had gone to inspect the new invention at Spezia, only to be shocked to learn that it had not yet yielded any results; a few days later another agent reported from Rome that Mussolini had witnessed an electric wave that shuts down motors and "sows death."[30] At the end of the month, there were renewed published reports about a "death ray." Marconi refused to make any statement whatsoever regarding his latest experiments, again only fuelling the stories.[31] One British newspaper, the *News Chronicle*, reported that Marconi had "revealed" that he was testing a wireless shortwave that would stop airplanes. He would not say any more until patents had been secured, except that the work was "well advanced."[32] It was also reported that Marconi had disclosed some of his experiments to both Mussolini and the king of Italy.

On August 28, the dramatized US radio news program *The March of Time* ran a clip recounting a fictionalized conversation between Mussolini, portrayed as a war-hungry tyrant, and Marconi at Santa Margherita. When Mussolini learns that Marconi had crafted a wireless ray that could stop an engine in its tracks, he asks: "It would mean that all motorized war machines would be useless, that foot soldiers and cavalry will again be the whole strength of an army?" And Marconi replies, "I believe the ray will also be invaluable in medical treatment." The presentation hints that Marconi's motivation may not be quite so virtuous, as he then says he would like to test the invention in the field in real war conditions.[33] This program appeared as Italy was building up to invade Ethiopia, and was a sign of growing awareness in the United States of Mussolini's true designs as well as of the ambivalent role of Marconi's research.

The outlandish stories—and Marconi's denials—kept coming.[34] The rumours came to a head in June 1936, when Marconi reportedly demonstrated a new discovery to Mussolini in broad daylight. The story had some basis in fact. It seems that one day, an operator working the microwave connection between the Vatican and Castel Gandolfo "happened to hear a queer noise, something like the sizzle produced by a person treading heavily on slushy ground," Marconi's secretary Umberto Di Marco later wrote.[35] Upon investigation, it was found that the strange noise had been caused by a gardener's cart passing through the beam of the microwaves. When this was reported to Marconi, he immediately recognized the military significance of the discovery—which was ultimately an embryonic form of radar—and dutifully wrote to inform Mussolini. "Mussolini showed great interest in the matter, expressed his desire to have the experiments repeated in his presence and turned up at the appointed place," an isolated section of the Rome–Ostia road, where it could be demonstrated that motorcars passing through a microwave beam created a distinctive signal. A communiqué was issued stating that Mussolini had attended some experiments "of military importance" by Marconi, but with no further detail.[36]

Marconi was of greatest use to Mussolini, however, in the area of international propaganda. In 1932 Mussolini asked for his advice. Mussolini had the idea of using radio to improve understanding of the Fascist regime outside of Italy. Marconi assured him that, technically, this could be done. Italian radio had the facilities to be heard in London, Paris, or Moscow. Mussolini was anxious about the impact of an embarrassing political failure and pondered the project carefully; finally, he gave the green light to foreign broadcasts. On Marconi's recommendation, a young journalist by the name of Lisa Sergio was hired to do the English-language program.[37] Marconi was also frequently on the air himself, especially in broadcasts aimed at the United States. Along with other personalities with international reputations, such as Enrico Fermi and Ezra Pound, he appeared in conversation on the short-wave program *The American Hour*, a production of the regime's press and propaganda department hosted by Sergio.[38]

Marconi needed little prompting at this time to sing the praises of fascism. On August 1, 1932, he sent a memo to Mussolini saying he was thinking of addressing a call to all Italians to contribute to scientific research, on the occasion of the upcoming tenth anniversary of the Fascist regime in October.[39] Mussolini's office took up the suggestion and tweaked it substantially, working with Marconi to craft an international, multilingual broadcast addressed to "the scholars, scientists, and artists of the world."[40] Marconi's intention to focus on scientific research, as head of the CNR, morphed into a panegyric to fascism, presented by the president of the Royal Academy. The tenth anniversary was marked by a number of important events celebrating the role of the new fascist intellectual.[41] Marconi's contribution was key to articulating the regime's official view of this role. Under the fascist prescription of the terms of intellectual discourse and practice, as aptly put by historian Mabel Berezin, "there was no life of the mind, or work of the mind, outside the state." Marconi's public performances certainly supported this ritual. With his unorthodox education and concrete practical accomplishments, Marconi was an excellent symbol for Mussolini's idea of the "new Italian fascist cultural hero."[42]

Marconi's address was set for broadcast on Saturday, October 15, 1932, the eve of a planned address by Mussolini to a mass rally in Piazza Venezia. The speech went through several iterations and the Marconi archive at the Lincei contains several drafts, some of them with notations in Marconi's hand. One draft was edited by Mussolini himself, with additions and modifications that Marconi either changed or validated with the notation "OK" in the margin. The original version was in Italian, obviously, and the English translation was Marconi's own, rendering it "more in conformity with the anglosaxon spirit"—not that Marconi had a clear idea of what he meant by that, but it was how he justified to Mussolini the changes he wanted to make. The English version was all Marconi's—for example, rendering Mussolini's self-characterization as "a man of genius" to "a great man"—and then, finally, to "a truly great man."[43]

The speech was eagerly anticipated in fascist circles, where it was known that Mussolini had a hand in it.[44] An EIAR (Italian national radio) hook-up transmitted the thirty-minute broadcast, in Italian, English, French, and German, across Europe and America beginning at 6:15 p.m. GMT. The speech was part travelogue, part encomium. Marconi issued an invitation "to visit Rome, the Eternal City, which thanks to Fascist rule and effort, is assuming once again something of the majesty and splendor which characterized it in the days of Augustus," a reference to Mussolini's archaeological renewal of the city centre. Throughout Italy, visitors will notice "the many other milestones set by Fascism along the path of civilization and progress...you will certainly appreciate above everything else the new spirit which the force of a new ideal, brought into action by a truly great man, has succeeded in arousing throughout Italy." In short, the speech was anything but a call for scientific research. It was one of the highlights of the tenth anniversary celebrations, eclipsing even Il Duce's address to the mass rally in Piazza Venezia the following day, which had to be cut short because of rain.[45] The fascist mainstream was delighted with Marconi's address—its simplicity, its eloquence, its message, and especially the fact that it went around the world; the invitation to visit Italy and see the progress that had been made would help "undo the calumny of a certain part of the foreign press," noted a Polpol agent in Milan.[46]

Marconi was still hoping to be more than a promoter of tourism for Mussolini. Only days before the tenth anniversary broadcast, on October 9, 1932, he made an important speech to the Italian Society for the Advancement of Science (commonly know by its Italian acronym, SIPS),[47] and two weeks later he published a long, sober, and statesmanlike article on "science and fascism," intended to rectify what he saw as an unfair myth circulating abroad that fascism had little sympathy for science and culture. Both as doctrine and as a system of government, he wrote, fascism was not averse to science and culture. To make his case, he evoked the role of the two great institutions he presided. The academy represented the centre of national culture and intellectual movement, while the CNR embodied the importance of scientific research. If ingenious ideas are almost always born in the head of an isolated individual, he wrote, drawing implicitly on his own life story, it is only patient research in well-equipped laboratories that can bring them to fruition. The Fascist government and the country's scientists were in harmony, and Mussolini had shown that he considered their work essential for the strength and growth of the Italian state.[48]

Read alongside Marconi's lobbying efforts, the article looks like wishful thinking, or the hope for a self-fulfilling prophecy. Marconi was trying to be strategic, however. Early in 1932, he oversaw another reorganization of the CNR.[49] He was now seeking to build a genuine research organization, while Mussolini still saw him as D'Annunzio's "magic hero" rather than a serious political figure. Marconi bombarded Mussolini with detailed memos on research policy throughout the year, culminating in a long brief in preparation for an important CNR meeting on November 19, 1932, which Mussolini himself attended to hear Marconi describe the role and accomplishments of science and technology in the service of the nation. For the Italian press, the example of a research council evolving under the direct patronage of Il Duce and the presidency of the world's greatest scientist was a symbol of the strength of fascism.[50] Marconi's position attracted attention outside Italy as well. "If we read Marconi right," the *New York Times* editorialized, "*Il Duce* has exercised his powers with such rare intelligence that Italy's scientific bureaus constitute an organic whole. The Research Council apparently initiates research, assigns problems to the proper scientific agency and exercises general scientific control." The technical and scientific paternalism of the state were long familiar in the United States, Germany, and the United Kingdom; what was new was *organization*.[51] Marconi was bringing credibility to Mussolini's model, without apologizing for it.

Busy as he was, Marconi also kept in touch with prominent fascist intellectuals like Giovanni Gentile, who was now the director of the massive Italian *Encyclopaedia of Science, Letters, and Arts*, published serially in thirty-six volumes between 1929 and 1936. (Known as "Treccani" after its initiator, senator Giovanni Treccani, the encyclopedia is still a major publication today.) Marconi was named president of the Enciclopedia Italiana in 1933. In this capacity he worked closely with Gentile, as the copious correspondence between them attests.[52] However, the relationship seems to have been mostly administrative, rather than political, and certainly not philosophical.

He also befriended the journalist Luigi Freddi (later Mussolini's director of film propaganda and founder of the state film studio Cinecittà[53]). Twenty years younger than Marconi, Freddi was a member of the futurist movement and had been a D'Annunzio legionnaire in Fiume; the two may have met during Marconi's celebrated visit to the rebel capital in September 1920. A few weeks later, Freddi published an effusive article on Marconi in the Trieste paper *Il Piccolo*, for which Marconi wrote a friendly letter of thanks (addressing Freddi as "Gigi"). They kept in touch during the 1920s, meeting in London or on the *Elettra* when their schedules allowed it, and Freddi was for a time first vice-secretary of the foreign branch of the National Fascist Party, of which Marconi was a member. In 1934, Mussolini named Freddi to head a newly created General Directorate of Cinematography within his Ministry for Press and Propaganda. One of the directorate's first major activities was to organize an international event on political cinema in September 1935. Freddi wrote a long letter to Marconi outlining the project, which would consist of a series of film screenings, a conference, and a publication on the uses and functions of political cinema. Questionnaires were sent to prominent figures in Italy, Germany, France, the United Kingdom, the United States, and the USSR—in addition to Italian intellectuals like Luigi Pirandello, Giovanni Gentile, and Marconi himself, the list was a compendium of the politicians whose delirium was about to embroil Europe: Italians Galeazzo Ciano (Mussolini's son-in-law, who was in charge of prop-aganda for the regime) and Achille Starace (first secretary of the Party); Nazi leaders Josef Goebbels, Hermann Goering, and Rudolf Hess; Soviet foreign minister Maxim Litvinov…One of the only non-fascist intellectuals on the list was George Bernard Shaw. Marconi remained in touch with Freddi until the end of his life; one of his last letters was to thank Freddi for sending him a new book.[54]

Because so much of Marconi's public persona at this time was based on hollow praise for the glories of fascism, the depth and complexity of his en-gagement with the regime was not apparent to outside observers—even to those whose job was to monitor such things. Mussolini—who, in the words of George Steiner, "spent lethargic days combing through bedroom gossip gathered for him by an indefatigable and prurient domestic-espionage ser-vice"[55]—clearly wanted to know what Marconi was getting up to. A brief report in Marconi's police file, dated November 18, 1932—the day before the CNR meeting mentioned earlier—provides an example. Here, the agent reports that it was being said in Rome that Marconi was in his dotage, his speeches were written by others, his experiments done by an army of assistants, especially an English engineer (Isted), that Marconi was like a great artist of yesteryear, like Raffaello, approving the work of others, his inven-tions the result of an organization that could afford to spend lots of money (that is, the British Marconi company).[56]

Not much bedroom gossip there, but the police reports give an indi-cation of how Marconi was perceived by the regime at that time. In one report, from September 1933, agent Virginio Troiani reported that Marconi was the pre-eminent political scientist of the period, among other things because of his relations with the Vatican, which protected him and provided him with the most modern studios, where he could work in secret, in full security and hospitality, far from the laws and indiscretion of others.[57] The police reports were often an amalgam of unfounded, unreliable rumours, in some cases picked up from sources that could be as casual as a conversation overheard in a theatre lobby. On February 11, 1934, a minor spy reported that it was rumoured Marconi was going to retire as head of the academy in order to have more freedom to do his research; there was absolutely no basis to the rumour. On other occasions, the police reports were clearly cribbed from the day's newspapers; thus, on July 23, 1934, an agent in Milan wrote that he heard Marconi would shortly be experimenting with a new invention, trying to have an unmanned vessel enter a port in darkness. On the report, someone has scribbled: "But this was in the newspapers! This informer is an imbecile."[58]

The police files occasionally recorded real but mysterious events that the reporting agent knew little about. On May 15, 1934, Beatrice Marignoli, formerly Marconi, met with Vatican Secretary of State Cardinal Pacelli. The meeting lasted an hour and it was said, the police agent reported, that *la Signora* spoke to the Cardinal about something political that could not be treated through official channels. It was also said that they may have spoken about the annulment of her marriage to Marconi. This was highly unlikely, as the annulment, awarded seven years earlier, was no longer in any way an issue. Far more likely, it was about Marconi's dwindling financial support. Beatrice was lobbying the government, and probably the Vatican as well, to try to influence her ex about his support and testamentary arrangements for their (now grown) children.[59]

In fact, the police reports were often the type of embellishment one hears when people think they know the truth behind a published story. When Marconi entered a London clinic on December 2, 1934, suffering from exhaustion and heart problems, an agent in Milan immediately reported to the police that it was being said that Marconi had gone completely mad. A few months later, there were reports that he had committed suicide "for family reasons"; an agent on the ball reported that this rumour was untrue.[60]

In September 1933, Marconi was invited to attend the Chicago World's Fair, appropriately themed "A Century of Progress." He told his daughter Degna, "Mussolini was very anxious I should go," but it seemed just as important that all his and Cristina's expenses were being paid.[61] He told Degna he expected to be in the United States about a month and even booked a return sailing in mid-October, but the trip turned into a three-month world tour that took him not only across the United States but to China and Japan as well. The publicity value for Mussolini and fascism was priceless; for many people on Marconi's itinerary, it was the first connection of fascism to an actual human face. In the United States, there was fascination and hope for the new ideology—progressives like *McClure's Magazine* founder Samuel McClure, for one, described Italian fascism as "a great step forward and the first new ideal in government since the founding of the American Republic"[62]—and the popular, glamorous Marconi couple were a tremendous promotional instrument. The trip also enhanced Marconi's political capital.

He was now someone who could carry delicate messages from one political leader to another.

Marconi's friends at RCA took charge of organizing (and paid for) the first leg of the trip. RCA and the fair officials were sufficiently sensitive to the political significance of the visit and to controversies surrounding Marconi's politics that there was talk of asking the Secret Service to assign a man to the party, and a police escort for all of Marconi's movements.[63] Marconi, Cristina, her maid Eugenia, Marconi's personal secretary (Umberto Di Marco) and his police bodyguard (Nunzio Lettieri[64]) sailed from Genoa on September 21, 1933, on the state-of-the-art *Conte di Savoia*.[65] It was the sort of vessel Marconi loved, "a magnificent ship with large salons and every possible comfort," according to Cristina.[66] When the party arrived in New York on September 28, RCA's George H. Clark, who had met Marconi before, found that he had "a quiet, almost sad, countenance," but by the time he left San Francisco a month later that "had changed to a permanent smile."[67]

Clark later prepared an hour-by-hour report of the trip. When the ship docked, Clark wrote, "Whistles blew; cheers from the assembled crowd mingled with the blare of automobile horns; New York's fireboats threw streams of Hudson River water skyward in honor of the arrivals. The party proceeded to the Ritz-Carlton Hotel, past cheering thousands."[68] It was reminiscent of Marconi's early triumphant visits to New York. On September 29, RCA hosted a dinner in Marconi's honour at the Ritz-Carlton. It was a friendly, informal event, and for once, Marconi seemed at ease, laughing and joking as the evening wore on. As observed by Clark, "the cloak of reticence with which he is wont to surround himself when in unfamiliar places fell away for the time, and he was genuinely and personally happy." Marconi also met with Orrin Dunlap, radio editor of the *New York Times*, who had just completed a manuscript on "his life and his wireless," and Marconi agreed to go over it upon his return from Chicago, to check it for accuracy and to supply additional material that he would like to see in the book.[69]

The next day, September 30, the Marconis left for Chicago by train, with thirty trunks and suitcases checked as baggage and as many more to be distributed throughout the compartments of their private car on the Pennsylvania Railroad line. They arrived in Chicago early the following morning, Sunday, October 1. A huge crowd had gathered, and as soon as the train was in the station, "the crowd went wild, and the air rang with 'Viva Marconi' for ten minutes or more," according to Clark. The Drake Hotel, where the party was staying, had to detail two Italian-speaking telephone operators to handle calls. The Marconis attended mass, then lunched as guests of an Italian-American association. The United States was still under Prohibition (it would be repealed in December), so dry toasts were made. It was the height of the Depression, yet the rest of the afternoon was spent sailing on the *Mizpah*, a yacht belonging to Eugene McDonald, founder of the Zenith Radio Corporation and one of Chicago's wealthier citizens. It was a perfect day, Clark remembered, a crisp breeze blowing off Lake Michigan, and the party grazed on "a typically American offering of food"— baked beans, hot dogs, cornbread, chicken, and potato salad.[70] At a banquet at the Drake that evening, Marconi spoke on the role of scientific research in bringing relief to the worldwide economic crisis (a suitable counterpoint to the day he had just spent sailing on the lake). Not only the physical but the social sciences could be marshalled "to make better use of the forces which have been placed at the disposal of man." The end of scientific research should be the more equitable, more remunerative, and more enjoyable distribution of work. "It is my belief and I might say my firm conviction that men will unite in their efforts and that better days are about to dawn," he said.[71] He couldn't have been more wrong.

Monday, October 2 was "Marconi Day" at the World's Fair. His company organized an elaborate stunt, lighting up the exhibition by moonlight from Italy, then issued an appropriately glowing press release: "A luminous impulse from the moon was picked up at the Observatory of Arcetri, in Tuscany, the last residence of Galileo, converted by a very sensitive photo-electric cell into an electrical impulse, relayed from Italy to Chicago, and there used to operate an apparatus which lit the electric lamps of the Exhibition."[72]

Marconi's first personal event of the day was a meeting with the American Legion. Looking pale and serious and dressed in a formal black cutaway, he appeared, according to Clark, "more like a distinguished statesman than an inventor." There was another distinguished visitor to the fair that day: President Franklin D. Roosevelt (inaugurated seven months earlier) had decided to make "a flying visit to Chicago" to address the Legion; FDR and Eleanor Roosevelt had left their home in Hyde Park, New York, at 5:00 p.m. the day before, and arrived in Chicago at 11:00 a.m. By noon they were at the fairgrounds.[73] While Marconi was having lunch, an aide to Roosevelt appeared, informing him that the president would appreciate the opportunity to see him. The Roosevelts and exhibition president Rufus Dawes were dining with an assortment of guests in the fair's administration building and Marconi arrived as dessert was being served. Dawes gave up his chair next to

the president and Marconi and FDR chatted for about twenty minutes, posing for a famous photograph.* FDR reminded Marconi that they had met at
a reception during Marconi's 1917 visit as part of Italy's wartime mission to
the United States, when FDR was assistant secretary of the navy and responsible for wireless. Marconi didn't remember the meeting. ("Where did I
meet that man?" he asked a colleague when he returned to his own party,
"For the life of me I cannot remember the occasion."[74]) Roosevelt invited
Marconi to lunch at the White House the following week, and by 4:00 p.m.
he was back on his train, to New York.[75]

   One might almost think that Roosevelt had made the trip precisely to
meet with Marconi. Certainly, it was one of his priorities; he was in Chicago
for all of four hours. FDR at this time had a serious interest in Italy's domestic politics; he had written to Mussolini a week after outlining his New
Deal program in a "fireside chat" radio broadcast on May 7, 1933,[76] and
Mussolini had acknowledged the affinities between his own policies and
Roosevelt's in a remark quoted in the *New York Times*, in June: "Your plan for
coordination of industry follows precisely our lines of cooperation."[77] More
important, FDR had hopes that Mussolini would provide a strong counterweight to the rising threat of Hitler (he had much less sanguine hopes for
the capacity of England and France to do the same) and was taking every
opportunity to reach out to prominent Italians who had Mussolini's ear (a
few months later it would be Margherita Sarfatti's turn to take tea at the
White House.)[78]

   America was fascinated by fascism, and with Marconi. The party left
the fair in a procession of cars inching its way through a massive crowd.
Again, as reported by Clark, "The air was filled with cheers, and cries of
'Viva Marconi,' and every hand was flung up in the Fascist salute. Marconi
smiled frequently and saluted in return."[79] Clark had an odd observation: he
found Marconi was touched, vindicated somewhat, by the demonstration.
Clark thought the reception made up for the pain and disappointment
Marconi may have felt on some of his previous visits—but they had been
happy ones. It seemed as though Marconi was being reflective about this
American display of adulation for fascism and his association with it.

   After three days in Chicago, Marconi and his party returned to New
York, via Niagara Falls. Then they went on to Washington, where they put

* An original vintage print of the newswire photograph was being sold on eBay as recently as
2008. Marconi is described in the caption as "Italian statesman and inventor of wireless."

up at the Mayflower Hotel. The official White House luncheon, for Marconi and fifteen guests, took place on October 11, 1933. It lasted all of an hour. By 2:00 p.m., FDR was in the Oval Office, reading and answering mail; by 2:45 p.m., he was in his office for a meeting with Dean Acheson.[80] There is no official record of what was discussed between Roosevelt and Marconi; FDR rarely had the substance of his meetings recorded.[81] Marconi later reported that the president had a good deal to say about wireless and expressed the view that it would now be difficult to imagine the world without it.[82] However, Marconi was clearly gathering and sharing diplomatic intelligence in Washington. Later that evening, at a formal dinner at the Italian embassy, he "had a most energetic talk with the British Ambassador [Sir Ronald Lindsay], in a secluded corner."[83]

Marconi was also a big hit with Eleanor Roosevelt, who was celebrating her forty-ninth birthday that day. She later told reporters: "I happened to speak to Senator Marconi of a certain new book I had just read, a book of rather revolutionary suggestions, and his instant query was 'Where can I get it?' I think that shows clearly the type of man he is, a man with an inquiring and youthful mind." Americans the First Lady had spoken to of the same book had been skeptical. It was men like Marconi, she said, "who are really going to contribute something to the solution of the problems of the future."[84] (She later sent him a copy of the unidentified book, for which he telegrammed from San Francisco, thanking her for it as well as for the White House reception.[85]) The Marconis were due to return to Italy a few days later but FDR proposed that they travel across the country and offered to put a train at their disposal. Cristina wrote in her memoir that she was anxious about prolonging her absence from her three-year-old daughter, Elettra, until Eleanor Roosevelt convinced her to extend the trip.

Marconi and his entourage spent seven days travelling across the United States, no longer guests of RCA but of the president. Marconi loved the days spent enjoying the Western scenery: the colours, the silence, the sunsets, the limitless space. "I felt that in those moments, faced with the beauty of nature, fresh ideas were springing into his mind," Cristina wrote. In Los Angeles their train was met by another huge crowd, including many Italian immigrants.[86] They were entertained by visits to Hollywood movie sets, and had lunch at RKO with Mary Pickford and John Barrymore. Pickford invited the Marconis to tea at Pickfair, the Beverly Hills home she shared with fellow actor Douglas Fairbanks. For Marconi, tea with Mary Pickford was a repeat performance; they had done it before when the actor visited Rome in April 1926.[87] They also met Charlie Chaplin and Paulette Goddard, who lived next door to Pickfair. Marconi and Chaplin had a long conversation, and Chaplin did an impromptu performance of his characteristic little tramp's gait.[88]

The highlight of the West Coast was unquestionably San Francisco. The party's train from Los Angeles stopped in Oakland and they took a ferry to the city, where their arrival was heralded by sirens and blasts from the whistles of ships in San Francisco Bay. They were greeted by an official reception committee led by San Francisco's Italian-American mayor, Angelo Rossi, and including representatives of the city government and business leaders. A band played "The Star-Spangled Banner" and the *Giovinezza*, the official hymn of the National Fascist Party (and Italy's unofficial national anthem during the Mussolini period). Again, Marconi looked grave and sombre, dressed in black and carrying a cane, a Fascist button in his lapel. "To the waiting cars they pressed, Mr Marconi bowing frequently and stretching out his arm in response to thousands of Fascist salutes," a local paper reported. But observers noticed that he rarely smiled. What was in his mind? Could it have been that he knew better than the crowd what support for fascism really meant? In any event, he felt compelled to respond to a reporter's question about the situation in Europe: "Mussolini and Hitler will not join forces. They are too different."[89]

While the Marconis were in San Francisco, they received another impromptu invitation, from the Japanese foreign ministry, and on November 2, 1933, they sailed for Japan. Cristina's memoir for this trip reads like a period travelogue.[90] After nearly a week at sea, they arrived in Hawaii, "still primitive and wild in 1933 ...there were very few white people." They were visited on board by the manager of the Honolulu Ritz, who had looked after them previously at the Savoy in London. When they arrived in Yokohama, they were whisked by car to Tokyo and received "at once" by Emperor Hirohito at the Imperial Palace. They stayed a few days at Frank Lloyd Wright's masterpiece Imperial Hotel, then did the standard travellers' tour of the country (Fujiyama, Kyoto, Osaka, Kobe ...), except that theirs was literally scattered with jewels, such as a lesson in the cultivation of pearls offered by Kokichi Mikimoto, the creator of the cultured pearl industry himself.

On November 13, 1933, London's *Evening Standard* reported that Marconi was en route from Japan to India, intending to join an Italian expedition to the Himalayas. He was, however, still enjoying Japan, where the emperor bestowed on him the Grand Cordon of the Order of the Rising Sun, an honour normally accorded only to Japanese royalty.[91] At the end of November, the Marconis sailed to Korea, crossed the country and entered China— travelling almost exclusively through Japanese-occupied territory. Eventually they reached Tientsin, then Peking, where Cristina noted that the only Europeans they met were missionaries and diplomats. The city was the seat of an important Vatican Legation as well as the Catholic University of Peking. Marconi was photographed looking tired and solemn, dressed in a heavy coat (loaned to him by the British ambassador) and leaning on his cane. He told an audience of a thousand people, including students, radio amateurs, and Chinese scientific leaders, that "he was very glad to say that the differences between the State and Church in Italy had all been settled, and that the Vatican City under the rule of His Holiness, and the Italian Government under the direction of Signor Mussolini, were in perfect agreement."[92]

The Vatican was indeed interested in Marconi's visit. The papal representative, Monsignor Ildebrando Antoniutti, reported to Secretary of State Pacelli that he had accompanied Marconi on several occasions during his stay in Peking. Marconi had visited the apostolic delegation and examined the equipment by which they received weekly broadcasts from Vatican Radio. Antoniutti proudly reported that the Catholic University was the only educational institution Marconi visited while in Peking, despite many invitations, adding that Marconi was vividly interested in missionary work.[93] Marconi wasn't ignoring secular concerns; he also held talks with the Italian embassy about setting up a direct radio link between Italy and China.[94] The Marconis continued surfing at the high end of the travel circuit. In Shanghai, where they spent six days, they were guests of Sir Victor Sassoon, owner of the Cathay Hotel; in Nanking, then the Chinese capital, they were received by Nationalist president Chiang Kai-Shek. They then left for Italy and home, from Shanghai, on an Italian ship; the trip took twenty-five days, via Hong Kong, the Straits of Malacca, Singapore (where they stayed at Raffles Hotel while Marconi spoke at the University of Singapore), Ceylon (where the Italian consul general was a nephew of Garibaldi), and Bombay. Having gone clear around the world, they arrived in Italy at the port of Brindisi on January 4, 1934. One of the first things Marconi did was send a telegram to Mussolini presenting his greetings and expressing the desire to meet as soon as possible to give him his devoted impressions of the trip.[95]

❈   ❈   ❈

Once back, Marconi's life was again bound up in Mussolini's regime. The CNR's annual plenary sessions were by now a regular occasion for both

Mussolini and Marconi to expound on the role of science. On March 8, 1934, the CNR met in Mussolini's balliwick at the Palazzo Venezia. Marconi spoke first, making a long speech detailing every aspect of the council's activity.[96] He recalled the importance of scientific research for both the economy and national defence, as well as the origins of the CNR: conceived in the aftermath of war, the council was now being adapted to peacetime needs. Mussolini then took the floor. Marconi had made a great speech, he said. But he took a different tack. Students of military affairs have wondered whether war is a science or an art; the preparation for war is a science, its execution is artistic. Scientific research thus had to be seen as a battle, and the CNR had to function like a general staff.[97] Mussolini's use of the military metaphor was a powerful message about his view of the role of research, as well as an ominous signal about the evolution of his thinking. In a few months he would have his first meeting with Hitler, in Venice in June 1934. For Marconi, the implications were multi-layered, not only for the organization over which he presided but also for his own ongoing research work, and—not a small thing—for his vision of the role of science.

On October 11, 1934 Marconi made an important address to the SIPS in Naples.[98] It was a short speech but one of the most political and least partisan speeches he ever made, on the relationship between science, technology, progress, and economics, and directed simultaneously at the Italian government, the scientific community, and the world outside. Reacting to the global economic depression then entering its third year, he highlighted the urgent necessity for Italy to organize scientific research broadly. "We are at a turning point in the history of humanity," he said. "Mechanical and technical progress and first of all the ever greater facility of means of communication and transportation have overturned the foundations of age-old economic traditions." One of the gravest signs of this upheaval was mass unemployment, "a phenomenon that can no longer be considered as passing and that needs to be attacked at its roots." The economic nationalism now favoured by many countries was a vain illusion that would only make the problem worse. Unfortunately, Italy too—for some reluctantly but out of necessity—was constrained to follow this path strewn with traps.

Meanwhile, through the zealousness of Secretary-General Giovanni Magrini, the CNR was involving itself more deeply in a different kind of fascist nation-building. In January 1935, the council gave its support to a project of one Carlo Magnoni, professor of ethnography at the University of Rome. Magnoni was proposing to create an "ethnic documentation centre" to conduct research of a "moral and social nature" in the area of racial studies. It took Magnoni until 1938 to get his centre off the ground, but the key foundational work was done between 1933 and 1935, under the auspices of the CNR. According to documents in the central state archives in Rome, the CNR board decided to support Magnoni's project in February 1934—at the height of Marconi's presidency. A letter from the president to the Ministry of National Education, dated January 29, 1935, attests to the CNR's support for Magnoni's research, insofar as it had determined that "the scope of the work proposed is in effect of serious scientific character." The only problem is that the letter was clearly not signed by Marconi, who was in a London clinic recuperating from another heart attack. On the letter, the signature of "Secretary-General G. Magrini" has been obliterated and replaced by the handwritten words "*Il Presidente.*"[99] Marconi was doing some work from his London hospital bed during that time. The question is whether he was a party to the projects Magrini was promoting.[100]

33

"Your Every Wish Is My Command"

There remains controversy to this day as to when exactly fascist Italy began moving toward a policy of racial purity. Mainstream historians hold that Mussolini was essentially indifferent to this issue until his alliance with Hitler, forged in 1936, drove him to adopt Nazi-style laws and regulations. A revisionist school sees antisemitism, in particular, as more deeply rooted in the DNA of fascism, and this is borne out by some of the documentation we have just seen, as well as in some of Marconi's activity in the 1930s. The controversy surfaced briefly in 2002, after one of these revisionist scholars, Annalisa Capristo, published an article in an academic journal that made news outside Italy as well. "Marconi blocked Jews from Il Duce's academy," headlined the *Guardian* on March 19, 2002.[1] The subtitle was even more damning: "Inventor of the radio helped in Italy's anti-semitic campaign." The article was based on research by Capristo, an independent Italian scholar, who scoured the archives of the Royal Academy, unearthing Marconi's personal notes on the long discussions around the naming of new members in 1932. Capristo noticed that Marconi had flagged Jewish candidates with the letter *E* (for *Ebreo*), without further commentary, on lists the academy was considering. This was a smoking gun, with which she constructed a plausible and convincing argument that led to the *Guardian* report and reopened a dormant debate in Italy on the buildup of antisemitism during the 1930s.[2]

Capristo was not interested in Marconi per se, but in the systematic exclusion (or non-inclusion, to be more precise) of Jews from the Royal Academy under his presidency between 1930 and 1937. She is part of a small cohort of scholars who have been unearthing evidence indicating a simmering buildup of institutional antisemitism in Italy in the 1930s, and that

Mussolini's race laws of 1938 were not an overnight aberration, as in the standard historical view (which is strengthened by the fact that it is the position taken by the dean of Italian historians of the period, the late Renzo de Felice, in his monumental *The Jews in Fascist Italy*). The standard view is a more comforting version of a discomforting story, with its idea that Italians were not inherently antisemitic like some of the other European peoples, that after the Risorgimento Jews were perfectly integrated into every area of Italian life, and that the race laws were strictly a cynical product of Mussolini's pandering to Hitler in the late 1930s.

The revisionist claim is far more disturbing and sinister: that antisemitic practices began creeping in to every area of Italian life much earlier, and that the race laws were a logical culmination of a process that started as much as a decade or so before. Italian Jews themselves are divided on this question, often depending on their own personal histories; some will tell you that their family felt no discrimination whatsoever prior to 1938, while others have stories of small and large indignities similar to those suffered in other parts of Europe: an unmotivated refusal of an apartment rental, an unexplained bypass for a promotion, a denial of a university application despite acceptable grades, a fiancé(e) whose family vetoed a match for no apparent reason....[3]

In January 1932, the Royal Academy began a process of nominating new candidates, aiming to refresh its membership in time for the regime's tenth anniversary celebrations in October. (The procedure called for the academy to recommend a maximum of ten new members per year to Mussolini for his consideration.[4]) In March and April 1932, Mussolini gave a series of interviews in which he told German writer Emil Ludwig: "Anti-Semitism does not exist in Italy." Mussolini emphasized that Italy's Jews were appreciated by their fellow citizens, that they had fought courageously for Italy, that they held important positions in Italian universities, the military, and banking. And yet, Ludwig countered, it is said that you would have excluded Jews from the academy. "Absurd," Mussolini replied. "We just haven't found the right person. At the moment, one of our most important scientists, the historian Della Seta, is a candidate."[5]* Alessandro Della Seta

---

* Emil Ludwig was an interesting example of a German whom Hitler turned into a Jew. He was born into a Jewish family but raised as a non-Jew. "Many persons have become Jews since Hitler," he once said. "I have been a Jew since the murder of Walter Rathenau [in 1922], from which date I have emphasized that I am a Jew." Ludwig became a Swiss citizen in 1932 and later emigrated to the United States (*The Sentinel* (Chicago), August 13, 1936).

was indeed one of Italy's most distinguished academics. An archaeologist and art historian specializing in Etruscan and Greco-Roman antiquity, he held the chair in Etruscan studies and Italian archaeology at the University of Rome. Della Seta was not only Jewish, he was also an outspoken supporter of fascism.[6] In the preface to one of his books, published in 1928, he referred to Italy's postwar political humiliation and the renewal of national spirit under the sign of fascism.[7] One finds this trope running throughout Italian intellectual thought at this time; and as we have seen, it characterizes Marconi's own adherence to fascist ideology.

The archives of the short-lived Royal Academy—a Mussolini creation that did not survive his regime—are located in a few cluttered rooms on an upper floor of the Corsini Palace in the old Roman quarter of Trastevere, across the road from the Villa Farnesina where the academy had its headquarters (and where Marconi had his presidential office). Thanks to Capristo's well-referenced article, it didn't take long to call up the relevant files.[8] Sure enough, there they were: the tell-tale E's, unmistakably in Marconi's handwriting. However, the lists themselves were a hodgepodge, some of them typewritten and very official-looking, others scribbled on small sheets of notepaper, the type of notes one makes to keep track of discussion during a long and meandering committee meeting. Some of the papers are dated, some are not, and they are not in any particular order (possibly the result of having been handled over and over down through the years). The lists include dozens of names, some of them recurring many times; three of them (Alessandro Della Seta, Cesare Vivante, and Giorgio Del Vecchio—all Jewish), sometimes but not always have an E beside them; in at least one case, the E's seem to have been added later. But the most important point is this: Marconi has made all sorts of notations beside the names on the lists. Not only E's, but also occupations, place names, X's, question marks, underlinings, totally inscrutable symbols, the occasional distracted doodle, and comments like: "possibly incompatible with D'Annunzio" (whose candidacy was also being considered at this time).[9] What kind of moral judgment can we make about a list that identifies some people as Jewish, some as "from Firenze," and some as likely to create friction with other members?

That said, given that the names marked with the letter E all belong to Jewish candidates, there can be no doubt that Marconi was identifying them—just as he was identifying others according to equally obscure criteria. But to what end? If one were to press playing devil's advocate to the extreme, one could argue that Marconi wanted to *favour* these candidates, and was reminding himself to give them a leg up when it came time for the vote. Was he identifying them for some purpose of his own, or as a service to Mussolini (however unlikely it was that Mussolini needed Marconi to tell him that Alessandro Della Seta was Jewish)? Capristo's research strongly suggests the latter. On March 18, 1932, a member of the academy, mathematician Francesco Severi, wrote to a Jewish friend, physiologist Carlo Foà, to say "The President [Marconi] has informed us that the exclusion of Senators and Jews is to be continued."[10]

The first stage for election to the academy took place in disciplinary sections. On March 8, 1932, at a meeting of the history and philosophy section, the candidacies of Vivante and Del Vecchio were narrowly defeated, while Della Seta's received a sound vote of support (the section considered thirty-two candidates in all, and recommended fourteen for advancement to the next level of the process). By the time the names were presented to a vote of the full academy, the list was down to nine, three candidates for each of the section's three vacant seats, presented in ranked order. Della Seta was the top recommended candidate for the section's third seat; after the vote by secret ballot was taken, he was still on top, with twenty-seven votes out of a possible thirty-five.

So Della Seta was recommended for nomination notwithstanding the fact that he was identified as Jewish in Marconi's notes. According to the statute creating the academy, the final decision belonged to the head of government—Mussolini. The academy forwarded its recommendations to the Council of Ministers, which approved them on March 16, 1932. There were no secrets in Mussolini's Rome, and word of the list leaked out almost at once. On March 18, Marconi and the academy's chancellor, Arturo Marpicati, met with Mussolini. There is no record of what was said at that meeting, but when the list of new academicians was announced on March 22, Della Seta's name was not on it—and the seat for which he had been nominated was left vacant. When Emil Ludwig questioned Mussolini a few days later about rumours that Jews were being excluded from the academy, Il Duce could reply with disingenuous indignation that the allegation was absurd; after all, Della Seta was a candidate.[11]

To hone in on Marconi's role, we need to speculate on what was said at his private meeting with Mussolini and Marpicati. As the meeting fell between the cabinet's approval of the nominees and Mussolini's final decision, it is safe to assume that the candidacy of Della Seta was discussed. Did Marconi, on the basis of his famous list, inform Mussolini that Della Seta was Jewish? That was, as noted, hardly necessary. Did Mussolini seek advice? Everything in Marconi's background indicates he would have advised Mussolini to go with the result of the vote; he was not fundamentally anti-semitic and it was the correct thing to do for the integrity of the institution over which he presided. Did Mussolini berate Marconi for having lost control of the academy after he had given a clear directive? Not likely, even if that were the case; Mussolini needed Marconi in this position more than Marconi needed him, and Marconi was one of the few people who could speak their mind to Il Duce without fear of repercussion, even if his counsel was rarely heeded. The most likely scenario is that the key player at this meeting was Marpicati.

Marpicati, chancellor of the academy and chief of its operations, was one of the most influential members of the National Fascist Party. His war record and political trajectory were similar to Marconi's, but he had been a much earlier adherent and stronger supporter of fascism, contributing to Mussolini's *Il Popolo d'Italia* as early as 1918, joining D'Annunzio in Fiume as a legionnaire, serving on the national bureau of the party in 1930 and 1931, and as vice-sec-retary from 1932 to 1934. He was an acclaimed poet and novelist in his own right and it was whispered that an important part of his job was to keep an eye on Marconi and see that the academy didn't stray from party policy.

If these were indeed still early days in the development of Italy's fascist racial policy, refusing to name the academy's first Jewish member was a good place to begin, and Marconi's *E*'s are a rare physical trace of a largely invisible process.[12] Mussolini was famously indifferent to Jews as a group (and, if anything, favourable to the individual Jews he liked, including an early string of Jewish girlfriends and his long-time mistress Margherita Sarfatti[13]); Marconi appreciated his extensive Jewish business relations, especially in Britain and the United States. But Marpicati, whether yet openly or not, was one of the strongest advocates for an exclusionary policy on race in Mussolini's immediate circle, and he was the sort of apparatchik who would always place the interests of the party above any moral or ethical considerations. There is no doubt what he would have advocated regarding the nomination of Della Seta.

The question then becomes, where did Marconi stand? As Capristo argued, even if Marconi was not at the origin of the policy, even if he was against it, he went along with it, and acquiescence seems to have been a fre-quently repeated theme of his relationship to fascism. Capristo was not par-ticularly interested in Marconi's moral position, but in the policies he objectively facilitated. But our interest *is* Marconi's moral position. Marconi had never been averse to going on record, to stating his views when necessary; at the same time, he was always careful never to burn a political bridge. There was always a calculation to be made, and eventually a trade-off. What he finally did, the choices he made in a given situation, provide the strongest clues we have to who he really was, what he was made of, and what he believed in. In this case, he chose to continue his support for the regime. He was now on a slippery slope.

The question of Marconi's moral stance on institutional antisemitism under fascism is even more acute considering that his personal connections were famously non-sectarian. In addition to his well-known close relations with Jewish businessmen, Marconi's family lawyer in Bologna was Eugenio Jacchia, who was not only Jewish but a prominent anti-fascist. Born in Trieste in 1869, Jacchia had to leave the city, then under Austrian control, after agitating for its annexation to Italy. He then moved to Bologna, where he became a partner of Giuseppe Marconi's lawyer Leonida Carpi, also marrying Carpi's sister. Jacchia was elected to Bologna city council on a left-radical ticket in 1902 and favoured Italian intervention in the First World War, but he became a strong early opponent of fascism and was frequently harassed by Mussolini's police in the 1930s. He was one of the executors named in Marconi's 1917 will and was still looking after Villa Griffone and Marconi's other interests in Bologna at the end of Marconi's life.[14]

The issue of Jewish membership in the academy may have been settled at that meeting on March 18, 1932, but it didn't go away. In any event, the unstated policy was soon evident. In 1933, Della Seta's name was put forward again, by writer and art critic Ugo Ojetti (a signer of the 1925 "Manifesto of the Fascist Intellectuals"); Ojetti lauded Della Seta's patriotism and recalled that Mussolini himself had mentioned him favourably in his interview with Emil Ludwig. In the discussion leading up to the secret vote, moral philosopher Francesco Orestano said the candidacy of Della Seta involved an issue on which only Mussolini could decide, and proposed placing Della Seta second to give Il Duce more room to manoeuver. But Della Seta came out on top again (with twenty-eight votes out of forty-four), and the academy maintained his top ranking. The results were transmitted directly to Mussolini on April 13, 1933, with an accompanying letter from Marconi. This time Mussolini filled the seat with the candidate who received the second-most votes, Raffaele Pettazzoni, a historian of religion. Marconi said nothing. Interestingly, the academy's third-place nominee was another Jewish candidate, law professor Cesare Vivante (also a party member). Needless to say, his candidacy was not considered.[15]

The outcome of the academy nomination process also reflected the situation of Italian intellectuals at this time. As we've seen, the fascist regime imposed a loyalty oath on academics in 1931; some, including a small number of Jewish university professors such as Vito Volterra, refused to take the oath and bore the professional consequences. Most went along for the sake of their careers if not out of conviction. The Della Seta votes of 1932 and 1933 showed clearly that whatever the regime's real policy on the accession of Jewish scholars to the academy, the membership did not make a link between loyalty to fascism and racial exclusion—although exclusion per se did not seem to trouble anyone; women were explicitly banned from academy membership; anti-fascists implicitly. Looking at the vote results, one cannot say the academy itself was antisemitic; if anything the academicians were bolder than one might have expected, actually creating a potentially embarrassing political situation for Mussolini by nominating Della Seta.

The regime's position was hardening. As for Marconi, while there were no E's anywhere on his lists in 1933, he couldn't possibly have been oblivious to the implications of the policies he was overseeing—Hitler had been sworn in as German chancellor less than three months before, on January 30, 1933, boosting the confidence of the right wing of the National Fascist Party. (Marpicati later proudly claimed he had been the first Italian to carry Mussolini's congratulations to Hitler.[16]) Again, Marconi maintained a total silence and just went about his business. A few days after sending the academy's April 1933 recommendations to Mussolini, he and Cristina were photographed at Rome's Campidoglio with Oswald Mosley, founder of the year-old British Union of Fascists, and Mosley's wife Diana Mitford.[17] London's Italian Fascists (whose honorary president was Marconi) were a bit perplexed as to how to deal with Mosley's group, after Mussolini instructed the foreign section of the party not to get involved in the internal politics of their host countries, but this didn't seem to pose a problem to Marconi.[18]† Marconi was also friendly at this time with Prince Philipp of Hesse, an early (1930) member of the Nazi Party whom Hitler often used as a courier of messages to Mussolini.[19]

Della Seta, Del Vecchio, and Vivante were on the lists again in 1934, and again there was no need for E's—although another Jewish candidate appeared in 1934, mathematician Tullio Levi-Civita. Levi-Civita's was one

† Mosley and Mitford were both interned during the war.

of several names beside which Marconi placed a question mark (but no *E*) on his lists. Levi-Civita—who the CNR's Magrini had once described as "a convinced communist and a crackpot [*squinternato*]," as well as a Jew—was indeed prominently anti-fascist.[20] This time, however, none of the Jewish candidates were even nominated.

❋   ❋   ❋

Throughout Marconi's time in the fascist hierarchy, he gave almost no sign of *active* engagement with the defining projects of the regime and its ideology. The one exception, the one piece of fascist policy that he unambiguously supported—and it was a huge one—was the colonization of Ethiopia.[‡] In the early 1930s, as we've seen, Italian fascism was still appreciated in the democracies as an apparently effective antidote to both the perceived inefficiencies of economic liberalism and the incapacity of modern states to contain the excesses of both capitalism and communism. This changed in 1935 after the Italian invasion and then annexation of Ethiopia.

Italy's colonial experience in East Africa began shortly after the opening of the Suez Canal in November 1869 (an event attended by one Giuseppe Marconi[21]), with the purchase by a Genoese shipbuilder and patriot of Italian unification, Raffaele Rubattino, of a strip of land thirty-six miles long and six miles wide on the Bay of Assab, at the southern entrance to the Red Sea in present-day Eritrea. In 1882 Rubattino ceded his property to the young Italian state (for which he is remembered today by a bronze statue in Genoa's Piazza Caricamento), and, encouraged by Great Britain, Italy began to develop interests in the region. An Italian expeditionary corps landed at Massawa on February 5, 1885, and Italy established relations with the emperor of Ethiopia, Yohannes IV, who was already beleaguered in conflict with British, Egyptian, and Sudanese forces. After a series of confrontations that fell just short of armed conflict, Italy and Ethiopia's new emperor, Menelik II, signed a treaty in 1889, granting Italy a portion of northern Ethiopia that would become Italian Eritrea in exchange for a promise of arms. Italy claimed the treaty gave it a protectorate over all of Ethiopia, but Menelik argued that there was no mention of such a thing in the Amharic version of the treaty.[22]

In 1891, Italy and Britain signed two agreements defining their respective spheres of influence. Italy "got" Ethiopia while Sudan was to be considered part of the British sphere. Further disputes between Italy and Ethiopia

---

[‡] I am using the country's modern name, Ethiopia, except in direct quotes from contemporary documents, which mostly (but not always) used the name Abyssinia.

ensued until Menelik's forces completely defeated the Italians at the Battle of Adowa on March 1, 1896, ending what came to be known as the First Italo-Ethiopian War. The defeat—the first of a European colonial power by an African nation—had a devastating effect on Italian national esteem (and made a lasting impression on twenty-one-year-old Guglielmo Marconi, who had just arrived in London and was following the news in the press[23]). On March 10, 1896, the government of Prime Minister Francesco Crispi collapsed. Italy's humiliating defeat at Adowa had a profoundly radicalizing effect on the rise of Italian nationalism—and eventually fascism—among young men of Marconi's generation and social class.

After the Battle of Adowa, Italy was forced to recognize the independence of Ethiopia. In August 1928, the Kingdom of Italy and the Ethiopian Empire signed a "friendship" treaty, providing for Ethiopian access to the sea at Assab, collaboration on the building of a modern road, and an agreement to settle further disputes through the League of Nations. Both Mussolini and the Ethiopian regent, Prince (Ras) Tafari (who later took the name "Haile Selassie I" when he became Emperor in 1930), had their own agendas to pursue. In 1930, Italy built an outpost at Walwal, in a poorly defined border area between Italian Somaliland and Ethiopia. After repeated confrontations, Ethiopia appealed to the League of Nations for arbitration on January 3, 1935, but the league failed to intervene. A few days later, following a meeting between Mussolini and French prime minister Pierre Laval, France and Italy signed an agreement under which, among other things, Italy felt it was given a free hand to deal with Ethiopia as it pleased. Emboldened by diplomatic signals from both France and Britain, which wanted to keep Italy as an eventual ally against Hitler, Mussolini decided to invade Ethiopia.

Ethiopia became a dividing line between the increasingly nationalist fascist regime and its critics. There was little opposition within Italy; outside, however, Ethiopia became a turning point in attitudes toward Mussolini's regime. Previously observed with watchful interest, fascist Italy was now feared by the dominant powers that had created the League of Nations as an expansionist potential ally of Hitler. The US position in foreign affairs in the early 1930s was generally anti-colonialist, but the attitudes of France and Britain were more complex and, from an Italian perspective, hypocritical. Why was it all right for them to have overseas colonies but not Italy? Italy presented itself as the liberator of the Ethiopian masses from oppressive tribal leadership and slavery; at the same time, the African war became a laboratory for burgeoning fascist theories of racial supremacy.

Italian patriots, nationalists, and, now, fascists had not forgotten—indeed, were determined to avenge—Adowa. The further humiliation of the treatment of Italian claims after the First World War, and a keen sense of the hypocrisy of the greater powers toward Italy's colonial aspirations (modelled, after all, on their own), provided the backdrop to Mussolini's claims. Marconi's Italian archive at the Lincei includes a 1935 booklet published in English in Rome that spelled out the justification for Italy's colonial entitlements: "Italy acquired unity and independence too late to compete for territories open to conquest beyond her seas. The best positions were conquered long before she could rise to the occasion."[24]

By the summer of 1935 it was clear that Mussolini was leading Italy to war. Marconi decided to support the cause unequivocally. On July 18, Mussolini wrote to Marconi, who was then in London, asking him to go to Brazil, where a new radio station was about to be inaugurated, on the government's behalf. On July 27, Marconi replied, in a handwritten letter from London: "Your every wish is my command." He accepted Mussolini's request, suggesting they meet so that he could present his ideas for doing "Italian and fascist propaganda" in South America, and stating his desire to "volunteer for the campaign in Africa" upon his return. (Mussolini's office sent back a boilerplate note stating that they had received his request to volunteer for East Africa and transmitted it to the military command headquarters, which replied in turn that it was unable to do anything with the request.)[25]

On August 28, Marconi told Reuters, "I shall probably offer my services. . . . If war breaks out between Italy and Abyssinia I shall place my services at the disposal of the *Duce*."[26] Although he emphasized that reports stating he had been called up were premature, it was reported in the press that he had "volunteered" and asked Il Duce to send him to any place in East Africa where he could be of use.[27] In later years, this became a marker of the extent of Marconi's support for fascism. Marconi's private secretary, close collaborator, and confidant Umberto Di Marco wrote, in 1945, that "Marconi first learned of his 'volunteering' upon reading [the] announcement" in the Rome press, whence it soon spread abroad. In addition to the correspondence cited here, which shows without any doubt Marconi offering to "volunteer," Marconi's Lincei archive—which remained in the care of Di Marco after Marconi's death—contains an exchange of telegrams between Marconi and Fascist Party secretary Achille Starace where Starace thanks Marconi on behalf of the party for asking Il Duce to send him to East Africa, and Marconi replies that he was only doing his duty. Although Di Marco tried to use this incident to indicate some of the "preposterous" press reports that appeared during Marconi's last years, his protestation was rather a fine example of the *ex post facto* efforts to sanitize Marconi's reputation.[28]

The trip drew the attention of the police spies keeping tabs on Marconi. "Many people have commented dubiously on the departure of Marconi for Brazil," reported Polpol agent Alfredo Bonati, one of the longest-serving of Arturo Bocchini's counter-espionage agents. "Many are asking: why exactly is he going, or even, will his work benefit the country?"[29] But they didn't need to take it any further; Mussolini had the best possible source for inside information about the trip: Arturo Marpicati was part of the entourage. Cristina's brother, Antonio Bezzi-Scali, kept a journal of the trip.[30] There was a lot of talk about politics; as the party left Genoa on September 12, the Italian-Ethiopian conflict was being discussed in Geneva (Italy presented its list of grievances to the League of Nations on September 4). Italy was standing firm, with Mussolini's position summed up by his aggressive slogan, *"Noi tireremo diritto"* (We will shoot straight). But the news from Geneva was disheartening. Britain was holding its ground as expected, but France's Laval, who had been a friend to Italy, now seemed to be abandoning his support. As the ship passed Gibraltar and left the Mediterranean, Marconi, observing the lush vegetation of Tangier, sent a telegram to Mussolini, encouraging him to continue. He now totally embraced Mussolini's struggle for the honour and grandeur of Italy, and whatever ambivalence he may have previously felt about fascism was gone.[31]

The *Augusto* arrived in Rio de Janeiro on September 24. Assailed by reporters, outdoors and bare-headed in a pouring rain, Marconi made an impromptu speech and received a great ovation before the party moved on to the luxurious art deco Copacabana Palace Hotel. Marconi received various honours in the coming days. On September 26 he visited the Brazilian Senate and attended a meeting of the Academy of Arts and Sciences. The mainstream press was full of favourable articles, while Rio's communist and Italian expatriate press condemned "the 'fascist' Marconi."[32] The Ethiopian conflict dominated the news. All of the world's anti-fascist forces seemed to be ganging up, and even friendly nations were letting go in order to follow England and its allies. Marconi and his party were discouraged by the reports, Bezzi-Scali wrote.

After Marconi made a radio speech to all of South America, the party moved on to San Paolo, where the visit was marked by fascist supporters demonstrating in the city with police protection and the approval of the authorities. On October 1, at a grand theatrical event, Marconi—wearing, along with the crowd, the fascist black shirt[33]—made another speech that was broadcast all over Brazil. As he went out on the balcony to salute the crowd that was unable to get into the theatre, it started to rain, and Cristina and her brother were concerned for Marconi's health. Sure enough, the following day he was struck with pharyngitis and had to cancel his activities for the rest of the trip. The party sailed from Rio on October 5, 1935. As far as the regime was concerned, it had been a great success.[34]

Two days earlier, on October 3, Italy had invaded Ethiopia. On the ninth, Baron Aloisi, head of the Italian delegation to the League of Nations, addressed the League Assembly in Geneva: "I shall not mention what western civilization owes to Italy," he began. Ethiopia is "a country where lawlessness and xenophobia hold sway," its state of internal disorder a constant menace to peace in East Africa. Ethiopia did not fulfill the league's requirement that every member state should possess a government capable of exercising authority over the whole of its territory. Nor was it a civilized country, he said. Ethiopia possessed its own colonies, which it ruled chiefly by atrocities and oppression, their native populations suffering slavery and extermination. Italy felt it was interpreting the true spirit of the league in Ethiopia; it was "a great proletarian nation demanding justice."[35]

Back from Brazil on October 17, 1935, Marconi began preparing two important radio broadcasts to the United States, using his connection with David Sarnoff, the head of NBC, to ensure maximum exposure.[36] On October 31 he opened a fifteen-minute address to a US radio audience by advising that he was not going to speak on his usual subject (wireless), but about the dispute between his country and "the so-called Empire of Ethiopia." When France and England pressed their peculiar colonial interests, he said, no one paid attention. But when Italy tried to protect itself against dangerous raids on its East African colonies and called for putting its existing treaties into effect, it was branded an "aggressor state." "Italy is trying to get what has been collectively but vainly promised to her for years and years by France, England and Abyssinia herself, namely the commercial and agricultural exploitation, the industrial development, the re-organization, in a word the civilization of an extensive territory ... first and foremost for the benefit of its own inhabitants," he said.[37]

Two weeks later Marconi addressed US radio listeners again, this time from a prestigious tribune offered by the Carnegie Endowment for International Peace. Carnegie had invited the government of Italy to designate a high-level personality to take part in an international Armistice Day broadcast on November 11, 1935. Mussolini asked Marconi to take on the task, invoking his worldwide reputation and mastery of English.[38] Although Marconi usually wrote his own speeches, the Ministry of Foreign Affairs took a special interest in this one, preparing an outline and sending it to Di Marco, who forwarded it to Marconi on November 4. Marconi informed Mussolini's office that he would submit the text on the morning of the broadcast "to have the opportunity to make the eventual changes."[39] Marconi's archive includes several handwritten drafts as well as revisions to typed versions of the speech in Marconi's handwriting; the ministry may have laid out the topics and vetted the text, but the eloquent turns of phrase and nationalistic hyperbole were definitely Marconi's.

Marconi began by stating that he had always maintained "that radio, which knows no frontiers, should be one of the greatest, if not <u>the</u> greatest instrument of international peace [his emphasis]." Italy was a founder of the League of Nations. The postwar peace treaties, however, had designed the world "for the exclusive benefits of those who had allotted to themselves the richest spoils of war." In the Far East, the League of Nations looked on, raised its voice, and did nothing when Japan invaded China, but now it was starting "a crusade against Italy," which was involved in "a merely colonial dispute." The League of Nations action "is considered by all the Italian people as an act of gross injustice, and we know only too well how a sense of suffered injustice of a whole people augurs ill for the peace of the world! There can be no lasting peace without justice."[40]

Notwithstanding these ardent appeals, the League of Nations denounced the invasion and imposed economic sanctions on Italy on November 18, 1935. Marconi was in Rome to inaugurate the academy's new season, and was one of the first international figures to decry the sanctions, pledging the academy's devotion "to Italy's cause for expansion in Africa" and calling on the nation to resist the "inhuman crusade" that had been launched against Italy from Geneva.[41] Italy was being punished for defending its colonies and seeking to bring civilization to a "backward and oppressed collection of tribes," and the academy had faith in the "final triumph of the sainted national cause."[42] The Italian press reported that Marconi's words expressed the feelings of all Italians. Later that day he put in a rare appearance and spoke at a late-night meeting of the Grand Council of Fascism called by Mussolini to report on the military and political situation.[43]

Marconi was due to visit England the following week and had accepted an invitation from the BBC to record a short radio broadcast for a

program commemorating his 1901 transatlantic signal.[44] Now he asked for
ten minutes of air time to explain the Italian viewpoint on the Ethiopian
crisis. This was, of course, not the first time he was intervening diplomatically in Italian-British relations. In July 1918, as head of the Italian delegation to the Inter-Allied Parliamentary Commercial Conference in London,
he had spoken eloquently about the "bond of sympathy and solidarity"
linking his two *patria*, England and Italy. "From time immemorial between
England and Italy there has never been any conflict or any cause of conflict," he had said in a dinnertime speech at the House of Lords, in 1918.[45]
By now, however, there was a conflict, and the day after the imposition of
sanctions, BBC director John Reith wrote to Marconi, courteously but
firmly refusing his request for air time. The reason given was that the BBC,
"wishing to protect its reputation for objectivity," permitted only officials of
the British government and the League of Nations to speak on the Italo-
Ethiopian controversy. "It was not going to be a hostile speech or made
with any sense of personal bitterness. I was simply going to give what I believe are the facts," Marconi told the press. "I suppose it might be called
propaganda, but I wanted to make a friendly speech, putting Italy's point of
view. I was going to make the point that no first-class nation could be expected to take sanctions lying down."[46]

Public and media response to the BBC refusal was far from harmonious. The first English-language biography of Marconi was published in
London on November 12, 1935, and author Doris Collier was besieged with
requests for comments from newspapers. The *Evening Standard* has not taken
sides in the dispute, she informed Marconi; the *Herald* is very anti-Italian
and has been asking for sanctions; the *Daily Mail* and *Sunday Dispatch* are so
pro-Italian that no one believes they give the truth. "I think the majority of
people feel as I do personally, very sorry that sanctions should be imposed
against the Italian people, but quite unable to appreciate Mussolini's aggressive attitude." In the *Sunday Pictorial* on November 24, 1935, Hugh Addison
wrote: "Here at last was a chance in which, between one civilized nation
and another, broadcasting would have done something in the cause of peace
and good-will.... The Marchese wanted to help, and should have been
allowed to try." An editorial in the *Morning Post* deplored that "the creator
of Broadcasting House" had "no access to his own invention" for an effort
that would have strengthened the amity between Great Britain and Italy.[47]

Criticism in London of the BBC refusal was noticed in Rome and at
the Vatican, where the office in charge of relations with foreign states was

compiling a huge file of news clippings on the Italo-Ethiopian conflict. Meanwhile, in Milan, an agent working under counter-intelligence spy Filippo Tagliavacche reported that Marconi's visit to London had the signs of an important mission: "Given Senator Marconi's great friendships in the English political world, hope is being nourished in some circles that the mission could help mitigate British intransigence towards our conflict in Africa and facilitate a rapid solution in our favour."[48]

Marconi and Reith had lunch together on November 29, 1935, and Reith wrote to him immediately afterwards confirming what they had discussed: "That if you would care to broadcast next week about anything (other than one thing) we should be delighted." He added, by hand, a friendly postscript: "I very much enjoyed our lunch today." Marconi replied, typically, reflecting Reith's tone back to him: "I am obliged for your offer of broadcast on any subject but the <u>one thing</u> [his emphasis]," adding his own thanks for lunch "and for the pleasant hour we spent together." Reith later wrote about the incident in his 1949 memoir, *Into the Wind*: "Marconi put me into some embarrassment. He wrote in November that he was coming to England and would very much like to broadcast a short talk on the Italian view of sanctions. I took him to lunch and we had a pleasant talk. He said he expected to be told that he could broadcast about anything under the sun except what he wanted to broadcast about. Quite right. I asked if Mussolini had put him up to this, and he did not deny it."[49]

It was all much cosier than the public perception let on: whatever their views, Reith and Marconi were respectively carrying out British and Italian policy on Ethiopia and sanctions—and they were both admirers of Mussolini. The editor of Reith's diaries, Charles Stuart, has written that the man identified as the founder of the BBC ethos of public service and independent journalism was slow to recognize the dangers of fascism and even Nazism, for that matter: "He admired Nazi achievement in spite of Nazi wickedness and as late as March 1939, when Prague was occupied, his only comment was that: 'Hitler continues his magnificent efficiency.'" Marconi certainly never went so far. Stuart wrote that Reith told Marconi when he was in England in November 1935: "I had always admired Mussolini immensely and I had constantly hailed him as the outstanding example of accomplishing high democratic purpose by means which, though not democratic, were the only possible ones."[50] If a respectable conservative like Reith (not to mention others, including Winston Churchill) could express admiration for fascism this late in the day without being tainted, it would later be said, why not Marconi?[51]

It may not have been necessary for Mussolini to instruct Marconi regarding propaganda on Ethiopia; of all the policies of Italian fascism, its colonial policy and Ethiopia in particular was the one Marconi supported most passionately. A stand against it, or even the sort of lukewarm acquiescence he was used to giving to many of Mussolini's policies, would have marked him as a progressive and cosmopolitan internationalist. Ethiopia had placed the issue of race on the Italian fascist agenda, opening the door to a dividing line between "pure" Italians and others that would shortly extend to Italy's Jews, who would find themselves slotted into the latter category no matter how well integrated they had been to Italian social, political, and cultural life since 1870.[52] As an outsider himself, Marconi could identify with the situation of European Jews. His position on Ethiopia recalled how profoundly his view of the world that he was seeking to improve through communication was that of the "white man's burden," as he had already demonstrated very early in his career.[53]

For Marconi, Ethiopia was a personal turning point as well; he was hurt and humiliated by rejection of his efforts to "explain" the Italian cause to Britain. Doris Collier wrote to him in January 1937 that the feeling against Italy had proven detrimental for her Marconi biography; there would be no American edition of the book.[54] At the same time, Marconi's interventions on Ethiopia marked the high point of his stature and prestige in Italy. As the country began to prepare for an even greater war, his importance as a bridge to the rest of the world, as well as the possible military value of his current research, took on heightened interest. Despite his unorthodox politics, Marconi was still very much *persona grata* in England; a few days before leaving, he received tickets from Buckingham Palace to attend the opening of Parliament.[55] He was leery of efforts to associate his rejection by the BBC with broad-stroke anti-British sentiment.[56] And despite (or perhaps encouraged by) the sting of the BBC refusal, he continued to enjoy the favour of large swaths of the British political sphere, especially on the right.[57] He received invitations to speak on the Ethiopian crisis from groups like the 1900 Club, a "private but extremely influential" association of Conservative Lords and MPs.[58] However, he never returned to England after this visit; in fact, he never left Italy again.

❧   ❧   ❧

Marconi, Cristina, and Di Marco embarked by train for Paris on December 6, 1935. The French prime minister, Pierre Laval, and British foreign secretary, Samuel Hoare, were meeting in Paris that weekend to hammer out a proposed solution to the Ethiopian crisis. Official diplomacy notwith-standing, Laval was sympathetic to Mussolini, and was in close contact with the new Italian ambassador to France, Vittorio Cerruti, in the days leading up to his meeting with Hoare.§ In London, Ambassador Dino Grandi, a seasoned and charming anglophile (like Marconi), was also in touch with the foreign secretary's office. But when details of the proposal—which was supposed to remain secret—were widely published on December 10, 1935, it was denounced by influential segments of both the British and French press. The plan proposed the partition of Ethiopia, on terms that Mussolini was apparently prepared to accept, but according to the London *Times*, Ethiopia would be left with little more than "a corridor for camels" as ac-cess to the sea.[59]

Laval, meanwhile, asked to meet with Marconi. From Paris, agent Guido Valiani reported to Polpol that it was being said that Marconi had paralyzed the electrical system of a British submarine and that Prime Minister Stanley Baldwin had telephoned Laval in the middle of the night and harangued him on the subject for an hour. Valiani reported that Laval had invited Marconi to the screening of a film about the French navy, which the entire government and diplomatic corps would be attending. Marconi apparently accepted only on the condition he be assigned a seat where he could enter and leave discreetly. Valiani, who headed a much-valued spy net-work working out of the Italian embassy in Paris,[60] added that he was filing these unusual reports to indicate the sort of strange things that were being said about Marconi.[61] But it seems that Marconi and Laval had more serious matters to discuss.

Multi-tasking as always, Marconi did a broadcast to Australia from his room at the Ritz, met with Laval, and informed Mussolini's office that he would be returning to Rome on December 17 and wished to be received by Il Duce the following day.[62] But on the last leg of his trip, on the Paris–Rome express train, just after leaving Genoa, he was stricken with a severe angina attack. Mussolini's office was immediately informed by a coded tel-egram from a police authority in Pisa that Marconi had been taken ill and was continuing his trip attended by two physicians.[63] When Di Marco called the next day to cancel Marconi's meeting, Mussolini was already aware of the situation; his agents were everywhere, even on the trains that he had fa-mously made to run on time.[64]

§ Cerruti had previously been Italian ambassador to Nazi Germany.

The League of Nations was due to meet about Ethiopia in Geneva on December 18, and Laval asked Italian delegate Renato Bova Scoppa to repeat the assurances he had given to Ambassador Cerruti and Marconi in Paris.[65] Whatever these were, an impression was being created that Marconi was a go-between, travelling between London, Paris, and Rome. The Grand Council of Fascism also met on December 18, and word was put out that Marconi had attended, although it was out of the question because of his heart condition.[66]

Marconi had suffered a life-threatening heart attack, as was clear from the detailed medical report of a physician, Giuseppe Paci, who happened to be travelling on the same train and seated at a neighbouring table when Marconi was stricken upon arriving for lunch in the dining car around 1:00 p.m. Paci managed to control the crisis with medication, and when the party arrived in Rome Marconi asked him to accompany them to his home to consult with his personal physician, Cesare Frugoni. Marconi was soon out of danger but the doctors agreed that he needed to rest and, above all, slow down the pace of his activities. For the rest of his life, he would have to be careful—more careful than he liked to be, and this itself caused additional strain. The state of Marconi's health, and the regime's determination to keep it secret, was clearly a matter of high-level concern in this delicate period. Di Marco later wrote that Marconi himself had "cooked" the report that he had attended the Grand Council of Fascism meeting, "lest any suspicion should arise in the mind of the public that the incident occurred in the train was of a serious nature."[67] This is not borne out by the evidence. Marconi was now targeted in the sights of the regime; his police file is full of often contradictory reports about the state of his health and the nature of his ailments; it also attests to renewed interest and concern about the possible military uses for his ongoing research, which had reached a new height in the preceding year.

In addition to his health and the fallout from his political commitments, Marconi also had to deal with a deteriorating state of relations with his company in 1935. The status of his personal research staff was a symptom. In May, Marconi's assistant Gaston Mathieu had received a notice of termination from London. Company chairman Lord Inverforth informed Marconi that the issue was cost: £9,000 for Mathieu's and Gerald Isted's work in Italy in 1934. Marconi, ever loyal and protective of his staff, replied that Mathieu had been his personal assistant for fifteen years and would find himself in a "peculiar position" if he were discharged, evidently alluding to Mathieu's precarious place as a foreigner in Fascist Italy. Mathieu prepared a convincing memo to Inverforth, justifying his contribution to the company's progress. Marconi, however, could not persuade the company to keep Mathieu in Italy. In June, Mathieu was offered the position of managing director of the company's Polish subsidiary and general manager of its agency in Warsaw. Marconi advised Mathieu to accept the offer and he did.[68]

Marconi's already strained relations with the company were further taxed a few months later when the company also requested that Isted return to the UK. This time, it was not only the cost but the possible risk to a British national of remaining in Italy that was invoked. Marconi wrote to managing director H.A. White on August 15, 1935: "You are under great misapprehension as to the present situation of foreigners generally, and Britishers in particular, in this country...a foreign engineer working anywhere in Italy under my orders, were he a 'poor islander' as you put it, or an Abyssinian, would be *ipso facto* above suspicion." Marconi's personal connections to Mussolini were sufficient to ensure the safety of anyone working under his direction, he said.[69]

Marconi was finding it difficult to strike an independent course. In his mind, his work was of interest to all parties, and he was trying to retain maximum control, playing all sides against one another: the British, the company, and the Italians. He saw the main benefit of his latest work as military, but did not seem to see that Britain and Italy might soon be on different sides in a military conflict. He was "evolving...an entirely new system of wireless communication which may prove extremely usefully especially in time of war," he told a company board meeting, then wrote to White that the Italian War Office was very eager that he should develop it under their aegis, and had offered him, for this purpose, "very substantial funds which I have, so far, refused in the belief that I could carry on with the Company's means as heretofore."[70] He recalled that during the Great War he had been able to work harmoniously with both the Italian government and the company at the same time. Of course, this was his way of pleading for London's continued support, but he failed to recognize that the present context was different.

White replied, somewhat irritated, as from his perspective the issue was fundamentally about Marconi's relationship with the company. When his agreement with the company was renegotiated in 1934, Marconi had given London the impression that it was his intention to gradually withdraw from research in order to devote more time to his public affairs in Italy. "There is nothing in the agreement which imposes any obligation to do anything except to illuminate the company with the luster of your great name," the company's chief operating officer wrote to its founder. Marconi persisted. He went to see Mussolini and mentioned the company's fears for the welfare of his British assistants in Italy. Mussolini "scoffed at the idea and confirmed that my men could be certain of the utmost consideration and protection," Marconi told White, warning him quite directly not to assume his complacency on the matter.[71]

The same day Marconi wrote this letter, the British press reported on his offer to join Italian forces in East Africa (as we've seen). White wrote to him: "We were all much interested in reading that you have volunteered for service at the Front."[72] All of Marconi's ventures were now bubbling over and mixing with each other—"evolving," like his research: his role as research council head, his personal research, his paternal concern for his staff, his political position in Italy, his relations with his parent company, his place in the firmament of British society and politics, his lustre in the media spotlight. . . . But with London pulling Mathieu and Isted out of Italy, Marconi was left alone, without technical or financial support, and his microwave research effectively ended. With his personal staff gone, Marconi had even less cause to remain in touch with London. Their correspondence shows him considered somewhat of an annoyance by the new corporate executives in charge of the company, although his aura still brought them valuable publicity and brand value. Nonetheless, he realized now that his practical future lay in Italy and, like it or not, in his relations with its leader. Marconi had made it his practice to personally keep Mussolini informed of the possible military importance of his latest research. Now this connection took on more importance, even as in his public pronouncements outside Italy he emphasized the benefits of the same research for peace.

From his corporate exile in Warsaw, Mathieu remained an unrepentant supporter of Marconi and Italian fascism, which he unequivocally linked. Mathieu ended up in Germany after the Second World War, remaining active well into the 1960s. After his death, his family left his papers, including much of his correspondence with Marconi, to the German Museum of Technology in Berlin, where it now forms part of the historical archive of Telefunken. Isted returned to England, and by November 1935 was settled in Chelmsford. He continued to work for the company until retiring in 1969, and believed that the reason he was called back to England was that

Marconi had been instructed by Mussolini to replace his foreign assistants with Italians.[73]

❖   ❖   ❖

As soon as he was back on his feet in January 1936, Marconi resumed his campaign against the League of Nations sanctions. In a speech at the academy on January 19, he recalled the refusal to allow him to speak on the radio in England, to "honestly explain Italy's reasoning," despite Britain's vaunted right to free speech.[74] Between his increasing alienation from the company and the reaction to his agitation for Italy's action in Ethiopia, there was now a palpable chill in Marconi's relations with England. On April 21, 1936, the *Daily Mirror* reported that he would not be paying his traditional spring visit. "Apparently he has not yet forgotten the refusal of the BBC to allow him to broadcast a message stating the Italian point of view."[75]

Mussolini, meanwhile, pursued his war in Ethiopia with renewed vigour and determination. After the failure of the Hoare-Laval proposals, Mussolini escalated the war, using chemical weapons, murdering prisoners, and pursuing "a politics of terror and extermination" against the Ethiopians that would be considered a crime against humanity, if not genocide, today.[76] On May 2, 1936, Haile Selassie fled to England with Ethiopia's gold; it was said that Mussolini allowed him to escape. Three days later Italy's general Pietro Badoglio entered Addis Ababa; by May 9 his forces controlled the country, and on May 15 Mussolini announced the annexation of Ethiopia and founding of the Italian "Empire." Broadcasting in fluent English from a huge rally in the Piazza Venezia, journalist Lisa Sergio—the Marconi protégé known abroad as "the golden voice of Rome"—told an international audience: "The people of Italy are proud to feel that whenever a momentous hour comes in their history, the *Duce* immediately calls out to them through this wonderful means which is the wireless, invented by an Italian, Guglielmo Marconi, who is now in the Grand Council and near the *Duce*." Moments later Mussolini proclaimed the decree placing Ethiopia under Italian sovereignty and the king of Italy as emperor.[77]

In the Italian Senate, Marconi was named rapporteur of a special commission created to convert the decree into law. On May 16, 1936, he made a glowing speech, laced with applause, declaring that a new and glorious era for Italy had begun, that the period started with the Risorgimento had ended, and that Italy would now take on greater tasks and responsibilities in the world.[78] Reviewing fourteen years of fascist government, he described Mussolini's "titanic" accomplishments with hallucinating hyperbole; under the guidance of "his infallible genius," Il Duce's revolution was creating the first example of a popular state founded on patriotism, authority, and labour. The opposition to Italy's "civilizing action" in Africa was "the most senseless and immoral coalition in history" as well as a threat to the cause of peace in Europe. "Thanks to you, o *Duce*, the Italian people have demonstrated to the world its strength, which is great, and its courage, which is unlimited."[79] A few days later, on May 21, Marconi's Wireless Telegraph Company held its thirty-eighth annual general meeting in London, chaired by Lord Inverforth; for the first time since 1897, there was no mention at all of Guglielmo Marconi.[80]

Mussolini's annexation of Ethiopia revealed the ineptitude of the League of Nations and heightened the growing crisis in international relations. Denouncing the world's inaction at the league on June 30, 1936, Haile Selassie declared prophetically, "It is us today. It will be you tomorrow."[81] US ambassador to Italy Breckenridge Long resigned and Italian ambassador Augusto Rosso was recalled from Washington.[82] (The United States was not a member of the league.) In July, faced with the Italian *fait accompli*, the League of Nations lifted its sanctions, while Japan and Germany recognized the Italian Empire. As Britain, France, and the United States had feared, Italy now entered increasingly closer relations with Hitler, forming an emerging international fascist "axis" that would be clad in iron within a few months.

Marconi's prestige abroad continued to be important for Mussolini's foreign policy. Discouraged by the attitudes he had encountered in Britain, Marconi altered his focus to swaying public opinion in the United States. On June 14, 1936, from Bologna, he broadcast again to America, again with David Sarnoff's support. He was launching a new radio station and began (as he almost always did) by recalling the history of his own early experiments. Radio was now important not only for entertainment and education "but also in the much more sensitive sphere of political propaganda." Shortwave broadcasting had annihilated the barriers of time; it knew no frontiers; it could promote understanding and peace. In the recent Ethiopian conflict, he said, "radio was profusely used, or rather abused, in spreading to the four corners of the world the most fantastic and absurd stories." However, peace and justice had prevailed in Italy's struggle against "savage African tribes and slave traders." Italy had "clear[ed] the path of progress and civilization of all the rubbish strewn upon it by the would-be champions of great ideals."[83] Cristina, too, was drawn into the propaganda campaign. On June 15, 1936, she also made a broadcast to the United States, addressed particularly to American women and intended to show how much fascism had done for the emancipation of their Italian sisters, declaring that Italy's half million working women were "on an absolute equality with men—equal work, equal pay."[84]

Despite the rupture in Italian-American relations, the month after these broadcasts, in July 1936, Marconi was invited by the National Advisory Council on Radio in Education to attend a conference in the United States on educational broadcasting. Italy's Ambassador Rosso advised the Foreign Office to draw the invitation to Mussolini's attention and note the political interest that Marconi should make the voyage: "The name and the personality of Guglielmo Marconi is immensely popular in this country in all levels of the population and his sole presence, even in a personal and unofficial capacity, would have favorable repercussions on American public opinion."[85] Marconi was too ill to travel, but on October 12, 1936, he made one more broadcast to the United States. It was Columbus Day, and, never one to miss a symbolic opportunity, he spoke from Genoa, Columbus's birthplace. This time he adopted a tone that his American audience had not heard before. "The might of right prevailed over the right of might, and a remote backward and aggressive corner of East Africa which so far had been grudged to us chiefly by the incomprehension of governments and ignorance of peoples, is at last thrown open to the civilising power of Italy under the unerring guidance of her great leader, Benito Mussolini."[86] Marconi was now in griplock with Mussolini and his fascist regime. Less than two weeks later, on October 25, 1936, Mussolini and Hitler signed the protocol forming the Rome–Berlin axis; Mussolini announced the alliance in a speech from the Milan cathedral on November 1.

As a prominent business, intellectual, scientific, and political figure, Marconi was constantly solicited to support causes, and the deteriorating situation of Europe's Jews in the 1930s was frequently brought to his attention in 1935 and 1936. In July 1935 he received a letter from a Zionist organization in Jerusalem, The Erez Israel (Palestine) Foundation Fund Keren Hayesod, a financial arm of the Jewish Agency, asking him for a brief message of support on its fifteenth anniversary. Addressing Marconi as a great friend of the Jewish national homeland (perhaps remembering his supportive comments about Zionism to an American Jewish newspaper in 1917), the fund reminded him of the masses of persecuted Jews whose only hope of survival was to seek refuge in Palestine. Umberto Di Marco replied on behalf of Marconi, stating that "His Excellency would be glad to accept your kind invitation were it not for the amount of work which takes literally every minute of his time. He asks me, however, to convey his best wishes for the success of your undertaking."[87] This was a standard Marconi blow-off—he implies that he supports the cause, but won't concretize that support (although it would not have taken much more effort to accede to the request than it took to dictate the letter of refusal). There's not much to be made of it except in the context of what came before and what was to follow. Whatever Marconi may have really thought of Zionism or immigration to Palestine, in 1935 he was not about to make a public statement in support of an organization involved in providing sanctuary to Jewish refugees. As the situation of Germany's Jews became more desperate, Marconi continued to be solicited for support. He was known to be tight with his money but generous with the use of his name to gain favours for people in difficult situations.

On January 26, 1936, Marconi received a letter from fellow Nobel laureate J.J. Thomson appealing for help on behalf of a daughter of Heinrich Hertz, who was considered partly of Jewish ancestry under the Nuremberg Laws (Hertz's father's family had converted to Lutheranism in 1834 and Hertz considered himself Lutheran). Mathilde Hertz, an animal psychologist, had escaped Germany and was now in England, working without pay at the Zoological Laboratory on Animal Psychology at Cambridge while hoping to find employment. She was living on an allowance of ten shillings a day, which would run out in June. Thomson was hoping to raise the sum of £900, which he estimated would support Hertz for three or four years and allow her to get on her feet. The Royal Society had agreed to contribute £150, Thomson himself was giving £100, and he was now "with extreme diffidence" approaching five or six wireless companies with a suggestion that they contribute £100 each.[88]

Marconi took nearly a month to reply, pleading that a persistent flu had kept him in bed and forced him to stop his activity almost entirely. "Now I hasten to assure you that I am in entire sympathy with your request in favour of Miss Hertz and I am writing without further delay to the Marconi Company submitting to them your proposal and giving it my full support." He did indeed write to company chairman Lord Inverforth the same day, describing Hertz even more starkly as "a refugee from nazist racial prejudice" and asking Inverforth to put Thomson's proposal to the company board. Marconi enclosed a copy of Thomson's letter, adding, "Personally I am in full sympathy with Sir J.J. Thomson's views."[89]

Another month went by before the board replied, rejecting Marconi's plea, and then Marconi wrote promptly to Thomson, on April 6, 1936: "I am terribly sorry to say that in spite of my strong recommendation to the contrary, the reply is a negative one; it appears that the Directors do not feel justified in acceding to appeals for assistance under present business conditions. My continued absence from England and the comparatively small share I take in the administration of the Company do not allow of my exercising any pressure on the subject; there is, consequently, nothing left for me to do but to express my regret that so deserving a cause should not have found a better response. If times were not so difficult owing to international and exchange complications [as a result of sanctions], I would have gladly made a contribution myself."[90]

Marconi could have made a personal contribution, even a small one. As in so many other circumstances, there was a complex mixture of motivations at play. Marconi took his time considering Thomson's request; he was indeed ill in January and February 1936, but not so ill that he ceased all activity; there are at least three letters in the archives where he or his secretary responds to correspondents during this period. His appeal to Inverforth was genuine. He may have been irritated by the board's response, and his letter to Thomson indicates that he is feeling somewhat alienated from the English company. One hundred pounds (equivalent to around $8,000 today) was a considerable sum as an individual contribution, even for Marconi. But could he not rise to the occasion if he indeed felt that Hertz's cause was worthy of support? Marconi's nominally positive response to Thomson's appeal is often cited by Italian scholars as evidence of Marconi's keeping a distance from fascist policies in 1936, although his failure to contribute concretely actually points in the opposite direction. The instance is a clear example of Marconi playing both sides of the fence; if necessary, he could make both claims—that he had supported Hertz and that he had not supported Hertz. This was undoubtedly more a question of character than one of an attitude toward Jews. Mathilde Hertz stayed in England, where she was eventually joined by her mother and sister; she is considered one of the first influential woman biologists and a pioneer in the field of comparative psychology, although mostly for her early work in Germany. She never managed to reestablish her career and died in poverty in 1975.

Marconi's complicated stance regarding the rise of antisemitism under fascism is highlighted by another case. In March 1936, Marconi received a desperate letter from Arthur Korn in Berlin. Korn was a physicist, mathematician, and inventor of the *Bildetelegraph*, a machine that transmitted photographs (today Korn's invention is considered the precursor of the fax machine). Korn, who was of Jewish descent, had been fired from his professor's position at the Berlin Polytechnic because, as he put it, "although of Protestant religion" he was "of non-aryan origin"—meaning that he had at least one Jewish parent or grandparent and was considered Jewish under Nazi race law; the Nazis didn't care about religion, what mattered to them was blood. Korn expected that Marconi was aware of his work in phototelegraphy, which was frequently written up in the popular and technical press—in 1923 he had successfully transmitted an image of the pope by wireless from Rome to Bar Harbor, Maine, with much media attention. Korn explained that he had no means of supporting himself and his family and pleaded for Marconi's help: "I think a single word from you would be enough to get me a position with the Marconi Company in England. I am sure I could be very useful in many respects with my knowledge of mathematics, physics and electrotechnics. I can attest that I speak English, French, Italian and a bit of practically all the European languages. I would be grateful for prompt action SOS."[91]

Marconi frequently replied himself to letters from acquaintances or even total strangers making requests of all sorts; the archives are full of examples where he uses his name to open a door—especially when there was no monetary cost to himself for doing it. And the company had benefited in the past from taking in a refugee scientist in difficult circumstances.[92] This time, Marconi had Di Marco reply politely to Korn, thanking him for his kind offer (!) and stating that he didn't believe his employment by the British Marconi company was likely "in view of the existing restrictions in the matter." Di Marco advised Korn to contact the company directly, and provided him with its London address.[93] How insensitive could Marconi have been to the plight of a German-Jewish scientist in 1936? Maybe he just didn't like Korn. Happily, Korn was able to move with his family to the United States, where he got a chair in physics and mathematics at the Stevens Institute of Technology, in Hoboken, New Jersey. He died in Jersey City in 1945.

Marconi's attitude toward Hertz and Korn strikingly contrasts with that of some of his scientific peers, such as Niels Bohr who, around the same time, was taking in refugees from Nazism at his Institute of Theoretical Physics in Copenhagen and helping many of them find positions at various institutions around the world. Bohr was half Jewish on his mother's side and thus possibly more sensitive to the plight of German refugees, but he provided an interesting counterpoint to Marconi. Einstein, of course, provided another.

All of this may just have been indifference, but then there is the story of Marconi's encounter with Hans Frank, a prominent Nazi lawyer who had been Hitler's personal counsel, was then the Nazi government's principal jurist and a minister without portfolio, and would later be governor-general of occupied Poland and a major war criminal; he was condemned to death for crimes against humanity and hanged at Nuremberg in 1946. (Frank was the only one of the high-ranking Nazis condemned at Nuremberg who tried to save his neck by repenting.[94]) Also president of the Academy of German Law, Frank came to Rome, ostensibly for a series of lectures and meetings, in early April 1936. The visit, the first by a German minister to the Italian capital since 1934,[95] was extremely important to both Hitler and Mussolini, as Hans Frank, who spoke Italian, was discreetly playing a key role in building the relationship between the two dictators.[96] By early 1936, Mussolini had decided that Italy and Germany shared "a community of destiny," and the Rome-Berlin axis was starting to take shape.[97] On March 7, 1936, Hitler's army occupied the Rhineland, violating the Treaty of Versailles, while Mussolini was vigorously pursuing his colonial war in Ethiopia. On April 1, Arturo Bocchini, head of Mussolini's political police, and Nazi SS chief Heinrich Himmler signed a secret protocol covering co-operation and collaboration against oppositional forces including "Bolshevism."[98] (Himmler was said to have modelled the organization of the Nazis' Gestapo on that of Bocchini's OVRA.) The following day, April 2, 1936, Frank arrived in Rome as Hitler's personal envoy to Mussolini.[99] He was greeted at the airport by Arturo Marpicati, in his capacity as director of the National Institute of Fascist Culture, which was officially hosting the visit.** On April 3, Frank placed a wreath at the tomb of the unknown soldier in Piazza Venezia, met briefly with Mussolini, and made a speech at the institute that was widely reported in the world press, justifying German race laws and sterilization policies.

The only Rome newspaper to criticize Frank's speech was the Vatican's official organ, *L'Osservatore Romano*. After quoting Frank—"We do not consider the Jews as Germans, because they have nothing to do with us, by blood and by race. They live happily in the bosom of their people and their

---

** Founded in 1926, the institute was intended to link the party to the sphere of culture. It was headed by philosopher Giovanni Gentile.

own civilization. To we Germans they are of no importance"—the Vatican daily added: "With this indignant introduction, Frank sought to explain the reasons for anti-Semitism, speaking of the 'incomprehension' of Nazism and defending the Nazis' anti-Jewish measures against international 'lies.'" The doctrines Frank tried to justify had no place in the civilized world, the paper said, adding that the Nazi practices of racial selection, like sterilization, were repugnant to all of humanity. The *New York Times* reported: "At the same time the Academy of Italy, presided over by Senator Guglielmo Marconi, a close friend of the Pope, received the German Minister. Senator Marconi praised the German in an address."[100] Indeed, on Saturday April 4, 1936, Frank was received warmly at the Royal Academy headquarters in the Villa Farnesina by Marconi and other officials, including Marpicati, justice minister Arrigo Solmi, and Giovanni Gentile.[101] Marconi presented cordial boilerplate greetings in the name of the academy before escorting Frank and the other dignitaries on a tour of the Farnesina grounds and offering the Nazi minister a *vermouth d'honneur*.[102]

The file on Frank's visit in the academy's historical archive tells a revealing back story. On March 21, 1936, the chief of staff of Rome's Ministry of Foreign Affairs wrote to Marconi informing him that Frank had been invited by the National Institute of Fascist Culture "to give a talk on a legal topic," suggesting that it would be good if the academy could offer the German minister some courtesy. Marconi was laid up with a bout of influenza and there is no sign that he actually received the letter himself. A handwritten notation on the letter says that Chancellor Marpicati replied to the ministry by telephone. On March 28, 1936, Marpicati wrote to Marconi informing him that Frank would visit the academy on Saturday, April 4, the date of its next regular general meeting, and that he would be accompanied by Italian justice minister Solmi and German ambassador Ulrich von Hassel. Marconi followed up with brief letters to Solmi and von Hassel confirming the visit.

The meeting took place as planned on April 4. The full text of Marconi's greeting to Frank, possibly drafted by Marpicati, distributed to the press and copiously cited in Italian, German, and US news reports, is in the archive. Aside from the usual cordialities, the key point was Marconi's affirmation that Frank's visit underlined the need for closer relations between "Hitler's Germany and Mussolini's Italy." He didn't have to say that, and certainly not that way. A press release issued by the academy summed up Marconi's greeting and Frank's response, and emphasized Marconi's points about the links between Italian and German law and the hope for closer relations with the Nazis. On his return home, Frank wrote a warm letter of thanks to Marconi, inviting him to visit Germany, which Marconi dutifully acknowledged.[103]

Once again, Marconi had publicly acquiesced with praise for a policy one would hope he privately found as repugnant as did the Vatican press. It was immediately noticed abroad. A Pittsburgh newspaper, the *Jewish Criterion*, headlined a brief notice datelined Rome, "Marconi Applauds Exposition of Nuremberg Laws," and reported the story as follows: "A painful impression has been created here by the fact that Guglielmo Marconi, world-famous scientist, six members of the Italian cabinet and many other high government officials who attended the series of lectures by Hans Frank, Nazi minister of justice [sic], applauded the Nazi official's laudatory references to the Nazi Nuremberg laws against the Jews."[104] (This was the entire text of the item.)

The world had still not fully digested the impact of the Nuremberg race laws when Hans Frank visited Rome in April 1936, and the speech may well have been a trial balloon to see how Italy—Germany's closest European ally—as well as the rest of the world would respond. The official Italian response (as opposed to the Vatican's) was enthusiastic, and Marconi's acquiescence was an important part of it. Marconi's public acceptance and the courtesy he displayed toward Hans Frank was more than an unfortunate obligation demanded by protocol. It was more than a desire not to rock the boat. Marconi provided a prestigious platform for one of the most unapologetic spokesmen for the Nazi regime's most offensive features. He did it without showing any sign, privately or publicly, that anything was wrong—not even a sign of discomfort. And he used the occasion to call for closer ties with the Nazi regime. If not quite a heartfelt endorsement, it was certainly more than a passive gesture.

Whatever it was, it was quickly forgotten. Marconi's assent in the barring of Italian Jewish academicians and his indifference to the emigration troubles of German-Jewish scientists were not revealed until decades after his death, and his role in entertaining Nazi minister Hans Frank while applauding his racial theories in Rome in April 1936 has never before been recalled. When Marconi died a little over a year later, an obituary in the *Jewish Criterion* praised him as a lover of peace and wondered whether the respect that Mussolini professed toward him was reciprocal. "It seems improbable that the man who did so much to bring the distant parts of the universe closer together relished the brutish attempts of an angular-jawed dictator to wedge them apart again. Marconi achieved, and it is we, his survivors, who will feed upon their fruits. Honors he received aplenty in his lifetime, but they will compare poorly with the glory that will be his as civilization continues to enjoy the benefits of his genius." Another Jewish-American publication went a step further, stating: "Marconi held the Jewish people in the highest esteem."[105] That may have been the case, and we have seen that antisemitism was anathema, even abhorrent, to Marconi. But he was, consciously or unconsciously, insensitive to its consequences, and when it arose in Italy he was certainly not prepared to stake anything important on the issue.

In this he was, unfortunately and uncharacteristically, typical. Hannah Arendt wrote in 1963 that Italy was "nearly immune to anti-semitism" before the 1930s, but that when it appeared, ordinary Italians didn't dare speak out on the subject.[106] This is probably closer to the truth than either of the more extreme positions one finds in Italy today. Ordinary Italians may have been "nearly immune to anti-semitism," but there was a dark undercurrent, present in some parts of the upper bureaucracy, in obscure corners of the Vatican, and in the hierarchy of the Fascist Party. When this current came to the forefront in the 1930s, ordinary Italians, as well as powerful ones like Guglielmo Marconi, did not speak out. Marconi's desperate need for acceptance translated into a deep, absolute, almost religious political conformity—however radical and uncompromising he was in scientific outlook, unconventional in his personal life, adventurous and risk-taking in business, he always, always situated himself on the side of political power, order, and authority. He was not just a common conformist. He had a charismatic pull on others less convinced than himself and he operated with ease across the conventional boundaries of his world. In Italy he appeared to be a voice of moderation, despite the rhetoric and bombast. In Britain he was a voice of reason. In America he preached for calm and understanding. Yet in 1936 he called for closer ties with Nazi Germany.

34

Controlling His Legacy

As 1937 dawned, an odd article appeared in the US magazine *Liberty*, entitled "Faith is man's bridge to infinity." The article was based on a pastiche of interviews Marconi had given over the years, synthesizing his views on science, politics, and—a topic he rarely broached in print—faith. Labelled "a devout Christian," Marconi was quoted as stating: "Science is hardly more than the dim light of a pocket lantern in the black forest through which mankind painfully gropes its way to God."[1]

The author of the article was George Sylvester Viereck, a German-American writer with some reputation as an early apologist for Nazism. Viereck had first interviewed Marconi in 1929, for an article where (among other things) he characterized Mussolini as "Nietzsche's Superman."[2] Viereck made several trips to Germany in the 1930s, ostensibly to visit his ailing mother, and became increasingly enamoured with what he saw there, sharing his observations in correspondence with Marconi. As he often did, Marconi revised Viereck's draft before it was published. Now, Viereck wrote to Marconi reiterating the suggestion that he ghostwrite Marconi's autobiography. Marconi replied: "I have had quite 'a dose' of biographies lately, and shudder at the idea of tackling another one."[3]

Marconi had, in fact, become obsessed with controlling his legacy. For years he had been constantly besieged by would-be biographers, some more serious than others. He directed much of his dwindling energy to dissuading them when he could, and bickering over details, correcting errors, and urging deletions when he couldn't. A typical response was the one he made after an Iowa writer wrote to US ambassador Breckenridge Long in Rome about how best to approach Marconi: "Nothing doing," Marconi scribbled on a letter from an embassy official, before passing it on to his secretary Umberto Di Marco to convey the regrets.[4]

There were already two Marconi books in circulation, both uncritical encomia. Marconi's long-time associate Luigi Solari published a breezy Italian account of Marconi's early glory years in 1928, with Marconi's blessing. The book was a paean to the Italian nationalism of the time, written in the style of fascist hagiography, and dressed up Marconi's story in full heroic mode. Solari was now patiently working on a full biography; he submitted some material to Marconi in 1933, but two years later, he still had not received any comments.[5] Doris Collier's "authorized" biography, published in 1935, was a bit more complicated. Marconi had tried to discourage Collier from publishing because of her wish to include some anecdotes regarding his first marriage and family life, and there was a long correspondence between them. Collier smartly cultivated Cristina's collaboration and Marconi grudgingly accepted to authorize the project in the end, on the condition that he be allowed to vet the manuscript. He didn't have time to revise the proofs and continued to grumble about little inaccuracies, tone, and nuance. He should have been pleased; the result was a wholly laudatory account of the key moments in Marconi's career, adding little to what could be gleaned from press archives.[6]

Another would-be biographer was Marconi's former personal secretary, Leon de Sousa, who had ghostwritten a four-hundred-page "autobiography" of Marconi under contract to the British company in 1919. De Sousa had been a hired gun; he received £750 for his manuscript and signed over all rights to the company. A contract was signed with Curtis Brown in 1918, but when the manuscript was completed, Marconi managing director Godfrey Isaacs would not agree to its publication, for unknown reasons. In 1930, Curtis Brown signed another contract for a book about Marconi—with Luigi Solari. De Sousa, now based in Paris as an agent for RCA, was still in touch with the publisher as well as with Marconi, and still hoped to publish a Marconi biography. De Sousa and Marconi met in Paris in March 1935, and when de Sousa tried to revive the idea of a biography, Marconi said, "But Solari has written one." De Sousa wrote immediately to Solari, and to Di Marco, proposing a collaboration that never took place.[7]

De Sousa also queried Marconi about a work in progress by Orrin E. Dunlap Jr., radio editor of the *New York Times*. A seasoned journalist and one of the earliest dedicated radio beat reporters, Dunlap too started working on his project around 1930; Marconi had been aware of it since meeting Dunlap in New York during his 1933 US visit. At Marconi's suggestion, Dunlap had approached RCA president David Sarnoff about writing a foreword. Sarnoff replied that he would agree to do it "if you would have Senator Marconi personally read your manuscript and give it his approval."[8] Whether an innocent suggestion or a thinly veiled bribe, Dunlap took it to heart; it turned into a nightmare for the author, delaying publication of the book for nearly four years. Dunlap sent Marconi a manuscript in February 1934, but Marconi had still not returned it a year later. Dunlap wrote to Marconi again in March 1935, informing him that he had rewritten the manuscript. Another three months went by without a reply, and then Marconi suggested they meet in London. Dunlap appealed to Sarnoff, who would soon be seeing Marconi in England: Could he try "to swing [him] into line"?[9]

Meanwhile, in New York, George Clark at RCA (who had organized Marconi's 1933 US visit) was also hoping to write a Marconi biography and even had an expression of interest from Oxford University Press. Clark wrote cagily to de Sousa in Paris, stating that he had been assigned by RCA to gather "all available data" on Marconi's life and achievements, and had been told that de Sousa had a large file of just such material. Would he loan it to them? Clark was more transparent with Marconi about what he was up to, but Di Marco informed him that "similar requests have come from time to time and not a single one has been acceded to." Clark then took another tack, writing to de Sousa and proposing that they join forces in order to eclipse Dunlap's book, which was almost ready for press: "I venture to say that my collection and yours contain much more actual facts about radio and its inventor than any other possibly could." Clark also tried to solicit Di Marco's collaboration, arguing that he possessed more "marconiana" than anyone else. Di Marco replied candidly: "There seems to be a plague, just at present, or, at any rate, the threat of a plague of such publications."[10]

Dunlap, meanwhile, persistently moved forward as he waited, patiently but in vain, for the promised input from Marconi. Reaching the end of his tether, he decided to go ahead and publish. In May 1936, he drafted a short preface and sent it to Marconi asking if he wanted to sign it and informing him that the book was going to press. On November 22, 1936, Marconi finally cabled Dunlap: "Hope your book yet unpublished as find many misstatements of fact which if published will compel my contradiction stop. . . . If still time can send you telegraphic statement of corrections which consider absolutely necessary." Dunlap replied immediately, complying with Marconi's request to stop the press while waiting for his corrections.[11]

Marconi really didn't want to see the book published at all. On November 29, 1936, he wrote Dunlap a long letter: "I realize how anxious you must be to go to press with this book. Let me tell you right away that I am equally anxious to go slow.... Let us both make the best of it.... The matter is too serious for both to be hastily settled. Let us abide by our time." Marconi asked Sarnoff to help him stop the publication; in addition to errors, the book contained "a lot of irrelevant information...which I am very reluctant to have divulged." He asked Sarnoff to try to persuade Dunlap "to give up the idea," but it was no use. Dunlap was anxious to please Marconi and would wait for his corrections, but the book was now in the hands of the publisher and there was no stopping it.[12]

In January 1937, Marconi sent Dunlap his corrections and agreed to sign the preface if Dunlap incorporated them. Dunlap made the changes, and the book was published in April 1937, with Marconi's preface (but no foreword from Sarnoff). There is a long list of Marconi's requested corrections along with all the related correspondence in the Lincei archive. At Marconi's insistence, Dunlap removed all reference to his one-time fiancée Josephine Holman and reduced the discussion of his first marriage to the simple mention that it took place, was not happy, and ended in divorce. Despite Marconi's objection, Dunlap kept thirteen pages on the Marconi scandal, but deleted a chapter on "financial trouble" that Marconi considered "undesirable" as well as inaccurate.[13]

Marconi's attitude toward Dunlap was in total contrast to his encouragement of Clark. In January 1937—as Marconi was still struggling with Dunlap's corrections—Clark sent him about "one-fifth" of his own biography, along with a sycophantic note. Marconi responded to Clark: "Allow me to congratulate you on the way in which you have tackled this rather arduous task." He had only a few changes to suggest, mostly concerning "certain myths which seem to have taken root in the imagination of historians," such as his having been a pupil of Augusto Righi or regarding the supposed adversity of his father to his early work.[14] Marconi typically bristled when reporters or would-be biographers (including Dunlap) brought up these "myths," but he was virtually bubbling with enthusiasm to Clark. Maybe he trusted him. More likely, he sensed that Clark would never complete his manuscript. The other four-fifths of Clark's biography never saw the light of day, while Dunlap's became the standard source on Marconi's life for the next twenty-five years.

❀   ❀   ❀

Marconi had been ruminating about his legacy for some time, certainly at least since a 1934 visit to Britain, when he expressed a melancholy thought in an exchange with the principal of St. Andrew's University in Scotland, where the students had just elected him Lord Rector: "Have I done the world good, or have I added a menace?"[15]

On the surface, Marconi's life seemed settled into a comfortable routine. Every day, he crossed the Tiber in a chauffeur-driven academy motorcar and went to his office in the Villa Farnesina, one of the most exquisite renaissance palaces in Rome and still very much the palace Goethe described in 1787 as having "the most beautiful decorations I ever saw."[16] With its frescoes by Raphael and Baldassarre Peruzzi, the sumptuous villa built for Sienese banker Augusto Chigi in the sixteenth century and then sold to Cardinal Alessandro Farnese was the headquarters of the Royal Academy and hence Marconi's workplace. Here, in a stately room to the left of the entrance, he sat behind a sixteenth-century desk in a high-backed chair, dictating correspondence to Di Marco, with his young daughter Elettra sometimes playing in the corner while her father ploughed through a pile of paperwork that kept regenerating itself.[17]

But he was still restless. Although he had not left Italy since his near-fatal heart attack on the train from Paris in December 1935, Marconi now hoped to make another trip to England. Early in 1937, he had the Italian ambassador in London, Dino Grandi, request an invitation for him and Cristina to attend the coronation of King George VI, which was scheduled for May 12. The reply from the Foreign Office was not edifying; there was unfortunately not enough room in Westminster Abbey and wives of foreign dignitaries were not being invited. Marconi protested that he had been invited to the coronation of George V; he even still had the invitation in his possession. Grandi took up the issue with Foreign Secretary Anthony Eden (who had succeeded Samuel Hoare after the failure of the Hoare-Laval Pact on Ethiopia) and Under-Secretary Robert Vansittart, but Marconi simply was no longer who he used to be. Grandi couldn't bear to transmit the unflattering reply to Marconi; he had foreign minister Galeazzo Ciano do it.[18]

Marconi still planned to go to England, although his plans became ever more vague. In early April, he wrote to a cousin, Helen Burn, in Folkestone: "We are all quite well and expect to be again in London if not exactly for the coronation, at any rate not much later." Around the same time, Di Marco wrote to an official at the British company headquarters: "A trip to London is constantly planned but just as constantly postponed: will it ever materialize? I sincerely hope so." The company was already reassigning Marconi's old office, however, refurnishing it and relocating his papers; certainly in their minds, he was not coming back.[19] As the year wore on, it became clear that there would be no trip. When Grandi invited Marconi to attend a minor diplomatic event in London, he replied that he would not be in England, although he still hoped to be there soon after the coronation. When the Marconiphone Company offered to install a state-of-the-art experimental television set in his hotel suite, Di Marco had to reply: "He will be delighted to see your suggestion carried out, but in view of the uncertainty of his arrival in London, he would ask you not to take any step in the matter before you hear further from him." All that said, other correspondence indicates that he was still contemplating a possible trip to London as late as the sixteenth of July.[20]

Marconi was endlessly receiving invitations and being solicited for causes, from progressives as well as reactionaries—one of the last was from Maria Montessori, sending him the program of the International Montessori Association's congress that was about to take place in Copenhagen, on the theme "Education for Peace." Marconi remained an honorary member of the association, despite Montessori's break with the fascist regime in 1934.* He was also invited to speak at an international conference on the fiftieth anniversary of the death of Heinrich Hertz, in Paris, scheduled for September 1937. Old friends like Margaret Sheridan still got in touch. Sheridan wrote from Milan in March, asking to meet him, but there is no sign that he answered. Leon de Sousa wrote from Paris in July, to say he was coming to Italy and hoped they would be able to meet.[21]

Marconi was now declining nearly every invitation, but in February 1937, encouraged by David Sarnoff, he agreed to deliver a broadcast on "the significance of modern communication" as part of a radio forum organized by the *Chicago Tribune*.[22] The broadcast took place on March 11, 1937. In the four years since his visit to Chicago it had become possible to send a signal around the world in one-seventh of a second (that is, at the speed of light; in October 1933, requiring relays, it had taken more than three minutes). "The science and art" of radio communication had reached the stage where an expressed thought could be transmitted almost instantaneously and received

---

* Montessori is another example of an activist intellectual who took an opposite trajectory under fascism to Marconi's. Initially attracted to Mussolini, whom she saw as a reformer who could improve Italy's education system, she set about organizing schools in the slums of Rome in the 1920s. As he did with Marconi, Mussolini sought to capitalize on Montessori's international reputation and named her inspector of schools. However, they later clashed over Mussolini's insistence that schoolboys wear the fascist uniform in class, and she broke with him in 1934 (Kramer 1976).

simultaneously by anyone, anywhere on the globe, who was equipped with the necessary apparatus, Marconi said. "Broadcasting, however, with all the importance it has attained . . . is not, in my opinion, the most significant part of modern communications, in so far as it is a 'one-way' communication."

As was his style, Marconi kept revising the talk up to the last minute. One of the five typed versions in the Lincei archive, the one that he actually read on the air, had the following addition: "A far greater importance attaches, in my opinion, to the possibility afforded by radio of exchanging communications wherever the correspondents may be placed. . . . It is only through radio, in fact, that we are capable—so far—of talking to each other with our own voice across the oceans as well as between the antipodes. The crowning reward to my work and effort lies exactly in this bare statement." This was well said. He then concluded: "With all our friction, jealousy and antagonism . . . and inspite of the bloody eruptions which from time to time rend it asunder, the ideal of peace and fraternity remains unabated in us: we all yearn for a better life, based on a better understanding of one another. In radio we have a fitting tool for bringing the people of the world together, for letting their voices be heard, their needs and aspirations to be mani-fested."[23] Here he was, of course, expressing an unrealized ideal. Radio was not able to prevent or give voice to the people in the global conflagration that was about to break out. It would, to the contrary, be used effectively by political elites on both sides in the coming calculus of conflict.

The dark side of fascism also continued to cloud Marconi's final months, as the implications of Mussolini's colonial policies and his alliance with Hitler came into the open. With the proclamation of the empire on May 15, 1936 (enthusiastically acclaimed by Marconi, as we've seen), Italians sud-denly found themselves co-subjects with millions of Africans. Prior to the annexation of Ethiopia, Italy had only a tiny non-Caucasian population, and a liberal tradition toward race prevailed; in another context, this could have been the seed of a multiracial society, but in Mussolini's empire, explicitly racist policies became the order of the day. Almost immediately, apartheid-style measures forbidding relations between metropolitan Italians and colo-nial subjects were institutionalized on the basis of race. There was now a dividing line between "pure" Italians and others. As Italy began to adopt the model of racial policies advocated by Hans Frank in his speech in Rome a year earlier, a fundamentalist antisemitism seeped into the Italian main-stream as well, from its previous confinement on the margins. In April 1937, an important fascist intellectual, Paolo Orano, published a treatise, *Gli ebrei in Italia* ("The Jews in Italy"), linking prevalent antisemitic stereotypes with Italy's new-found theories of racial purity. Colonial racism and antisemitism began to merge in an overall national policy, where blacks and Jews were equally not Italian.[24]

Marconi couldn't possibly have missed these developments. His attitudes toward race were a corollary of his total, uncritical support for European colonialism and twentieth-century imperialism. It is unlikely that he had a single personal relationship with a non-Caucasian in his entire life, and in this he was typical of his time and social class. But he was also a perpetual outsider—he hated it, but that is what he was—an "anglo-saxon" in Latin Italy, a "foreigner" in imperial Britain, a baptized Catholic brought up Protestant and then reconfirmed Catholic at the age of fifty-three, a "cosmopolitan." In this respect, he had much in common with many of his Jewish business associates. Yet when they came under attack as a racial group he remained silent. His lifelong friend, Margherita Sarfatti, had seen the writing on the wall and had herself baptized as early as 1928 (followed by her two children a few years later). Marconi's own nephew, his half-brother Luigi's son Pietro, married a Jewish woman, Giselda Fano, in August 1938; they emigrated to Addis Ababa after the racial laws were promulgated a few months later.[25]

As Mussolini's position on an Italian-German alliance hardened, Marconi found himself in an increasingly uncomfortable position. On September 5, 1936, the chancellor of the Royal Academy, Arturo Marpicati, wrote to Marconi: "Dear President, I inform you that I am invited personally by Hitler to the Nazi congress in Nuremberg as guest of the German Government. *Il Duce* has decided favourably for my voyage which will last seven days. I will return to Rome on the 15th of September." Marconi replied promptly, from Santa Margherita Ligure where he was staying on the *Elettra*: "I congratulate you for the honour done to you by the German Government with the personal invitation of the Head of the Reich to the National Congress at Nuremberg; I am also very curious to know if the invitation is addressed to you in your quality as Chancellor of the Royal Academy or as Hierarch of the PNF [a leader of the National Fascist Party]. Anyhow, I wish you *bon voyage* and will be happy to have news from you directly on your return to Rome on the 15th."[26]

This exchange is fascinating: Marpicati informs Marconi, nominally his boss at the academy, that Mussolini has approved his trip to the Nazi congress at Nuremberg; Marconi has no choice but to accept, but he is curious to know in what capacity Marpicati has been invited—was Hitler reaching out to the academy or to the party, and if it was to the academy, why was the invitation not offered to Marconi? It has been suggested that Marconi was invited as president of the academy and asked Marpicati to make the trip because he did not want to shake the hand of Hitler. The correspondence does not back up this view; Marconi may even have been miffed not to have received the invitation himself. And had it been offered to him, what would he have done?

Marpicati replied immediately: "The invitation from Hitler was transmitted to me by von Ribbentrop not as a hierarch (which I am no longer) but as a political person, certainly remembered in Germany for having carried the first wishes from Mussolini after Hitler's coming to power." (Marpicati had attended the previous Nuremberg Congress in 1933.) Marconi's intuition was correct: Hitler was seeking closer relations with the Italian Fascists, and they with him. Marpicati was an important figure in this regard; his presence at Nuremberg was desired both by Itay's Fascist Party and by the Nazis, "as a political person." Marconi took a few days to digest this before replying, with tongue in cheek: "Dear Marpicati, Many thanks for your information from which I gather with pleasure that your qualities as an orator are appreciated in high places."[27]

Marpicati filed a detailed report when he got back from Nuremberg. The original was submitted to Mussolini, and Marconi was provided with a copy.[28] The report was a mixture of adulation and prudent bureaucratese. Hitler had made "immense progress" in his three years in power. He was more secure and had an almost mythical impact on the crowd, which listens to him "with religious stupor" as to an oracle. Marpicati chose his highlights carefully: one of the key developments was Hitler's insistence that German science was no longer dependent on "eternal and infamous Hebraic pollution." Hitler was to be taken seriously. That said, Marpicati was annoyed at Nazi indifference and ignorance toward Italy, and especially the lack of recognition of the role of Italian fascism as an antecedent to Nazism; not a word had been said about Italy's victorious war in Africa, despite what it meant as an international success for fascism. Marpicati's report left an ominous sense that the Nazis might turn out to be difficult, grasping allies.

Apart from Marconi's wisecrack about the reason for his invitation to Marpicati, there is no direct comment anywhere in the documents to indicate Marconi's views or feelings about the Nazis (and, as we saw, he participated in public pro-Nazi displays when called for in his official functions). And yet, later commentators always take care to state that Marconi was critical of Italy's alliance with Germany. Cristina, for example, wrote: "Guglielmo was against Nazism. He was worried by the understanding between Mussolini and Hitler and thought it could have serious consequences." According to Cristina, Marconi was concerned about the prospect of war and asked to see Mussolini at the Palazzo Venezia. "He was received immediately in the Map Room. The conversation was very forthright. Guglielmo bravely warned Mussolini against allying with Hitler." Mussolini was not interested in Marconi's advice, however. "You say these things because your mother was English," Cristina reports Mussolini saying to Marconi.[29] Marconi never said a word in public about the alliance.

❀   ❀   ❀

In September 1936, Marconi oversaw yet another reorganization of the CNR. Tired of perpetually arguing with the Ministry of Education, Marconi proposed that the council be attached directly to Mussolini's office. The council's basic law, dating from May 26, 1932, would now state that the CNR was "the supreme technical Council of the State *and is placed under the direct orders of the Head of Government* [emphasis added]."[30] At this late date, notwithstanding his growing skepticism toward Mussolini, Marconi still preferred to deal directly with the summit of power. Mussolini, at his end, still preferred dealing with Marconi as well; despite their sparring, he renewed Marconi's tenure as head of the CNR on February 23, 1937.

The CNR was by now a shell of its earlier self, and certainly a far cry from what Marconi had thought it might be. The board didn't meet at all for eleven months, between July 1936 and June 1937, although Marconi himself continued to intervene on relevant questions, and some of his collaborators, like Enrico Fermi, continued making research proposals that fell on deaf ears. Other official institutions also began noticing the sluggishness of research and innovation, and expressing it. On February 19, 1937, the Fascist Federation of Industrialists wrote to Mussolini, copying their letter to Marconi, calling for a less bureaucratic approach and more support for pure research.[31] On March 1, Marconi spoke at the Grand Council of Fascism (along with other party hierarchs like Achille Starace, Paolo Thaon di Revel, and Dino Grandi) regarding the role of science and technology in achieving economic "autarchy" as Italy underwent military preparation.[32] On March 23, Mussolini's chief of staff had to remind him that he had been promising Marconi a meeting for some time.[33]

The regime's lack of support and eventually the constraints on their intellectual freedom severely hampered Marconi's cherished physicists. In

June 1938, Enrico Fermi received less than half the funds he had requested for his team's research on slow neutrons (the basis for the nuclear reactor). A few months later, at the age of thirty-seven, he was awarded the Nobel Prize. (Marconi won his at thirty-five.) It is difficult to say how much longer Fermi might have remained loyal to Mussolini, but another issue, which had nothing to do with scientific research, settled the question. Fermi's wife, Laura, was Jewish, and the Italian race laws adopted in November 1938 made it untenable for them to remain in Italy. After travelling to Stockholm to receive the Nobel, Enrico and Laura Fermi did not return to Italy, sailing instead for the United States, where Fermi was immediately offered research chairs at no less than five universities. The American government was more generous in support of Fermi's research, which, of course, became the key component of the Manhattan Project and the atomic bomb.

Ultimately, the CNR was more like a ministry of research administration than an actual catalyst of research. Marconi managed to maintain its independence in the last months of his life, so far as possible in a fascist regime. After his death, it passed into the hands of the military and the politicians. Marconi was succeeded as president by Marshall Pietro Badoglio, the hero of Ethiopia (and later head of government after the overthrow of Mussolini in 1943). The council's principal meeting room was renamed Aula Marconi. Today, a smaller room beside it commemorates Vito Volterra.[34]

After forty years in the public eye, it was expected that anything Marconi did would be spectacular and game-changing, that he would always be working on something "revolutionary." In 1937, he was indeed thinking about things like transmitting electrical power by radio waves, the possibilities of interplanetary communication, extraction of gold from the sea, radar, and the communicative possibilities of broadcasting. He was also keenly aware that his inventions could easily be hijacked from the intentions he set for them, especially in a context of war. The newspapers kept talking about his "death ray," but while the idea of Marconi devising such a device for Mussolini is intriguing, and, as we've seen, he supported Italy's colonial wars, he was terrified of the idea of war between Italy and Britain and sensitive enough to see that that was where Mussolini's fateful alliance with Hitler was leading.

In May 1937, Marconi gave his last press interview, to the journalist and propaganda specialist Lisa Sergio. The interview was carried around the world, by the Hearst syndicate in the United States, the *Daily Express* in the

UK, and the fascist *Il Popolo d'Italia* in Italy. Sergio reported that Marconi told her he had indeed discovered and tested a death ray, and then dropped the idea. The way he told it, it was a practical problem, not a moral one, but in any case, he was no longer looking for new military applications for wireless. This may have been Marconi's way of signalling the news to the Palazzo Venezia. It was certainly not the first time Marconi used the media to get a message out, but the story of his relationship with Lisa Sergio is another piece of the puzzle that was Marconi's last year in Rome. Born in Florence in 1905, she was the daughter of Agostino Sergio, a baron, landowner, and acquaintance of Marconi's, and Marguerite Fitzgerald, an American from a socially prominent family in Baltimore.[35] Perfectly fluent in colloquial English, Sergio worked as a journalist with the Florence-based English-language *Italian Mail* before moving to Rome in 1929. When Mussolini sought Marconi's advice on setting up international propaganda broadcasts in 1932, and asked if he knew a convincing English-language broadcaster, Marconi recommended Sergio, who had impressed him with some editing and translation work she had done for him.[36]

Sergio met with Mussolini, and a few days later received a call inviting her to come in to see the head of the press department of the Italian Ministry of Foreign Affairs, Gaetano Polverelli (later Mussolini's last minister of popular culture). After explaining what they had in mind for a foreign-language news broadcast, Polverelli told Sergio that Marconi had suggested her for the job, and asked if she was an ardent fascist. "Ardent Fascist?" she replied. "Everybody is."[37] Sergio later wrote that she initially refused the offer, but when she consulted Marconi he urged her to take the position. The combination of "flattery, il Duce [and] Marconi" was irresistible, she said, and she took the job.[38] A file in the archives of the Italian Ministry of Popular Culture tells the story a bit differently. Sergio had been known to the regime since applying to do freelance work in 1930, shortly after arriving in Rome.[39] In July 1933, Polverelli hired her to do translation and an English-radio news broadcast, and she was assigned directly to the foreign affairs branch of Mussolini's personal press office. On December 2, 1933, the appointment was confirmed in a brief note signed simply "Ciano" (Mussolini's son-in-law, Galeazzo Ciano, who was then head of the government's office for press and propaganda).[40] There is nothing further in the file, but the Fascist political police also had a file on Sergio, started with an entry dated August 30, 1933, in which the agent says he was told that Sergio is "a member of our organization." Had she also been recruited by OVRA?[41]

Sergio claimed, inflatedly, that she was Europe's first woman radio announcer; in fact, she wasn't even Italy's (that honour goes to Maria Luisa Boncompagni, who was the host of Italian radio's first broadcast in October 1924[42]). Nonetheless, Sergio became a fixture on the English program of Italian international shortwave radio between 1932 and 1937 (where she styled herself "the English speaker"). She was noticed by the European and US press, London's *Daily Telegraph* commenting that, "so perfect are her accent and inflexions that only when she comes to place names does any suspicion arise that she is not English." Janet Flanner, in the *New Yorker's* "Paris letter" of October 28, 1936, said Sergio's "high-powered sales talks for the Fascist point of view...have constituted a new propaganda departure."[43] Sergio also served Mussolini as an English interpreter, and claimed to have coined the term "axis" when translating the infamous speech of November 1, 1936, in which Il Duce announced Italy's alignment with Germany.

At first, Sergio had no difficulty marching to Mussolini's drumbeat; for example, describing Ethiopians, when it was necessary for an "ardent fascist" to do so, as "a bunch of black savage tribes uninterested either in progress or prosperity."[44] Eventually, she later wrote, her association with the propaganda designed to cover up Italian brutality in Africa soured her on Mussolini and on fascism. She began to "skip a phrase or two" from the dictated scripts and use "circumlocutions" in her broadcasts. Early in 1937 she was in trouble after Mussolini discovered she had been tampering with his speeches and, she said, she was ordered to resign.[45] She later wrote that she refused and was confronted by Ciano—now minister of foreign affairs, as well as, according to an FBI report, Sergio's lover[46]—who apparently lashed out: "If I were in *il Duce's* place I would have had you shot a long time ago"— to which she says she retorted: "It will be you that ends up shot, not I." This is almost certainly an apocryphal tale. For one thing, it is just too good to be true—Ciano was indeed shot by a fascist firing squad in Verona in January 1944.

After that, Sergio later wrote, "my telephone was permanently out of order, the mail did not reach me and a man, posted outside my door night and day, shadowed me wherever I went."[47] She learned that her arrest was imminent and that she was earmarked for deportation to a prison island for hard labour. Then, after nearly three months without work, she received the assignment to interview Marconi. John Whittaker, the Rome bureau chief for the *New York Herald Tribune*, suggested she ask Marconi for help. She did, and as she later told the story, Marconi provided her a passport with a visa for the United States, a first-class steamship ticket to New York, $500 in cash, and a letter of introduction to David Sarnoff.[48] Not bad for someone who had just been exposed as an enemy of the regime. She sailed from Naples on June 27, 1937, on a reservation made in Marconi's name. "The name was magic," she later wrote. "Nobody troubled me and everything was done to make me comfortable." In New York, on the basis of Marconi's recommendation, Sergio was hired by Sarnoff and enjoyed a long and successful US career as a radio broadcaster. Sergio's 1989 obituary in the *New York Times* described her as "a slender, fiery contralto once known as 'the Golden Voice of Rome,'" and embellished the story of her flight from Italy, stating that she was "smuggled aboard" a transatlantic liner by Marconi, something that even she never alleged.[49]

One wonders why Marconi helped Lisa Sergio escape. He might have been helping the regime rid itself of a thorny opponent. Fascist Rome was still not Nazi Germany, and troublemakers with friends in high places could not just be made to disappear. Perhaps he believed she was in genuine danger. Marconi did see himself as a fixer; he was important and powerful enough to have access to exit papers. Did he now share the growing antipathy of many "ardent fascists" like Lisa Sergio for the corrupt disaster course on which the regime was set? Or was it just one more instance of Marconi helping out an attractive young woman he liked?

There are small but revealing indications that Sergio was not as *persona non grata* as she pretended at the end of her time in Rome. Her May 1937 interview with Marconi was carried by the fascist mouthpiece *Il Popolo d'Italia*, hardly a sign that the regime was trying to isolate her professionally. The writer Niccolò Tucci, a long-time friend of Sergio's, was quoted years later as stating that her political discomfort in Italy was genuine if perhaps not wholly anti-fascist. "Sergio became anti-fascist in 1937, but not completely. When she left Italy she was confused," Tucci said in a 1990 interview.[50] With the combination of insecurity (whether due to political or romantic difficulties) and increasing unease over her ambivalence toward fascism, "her future in Italy must have looked bleak," writes the American scholar Stacy Spaulding, author of a doctoral thesis on Lisa Sergio.[51]

It may have been this very ambivalence and melancholy forecast, mirroring his own, that endeared Marconi to Sergio. (She certainly had more success at attracting his support than, say, Arthur Korn.) It is highly unlikely that Marconi and Sergio were romantically involved. They maintained a certain formality and, anyway, those days were well behind him.[52] But Sergio perfectly fit the profile of the women to whom Marconi was attracted: young, glamorous, strong-headed, interesting, accomplished, and well bred. And Marconi, too, was adept at inventing and reinventing versions of his past, and even his present, for public consumption. So we don't know and may never know the answers to these questions. But there are some facts. In her apartment in Rome, Sergio kept signed photographs of two prominent countrymen: Galeazzo Ciano and Guglielmo Marconi. The Marconi photograph, which she brought with her to the United States and kept for the rest of her life, is inscribed, in Italian: "To Lisa Sergio, in admiration of her work of Italian propaganda. From Guglielmo Marconi, Rome, 20 June 1936."†

One of the last foreign political figures to meet with Marconi was the British Labour MP George Lansbury, who saw him in Rome in early July 1937. Lansbury, a principled pacifist, critic of European rearmament, and advocate for dismantling the British Empire, believed that peace was tied to international economic co-operation; he had broken with his party's establishment and resigned the leadership in 1935 over economic sanctions against Italy, which he opposed. As a result, he was considered soft on fascism and politically naive. Lansbury travelled to the United States in 1936, meeting with Roosevelt to propose a world peace conference. In 1937 he toured Europe, to try to convince European leaders, including Hitler and Mussolini, of his cause. He later wrote that his most encouraging talk was with Marconi. "Marconi could not be placed in the same category as ordinary politicians.... He was a Fascist and had accepted that form of government because he thought it best for Italy. All the same he had a deep abiding affection for Great Britain and...was emphatic in his declaration, made again and again, that war in these days of potential abundance for all was a terrible blunder." Remembering Marconi's sad demeanour, Lansbury wrote: "I think he felt himself a doomed man...he seemed to be unable to bear the thought of war between Italy and England."[53]

The Lansbury interview was another marker of the enigma that shrouded Marconi's final months—an enigma that may have been one to him as well. Although Marconi characteristically adapted his stance to suit the person to whom he was speaking, his frank comments to Lansbury were a strong sign that he was genuinely shifting away from blind support for

† The inscribed photograph of Marconi, seated at an ornate desk, stiffly dressed, his glass eye staring vacantly off at an angle while his good eye looks at the camera, typically unsmiling, is in the Lisa Sergio Papers, Georgetown University Library (14/41).

Mussolini to a higher concern for the state of world affairs. Reflecting on the role of science as an instrument of peace not war, Marconi anticipated the ambivalent role of the atomic scientists during the Second World War, not to mention today's computer technologists, security state specialists, and drone developers.

Mussolini, however, never stopped believing in Marconi's death ray, and the failure to get his hands on the elusive weapon haunted him in his final days. As the Allies were closing in on him in March 1945, in his last interview, he "mentioned with regret a 'death ray' that Marconi had invented but kept secret."[54]

❁   ❁   ❁

Marconi continued to worry about money. With the future of his British income increasingly insecure, and London cutting its support for his research, he put the word out that he was thinking about selling his beloved yacht. He felt it was worth a million lire, but the best private offer he received was 650,000.[55] (At this time the yacht was insured in England for £21,000, or the equivalent of just over two million lire.[56]) He then began negotiating an eventual sale to the Italian government and, in early July, asked the *Elettra*'s captain, Gerolamo Stagnaro, to make an inventory of material on the yacht. The sale was so likely that Stagnaro decided not to refill the yacht's coal supply. On July 13, 1937, Di Marco told Stagnaro to expect a visit from government officials imminently.[57] (Marconi never considered selling his other major asset, Villa Griffone, which was worth approximately the same as the yacht. Although he no longer used Villa Griffone, it was the legacy of his father.)

Marconi's relations with his older children, and especially their financial affairs, were still a great concern to him. He had been increasingly alienated from them during the past few years, as he came to rely more and more exclusively on Cristina for a structure to his personal life, but also because of the ongoing struggles with Beatrice and Marignoli about money. He felt that they were all colluding to get as much as they could out of him, while in his mind his first duty was to ensure the future of his new family. He had always been absent but loving toward his older children; now he withdrew his affection as well.[58] Gioia, the youngest, especially felt this. She was still living with her mother and got to see the drama of her divorced parents from Beatrice's side. Of the three children from his first marriage, she was also the one who had had the least to do with her father when she was growing up. His absence stayed with her forever. Gioia's son, Michael Braga, believes that to the end of her life, Gioia could never have enough of Marconi.[59]

Around the same time that he was trying to help Lisa Sergio relocate to New York, Marconi was also soliciting Sarnoff's help in arranging the future of his son, Giulio. Giulio had joined the international Marconi company on January 1, 1934, after a two-year stint in the Italian navy, from 1931 to 1933. But by the end of 1934 he was back in a rest home.[60] From 1934 to 1937 he did various jobs for the Marconi companies but was unable to apply himself fully to work.[61] On January 12, 1937, Marconi wrote to Sarnoff asking him to consider employing Giulio at NBC. He was able to put a positive spin on his son's professional itinerary: after studying at the Royal Navy Academy, Marconi wrote, Giulio had left the navy in 1933 "as he did not feel suited for a sea-faring life." He then worked for the Marconi companies in London until 1936, when "for political reasons they did not wish to employ Italian nationals in England." He was now working for the Italian Marconi company in Rome. "I know you too might have difficulties over employing a person of foreign nationality in America, but think you could get over them." Sarnoff replied that he would be happy to employ Giulio, and take him under his wing. He told Marconi how Giulio should go about obtaining his visa, what to say to the State Department if necessary, and that he would meet him at the port when he arrived. Marconi asked Sarnoff to help Giulio find appropriate lodgings but not to go too softly with him: "I should like him to feel that he must rely on his own income and to be stimulated by the wish to improve his position through sheer work and steadfastness of purpose."[62] Giulio sailed from Rome on April 7, 1937, and, in New York, was well looked after by Sarnoff. For the first time in his life, he flourished, becoming "production director for sustaining programmes and later assistant to the director of television programmes, then in an experimental stage," as he later put it in his curriculum vitae. One of Marconi's last family letters, heartfelt and full of fatherly advice, dated May 17, 1937 (and written in Italian), was to his son.[63]

Marconi continued to niggle with Degna over financial issues. In one of his last letters to her, on March 11, 1937, he mentioned a recent operation she had had in England. "You ask me to help you with the doctors' and nursing home bills. I am doing this by sending you cheques.... The other smaller ones you can pay out of your allowance which is now not inconsiderable."[64] He was still nickel-and-diming his children. Marconi's financial relations with the family of his first marriage had been bubbling on a high heat for years. On January 31, 1934, they spilled into the political realm when Beatrice wrote to Mussolini's office requesting a meeting to discuss "a delicate question regarding the children of myself and Sen. Guglielmo Marconi." Someone has written on Beatrice's letter, in Mussolini's archive: "delicate topic" ("*argomento delicato*").[65]

There is no trace of a meeting, but Mussolini's files contain a sheet of paper dated April 7, 1934, initialled "GM," with the notation "seen by Il Duce, 29 April 1934." The document precisely lists the details of the payments Marconi had committed to make for the support of his and Beatrice's three children, the British trust he had set up to their benefit at the time of his marriage settlement of 1905, and the divorce settlement.[66] Whether this was furnished by Beatrice, the political police, or another source is not known. As we've seen, on May 15, 1934, Beatrice met with the Vatican secretary of state, Cardinal Pacelli, presumably on the same topic.[67] Around the same time, Marconi's lawyers drafted a brief, a copy of which also found its way into Mussolini's files, arguing that Marconi's finances were in a sorry state, that he had been generous to Beatrice and the children, that they were ungrateful, mean, and threatening toward him, et cetera.[68] (Not noted in any of these documents was that Marconi was negotiating a new agreement with the British company at this time, under which he would receive a one-time payment of £40,000 in lieu of future remuneration.[69])

There is no record of any follow-up to this information by Mussolini's office. But in November 1934, Degna Marconi picked up the flame, writing to Mussolini's office as well, laying out the details of her mother's financial situation and also asking for a meeting. "I do not intend to attack my father, I only wish to seek help for a grave situation," she wrote. Degna's request was shuffled from desk to desk; it is covered with notations like "She insists on speaking with you" and "You see her." On November 13, 1934, Degna was received by Galeazzo Ciano.[70] She also evidently appealed to her father, who wrote to her from his London hospital bed on December 23, 1934: "I am sorry of what you say about your financial difficulties and cannot but regret your refusal of my offer to assist you all, with the expenses of running the house."[71]

Beatrice's problems stemmed from the serious financial difficulties of her husband, Liborio Marignoli. Beatrice had given Marignoli the proceeds of her three million lire divorce settlement in exchange for his villa at Posillipo; now Marignoli's land in the region of Montecorona had been sequestered in a complicated legal action with the state agricultural credit agency, and he was petitioning Mussolini to intervene. (He was political secretary of the local fascist party branch.) In 1931, Marignoli had used his and Beatrice's property as collateral for a bank loan and the debt had grown to sixteen million lire.[72] Their affairs got tangled up in Marconi's efforts to control his financial commitments to his children; on May 10, 1935, Marignoli wrote a note to Mussolini's office, complaining that Marconi had told his youngest daughter, Gioia, that the Marignolis were "flourishing," financially, and seeking to set the record straight, lest this false information prejudice his petition. On September 1, 1935, spunky and showing no lack of her father's chutzpah, Degna pleaded again to Mussolini's chief of staff: please do something. Despite the fact that this type of petitioning was the only way to get things done under the fascist regime, she received a reply that, regretfully, Mussolini's office could not intervene in a judicial matter.[73]

Degna, meanwhile, had her own financial issues with her father. A thick file in the Lincei archive contains thirty-three documents related to Degna's unruly finances in 1934 and 1935: letters about unpaid bills, overdrawn accounts from banks and retailers (for example, Harrods and Selfridges), imploring letters and telegrams from Marconi. On May 28, 1935, the manager of the Westminster Bank in London wrote to inform Marconi that Degna had come to see him, appearing very distressed about her situation; she had already completely overdrawn her next month's allowance. She was facing a summons and to save Marconi's name from coming up in court, the bank had settled the amount, a relatively small sum of £8-6 (a little over $40). Marconi made good the account, thanking the bank for its courtesy, but insisting that he declined to assume any responsibility whatsoever for his daughter's debts. To a lesser extent, the files tell a similar tale about Gioia's and Giulio's spending habits.[74]

Totally exasperated, Marconi wrote a long, pleading letter on June 7, 1935, to the children's uncle, Beatrice's brother Donough O'Brien, who was one of the marriage settlement trustees. "Dear Donough, You are undoubtedly aware of the reckless manner in which my daughters Degna and Gioia have been spending, and I might say squandering money during the last few years, notwithstanding my repeated exhortations and warnings not to exceed the allowance which I have been issuing to them and which is ample considering my means....‡ I cannot possibly allow this to go on any longer,

---

‡ Here he details the amounts, which are exactly as indicated on the list dated April 7, 1934, in Mussolini's files: Degna, allowance £600 per year plus bills paid totalling £728 since March 1, 1934; Giulio, salary £500, allowance £400 per year plus £196 paid since March 1, 1934, mostly for medical bills; Gioia, allowance plus £36 in bills, for total £960 in bills alone, in addition to allowance.

and it is because of this that after consulting my lawyers I am making an appeal to you to help me in the matter." Marconi enclosed a cheque for £600 and asked O'Brien to sit down with his nieces and decide which of their debts were the most urgent to pay.[75] The day after writing this letter, Marconi sent his lawyers the new holograph will he had written in April.

The financial stresses placed a strain on Marconi's already fraught relationship with his children. Degna hardly saw him for two years, then renewed contact when she spent some time in Rome in 1936. She found him a different person, worn down by ill health, indifferent to business battles that would have stimulated him in the past, and pessimistic about the darkening political situation.[76] Degna returned to Rome in the winter of 1937. She was troubled to see how much her father had aged, although she had no idea how ill he was. She was happy that "the old affection had returned to replace our bruised relationship and the kindness I had so desperately missed was mine again. Misunderstandings vanished and by mutual consent we forgot them." As they spoke, "His mind turned again and again to London and, incredulous though I was at first, I began to be sure that he intended to leave Rome and move there, so that he could once more enjoy the old independence to work. His plans, as he unfolded them, were clear and concise and centered on taking a house where Giulio and I were to live with him." But what about his new family? "Only one thing he did not specify—where Cristina and Elettra fit into this pattern."[77]

Marconi was indeed boxed in. On the one hand, he had little left in England besides bittersweet memories and lost hopes; on the other, he was becoming increasingly disillusioned with the way things were going in Italy. His family life was both his only solace and his greatest constraint. He had traded a life of wandering freely for one where his wife rarely left his side. Some old friends looked askance at the new relationship. Marconi's former sister-in-law Lilah, one of the people who knew him best, described running into Guglielmo and Cristina at a British embassy reception in Rome, "she fanning and preening herself and he with a stick, hobbling after her with a seraphic air of satisfaction."[78] At sixty-three, Marconi was living with his in-laws. He liked them, especially his mother-in-law, who was five years younger than him and a lively socialite. But as he confronted the inevitability of giving up his yacht, he had to face a future where his only private personal space would be a tiny study in a cubicle of the apartment. Marconi's grandson, Degna's son Francesco Paresce, has a good word for it: Marconi was "cloistered." It was by choice. "He had fought his whole life so when he could finally just sit back and be adulated, he just did that."[79] And no one adulated him more than Cristina. An exemplar of wifely devotion, she told a Chicago newspaper reporter: "I realize that my husband is an extraordinary man and so I listen quietly and never disturb him with foolish questions."[80]

Marconi was indeed hoping to visit England in the months before he died, as we've seen, but aside from Degna's impression, there is no indication that he was contemplating leaving his young family behind. Rather, his family situation had created a new dilemma. He adored his little daughter, Elettra. Cristina's Roman roots ran deep and she had no reason to feel a need to expatriate. He, on the other hand, could easily give up Rome, which had become more of a burden than a comfort to him. Elettra Marconi says that as the situation in Italy darkened, her father was thinking of taking her and her mother away—not to England, nor to America, but to South Africa, Canada, or Australia.[81]

Degna's view, that Marconi was considering moving back to England, is contradicted by one of the last pieces of business Marconi was working on: a new agreement with the British Marconi company. The Oxford Archives contain a draft document updating Marconi's existing 1934 contract, under which he was engaged as "Director of Research to the Company and when requested to other Companies" for the remainder of his lifetime. According to this document, Marconi was asking to be relieved of his obligations to the company "in order to devote more time to the high offices entrusted to him by the Italian Government and to his scientific researches in the exclusive interest of Italy." The company was prepared to release him from these obligations as of August 1, 1937, and nonetheless to continue paying him £5,000 per annum for life, as per the existing contract. Under the terms of a second, parallel agreement being negotiated at the same time, Marconi would resign from all his positions with the Marconi companies, except for the presidency of its Italian subsidiary, for which he would receive an expense allowance of £800 per annum.[82] A final draft of this latter document, dated July 16, 1937, was sent from London to Solari's home in Rome, but there is no sign that it arrived in time for Marconi to act on it.[83]

❊   ❊   ❊

Marconi now had less compelling personal ties to England as well. On April 25, 1936, Marconi's sixty-second birthday, his brother Alfonso died in London, of heart failure. He was seventy-one and had been living in hotels for the past sixteen years, since Annie Marconi passed away in 1920. In recent years, he and Guglielmo were only Christmas-card close but the connection was strong, and recalled the Marconis' deep family ties. Degna and

Giulio were in London and were called by the manager of the Hotel Splendide to say Alfonso was ill, but by the time they arrived he was dead. By some "alchemy" that Degna could not quite comprehend, their cousin Henry Jameson Davis turned up moments later and took charge with his usual *savoir faire* and "the same great kindness which he had extended when Father and Grandmother first arrived in England."[84] Alfonso was described in an obituary as "a gifted violinist, an artist of no mean ability, and a collector and connoisseur of old stringed instruments and pictures."[85]§ He was buried next to Annie at Highgate Cemetery, in the plot Davis had provided for Annie back in 1920. Marconi was too ill to travel to attend the funeral.

A few months later, on Christmas Day 1936, Henry Jameson Davis, too, passed away, at his home in Surrey at the age of eighty-two. A tribute in the *Miller*, organ of the British milling trade, described him as "the aristocrat of the party, [he] generally assumed an air of superiority and aloofness... [and] was an extremely courteous and friendly competitor.... In the most trying conditions he always acted the gentleman."[86] Not unlike his "affectionate cousin" Guglielmo. Marconi and his oldest collaborator had barely seen each other in the past twenty-five years.

Since the December 1935 incident on the train from Paris, Marconi had been living on borrowed time. His world was closing in on him. His heart attacks became more frequent, and were considered so routine that he even joked about them. When it was necessary to make public excuses for his absence from events, it would be said that he was having digestive problems. In May 1937 he had the most severe attack yet. When Di Marco went to his house the next morning he found him "lying in bed, though quite cheerful, and ready to go through the usual (day's) work." Marconi showed him some blood stains on his pyjama sleeve and said, with a smile: "See what this blood-thirsty doctor of mine has done with me last night. If it goes on like this, one of these days I shall really and truly give up my ghost."[87]

Di Marco portrayed Marconi at this time as an austere figure who stuck to a rigid, sober routine, rising at seven, having an unvarying breakfast at eight of two boiled eggs, a cup of tea, and a couple of slices of bread and butter. After breakfast he read the newspapers and dealt with correspondence; he came to the office around eleven and always left for lunch punctually at one. He ate frugally, almost abstemiously, took a few minutes rest, then worked again until

---

§ A Stradivarius from Alfonso Marconi's collection sold at Sotheby's in London for more than £670,000 in October 1914.

five when he broke briefly for tea and biscuits, then signed the day's correspondence. At eight on the dot he sat down to dinner, as frugal as his midday meal, and was in bed soon after. Marconi didn't like to go out or enjoy walks, to the point where one rarely met him in the streets of Rome.[88] He was basically solitary, melancholy, and dull, a sad shadow of the one-time ladies' man—yet consistent with the depressive described by Beatrice in her memoir.

Marconi was ill with a draining spring flu in June and early July, but he came to the office every day nonetheless.[99] He hoped to travel that summer but his doctors were insisting he stay close to Rome. Pressured by his family, he agreed to holiday at the Hotel Astor in Viareggio and put off the long-planned trip to London once again. Elettra and her grandmother left for the seaside in early July but Marconi and Cristina remained in the city; he was embroiled in yet another financial squabble with the London company and was hoping to settle it before going on vacation.[90] He also had some other important business to attend to. On July 12, 1937, Mussolini received a note from his private secretary, advising that Marconi had requested a meeting, during the following week if possible.[91] The purpose was not specified, but a meeting was set for Monday July 19, 1937, at 6:00 p.m.

Marconi also requested a meeting with the pope and was received at Castel Gandolfo on Saturday, July 17. What was said there has never been revealed. Many pundits (as well as Mussolini police agents) later speculated that Marconi sought counsel as to what to do to avoid having his "death ray" end up in Mussolini's hands, but there is no evidence for this hypothesis. Degna was convinced that Marconi's motive was personal: "He was on the threshold of a new life ... he had decided to live alone and to transfer himself and his work to England, despite his abiding love of his own country. At such a juncture he would have turned to a man of elevated spiritual insight who had repeatedly shown him understanding and friendship."[92] However, there is no evidence that this was the purpose of the meeting either. It was far more likely to do with business or politics, the topics Marconi usually discussed with the pontiff—especially on the eve of an anticipated meeting with Il Duce.

Cristina rarely left Marconi's side in those days, but now, "seeing how well he was looking and trusting that the warm season would make another heart attack less likely," she left for Viareggio on July 19, in order to be there for Elettra's seventh birthday on the twentieth. Marconi was to join them on the twenty-first. He saw Cristina off at the railway station at noon on Monday the nineteenth and then went, as usual, to his office at the Farnesina, where he met briefly with senior officials of the academy. Di Marco said, "I noticed nothing unusual in his appearance nor in his demeanor."[93]

One of the people Marconi met at the Farnesina on July 19 was Arturo Marpicati. Marconi was "pale, tired, and smoking an inoffensive cigarette of dry sage smelling of incense" (a concoction of his doctors, to get him to cut down on tobacco), Marpicati remembered. They discussed the theme of the next Volta conference—Africa—and Marconi said he was looking forward to lunch. Then he said: "Did you know I saw the Pope the day before yesterday at Castel Gandolfo? We had a very cordial conversation. He's in good humour and good shape, better than us. I asked him for a blessing for my little Elettra, it's her birthday tomorrow. . . . I asked him to give me a special blessing too, in case I would soon need it." As though to underline that he was being ironic, he added: "If you're on vacation, promise me you'll come back for my funeral." With that Marconi grabbed his hat and ubiquitous ivory-headed walking stick and he was off.[94]

Marconi left the Farnesina at 1:00 p.m., taking Di Marco with him in his car, as usual. "He was silent throughout the ride," Di Marco said.[95] Di Marco left him at the office of the Italian company, where Marconi wanted to talk to Solari (who also wrote about the meeting in his 1940 biography). Rome was having a heat wave and Marconi looked pale and tired. "You're doing yourself a lot of harm by being in Rome in this heat," Solari told him. "You're right, but I must fix so many things," Marconi replied, looking at Solari sadly, with a melancholy smile. They briefly discussed Marconi's current efforts to pursue his microwave research and then Marconi left for lunch, saying he would see him later.[96]

At five o'clock Di Marco arrived at the house as planned, with papers for Marconi to sign. When he arrived, Marconi's father-in-law was about to send for the doctor. Di Marco was asked to call the Palazzo Venezia and cancel the meeting with Mussolini, which he did (Solari wrote that Di Marco called *him* at 5:00 p.m. to cancel their meeting). About an hour later the doctor, Arnaldo Pozzi, arrived. Di Marco stayed on for a while, "without, I must confess, any special anxiety, used as I was to see those crises happily overcome." Pozzi confirmed Di Marco's optimistic feeling; things were not as bad as they had been in May. Pozzi nonetheless informed his chief, Cesare Frugoni, Marconi's chief physician, who was out of town. Sitting in an adjoining room, listening to Marconi chatting calmly with Pozzi and making pleasantries, Di Marco was reassured and left for his home in the Alban Hills. However, things took a turn for the worse; around 9:00 p.m., Frugoni was picked up at the train station and rushed to via Condotti.[97]

By then he could not do much. "It was very warm in the room where father lay with hardly a breath of air filtering through the open window," wrote Degna. "On the hour, the bronze notes of the bell of the old church of Trinità dei Monti hung in the air, suspended in the heat and silence."[98] Frugoni later described to Degna his last conversation with her father. "Reclining and looking very pale he lifted his forearm and saw that the blood in the artery was not beating any more. Turning to me he said in a low voice, 'How is it, Frugoni, that my heart has stopped beating and I am still alive?' To which I replied: 'Don't ask such questions, it is only a matter of position, because your forearm is raised.' With a little wry smile, he said: 'No my dear doctor, this would be correct for the veins but not for an artery,' ... He then frowned slightly and said, 'But I don't care' and again, while his forehead slowly relaxed, 'I don't care at all.'"[99] At 3:45 a.m., on July 20, 1937, Guglielmo Marconi died.

35

The Heritage

Mussolini arrived at Via Condotti at 7:00 a.m., accompanied by his undersecretary of state, Giacomo Medici del Vascello, and academy chancellor Arturo Marpicati. The academy took over the organization of the funeral arrangements, at the expense of the state.[1] Marconi's body lay exposed in the Galatea room of the Villa Farnesina, beneath the frescoes of Raphael, for the next twenty-four hours, while more than fifty thousand Romans filed by solemnly. One of them, there incognito, was Beatrice, who reported to their daughter Degna: "The ambiance was simple and austere...no flowers, no candles, just one continuous flow of people of all classes, all nationalities.... Those who saw the procession say they have never in any occasion in Rome seen anything to equal it."[2]

The state funeral was set for 6:00 p.m., Wednesday, July 21, 1937, at the Church of Santa Maria degli Angeli in Piazza Esedra (now Piazza delle Repubblica), near the Termini train station, where Marconi and Cristina were married in 1927. The basilica is one of the most majestic in Rome. Built in the sixteenth century within the ancient complex of the Baths of Diocletian (in other words, initially a pagan site), according to plans designed by Michelangelo, it was often used for state funerals, going back to the Medici Pope Pius IV in 1565. There was also a scientific connection Marconi must have appreciated. A 150-foot sundial on the floor, inaugurated by Pope Clementine XI in 1702, indicates the spring equinox and hence helps determine the exact date of Easter Sunday. An old marble inlay with the signs of the zodiac runs alongside the sundial; every day, at noon, sunlight entering through the roof touches a different point of the line. The Church is said to contain the only meridian in the world for tracking the polar star.[3]

Around 4:30 p.m., Marconi's body was enclosed in a rich oak casket, finely carved with laurel wreaths and lions' heads, and decorated with brass.

The casket was covered with gold-fringed yellow draperies and placed on a horse-drawn hearse along with four wreaths: one each from the emperor, Il Duce, the Fascist government, and the largest, adorned with a red ribbon and Nazi swastika, sent by Adolf Hitler.* A few minutes before six, under a blue sky and in bright sunshine, the hearse left the Farnesina drawn by six horses, attended by thirty-six footmen, and followed by twenty-five carriages and nine army trucks filled with more wreaths. The procession, described by the *New York Times* as "one of the most imposing ever seen in Italy," was nearly a mile long. Over half a million people lined the streets.[4]

The *New York Times* correspondent described the scene: "The funeral procession was led by a detachment of mounted carabineers, followed by a battalion representing all branches of the Italian fighting service. Then came two long rows of monks carrying lighted candles, and the hearse, flanked by thirty-six employees of the Rome municipality in magnificent medieval costumes."[5] Mussolini joined the procession on the Via Nazionale, for the last stretch to the Piazza Esedra, where an enormous crowd was gathered. When the procession arrived in the piazza, in front of the church, "the funeral car was placed as if in the centre of a Black Shirt parade, flanked with a Fascist guard of honour… every arm within radius of vision was raised in the Roman salute.… The Fascist flag and the flag of his native Bologna were carried at the head of the procession."[6] The Fascist funeral rite was then performed. Party secretary-general Achille Starace shouted "Comrade Guglielmo Marconi!" and the assembled *fascisti*, led by Mussolini, replied "Present!"[7]

At precisely 6:00 p.m., Rome time, telegraph and radio stations in Italy, Britain, the United States, and Canada went silent, along with the thirty-one beam and wireless stations of the Cable and Wireless network around the world, as well as others in China, Japan, the Near East, and Europe.[8] By order of governments on all sides of the brewing international conflict, for a brief moment the world was as it had been forty-two years earlier, before Guglielmo Marconi burst into view and changed it, irreversibly, forever.

The British and American press reported Marconi's death in a whirl of hyperbole. The *Manchester Guardian* said Marconi had been inspired by a vision of a world without barriers, a federation of mankind; to the *Yorkshire*

---

* Hitler also sent a telegram of condolences to Mussolini, from Berchtesgaden on July 20, 1937 (Hitler to BM, telegram, July 20, 1937, ACS, SPD CO GMF, 197.598/1); "Condolence from Hitler. Message to Mussolini and Widow Sent by Chancellor," *NYT*, July 21, 1937.

Post he was "a citizen as cosmopolitan as Edison and Einstein. The *Observer* wrote, "No man has made a more visible difference to his age than Marconi." The *New York Times* called him "a citizen of the world," a "master magician," and "ruler of the electric waves . . . a great gentleman as well as inventor." In Ireland, he was hailed as "a modern magician" and "the great Marconi . . . one of the greatest men of all times." To the *Irish Times* his name was "a household word."[9]

As had often been the case during Marconi's lifetime, the standard was set by the *Times* of London. A lead editorial, titled simply "Marconi," said he could be regarded as "the supremely significant character of our epoch, the name by which the age is called." No one had transformed the world as he had; combining "diligent study with an intuitive impulse," he had accomplished what others were content to deem impossible. The *Times* also highlighted Marconi's "friendship" with Mussolini, in a separate section of its long obituary. "The Duce loses in Marconi a devoted admirer, and a trusted friend, and the Fascist Party one of its staunchest supporters." Marconi had joined the Party "at a time when other prominent Italians were either opposed or lukewarm towards the government of Mussolini. The Duce greatly appreciated Marconi's gesture, and besides bestowing numerous honours and distinctions upon him, treasured a friendship which grew deeper and deeper with their increased contacts."[10]

A regional British newspaper, the *Eastern Daily Press* (Norwich), offered some rare reflection, suggesting that Marconi's "very genuine regard" for Mussolini provided a window into the differences between Italian Fascism and German National Socialism. Where Hitler had alienated important parts of German life through a policy of repression, it said, Mussolini had rallied the cultural life of his country. As evidence of this claim, the paper contrasted Marconi's attitude to that of Einstein. Unlike Einstein, Marconi "was content to be a proud and patriotic son of his own fatherland, with no desire to be an internationalist."[11] The article elided the obvious fact that Einstein, as a Jew, was inarguably a member of the main group being "alienated" by Hitler, while Mussolini was able to use Marconi as a shield for much of what he was doing. Marconi was both a cosmopolitan internationalist and a patriotic Italian nationalist, and this was part of the tragedy that caused him so much turmoil at the end of his life.

Mussolini was intensely interested in the international press coverage of Marconi's death. The archives of the propaganda office of the Ministry of Foreign Affairs in Rome contain a file, literally bursting with dozens of cuttings, sent by Italian legations around the world. A special red folder, labelled "*Stampa tedesca*," contains summaries in Italian of the German press reports, prepared for Mussolini personally. One of the first German papers to report the news, the *Berliner Börsen-Zeitung*, noted in a brief, sober report that Marconi was "one of the first followers of Fascism and a loyal soldier of Mussolini." His death was a loss not only for his country but for the entire world and for all of science. With Marconi's passing one of the most "representative personalities of our era" disappeared. Another, the *Berliner Tageblatt*, recalled that Marconi never claimed to be a stateless "citizen of the world," but always remained an Italian patriot.[12]

Marconi's death also launched, or relaunched, some persistent urban legends about his life. The *New York Times* embellished an Associated Press obituary by adding a single line to the end regarding Marconi's second marriage: "At this ceremony his best man was Mussolini." It was absolutely false, but this factoid has been reproduced dozens of times down through the years. The AP also circulated a fantastic story that Marconi had a "premonition" of his death, based on the fact that he asked to meet the pope before July 19 "because on the 20th I am going away."[13] The innocent truth was that Marconi was planning to leave Rome to meet his family at the Italian seaside on July 20.[14]

The Vatican, too, was keenly interested in the event and, indeed, the most fabulous story of all appeared in the official Vatican daily *L'Osservatore Romano*, reporting that Marconi had died "a Christian death." Around midnight, it said, Father Paolino Rapa, a local parish priest, was called to Via Condotti. Observing that Marconi's condition was not grave, he left. A few hours later, the report continued, the religious nurse attending Marconi asked if he wanted to receive last rites and he said yes. Rapa was called to return and administered the last rites. Around 3:45 a.m., the paper reported, while Rapa recited the Lord's Prayer, Marconi expired serenely.[15] This was totally false.

Thanks to the church's obsession with having its own secret record of the true story, we know what really happened. A few weeks after Marconi's death, Rapa and the nurse who attended Marconi in his final hours, Sister Agnese Bengoa, were called in to the Vicariate of Rome and asked to confirm what they had witnessed. Their accounts differed from the newspaper report, and concurred with each other in the one crucial detail: Marconi was already dead when Rapa arrived at Via Condotti for the second time at 3:45 a.m. and had died without receiving the last rites.[16]

Bengoa attested that she arrived at the house around 8:00 p.m. and found Marconi in good shape. Around midnight, however, she noticed that he had fluid in his lungs and was having difficulty breathing. "I thought it my duty to inform his father-in-law so that he could call a Priest," Bengoa said. "While the Priest was called, I tried to prepare Senator Marconi to receive the Sacraments. The Senator replied, smiling, 'I don't think it's necessary.' Upon my insistence, he replied energetically: 'Sister, no one is entering here without my invitation.'" Knowing Marconi's character, she said, "I tried to influence him indirectly." Soon afterwards, she said, Marconi asked her to recite a prayer and kissed the crucifix with devotion. Meanwhile, the priest arrived, and having heard that Marconi was not ready for him, returned to his parish, with the assurance that he would come again immediately as soon as he was called.

Bengoa remained alone with Marconi while the doctor, Arnaldo Pozzi, stood by in a neighbouring room. Marconi was lucid and talkative, repeating his prayers every so often, until around 3:00 a.m., when he was given an injection of morphine and began talking to himself. "I took the opportunity to reiterate my insistence and he replied that he still wanted to wait," the nurse testified. Soon Marconi had a coughing fit that provoked vomiting and the doctors intervened. Around 3:30 a.m. his condition seemed to improve a bit, and the doctors returned to their room to rest. "Seeing that I was still standing, the Senator invited me to sit. Suddenly he said: 'I feel sick.' Dr Pozzi heard him and entered the room at once. Marconi repeated to Dr Pozzi that he felt ill, and while still talking, bent his head in pain and lost consciousness." Bengoa immediately told an attendant to call Rapa and also sent Marconi's valet to fetch him. "The pastor arrived ten minutes later.... Unfortunately the Senator had already expired," she said. Rapa testified that he had rushed to the house as soon as he was called but that it was too late. Marconi had been in no hurry to receive the last rites of the Catholic Church. Or at least he was not worried about dying without them. He prayed, in his own way, but he didn't care to see a priest. In his dying moments, Marconi kept his own counsel, as he always had.

Certified copies of the two affidavits are in Secretary of State Pacelli's files in the Vatican Secret Archives, buried amidst a handful of condolences and a cutting of the bogus *L'Osservatore Romano* article. There is no record of church reaction to this sign of Marconi's peculiar approach to the prescribed practices of his faith but in the secular world another report was filed for Marconi's police dossier by Virginio Troiani or one of his sub-agents on

September 7, 1937, recording that the Vatican was abuzz about "the published lie around the death of Senator Marconi, according to which he received religious consolation. It is positively to the contrary that this news was false, since Senator Marconi died without the Sacraments." The police agent allowed himself to editorialize in conclusion: "And that's the truth that was published in the newspapers!"[17]

❋   ❋   ❋

Immediately upon Marconi's death all the interests invested in him in one way or another staked a claim to his legacy. Luigi Solari's role in the days following Marconi's death was pivotal; as the crucial link between Marconi's Italian affairs and the British company—which still had a large stake in the Marconi brand—he became a linchpin for both the company and the fascist regime in establishing control over Marconi's legacy. On July 22 he wired London, in code: "Imposing funeral government account took place last night with intervention all Italian government diplomatic body senate parliament all scientific associations and all roman people. Tomorrow body will be carried Bologna where I shall go also as your representative. Upon my return Rome shall ask audience from the Head of Government to express personally your company condolences." He proposed that the company donate a year of Marconi's salary to a Rome charity (a far cry from anything Marconi himself had contemplated, as we shall see) "as homage to (the) great departed."[18]

On July 23, Solari read a "posthumous message from Guglielmo Marconi" on Bologna radio, that Marconi had purportedly been planning to deliver to mark the launch of the station. The message evoked the emotional power of radio, the impact for Italians, especially in the diaspora, of hearing the voice of Il Duce assert "our country's firm will to progress continually through work, through peace, and, if necessary, through force." The dedication hoped the station would "spread throughout the world news of fresh victories and of spiritual and material conquests made by Fascist Italy." The underlying bellicose tone of the message belied everything Marconi said *in private* during the last months of his life, where he rather deplored the talk of force, especially in any connection to radio.[19]

The political police were unusually busy in the days following Marconi's death, and the reports began coming in right after the funeral. Old themes like the mysterious nature of Marconi's recent research came up, sparking rumours that his death was not the result of a heart attack at all, but somehow related to the famous "death ray."[20] There were nearly thirty reports in all, the most interesting, more critical ones coming from agents working inside the Vatican or close to the inner sanctum of the Fascist Party. "Some considerations are being raised about the honours that have been bestowed" on the deceased, and there is reason for disapproval and criticism by the fascist regime, Troiani's sub-agent Umberto Purinan (aka "Puberto") reported on July 22.[21]

When Marconi's will was read on July 23, 1937, it was noted with some surprise that he had provided nothing at all outside his family, and, especially, nothing for either the church or the party.[22] The two-page holograph will was made in 1935 (an accurate typewritten copy is in the Polpol file). The fact that Marconi left not a *centesimo* to either religious or political organizations, from an estate rumoured to be as great as 500 million lire (£5 million or $25 million at the time—$400 million today), was the object of ongoing amazement, "Puberto" stated in a long report filed on July 30. Observers were astonished that after showing so much attachment to the fascist regime and the Vatican in his lifetime, Marconi left them nothing. The press was perplexed as to how to interpret this and was wondering whether Marconi hadn't left some kind of "moral testament" somewhere.[23] It was stupefying that Marconi had not even left anything for a scientific foundation in memory of his name or to support his favourite research.[24] Inside the Vatican it was said that if Marconi had been as fervent a fascist as he wished to appear, he would have left something to the party or to charity.[25]

The will, dated April 27, 1935, named Marconi's youngest daughter Elettra his "universal heir," leaving the three older children from his first marriage only what they were entitled to under Italian law. Degna, Giulio, and Gioia would each get one-eighth of their father's estate and Elettra would get the rest, although her mother would enjoy the usufruct during her lifetime. Marconi justified this allocation with the remark that the three older children had already received a total of over one million Italian lire from him during his lifetime, in addition to the fruits of the marriage settlement he made when he married Beatrice in 1905.

The size of the estate was the object of wild speculation. It included Villa Griffone, the yacht and whatever monies, stocks, and other considerations Marconi had accumulated. The figure 500 million lire continued to be repeated in press reports, although that was a stupendous exaggeration. In September 1937, two months after Marconi's death, David Sarnoff visited Cristina in Rome and issued a press release on her behalf stating that

Marconi's estate was worth only $150,000 ($2.4 million in today's money), leading him to quip that an inventor was someone who made other people rich.[26] The true figure lay somewhere in between—Marconi's declared estate in England alone was a little over £48,000, or £36,000 ($180,000) after taxes.[27] The fact was that Marconi left relatively few liquid assets. According to a balance sheet prepared at the time, now in the possession of Francesco Paresce, the assets of Marconi's estate in Italy—excluding Villa Griffone—totalled just over 1.2 million lire ($60,000), of which more than half (665,000 lire, or $33,000) was represented by the yacht and less than 200,000 lire ($10,000) was cash. The value of Villa Griffone and its surrounding property was estimated at around one million lire ($50,000).[28]

One reason for the speculation was the widely held assumption that Marconi owned a large part of his company and was entitled to receive royalties on his patents. In fact, he owned a relatively modest number of shares, and whatever income he received as president and technical advisor ceased with his death.[29] Furthermore, he received no royalties. Marconi was cash poor, using his income for current expenses and, as we saw, even facing the sale of his beloved yacht in the last weeks of his life.[30] There was, however, a safe containing some of Marconi's papers at company headquarters in London. Cristina had the only key but Degna and her siblings insisted that they or their representative be present when it was opened. In addition, the company secretary pointed out to the estate's lawyers, Marconi had deposited in a basement store room "various travelling trunks, cases, and filing cabinets" containing both company and personal papers. The company official estimated that it would take at least a month to do a careful investigation of the contents.[31]

A preliminary examination of the effects held by the company in London was completed in November 1938. The contents included about twenty packing cases "filled with nothing but old periodicals, of no possible value whatever." These were destroyed. Another twenty trunks contained "a confusion of papers" including cheque book stubs, copies of typewritten letters, and so on. A number of small boxes contained various personal documents including letters from Marconi to Beatrice and papers relating to their divorce. There were also three bibles (one belonging to Annie Marconi, one that Annie had presented to Guglielmo, and one that had been Alfonso's), and, finally, "a Court dress, hat and sword," that Marconi had apparently worn to the coronation of Edward VII in 1902.[32] Aside from the bibles and court outfit, this is a rough inventory of the contents of the Marconi archive that eventually found its way to the Bodleian Library in Oxford.†There was more correspondence with the two branches of Marconi's beneficiaries, but before anything could be decided about the disposal of the estate, Italy and Britain were at war and Marconi's assets in England were frozen as enemy property.

❋   ❋   ❋

The rest of the world had other things to think about, but in Italy, Marconi's presence continued to serve the fascist regime while casting a long shadow over public life and the lives of those closest to him. On December 9, 1937, four days before Italy withdrew from the League of Nations, Mussolini memorialized Marconi in a speech in the Senate. He recalled that just before his death, Marconi had been doing experiments of "a military character," and that these had taken Italy "to the beginning of a road which will be followed," Mussolini said cryptically. Marconi's unfinished work would be taken up again and his memory honoured in this "most typical fascist way."[33] A few days later, on December 18, Marconi's birthday was declared a national holiday. A few weeks after that, in January 1938, Mussolini received Marconi's widow, who presented him with her husband's Fascist uniform, which Il Duce announced the state would display.[34]

Beneath the surface, speculation over Marconi's true colours continued to bubble. The gap between the public adulation and the private skepticism surrounding Marconi's loyalty mirrored the ambivalence that he had fostered in his own relations in the years before he died. The issue remained a concern of the political police and wild rumours continued to fill the Polpol file on Marconi in 1938 and 1939. From St. Moritz, Switzerland, an unidentified agent reported that, shortly before his death, Marconi had left the pope a coffer containing his "death ray" invention, to avoid its falling into Mussolini's hands. The Vatican was maintaining total silence on the matter. Another report alleged that Marconi had left a secret package with his daughter to be delivered to Mussolini only in case of war, or ten years after his death.[35] One of the most sensational reports, from an agent in Bologna, questioned the official cause of Marconi's death. "Everyone thinks Marconi died of a heart attack," the agent wrote, "but that's not the truth; he died victim of his knowledge! He knew things that he took to the grave." And on and on.[36]

❋   ❋   ❋

---

† The cocked hat and sword are now held by the Museum of the History of Science at Oxford, where they were displayed in 2011 as part of an exhibition called "Eccentricity: Unexpected Objects and Irregular Behaviour." The bibles, presumably, were returned to the family.

Marconi's survivors continued to solicit Mussolini's support, and he continued to take an interest in their affairs. Barely two weeks after Marconi
died, Cristina requested a meeting with Mussolini.[37] Giulio Marconi wrote
a few days later, thanking Il Duce for his kindness on the death of his father.
In September, Beatrice wrote asking if the state would purchase her villa at
Posillipo. Her husband Liborio Marignoli made a similar request, asking
that the state acquire a historic property he owned, the Hermitage of Monte
Corona.[38] The financial affairs of Marconi's heirs continued to occupy the
agenda of Il Duce and his senior ministers, despite their myriad concerns.
On December 20, 1937, Galeazzo Ciano told his diary: "Marconi's daughter
[Degna] came to ask me for help for herself and her brother. The hostility
of the stepmother has driven them to poverty. Without money and without
circumstance. He may have been a great genius, but I only knew him when
he was already very much over the hill."[39]‡

The financial situation of Marconi's first family kept getting worse. In
April 1939, Degna wrote to Mussolini, again exposing the sad state of relations between the children and their stepmother, and how they had been
progressively estranged from their father after his second marriage and now
found themselves in a painful situation. Nonetheless, wrote Degna, she and
her siblings wished to gift Villa Griffone to the state if Cristina would agree.
Degna sent a draft of her letter to Mussolini's daughter Edda, who passed it
up the chain until it eventually reached Mussolini, who passed it back down
to Luigi Federzoni, president of the new Fondazione Guglielmo Marconi
(created in 1938), as well as head of the Royal Academy and president of
the Senate.[40] On July 21, 1939, Cristina wrote to Mussolini saying she was
pleased with the idea. The acquisition of the villa was reported in the press
the same day.[41]§

Gioia Marconi's financial problems also crossed Mussolini's desk. In
December 1939, Ciano informed Mussolini that Gioia, now twenty-three, was
employed by the Ministry of Popular Culture; she was a conscientious worker

‡ Ciano's lack of respect for Marconi had already been noted in the July 30, 1937, Polpol report by
agent Puberto cited above: Ciano apparently went to the beach on the afternoon of Marconi's
funeral, and arrived at the church dressed in a military uniform whereas everyone, including his
father-in-law, Il Duce, was in formal dress. "It is a detail that has been widely noticed and that
has not failed to be underscored, as we point out, by unfavorable comments" (Agent 40
["Puberto"], report, July 30, 1937, ACS, PP, b.60/A).
§ Villa Griffone remains the one constant connecting all the Marconis to this day. After its gifting
to the Fondazione, the surrounding land remained the property of Marconi's estate and is today
rented to neighbouring farmers by Elettra Marconi.

and able to earn her living. However, she had accumulated around 50,000 lire of debts that she could not clear. Ciano believed there was a case for helping her. Mussolini paid Gioia's debts.[42] Not only that, he decided to set up official state pensions for Cristina and Marconi's three adult children. Perhaps he tired of being solicited every time they needed something, but that still didn't end the solicitation. On June 25, 1941, Degna asked for, and was granted, a meeting with Mussolini, after which she wrote thanking him for his "generous decision" in her favour. On October 25, 1941, Gioia wrote, addressing Mussolini as "you whom I consider like a father," pleading for a larger stipend (she had been awarded an annual allowance of 36,000 lire but felt she should not get less than Degna, who was now married and to receive 60,000). Mussolini also met with Cristina to inform her of her pension.[43] Despite the pressures of war, he had both time and money for Marconi's survivors.

A life annuity for Marconi's three adult children was approved by the Italian Senate and announced in November 1941. The amounts were not mentioned but there is a receipt in Mussolini's archive for 300,000 lire to be paid to Maria Cristina Marconi, from Il Duce's discretionary administrative funds.[44] The pensions immediately attracted the attention of Polpol, and an agent in Virginio Troiani's team commented: "The award of a state pension to the millionaire children of the late Marconi has provoked sharp surprise in intellectual circles." The news was the object of chatter inside the Vatican, still smarting from Marconi's testamentary snub.[45]

Mussolini also took care to look after Marconi's remains. In 1939, soon after Villa Griffone was gifted to the Italian State, a plan for a mausoleum on its grounds was announced. Construction started in April 1940 and the mausoleum was completed in July 1941. It was designed by Marcello Piacentini, an official architect of the regime.[46] Piacentini, whose style was known as "simplified neoclassicism," was one of Rome's most distinctive fascist architects; he designed much of the administrative EUR district, the University of Rome La Sapienza, the broad Via della Conciliazione connecting central Rome to Vatican City, as well as many works in Italy's colony in Libya. On October 6, 1941, in the presence of Mussolini as well as a bevy of uniformed Fascist and visiting Nazi officials, Marconi's remains were interred in the mausoleum.[47] Astonishing as it may seem, this monument of faded (not to say decadent) fascist grandeur still stands untouched today, proudly and untroubled, in the Pontecchio countryside.

The destiny of the *Elettra* is almost an allegory for Italy's wartime and postwar trajectory. After a period of confusion, where it was first reported that the yacht was being sold, and then that it would not be sold but gifted to the Italian government and exhibited at the New York World's Fair of 1939, Mussolini finally bought it for the Italian state, after refusing an American offer of $1 million for it. After the collapse of the fascist regime in 1943, the yacht was commissioned by the Germans and retooled as a warship. Marconi's scientific apparatus was put in storage in Trieste. In 1944, the yacht was sunk in shallow water by an Allied torpedo during a naval action off the Dalmatian coast. Immediately after the war, the Italian government applied to have Marconi's research equipment returned to the CNR; it was back in their hands by the end of 1946. The allied authorities agreed with amazing alacrity and detailed attention to what must have seemed a relatively trivial issue at the time. But the ship itself got caught up in the postwar negotiations over the status of Trieste; it ended up in the hands of Yugoslavia and was only raised from the seabed and returned to Italy in 1961. It then lay in the Bay of Muggia, near Trieste, until 1977, when it was broken up and its sections dispersed among various institutions. The keel, for example, now resides on the grounds of the Fondazione Marconi at Villa Griffone. In June 2015, an Italian amateur radio club organized an event called "*Elettra* back on air!" and broadcast to some American counterparts in New Jersey from the yacht's relic at Pontecchio.[48] The efforts to keep the *Elettra's* memory afloat seem indefatigable.

On July 20, 1937, the day of Marconi's death, Mussolini was quoted in the Italian-American newspaper *Il Progresso* as stating that Jews were no different from any other Italian citizens. The statement was widely circulated in the US press and read out at synagogues and Jewish-American meetings.[49] As late as February 1938 there was still no *official* anti-Jewish policy in Italy, and Mussolini declared that "the Fascist government had no intention of ever taking political, economic or moral action against the Jews."[50]

On July 14, 1938, nearly a year to the day of Marconi's death, the Manifesto of the Racist Scientists (*Manifesto degli scienziati razzisti*) was published in *Il Giornale d'Italia*, stating that "It is time that the Italians proclaim themselves frankly racist." The manifesto was signed by some of Italy's most prestigious scientists. Later that month, an article appeared in the journal *La Vita Italiana* claiming that there were "too many Jews" on the staff of the CNR, which Marconi had still presided at the time of his death; a flurry of notes between the council and Il Duce's office provided a precise count and arguments for dismissing or retaining those who were deemed "replaceable" or "irreplaceable."[51] On October 6 and 7, 1938, Mussolini's program for racial discrimination was adopted by the Grand Council of Fascism. Marconi would have been forced to take a stand. (Only one member of the council, Cesare Maria DeVecchi, was absent for the vote; three members, Italo Balbo, Emilio De Bono, and Luigi Federzoni, opposed the measures.[52]) On November 17, 1938 (a week after the German Kristallnacht), a royal decree transformed the main elements of Italy's fascist program on race into law. Specific acts of legislation then followed.[53] A handful of life senators, holdovers (like Marconi) from the liberal period "who had nothing to fear from opposing the regime," refused to acquiesce.[54] Marconi was joined at the hip to Mussolini and his friends were not among the liberals. But he wouldn't have been happy with how things were going. The racial question was the tip of a very big iceberg and at the end of his life, Marconi may have felt like he was on the *Titanic* after all.

It is pointless to speculate on what he would have done, but the pressure on him would have been tremendous, the consequences enormous. In this respect, Marconi died a convenient death. In the immediate aftermath, Marconi's memory held an important symbolic place in fascist and even Nazi mythology. The photograph of Hitler's wreath next to Mussolini's in the Fascist archives attests to that.[55] When Hitler visited Italy in May 1938, his train passed through Bologna on its way to Rome; here, at Marconi's birthplace, "an allegorical wireless station formed by 12 aerials had been erected near the platform," the London *Times* reported. "The entrance of the Führer into the station was greeted by the whistling of 200 engines."[56]

As it turned out, Marconi's ambivalence regarding the world situation toward the end of his life was the starting point for his posthumous rehabilitation. By the 1960s, Marconi's reputation was being sanitized in Italy, occasionally by former associates like Adelmo Landini, who had worked on the *Elettra* in the 1930s. Landini claimed that in October 1935, when Mussolini was thinking of blockading the Suez Canal against the British fleet, Marconi told him that war with Britain would be folly and that he, Marconi, would speak against it openly in the Senate. Mussolini, after a moment's hesitation, said, "Certainly you know England better than me. I accept your advice, I will not attack first." Had Marconi been alive in 1940, things would have surely gone differently, Landini wrote in 1963: "Who better than he could have enlightened and saved Italy from entering the war? He enjoyed enormous prestige with the King as well as Mussolini.... Plus he had a rare gift, demonstrated many times in the course of his life: the gift of intuition and clairvoyance." Twenty-five years after Marconi's death, his reputation was feeding off the self-interest of former associates like Landini.[57]

Today, in Italy, Marconi's rehabilitation is well on its way to completion. In 2009, Pier Ugo Calzolari, president of the Italian national committee for the Marconi Nobel Prize Centenary Celebrations, was able to state that Marconi is now remembered "under a better light than that which had darkened his illustrious career after his death."[58] Yet regimes like Mussolini's fascism are held in place and kept alive by well-meaning charismatic believers like Marconi, whose enthusiasm inspires hope and confidence in masses of ordinary followers.

❁   ❁   ❁

Outside Italy, Marconi's thorny relation to fascism started to be taboo even before his death. It was not mentioned at all in Jacot and Collier's 1935 biography, and Dunlap skirted the issue, stating that Marconi was a member of the "Italian Fascisti," but not the Party (which was, of course, incorrect): "He never belonged to any political party, but when he saw Fascism saving Italy he said: 'I am a Fascista by conviction. Fascism is a regime of strength necessary for the salvation of Italy.'"[59] That's all Dunlap said about it in his book, published in April 1937, a few months before Marconi's death. In his US presidential radio broadcast on the fall of Rome in June 1944, Franklin Delano Roosevelt remembered the four "great sons of the Italian people—Galileo and Marconi, Michelangelo and Dante."[60] All the others were from the seventeenth century or earlier, and there was no hint that Marconi had been an influential supporter of the fascist regime whose fall was being celebrated. A few months later, a reporter in the *Stars and Stripes* wrote: "Marconi died two years before World War II began, and it is of little consequence to discuss his connections with Fascism."[61]

The British Marconi company made a particularly heroic effort to keep the darker side of Marconi's last years buried when questions surfaced in the 1960s. In 1961, a writer by the name of Leslie Reade contacted the company to ask about Marconi's relations with Italy's Fascist Party.[62] Reade had been commissioned by Faber and Faber to write a life of Marconi aimed at an adolescent audience, and he was interested in this part of the story. The company asked three old hands who had known Marconi when he lived in Italy to comment on the question.[63] A spokesman wrote back to Reade on March 21, 1961: "I ought to say how very much disturbed we are here over your preoccupation with Marconi's dealings with the Italian Fascist Party."

The three men the company consulted were all convinced "that Marconi was no more than a 'Fascist de convenance.'"[64]

MWTC secretary L.J. King confirmed the obvious fact that Marconi was a party member, but couldn't state that he was an "ardent supporter"; his impression was that "Italians living in Italy who had a public standing were virtually obliged to become members of the Party in order to retain their social standing and public recognition. My personal opinion is that Marconi's interest in the Fascist Party never went beyond that, although I am aware that he allowed his name to be used for some of the important activities of the Party."[65] Company engineer E.A. Payne, who spent part of the 1920s seconded as personal assistant to Marconi on the *Elettra*, wrote a more elaborately apologetic note. Marconi "showed no interest in politics, in the same way that he never attended church or showed the slightest interest in religion.... The fact that Marconi wore the Fascist badge in his buttonhole and made broadcasts over the Italian network on politics, I ascribed to the exigencies of the day as he always seemed to me to be utterly bored with the whole business."[66] Marconi's former assistant Gerald Isted, who had been with him in Italy until 1935, met with Reade and understood that the author was going down a dangerous path. If Reade went ahead, Isted wrote to the company's publicity manager, it would do untold harm to the name of Marconi. "The main idea he expressed to me, which I must confess gave me quite a shock, was that not only was Marconi a Fascist but he was an obnoxious kind of Fascist. Of course he was a Fascist, but I am quite sure that, like 80% of other eminent men in Italy at that time, he gave only lip service to the movement because it was politically expedient to do so." Isted also recalled "many occasions on which he privately ridiculed the Fascist 'popinjays.'"[67]

The company officials were learning about a side of their founder that they had been eager not to know. Reade wrote on March 23, 1961, enclosing an extract from the *Times* obituary (cited above) reporting that Mussolini had lost a devoted admirer and the party one of its staunchest supporters. Significantly, he said, "I could not find one word of dissent or contradiction in the form of Letters to the Editor or otherwise from Marconi's intimates about the passage quoted. I might also say that I have met at least one person who knew Marconi, who, in response to a straight, and not a leading, question ('Do you know if Marconi had any political opinions?'), answered instantly, 'Yes, he was a Fascist.'"[68] The company concluded that the "evidence" was overpowering. After all, it had been in the *Times*. One had to assume that if there had been anything to present in Marconi's defence against "these very firm statements," it would have been done at the time. "Faced with this, and with Leslie Reade's other 'evidence for the prosecution,' I think that if I were on the jury and completely unbiased in the matter, I might tend to lean towards his side of the case," a company public relations official wrote in an internal memo.[69]

Reade produced a balanced but clearly positioned account of Marconi's relationship to Mussolini and fascism, giving voice to those who argued that Marconi was at most a "fascist of convenience." It was the first published acknowledgement of this role to appear in English in the twenty-five years since Marconi's death. "Altogether, the most charitable explanation, and still perhaps the most likely, of Marconi's long association with the unsavoury movement of Fascism is not that now suggested by his old colleagues, that he was actuated solely by his own personal materialistic interests, but that he thought, however mistakenly, that he was acting for the good of his country. In the light of the heroism of many poor and humble opponents of Fascism, it is an unsatisfying excuse, but it is better than none. Some of Mussolini's strongest supporters, especially in countries outside Italy, had even less."[70]

The company historians and public relations people had to be wearing pretty strong blinders to be so oblivious to their founder's last years. However, the idea that Marconi's connection to fascism was merely casual if not non-existent became an integral part of company lore. In W.J. Baker's semi-official 1970 *History of the Marconi Company*, a generally reliable and very good source on company matters, Baker wrote rather disingenuously: "There is no evidence that he made the slightest contribution to Fascism."[71] Yet, Baker was employed by the company as "Technical Editor (Research)" and had access to all its files in order to write his book. Marconi's close associate H.M. Dowsett, whose unpublished company history formed the framework for Baker's book, was more nuanced: "It is fair to say that his mind...was not furnished for mature political judgments."[72]

36

He Only Cared About Wireless...

On September 14, 2012, I was working in the library at the Lincei in Rome, beneath the frescoes by Raphael, when I was approached by a man who reminded me of the actor Geoffrey Rush. "Excuse me, where are you from? I understand that you are working on Marconi. Me too, can we speak for a few minutes...." Riccardo Chiaberge, a retired journalist with *Corriere della sera*, was a friend and colleague of the late Giancarlo Masini, author of the last Italian biography of Marconi, which appeared in 1976. Chiaberge told me he was writing a book on the life of Marconi: "I have not decided what it will be, something between fact and fiction," he said.

Chiaberge's book appeared in November 2013. It is a lively read, and exposed many aspects of Marconi's life that had never before been revealed. The book generated a flurry of media interest in Italy; as the first work in Italian to candidly acknowledge Marconi's relationship to fascism, it rekindled a cathartic discussion about Marconi's legacy.[1] When Chiaberge and I met again in Rome in May 2014, he asked me a pointed question: "Do you think Marconi was Italian or British?" I answered unequivocally: "Italian." He shook his head and said, "I think he was British." Marconi's otherness remains his defining feature. No matter what he did, he remained an outlier.

As I worked on this book, I felt as though I had become obsessed with this connection between Marconi's sense of otherness and his attraction to power and, ultimately, a particular form of power—fascism. This is what made Marconi interesting (to me), eclipsing the accomplishments that make him an important historical figure. But Marconi's accomplishments continue to stand as a benchmark for innovation in science and technology. Is that beyond, aside from, or integrally connected to his political and existential stance, I continue to wonder.

Marconi is important because he was the first person to imagine a practical application for the wireless spectrum, and to develop it successfully into a *global* communication system—in both terms of the word; that is, worldwide and all-encompassing. He was able to do this because of a combination of factors—most important, timing and opportunity—but the single-mindedness and determination with which he carried out his self-imposed mission was fundamentally character-based; millions of Marconi's contemporaries had the same class, gender, race, and colonial privilege as he, but only a handful did anything with it. Marconi needed to achieve the goal that was set in his mind as an adolescent; by the time he reached adulthood, he understood, intuitively, that in order to have an impact he had to both develop an independent economic base and align himself with political power. Disciplined, uncritical loyalty to political power became his compass for the choices he had to make.

At the same time, Marconi was uncompromisingly independent intellectually. A few months before leaving Italy in 1938, Enrico Fermi was asked to write the preface to a commemorative publication on Marconi's work. Marconi proved that theory and experimentation were complementary features of progress, Fermi wrote. "Experience can rarely, unless guided by a theoretical concept, arrive at results of any great significance ... on the other hand, an excessive trust in theoretical conviction would have prevented Marconi from persisting in experiments which were destined to bring about a revolution in the technique of radiocommunications."[2] In other words, Marconi had the advantage of not being burdened by preconceived assumptions.

As Fermi noted, and as we've seen, Marconi was not deterred by conventional wisdom. Nor was he deterred by institutional obstacles. In June 1943, the US Supreme Court finally ruled on a patent infringement suit taken by the Marconi Wireless Telegraph Company of America against the United States government in 1916. The company claimed that the government had infringed on Marconi's 1904 US "tuning" patent. When the Marconi company sold its US assets, including its patents, to the new Radio Corporation of America in 1919, it reserved this unresolved claim for its own prosecution. It was Marconi's last commercial interest in the United States. The infringement suit claimed that the US government was using Marconi's patent without paying royalties. The government argued that the patent was not original and hence invalid.[3] (There is some irony that while the case was still pending, the US Congress voted to erect a monument to

Marconi in Washington, recognizing him as the inventor of wireless telegraphy. President Roosevelt approved the resolution on April 13, 1938.[4])

The 1943 Supreme Court ruling, a 5–3 decision written by Chief Justice Harlan F. Stone (one justice did not take part) stated: "Marconi's reputation as the man who first achieved successful radio transmission rests on his original patent...which is not here in question. That reputation, however well-deserved, does not entitle him to a patent for every later improvement which he claims in the radio field."[5] While that ended the Marconi company's claim against the United States government, it brought a scathing dissent from Justice Felix Frankfurter, the most eloquent voice on the court:

> The inescapable fact is that Marconi in his basic patent hit upon something that had eluded the best brains of the time working on the problem of wireless communication.... To find in 1943 that what Marconi did really did not promote the progress of science because it had been anticipated is more than a mirage of hindsight. Wireless is so unconscious a part of us...that it is almost impossible to imagine ourselves back into the time when Marconi gave to the world what for us is part of the order of our universe....nobody except Marconi did in fact draw the right inferences that were embodied into a workable boon for mankind.[6]

Frankfurter got the main point: regardless of the validity of a particular ancillary patent in hindsight, it was Marconi who changed the world. The 1943 US decision has become a set piece in the arsenal of Marconi doubters, but Justice Frankfurter's dissenting opinion was more in line with the common-sense view that Marconi did what others *might* have done or *could* have done but didn't do. The decision also elided the fact that in 1943 the United States was at war with Italy, and the US government was in no mood to pay a settlement to a British company over a forty-year-old dispute involving the inventions of a dead "Italian scientist" (as Marconi was called in a lower court ruling). It highlighted the continuing political resonance as well as the mystique associated with Marconi.

The Marconi name continued to be a recognizable and valuable asset as the company passed from corporate hand to corporate hand through a labyrinth of mergers, breakups, and spinoffs. Marconi's Wireless Telegraph Company was taken over by the English Electric Company Limited in 1946 and renamed the Marconi Company Limited in 1963.[7] Following the merger of English Electric with the (British) General Electric Company in 1968, the company was rebranded as GEC-Marconi Electronics Limited. It was reorganized as Marconi Electronic Systems Limited in 1998. By then it was a diversified conglomerate dealing mainly in telecommunications equipment. After parts of the company were spun off and merged with British Aerospace in 1999 (becoming BAE Systems and thus losing its distinct identity for the first time in a hundred years), GEC retained the brand and renamed itself Marconi plc. In 2003, a new entity was formed, Marconi Corporation plc, and in 2006, the bulk of the Marconi assets, including the name, were sold for £1.2 billion to Ericsson, the Swedish electronics transnational that bills itself as the world leader in global mobile communication, with some thirty-seven thousand patents.[8]

In the course of these changes, the company archives—which held most of Marconi's personal papers for the British period of his career, from 1896 to 1927—were nearly lost. In January 1997, Christie's announced an unprecedented auction of "the remarkable and little known archives of the Marconi company, the pioneers of wireless communication." The sale was set for April 24 and 25, 1997, coinciding with Marconi's birthday as well as the company's centenary. News of the impending auction sparked an outcry in England, with calls for the archives to be preserved as a national asset. From Rome, Marconi's daughter, Elettra, said she was "shocked and heartbroken," and expressed her wish that the collection remain in England. British media historian Lord Asa Briggs wrote to the *Times* that dispersal of the archives would be "thoroughly irresponsible." The company responded that there were good reasons to sell off the archives; they were "little-known, rarely used and inaccessible," housed in an obscure location at Great Baddow, Essex, near Chelmsford. But, clearly embarrassed and facing a PR disaster, it halted the sale and an arrangement was made with London's Science Museum, by which the collection would be donated "to the nation."[9] In 2004, the archives were given by Marconi plc to the University of Oxford, along with a collection of historic equipment that is now housed in Oxford's Museum of the History of Science.[10]

Among scholars and aficionados, there are still many controversies surrounding Marconi. Was he self-taught or formally educated? Did he have a flash of insight or did his idea grow incrementally? When did he make his transformational discovery and what were the influences on him? It would be nice—if it were possible, from this distance—to answer these questions, and indeed, recent research allows us to gain greater clarity if not quite full closure. More interesting is the fact that, over a century after the fact, such questions still exist at all. There is a minor Marconi industry that survives on and feeds debates the rest of us might prefer to see either buried or resolved.

So what did Marconi actually invent? An anecdote may be helpful here. There is an apocryphal story about the competition to build the cupola of the basilica of Santa Maria del Fiore in Florence, known as the Duomo. After receiving proposals from some of the most famous designers of the early Renaissance, the planning committee announced that whoever could make an egg stand on end would win the commission. After all others failed the test, the architect Filippo Brunelleschi simply cracked the egg and stood it upright. When his rivals protested that they could have done the same thing, Brunelleschi retorted that they would be able to vault the cupola, too, if they only knew how he planned to do it. Brunelleschi got the commission.[11]* Marconi's early benefactor, William Preece of the British Post Office, used to tell a version of the Brunelleschi story in which the architect has morphed into Christopher Columbus. "Columbus did not invent the egg. He showed how to make it stand on end," Preece would say, to make the point that Marconi may have been working with what was already known but he turned it into something previously unimaginable.[12] In modern terms, this story is reminiscent of the scene in the 2010 Hollywood film *The Social Network* where the Mark Zuckerberg character looks coolly across a boardroom table at his archrivals, the Winklevoss twins, and says: "If you guys were the inventors of Facebook"—pause—"you'd have invented Facebook." The moral of the story, of course, is that anything looks easy once one knows how it is done.

Regardless of who actually did what in connection with the invention of wireless, Marconi was undoubtedly the one who made the egg stand on end, and as a result his is the name most closely associated with the advent of wireless communication technology—or *radio*, as it became known around 1910, then *broadcasting, telecommunication, mobile telephony, Wi-Fi, social media,* and *cloud computing* today. As Marconi himself enjoyed saying in later years, it really didn't matter who was the first to discover the techniques of long-distance wireless telegraphy; what was important was that he was the first to discover a *use* for it, to channel the energy and focus on that use, to organize a system for exploiting it, to mobilize powerful commercial and political forces to harness it, and to use it to change the world. In this re-

---

* Brunelleschi is also said to have received the world's first patent, in 1421 (M. Frumkin, "Early History of Patents for Invention," *Transactions of the Newcomen Society* 26 [1947–49]: 48).

spect, he was more like Edison or Franklin than Galileo or Einstein. Marconi was without a doubt the dominant—as well as the most enigmatic and controversial—figure in the pioneering stage of the information age. After a certain point, it does not really matter who did what; it is impossible to speak about the history of modern communication—from the wireless telegraph to radio to radar, the cellphone, GPS, and the Internet—without paying close attention to Marconi and his career. Or as Marconi was too polite to say but could have said to any one of his detractors: if you had invented wireless, you would have invented wireless.

Having "invented" wireless, Marconi also had to develop the conditions for exploiting it. Again, character played an enormous role in his success in this regard. Marconi's close collaborator, Charles S. Franklin, whose research led to the "Marconi beam system" in the 1920s, wrote: "You can't call him a scientist but perhaps because he was not a scientist he had intuitions which he gambled on and which in many cases confounded the scientists. . . . His scientific knowledge was weak, his engineering knowledge was weak, but he had a damned lot of intuition and common sense. He may have initiated the beam system but he didn't know a thing about it."[13] To some, Marconi was a virtuoso who created a vast *trompe l'oeil scientifique*. Historian of technology Sungook Hong recalls: "Marconi's inventions, modifications, and improvements fit into a small box, at that time dubbed Marconi's 'secret-box' or 'black-box.' When Marconi 'opened' this 'black-box' by publicizing his first patent in 1897, people were amazed and intrigued by its simplicity. The solutions appeared so simple and so obvious that many began to wonder why no one else had come up with them."[14] Marconi was the first geek, his followers the first hackers; the entrepreneurial structure he created in 1897 was also the prototype for the dot.com start-up.

At the beginning of this book, we surveyed an array of protagonists, Marconi could-have-beens had they enjoyed similar circumstances or been able to muster comparable resources. Their likes are still around today. In August 2010, a Brooklyn cinematographer by the name of Luke Geissbühler, assisted by his seven-year-old son, Max, placed an Apple iPhone and a high-definition video camera in a fast-food takeout carton and propelled it one hundred thousand feet into space using a weather balloon. They recorded the journey on video, and, tracking the iPhone by GPS, recovered the device after the balloon exploded and the iPhone came back to earth by parachute, landing thirty miles from the launch site near Newburgh, New York. It's not clear what the Geissbühlers were hoping to achieve by this feat—besides having a good time—but a week later, the *Economist* reported, NASA spent millions doing essentially the same thing.[15] Technological development is an apparently boundless field that continues to occupy both personal tinkerers as well as the corporate/military/state-based establishment.

What enabled Marconi to stare into the eyes of that establishment was the legal and corporate structure that he put into place (with just a little help from his friends in London and elsewhere). In business, too, he was an outlier, arriving out of nowhere to stir the placid (and profitable) pot of the cable companies. His inclination, indeed preference, was to work with governments rather than the market, but he chose to use the favourite instrument of capitalism, the private corporation. A precursor of the digital economy, Marconi's company pioneered the practice of negotiated showdowns between capital and the state. Capitalism has arguably proven to be more robust than colonialism or imperialism; the multinational firm has never been as powerful as it is today, but this power is based on the alliances it has been able to make with governments. In 2012, the CEO of Google sat down with the president of France to negotiate royalty payments to French newspaper publishers, to give but one small example;[†] a century earlier, Marconi was making comparable deals with the world's most powerful governments. The dealing went both ways. The role of today's information service providers in supplying big data to governments for surveillance purposes recalls Marconi's collaboration with Scotland Yard and British security in the 1920s. But less than ten years after he debuted, the leading powers of the day decided to put in place an international regulatory regime in order to block Marconi from achieving a private global monopoly on the use of the newly discovered radio spectrum. No single individual before or since has ever amassed such power in the field of communication.

There was a price, as there always is. Only the people closest to him realized that Marconi clearly suffered from cyclical bouts of depression. He regularly spent long periods out of circulation, taking to his bed for unspecified reasons. His wife Beatrice said he would often spend days in a state of oblivion, with a faraway expression in his eyes. Her sister Lilah wrote that twice, in moments of despair over the company's financial situation, he voiced the possibility of suicide to Beatrice.[16] None of this was ever diagnosed, let

---

[†] The French government successfully insisted that the global Internet giant pay a small royalty fee every time its search engine referred a user to a French newspaper's website ("*Le ton monte entre la France et la firme Google*," *Le Devoir*, October 30, 2012).

alone treated. And he could be merciless. The two or three times he had major clashes with business associates, such as Cuthbert Hall and Jameson Davis, he crushed them. The same with patent rivals such as Reginald Fessenden and Lee de Forest. In his public scientific disputes and comments to the press he could be brutal. With subordinates, especially his technical associates, though, he was usually courteous, respectful, and supportive, winning him their absolute loyalty. Marconi's assistants found him a good chief to work for, provided one quickly came to terms with the idea that one was expected to be on duty twenty-four hours a day, seven days a week. He certainly drove his assistants hard, but never so hard as he drove himself.

Marconi's monomaniacal focus played out in every aspect of his life. His daughter Gioia, in a 1985 lecture, told a story about a rare outing with her dad, which had seemed to begin well until he realized that their driver was a ham radio operator. "So off we went to the poorest part of town and up the stairs we climbed to the driver's little top floor room. I remember that father became so engrossed with what he was doing that he completely forgot about me.... As I look back to the years with father, they weren't easy.... The thrust of his vision was so great that it swept all else from his path."[17] The people closest to him concurred. P.W. Paget, the assistant who had been with Marconi and George Kemp in Newfoundland in 1901, told the *Times* of London the day Marconi died: "He had little interest in anything outside wireless."[18]

Marconi's accomplishments were immersed in the muck of his messy life, and his personal relations took a back seat to his work. Among men, he didn't really have friends, only associates. He desired female attention and companionship, and was constantly infatuated, often with very young women. He wanted more than anything to be loved. He courted women who challenged him but married two who would have never dreamed of doing so. His lovers (if that's what they were) were exciting, adventurous, ambitious, and interested in politics; his wives were expected to be appendages and to look after him, as his mother had. Something was missing in his emotional development, perhaps a result of his mother's unconditional devotion; he always, painfully, sought and was never able to find that in a companion. It didn't work with Beatrice—who probably knew him better than anyone else but wanted more—and his first family was totally dysfunctional. His relationship with Cristina, who was prepared to play the role, was perhaps his most successful, but by then he was tired, ready to settle down, and had lowered his expectations.

"My father had a foot in both centuries," his daughter Degna wrote. "By tradition and breeding, he was a man of the nineteenth century. It showed in the formality of his manners, the impeccability of his dress, and the reticence behind which he walled his personal self. Intellectually and temperamentally, he was of the twentieth century. . . . His genius shaped it and it was very much to his taste."[19] I would go further. Marconi was also of the twenty-first century. His "genius," if that's what it was, lay in *making connections*—through technology, personal networking, and harnessing the potential of existing knowledge. From the time he came out to the press, following the demonstration of his "magic box" in London in 1896, Marconi was surrounded by an aura of myth, which he cultivated carefully and made his greatest asset. At the height of his fame he was so well known that he literally needed no introduction; when he received an honorary degree from Oxford in 1904, he was presented simply as "Signor Marconi," with no further qualification.[20]

That aura continues to shine. Today we would consider someone like Marconi part of a scientific-technological elite, but what really set him apart in his day was the vision he articulated for the benefits of global communication. Decades before McLuhan coined the phrase, Marconi was living in a global village of his own making, personifying a utopian idea of the liberating power of communication technology. He was perhaps the first to believe in, understand, and *express* the power of communication. (McLuhan himself stated in 1969: "The Gutenberg Galaxy [print] is being eclipsed by the constellation of Marconi."[21])

Alongside the Marconi aura, there was the rhetoric that went with it. When he spoke about wireless, Marconi presented a discourse of possibility.[22] Marconi's rhetoric was camped in the ideology of modernism, and it became a central part of that ideology. In Marconi's hands, wireless was not only a practice, it was a way of life. Marconi's vision remained consistent, and he kept restating it in a range of artful interventions, from his 1896 interview in *La Tribuna* to his testimony at the *Titanic* inquiries, from the 1920 paper in which he described the cellphone to his last radio broadcast in 1937: communication between people widely separated by space and time could be a force for good.

People believe what they want to believe, and Marconi is interesting today because we believe we are living in the world he imagined, empowered by mobile, instantaneous, wireless communication. But as we saw, by the end of his life, Marconi was wondering whether the uses begat by his technology were a good or a bad thing for humanity. The true extent of the emancipatory character of modern communication remains to be understood. Marconi thus personifies the paradox of communication. His ambivalence is ours. How does a technology that promotes and facilitates contact, openness, and human potential become an instrument for domination, manipulation, and control? That is the question that Marconi's story asks and, maybe, begins to answer.

Postscript

Beatrice O'Brien Marconi Marignoli, Marchesa of Montecorona, died in Rome in 1976, at the age of ninety-five. The newspaper *Il Tempo* devoted a full page in tribute to her life.

Degna Marconi married the press officer of the Italian embassy in London, Gabriele Paresce, in 1938, and their son Francesco was born in 1940. The Paresces spent the war in Rome and then lived as a diplomatic family in various parts of the world, most notably in Washington, DC, and Seoul, where Gabriele Paresce was the Italian ambassador to South Korea. Degna spent the better part of a decade researching and writing her memoir, *My Father, Marconi*, which is in many respects the most candidly interesting book about Marconi. She died in 1998.

Giulio Marconi returned to Rome after the war and worked for the Italian Marconi company until his death in 1971. He never married.

Gioia Marconi worked in radio in Italy and the United States for some years, and married George Atkinson Braga, an American sugar broker with interests in Cuba, with whom she had two children, Allegra (born 1958) and Michael (born 1961). In 1974, the centennial of her father's birth, she created the Guglielmo Marconi International Fellowship Foundation, now known as the Marconi Society, which awards an annual prize "to individuals whose scope of work and influence carry on the legacy of Guglielmo Marconi." Laureates have included Google founders Sergey Brin and Lawrence Page, Sir Tim Berners-Lee, who conceived the World Wide Web, Internet pioneer Vint Cerf ("Chief Internet Evangelist" at Google), and Martin Cooper, inventor of the cellphone. Gioia died in 1996.

Maria Cristina Bezzi-Scali Marconi was Marconi's widow for fifty-seven years, until her death in 1994. She continued to receive company emissaries and other Marconi interestees well into her eighties. After her death, her daughter Elettra completed and published Cristina's memoir, *Marconi My Beloved*.

Elettra Marconi married Prince Carlo Giovanelli in 1966, and their son Guglielmo was born the following year. She has devoted her life to promoting the heroic version of her father's memory and is solicited to take part in Marconi commemorative activities around the world. Elettra lives in Rome, in the grand house on Via dei Condotti where she grew up and where her father died, receiving callers in a dark, funereal living area that looks and feels as if it hasn't changed since 1937. The apartment is crammed with Marconi memorabilia: busts and photographs, a scale model of the *Elettra*, a bottle of Jameson whiskey on a sideboard. The furniture is ancient; an elaborate Murano glass chandelier, white with cranberry-pink highlights, is the only touch of playfulness in the room. Elettra keeps the shutters drawn, and explains that she must continue to carefully guard her father's last letters, away from inquisitive eyes.

Acknowledgements

This project was a massive undertaking and would have been unthinkable without the enthusiastic support and assistance of an international army of friends, volunteers, professionals, people with jobs and families, and Marconi buffs who became as passionate about the project as I was and put their own priorities aside at least momentarily to help make it happen.

My agent, John Pearce, the best a writer could hope for, has been a stalwart anchor of support and wisdom, helping me to transform the project from a dreamy idea into a book that will hopefully bring Marconi's intricate story into wider view. My sharp-eyed commissioning editor at OUP, Tim Bent, had his patience tried in too many unspeakable ways but never flinched. Without the constant encouragement of these two, *Marconi* would still be a work in progress. Also at OUP, Alyssa O'Connell tracked down illustration material and the requisite permissions while performing a plethora of background editorial tasks with efficiency and aplomb. Richard Johnson deftly shepherded the manuscript through the production process, and Linda Pruessen copy-edited it with the care and precision of a brain surgeon.

A number of academic institutions did what institutions do best—provide a discreet framework of support for research, reflection, and writing. The Social Sciences and Humanities Research Council of Canada provided a grant that allowed me to do my research properly. Thanks to them, if I could find the stone it was not left unturned. My home base at McGill University provided a secure and supportive environment as well as additional funding from the Faculty of Arts and the Beaverbrook Fund for Media@McGill, and, crucially, sabbatical leaves to kick-start and then complete this project. The Department of Media, Culture, and Communication at New York University and the Department of Media and Communications at the London School of Economics and Political Science (LSE) were welcoming havens during these leaves. In addition to test-driving some of the material in informal talks at McGill and the LSE, I benefited from presentation platforms and discussions with numerous colleagues within the European

Communication Research and Education Association and the International Association for Media and Communication Research.

One group of people, more than any other, made it possible to push this project to its full potential: the librarians and archivists, bless them all. At the University of Oxford, the Bodleian Library's senior archivist Michael Hughes made the Marconi Archives accessible. The online publication of Hughes's exhaustive catalogue, supplemented by his personal navigation tips, made it feasible to organize several research furloughs in Oxford between 2010 and 2012. Colin Harris and the staff of the Bodleian's Special Collections Reading Room (SCRR), especially Rebecca Wall and Julia Wagner, made this a pleasant and productive experience. A cast of anonymous scribblers and scholars in the SCRR provided colourful company during the months I was poring over boxes of Marconi files. To top it off, the Bodleian waived its usual permission fees to use images from the Marconi Archives in this book. Thank you, Bodley!

Negotiating the vaults and corridors of Italian archives requires a special combination of patience, connections, and good humour, and I was tremendously fortunate to have expert guidance in this regard. Carlotta Darò did groundbreaking preliminary research in Rome, finding and introducing me to the principal gatekeepers of the copious Marconi archival sources in Italy. Alessandro Romanello, of the Accademia Nazionale dei Lincei, helped me open many doors (not only to archives but to some of the lesser known bistros of Trastevere) and was a steadfast friend and companion. Mauro Canali, one of the leading experts on the history of Italian fascism, steered me through the intricacies of the central state archives (Archivio Centrale dello Stato) and introduced me to the astonishing files of the Fascist political police which, to my knowledge, have never before been scrutinized by Marconi researchers. Giovanni Paoloni, the dean of Italian Marconi scholars, shared his vast knowledge and powerful insights about the custodians of Marconi's legacy. Alessandro Visani paved the way for my incursion to the Vatican Secret Archives (and would have done more had he not been sidelined by a motorcycle accident). Margherita Martelli (Archivio Centrale dello Stato), Rita Zanetti (Reale Accademia d'Italia), Paola Cagiano de Azevedo (Accademia Nazionale dei Lincei), Stefania Ruggeri (Archivio Storico del Ministero degli Affari Esteri), and Chiara di Vecchis (Biblioteca del Senato) were all helpful and effective archival facilitators at their respective institutions in Rome.

In addition to the Oxford and Italian archives, numerous library and archival sources were tapped for this research. Thank you to Jane Harrison, who opened the doors of the Royal Institution in London; Wendy Shay and Kay Peterson of the Smithsonian Institution in Washington; Lisette Matano and Nicholas B. Scheetz of the Georgetown University Library; Dan Lewis and Catherine Wehrey of the Huntington Library in San Marino, CA; William Baehr of the Franklin D. Roosevelt Presidential Library and Museum in Hyde Park, New York; Anne Macneil and Mariah Hudec of the Beaton Institute at the University of Cape Breton; Jocelyn Bethune of the Alexander Graham Bell Museum in Baddeck, Nova Scotia; Claryn Spies of Yale; and the staffs of the Ticonderoga Historical Society, The Rooms Provincial Archives in St. John's, Newfoundland, the Library of Congress in Washington, the Friends of Highgate Cemetery, and Library and Archives Canada in Ottawa. Sara Bannerman mined the National Archives of South Africa on my behalf, and Christian Herzog did the same at the Deutsches Technikmuseum in Berlin. At McGill, librarians Jennifer Garland, Francisco Uribe, and Lonnie Weatherby chased down dozens of obscure sources from around the world, and my able research assistant (and PhD supervisee) Errol Salamon kept track of it all.

The specialized world of Marconiana is a microcosm of its own, and no one knows more about it than Barbara Valotti of the Fondazione Guglielmo Marconi. Barbara became my "go-to" person for dates and details, sources, resources, and fact-checking. Two of Marconi's grandchildren, Francesco Paresce and Michael Braga, generously opened their family archives to me and spent hours answering my probing questions with candour and genuine interest in setting the record straight. Riccardo Chiaberge shared precious documents, and insights, from his own research, as did Annalisa Capristo, whose work has shaken some academic certainties about the origins of Mussolini's racial policies. Claudia Padovani and Elena Pavan performed some essential tasks that required an Italian chequing account and postal address.

In Clifden, Ireland, the delightful Marconi enthusiast Shane Joyce walked me around the ruins of Marconi's station on the bog of Derrigimlagh and later answered dozens of queries on some of the finer points of Marconi's Irish connection. Shane is a fount of information about Marconi, Ireland, and other matters, and I am happy to report that his project to redevelop the Clifden site has captured the imagination and loosened the purse strings of the Irish Ministry of Tourism. Elsewhere in Ireland, the burden of research was lightened by folks like Pedro, the manager of Daddy Nottage's pub in Crookhaven, and Sean Kilkenny, the master of pony and trap rides at Dromoland Castle. Grania Weir opened the Dromoland guestbook to my inquiries; Jarlath Glynn of the Enniscorthy Library provided clues to the mysteries that still surround some of Marconi's Irish antecedents, as did Dorothy Davis, Ken Hemingway, Kristin Jameson, Robert Somerville-Woodward, Marianne Young, and the librarians of the Church of Ireland in Dublin. Seán o Siochru and Aine Casey provided a retreat in the Beara peninsula when I needed one.

For quickly answering queries that arrived out of nowhere, I am grateful to Ruth Teer-Tomaselli, Donal McCracken, Maria Rikitianskaia, Leonard Laborie, Garth Bulmer, Alexander Rufus-Isaacs, Michael Fisher, Richard Collins, Alban Webb, Jeffrey Lack, and Grace Westcott, as well as fellow authors Susan J. Douglas, Linda Lumsden, Denis Judd, Keith Jeffery, and Jonathan Sterne. I especially appreciated several exchanges with Charlie Foran and the participants in a workshop on writing biography that Charlie led for the Quebec Writers' Federation in April 2013. In addition to the work of the aforementioned Barbara Valotti, Mauro Canali, and Shane Joyce, parts of the manuscript were also graciously read and commented on by Italian telecom historian Gabriele Balbi, Pulitzer Prize–winning author David Kertzer, and documentary filmmaker Alan Mendelsohn. *Mia cugina* Lillian Arsenault read the entire first draft of the manuscript and provided priceless advice.

There are not many people left alive who knew Marconi, and those who did knew him when they were small children at best, or by association with other family members. Foremost among these, of course, is Marconi's daughter, Princess Elettra Marconi Giovanelli, who was seven years old when her father died. I had six meetings with Elettra, three of them in her home (which was also Marconi's last residence). Others who shared family anecdotes include Marconi's grandchildren Francesco Paresce, Michael Braga, and Guglielmo Marconi Giovanelli; John Jameson-Davis, a grandson of Marconi's Irish cousin and business associate, Henry Jameson Davis; the Honorable Peter Smith, grandson of Marconi's first fiancée, Josephine Holman, and his wife Betty Smith; John Tepper Marlin, a cousin by marriage of Marconi's second fiancée, Inez Milholland; Doug Cunningham, whose father bought Marconi's station house at Glace Bay; and Ambassador Luigi Solari Jr., grandson of Marconi's collaborator and earliest biographer.

Finally, numerous dear friends and family members provided moral support, wise counsel, and conviviality during the years I worked on this project. Their roles were unequal but at this moment I remember them all, especially Elaine Arsenault, who was there when it started, and Lucie Rodrigue, who is watching it end.

Sources and Abbreviations

ARCHIVAL SOURCES

AAS = Acta Apostolicae Sedis [official gazette of the Holy See] (online)

ACS = Archivio Centrale dello Stato [Central Archives of the State], Rome

   AEI = Archivio dell'Enciclopedia Italiana

   CNR = Consiglio Nazionale delle Ricerche

   FSN = Archivio Francesco Saverio Nitti

   MRC = Ministero della Real Casa

   PCM = Presidenza del Consiglio dei Ministri

   PP = Ministero dell'Interno, Direzione Generale della Pubblica Sicurezza, Divisione Polizia Politica, Fascicoli Personali, Serie A 1927–1944

   SPD = Segreteria Particolare del Duce

      SPD CO CNR = Carteggio Ordinario 550.752 (former 129.107), "Consiglio Nazionale delle Ricerche"

      SPD CO GM = Carteggio Ordinario 523.201, "Marconi, Guglielmo"

      SPD CO GMF = Carteggio Ordinario 197.509 (formerly 197.598), "Marconi. Sen. March. Guglielmo. Famiglia"

      SPD CR = Carteggio Riservato Categorie 480/R ("per la questione ebraica")

AGBFP = Alexander Graham Bell Family Papers, Manuscript Division, Library of Congress, Washington, DC

AGBMA = Alexander Graham Bell Museum Archive, Baddeck, Nova Scotia

AIL = Archivio Storico Istituto Luce, Rome (online)

ASMAE = Archivio Storico del Ministero degli Affari Esteri, Rome

   AI = Ambasciata d'Italia in Washington

   AL = Ambasciata Londra 1861–1950

   CP = Conferenza della Pace 1919–1922

   GI = Carte Imperiali

   MCP = Ministero della Cultura Popolare

ASS = Archivio Storico del Senato della Repubblica, Rome

   ACG = Alta Corte di Giustizia

ASV = Archivio Segreto Vaticano [Vatican Secret Archives], Vatican City

ARRSS (AA.EE.SS.) = Segreteria di Stato, Sezione per i Rapporti con gli Stati, Archivio Storico (Affari Ecclesiastici Straordinari)

BEA = Guglielmo Marconi Collection, Beaton Institute, University College of Cape Breton, Sydney, Nova Scotia

BS = Biblioteca del Senato, Rome

DTM = Deutsches Technikmuseum, Berlin

EMH = Edward Mandell House Papers, Yale University Library Digital Collections (online)

FDR = Franklin D. Roosevelt Presidential Library, Hyde Park, New York

AER = Anna Eleanor Roosevelt Collection

ASN = Papers as Assistant Secretary of the Navy

DBD = "FDR Day by Day" (online)

GTC = Grace Tully Collection (online)

FGM = Fondazione Guglielmo Marconi, Bologna (online)

FP = Francesco Paresce Collection, Bologna

GCR = George H. Clark Radioana Collection, Archives Center, National Museum of American History, Smithsonian Institution, Washington, DC

HUN = Guglielmo Marconi Correspondence, The Huntington Library, San Marino, California

HWL = Henry Willard Lende Collection, San Antonio, Texas

IMP = Inez Milholland Papers, Arthur and Elizabeth Schlesinger Library on the History of Women in America, Radcliffe Institute for Advanced Study, Harvard

ITU = International Telecommunication Union, Library and Archives, ITU Conferences Collection (online)

LAC = Library and Archives Canada, Ottawa

LDF = Lee De Forest Papers, Manuscript Division, Library of Congress, Washington, DC

LIN = Archivio Guglielmo Marconi, Accademia Nazionale dei Lincei, Rome*

LSP = Lisa Sergio Papers, Special Collections, Georgetown University Library, Washington, DC

MB = Michael Braga Collection, Sarasota, Florida

MFP = Milholland Family Papers, Ticonderoga Historical Society, Ticonderoga, New York

---

* The inventory of the Marconi Archive at the Accademia Nazionale dei Lincei was reclassified in 2014. Except where explicitly noted, the references in the endnotes follow the earlier 1996 classification.

NA = The National Archives [UK], Kew (online)

NARA = The U.S. National Archives and Records Administration, Washington, DC

NARS = National Archives and Records Service of South Africa, Cape Town/
Pretoria, South Africa

NP = Nobel Prize Nomination Archive (online)

OX = Marconi Archives, Bodleian Library, Oxford

RAL = Archivio Storico, Reale Accademia d'Italia, Accademia Nazionale dei
Lincei, Rome

RI = Marconi Papers, Royal Institution, London

RPA = The Rooms Provincial Archives, St. John's, Newfoundland

RSB = Ray Stannard Baker Papers, Dept. of Rare Books and Special Collections,
Princeton University Library (online)

TEP = The Thomas A. Edison Papers, Rutgers (online)

CORRESPONDENTS

AJM = Annie Jameson Marconi

AM = Alfonso Marconi

AS = Adolf Slaby

BM = Benito Mussolini

BOM = Beatrice O'Brien Marconi

DM = Degna Marconi

DP = Daisy Prescott

DS = David Sarnoff

FGK = Frederick G. Kellaway

GAM = Gaston A. Mathieu

GCI = Godfrey Charles Isaacs

GIO = Gioia Marconi

GIU = Giuseppe Marconi

GLO = Giulio Marconi

GM = Guglielmo Marconi

HAW = H.A. White

HCH = Henry Cuthbert Hall

HJD = Henry Jameson Davis

IM = Inez Milholland

JAF = James Ambrose Fleming

JBH = Josephine Bowen Holman

JEM = John Elmer Milholland
JHH = John Henniker Heaton
LDF = Lee de Forest
LOB = Lilah O'Brien
LS = Luigi Solari
MCM = Maria Cristina Marconi
OED = Orrin E. Dunlap
RAF = Reginald A. Fessenden
RNV = Richard N.Vyvyan
SFP = Samuel Flood-Page
UDM = Umberto Di Marco
WP = William Preece

NEWSPAPERS

LT = *Times* (London)
NYT = *New York Times*

COMPANIES

MIMCC = Marconi International Marine Communication Company
MWTC = Marconi's Wireless Telegraph Company
MWTCA = Marconi Wireless Telegraph Company of America
MWTCC = Marconi Wireless Telegraph Company of Canada
WTSC = Wireless Telegraph & Signal Company

Notes

Unless otherwise noted, translations of documents in the Marconi Archives in Oxford were completed by the Marconi company, and all other translations were completed by the author.

PROLOGUE

1. M. Gladwell, "Creation Myth," *New Yorker*, May 16, 2011, 44–53.
2. Daily Express [Lisa Sergio], "Marconi Drops Deathray," *Daily Express,* May 18, 1937.

CHAPTER I

1. This section on the origins of the Marconi family is based on Giacomelli and Bertocchi 1994.
2. Museum of the History of Bologna, Palazzo Pepoli, texts on display (Bologna: Bononia University Press, 2011). For an interesting portrait of politics and life in Bologna in the 1850s, see David I. Kertzer, *The Kidnapping of Edgardo Mortara* (New York: Vintage, 1998).
3. Solari 2011, 21.
4. GM, typescript draft for *Encyclopedia Britannica*, 1926, OX 1772.
5. Comitato Guglielmo Marconi International, *Documenti rari su Guglielmo Marconi*, www.radiomarconi.com/marconi/gm_docum.html. Marconi told biographer Orrin Dunlap in the 1930s that his father did not speak English (LIN 3/124), although of course he could still have been a partner in a London business.
6. LIN (2014) B.59 contains some souvenir material from the trip. (The inventory of the Marconi Archive at the Accademia Nazionale dei Lincei was reclassified in 2014, and some new material was added. Except where explicitly noted, the LIN references in the endnotes follow the earlier 1996 classification.)
7. Solari 2011, 39.
8. Jameson is now part of Irish Distillers, which was acquired by the French liquor conglomerate Pernod Ricard in 1988. There is some controversy surrounding the founding of the Jameson distillery. According to the most likely version, John Jameson was the manager of the Dublin Bow Street company actually founded by John Stein of Kennetpans, Scotland, which Jameson (or his son, also named John) bought from Stein in 1805 and renamed John Jameson and Son. As an aside, Irish whiskey takes an *e*; Scotch whisky does not. (*The*

History of Kennetpans, http://www.kennetpans.info/; see also "Signor Marconi's Irish Lineage. His Mother a Wexford Lady," *Irish Times,* January 15, 1898.)

9. Today, Enniscorthy (about two hours south of Dublin) is easily accessible by road, bus, and train. Daphne Castle no longer exists but the former distillery still stands, across the road from a popular pub in an area known locally as "The Still." See Walsh [1983?], and Walsh 2010.

10. See Ireland, National Archives, Bureau of Military History 1913–21, www .bureauofmilitaryhistory.ie/reels/bmh/BMH.WS1373.pdf, for nationalist activity in Enniscorthy leading up to and during the 1916 rising.

11. See Scotlands People, "Old Parish Records, Banns and Marriages," at scotlandspeople.gov.uk; Prescott 1910, 20.

12. "The Mother of Guglielmo Marconi," undated typescript, OX 1758.

13. Walsh 2010.

14. *Wexford Independent,* April 6, 1859. According to the Griffith's Valuation—a property valuation done between 1847 and 1864 and considered one of Ireland's most important genealogical sources— Andrew Jameson had several tenants at Fairfield in the 1850s (Ask about Ireland, "Griffith's Valuation," www .askaboutireland.ie/griffith-valuation/).

15. D. Marconi 1962, 6.

16. D. Marconi 1962, 7. See also "Town's Connection with Marconi Recalled," *Enniscorthy Echo,* July 9, 1976, OX 1758.

17. Valotti 1995, 14.

18. Enniscorthy telephone directory, consulted in July 2013.

19. See Fondazione Guglielmo Marconi, www.fgm.it/en/archive.html. The documents now form part of the Marconi Archive at the Accademia dei Lincei, across the road from the Villa Farnesina, and have been analyzed most notably by historian Barbara Valotti, whose unpublished thesis (1994–95) convincingly picks apart the inconsistencies in the standard accounts of Marconi's life up to 1895. The most reliable single source for detail on the young Marconi, Valotti's thesis is also an interesting case study of the use of mythical narrative in constructing the history of science and technology (especially where the figure of the "heroic inventor" is concerned). The Marconi Archive at the Lincei includes a separate section covering Marconi's correspondence and other documents from the 1930s, also curated by Giovanni Paoloni. These papers were donated to the academy in the 1970s by the family of Umberto Di Marco, Marconi's last personal secretary, who took custody of them at the time of Marconi's death in 1937, apparently on his instructions (G. Paoloni, interview with author, July 7, 2011). See also Valotti 1995, Bettòlo 1986, and Bigazzi 1995.

20. Marconi's eldest grandson, Francesco Paresce, has described his grandfather as "a rather special but otherwise ordinary mortal striving against long odds in showing the way to a sympathetic but mostly uncomprehending public" (Paresce n.d.). According to Paresce (n.d.), Marconi was above all a man in the right place at the right time. Giovanni Battista Marini Bettòlo, who knew

Marconi toward the end of his life, has referred to the first period of Marconi's career as "heroic" (Bettòlo 1986). The notion of Marconi as hero is an almost unavoidable trope recurring in the literature about him, as well as in journalism, poetry, and literary fiction.

21. Giuseppe Pession states that the family occupied the apartment at the end of 1873, and gave it up around a year after the birth of Guglielmo (Pession 1941).

22. Province of Bologna, "Marconian Itineraries," n.d.; M.C. Marconi 1999, 18.

23. G. Bonuzzi, "The Father of Radiotelegraphy," *Corriere dei Piccoli*, July 4, 1926. Translated typescript, OX 1772.

24. Sources mention both 1876 and 1877, but I have not seen any original documentation establishing the date.

25. R. Rodwell (company historian), "From Miss Bell, County Archivist, Bedford, 20 January 1984," OX 1771.

26. Archivist, Bedfordshire County Hall to MWTC, 1984, OX 1757; G. Marconi 1913, OX 54 (this is the most reliable of Marconi's many autobiographical accounts of his early life and experiments). Marconi's cousin Daisy Prescott (1910) also states that he received some schooling in England. In 1903, Marconi told the *Giornale d'Italia* that he had studied in England between the ages of five and ten. ("*Salutando Guglielmo Marconi. Intervista al momento della partenza da Roma*," *Il Giornale d'Italia* (Rome), May 20, 1903.)

27. Giacomelli and Bertocchi 1994.

28. A letter cited by Barbara Valotti refers to Giuseppe as "an enemy of the Government of Italy, having remained attached to the cause of Papal temporal power" (Valotti 1994–95, 12). He never regained his Italian citizenship (Giacomelli and Bertocchi 1994).

29. AJM to GIU, February 19, 1873, and June 4, 1873, LIN 37 (cited in Valotti 1994–95, 13 and Valotti 2014). Annie was in Ireland with Alfonso at this time, before Guglielmo's birth.

30. W. Stimson to AJM, April 21, 1881, LIN 39/A5. See also Giacomelli and Bertocchi 1994.

31. Prescott 1910, 1. Daisy wrote this text in January 1910 with the intention of having it published and asked her aunt to vet it with Guglielmo (DP to AJM, January 16, 1910).

32. Sullivan 2014, xxi.

33. *Giornale della Domenica* (Rome), January 17, 1937 (cited in Pession 1941). In a similar vein, Marconi told a US newspaper interviewer on April 14, 1912: "I always believed in myself, dreamed I was going to be somebody—make the world talk." He said he did not imagine wireless at first, but "had in mind always the idea of bringing countries closer in touch with each other, uniting remote spots and centres of life, but it was all so vague" (K. Carew, "Attuned to Wireless Waves, Gets Marconi's Message," *New York Tribune*, April 14, 1912, cited in Valotti 2014).

34. Pession 1941.
35. B. Valotti, interview with author, February 24, 2015.
36. "*Il maestro di Marconi,*" *Corriere delle sera* (Milan), May 9, 1903.
37. Valotti 2014.
38. B. Valotti, interview with author, February 24, 2015.
39. Solari 2011, 24; D. Marconi 1962, 88.
40. Prescott 1910.
41. Solari 1942, 2. Solari met Marconi professionally in 1901 and became his closest Italian collaborator, a relationship that continued until the day Marconi died. He wrote no less than four books about Marconi; unfortunately, the details they report are not always reliable. Here, for example, he dates the meeting he describes in 1886, but the Marconis left Florence in 1885.
42. M. Twain, *The Autobiography of Mark Twain,* ed. Charles Neider (New York: Harper & Row, 1959), 412.
43. Giacomelli and Bertocchi 1994. Luigi and Letizia had two children, Giovanni (b. 1888) and Pietro (b. 1891). In May 1891, Luigi was elected to the Bologna municipal council on a liberal-conservative ticket, but a year later he and his family also moved to Villa Griffone, now the real centre of the Marconi clan.
44. Vaccari 2006.
45. Valotti 1994–95, 12. The original *Valdese,* or Waldensians, were followers of a movement initiated by Valdo di Lione (Waldo of Lyon) in 1173. In medieval times they were pauperist, pacifist, and literalist with regard to biblical interpretation, and persecuted as heretics by the Inquisition. The Waldensian Evangelical Church was founded in northern Italy in 1532 by Calvinist refugees from France, and claims to be the only Italian Protestant church to have survived the religious wars of the Counter-Reformation (information taken from posters at Chiesa Valdese di Livorno, September 29, 2012.)
46. D. Marconi 1962, 277.
47. According to Giotto Bizzarrini (letter to UDM [n.d.], cited in Bettòlo 1986), Marconi registered for the technical course on November 4, 1885.
48. G. Marconi 1913, 1, OX 54.
49. Prescott 1910, 2–3. Valotti maintains that Marconi wanted to "be an inventor" and was searching for an idea long before he read of Hertz's discovery (interview with author, February 24, 2015).
50. Little is known about Marconi's life between 1889 and 1891. According to Bizzarrini (cited in Bettòlo 1986), the family spent most of that time in Bologna. According to Mario Giorgi and Barbara Valotti (2010), Rosa taught Marconi starting in autumn 1891.
51. "*Il maestro di Marconi,*" *Corriere delle sera,* undated cutting [May 9, 1903], LIN 37. Marconi began his 1909 Nobel lecture by stating that he never studied physics "in the regular manner" but did "attend one course of lectures on physics under the late Professor Rosa at Livorno" (Commendatore G. Marconi, "Wireless Telegraphic Communication" [Nobel lecture, December 11, 1909],

www.nobelprize.org/nobel_prizes/physics/laureates/1909/marconi-lecture.pdf;
Guglielmo Marconi. Nobel Lecture. Original Manuscript, Bologna: Fondazione
Guglielmo Marconi, 2009, 2).

52. G. Bizzarrini to UDM, n.d., cited in Bettòlo 1986. Also cited in Valotti 1995, 17,
and Valotti 1994–95, 43–44.

53. Pession 1941.

54. Solari 1942, 2–3.

55. AJM to GM, n.d, LIN 37/A1.

56. Pession 1941.

57. B. Valotti, interview with author, February 24, 2015. *L'Elettricità* was founded in
Milan in 1882 and was concerned with practical applications of electricity and aim-
ing to reach a target audience of specialized tinkerers—a cross between *Popular
Mechanics* and *Wired* (maybe more like the former), if we compare it to today's field.

58. Valotti 2014.

59. "Expenses incurred by Giuseppe Marconi for the experiments of his son
Guglielmo and for his maintenance in England in 1896–1897," LIN 39/A5.

60. Prescott 1910, 18.

61. G. Marconi 1937, opening sentence of document prepared for his biographer
Orrin Dunlap, LIN 3/124.

62. *Giornale d'Italia*, May 20, 1903, cited in Valotti 2003, 11.

63. GM to G.H. Clark, March 8, 1937, LIN 2/77.

64. GM to AM, n.d., LIN 37/A1. Known as the "blue notebook letter," this undated letter is considered by Paoloni to date from spring 1892 (Paoloni and
Simili 1996, 35). Marconi also mentioned in this letter that he was taking three
lessons a week from Bizzarrini, further situating it in 1892 or 1893.

65. Recognized in his time as a great experimental physicist, Righi is an important
figure in the history of science in his own right. He was nominated for the Nobel
Prize in Physics, unsuccessfully, every year from 1905 until his death in 1920 (Giorgi
and Valotti 2010). A prominent street in the centre of Bologna is named after him.

66. Unpublished memoir of Antonio Marchi, OX 1757.

67. GM to OED, n.d., LIN 3/124. Solari (2011) also quotes Marconi saying he rode
between Bologna and Griffone on a donkey.

68. Testifying in a US patent case in 1903, Marconi declared: "I was by no means a
regular attendant at his [Righi's] lectures, but only attended occasionally. I knew
him personally, and I don't believe I have met him privately more than three or
four times previous to 1897. ... I did not meet him privately after 1897, but I
met him at a public reception, I believe in July, 1897, and I met him at another
public reception in September, 1902" (Marconi testimony, September 28, 1903,
US Patent Office, Minutes of evidence of an investigation of interference
between applications of Marconi, Fessenden and Shoemaker, 1904, OX 533).

69. Dragoni 1995.

70. On May 28, 1897, Righi told the Bologna newspaper *Il Resto di Carlino* that
Marconi was not his student, and Marconi himself was forever trying to set the

record straight. In an interview in *La Tribuna* on July 4, 1897 (translated in OX 1757): "Q: 'Did you study physics at Bologna under Prof Righi?' A: 'I studied under the renowned Professor Vincenzo Rosa at the Livorno Lyceum and would be most happy that it be known that he was my only physics master.'"

71. M. Bigazzi, "Young Marconi's Experiments," *Universitas* 7 [January 1995]: 1–5. Bigazzi was a radio technician and scientific advisor to the Fondazione Guglielmo Marconi in Bologna.

72. "*L'Avvenire dell'Elettricità*," *L'Elettricità* 12, no. 44 (October 29, 1893): 697–98, cited in Valotti 2014, 11 (translated by Francesco Paresce). Valotti also points out that issue 32 of the 1894 issue included an article by Righi as well as a report of Lodge's famous 1894 lecture to London's Royal Institution.

73. Paresce n.d.

74. Hertz died in January 1894 and Marconi would have learned about his death in the February 1894 issue of *L'Elettricità* (B. Valotti, interview with author, February 24, 2015).

75. GM to DP, n.d., LIN 37/A1.

76. Marconi's British patent agent, Edward Carpmael, maintained that he was told Marconi conceived his invention in November 1894 and put it into practice in July 1895. If this were indeed the case, Marconi would have had a more difficult time establishing the priority of his invention with respect to others, like Lodge, whose work was publicized during the summer of 1894 (Carpmael memo, March 14, 1900, OX 530).

77. di Benedetto 1974b, 136 (the original Italian version is on page 94).

78. G. Marconi, *Le Vie del mare e dell'aria*, vol. 1 (1918): 6–7; see also Giuseppe Pelosi, "Where was Marconi in the Summer of 1895," *IEEE Antennas and Propagation Magazine* 50, no. 4 (August 2008): 218–19.

79. O. Malagodi, "*Il telegrafo senza fili. Importante invenzione d'un Italiano. Intervista coll'inventore*," *La Tribuna*, December 27, 1896.

80. G. Marconi 1913, 3, OX 54. In his Nobel lecture, Marconi states that he was "fairly well acquainted with the publications of that time dealing with scientific subjects including the works of Hertz, Branly and Righi," but he doesn't say *when* he became acquainted with these publications, lumping the statement in with the previously cited declaration that he had studied with Rosa in Livorno, which is known to have been in 1891–92, before Hertz's work had been widely publicized.

81. As suggested by Tabarroni 1974, 53.

82. G. Marconi 1937, LIN 3/124. Marconi's timeline here is consistent with his 1903 US patent litigation testimony, where he states that he first took up investigation with Hertzian waves in the spring of 1895 (OX 533). What Marconi here calls his "attic laboratory" is often referred to in the literature as the "granary," or the "silkworms' room" on the top floor of Villa Griffone. Elsewhere, Marconi wrote that it was his father who had suggested he use the

Morse symbol for *s* as his main signal, because of its simple clarity. Marconi stuck with this practice throughout his career. For example, his 1901 transatlantic signal also transmitted the Morse letter *s*. Valotti (2014) dates the gun experiment to November 1895.

83. Ibid.

84. G. Marconi 1913, 3 and 13, OX 54.

85. In the early 1970s, Manfredo Gervasi searched the Italian State Archives, the archives of the Ministry of Posts and Telecommunications, and the archives of the Ministry of Foreign Affairs, looking for evidence of correspondence between the Marconi family and the Ministry of Posts and Telegraphs before Marconi left for London in February 1896. He found no trace of any such correspondence (Gervasi 1973; copy in OX 1756).

86. *Giornale d'Italia*, May 20, 1903.

87. GM to BM, October 17, 1923 (English translation), OX 212.

88. The myth usefully outlived Marconi; see, for example, Marcello Manni, "*Marconi e l'Italia*," *L'Arena* (Verona), April 22, 1943.

89. G. Marconi 1913, 13–14, OX 54. According to Daisy Prescott, the Marconis had intended to leave for London as early as May 1895, "but just about that time my Mother was taken very seriously ill, and instead of going to London, Guglielmo and his Mother came direct to Florence" (Prescott 1910, 16).

90. di Benedetto 1974b, 136. This was also more or less the position maintained by Marconi in his May 1903 interview with the *Giornale d'Italia*, where he asserted that it was false that he had first offered his invention to the Italian Ministry of the Marine and they had refused it.

91. See, for example, Savorgnan di Brazzà 1939, 394.

92. D. Marconi 1962, 33.

CHAPTER 2

1. The Canadian political economist Harold Innis (1950, 1951) was the first to point this out.

2. See, for example, Mumford 1966.

3. R. Williams 2003, 134. The exact quote is: "there is no way to teach a man to read the Bible which did not also enable him to read the radical press."

4. Mackinnon 1971. Mackinnon was also one of the first advocates of a "Continental States of Europe."

5. On this point, Marconi provides an instructive example of the take-up of innovation: "Reuter, a specialist in news gathering and distribution by means of carrier pigeons, before the adoption of wire telegraphy, was one of the first to realize the importance of the discovery when it came. He did away with carrier pigeons where possible, but for months he was unsuccessful in his attempts to persuade London editors to use his telegraphic news. Finally, by supplying a month's service free, so that comparisons might be made . . . he succeeded in

introducing his system. It illustrates the common reluctance to take up improvements" (G. Marconi 1899, 52, OX 48).

6. Magder 2011.
7. See, for example, Wu 2010.
8. Magder 2011.
9. Rowe 2009.
10. Sterling 2007, 152–53.
11. Winseck and Pike 2007.
12. Headrick 1988, 102.
13. Ibid.
14. Headrick 1988, 98. See also Satia 2010.
15. Magder 2011.
16. ITU, International Telegraph Conference (Paris, 1865), Convention télégraphique internationale, Article 4.
17. Magder 2011.
18. ITU, International Telegraph Conference (St. Petersburg, 1875), Convention télégraphique internationale.
19. Ibid.
20. G. Marconi, 1899, 4–5, OX 48.
21. The manuscript is remarkably clean, clearly retyped after editing, and with a few corrections made in Marconi's hand directly on the text. An early archival notation suggests that there had been some uncertainty as to the authorship of the manuscript (it was said to be "ostensibly written by Marconi…Actual authorship not confirmed") but a second, 1972 notation from company staff states unequivocally: "There can be no doubt that Marconi himself dictated these chapters." The style and tone, as well as the handwritten notations, support this view.
22. Except as otherwise mentioned, what follows is largely based on Sarkar et al 2006, 1–50.
23. Lindell 2006, 247.
24. Ibid., 258.
25. The obvious line connecting Galvani, Volta, and Marconi provided the utopian novelist H.G. Wells an opportunity to have one of his fictional characters, the Edwardian Mr. Britling, say: "The British mind has never really tolerated electricity; at least not the sort of electricity that runs through wires. Too slippery and glib for it. Associates it with Italians and fluency generally, with Volta, Galvani, Marconi and so on…" (H.G. Wells, *Mr. Britling Sees It Through* [London: Cassell, 1916], cited in Weightman 2007, 325).
26. Sarkar et al 2006, 7.
27. Sir William Bragg, president of the Royal Society, thanking J.A. Fleming following his Marconi memorial lecture on November 26, 1937. Published in *Journal of the Royal Society of Arts*, November 26, 1937, 63.

28. G. Marconi, "The Development of Wireless Telephony," typescript, OX 309. A handwritten note on the typescript indicates that this was the text for a BBC radio broadcast Marconi was due to make on September 19, 1931. The broadcast actually took place on September 21.

29. G. Marconi, "Development of Wireless Communication," *LT*, special centennial commemorative supplement on Michael Faraday, September 21, 1931.

30. Sarkar et al 2006. See also Howe 2007.

31. Catania 1990, 1992.

32. C. Gray 2006, 198.

33. Catania makes a strong case for Meucci's priority, and recalls US government support at the time for his claim (Catania 2002, 426–42). In 2002, the US House of Representatives passed a resolution recognizing Meucci's contribution to the invention of the telephone. A week later, the Canadian Parliament unanimously affirmed that Bell—who during part of his life was Canadian—was the sole inventor of the telephone. Bell himself said he had "conceived" the telephone in Brantford, Ontario, in 1874 and "invented" it in Boston in 1876 (Costain 1960).

34. Briggs and Burke 2009, 142.

35. G. Marconi 1899, OX 48. The story of the snail telegraph was reported by Jules Allix in *La Presse* (Paris) on October 12, 1851, and in Baring-Gould 1889.

36. Ibid., 11.

37. C. Gray 2006, 25.

38. "The invention of wireless—the case for Marconi v. others," OX 1756.

39. Sarkar et al 2006, 68.

40. Belrose 2006, 352. In 1964, the US Congress passed a resolution recognizing "the foresight, ingenuity, and outstanding achievement of Dr Mahlon Loomis in being the first person to invent and demonstrate a system of wireless communication" (US Congress Joint Resolution No. 1181, September 24, 1964).

41. Worrall 2007.

42. Aitken 1976, 103.

43. J.A. Fleming, "The Centenary of the Electric Current," lecture to the British Association, 1899, *Electrician*, September 22, 1899, 765. Copy in OX 153.

44. A.A. Campbell Swinton, "Some Early Electrical Reminiscences," *Journal of Electrical Engineers* (1922): 492. Cited in Bailey and Landecker 1947. See also Worrall 2007.

45. Sarkar et al 2006.

46. C. Kinnaird, "Your America," *New York Journal American,* June 1, 1962. Murray's claim regarding its billing as "the birthplace of radio" can be found at http://www.nathanstubblefield.com.

47. This section on Maxwell is based on Sarkar et al 2006, 29ff.

48. Ibid., 224.

49. Ibid., 46.

50. Ibid., 224.

698 NOTES TO PAGES 49–54

51. Following based on Lindell 2006, 247–66; in turn based on Fahie 1899.
52. G. Marconi 1899, OX 48.
53. T.A. Edison, Menlo Park, New Jersey, "Means for transmitting signals electrically." US Patent Office 465,971, December 29, 1891 (filed May 23, 1885). Copy in OX 415.
54. Cited in Sarkar et al 2006, 259; also Orrin E. Dunlap, Jr., "Edison Glimpsed at Radio in 1875," *Scientific American* (December 1926): 424–25.
55. "Edison Enters Marconi Company," *NYT,* May 28, 1903; "Edison Becomes Marconi's Ally. Joins Wireless Telegraph Company as One of its Technical Directors," *New York World,* May 28, 1903. Historians agree that Edison's greatest invention was not the light bulb or the phonograph but the industrial research laboratory. By the 1920s, labs based on the prototype he established at Menlo Park, New Jersey, in 1876 had become common. Most of Edison's patents were developed in the corporate research environment that he pioneered. Marconi unwittingly adopted a similar model (Douglas 1987).
56. The Royal Commission issued five reports between 1892 and 1898 (Great Britain, Parliament, House of Commons, *Reports of Commissioners.* House of Commons Parliamentary Papers: 19th century, 1801–1900, London: HMSO).
57. Lindell 2006, 264–65.
58. Hong 2001, 12.
59. Lindell 2006, 264–65.
60. G. Marconi 1899, OX 48.
61. The term is Hong's (2001, 2).
62. Dragoni 1995.
63. Commendatore G. Marconi, "Wireless Telegraphic Communication" [Nobel lecture, December 11, 1909], www.nobelprize.org/nobel_prizes/physics/laureates/1909/marconi-lecture.pdf; *Guglielmo Marconi. Nobel Lecture. Original Manuscript,* Bologna: Fondazione Guglielmo Marconi, 2009, 5.
64. Sarkar et al 2006, 87, 309.
65. A. Popov, "Apparatus for the Detection and Registration of Electric Vibrations" (translation), *Zhurnal Russkogo fiziko-khimicheskogo obshchestva* 28, 1896.
66. "Questions to be submitted to counsel concerning the claims in Marconi's British specification No. 12309 of 1896. Prepared by Dr. J.A. Fleming, F.R.S," n.d. [November 6, 1899?], OX 416.
67. Betts, Betts, Sheffield & Betts, New York. "Opinion in re validity of Marconi Patents Nos. 586,193, 624,516 and 627,650," January 2, 1900, OX 530; and Carpmael & Co., untitled, March 14, 1900, OX 530.
68. A. Popov, letter to the editor, *Electrician* 40 (December 10, 1897), 235. Mentioned by Bailey and Landecker 1947 and Gervasi 1973; Belrose 2006, 349–420; D. Marconi 1962, 169. The one time Marconi contested the claim of Popov's priority was indirectly, in an acidic rebuttal to a new edition of Lodge's *The Work of Hertz* in 1899 that referred to Popov's "telegraphic application of 1895" (G. Marconi, "Comparison between the 1st and 3rd edition of 'The Work of

Hertz' by Professor Lodge F.R.S.," carbon typescript, February 15, 1900, OX 416). A raft of articles for and against the claims to Popov's priority over Marconi are assembled in "Additional notes" to S.T. Cope, undated typescript, "Notes on the history of the first twenty-five years of the Marconi Company," in-house document, OX 1838.

69. Seifer 1996, ix. Tesla rarely mentioned Marconi by name, but referred to him and his ilk indirectly and disparagingly. In his autobiography, Tesla characterized those who sought to thwart his efforts as "nothing more than microbes of a nasty disease" (Tesla 2005, 80. Originally published in *Electrical Experimenter* [February–July 1919]).

70. Seifer 1996, 88.

71. Ibid., 107–9. See also T.C. Martin, "Tesla's Oscillatory and Other Inventions," *Century* 49, no. 6 (April 1895): 916–33, quoting his Philadelphia lecture of February 1893, in which he expressed confidence in the possibility of wireless telegraphy.

72. Seifer 1996, 142, 183.

73. Ibid., 172.

74. G. Marconi 1899, 79, OX 48.

75. On this basis, one of the most important contemporary proponents of this view, Sungook Hong, states unequivocally and convincingly that "any claim for Lodge's priority is incorrect" (2001, 26). Hugh Aitken, author of the authoritative *Syntony and Spark*, on the other hand, comes down on the side of Lodge.

76. Guagnini 2002, 192; see Lodge-Righi correspondence cited in chapter 4, and GM to WP, November 16, 1897, OX 128/79; Oliver Joseph Lodge, "Improvement in Synchronized Telegraphy without Wires, Syntonic Wireless," UK patent 11575/97, May 10, 1897. In 1911, Lodge sold his patents and became a scientific advisor to the Marconi company (see chapter 4); in 1925 he nominated Marconi for membership in Britain's Royal Society (see Lodge-GM correspondence, OX 28); and in 1931 Marconi warmly invited Lodge to a conference he was organizing in Italy (see correspondence in LIN 18/2).

77. Hong 2001, 22–23.

78. J.A. Fleming, "A few notes on the method of presenting the Marconi case," typescript, April 26, 1904, OX 1756. Marconi won the groundbreaking case, which is described in chapter 13. Fleming changed his tune after an acrimonious severance from Marconi and his company in 1931. In a memorial lecture following Marconi's death, Fleming (1937) revised the position he had defended for decades and said that priority for the conception of wireless rightly belonged to Lodge. See Hong 2001, 26–28.

CHAPTER 3

1. The exact date has been reported by various sources as late January 1896; February 2, 1996; February 12, 1996; or simply February 1896. A 1922 attestation

from the Italian embassy in London in the archives of the Italian Senate states that Marconi established residency in London at the end of the first month of 1896 (ASS, ACG 279/16bis/15). However, Marconi's Italian passport (http://www.radiomarconi.com/marconi/gm_docum.html) bears the date of issue February 10, 1896, arguing for the later departure date—and indicating that Marconi's initial rise to prominence in England was even more meteoric than had been thought. (According to this website, Marconi only requested his passport on January 24, 1896, stating that he would return to Italy to do his military service.) The story of how Marconi's instruments were damaged by customs officers zealously looking for terrorist bombs has been told many times, for example by Weightman 2003. See "In the Witness-Box. Wireless Telegraphy. Interview with Signor Marconi," *Sunday Times*, October 9, 1898, for a contemporary account.

2. Cecil 1969, 9, 145.
3. Ibid., 194.
4. Ibid., 183.
5. The Marconis had another address in the area, 21 Burlington Road, which also no longer exists.
6. HJD to GM, September 27, 1912, OX 581.
7. de Sousa 1919, 25–6, OX 55.
8. GM to GIU, March 4, 1896, HWL, copy in OX 1849. In the mid-1990s, a Texas collector, Henry Willard Lende, acquired thirteen letters written by Marconi to his father between March 11, 1896, and September 7, 1898, that had been preserved by the Franceschini family, former caretakers at Villa Griffone. The Lende collection, in San Antonio, Texas, is the most complete source for Marconi's activities and state of mind during his first year in England. Transcribed copies of the letters were given to the Accademia dei Lincei in Rome, and an English translation was made by the Marconi company. See OX 1849; LIN 35/27. Citations are from the English version. The letter cited here was acquired later by Lende, not as part of the original collection.
9. GM to GIU, March 4, 1896, HWL, copy in OX 1849. Marconi also showed his interest in politics in this letter with a comment on the news of the terrible Italian losses at the Battle of Adowa (Ethiopia) a few days before: "I hope that they will once and for all decide to abandon that country which in everyone's opinion will be the ruin of Italy" (see chapter 33).
10. Guagnini 2002, 173.
11. *Improvements in telegraphy and in apparatus therefor*. Published in "Provisional Patents, 1896," *Electrical Engineer*, March 13, 1896, 308. Crucially, under Section 4.1 of the Patents, Design and Trade Marks Act, 1883, "Any person, whether a British subject or not, may make an application for a patent."
12. The original notebook is in OX 127.
13. Guagnini 2002.
14. Aitken 1976.

15. Hong 2001, 53; Guagnini 2006.

16. G. Marconi, "Improvements in transmitting electrical impulses and signals and in apparatus therefor." UK Patent No. 12039/96, OX 411, OX 414. The provisional specification was filed June 2, 1896, the complete specification on March 2, 1897, and the patent was granted on July 2, 1897.

17. Fahie 1899, xi (Fahie includes in an appendix the full text of the patent with diagrams).

18. Solari 2011, 33.

19. Hong 2001, 22.

20. Aitken 1976, 211.

21. This, one of the iconic documents of Marconi historiography, is frequently quoted in full in the standard biographies. The original is in the British Postal Museum & Archive in London, and there are early photostatic copies in OX 1774 and other collections.

22. GM to GIU, April 1, 1896, HWL 3, OX 1849. The standard Marconi literature is confusing on this point, with some authors stating that the first meeting with Preece only took place in July, as later claimed by Marconi's first assistant, George Kemp. However, Marconi mentions Preece in at least three earlier letters to his father, on April 1, 1896, April 8, 1896, and May 6, 1896 (HWL 3, 4, and 5; OX 1849).

23. Aitken 1976.

24. "Account by P.R. Mullis Esq. of his early recollection of Signor Marconi in 1896," 1940, OX 1849. This is one of the sharpest indications we have of the importance of the early experiments at the GPO for the perfection and development of Marconi's embryonic system. A truncated version of Mullis's report was published in the *Post Office Electrical Engineers Journal,* vol. 37, 1944–45. Mullis's report is the only detailed account of Marconi's first meeting with Preece.

25. Undated document on stationery of Urquhart & Small, Consulting Engineers, London, OX 168.

26. GM to GIU, April 1 and 8, 1896, HWL 3 and 4; also in OX 1849.

27. GM to GIU, April 8, 1896, HWL 3 and 4; OX 1849; GM to GIU, May 12, 1896, HWL 6, OX 1849.

28. B. Valotti, interview with author, February 24, 2015.

29. GM to Secretary of State for War, May 30, 1896, Public Record Office, London. Cited in Valotti 1994–95, 101.

30. Inspector General of Ordnance, War Office, to GM, June 15, 1896, OX 180.

31. One of Marconi's biographers, Giancarlo Masini, suggests mischievously that Preece may have appointed Kemp "at least in part, to keep an eye on Marconi and report what he was doing" (Masini 1995, 74–75).

32. D. Hay (Archives Manager, British Telecom) to J. Greenwood (Operator Services, Cheapside House), January 12, 1989, OX 1774; Solari 2011, 35.

33. G. Kemp, Experimental Room Notebooks, 1896, OX 112/91–93; OX 113/2–14.

34. Sarkar et al 2006.

35. H.B. Jackson to GM, September 15, 1896, OX 129/1; H.B. Jackson to GM, September 17, 1896, OX 180.
36. WP to J.C. Lamb, September 20, 1896, OX 129/43–45.
37. In the interim, Kemp assisted Marconi with experiments at Cardiff in November 1896, and again at Salisbury in March 1897. See Kemp experimental notebooks, OX 114; GM to WP, n.d., OX 127/49; GM to WP, OX 128/45; GM to WP, March 18, 1897, MB.
38. Jacot and Collier 1935, 233.
39. Catalogue of the Marconi Archives, Bodleian Library, University of Oxford, 2008, section A.2, and OX 684. The original of the complete diary as well as a series of "expanded extracts" Kemp compiled around 1930 are in OX 56–88 and OX 89–111. The company's unofficial corporate memory was rather unkind to Kemp. Publicity Manager R.P. Raikes told Degna Marconi in 1954: "He was a man of fanatical fidelity and a prodigious worker, but it seems very doubtful that he was ever a closer confidante of your father than would be, shall we say, a favourite horse" (R.P. Raikes to DM, October 25, 1954, OX 1771).
40. Solari 2011, 35, translation in D. Marconi 1962, 46.
41. "The British Association," LT, September 23, 1896. Also, "Telegraphy without Wires," Electrical Engineer (London), September 25, 1896, and "New Kind of Submarine Cable," New York Tribune, November 22, 1896.
42. GM to WP, December 5, 1896, OX 128/23–24.
43. "Telegraphy without Wires," St. Martin's-Le-Grand, January 1897. The lecture was reported in the Westminster Gazette, December 12, 1896; the Daily Chronicle and other London newspapers, December 14, 1896; the Manchester Guardian, December 15, 1896; the Electrical Engineer, December 18, 1896; the Weekly Irish Times, December 26, 1896; in the European continental press, and, in the United States, in the Chicago Daily Tribune, January 4, 1897; San Francisco Chronicle, January 23, 1897; and Los Angeles Times, January 24, 1897. The thrust of these reports, which all marvelled at Marconi's youth and apparent ingenuity, was summed up in Scientific American on January 23, 1897 ("Telegraphy without Wires"), reporting on Preece's lecture and introduction of Marconi and his "invention which promises to be of the greatest practical value in the world of telegraphy."
44. "Telegraphing without wires," Westminster Gazette, December 12, 1896.
45. GM to WP, December 13, 1896, OX 128/25/27.
46. GM to GIU, January 9, 1897, HWL 7, OX 1849.
47. Gervasi 1973, 280–86. Gervasi discovered the letter (GM to A. Ferrero, December 20, 1896) in the archives of the Italian Ministry of Foreign Affairs in Rome (ASMAE, AL, "Londra 1896," b.227). In this letter, Marconi recalls having informed Ferrero some ten months earlier (i.e., February 1896) "that I had discovered a system which permits telegraphy between two locations without the need for conductor wires." Ferrero replied on January 4, 1897, inviting Marconi to lunch at the embassy. (Ferrero's reply is misdated January 4, 1896, an obvious

error as he refers explicitly to Marconi's of December 20, 1896, and Marconi only arrived in England in February 1896.)

48. GM to GIU, January 9, 1897, HWL 7, OX 1849; GM to F. Wynne, n.d., OX 127/47.

49. GM to GIU, January 20, 1897, HWL 8, OX 1849.

50. "To Guglielmo Marconi re Sale of your patent," statement for January–July 1897, Morten, Cutler & Co., 99 Newgate Street E.C., OX 168. Except as otherwise noted, references in this section are to this document.

51. Marconi was also in close contact with Preece at this time, see OX 128/5, January 28, 1897.

52. F.H. Bowman, Consulting Engineer, London, "Report on Marconi's Receiver," February 27, 1897, OX 530, cited in the draft prospectus (OX 168). Marconi considered this report important enough to include it in a list of reference documents that he compiled for his 1904 US court case.

53. The opinions of Edward Carpmael (March 5, 1897), J. Fletcher Moulton (March 6, 1897), and J.C. Graham (March 8, 1897) are in OX 530. See also Carpmael report, February 13, 1900, OX 530.

54. Fleming would become the first prominent British scientist to publicly endorse Marconi's work. He had been associated with the introduction of the incandescent lamp to Britain, as an advisor to the Edison and Swan Electric Lighting Company. In 1879, also acting for Edison, he had fought the contention of the British Post Office that the telephone was a telegraph and thus fell under the scope of existing legislation. Fleming worked for the Marconi Company from 1900 to 1931, notably providing expertise in numerous patent litigations.

55. As another example of interconnections in the small world of late Victorian politics, business, and technology, Hozier's daughter Clementine later married Winston Churchill, who was born the same year as Marconi.

56. GM to WP, March 18, 1897, MB, discussing details of the latest experiments, is brilliant testimony to Marconi's composure and coolness of mind and to the ability to compartmentalize that was one of his great strengths. There is no hint in this letter of the ongoing tension surrounding his financial future and the future of the invention.

57. AJM to GM, March 20, 1897, RI 12.

58. J.C. Graham to WP, April 9, 1897, OX 168.

59. GM to WP, April 10, 1897, OX 168.

60. HJD to Morten et al, April 14, 1897, OX 168.

61. J.C. Graham to GM, April 19, 1897, OX 168.

62. J.P. Smith to HJD, April 26, 1897 (emphasis in original), OX 168. The brother referred to, William Smith, became one of the first directors of the company.

63. "Wire-less Telegraphy. Government trials. Inventor's explanation," *Daily Mail*, April 26, 1897.

64. This was Davis's usual salutation and signature in letters to his cousin. Marconi, typically adapting his tone and level of discourse to whomever he was speaking,

addressed and signed his letters to Davis the same way. HJD to GM, April 27, 1897, OX 168. There is a more intense, unsigned, and undated version of this note, or a note expressing same, in OX 168 (see "82 M L" to GM).

65. These last details are not actually mentioned in Marconi's note dated April 30, 1897, but were part of the deal and appear in all subsequent documents, as well as in Marconi's letters to Giuseppe.

66. GM to HJD, April 30, 1897, OX 168. The list nicely mirrors British colonial and trade interests of the time. To be taken out at once: Argentine Republic, Malta, Transvaal, Egypt, Queensland, Natal, Orange Free State, Tasmania, New Zealand, Switzerland, Brazil, New South Wales, Japan, Norway, South Australia, Newfoundland, Sweden, Western Australia; and later on: Gibraltar, Ceylon, British North Borneo, Victoria, Canada, Hong Kong, Zululand, Straits Settlements.

67. Davis memo to file, May 4, 1897, OX 168.

68. *NYT,* May 26, 1897.

69. G. Kemp, Diaries, exp. extracts, OX 89; see also OX 1775.

70. G. Marconi 1923, 1, OX 52.

71. W.H. Preece, "Signalling through Space without Wires," Friday evening discourse delivered before the Royal Institution, June 4, 1897; see "Signalling through Space," *LT,* June 7, 1897; "Telegraphy without Wires," *LT,* June 11, 1897.

72. Italian Naval Attaché Bianco to GM, May 21, 1897, and June 1, 1897, OX 212. See also Solari 2011, 38.

73. J.F. Moulton, "Opinion," May 21, 1897, OX 530.

74. GM to C. Maynard Owen, power of attorney, June 24, 1897, OX 168.

75. GM to C. Maynard Owen, June 24, 1897, OX 168.

76. O. Lodge, "Telegraphy without Wires," *LT,* June 22, 1897.

77. A. Righi to O. Lodge, June 18, 1897 (Lodge Collection, University College, London, cited in Valotti 1994–95, 50). The interview with Righi appeared in *Il Resto del Carlino* (Bologna), May 28, 1897; O. Lodge to A. Righi, June 21, 1897, Archives of the Accademia Nazionale delle Science detta dei XL—Fondo Righi, cited in Bettòlo 1986, 448; A. Righi to O. Lodge, June 25, 1897, cited in Valotti 1995 and Valotti 1994–95, 54.

78. GM to GIU, June 29, 1897 [misdated June 29, 1899], LIN 35/27. Some details related here are taken from press reports and Marconi's own retrospective accounts.

79. "*Esperimenti del telegrafo senza fili al Ministero della marina,*" *Il Resto del Carlino* (Bologna), July 3, 1897. Marconi set up a transmitter on the third floor of the ministry's headquarters, a receiver on the first floor, and telegraphed the words "*Ministero della marina*" some four hundred times. Marconi's Italian visit received blanket coverage in the Italian press, with at least twenty-five articles appearing in the main Rome, Milan, and Bologna newspapers in July 1897.

80. See, for example, "Copy of extract from the '*Morning Leader*' of July 7th, 1897. The New Telegraphy. Rome, 6th July," typescript on letterhead of H. Jameson

Davis, London, OX 133/4: "Signor Marconi to-day gave a series of experiments illustrating the working of his apparatus for telegraphing without wires before the King and Queen at the Quirinal. Their Majesties heartily congratulated the inventor, and were much interested to hear of the support which he had received in England." Note that he is already simply "Signor Marconi" to the British press.

81. W. J. Baker 1970, 173.

82. HJD to GM, July 9, 1897, OX 168; HJD to GM, n.d., OX 168. The first page of this letter has been lost so the date is unknown, but it appears to have been written later than the one dated July 9, 1897. The initial directors of the company were James Fitzgerald Bannatyne, William Smith, Edgar Appleby (chair), Davis, and Marconi (OX 392).

83. G. Marconi 1923, 1, OX 52.

84. AJM and GIU to GM, July 7, 1897, RI 9, copy (with translation) in OX 1849. Giuseppe wrote in Italian, Annie in English. See also AJM to GM, "Sunday" (July 11, 1897) and "July 1897," RI 10 and 11.

85. See "*Documenti rari su Guglielmo Marconi*," Comitato Guglielmo Marconi International, www.radiomarconi.com/marconi/gm_docum.html.

86. Solari 2011.

87. D. Marconi 1962, 48; de Sousa 1919, OX 55.

88. HJD to GM, July 21, 1897, OX 168.

89. "*Guglielmo Marconi a Bologna*," Il Resto del Carlino, July 25, 1897; "*Per Guglielmo Marconi*," Il Resto del Carlino, July 28, 1897.

90. WTSC directors' meeting minutes, July 22, 1897, OX 393.

91. Seven of the nine original subscribers were Irish "corn factors" (grain merchants), see OX 392, OX 682. The other two were James Fitzgerald Bannatyne of Limerick, whose occupation was listed as "Gentleman," and Davis.

92. HJD to investors, "Re Wireless Telegraph & Signal Company, Ltd.," July 24, 1897, OX 168.

93. GM to WP, July 27, 1897, OX 168.

94. WP to GM, August 6, 1897, OX 168.

CHAPTER 4

1. *Neue Freie Presse* (evening edition), December 17, 1896.

2. *Il Resto del Carlino*, December 22, 1896. The report noted that the inventor was the half-brother of the former Bologna city council member, Luigi Marconi. Luigi was elected to the Bologna municipal council in May 1891 on a liberal-conservative ticket (Giacomelli and Bertocchi 1994).

3. *La Tribuna*, December 27, 1896.

4. *Chicago Daily Tribune*, February 8, 1897.

5. Dam 1897.

6. The term is from Douglas 1987.

7. LIN 22/28. The copy contains the hand notation: "Mr Dam told me that Marconi himself corrected the proofs of this article."

8. *Electrical Review,* July 16, 1897; *Daily Chronicle,* July 27, 1897; *Globe,* July 27, 1897; *Morning,* July 28, 1897; *Daily Mail,* July 29, 1897.

9. *NYT,* May 26, 1897; *New York Herald,* July 4, 1897; *New York World,* August 3, 1897; *Press* (Philadelphia), August 4, 1897; *New York Herald,* August 8, 1897; *New York Sun,* September 5, 1897.

10. The labels have stuck. See recent biographies of Edison (Stross 2007) and Tesla (Seifer 1996).

11. Brandon 2003; Sandford 2011; Rapaport 2010.

12. The term was coined by Society for Psychical Research secretary Frederick Myers (J. Gray 2011).

13. Dam 1897.

14. J. Gray 2011.

15. Doyle's troubled friendship with Houdini and thwarted effort to contact Marconi in the 1920s are discussed in chapter 27.

16. Haynes 1982, xiii. The Society for Psychical Research's goal today is not far off the original: "The scientific investigation of the ways that organisms communicate and interact with each other and with the environment, that appear to be inexplicable within current scientific models" ("Overview of Psychical Research," Society for Psychical Research website, www.spr.ac.uk/main/page/overview-psychical-research-parapsychology).

17. *LT,* September 8, 1898.

18. W. Crookes, "Presidential Address," *Electrical Engineer,* September 9, 1898, 327–32.

19. W. Lynd, "What Is a Brain Wave? Facts about Mind-Reading," *Christian World,* September 22, 1898.

20. GM to GIU, HWL 9, August 8, 1897, OX 1849; GM to GIU, August 14, 1897, cited in Masini 1995, 106, as well as Solari 1942.

21. WP to J. Gavey, August 24, 1897, OX 128 /49 and 54.

22. For example, in January 1898 Preece was talking about giving Marconi a contract to fit up the *Goodwin* lightship (January 1, 1898, OX 128 /11); a few months later, on May 3, the *Daily News* (London) reported that Marconi had severed his connection with the GPO. And so it went. Marconi later claimed that in September 1899 the company offered the Post Office the use of its patents for the neat sum of £50,000 a year, but the offer was rejected (Minutes of Evidence from the Select Committee on Marconi's Wireless Telegraph Company Limited, Agreement (1913), 509, May 7, 1913).

23. WTSC directors' meeting minutes, August 18, 1897, OX 393.

24. "Drawings of early equipment, 1897–8," OX 365. Eleven of these drawings, some of them more than six square feet in size, are preserved in OX 365. Seven of the drawings are of oscillators (the key component of Marconi's wireless transmitter) in various arrangements, and one depicts the coherer, Marconi's earliest wireless receiver.

25. WTSC directors' meeting minutes, August 18, 1897, OX 393.

26. Ibid.

27. H.M. Hozier to GM, August 20, 1897, OX 248.

28. WTSC directors' meeting minutes, September 8, 1897, OX 393.

29. GM to WP, September 9, 1897, OX 128; and September 12, 1897, MB (copy in OX 128); WTSC directors' meeting minutes, September 8, 1897, OX 393. Baden Baden-Powell, who became a friend as well as supporter of Marconi, is considered a pioneer of military aviation. He was the youngest brother of Major Robert Baden-Powell, later famous as the defender of Mafeking in the second Anglo-Boer War and founder of the international boy scout movement.

30. H.W. Allen, cited in "Death of Marconi's Collaborator," MWTC press release, December 29, 1936, OX 682.

31. WTSC directors' meeting minutes, September 30, 1897, OX 393.

32. *Investors Review*, October 7, 1897, 484, cited in Jome 1925, 21 (page 2 of typescript in OX 581).

33. *Sunday Call* (Newark), October 3, 1897, OX 1434.

34. *Daily Chronicle*, September 20, 1897, 36, OX 1434.

35. WTSC directors' meeting minutes, November 18, 1897, OX 393.

36. The station operated until May 26, 1900, and was then dismantled and moved to a stone cottage at Niton, Isle of Wight, which became a permanent station and was later taken over by the GPO.

37. G. Marconi 1923, 2, OX 52.

38. *McClure's* 13 (June 1899): 99–112.

39. G. Marconi 1923, 2–3, OX 52; also, "Some Facts about Wireless Telegraphy," *Electrical Review* 42, no. 1065 (April 22, 1898): 538.

40. "Notes by Kemp on Experimental Transmissions from a Tug in Alum Bay to The Needles Station, Dec. 1897," OX 134.

41. GM to GIU, HWL 11, December 24, 1897 (author translation of Italian transcript in LIN 35).

42. GM to GIU, HWL 10, December 1, 1897, OX 1849.

43. GM to GIU, HWL 11, December 24, 1897, LIN 35.

44. GM to GIU, HWL 12, February 15, 1898, OX 1849.

45. HJD to GM, January 27, 1898, OX 581.

46. WTSC directors' meeting minutes, February 19, 1898, OX 393.

47. "Agreement as to Loan and Option re MWTC Ltd," August 8, 1901, OX 581.

48. W.J. Baker 1970, 37.

49. AM to GM, May 17, 1898, LIN 37.

50. Rowe 2009.

51. G. Isted, draft manuscript, 1981, OX 3991; repeated by D. Marconi 1962, 45, and many other sources.

52. WTSC directors' meeting minutes, July 26 and August 9, 1898, OX 393.

53. "Demonstrations to Lord Kelvin and Lord Tennyson, with related items, 1898," OX 142.

54. AM to GM, June 13, 1898, LIN 37.

55. "The Eastern Telegraph Companies and Their Charges," *LT*, July 21, 1899.

56. HJD to First Lord of the Admiralty, May 9, 1898, OX 245.

57. "On Board the 'Flying Huntress,'" and "With Marconi in Dublin Bay," *Daily Express*, July 21, 1898; both in *Wireless Telegraphy and Journalism*, commemorative booklet published by the Dublin *Daily Express* (OX 147). See also G.F. Fitzgerald, "The Meaning and Possibilities of Wireless Telegraphy," *Dublin Evening Mail*, July 21, 1898.

58. WTSC "Confidential Circular. Wireless Telegraphy," New York, October 1899, OX 180.

59. G. Marconi 1899, 132, OX 48.

60. "Marconi Telegraphy," *Electrical Review* 43, no. 1082 (August 19, 1898): 264-65.

61. "Wireless Telegraphy at Cowes," *Daily Express* (Dublin), August 3, 1898; GM to GIU, September 7, 1898, HWL 13, OX 1849.

62. G. Kemp, Diaries, February 10, 1902, OX 58.

63. "Haven History," hotel handout, 2011.

64. "The Cradle of Wireless," *Bournemouth Echo*, January 16, 1926; "How We Made the First Radio Signals. Senatore Marconi's Assistant Looks Back," *Evening News*, January 16, 1926.

65. W.J. Baker 1970, 43.

66. Aitken 1976, 192.

67. G. Marconi 1923, 4, OX 52.

68. Ibid. Marconi's early GPO experiments were with waves about 120 centimetres in length (W.H. Preece, "Signalling through Space Without Wires," Friday evening discourse delivered before the Royal Institution, June 4, 1897).

69. GM to GIU, September 7, 1898, HWL 13, OX 1849. See also other correspondence from this period (GM to GIU, December 23, 1898; GM to GIU, March 5, 1899; GM to GIU, April 3, 1899), LIN 37.

70. GM to AJM, n.d. [1898], OX 17.

71. AJM to GM, October 8, 15 and 17, 1898, RI 13-14, 15, and 16.

72. Twelve letters from AM to GM, April 5-24, 1899, and ten letters from AJM to GM, April 28, 1898-June 28, 1899, all in LIN 37. Cited here are AM letters dated May 15 and 17, and June 13 and 23, 1898; and AJM letters dated April 29, 1898; March 22, 1899; April 12, 1899; and June 14 and 28, 1899.

73. "London Letter. Shipping, Finance, Commerce," *Journal of Commerce* (Liverpool), October 8, 1898. Officially, it was the company's second general meeting; a *pro forma* first one had been held on November 18, 1897, a few months after incorporation.

74. "The Wireless Telegraph & Signal Company, Limited," October 7, 1898, OX 400. The typescript of Davis's report includes edits in Marconi's hand.

75. O.J. Lodge, "Improvement in Synchronized Telegraphy without Wires, Syntonic Wireless," UK patent 11575/97, May 10, 1897.

76. WTSC directors' meeting minutes, July 26, 1898, OX 393. See also correspondence between HJD and Lodge, September to October 1898, OX 416, where Davis says that notwithstanding their failure to arrive at a deal, the company would like to employ Lodge as a consultant.

77. "Patents originally granted to Lodge and Muirhead, together with a deed of assignment of patents to the company, 1897–1912," 1911, OX 412.

78. Slaby 1898.

79. For example, Kittler 1986.

80. Extract from Slaby pamphlet, November 1897, OX 417. Also cited in "Some Facts about Wireless Telegraphy," *Electrical Review* 42, no. 1065 (April 22, 1898): 537.

81. Slaby 1898.

82. AS to GM, May 20, 1898, OX 18/5.

83. AS to GM, August 14, 1898, OX 18/6.

84. GM to AJM, n.d. [1898], OX 17.

85. AS to GM, August 14, 1898, OX 18/6.

86. See especially AS to GM, October 13, 1898, OX 18/7.

87. AS to GM, November 6, 1898, OX 18/8.

88. Ibid. (emphasis in original).

89. AS to GM, January 31, 1899, OX18/9.

90. AS to GM, February 9, 1899, OX 18/10.

91. Friedenwald 2000, 441–63; Dunlap 1937, 91.

92. Friedenwald 2000.

93. WTSC directors' meeting minutes, September 30, 1897, OX 393.

94. WTSC directors' meeting minutes, March 31, 1898, OX 393.

95. Rollo and Queiroz 2007.

96. A.J. Young, English Electric Valve Co., Chelmsford, to R.J. Bell, Historian, Publicity Dept., Marconi House, Chelmsford, June 15, 1965, "Early Marconi History," OX 1773.

97. G. Marconi 1923, 5–6, OX 52.

98. "Signor Marconi on Wireless Telegraphy," *LT,* June 3, 1899; G. Marconi, "Wireless Telegraphy," *Electrical Engineer,* March 24, 1899.

99. de Sousa 1919, OX 55.

100. *LT,* March 29, 1899; *NYT,* April 9, 1899.

101. A. Baden-Powell to GM, n.d., OX 153.

102. WP to GM, April 8, 1899, cited in G. Marconi 1913, OX 54.

103. W.W. Goodbody, WTSC "Confidential Circular. Wireless Telegraphy," New York, October 1899, OX 180.

104. Ibid.

105. JAF to HJD, August 19, 1899, OX 153.

106. W.J. Baker 1970, 48. See also J.A. Fleming, "The Centenary of the Electric Current. 1799–1899," *Electrician,* September 22, 1899, 764–68 (abstract of a lecture delivered to the British Association, Dover, September 18, 1899), OX 153.

107. C. Moffett, "Marconi's Wireless Telegraph," *McClure's* 13 (June 1899): 99–112. All references to the Moffett article in the following paragraphs are taken from this source.
108. "Court Circular," *LT,* June 10, 1899.
109. Rittenberg 1914, 9, 11, and 41.
110. Cecil 1969.
111. JAF to GM, August 23, 1899, OX 153; *Daily Mail,* reported in *NYT,* August 21, 1899.

CHAPTER 5

1. de Sousa 1919, 100–1, OX 55.
2. T. Edison to J. McMillin, May 12, 1899, TEP.
3. Alfred Stieglitz Collection, The Metropolitan Museum of Art, New York.
4. "The Hoffman House Sold for $3,500,000," *NYT,* February 24, 1915.
5. "Art: Tales of the Hoffman House," *Time,* January 25, 1943.
6. "The Hoffman House sold for $3,500,000," *NYT,* February 24, 1915.
7. W. Densham, Diaries, September 21 to November 15, 1899, OX 4007.
8. Douglas 1987, 9. Marconi describes the deal and his experience in G. Marconi 1913, OX 54, which also includes extensive press citations. Here he says the *Herald* paid the expenses of him and his assistants.
9. Howeth 1963, chapter III. Howeth's study, considered the authoritative US source for early radio history, includes much primary source detail unavailable elsewhere.
10. de Sousa 1919, 104, OX 55.
11. "Message from Parade to Navesink without Wire," *New York Herald,* September 30, 1899, OX 154/30.
12. Ibid.
13. *NYT,* October 4, 1899.
14. *New York Herald,* October 4, 1899, OX 154/31.
15. W. Densham, Diaries, October 3, 1899, OX 4007.
16. Howeth 1963, chapter III.
17. de Sousa 1919, 108, OX 55.
18. *NYT,* October 17, 1899.
19. Howeth 1963, chapter III.
20. "Report of the Chief Signal Officer to the Secretary of War," Washington, for the year-end June 30, 1899, OX 154/1 (the report was actually dated September 30, 1899, the first day of Marconi's New York experiments).
21. "Not to Use Marconi's System. US Army Will Cling to Its Own Wireless Telegraphy," *NYT,* November 5, 1899.
22. Marconi Wireless Telegraph and Signal Co., Ltd. to Lt. Comdr. J. C. Colwell, USN, Bureau of Equipment, September 2, 1899, in NARA, cited in Howeth 1963.

23. Secretary of the Navy to Lt. Comdr. John T. Newton, October 23, 1899, Bureau of Equipment, NARA, cited in Howeth 1963.

24. "Topics of the Times," *NYT*, October 29, 1899.

25. "Wireless Telegraphy Tests," *NYT*, November 2, 1899.

26. W.W. Goodbody, WTSC "Confidential Circular. Wireless Telegraphy," New York, October 1899, OX 180.

27. Howeth 1963, citing document in the GCR (SRM 100, p. 2).

28. GM to "Wireless Board," October 29, 1899, cited in Howeth 1963, chapter III.

29. GM to A.G. Bell, November 6, 1899, AGBFP, Box 132. Marconi privately stated a few weeks later that the government wanted him to go to South Africa but that he was not keen to do so; in fact, he never went (GM to JBH, November 25, 1899, HUN 1/1). Bell invited Marconi to visit him many times over the years, but Marconi never did.

30. "Wireless Telegraphy in War," professional notes, The Proceedings of the United States Naval Institute 26 no. 3 (September 1900): 535. Cited in Howeth 1963.

31. See B.A. Austin, "Wireless in the Boer War," in *International Conference on 100 Years of Radio, IEE Conference Publication No. 411*, September 5–7, 1995, Savoy Place, London, 1995; D.C. Baker 2006, 421–54; Tim Wander, "Wireless Goes to War—The First Time," *Practical Wireless* (May 1989): 34–36, OX 1809; and Sterling 2007, 65–66.

32. C.K. van Trotsenburg to Messrs Siemens Bros & Co, February 28, 1898. Except as otherwise indicated, the correspondence cited in this section is in NARS and cited in Austin 1995, in D.C. Baker 2006, or in OX 1809.

33. C.K. van Trotsenburg to L.W.J. Leyds, March 2, 1898, OX 1809, translation provided in Austin 1995.

34. A. Siemens to C.K. van Trotsenburg, March 26, 1898, unreferenced article, OX 1809.

35. L.W.J. Leyds to C.K. van Trotsenburg, April 20, 1898, NARS, cited in D.C. Baker 2006, 429; A. Siemens to C.K. van Trotsenburg, June 11, 1898, OX 1809.

36. D. C. Baker 2006, 436.

37. HJD to van C.K. Trotsenburg, July 1, 1899, OX 1809.

38. Unsigned memo in NARS, August 24, 1899, cited in D.C. Baker 2006, 440.

39. D.C. Baker 2006, 442–43.

40. WTSC report for year ending August 31, 1899 (dated December 19, 1899), OX 401.

41. GM to JBH, November 25, 1899, HUN 1/1.

42. G.L. Bullocke to H.W. Allen, December 17, 1899, OX 334.

43. Lecture, Royal Institution, London, February 2, 1900. The chair of the meeting was none other than Alexander Siemens (*LT*, February 5, 1900).

44. de Sousa 1919, 121, OX 55; GM to JBH, February 14, 1900, HUN 1/8.

45. Lecture, Royal Institution, London, February 2, 1900. Marconi was still considering the possibility of going to South Africa himself at this time (GM to JBH, February 14, 1900, HUN 1/8).

46. The Company's South African contract was announced by managing director Samuel Flood-Page in a letter to the *Times*, July 18, 1900 (Satia 2010).
47. D.C. Baker 2006, 450.
48. Howeth 1963.
49. Ibid.
50. Henry S. Saunders to Frederick Hamilton, August 21, 1899, OX 180.
51. W.W. Goodbody, WTSC "Confidential Circular. Wireless Telegraphy," New York, October 1899, OX 180.
52. Douglas 1987, 64–65.
53. Howeth 1963, chapter II, n.28. Both the *NYT* and *New York Herald* reported the company formation on November 23, 1899.
54. *Transatlantic Times* 1, no.1, November 15, 1899, OX 250.

CHAPTER 6

1. D.E. Holman 1909, xxvii. Biographical details composed from various newspaper sources in April and December 1901. See also, D. Bodenhamer and R. Barrows, eds., *The Encyclopedia of Indianapolis* (Indiana University Press, 1994).
2. According to US Census records for 1880, Josephine was born in 1875, but later census data give the date as 1874. Her birthday was in October (JBH to GM, October 26, 1900, OX 18/27).
3. "Marconi and His Fiancée. Chance Brought about Their Meeting. Miss Holman a Typical American Girl," undated newspaper clipping in HUN 2/40, provenance not indicated, but evidently from April or May 1901, as the couple's engagement is announced.
4. Ibid.
5. Marconi was still in touch with Sita Camperio in 1925, reminiscing about the old days in Livorno (GM to "Cara Sita," January 24, 1925, MB). She was the godmother of Marconi's son, Giulio, who was named after her late brother.
6. Solari 2011, 50.
7. Holman's letters to Marconi are in the Marconi Archives in Oxford; Marconi's to Holman and Holman's other documents are in the Huntington Library in San Marino, California.
8. GM to JBH, November 25, 1899, HUN 1/1.
9. JBH to GM, December 31, 1899, OX 20/23.
10. GM to JBH, January 15, 1900, HUN 1/6.
11. Masini 1995, 142–43. Annie and Giuseppe then went on to Dublin, and spent at least a month in Ireland before returning to Italy. Annie's letters detail Giuseppe's health problems during this trip; he has a cold; he has a cough, et cetera (AJM to GM, February 16, 1900, RI 31 [incorrectly dated February 16, 1901]; AJM to GM, March 26, 1900, RI 19; AJM to GM, April 20, 1900, RI 20–21).
12. G. Marconi and MWTC, "Improvements in apparatus for wireless telegraphy," UK Patent No. 7777/00, provisional specification filed April 26, 1900, complete

specification filed February 25, 1901, patent accepted April 13, 1901 (OX 414). The patent explicitly built on Marconi's original patent, 12039/96; the specification read: "The object of this invention is not only to increase the efficiency of the apparatus hitherto employed, but also to so control the action as to cause intelligible communications to be established with one or more stations only out of a group of several receiving stations."

13. See, for example, Marconi's testimony in a 1913 court case, where he states: "My Patent No. 7777 of 1900 played an all important part in the development of wireless telegraphy. It represents what was originally known as my syntonized system...." ("In the Central Criminal Court. Rex v. Cecil Chesterton. Proof of Commendatore G. Marconi," n.d. [May 1913], OX 224).

14. Bill of Complaint. City Court of the United States Southern District of New York. Lyman C. Learned v. Guglielmo Marconi, October 16, 1899, OX 532.

15. SFP to E.H. Moeran, March 23, 1900, OX 532.

16. G. Marconi testimony in Bill of Complaint. City Court of the United States Southern District of New York. Lyman C. Learned v. Guglielmo Marconi, n.d. [June 1900], OX 532; Learned v. Marconi, "Defendant's Record on Final Hearing," n.d., OX 532.

17. "Suit against Marconi dismissed," *New York Tribune*, March 23, 1901. The company immediately used this outcome, and the recent awarding of its German patents, to bolster the strength of its claims in the UK (SFP to "My Lord," April 19, 1901, OX 204).

18. "A few notes on No. 7777 of 1900 by J.A. Fleming," undated typescript, OX 416. Thought to be written in 1906 or 1908.

19. GM to JBH, December 20, 1899, HUN 1/3, and January 15, 1900, HUN 1/6.

20. R. Kipling, "Wireless," *Scribner's*, August 1902.

21. R. Kipling, "White Man's Burden," *McClure's*, February 12, 1899.

22. Morel coined the term in 1903; his book of the same title was published in 1920.

23. de Sousa 1919, 125, OX 55.

24. GM to JBH, February 9, 1900, HUN 1/7. The letter is dated in Marconi's hand, but evidently in error. According to Densham's diary (OX 4007), Marconi was in Brussels from March 7 to 12, 1900, and the demonstration took place on March 10. From the context, the letter appears to have been written on March 9. See also, "Various News Items. In Brussels the Inventor Marconi at the Palace of the King, Wireless Telegraph Experiments," typescript, translated extract from *Étoile Belge*, March 11, 1900, OX 166/21.

25. Hochschild 1998.

26. This paragraph based on Hochschild 1998.

27. "Treatment of Natives on the Congo," *LT*, May 13, 1897.

28. "The Congo Free State," *LT*, February 23, 1900.

29. "House of Commons," *LT*, February 28, 1900.

30. Hochschild 1998, 175.

31. "Various News Items," *Étoile Belge*, March 11, 1900. See also "Leopold and Wireless Telegraphy," *NYT*, March 12, 1900. Flood-Page reported the results of the successful meeting with Leopold to the board on March 14, 1900 (MWTC directors' meeting minutes, March 14, 1900, OX 393).

32. "Wireless Telegraphy," *LT*, November 5, 1900. See also "The Marconi Experiments on the 'Princess Clementine,'" *British Underwriting and Shipowners' Gazette*, no. 11 (December 1900), OX 249.

33. MWTC, Directors' Report for year ending September 30, 1900, dated December 8, 1900, OX 401.

34. HCH to GM, July 4, 1901, OX 175.

35. HCH to GM, June 28, 1901, July 3, 4 and 10, 1901, all in OX 175.

36. GM to SFP, October 25, 1901, OX 171.

37. "The Congo Free State," *LT*, May 23, 1903; "The Congo State," *LT*, July 29, 1903.

38. Hochschild 1998, 283.

39. Hawaii, an independent kingdom from 1801 to 1893, was then briefly a republic before being annexed by the US in 1898.

40. F.J. Cross to GM, July 5, 1899, OX 179.

41. "Wireless Telegraphy in Hawaii," *NYT*, July 16, 1899.

42. H.W. Allen to F.J. Cross, August 2, 1899, OX 179.

43. A. Gray to F.J. Cross, November 27, 1900, OX 179.

44. A. Gray to F.J. Cross, December 29, 1900, OX 179.

45. A. Gray to F.J. Cross, March 28, 1901, OX 179.

46. V. Pletts to SFP, n.d., OX 179.

47. "Marconi's System in Hawaiian Islands," *Los Angeles Times*, February 18, 1901.

48. Marconi's address to MWTC shareholders, 5th AGM, February 20, 1902, original typescript in OX 400; in *Electrical Review* 50, no. 1266 (February 28, 1902): 354–56. The subsequent edition (vol. 50, no. 1267, March 7, 1902) chastised Marconi for blaming the collapse of the Hawaiian operations on "ill-paid operators." The February 21 issue also carried the report of Marconi's speech, identical except that the offensive parenthetical phrase is absent. It also appears, however, in de Sousa 1919 (OX 55).

CHAPTER 7

1. GM to SFP, June 5, 1900, OX 532; MWTC directors' meeting minutes, July 2, 1900, OX 393.

2. GM to JBH, June 3, 1900, HUN 1/10.

3. "Signor Marconi's Visit," HUN 2/40.

4. GM to JBH, June 13, 1900 [misdated July 13, 1900], HUN 1/17. The envelope is postmarked June and it is obvious from the context that the letter was written in June—Marconi returned to England on June 20. Marconi was accompanied on this US trip by his assistant William Densham. From Densham's diary, it

appears that Marconi spent less than forty-eight hours at Cragsmoor; he had lunch with Densham in Manhattan on June 16 and again on June 18 (W. Densham, Diaries, OX 4007).

5. GM to JBH, June 18, 1900, HUN 1/12.
6. GM to JBH, June 20, 1900, HUN 1/13. The letter was probably posted from sea, on the last post brought back by an accompanying tugboat.
7. GM to JBH, June 21, 1900, HUN 1/14.
8. GM to JBH, June 21, 1900, HUN 1/14. The last bit was dated London, July 4, 1900, but was still part of the long shipboard letter. The reporters he refers to were from the *Irish Daily Independent* ("Signor Marconi Interviewed," June 28, 1900, HUN 2/40). Marconi told the paper his US trip was strictly pleasure.
9. GM to JBH, July 17, 1900, HUN 1/15.
10. GM to JBH, July 31, 1900, HUN 1/18. A year later he would be sending her condolences on the assassination of US president William McKinley (GM to JBH, September 17, 1901, HUN 1/47).
11. AJM to GM, August 9, 1900, RI 22–23; AJM to GM, August 31, 1900, RI 26–27.
12. AJM to GM, August 26, 1900, RI 24–25.
13. JBH to GM, October 26, 1900, OX 18/27.
14. GM to JBH, November 8, 1900, HUN 1/22.
15. JBH to GM, November 18, 1900, OX18/20.
16. JBH to GM, n.d., OX 18/40.
17. GM to JBH, December 20, 1900, HUN 1/24.
18. GM to JBH, December 3, 1900, HUN 1/23; AM to GM, November 11, 1900, RI 4; AJM to GIU, December 16 and 29, 1899, LIN 37; AJM to GIU, December 30, 1900, LIN 37. See also AJM to GIU, January 18 and 26, 1901, LIN 37; AJM to GIU, December 5, 1900, LIN 37.
19. HJD to members of the WTSC board, August 22, 1899, OX 581.
20. Philip Geeves, "The Switched-On Generation," *Newsletter of the Royal Australian Historical Society* (August 1974): 3–4, OX 683.
21. At first as a director, taking on the managing directorship two months later.
22. SFP report to GM, October 6, 1899, OX 417.
23. Slaby had most recently written to Marconi in February 1899.
24. SFP report to GM, October 6, 1899, OX 417.
25. E. Hoffmann to SFP, October 4, 1899; SFP to E. Hoffmann, October 4, 1899 (translated from German), OX 417.
26. SFP to E. Hoffmann, October 4, 1899 (translated from German), OX 417.
27. SFP report to GM, October 6, 1899, OX 417.
28. "Marconi's Wireless Telegraph Company (Limited)," *LT*, December 19, 1900.
29. SFP to GM, January 3, 1901, OX 417.
30. SFP to GM, January 16, 1901, OX 417.
31. Inspired by T.P. Hughes 1983.
32. W.R. Elliott to HJD, July 13, 1899, OX 180.

33. H. Saunders, weekly report to WTSC directors, August 25, 1899, OX 178.

34. MIMCC, notice of share offer (prospectus), June 18, 1900; MIMCC, *General orders*, February 15, 1902, OX 243.

35. "Imperial Telegraphic Communication," Report of the Debate of the House of Commons on Tuesday, May 22, 1900 (booklet), OX 218. See Great Britain, House of Commons, *Debates*, May 22, 1900, vol. 83, cc969–1,011. All references to the House debate are taken from this source.

36. The Liberal Unionist Party was formed in 1886 by a faction that broke away from the Liberal Party in opposition to that party's support for Irish home rule; it merged with the Conservative Party in 1912. Another MP who took up the cause of cheap cable rates was John Henniker Heaton. The reformers included men such as Sir Sandford Fleming, originator of the idea of standard time, and engineer Charles Bright. They advocated a state-owned cable system run as a public service (Winseck and Pike 2007). All of these men were associated in one way or another with Marconi.

37. A departmental committee on imperial telegraphic communication was established later in 1900, consisting of Balfour, Hanbury, the postmaster general, the undersecretaries for India and the colonies, and two members from the intelligence branches of the Admiralty and War Office (Great Britain, House of Commons, *Debates*, July 23, 1900, vol. 86, c867).

38. W.J. Baker 1974, 63.

CHAPTER 8

1. Bettòlo 1986.

2. de Sousa 1919, 134, OX 55.

3. See, for example: GM to JBH, January 15, 1900, HUN 1/6; March 9, 1900 [misdated February 9, 1900], HUN 1/7; February 14, 1900, HUN 1/8; July 4, 1900, HUN 1/14; August 20, 1900, HUN 1/19.

4. SFP to Viscount Clifden, August 10, 1900, OX 187; SFP to GM, August 18, 1900, OX 389; SFP to MWTC directors, August 30, 1900, OX 187; Bussey 2000, 15.

5. SFP to Clifden, August 10, 1900, OX 187.

6. H. Williams 1999, 39.

7. AJM to GIU, January 26, 1901, LIN 37; GM to JBH, September 7, 1900, HUN 1/20, and October 8, 1900, HUN 1/21.

8. de Sousa 1919, 135, OX 55.

9. Bussey 2000, 16–17.

10. D. Marconi 1962, 92.

11. After Poldhu, Vyvyan also built the Marconi receiving stations at South Wellfleet, Massachusetts, and Glace Bay, Nova Scotia, where he was in charge for many years. He remained with the company into the 1930s, and published a lasting record, *Wireless over Thirty Years* (Vyvyan 1933; see also OX 688).

12. For example, JAF to GM, November 15 and 26, 1900, OX 174.

13. SFP to GM, November 29, 1900, OX 654. Fleming had expressed his concerns privately to Flood-Page days before (JAF to SFP, November 23, 1900, cited in Bondyopadhay 1994, 57–58).

14. MWTC directors' meeting minutes, November 30, 1900, OX 393; SFP to JAF, December 1, 1900, MS Add 122/47, Fleming Collection, University College London (cited in Hong 2001, 54 and 68, and in Sarkar et al 2006, 318). See also Bondyopadhyay 1994.

15. A seasoned corporate consultant, Fleming understood the rules of the game. In his original May 1899 offer of service to the Marconi company, he proposed "that you would possess in return my thoughts and any inventions or suggestions I may make in your business as your exclusive property" (JAF to HJD, May 2, 1899, cited in Bondyopadhyay 1994, 55).

16. GM to JAF, December 10, 1900, MS Add 122/47, Fleming Collection, University College London (cited in Hong 2001, 69); JAF to GM, December 13, 1900, OX 187.

17. See undated memo renewing Fleming's contract as scientific advisor to the company, first made December 1, 1900. Attached list of patents taken out by Fleming for MWTC, 1900–03, and those taken out by JAF and yet to be transferred "on re-establishment of the Agreement," 1904. Re 24850/1904, "Improvements in Instruments for Detecting and Measuring Alternating Electric Currents," the memo says: "This is the patent for the Oscillation [thermionic] Valve likely to be very valuable." All in OX 174.

18. See Hong 2001, 26–28.

19. Fleming 1937, 41–64, OX 1756.

20. GM to JBH, January 6, 1901 (emphasis in original), HUN 1/26; GM to JBH, January 21, 1901, HUN 1/28.

21. GM to JBH, January 21, 1901, HUN 1/28. (The Queen died on January 22, 1901. Marconi must have started the letter the day before or mistaken the date.)

22. JAF to GM, February 20, 1901, OX 174; JAF to GM, February 27, 1901, OX 174; J.A. Fleming, "Private & Confidential. Recommendations with regard to the Marconi electric power station in the US," typescript, February 20, 1901, OX 191.

23. See, for example, E.H. Moeran to MWTC, April 16, 1901, OX 381.

24. J.F. Bannatyne to GM, March 6, 1901; GM to SFP, March 6, 1901; OX 245.

25. GM to JBH, March 1, 1901, HUN 1/30; GM to JBH, March 14, 1901, HUN 1/31. Annie gets into the act, writing to her son: "I hope you take the cod liver oil and keep plenty of warm clothes on…" (AJM to GM, March 24, 1901, RI 33–34)

26. "Mr. Marconi and Wireless Telegraphy," *Sketch,* March 20, 1901, HUN 2/40.

27. G. Marconi, Diaries and Notebooks, March 1901, OX 1; telegram (unsigned) to JBH, March 20, 1901, HUN 2/33.

28. GM to JBH, March 18, 1901, HUN 1/32.

29. GM, "Report from G. Marconi on his recent visit to America," n.d. [1901], OX 191. Fleming had laid out the criteria for a site in his memo of February 20,

1901 (J.A. Fleming, "Private & Confidential. Recommendations with regard to the Marconi electric power station in the US," typescript, OX 191).

30. Whatley 1987, 11.
31. "Mr. Marconi Comes to South Wellfleet," January 16, 2013, http://southwellfleet .wordpress.com/2013/01/16/mr-marconi-comes-to-south-wellfleet/.
32. Whatley 1987, 11.
33. Henry David Thoreau, cited in Whatley 1987, 11.
34. de Sousa 1919, 204, OX 55.
35. E.H. Moeran to MWTC, April 16, 1901, OX 381.
36. "Wireless Telegraph System to Be Installed at South Wellfleet," *Barnstable Patriot*, May 5, 1901.
37. RNV to JBH, July 25, 1901, HUN 2/9. Vyvyan was one of the few Marconi associates to develop his own relationship with Holman. See RNV to JBH, twelve letters and one cable, May 18–November 2, 1901, HUN 2/4–2/16.
38. Whatley 1987; "Mr. Marconi Builds His Station in South Wellfleet," January 16, 2013, http://southwellfleet.wordpress.com/2013/01/16/mr-marconi-builds-his-station-in-south-wellfleet/.
39. Ralph Hamilton Beach to Francis Jehl, January 20, 1937 and November 12, 1939, TEP. See also, "Marconi the Guest of Edison," *NYT*, April 17, 1901.
40. GM to JBH, April 20, 1901, HUN 1/34.
41. GM to JBH, April 21–26, 1901, HUN 1/34.
42. GM to JBH, telegram, April 28, 1901, HUN 1/35.
43. "Inventor Marconi to Wed," *NYT*, April 27, 1901.
44. GM to JBH, April 30 and May 3, 1901, HUN 1/36, HUN 1/37.
45. AJM to GIU, April 28, 1901, LIN 37.
46. AJM to GIU, May 11 and 20, 1901, LIN 37.
47. GM to JBH, May 21, 1901, HUN 1/39; AJM to JBH, May 24, 1901, HUN 2/20; also AJM to GM, May 26, 1901, RI 36; AJM to GM, July 4, 1901, RI 40; and AJM to GM, August 31, 1901, RI 43–44.
48. AJM to GM, July 3, 1901, RI 38–39.
49. Barbara Valotti asked me in October 2012 whether I had uncovered any evidence that Marconi had really been engaged to Holman. Before the discovery of Marconi's letters in the Huntington collection, this was a reasonable question.
50. GM to JBH, July 9 and 24, 1901, HUN 1/43, HUN 1/44.
51. JBH to GM, June 18, 1901, OX 18/33.
52. Memo, SFP to MWTC board of directors, October 1900, OX 581; "Report," March 20, 1901, OX 177.
53. "Queen's Bench Division," *LT*, January 30, 1896.
54. MWTC directors' meeting minutes, March 21, 1901, OX 393. Hall would be named managing director on July 24, 1901 (OX 393).
55. A few photos of Hall survive in London's National Portrait Gallery.

56. G. Marconi, "Syntonic Wireless Telegraphy," *Journal of the Society of Arts* 49, no. 2530 (May 17, 1901): 506–17 plus figs.

57. HCH to GM, May 23, 1901, OX 249.

58. GM to HCH, May 26, 1901, OX 249; HCH to GM, June 17, 1901, OX 249.

59. Marconi testimony, September 28, 1903, US Patent Office, "Minutes of evidence of an investigation of interference between applications of Marconi, Fessenden and Shoemaker, 1904," OX 533.

60. "Report on Negotiations in Progress," n.d. [April 2, 1901], OX 180.

61. "Memorandum. Our Position with the Belgian and National Companies," n.d., OX 389. Also, "Proposal," September 23, 1899, OX 389.

62. C.E. Rickard, "Report re Visit to Malta," October 9, 1900, OX 180.

63. A.D. Raine to MWTC, February 25, 1901, OX 180.

64. SFP to GM, November 18, 1901, OX 389.

65. Sexton 2005; Clarke 1995.

66. GM to JBH, July 9, 1901, HUN 1/43.

67. H.W. Leach, cited in Sexton 2005, 69.

68. GM to JBH, July 24, 1901, HUN 1/44.

69. SFP to GM, July 21, 1901, OX 249.

70. "Marconi's Own Story of Transatlantic Signals," (extract) *Weekly Marconigram*, July 25, 1903, OX 1779.

71. HCH to GM, October 28, 1901, OX 175.

72. Sexton 2005.

73. de Sousa 1919, OX 55.

74. See chapter 1.

75. See "Luigi Solari," http://www.carlobramantiradio.it/solari.htm. See also B. Valotti's preface and Luigi Solari Jr.'s biographical note in the 2011 reissue of Solari's 1940 biography of Marconi (Solari 2011); Marconi's deposition in LIN 19/9; and Luigi Solari Jr., interview with author, October 5, 2012.

76. Solari 2011. Retold in Masini 1995, 178–79.

77. Marconi was always ambivalent about his treatment by the Italians. After his Newfoundland success in December 1901 he personally informed the Italian government before announcing the news to the press; at the same time, he wrote in a cable to his London office: "Italy made great number of promises in order to see my first experiments in England but nothing came of it" (OX 157/3).

78. C.E. Rickard, a Marconi engineer, described the Italian navy coherer as "a glass tube, with an iron and a carbon electrode and between them finely adjusted was a drop of mercury. The iron end was connected to the aerial and the carbon to the earth" (C.E. Rickard to E.A. Payne, February 24, 1950; A. Gray to E.A. Payne, February 26, 1950; both in OX 1778).

79. HCH to GM, June 26, 1901, OX 245; HCH to GM, July 1, 1901, OX 187; GM to JBH, July 9, 1901, HUN 1/43; GM to JBH, July 24, 1901, HUN 1/44; GM

to JBH, September 10, 1901, HUN 1/46; GM to JBH, September 17, 1901, HUN 1/47.

80. Bussey 2000, 35; W.J. Baker 1970, 65; G. Kemp, Diaries, September 17, 1901, OX 57.
81. HCH to GM, October 3, 1901, OX 187.
82. GM to JBH, October 15, 1901, HUN 1/48.1.
83. G. Marconi 1923, with archival note: "File carefully," OX 52. See also MWTC directors' meeting minutes, September 19, 1901, OX 393.
84. HCH to GM, October 3, 1901, OX 155/1–2.
85. Telegram, November 3, 1901, from [GM?] Crookhaven to Marconi Expanse London. "Goodbody agrees advisable make experimental connections with Newfoundland." See also "Newfoundland Agreement," signed by Governor Bond of Newfoundland on October 8, 1901, and by three directors of the company on November 7, 1901 (LAC, Canadian Marconi Company MG 28 III 72, Vol. 6).
86. Rowe 2009.
87. J. Bottomley to MWTC, n.d., OX 155 /9–10; extract from Anglo-American Charter (Act of the Newfoundland Legislature passed April 15, 1854), OX 155.
88. MWTC directors' meeting minutes, October 17, 1901, OX 393; HJD to GM, October 27, 1901, OX 381.
89. GM to JBH, October 15, 1901, HUN 1/48.1.
90. Telegram, GM to G. Kemp, November 4, 1901, OX 91 (G. Kemp, Diaries, exp. extracts).
91. University of Oxford, Bodleian Library, Marconi Archives, Catalogue description for OX 155.
92. "Wireless Telegraphy for the Canadian Coast. A Chat with Signor Marconi," *Journal of Commerce* (Liverpool), November 26, 1901. On November 18, 1901, Flood-Page reported to Marconi that he told the *Herald's* man to come to see Marconi "so that you might tell him whatever you wish him to say. He was quite willing to accept this arrangement" (SFP to GM, OX 155). To which Marconi added by hand: "see him." See also McGrath 1902.
93. GM to JBH, November 18, 1901, HUN 1/49.
94. GM to W.S. Entwistle, November 22, 1901, OX 171.
95. GM to H.W. Allen, November 23, 1901, OX 171.
96. Allan Bros., London, to Capt. Daly, MWTC, November 11, 1901, OX 155; "Correspondence concerning preparations for the transatlantic signal attempt, Oct.–Dec. 1901," OX 155.
97. G. Kemp, Diaries, November 26, 1901, OX 57.

CHAPTER 9

1. Newfoundland Historical Society, http://www.nlhistory.ca.
2. GM to "Jumbo," handwritten draft, n.d., OX 157/1.
3. G. Kemp, Diaries, December 6, 1901, OX 57. Cochrane House, rebuilt after its destruction in the great fire of 1892, was St. John's top hotel. It burned again in

1985 and was rebuilt as a seniors' residence. A plaque on the building commemorates Marconi as one of its famous guests, along with Leon Trotsky, who
may have stayed there on his way to join the Bolshevik Revolution in 1917 (see
The Scope, "Leon Trotsky? At Cochrane House?", http://thescope.ca/nooks/
leon-trotsky-at-cochrane-house), and aviators John Alcock and Arthur Brown,
who set out from St. John's on their pioneering transatlantic flight in June
1919—a flight that ended in the bog beside Marconi's station at Clifden, Ireland.

4. Candow 2011, 52.

5. de Sousa 1919, 142–43, OX 55. This passage is almost identical to a feature
article in the *Times* of January 3, 1902 ("Marconi in Newfoundland"), based on
an interview with Marconi, and is an indication of de Sousa's method.

6. GM to "Jumbo," handwritten draft, n.d., OX 157/1.

7. GM to JBH, December 8, 1901, HUN 1/50.1.

8. GM to "Jumbo," handwritten draft, n.d., OX 157/1. Marconi signed the letter
"Markie," a nickname he sometimes went by in England.

9. One of Marconi's biographers, Giancarlo Masini, writes that Marconi "consoled himself" after breaking up with Josephine by immediately starting another affair, "a habit he would follow throughout his life" (Masini 1995, 139).

10. HCH to GM, November 27, 1901, OX 175. See also SFP to GM, December 9,
1901, OX 187.

11. Canadian Marconi Company, "Marconi—The Man," LAC, Canadian Marconi
Company, MG 28 III 72, vol. 7.

12. "Marconi's Own Story of Transatlantic Signals," extract from *Weekly Marconigram*,
June 25, 1903, OX 1779.

13. G. Kemp, Diaries, December 11, 1901, OX 57.

14. GM to MWTC chairman and directors, December 23, 1901, OX 158 /1–6.

15. G. Kemp, Diaries, December 12, 1901 (emphasis in original), OX 57; exp. extracts,
December 12, 1901, OX 91.

16. Marconi's diary for 1901 is the first of more than four thousand boxes in the
Oxford Marconi Archives (OX 1). In 1904 testimony in a US patent case involving Reginald Fessenden and Harry Shoemaker, after referring to his and Kemp's
diaries to establish a timeline of events, Marconi specified: "I wish to state that
the notebook was not kept as a regular diary, but only as a convenient means of
noting and ascertaining certain facts as they occurred" (US Patent Office, 1904,
OX 533). Marconi said his recollection was that the pencil notations were made
prior to the dates under which they were noted and referred to unrelated tests
(Interference Proceedings No. 22505, Marconi v. Shoemaker v. Fessenden in
USA, OX 1778). Even within the company, controversy about the interpretation
of Marconi's diary continued as late as 1950 (see correspondence between
MWTC publicity manager W.G. Richards and former chief engineer C.E.
Rickard, April 26–27, 1950, and May 1, 1950, OX 1778). Marconi claimed he and
Kemp heard the Morse *s* perhaps twenty times on the twelfth; and on the fol-
lowing afternoon, "having succeeded in flying another kite, we received another

series of signals, but they were less distinct and not renewed after a brief period....On Saturday we tried again, but the weather conditions were still more adverse...and I therefore ceased my experiments" ("Marconi's Own Story of Transatlantic Signals," extract from *Weekly Marconigram*, June 25, 1903, OX 1779).

17. de Sousa 1919, 143–44, OX 55.

18. Ibid., 144–45, OX 55. Solari later confirmed, based on conversations with Marconi as well as photographic evidence, that Marconi had used a telephone receiver connected to the mercury "Italian navy coherer" he had given him in England. According to Solari, Marconi felt his usual receiver would not be sensitive enough to capture the signals from Poldhu ("Extract of letter from Marquis Solari, Rome," January 1, 1948, OX 1778). Marconi had several coherer receivers with him in St. John's. Marconi sometimes claimed he himself had developed the new, more sensitive coherer ("Marconi's Own Story of Transatlantic Signals," extract from *Weekly Marconigram*, June 25, 1903, OX 1779).

19. GM to MWTC chairman and directors, December 23, 1901, OX 158 /1–6.

20. "First Radio Signal across the Atlantic. Thirtieth Anniversary of Historic Poldhu-Newfoundland Experiment," draft article submitted to *NYT*, December 9, 1931, OX 1779.

21. Telegram, GM to MWTC, n.d., OX 157. Marconi cautioned London not to show the system to the German navy unless on a firm business proposal.

22. "Wireless Signals across the Ocean. Marconi Says He Has Received Them from England. Prearranged Letter Repeated at Intervals in Marconi Code," *NYT*, December 14, 1901.

23. "Mr Edison Believes Mr Marconi Did It," *New York Herald*, December 18, 1901. Also "Edison Believes in Marconi's Success," *NYT*, December 19, 1901; GM to Edison, December 23, 1901, TEP.

24. "Professor Pupin Praises Marconi's Great Feat," *NYT*, December 16, 1901; "Believes in Marconi," *NYT*, December 17, 1901.

25. H.J. Powys, "An Eye-Witness Account of a Great Day," *Marconi Review*, supplement to vol. XIV, no. 103 (1951): 29–30, OX 1779.

26. G. Kemp, Diaries, December 17, 1901, OX 57; "Marconi's Experiments and the Anglo-American Telegraph Company," *Evening Telegram* (St. John's), December 17, 1901. The original Vey photographs are in the Newfoundland provincial archives in St. John's (RPA).

27. Lodge 1925 (cited in draft article submitted to the *NYT*, December 9, 1931, OX 1779).

28. Aitken 1976, 243. Arthur Kennelly, an American electrical engineer, and Oliver Heaviside, a British Maxwellian physicist (see chapter 2), were working independently, but their hypotheses were so similar and close in time that the area of the atmosphere they refer to has come to be known as the "Kennelly-Heaviside layer."

29. Morine & Gibbs to GM, December 16, 1901, OX 156 /1.

30. W. Smith, "How Marconi Came to Canada," LAC.

31. GM to Morine & Gibbs, December 16, 1901, OX 156 /2–6; Morine & Gibbs to GM, December 17, 1901, OX 156 /7; J.S. Winter (counsel for GM) to Morine & Gibbs, December 18, 1901, OX 156/8.

32. "To Restrain Marconi" (editorial), *NYT*, December 18, 1901; "Marconi and the Anglo-American" (editorial), *NYT*, December 19, 1901; *LT* (editorial), December 21, 1901; telegram, SFP to GM, December 17, 1901, OX 157/21. The company was able, as always, to see the positive side of a situation: the Anglo-American legal intervention "at least gave the satisfactory assurance that one of the great Telegraph and Cable Companies not only believed but also feared the possibility of Wireless Transatlantic communication" (MWTC press release, "First Wireless Signal across the Atlantic. 25th anniversary of Senatore Marconi's notable achievement. Senatore Marconi describes his experiments," December 10, 1926, OX 1779); HJD to GM, December 19, 1901, OX 187.

33. Telegram, A.G. Bell to GM, n.d., OX 157/23. Marconi replied on December 19, 1901: "Thanks for most generous offer hope visit Cape Breton next week shall telegraph more fully tomorrow." Like so many compulsively busy and self-absorbed people, Marconi liked to keep all his options open until the very last minute and not commit, even in the face of a fabulous offer. He kept Bell dangling until he was sure he had a deal from the Canadians and only declined the invitation after he got to Montreal on December 29, 1901: "Sorry cannot accept your kind invitation as it is essential that site should be on ocean…" (AGBFP, Box 132). Bell was fascinated with Marconi and kept a thick scrapbook of news clippings, chronicling his exploits between December 17, 1901, and March 30, 1902 (AGBFP, Box 264).

34. "The nineteenth century was the century of the United States. I think we can claim that it is Canada that shall fill the twentieth century" (Sir Wilfrid Laurier, speech to the Canadian Club, Ottawa, January 18, 1901, cited in King 2010, 15).

35. Prior to Confederation, there was a gap in rail service between Rivière-du-Loup, east of Quebec City, and Truro, Nova Scotia. The new Dominion government hired Sir Sandford Fleming, the inventor of standard time, to oversee construction of a new line to fill the gap.

36. "By 1900 there were more than 56,000 kilometres of telegraph line in the country" (Bird 1988, 3).

37. See Gagné 1976, 9–15.

38. It was reported as early as September 1899, during Marconi's first US visit, that the governments of Canada and Newfoundland were negotiating for the installation of his system ("Negotiating for Marconi's System," *NYT*, September 29, 1899).

39. Bird 1988, 7.

40. Canada, Reports of the Deputy Minister of Public Works, *Sessional Papers*, December 24, 1900, and January 15, 1902, cited in Bird 1988, 8–9. The company signed a contract with the Canadian government in June 1901 to build two

stations for communicating twenty miles across the Belle Isle Straits, at Chateau Bay and Belle Isle; the stations opened in September 1901 ("Particulars and information with regard to the Marconi Wireless Telegraph System, showing its applications and utility," March 1905, OX 180). An agreement was signed in November (see chapter 8), and Marconi used this as the public pretext for his trip to Newfoundland in December. (The Canadian government maintained and operated lighthouses and stations on the coast of Newfoundland and Labrador that were deemed to be a part of Canada, so Canada as well as Newfoundland could both negotiate agreements with Marconi; presumably the Anglo-American monopoly did not apply to the Canadian stations, even when those were located in Newfoundland.)

41. W. Smith, "How Marconi Came to Canada," LAC.

42. W. Mulock to W. Smith, April 23, 1923, LAC, Fonds William Smith, MG30-D18, Vol. 1. See also E. Roberts, "Why Marconi Left Newfoundland," *Compass*, October 2, 2012, and Tarrant 2001.

43. Fielding was a quintessential Canadian politician, moving adeptly between provincial and federal levels of government by pitting interests against one another. Under his premiership, the US-owned Dominion Coal Company was given a ninety-nine-year lease on Nova Scotia coal mines in exchange for a royalty of twelve cents a ton. A masterful politician, Fielding was a key builder of the federal Liberal Party, which became Canada's "natural governing party" in the twentieth century (Whitcomb 2009, 37–39).

44. Telegram (ironically sent via Anglo-American Telegraph Company cable), W.S. Fielding to GM, December 20, 1901, OX 157/9. Marconi's reply is cited in GM to MWTC chairman and directors, December 23, 1901, OX 158/1–6.

45. "Transatlantic Wireless Telegraphy," *LT*, December 28, 1901.

46. See chapter 10.

47. See W. Mulock to W. Smith, February 2, 1901; W. Smith to W. Mulock, March 8, 1902; W. Smith to GM, December 18, 1902; all in LAC, Fonds William Smith, MG30-D18, Vol. 1.

48. "To Wed Marconi," *New York Sun*, December 18, 1901; "Cupid Waits on Wireless Telegraph," source not indicated, dateline Indianapolis, December 19, 1901, cutting in HUN 2/40; GM to JBH, December 22, 1901, HUN 2/1. The telegram, addressed to Holman's mother, indicates that Josephine is apparently ill.

49. "Marconi Speaks at a Luncheon in His Honour," *NYT*, December 19, 1901.

50. "Possible Developments of Wireless Telegraphy," *NYT*, December 29, 1901. This meant, for example, that a permanent connection between New York and England, via Newfoundland or Cape Breton, could be established for $180,000. Despite his commitment to the Anglo-American to cease his activities in Newfoundland, Marconi continued talking to Newfoundland business interests and scouting locations for a possible station. On December 22, 1901, the *New York Herald* reported an inspection visit for that purpose to Quidi Vidi village in the outskirts of St. John's ("Canadian Capitalists to Aid Mr Marconi in Experiments," *New York Herald*, December 22, 1901).

51. GM to MWTC chairman and directors, December 23, 1901, OX 158/1–6.
52. "Report on Negotiations in Progress," n.d. [believed to be April 2, 1901], OX 180.
53. G. Kemp, Diaries, December 24 and December 25, 1901, OX 57. Within two years, Marconi's did have a Newfoundland contract, see OX 180.
54. MacLeod 1992.
55. W. Smith, "How Marconi Came to Canada," LAC; "Marconi Arrives in Cape Breton," *Sydney Daily Post*, December 27, 1901.
56. "*L'homme du siècle. Signor Guglielmo Marconi, le célèbre inventeur de la télégraphie sans fil à longue distance de passage à Montréal*," *La Presse* (Montreal), December 30, 1901, OX 160/55. Quebec historians date the history of radio in the province from this moment (Pagé 2007, 15 and 28–29).
57. W. Smith, "How Marconi Came to Canada," LAC.
58. "Marconi Entertained in Ottawa," source not indicated, dateline Ottawa, January 1, 1902, cutting in HUN 2/3; "Marconi Got Electric Shock," *New York Herald*, January 5, 1902 (this item was also in Alexander Graham Bell's scrapbook in AGBFP; Marconi was considering an invitation to visit Bell in Washington at this time); GM to JBH, n.d., HUN 2/3.
59. G. Marconi, Diaries and Notebooks, 1902, OX 2.
60. "Mr. Marconi and Telegraphic Communication with Canada," *LT*, February 10, 1902.
61. GM to Sir Wilfrid Laurier, January 6, 1902, LAC, RG13-A-2, Vol. 1903 (1902–129). The typed letter was accompanied by a handwritten note to Laurier dated January 7, 1902, LAC, MG26-G, Microfilm C-790.
62. Lord Minto, "Memoranda of conversations," LAC MG27-IIB1, R4688-0-9-E. Microfilm reel C-3113 (Vol. 4, p. 18–20), January 7, 1901 [sic, the date was January 7, 1902]. Typed on stationery of Government House.
63. L.H. Officer, "Exchange Rates Between the United States Dollar and Forty-One Currencies," *MeasuringWorth*, 2015 (http://www.measuringworth.com/exchangeglobal/).
64. See Bird 1988.
65. See Hardin 1974. In this work, one of the classics of the new Canadian nationalism that emerged in the 1970s, Hardin comments wryly that the Canadian nation itself was seen by its founders as an enterprise.
66. See "Wireless Telegraphy," *LT,* January 6, 1902.
67. de Sousa 1919, 152, OX 55.
68. H.H. McClure to JBH, December 31, 1901, HUN 2/28.
69. M.W. Sewall to JBH, January 10, 1902, HUN 2/30.
70. HCH to GM, January 10, 1902, OX 175.
71. *LT,* January 10, 1902; G. Kemp, Diaries, January 1–12, 1902, OX 58.
72. *New York World,* January 13, 1902.
73. "Marconi Predicts," New York evening newspaper (source unclear), January 13, 1902, cutting in HUN 2/40.
74. *NYT,* January 13, 1902.

75. J.B. Holman, Diary, HUN 2/38.

76. The original autographed menu is in OX 49.

77. Mina Miller Edison to Mary V. Miller, January 12, 1901, TEP; N. Tesla to AIEE, *New York Herald*, January 14, 1902, AGBFP and OX 54.

78. *Electrical World and Engineer* 39, no. 3, p. 107, January 18, 1902, OX 49 /154.

79. "Marconi Tells of Wireless Message Tests," *NYT*, January 14, 1902.

80. "Marconi," *NYT*, January 15, 1902.

81. J.B. Holman, Diary, HUN 2/38.

82. "Marconi Sets Fiancée Free . . . Says He Is Sorry. Rumor Has it that Woman Was Piqued at Inattention," *Inter Ocean* (Chicago), January 21, 1902, HUN 2/40. Marconi made a gesture to McClure in one of the few ways he knew to express his gratitude; he presented him with a handsome studio portrait photograph of himself, made a few days earlier in Ottawa, and autographed "To Mr. Harry McClure, from his sincere friend, Guglielmo Marconi, 21st Jan., 1902" (sold by RR Auction, Amherst, Mass., January 17, 2013, http://www.artfact.com/auction-lot/guglielmo-marconi-177-c-986e1e49f6).

83. "No Loveless Marriage for Wireless Inventor. Marconi Freed by His Fiancee," *New York Herald*, January 22, 1901, HUN 2/40.

84. "Signor Marconi Sails. Will Say Nothing Further about Broken Engagement," *NYT*, January 23, 1902.

85. Undated unsigned typescript, HUN 2/36; *NYT*, January 23, 1902.

86. J.B. Holman, Diary, January 27 1902, HUN 2/38.

87. "Miss Holman to Wed Soon. Ex-fiancee of Signor Marconi to Marry a Hungarian," *NYT*, May 3, 1902; "Miss Josephine Holman Weds. Marries Eugen Boross of Budapest—Was Formerly Engaged to Marconi," *NYT*, May 23, 1902.

88. Bryn Mawr Alumnae Bulletin, vol. XVI, no. 1, January 1936. Various Internet sources.

89. http://www.oac.cdlib.org/findaid/ark:/13030/c8x06chf/.

CHAPTER 10

1. Quoted material in the next few paragraphs is taken from Ray Stannard Baker, "Marconi's Achievement: Telegraphing across the Ocean without Wires," *McClure's* 18, no. 4 (February 1902): 291–99. An identical version appeared in *Windsor Magazine*, May 1902.

2. See, for example, "Cable Men Do Not Fear Marconi," *NYT*, February 16, 1902; "Marconi vs. Cable Stocks," *NYT*, March 23, 1902.

3. G. Marconi, Diaries and Notebooks, January 22, 1902, OX 2; G. Kemp, Diaries, January 22 and 29, 1902, OX 58.

4. A. Gray to GM, January 31, 1902, OX 249; HCH to GM, February 3, 1902, OX 175; G. Kemp, Diaries, February 10, 1902, OX 58.

5. Telegram, W.S. Fielding to GM, January 29, 1902, in LAC MG26-G; MWTC directors' meeting minutes, February 6, 1902, OX 393; LAC RG13-A-2, Vol. 1903

(1902-129) [also in MG26-G Microfilm C-790]; Canada, 9th legislature, 2nd session, February 13, 1902.

6. MWTC, Directors' Report for Year Ending September 30, 1901 (dated February 11, 1902), OX 401.

7. MWTC, Directors' Report for Year Ending September 30, 1902, OX 400.

8. Marconi's statement to MWTC AGM, February 20, 1902, original in OX 400. See also *Electrical Review* 50, no. 1266 (February 28, 1902): 354–56; Hong 2001; *LT*, February 21, 1901.

9. Ibid.

10. Ibid.

11. "Prince Henry Dies; Brother of Kaiser," *NYT*, April 21, 1929.

12. "Ready for Prince Henry," *NYT*, February 19, 1902; "Prince Henry's Visit to America," *LT*, February 22, 1901, and "Waiting for the Kronprinz Wilhelm. Prince Henry's Arrival May Be Retarded by Storms," *NYT*, February 22, 1902; "Prince Welcomed by Chiefs of Industry," *NYT*, February 27, 1902.

13. "Prince's Trip Across," and "Liner's Wireless Messages," both *NYT*, February 24, 1902.

14. "Prince Henry Near Home," *NYT*, March 18, 1902; "Germany," *LT*, March 19, 1902.

15. GM to HCH, September 16, 1901, OX 169; HCH to GM, October 4, 1901, OX 247.

16. de Sousa 1919, 274, OX 55.

17. "Marconi Discusses the Slaby-Arco System," *NYT*, March 30, 1902. See also "Marconi in Defense of His Inventions," *NYT*, March 28, 1902; "Tesla's Work and Marconi's: The Former Is the 'Father of Wireless Telegraphy,'" *New York Sun*, March 30, 1902.

18. J. Loewe to SFP, April 9 and 11, 1902, OX 417.

19. "Fears a Marconi Monopoly," *NYT*, March 22, 1902.

20. "To Call a Telegraph Congress. German Plan for International Action to Prevent Wireless System Monopoly," *NYT*, March 30, 1902; "Wireless Telegraphy," *LT*, March 31, 1902.

21. Memorandum of Agreement between MWTC/MIMCC and HM King Edward Seventh represented by RH Sir Wilfrid Laurier, March 17, 1902. Signed by GM and Laurier, OX 180 (also in LAC, RG12-A, R184-275-X-E. Files related to the Canadian Marconi Company, vol. 6 and vol. 396); GM to HCH, March 23, 1902, OX 171.

22. de Sousa 1919, 171, OX 55.

23. MWTC directors' meeting minutes, July 25, 1902, OX 393.

24. "Marconi May Change His Stations Abroad," *NYT*, March 5, 1902.

25. "Topics of the Times" (editorial), *NYT*, March 6, 1902.

26. Memorandum, attributed to SFP, February 25, 1903, OX 248; Memorandum of Agreement between the Marconi International Marine Communication Company Limited and the Society incorporated by Lloyd's Act 1871 under the

name of "Lloyd's," September 26, 1901, OX 248. See also MWTC, Directors' Report for Year Ending September 30, 1901 (dated February 11, 1902); HCH to GM, August 29, 1901, OX 175.

27. H.M. Hozier to HCH, n.d., OX 248; HCH to GM, July 25, 1901, OX 175.
28. *Electrician*, October 25, 1901.
29. The contract was especially good for Sir Henry Hozier, who himself received £3,000, five hundred MWTC shares, and a seat on the MIMCC board as part of the deal. Marconi didn't do as well. MIMCC had to buy five hundred of his personally held MWTC shares below market price and issue a new offering in order to meet its obligation to Hozier. The company's cash flow was operating from month to month at this point (HCH to GM, October 1, 1901, OX 175).
30. See for example, HCH to GM, November 13, 1903, detailing a clash between Hall and Hozier; and HCH to GM, November 27, 1903, about contract interpretation issues; both OX 248.
31. See Bruton and Gooday 2010.
32. J.C. Lamb to MWTC, March 16, 1901, OX 180.
33. The letter is addressed to "My Lord," April 19, 1901, OX 204.
34. HCH to GM, July 10, 1901, OX 204.
35. W. Blaydes to HCH, May 28, 1902, OX 171. Also GM to HCH, May 30, 1902, OX 171; HCH to A. Chamberlain, October 10, 1902, OX 204; GM to HCH, November 18, 1902, OX 172.
36. Memo, HCH to [name illegible], December 17, 1902, OX 204.
37. J.C. Lamb to HCH, December 31, 1902, OX 204.

CHAPTER 11

1. de Sousa 1919, 179, OX 55.
2. Consul General Mazza to Foreign Affairs Minister Giulio Prinetti, February 18, 1902, DDI, Terza Serie, VI, doc. 155, 113, February 18, 1902.
3. GM to LS, April 12, 1902 (Solari 1942).
4. GM to HCH, June 29, 1902, MB.
5. MWTC directors' meeting minutes, July 3, 1902, OX 393.
6. Landini 1960.
7. HCH to *LT*, September 11, 1902.
8. de Sousa 1919, OX 55. MWTC directors' meeting minutes, September 11, 1902, OX 393.
9. GM to HCH, September 3, 1902, MB.
10. de Sousa 1919, 196, OX 55.
11. Solari 1942.
12. Telegram, GM to HCH, September 10, 1902, reported by Hall in MWTC directors' meeting minutes, September 11, 1902, OX 393.
13. Masini 1995.
14. *LT*, September 22, 1902.

15. Marconi said he had initially laid out these conditions for Italian military use of his system in a letter dated February 8, 1902; Solari would refer to this letter as the basis for Italy's position at the Berlin conference in August 1903. I have never seen this letter in any of the archives.
16. MWTC directors' meeting minutes, September 26, 1902, OX 393.
17. GM to HCH, October 7, 1902, OX 171.
18. GM to HCH, October 1, 1902, MB (copy in OX 171).
19. HCH to W. Laurier, September 26, 1902, LAC, MG26-G, Microfilm C-795, p 67328-9.
20. GM to *LT*, October 10, 1902, OX 171; HCH to GM, October 17, 1902, OX 161/2. Details of the arrangement were spelled out in a letter from Fred Cook of the *Times* in Ottawa, to Marconi in Glace Bay, November 1, 1902, OX 161. (Cook was an interesting figure. A well-connected local journalist and principal Canadian correspondent for the *Times*, he was also the mayor of Ottawa.)
21. Material cited over the next several paragraph is, unless otherwise noted, taken from MacLeod 1992, 65–72.
22. MacLeod 1992. See also MacLeod 1982, BEA, MG 12.214.
23. GM to HCH, November 26, 1902, MB.
24. GM to W.H. Bentley, chairman, MWTCA, January 18, 1903, OX 172. Bentley brought Marconi's request to the US board immediately, and replied with approval three days later (W.H. Bentley to GM, January 21, 1903, OX 212).
25. MWTC directors' meeting minutes, December 18, 1902, OX 393.
26. F. Braun to SFP, April 7, 1902; SFP to F. Braun, April 11, 1902; OX 417.
27. GM to HCH, October 10, 1902, OX 171.
28. GM to Minto, November 7, 1902, OX 161/6. Also in LAC, Minto Papers, Vol. 7 XXXI, 10–11.
29. GM to H.H. McClure, November 7, 1902, OX 171; "Marconi has Triumphed," *Mail and Empire*, November 24, 1902, OX 161/13. Marconi was characteristically annoyed by the report, especially as it was attributed to his friend, Member of Parliament John Henniker Heaton, and asked Hall to investigate whether it was true that Heaton had made such an announcement (GM to HCH, December 3, 1902, OX 171).
30. The date is often given erroneously in the literature as December 15, 1902. An attempt was made on that date but it was unsuccessful (see LAC, MG30-D77, Marconi, Guglielmo—His message from Canada to the *Times*).
31. "Parkin's Account of Marconi Feat," *NYT*, December 12, 1926. After the initial message from Minto to the King, Parkin messaged the *Times* attesting to what he had just witnessed. His full account, however, was delayed for several weeks because of insufficient postage on the article he sent to the *Times*. The whole story is told in Hilary Brigstocke, "For the Want of a Dime a Scoop was Lost," *LT* "House Journal," April 1964. An extensive file of original notes and correspondence detailing Parkin's adventure is in LAC MG30-D77, Marconi, Guglielmo—His message from Canada to the *Times*. See also "Marconi Messages

Cross the Ocean. Inventor Announces He Has Established Communication Between Canada and England," *New York Herald*, December 22, 1902, OX 161/17.

32. See GM to Minto, December 26, 1902, LAC, Minto Papers Vol. 7 XXXI, Microfilm reel C-3116, 12–14; Minto to King Edward, January 18, 1903, LAC, Minto Papers Vol. 25, Microfilm reel C-3115, 57–61; GM to W. Laurier, February 11, 1903; HCH to W. Laurier, February 11 and 14, 1903. Laurier referred the matter to his finance minister, Fielding (February 23, 1903), who replied (February 24, 1903): "I do not quite see what we could do to influence the Imperial Government in the manner suggested by the writer of these letters." Finally, Laurier replied to GM on February 27, 1903, "that it does not seem expedient that the Canadian Government should take the action which you ask for." LAC MG26-G. Correspondence (Sir Wilfrid Laurier), Microfilm C-798. Laurier and his government felt that they had already done enough.

33. GM to L. Clow & Co., August 29, 1901, OX 171; Ochs Brothers to E. Geoffrey Chubb, re: "MWCo £30,000 loan (1901)," December 23, 1902, OX 581; various telegrams (GM, HCH, HJD and J.F. Bannatyne), December 23, 1902, to January 7, 1903, OX 161/84–141; GM to HJD, January 5, 1903, OX 161/136; GM to HJD, January 8, 1903, OX 581.

34. GM to Willard R. Green, January 13, 1903, OX 161/142.

35. W.J. Baker 1970, 80.

36. *Barnstable Patriot,* January 26, 1903.

37. Whatley 1987, 27–29; "Mr. Marconi Builds His Station in South Wellfleet," January 16, 2013, http://southwellfleet.wordpress.com/2013/01/16/mr-marconi-builds-his-station-in-south-wellfleet/.

38. GM to J.D. Taylor, January 12, 1903, OX 191.

39. *Boston Daily Globe,* January 20, 1903.

40. Whatley 1987, 20; *NYT,* January 20, 1903.

41. *Barnstable Patriot,* October 5, 1903.

42. *Hyannis Patriot,* September 21, 1908 (Marconi and his party visited on September 13 and 14, 1908).

43. *Hyannis Patriot,* April 16, 1917; *Hyannis Patriot,* February 4, 1918.

44. GM to LS, March 7, 1903 (Solari 1942).

45. LS to GM, February 16, 1903, OX 161/146; translation of "Deliberations in the Italian Parliament re Bill for erection of High power station," Ministry of Posts and Telegraphs, *The Marconi Radio-Telegraphy in Parliament,* April 5, 1903, OX 212.

46. Quoted material from the Chamber of Deputies debate over the next few paragraphs is taken from "Discussion of the Bill in the Lower House at the Sitting of the 20th February 1903," 21–36, translated transcript in OX 212.

47. The final document was dated April 5, 1903.

48. de Sousa 1919, 216–17, OX 55.

49. D. Marconi 1962, 148; slightly different in Jacot and Collier 1935, 78; Dunlap 1937, 148.

50. Solari 1939. The full speech (Guglielmo Marconi, "*La telegrafia senza fili*") was published as a special supplement of *L'Elettricista,* May 15, 1903. Regarding the British War Office translation, see G. Marconi 1913, 106, OX 54.

51. *NYT,* May 9, 1903.

52. AM to GIU, various, May 1903, LIN 37.

53. AM to GIU, n.d., along with cutting, "*Il maestro di Marconi,*" May 9, 1903, LIN 37. See also Marconi's interview with the *Giornale d'Italia* ("*Salutando Guglielmo Marconi. Intervista al momento della partenza da Roma*"), May 20, 1903.

54. See Hong 2001.

55. J.N. Maskelyne Jr., "Improvements in Electrical Signaling," 7983/00, April 30, 1900, complete specification dated February 28, 1901, OX 415.

56. H. Hozier, transcript of lecture "About 'Aetheric Signalling' and Wireless Telegraphy, with Particular Reference to Experiments Conducted by Nevil Maskelyne," 1901, OX 166/24; extract of unsigned typescript dated August 9, 1901, saying that Lloyd's' arrangement with Maskelyne was coming to an end because they were going to adopt the Marconi system, OX 166/52.

57. Brandon 2003, 85. Maskelyne, like Houdini (and Marconi), was also a vigorous opponent of spiritualism. Although he was often considered a medium, he would protest that in fact the opposite was the case, that mediums were conjurors (Brandon 2003, 236).

58. Maskelyne 1936, 170–71.

59. Hong 2001, 104.

60. G. Marconi, "Address," *Electrician* 50 (April 3, 1903): 1,001–2; see also Hong 2001, 107.

61. Hong 2001, 108.

62. This account is based on an eyewitness report, Arthur Blok, "Some Personal Recollections of Sir Ambrose Fleming," in *The Inventor of the Valve: A Biography of Sir Ambrose Fleming,* ed. J. MacGregor-Morris (London: The Television Society, 1954). Cited in Hong 2001, 110. Hong devotes an entire chapter to the Maskelyne Affair (89–118).

63. "Who Was the Hooligan?" *Electrical Review* 52, no. 1334 (June 19, 1903), OX 166/49. See also *LT,* June 11, 13, 16, and 20, 1903; *Morning Leader,* June 12 and 14, 1903; *Daily Express,* June 12 and 13, 1903; *St. James Gazette,* June 13, 1903; and *Star,* June 16, 1903. Fleming gave his side of the story in two letters to Marconi on June 5 and 6, 1903, OX 166. See also OX 1780.

64. Hong 2001, 166.

65. G.H. Murray to HCH, January 31, 1903; HCH to G.H. Murray, February 4, 1903; both in OX 204.

66. GM to A. Chamberlain, March 2, 1903, OX 204.

67. A. Chamberlain to GM, March 4, 1903, OX 204; see, for example, GM to A. Chamberlain, March 15, 1903, OX 173.

68. de Sousa 1919, 224, OX 55; H. Kershaw to A. E. Heming, July 2, 1903, OX 173.

69. HCH to GM, July 31, 1903; F.H. Villiers (Foreign Office) to Secretary, MWTC, July 30, 1903; both in OX 204.

70. Memo "re Berlin Conference," August 1, 1903, OX 204.

71. H. Cuthbert Hall, "The Berlin Wireless Telegraph Conference," *LT*, August 1, 1903.

72. "Wireless Companies' Quarrel. German Concern Says It Would Be Easy to Interrupt Marconi Messages," *NYT*, July 30, 1903.

73. ITU, Preliminary Conference on Wireless Telegraphy (Berlin, 1903), *Documents*, "Suggestions Submitted by the German Government for Discussion at the Conference," 1 (Art. 1, Para. 1).

74. All quoted material in the following paragraphs is taken from ITU, Preliminary Conference on Wireless Telegraphy (Berlin, 1903), *Documents*, "Proceedings at the Conference."

75. Historian Susan J. Douglas (1987, 110-19) argues that this stance was partly rooted in groundless prejudice. Douglas states that Marconi was thought by some navy officials to be part of a Jewish corporate cabal. Barber, for one, doubted Marconi's claim to have signalled across the Atlantic and wrote to a colleague in April 1902 that Lloyd's secretary Henry Hozier (now a Marconi director) had told him "the whole thing was a stock-jobbing operation worked in the interest of 'a lot of Jews'" (F.M. Barber to Admiral R.B. Bradford, Chief, US Bureau of Equipment, April 22, 1901, NARA, cited in Douglas 1987, 115; also F.M. Barber to R.B. Bradford, June 22, 1906, Douglas 1987, 113).

76. ITU, Preliminary Conference on Wireless Telegraphy (Berlin, 1903), *Documents*, "Proceedings at the Conference," 39–43, Annex to the minutes of the fourth sitting. Memorandum by Captain Quintino Bonomo del Casale, Italian delegate.

77. GM to HCH, August 12, 1903, MB.

78. ITU, Preliminary Conference on Wireless Telegraphy (Berlin, 1903), *Final Protocol*, Art. 1, Para. 2; Art. 4; Art. 3.

79. See Wu 2010, and T. Wu, "Does a Company Like Apple Need a Genius Like Steve Jobs?" *New Yorker*, February 19, 2013.

CHAPTER 12

1. MacLeod 1992, 80. MacGilivray admitted to a friendship with Marconi but denied reports of an engagement ("'Wed Marconi?' It Makes Miss MacGilivray Smile," cutting in BEA, source unclear, dateline Boston, January 14 or 16, 1903).

2. Solari 2011, 87.

3. Ibid., 45 and 48.

4. Ibid., 48.

5. Ibid., 111.

6. de Sousa 1919, 227, OX 55.

7. "Edison Enters Marconi Company," *NYT*, May 28, 1903.

8. Jacot and Collier 1935, 92; also told by Solari 2011, 116, and others.

9. Marconi v. de Forest, and Fessenden and Shoemaker v. Marconi (discussed in chapter 13).

10. "Wireless telegraphy," *LT*, September 1, 1903; GM to HCH, September 4, 1903, MB; de Sousa 1919, 227–32, OX 55; MWTC directors' meeting minutes, October 20, 1903, OX 393.

11. Solari 2011, 117.

12. Lumsden 2004, 9–18.

13. Solari 2011, 111; Lumsden 2004, 26 and 128.

14. J.E. Milholland, Diaries, October 18, 1903, MFP; IM to JEM, October 24, 1903, cited in Lumsden 2004, 26.

15. Solari 2011, 111.

16. John Tepper Marlin (great-nephew of Eugen Boissevain), correspondence with author, July 26, 2012.

17. D. Marconi 1962, 149.

18. GM to James G. Lambeth, June 30, 1903, OX 173; AJM to H. Kershaw, August 21 and 26, 1904, September 4, 1904, all in OX 17.

19. Cecil 1969, 80.

20. *LT*, March 29 and December 27, 1902.

21. "Papers concerning motoring, 1901–12," OX 27.

22. Giacomelli and Bertocchi 1994; documents in OX 22.

23. "Hand written official publication of Giuseppe Marconi's will by the *Notario* (comm. of oath)," OX 22/31.

24. Solari 2011, 121 and 125.

25. G. Marconi 1913, 123, OX 54. See also *LT*, June 23, 1904; *Oxford Chronicle*, June 24, 1904; *Guardian*, June 29, 1904.

26. Brownsea was also the location of Major Robert Baden-Powell's first scout camp, set up in 1907. Charles van Raalte, who also served as mayor of Poole, died unexpectedly while on a trip to India in 1908 and is buried on the island. Florence van Raalte maintained the Brownsea traditions until 1925, when she sold it to Mary Bonham Christie for £125,000. In 1961 the island was taken over by Britain's National Trust, in lieu of death duties, and has been operated as a tourist site and nature reserve since that time. According to the National Trust brochure, "Brownsea Island," Marconi was "one of the more unusual visitors."

27. D. Marconi 1962, 162.

28. B. Marconi, Memoir, FP.

29. Beatrice Marconi (Memoir) and Degna Marconi (1962) do not give dates, but according to archival records, Marconi was at the Haven from July 6–11, 1904, then at Bath and Liverpool for a few days on business, then in London around the fifteenth.

30. D. Marconi 1962, 163.

31. Ibid., 164.
32. According to the Dromoland guest book, Marconi was there on September 4–5, 1904 (Grania Weir, correspondence with author, August 23, 2013).
33. GM to AJM, September 8, 1904, MB.
34. *Halifax Herald*, October 7, 1904.
35. GM to AJM, November 16, 1904, MB; GM to HCH, October 27, 1904, MB; GCR, 30/2.
36. GM to HCH, December 13, 1904, MB.
37. BOM to LOB, December 21, 1904, cited in D. Marconi 1962, 165.
38. GM to AJM, December 21, 1904, MB.
39. D. Marconi 1962, 166.
40. GM to AJM, January 14, 1905, MB.
41. The story made it from Rome to the *New York Times* in a matter of hours. "Marconi Engaged Again? His Fiancée Said to Be Princess Giacinta Ruspoli," read the January 17, 1905, headline, citing Rome's *La Patria*; the next day, the *NYT* picked up a dispatch from *L'Italia* reporting that Marconi and the princess were *not* engaged.
42. "Marconi Romance. 'Wireless' Wizard Engaged to a Peer's Daughter. Beauty and Wealth," *Daily Mirror*, January 24, 1905.
43. Lumsden 2004, 26.
44. Marconi was, of course, the principal shareholder in MWTC and its subsidiaries, but, as we've seen, the company's financial situation was chronically precarious, and Marconi was constantly buying up shares or lending it money, using his personal capital (largely, the proceeds from the sale of his original patent). See chapter 14.
45. L. O'Brien 1952, FP.
46. The O'Brien family history that appears in the next several paragraphs is taken from the following sources: Joe McCarthy, "Dromoland: A Brief History of Dromoland Castle." Brochure, 1971; http://www.dromoland.ie/index.html; "O'Brien Family Gathering," *LT*, March 18, 1936; thepeerage.com; personal communications and observations during my own visit to Dromoland, July 2013.
47. D. Marconi 1962, 157.
48. Ibid., 155.
49. de Sousa 1919, 318, OX 55.
50. GM to AJM, February 1, 1905, MB.
51. A copy of the settlement and the original holograph will are in OX 22/36 and /42. Marconi wrote a more extensive holograph will on April 8, 1906, leaving Beatrice the total income from all his "stocks, shares, loans and investments"; the property inherited from his father in Bologna to his mother; six thousand MWTC shares to Alfonso; two thousand shares to each of his two nephews, sons of his half-brother Luigi, who died in February 1906; three hundred shares each to Kemp and William Bradfield (his chief American engineer); and 150

shares to Kershaw. The executors of the 1906 will were Hall, Lowry-Corry, and the Marconis' Bologna lawyer, Leonida Carpi (OX 22/44–45).

52. D. Marconi 1962, 170; *NYT,* March 17, 1905; *LT*, March 17, 1905.
53. The Dromoland guest book shows Marconi, Beatrice, Annie, and Alfonso in residence between March 19 and March 22, 1905. According to Beatrice (Memoir), the newlyweds arrived on March 17, 1905, and stayed a week.
54. BOM to Lady Inchiquin, March 18, 1905 (emphasis in original), RI 91.
55. In 1908, shortly after the birth of Degna; possibly once more in 1909 (Marconi addressed a letter to her there on their anniversary in March 1909, but if she was there she did not sign the visitors book); and then not until August 1954, when she was accompanied by her son Giulio. Marconi visited briefly in December 1905, probably on his way to Clifden, and again in 1913 (Dromoland guest book, courtesy of Grania Weir, correspondence with author August 23, 2013).
56. B. Marconi, Memoir, FP.
57. BOM to G.S. Kemp, n.d., OX 123.
58. B. Marconi, Memoir, FP.

CHAPTER 13

1. D. Marconi 1962, 177.
2. F.H. Betts to GM, April 11, 1905, OX 532.
3. Marconi Wireless Telegraph Co. of America v. De Forest Wireless Telegraph Co., Circuit Court, S.D. New York. 138 F. 657; 1905 U.S. App. LEXIS 4624. April 11, 1905.
4. MWTC, Report of Proceedings at the Sixth Ordinary General Meeting, March 31, 1903, OX 401.
5. Dunlap 1944, 175.
6. G. Marconi, "Improvements in Transmitting Electrical Impulses and Signals and in Apparatus Therefor," UK Patent No. 12039/96. Copies in OX 411 and OX 414.
7. There is no biography of Fessenden other than an apologetic hagiography by his wife, Helen, which has been described as both "bitter and defensive" (Douglas 1987, 329 note 31), and "in a sense a love story, cataloguing inventions stolen from and injustices visited upon an innocent and long-suffering genius" (Zuill 2001). This section is based on various sources, including H. Fessenden 1940; Douglas 1987; Belrose 2006; and Zuill 2001; as well as sections of http://earlyradiohistory.us/sec007.htm and GCR.
8. R. Fessenden 1925, part VI (June), 2218.
9. Douglas 1987, 45.
10. Belrose 2006, 411–12.
11. R. Fessenden, "High Speed Telegraphy in Connection with an Alaska-Russian Telegraph Line," *Electrical World*, August 7, 1897, OX 1434.
12. RAF to GM, October 6 and 16, 1899, OX 18.
13. RAF to GM, October 6, 1899, OX 18.

14. Douglas 1987, 45; Aitken 1985, 25.

15. HCH to GM, June 12, 1901, and encl.; HCH to GM, July 25, 1901; both in OX 175.

16. "Government Test of Wireless Telegraphy," *NYT*, April 27, 1902.

17. Douglas 1987, 112.

18. Fessenden's backers were Thomas H. Given, president of the Farmers Deposit National Bank, and Hay Walker Jr., a soap and candle manufacturer.

19. US Patent Office 706,735 to 706,747, all dated August 12, 1902, OX 418.

20. C.S. Franklin to MWTC, December 18, 1902, OX 416; GM to HCH, October 4, 1902, OX 171; GM to HCH, October 12, 1902, OX 171.

21. Lee de Forest published an autobiography (de Forest 1950). Also, Samuel Lubell, "Magnificent Failure," *Saturday Evening Post*, January 1942; LDF; GCR; and OX.

22. Lee de Forest, "Reflection of Electric Waves of Very High Frequencies at the Ends of Parallel Wires" (doctoral thesis, Yale 1899).

23. L. de Forest, *Diaries 1891–1903*, April 11, 1899, LDF.

24. LDF to GM, September 22, 1899, OX 381.

25. "Brief Company Histories from the Radio Industry, 1900-1930s," GCR (online catalogue).

26. L. de Forest, *Diaries 1891–1903*, January 13, 1902, and March 27, 1902, LDF.

27. "A New York Rival of Marconi," *New York World*, April 1902, cited in de Forest 1950, 134.

28. Ibid, 135.

29. Douglas 1987, 94.

30. GM to HCH, October 12, 1902, OX 171.

31. LDF to RAF, January 27, 1903; RAF to LDF, January 28, 1903; RAF to LDF, January 31, 1903; all in GCR, 57/4.

32. RAF to LDF, March 28, 1903, GCR 57/4.

33. A. Righi and B. Dessau to *Electrician*, April 1, 1903, OX 18. The review was published in *Electrician* no. 1295 (April 1, 1903): 857.

34. GM to *Electrician,* April 7, 1903, OX 18.

35. A. Righi to GM (translation), May 2, 1903, OX 18.

36. F.H. Betts to MWTC, May 22, 1903, OX 416.

37. US Patent Office, "In the matter of the Interference between an Application of Guglielmo Marconi and an application of Reginald Fessenden. Interference No. 22,360. Preliminary Statement of Guglielmo Marconi," n.d. [September 1903], OX 416.

38. LDF to RAF, September 9, 1903; RAF to H. Walker, October 9, 1903; both in GCR 57/5.

39. HCH to GM, October 26, 1903, OX 418.

40. US Patent Office to GM, December 22, 1903, OX 418.

41. F.H. Betts to MWTC, December 31, 1903, OX 418. The evidence submitted included Marconi's diaries for 1901 and 1902 (OX 1, OX 2) as well as Kemp's diary for 1898 (OX MS. Photogr. d. 43).

42. Seifer 1996, 280.

43. MWTC directors' meeting minutes, February 23, 1904, OX 393.

44. *NYT*, July 3, 1904.

45. GM to HCH, July 11, 1904; HCH to GM, July 12, 1904; both in MB.

46. Douglas 1987, 97–98.

47. G. Marconi, Diaries and Notebooks, September 1904, OX 13.

48. "Minutes of evidence of an investigation of interference between applications of Marconi, Fessenden and Shoemaker, 1904," OX 533.

49. LDF, Reel 2, October 25, 1904.

50. See HCH to GM, July 12, 1904, OX 532.

51. "Minutes of evidence of an investigation of interference between applications of Marconi, Fessenden and Shoemaker, 1904," OX 533.

52. See JAF to HCH, July 8, 1904; HCH to GM, July 8, 1904; HCH to JAF, July 8, 1904; all in OX 418.

53. Marconi Wireless Telegraph Co. of America v. De Forest Wireless Telegraph Co. Circuit Court, S.D. New York. 138 F. 657; 1905 U.S. App. LEXIS 4624. April 11, 1905. One piece of evidence submitted by Marconi that may have influenced the decision was the erroneously dated letter from Ambassador Annibale Ferrero (mentioned in chapter 3) suggesting that Marconi had demonstrated his apparatus in London in 1895 and not 1896.

54. Carpmael & Co., "Report on Dr Marconi's Patent No. 12309 of 1896," November 30, 1906, OX 530. Also Sheffield & Betts to MWTC, December 27, 1916, OX 538.

55. "Decision in Wireless Suit...Both Claim Victory," *NYT*, April 13, 1905.

56. W. Bradfield to GM, April 11, 1905, OX 381.

57. MWTC director's meeting minutes, June 21, 1905, in OX 393.

58. J.A. Fleming, "Memorandum on the ruling of Judge Townsend in the Marconi vs De Forest action," June 26, 1905, OX 532; GM to HCH, June 30, 1905, OX 532;

59. HCH to GM, October 1 and 13, 1905, OX 418.

60. LDF to GM, October 30, 1905; GM, unaddressed, October 31, 1905; both in OX 249.

61. GM to HCH, November 7, 1905, in MB; HCH to Departments, n.d., OX 416.

62. HCH to GM, June 12, 1906, OX 418; HCH to The London & South African Agency, Ltd., January 3, 1906, OX 205.

63. Zuill 2001.

64. J. Bottomley to HCH, May 21, 1906; HCH to GM, May 28, 1906; both in OX 391.

65. Typescript, unpublished letter from RAF to *LT*, February 25, 1907, OX 391.

66. Clark 1936, OX 36.

67. *NYT*, July 21, 1937; de Forest 1950, 422.

CHAPTER 14

1. See "Wireless Telegraphy," *LT*, September 3, 1903; HCH to GM, September 5 and 9, 1903, OX 175; MWTC directors' meeting minutes, September 8, 1903, OX 393.
2. GM to HCH, September 4, 1903, MB.
3. Italy had agreed, in the final version of the Berlin protocol, to try to negotiate a modification of its contract with Marconi. See Regnandi to GM, November 11, 1903, OX 212.
4. HCH to GM, September 26, 1903, OX 204.
5. HCH to GM, December 16, 1903, OX 204.
6. HCH to GM, December 18, 1903, OX 204.
7. G. Marconi, "*A Sua Eccellenza, Il Ministro della Marina, Roma*," translation dated January 7, 1904, OX 204.
8. "The Marconi Company and the USSR," information sheet, n.d., OX 1821.
9. A Russian naval commander, Wladimir Semenoff, vividly described an incident that occurred during the blockade of Port Arthur in the early stages of the war, where the German-equipped Russian ships were unable to communicate with one another: "And so the Russian battle fleet, the last card in our game, was placed at the disposal of Herr Slaby-Arco for experimental purposes" (W. Semenoff, *Rasplata (The Reckoning)* [London: John Murray, 1909], 282).
10. "Russia and Wireless Telegraphy," *LT,* April 20, 1904.
11. Sato and Sato 2006, 461.
12. Headrick 1991, 123.
13. "Aleksandr Stepanovich Popov," in Giorgi and Valotti 2010, 85.
14. "The Month with Marconi," *Munroes' Marconigram* 3, no. 8 (November–December 1905), GCR 16/5.
15. Sato and Sato 2006, 461.
16. GM to AJM, December 10, 1904, MB.
17. HCH to GM, January 8, 1904, OX 176.
18. Faleni and Serafini 1904, OX 41 /33–73. The *Album Marconi* is an impressive cultural artifact, including ads for branded products such as a Marconi homburg by Borsalino, an aperitif called Amaro Marconi, and a line of Marconi havana cigarillos. On the front cover, framed in an art nouveau motif, two angels emerge from a pair of antenna towers to float over the globe, connected by an electrical spark, as Marconi and his famous transmitter look on. The back cover is a colourful display ad for the Buenos Aires Teatro Marconi, a neoclassical structure that opened in 1903 and was used for opera well into the 1950s.
19. de Sousa 1919, 227, OX 55; Solari 1942; GM to LS, June 3, 1904, and July 9, 1904, in Solari 1942; Pession 1941, chapter 11.
20. MWTC directors' meeting minutes, January 12, 1904, OX 393; *LT*, May 31, 1904; *NYT,* August 21, 1904.
21. de Sousa 1919, 303–5, OX 55.

22. GM to AJM, August 14, 1904, MB.

23. Ibid.

24. de Sousa 1919, 306, OX 55.

25. HCH to GM, March 17, 1904, OX 204.

26. HCH to GM, March 25, 1904, OX 204.

27. "MWTC Limited," *LT*, March 31, 1904.

28. "Transatlantic 'Wireless' Service. Mr Marconi to Start it Almost Immediately," *St. James Gazette,* April 15, 1904, OX 218.

29. "Wireless Girdle of the Globe Marconi's Aim," *NYT*, May 15, 1904.

30. MWTC directors' meeting minutes, June 9, 1904, OX 393.

31. Bills (UK), Wireless Telegraphy. A Bill to Provide for the Regulation of Wireless Telegraphy. Presented by Lord Stanley, July 18, 1904.

32. Public General Acts (UK), Wireless Telegraphy Act, 1904. 4 Edw. 7, c. 24. See also Lord Stanley, "Memorandum Explanatory of the Wireless Telegraphy Bill," House of Commons, July 19, 1904, OX 204.

33. "State Control of Wireless. Canada Likely to Follow the British Government's Example," *NYT*, November 7, 1904; Revised Statutes of Canada, Wireless Telegraphy Act*,* 4&5 Edw. 7, c. 49, July, 20, 1905; Australia. Laws, Statutes, etc., Wireless Telegraphy Act*,* Act 8 of 1905, October 18, 1905.

34. The Government of New Zealand passed the world's first wireless legislation in 1903, although there were as yet no wireless operations at all in the colony. The legislation gave the government an absolute monopoly on wireless transmission and reception in order to *prevent* the establishment of wireless in New Zealand, so as to protect the wired telegraphy system and the tax revenue it generated. (Marconi had no stations at all in the southern hemisphere, let alone New Zealand, in 1903). The New Zealand Wireless Telegraphy Act of 1903 is a rare example of national legislation *in anticipation* of a new communication technology. See Radio Spectrum Management, "A brief history of regulation of radiocommunications in New Zealand," http://www.rsm.govt.nz/cms/customer-support/about-us/a-brief-history-of-regulation-of-radiocommunications-in-new-zealand/multipagedocument_all_pages.

35. HCH to GM, July 30 and August 5, 1904; JHH to Lord Stanley, August 9, 1904; both in OX 204.

36. HCH to GM, August 11, 1904, OX 204.

37. Heads of Agreement between the Postmaster General, Marconi's Wireless Telegraph Company, Limited, and the Marconi International Marine Communication Company, Limited, August 11, 1904, OX 204.

38. MWTC directors' meeting minutes, September 6, 1904, OX 393.

39. Ibid.

40. *LT*, October 25, 1905.

41. HCH to H. Babington Smith, August 20, 1905, OX 204.

42. HCH to A.J. Balfour, "Imperial Defence and Wireless Telegraphy," November 10, 1905, OX 204.

CHAPTER 15

1. MacLeod 1982. The site is often said to have been in the nearby town of Port Morien, but it is actually just outside the city limits of Glace Bay.

2. B. Marconi, Memoir, FP.

3. GM to AJM, "Monday" [March 5, 1906], MB.

4. In 1942, the Marconi company was interested in learning the details of Lucia's burial and wrote to the Westminster City Hall for a copy of her grave certificate. They received a rather ghoulish note from the town clerk informing them that Lucia's grave had been dug to a depth of nine feet and there was still room in it for two more interments (Town clerk, City of Westminster, to W.G. Richards, Chelmsford, re City of Westminster cemetery, Hanwell, Grave No. 5095, September 15, 1942, RI 163). According to this document, Lucia was twenty-six days old when she died, dating her birth at February 4, 1906.

5. B. Marconi, Memoir, FP.

6. AJM to H. Kershaw, April 22 and 28, 1906, OX 17; RNV to GM, April 30, 1906, OX 176; "Court Circular," LT, May 5, 1906; HCH to GM, May 24, 1906, OX 176; HCH to GM, May 28, 1906, OX 248.

7. B. Marconi, Memoir, FP; D. Marconi 1962, 181–82.

8. HCH to GM, June 18, 1906, OX 205.

9. GM to HCH, June 19, 1906, MB; HCH to GM, June 20, 1906, OX 205.

10. GM to HCH, June 25, 1906, OX 814 and MB; GM to HCH, June 30, 1906, OX 814.

11. HCH to GM, June 19, 1906, OX 177; GM to HCH, August 6, 1906, MB.

12. NYT, August 16, 1906; NYT, September 12, 1906.

13. HCH, "Wireless Telegraphy. Note on the Berlin Conference Proposals," February 20, 1906, OX 205; see also Hall 1906, OX 205.

14. ITU, International Radiotelegraph Conference (Berlin, 1906).

15. "Wireless Telegraphy Troubles the Nations. Conference to Be Held in Berlin Wednesday to Consider It. Marconi Company Opposed. It Refuses to Co-operate with Other Systems and Germany Would Bring Pressure on it," NYT, October 1, 1906. In 1901, the New York Herald installed Marconi-equipped stations on US Navy lightships operating off Nantucket, allowing US-bound transatlantic passenger ships to announce their impending arrival. After the German government complained about Marconi's ongoing refusal to allow communication with differently equipped stations, the US government had the Herald stations removed in 1904. The navy then installed its own station, which was boycotted by Marconi operators because it was not using Marconi equipment (see Report of the Inter-Departmental Wireless Telegraphy Board, 1904, in earlyradiohistory.us; "Why the Herald Stopped Wireless Nantucket News," New York Herald, July 14, 1904; "Ignored by Marconi. No Communication From Steamer by Wireless off Nantucket," NYT, October 28, 1904).

16. NYT, October 1, 1906.

17. "Hope for Wireless Accord. Attitude of British Delegates More Favorable than Expected," *NYT,* October 5, 1906.
18. GM to HCH, October 14, 1906, MB.
19. Ibid.
20. HCH to GM, October 12, 1906, OX 205.
21. See, for example, "Wireless Telegraphy Conference. German Rejoicings at British Surrender," *The Morning Post*, October 12, 1906; *Electrical Industries* 286 (October 17, 1906): 1, 396, OX 205.
22. HCH to GM, October 17, 1906, OX 205.
23. Notes from Sir Charles Euan-Smith, chairman of MWTC, on his meeting with the prime minister at 10 Downing Street, October 29, 1906, OX 205.
24. MWTC booklet, press review, December 17, 1906, OX 207; ITU, International Radio Telegraph Conference (Berlin, 1906), Convention radiotélégraphique internationale.
25. "The Wireless Telegraphy Conference," *Electrical Review* 59, no. 1512 (November 16, 1906): 805–6.
26. H.C. Hall, "Wireless Telegraphy," *Electrical Review* 59, no. 1514 (November 30, 1906): 865–66.
27. *NYT*, November 4, 1906.
28. See, for example, MWTC, "Wireless Telegraphy Convention," December 17, 1906, OX 207.
29. HCH to GM, November 8, 12, and 19, 1906; all in OX 205.
30. HCH to GM, November 21 and 22, 1906, OX 205.
31. HCH to GM, November 30, 1906, OX 205.
32. All references to the Select Committee hearings and report in the following paragraphs are based on *Report of the Select Committee on the 1906 Radiotelegraphic Convention,* July 8, 1907, OX 210.
33. G. Marconi, "General remarks," handwritten notes, and typescript addendum, n.d., OX 205.
34. Unsigned [HCH to GM?], n.d., OX 205.
35. *Report of the Select Committee on the 1906 Radiotelegraphic Convention,* July 8, 1907, OX 210.
36. HCH to GM, July 3, 1901, OX 175.
37. GM to HCH, August 4, 1902, OX 175.
38. HCH to GM, March 27, 1903, OX 581.
39. HCH to GM, October 25, 1904, OX 187.
40. See, for example, HCH to W. Laurier, July 11, 1907, LAC, MG26-G.
41. H.W. Allen to GM, February 7, 1908, OX 177.
42. GM to C. Euan-Smith, February 12, 1908, OX177.
43. "Copy of a Resolution passed at a Special Meeting of the Board of Directors of MWTC Ltd.," February 28, 1908, OX 177.
44. "Meeting of the Board of MIMCC Ltd.," March 3, 1908, OX 177.

45. *LT*, March 4, 1908. The significance of the change in regime was not lost on the company's overseas correspondents. In Ottawa, superintendent of telegraphy Cecil Doutre wrote to Laurier's private secretary on April 16, 1908, regarding an outstanding matter: "The fact that Mr Cuthbert Hall…who was largely responsible for the attitude adopted by the Parent Company as well as its subsidiary companies, toward those with whom they were doing business, has been recently forced to resign his position and has been replaced by Mr Marconi himself, will have, I think, quite a bearing on this action…" (LAC, MG26-G, microfilm reel C-861).

46. H.W. Allen to Messrs Hollams, February 29, 1908; see also undated agreement between H.C. Hall and the two Marconi companies, OX 177.

47. Hollams to chairman and directors of MWTC, March 3, 1908, OX 177.

48. HCH to C. Euan-Smith, April 10, 1906, OX 177.

49. C. Euan-Smith to H. Babington Smith, April 13, 1908, and C. Euan-Smith to "B.F.," April 13, 1908; both in OX 177; and "Public companies…Marconi's Wireless Telegraph Company," *LT*, April 23, 1908.

50. HCH to MWTC, April 13, 1908, OX 177.

51. See chapter 21.

52. Maxse and his journal were among those scandalized by the awarding of the Marconi contract and insider trading by ministers in Marconi company shares in 1912. See chapter 21.

53. Hall 1915, 877.

CHAPTER 16

1. It was a Sunday morning and one of the first witnesses was a local minister, Reverend O'Shea, on his way by motorcycle on a narrow track to conduct prayer service for the staff. The incoming plane frightened the donkeys carting turf to the station. After a walk over the bog that Brown described as "a dragging discomfort," the aviators were treated to a full Irish breakfast in the station's mess hall. A monument on the main road points to the historic spot (Lynch 2009, 176).

2. HCH to GM, September 1, 1904, OX 193; GM to AJM, December 10, 1904, MB.

3. HJD to HCH, August 5, 1905, OX 193; Shane Joyce, "Marconi in Connemara" (correspondence with author, 2013). According to Joyce, a Clifden civil engineer who has done extensive research on the station, the vendors were a family named Kendall, one of a dozen or so large landed families in West Connemara, and the deed of sale was dated December 30, 1905. Also, Paddy Clarke, "The Marconi Wireless Station at Derrigimlagh—Clifden and Letterfrack," undated article in OX 1845; Clarke 1995.

4. Shane Joyce, "Marconi in Connemara" (correspondence with author, 2013).

5. Clarke 1995; W.S. Entwistle, *Yearbook of Wireless Telegraphy & Telephony*, 1922, 1,251 (courtesy of Shane Joyce).

6. Heather 1993, 73. The Walls had roots in Wexford, but Henrietta's father, Richard Henry Wall, had settled in Connemara and bought Errislannan in 1793. The house is now owned by Stephanie Brooks, who graciously offered two visitors an ample glass of single malt Scotch when they dropped in unannounced in July 2013.

7. Shane Joyce, "Marconi in Connemara" (correspondence with author, 2013).

8. P.H. Bodlein, Acting Clerk, Poor Law Office, Clifden, to GM, February 17, 1906, OX 193.

9. "Opening of the Service," NYT, October 18, 1907; "Bernhardt on Wireless," NYT, October 23, 1907.

10. "Our Marconi Wireless Service" (editorial), NYT, November 4, 1907; "Marconi's Triumph" (editorial), NYT, December 2, 1907.

11. "Wireless Wonders Are Yet to Come," NYT, December 19, 1907.

12. Shane Joyce, "Marconi in Connemara" (correspondence with author, 2013).

13. "Industrial Canada," LT, August 15, 1908.

14. William Appleton, "The Marconi Company in Canada 1906," 6–7, BEA, MG 12.214, section F.1.

15. "Marconi—The Man," Canadian Marconi Company, unpublished typescript, October 1956, in LAC, MG 28 III 72.

16. Vyvyan 1933, 48.

17. I was shown around in July 2013 by the current owner, Russell's son Doug Cunningham, who told me: "Everyone loved Marconi. These were good jobs."

18. E. O'Brien, Diary, June 28–July 27, 1907, OX 192. All material referring to Eileen O'Brien's diary is from this source.

19. D. Marconi 1962, 182.

20. "Marconi Station, Glace Bay, Nova Scotia, Cape Breton, Canada," July 9–17, 1907, depicting Marconi, Vyvyan, and Beatrice, Eileen, and Barnaby O'Brien, OX b.62.

21. E. O'Brien, Diary, June 28–July 27, 1907, OX 192.

22. "Wireless across the Atlantic Soon," NYT, September 24, 1907.

23. Charles Moberly Bell, editor of the London Times, was the first to suggest the term marconigram to describe a wireless telegram. Marconi was asked what he thought and was noncommittal; he said he had "no personal objection...but he does not care to give it his personal recommendation" (W. Blaydes to HCH, March 27, 1903, OX 173).

24. B. Marconi, Memoir, FP.

25. Lady Inchiquin to GM, February 2, 1908, OX 17. Beatrice was in Glace Bay July 10, 1907, to January 8, 1908; Marconi was there July 10 to July 17, 1907; September 13 to November 8, 1907; and early December 1907 to January 4, 1908.

26. "Somborne Park," Hampshire Gardens Trust Research, http://hantsgtrg.pbworks.com/w/page/16630124/Somborne%20Park.

27. B. Marconi, Memoir, FP.

28. "Marconi Arrives to Inspect Wireless," *NYT*, September 10, 1908.
29. GM to LOB, September 1, 1908, RI 153.
30. SFP to GM, September 26, 1908, OX 581; *Barnstable Patriot*, September 20, 1908; *LT*, September 30, 1908.
31. GM to BOM, October 9, 1908, MB.
32. D. Marconi 1962, 186.
33. GM to BOM, November 13, 1908, MB.
34. GM to BOM, December 10, 1908, MB.
35. GM to BOM, December 11 and 14, 1908, MB.
36. GM to BOM, December 23, 1908, MB.
37. B. Marconi, Memoir, FP.
38. GM to BOM, [December 14, 1908], MB; GM to BOM, November 13, 1908, MB.
39. GM to BOM, December 1, 1908, MB.
40. "Draft of proposed agreement between MWTC Ltd and his Grace the Duke of Manchester," n.d. [1913], OX 193. Also, Manchester to GM, October 7, 1911, OX 193.
41. "Kim" to GM, n.d., OX 18/180.
42. GM to BOM, December 7, 1913, MB; G. Marconi, Diaries and Notebooks, 1919, OX 8.
43. Manchester 1932. Villiers-Tuthill (2002) writes that Manchester's memoir portrays "a conceited snob [with] delusions of grandeur" and no regard for his Irish tenants.
44. Manchester to GM, November 27, 1934; GM to Manchester, November 30, 1934; both in LIN 4/227.
45. John R. (Jack) Binns, testimony to US Senate inquiry on the *Titanic* disaster, Day 13, May 3, 1912, http://www.titanicinquiry.org/.
46. See "Papers relating to the sinking of the SS *Republic*," OX 296; "The Republic-Florida Collision," Jack Binns—Wireless Hero, http://jackbinns.org/the_republic-florida_collision.

CHAPTER 17

1. Crawford 1984.
2. "Alfred Bernhard Nobel," Encyclopaedia Britannica, http://www.britannica.com/biography/Alfred-Bernhard-Nobel; "Full text of Alfred Nobel's Will," Nobelprize.org, http://www.nobelprize.org/alfred_nobel/will/will-full.html.
3. JAF to GM, October 4, 1901, OX 37/1.
4. Regarding Marconi's 1903 nomination, see W. Blaydes to Monsieur l'Administrateur Délégué, Cie de Télégraphie sans Fil, Bruxelles, March 17, 1903, OX 173; *NYT*, October 22, 1903; Feldman 2000.
5. Feldman 2000, 7.
6. Dalle Donne and Valotti 2010.
7. Feldman 2000, ix.

8. Thor Lutken to HCH, February 2, 1904; HCH to GM, February 6, 1904; both in OX 37.

9. Crawford 1984, 64.

10. See GM to BOM, November 13, 1908, MB, cited in chapter 16.

11. C. Sunnelius to GM, November 10, 1909, OX 37/42r-v.

12. "Guglielmo Marconi—Facts," nobelprize.org, http://www.nobelprize.org/nobel_prizes/physics/laureates/1909/marconi.html.

13. Barbara Valotti and Mario Giorgi, *Marconi—His Life with Images*, FGM, 2010 (online); "Karl Ferdinand Braun," in Giorgi and Valotti 2010, 89.

14. Marconi supposedly told this story at a lunch at Caltech during his visit there in October 1933 (Clark 1936, OX 36). Braun spent the last years of his life in the United States, after coming to testify in a patent case early in the First World War and staying until his death in Brooklyn in 1918. In Germany today, he is often confused with rocket scientist Werner von Braun (Christian Herzog, correspondence with author, December 17, 2013).

15. Oliver Lodge was never nominated; he was one of several nominators of Augusto Righi in 1908 (Crawford 1987).

16. G. Mittag-Leffler to P. Painlevé, December 9, 1908; P. Painlevé to G. Mittag-Leffler, December 28, 1908; both cited in Crawford 1984, 141.

17. GM to BOM, September 26, 1909, MB.

18. Crawford 1984, 143.

19. Mittag-Leffler continued his campaign according to plan, soliciting nominations for Poincaré from some fifty eligible nominators, including all the Nobel laureates in physics, in 1910; Marconi was one of the thirty-four who signed, giving Poincaré the highest number of nominations ever received by a candidate in a single year to that time, but Poincaré never got the prize; he died in 1912.

20. Crawford 1984, 166.

21. Nobel physics committee minutes, February 2, 1901, cited in Crawford 1987, 160.

22. Thor Thörnblad to GM, November 29, 1909, OX 37/65–66.

23. In her memoir, Lilah O'Brien mentions meeting Hedin and Lagerlöf (L. O'Brien 1952, FP).

24. "Nobel Prize in Physics 1909—Presentation Speech," Nobelprize.org, http://www.nobelprize.org/nobel_prizes/physics/laureates/1909/press.html.

25. "Ferdinand Braun—Nobel Lecture: Electrical Oscillations and Wireless Telegraphy," Nobelprize.org, http://www.nobelprize.org/nobel_prizes/physics/laureates/1909/braun-lecture.pdf.

26. "Guglielmo Marconi—Nobel Lecture: Wireless Telegraphic Communication," Nobelprize.org, http://www.nobelprize.org/nobel_prizes/physics/laureates/1909/marconi-lecture.pdf. Marconi's original draft is in OX 50.

27. Cited in Giorgi and Valotti 2010, 45. Regarding his stop in Berlin: on January 6, 1910, Marconi reported to his board that in passing through Berlin, on his

way back to London, he met with Georg von Arco and discussed a possible "business understanding" between the Marconi company and Telefunken for countries where they were in severe competition (MWTC directors' meeting minutes, January 6, 1910, OX 394). This was his only known visit to Berlin. Marconi and Telefunken negotiated a series of non-competition agreements covering various countries in 1912–13 (see OX 417 and chapter 20 below).

28. Toscano 2012.
29. "*Manifeste du futurisme*," *Le Figaro*, February 20, 1909.
30. Marinetti 1913, cited in Andrews 1991, I.
31. Andrews 1991, 1–2; see also Lyttleton 2014.
32. Hellemans 1995, 293. See, for example, "Wells and the Marconi inventions," *NYT*, January 31, 1903. ("No sooner does Mr H.G. Wells dream something up than along comes that practical thinker Marconi and makes it real....")
33. Toscano 2012.
34. Campbell 2006.

CHAPTER 18

1. MWTC directors' meeting minutes, April 7 and 10, 1908, OX 581.
2. HJD to GM, October 29, 1909, OX 581.
3. GM to HJD, October 29, 1909, OX 581.
4. MWTC directors' meeting minutes, January 25, 1910, OX 394.
5. GM to LS, January 29, 1910, in Solari 1942, TAV. XXXI.
6. The Isaacs Family and the Marquesses of Reading, from c. 1659, www.tzorafolk .com/genealogy/history/Reading.htm.
7. T.P. O'Connor, "Late Mr. Godfrey Isaacs: A Personal Memoir," *Daily Telegraph*, April 18, 1925.
8. "Mayfair Gallery: Men of the Day," *Mayfair* newspaper supplement, October 9, 1915, 7.
9. "In the Central Criminal Court. Rex v Cecil Chesterton. Proof of Commendatore G. Marconi," n.d. [May 1913], OX 224.
10. Draft agreement between MWTC and GM, 1910, OX 814.
11. *NYT*, July 28, 1910.
12. Posthumous list of Marconi's positions with the company, OX 654.
13. See OX 369, OX 413, and OX 415 for patents in names of various company employees (Entwistle, Gray, Franklin, Round, et cetera), some in their own names, some jointly with the company.
14. A. Gray to GCI, August 29, 1910, OX 814.
15. Chronological notes on company history, OX 1866.
16. *Argentine Radio: Over 60 Years on the Air*, Argentine Information Secretariat, 1981.
17. London Borough of Richmond Upon Thames, "Local History," http://www .richmond.gov.uk/local_history_trumpeters_house.pdf; see also B. Marconi, Memoir, FP.

18. *LT*, April 1, 1911; May 19 and 26, 1911; July 6 and 9, 1911; B. Marconi, Memoir, FP.

19. B. Marconi, Memoir, FP.

20. GM to BOM, May 6 and 8, 1911, MB.

21. BOM to GM, n.d., RI 93 (the reference to Giulio's birthday dates the letter as May 21, 1911); according to Kemp's diary, the Marconis were in Clifden on May 17–18, 1911.

22. Marriage Records for Charles Clover and Florence Bell, http://www.ancestry .ca/genealogy/marriage-records/charles-clover-and-florence-bell.html.

23. GM to C. Clover, July 28, 1911, OX 18; GM to F. Clover, July 31, 1911, OX 18.

24. GM to C. Clover, August 1, 1911, OX 18.

25. GM to F. Clover, August 2 and 4, 1911, OX 18.

26. Vita Sackville-West, *The Edwardians*, London: Virago Press, 1983 (reprinted 2008), 171–73. Originally published in 1930 (London: Hogarth Press).

27. Marconi's sister-in-law Lilah wrote: "One always felt one could go to him about a personal affair and receive sympathy and help. He enjoyed being consulted" (L. O'Brien 1952, FP).

28. For example: The Clovers' address in Wales (n.d., 1914, OX 5); "Cartier Betty Clover" (January 24, 1925, OX 11); "Degna – Giulio – Gioia – Betty Clover – Gina" (January 31, 1925, OX 11); "dinner Mrs Clover – Betty C" (November 18, 1925, OX 11); "scritto Degna Mrs Clover" (January 20, 1926, OX 12).

29. G. Marconi, Diaries and Notebooks, January 31, 1925, OX 11.

30. G. Marconi, Diaries and Notebooks, January 24, 1925, OX 11.

31. Charles and Florence Clover remained together and had two more daughters. In June 1935, Betty Clover married Geoffrey Adams, an Australian cricket champion, in Cheshire, England. Marconi was in London at the time, but he does not seem to have attended the wedding. Betty was described in a 1998 obituary of Geoffrey Adams as "a noted beauty with a love of parties and one of three sisters known as 'the belles of Cheshire'" (*LT*, March 13, 1998). In 1947, Florence Clover was in London's Chancery Court claiming a share in the estate of Sir Edmund Fleming Bushby, a Liverpool cotton magnate with whom she said she had had an intimate relationship for over twenty years beginning in 1921. She testified that she had met Marconi in Canada as a child (through her family's connection to Alexander Graham Bell) and remained in touch with him after she married and moved to England. In 1914, she said, Marconi gave her six thousand New York telephone bonds as commission for facilitating a business deal; she gave Bushby the bonds to hold for safekeeping in 1922 and later allowed him to sell some of them. Some time after that, she said, she asked Marconi to give her £5,000, "which he did when I saw him breakfasting in bed after dining with him at the Savoy." She gave the money to Bushby, who used it for gambling, she said in court (*The Advertiser* [Adelaide], October 24, 1947).

32. G. Kemp, Diaries (exp.), April 11, 1911, OX 100.

33. GM to BOM, "Tuesday" [possibly April 11 or 18, 1911], MB. See also Shane Joyce, "Marconi in Connemara" (correspondence with author, 2013).
34. Clarke 1995, 23.
35. G. Kemp, Diaries (exp.), May 21, 1911, OX 100.
36. Shane Joyce, correspondence with author, September 23, 2014.
37. G. Kemp, Diaries (exp.), May 26, 1911, OX 100. Kemp referred to Marconi as "Dr." in his diary after Marconi was awarded an honorary degree from Oxford in 1904.
38. Ibid., July 15 and 16, 1911, OX 100.
39. *LT*, July 21, 1911.
40. Congested Districts Board for Ireland, Dublin, to GM, July 21 and August 3, 1911; both in OX 193.
41. G. Kemp to GCI, September 26, 1911, OX 193; see also G. Kemp to GM, September 21, 25, and 26, 1911, all in OX 193.
42. MacLeod 1992, 116.
43. AGBMA, July 26, 1914, HN vol. 84, 136.

CHAPTER 19

1. Mango 1999.
2. GM to GCI, n.d. [December 15, 1911], OX 19.
3. *LT*, November 20, 1911.
4. W.J. Baker 1970.
5. *LT*, January 6, 1912.
6. "Agreement between the Italian Government and Guglielmo Marconi for the use of Radiotelegraphy and Radiotelephony in Italy and in the Colonies," 1915, Annex 6: "Working of the Coltano Station" (L. Cattolica to MWTC, Rome, December 29, 1912), OX 212.
7. "The Tripoli Campaign: Review of the Situation," *LT*, January 2, 1912.
8. See, for example, "The Recent Progress of Radiotelegraphy," lecture given by G. Marconi, Rome, March 3, 1914, translation, OX 51.
9. "Agreement between the Italian Government and Guglielmo Marconi for the use of Radiotelegraphy and Radiotelephony in Italy and in the Colonies," 1915, Annex 5: "Summary of a Report Regarding Comparative Experiments Executed between Marconi and Telefunken Transportable Stations [May 1911]," OX 212.
10. W. Blaydes to HCH, October 8, 1902, OX 171.
11. de Sousa 1919, 289, OX 55.
12. de Sousa 1919; Solari 1928, 230.
13. Minutes of Evidence from the Select Committee on Marconi's Wireless Telegraph Company, Limited, Agreement (1913), May 7, 1913, 513, OX 222; de Sousa 1919, 289, OX 55.
14. *LT*, January 6, 1912.

15. This and following, GM to GCI, n.d. [December 15, 1911], OX 19. In contrast to Marconi's updates, the *Times,* in fine ethnocentric fashion, reported: "There is little doing at the moment and 40,000 Italians are eating the *polenta* of idleness" (January 2, 1912).
16. GM to BOM, December 10, 1911, MB.
17. *NYT,* March 24, 1912.
18. GM to AJM, n.d. [December 11, 1911], OX 19; G. Kemp, Diaries (exp.), December 24, 1911, OX 100.
19. *LT,* January 6, 1912.
20. McMeekin 2011, 80.
21. de Sousa 1919, 288, OX 55.
22. Sterling 2007, 369.
23. GM, draft letter fragment, n.d., OX 19.
24. V.I. Lenin (signed "T"), "The End of the Italo-Turkish War," *Pravda* 129, September 28, 1912, Lenin Internet Archive.
25. "Mr. Marconi's New House," *LT,* February 26, 1912. Eaglehurst was on the housing market in 2014 for an asking price of £6.5 million. One of its featured attractions was that it had once been the home of Guglielmo Marconi.
26. D. Marconi 1962, 198.
27. Ibid.
28. "Folly Which Inspires Great Ideas," *Southern Evening Echo,* November 28, 1986; "Luttrell's Tower and Marconi," information sheet; both in OX 1759.
29. D. Marconi 1962, 197.
30. British Post Office, "License to use Wireless Telegraphy for Experimental Purposes," issued by The Right Honourable Herbert Louis Samuel MP, Postmaster General, to Guglielmo Marconi, Esquire, March 20, 1912, OX 24/98.
31. B. Marconi, Memoir, FP.
32. GM, unaddressed fragment, n.d., OX 18 / 190–91.
33. HJD to GM, December 19, 1911, OX 581.
34. HJD to GM, August 7, 1912, OX 581. Marconi's letter of August 6, 1912, is lost.
35. HJD to GM, August 27, 1912, OX 581.
36. HJD to GM, September 27, 1912, OX 581.
37. Ibid.
38. Davis's son Ronald Jameson-Davis, ninety-one years old in 2013, described meeting Marconi with his father on two occasions as a boy, once at the Ritz Hotel in London and once in Brighton (John Jameson-Davis, correspondence with author, September 3, 2013).
39. Solari 2011, 173–76.
40. Marconi had left his Rolls-Royce behind in England; it was a conspicuous symbol of wealth. A worker at the Fiat plant would have had to work one hundred years to buy one (Chiaberge 2013).
41. Chiaberge 2013, 162–67; Jacot and Collier 1935, 137–39; Masini 1995, 265–69; B. Marconi, Memoir, FP.

42. "Inventor and Wife in Auto Crash," *Philadelphia Inquirer*, September 26, 1912; "Marconi's Hurts Painful," *Chicago Post*, September 26, 1912; *Philadelphia Public Ledger*, September 27, 1912.

43. "Marconi Slightly Improved," *New York Sun*, September 27, 1912; "The Inventor, In Spite of His Injuries, Maintained His Habitual Calm," *New York Herald*, September 27, 1912; "Marconi Progressing Well. But Doctors Are Still Unable to Tell How Much His Eye Is Hurt," *NYT*, October 1, 1912; "Marconi's Eye Badly Hurt. Physicians Can't Say If Sight of It Can Be Saved," *New York Sun*, October 3, 1912; "Mr. Marconi's Sight. Removal of Injured Eye by Vienna Specialist," *The Star*, October 18, 1912.

44. AM to Emma Davis, November 13, 1912, MB; AJM to GM, November 19, 1912, RI 69.

45. Solari 2011, 176.

46. GM to BOM, n.d., MB; GM to BOM, October 21, 1922, MB; GM to LOB, February 28, 1923, FP; GM to DM, December 10, 1924, RI 110; GM to BOM, April 5, 1927, FP.

47. *Glasgow Evening Times*, December [?], 1897, cutting in OX 1435. Examples of photographs in which Marconi's glass eye is apparent include his 1915 Italian passport photo (OX 20/5) and the frontispiece to Solari 1928. Paddy Clarke (1995, 23) writes that people in Connemara remembered Marconi wearing an eye patch during one of his visits there, indicating that he may have occasionally used it to give the glass eye a rest.

CHAPTER 20

1. Minutes of Evidence, 1913, May 7, 1913, OX 222.

2. GCI to Colonial Office, March 10, 1910, in "Imperial wireless installation. Copies of correspondence relating to the Contract for Imperial wireless stations," Great Britain, Parliament, House of Commons, Select Committee on Marconi's Wireless Telegraph Company Agreement, *Report,* 1913, OX 223; and GCI testimony in case of Isaacs v. Hobhouse, July 17–18, 1918, High Court of Justice, King's Bench (copy in OX 540).

3. de Sousa 1919, 363, OX 55. Telefunken would build stations at Dar es Salaam (German East Africa, now Tanzania), Yap (Caroline Islands), Kamina (Togoland, now Togo), Samoa, Nauru (Marshall Islands), Herbertshöhe (German New Guinea, now Kokopo, Papua New Guinea), Duala (Cameroon), Kiao-Chau (now Jiaozhau, China), and Windhoek (German Southwest Africa, now Namibia). All would be captured or destroyed in the early months of the war.

4. W.J. Baker 1970, 131.

5. GCI testimony in Isaacs v. Hobhouse, July 17–18, 1918, OX 540.

6. Donaldson 1962, 54.

7. "Marconi Describes Imperial Wireless," *NYT,* March 16, 1912.

8. GCI testimony in Isaacs v. Hobhouse, July 17–18, 1918, OX 540.

9. *"40 Jahre Groß-Sendestation Nauen,"* *Funk- und Fernseh-Illustrierte* 47 (1956), 18–19. Telefunken Unternehmensarchiv I.4.NL.048 G. Goebel 0300.

10. Partheil 1907, 102.

11. Partheil 1910, 13.

12. GM to J. Eustace Jameson, April 1908, OX 581.

13. JAF to GM, February 7, 1910, and March 16, 1910, both in OX 417.

14. "Wireless Telegraphy and German Shipping," *LT*, January 16, 1911.

15. *LT*, November 12, 1912, and November 13, 1912.

16. Contract between MWTC and Telefunken, March 6, 1913, OX 417.

17. Headrick 1988, 128.

18. Correspondence between GCI and H. Bredow, November 23, 1912, and following, OX 417.

19. GCI testimony in Isaacs v. Hobhouse, July 17, 1918, OX 540.

20. MWTC directors' meeting minutes, June 20, 1911, OX 394.

21. Donaldson (1962) is the main reference for otherwise unsourced detail on the contract and the ensuing "Marconi scandal."

22. "Wireless to Link All British Empire," *NYT*, February 1, 1912.

23. See GCI to GM, April 10 and 13, 1911, and GCI to M. Nathan, April 13, 1911, all in OX 219.

24. Letter annexed to John Bottomley affidavit in United Wireless v. NESCO suit, April 15, 1912, OX 538.

25. "Accepts Wireless for British Empire. Government Agrees to Marconi Company's Terms for Inter-Imperial Communication," *NYT*, March 9, 1912.

26. "In the Central Criminal Court. Rex v Cecil Chesterton. Proof of Commendatore G. Marconi," n.d., [May 1913], OX 224.

27. G. Kemp, Diaries (exp.), March 1–3, 1912, OX 101.

28. GM to BOM, March 15, 1912, MB.

29. B. Marconi, Memoir, FP; D. Marconi 1962, 198.

30. G. Kemp, Diaries, March 8, 1912, OX 101.

31. See chapter 13.

32. GCR, online catalogue; Douglas 1987, 177–85; Maclaurin 1949, 79–81; Howeth 1963, chapter XI; "Government Raids United Wireless," *NYT*, June 16, 1910; "Five Wireless Men are Found Guilty," *NYT*, May 30, 1911.

33. "Marconi Describes Imperial Wireless," *NYT*, March 16, 1912.

34. GM to BOM, March 22, 1912, MB. Also, GM to BOM, mistakenly dated April 29, 1912, and GM to BOM, February 9, 1910, MB.

35. Lumsden 2004, 52.

36. Maria Montessori's book, *The Montessori Method*, appeared in 1912. Beatrice remembered that Inez had got her interested in the "method" and persuaded Marconi to send her and the children to Rome to try the system. "It was a great success and after that I had always with them a Montessori teacher" (B. Marconi, Memoir, FP). Marconi, who was very much involved in the children's education, also took some credit for the decision. He wrote to Beatrice

on August 15, 1913: "I have made inquiries about the Montessori method from an Italian I met here the other day and hope to soon be able to give you more information" (GM to BOM, August 15, 1913, MB).

37. "10-minute wireless, London to New York," *NYT*, March 17, 1912.

38. "The Marconi Wireless," (editorial), *NYT*, March 18, 1912.

39. Briggs and Burke 2009, 153.

40. "Ocean-Wide Gossip of *Times*'s Wireless," *NYT*, March 20, 1912.

41. Massie and Perry 2002.

42. "Pioneering Amateurs (1900–1917)," http://earlyradiohistory.us/sec012. htm. The word *amateur* was dropped from *Radio Amateur News* in September 1920, and under Gernsback's editorship *Radio News* became the most important of the new radio magazines in the 1920s. Gernsback remained as publisher and editor-in-chief until 1929, and the magazine continued to publish until 1958.

43. "Boy Experts in Wireless Telephoning," *New York Herald*, October 8, 1911.

44. See "The Autobiography of a Girl Amateur," *Radio Amateur News*, March 1920, http://earlyradiohistory.us/1920auto.htm; "The Feminine Wireless Amateur," *Electrical Examiner*, October 1916, http://earlyradiohistory.us/1916fem.htm.

45. H. Gernsback, letter to the *NYT*, March 29, 1912.

46. Douglas 1987, xxiii.

47. "United Wireless Offer Is Too Late," *NYT*, March 17, 1912; "Accept Plan for United Wireless," *NYT*, March 18, 1912; "Waives Wireless Profits," *NYT*, March 20, 1912; "Marconi May Absorb United Wireless," *NYT*, March 25, 1912; "United Wireless Arrangements," *Wall Street Journal*, March 26, 1912. Also, Douglas 1987 and documents regarding UWTC bankruptcy in OX 391 and OX 538.

48. Cited in Douglas 1987, 184–85.

49. GM to BOM, March 22, 1912, MB.

50. GM to BOM, mistakenly dated April 29, 1912, MB.

51. "Marconi Plans New Inventions as Useful as Wireless," *NYT*, March 24, 1912.

52. "Wireless Compass Ends Sea Fog Perils," *NYT*, April 4, 1912.

53. "Boom in Marconi Shares," *NYT*, March 22, 1912.

54. H. W. Allen, undated typescript, OX 1767.

55. Marcus Goodbody affidavit, in UWTC v. NESCO, April 15, 1912, OX 538.

56. Donaldson 1962, 179.

57. "Marconi Head Calls Us Slow in Business. We Waste Time Telephoning and Deliberating, Says G.C. Isaacs, Sailing Away. But He Likes Us Immensely. Our Hospitality Unsurpassed—Enthusiastic Over Wireless Here—Hotels Too Democratic," *NYT*, April 3, 1912.

58. GM to BOM, mistakenly dated April 29, 1912, MB.

59. GM to BOM, April 4 and 5, 1912, MB.

60. "In the Central Criminal Court. Rex v Cecil Chesterton. Proof of Commendatore G. Marconi," n.d. [May 1913], OX 224. Also, Select Committee Report, May 7, 1913.

61. GM to BOM, April 4, 1912, MB.

62. "Parliament," and "The Debate in the Commons," *LT*, April 4, 1912. "Favors Wireless, Not State Cable," *NYT*, April 4, 1912, and "Wireless Merger Is Now Assured," *NYT*, April 6, 1912.

63. Hughes and Bosworth 2012.

64. *NYT*, April 15, 1912.

65. R.M.S. *Carpathia*, reconstructed *procès-verbal*, in Hughes and Bosworth 2012, 84. The original *procès-verbaux* of the ships and shore stations that received messages from the sinking *Titanic* are in OX 260–62.

66. Hughes and Bosworth 2012; Titanic Inquiry Project, http://www.titanicinquiry .org.

67. "False Titanic News to Be Investigated," *NYT*, April 18, 1912.

68. "Biggest Liner Plunges to the Bottom at 2:20 a.m.," *NYT*, April 16, 1912; "Partial List of the Saved," *NYT*, April 16, 1912.

69. GM to BOM, April 16, 1912, MB.

70. "Topics in Wall Street," *NYT*, April 17, 1912; "Plead with Sable Island," *NYT*, April 17, 1912; "Carpathia Not Expected Until Thursday Night with Survivors. Not in Wireless Touch . . . ," *NYT*, April 17, 1912.

71. "Marconi Cheered for Wireless Feats," *NYT*, April 18, 1912; G. Marconi, "The Progress of Wireless Telegraphy," typescript hand-dated April 17, 1912, OX 51.

72. "Carpathia Not Expected Until Thursday Night with Survivors. Not in Wireless Touch . . . ," *NYT*, April 17, 1912.

73. "Marconi Stock Booms," *NYT*, April 18, 1912.

74. "Marconi stock $10,000,000. Brisk Trading in New Shares Which Present Holders May Buy," *NYT*, April 19, 1912.

75. J.E. Milholland, Diaries, May 4, 1912, MFP. Inez's father happened to see the newsreel in a London theatre, and commented on Marconi's demeanour. See http://www.britishpathe.com/video/titanic-disaster-1/query/titanic+disaster; see also, "Marconi Pays Visit to the Rescue Ship," *NYT*, April 19, 1912.

76. "Marconi Pays Visit to the Rescue Ship," *NYT*, April 19, 1912.

77. "Topics of the Times: The World's Debt to Marconi," *NYT*, April 20, 1912.

78. "Throng Honors Dead at Theatre Service," *NYT*, April 22, 1912.

79. "Western Union with Marconi Wireless," *NYT*, April 20, 1912.

80. "Marconi Stock Boom Suddenly Collapses," *NYT*, April 21, 1912; "Topics in Wall Street: Placid Market in Marconi," *NYT*, April 23, 1912.

81. Donaldson 1962, 53.

82. "Marconi Finance. Excessive Speculation in the Shares of the Companies," *LT*, April 23, 1912.

83. All information cited from the hearings is taken from US Senate Report, "Investigation into loss of S.S. '*Titanic*'," OX 290. Also US Senate, "*'Titanic'* Disaster," online at http://www.titanicinquiry.org/.

84. Alexander Graham Bell, "Home Notes" for 1912, L.B., vol. 37, 41, AGBMA.

85. *New York Herald*, August 17, 1912.

86. The amateurs' opposition to the 1912 Radio Act was mirrored nearly a century later by the efforts of cyber-libertarians to keep the Internet free of government regulation. See, for example, Barlow 1996.

87. H.T. Cottam, "Titanic's 'CQD' Caught by a Lucky Fluke," *NYT*, April 19, 1912; H. Bride, "Thrilling Story by Titanic's Surviving Wireless Man," *NYT*, April 19, 1912.

88. See also, "Marconi Pays Visit to the Rescue Ship," *NYT*, April 19, 1912.

89. "No Order to Hold Up Titanic News. On the Contrary, Marconi Tells Senate Committee, Every Effort Was Made To Get It," *NYT*, April 26, 1912; "The Preposterous Smith" (editorial), *NYT*, April 27, 1912; "Threaten to Quit the Titanic Inquiry. Senators Dissatisfied with the Way Chairman Smith Is Conducting It," *NYT*, April 29, 1912.

90. "No Wireless Order to Hold Back News. Sea Gate Operator Explains the Messages to Bride and Cottam on the *Carpathia*. Ship Then in the Harbor. 'Keep Your Mouth Shut' Not Official, but Friendly Words of One Operator to Another," *NYT*, April 27, 1912.

91. David Sarnoff, "Radio," as told to Mary Margaret McBride, *Saturday Evening Post*, August 7, 1926, 141–42.

92. "The Tragedy of the *Titanic*—A Complete Story" (special supplement), *NYT*, April 28, 1912.

93. "Marconi Sails on Safeguarded Liner," *NYT*, May 1, 1912.

94. "Minutes of Evidence of the British Enquiry, 1912," OX 291.

95. MIMCC (unsigned) to Sir R. Ellis Cunliffe, Solicitor to the Board of Trade, May 1, 1912, OX 258/13–18; MIMCC (unsigned) to R.E. Cunliffe, May 7, 1912, OX 258/67.

96. MIMCC, "Notes re Interview with Sir R. Ellis Cunliffe at '*Titanic*' enquiry to-day 7th June, 1912," OX 258/62–63; Memo, G. Turnbull to GM, June 13, 1912, OX 258/157–162.

97. "Minutes of Evidence of the British Enquiry, 1912", G. Marconi testimony, Day 26, June 18, 1912, 713–17, OX 292.

98. *Report on the Loss of the 'Titanic' (S.S.)*, July 30, 1912, paragraph 20, OX 293.

99. ITU, International Radiotelegraph Conference (London, 1912), International Radiotelegraph Convention, Art. 3.

100. *LT*, June 6, 18, and 22, and July 2 and 6, 1912; *NYT*, June 13 and July 4 and 10, 1912. Regarding the 1914 conference, see OX 295.

101. "Spanish and General Wireless Trust (Limited)," *LT*, May 10, 1912.

102. Rollo and Queiroz 2007.

103. "The Pulse of the Empire," *LT*, May 24, 1912; see also "Special Article: The Situation on the Stock Exchange," *LT*, May 30, 1912.

NOTES TO PAGES 365–373   755

104. "The New Home of Wireless. Opening of 'Marconi House,'" *Marconigraph*, June 1912, typescript, OX 201; "The Passing of Marconi House," *Zodiac*, n.d. [1933], 438–41, OX 201; G. Marconi 1923, OX 52; H.A. White, "Moving House on a Large Scale," typescript, September 1933, OX 201.

CHAPTER 21

1. "Political Notes," *LT*, July 11, 1912.
2. W.R. Lawson, *Outlook*, July 20, 1912, and August 31, 1912, the latter cited in Donaldson 1962, 27–28.
3. In his antisemitic treatise *The Jews* (London: Constable, 1922, 3), Belloc wrote that "the continued presence of the Jewish nation intermixed with other nations alien to it presents a permanent problem of the gravest character." See also Searle 1987.
4. *Spectator*, September 14, 1912, cited in Donaldson 1962, 28.
5. Cited in Searle 1987, 197.
6. Maxse owned and edited the *National Review*; in 1913 he published *The Great Marconi Mystery* (London: The National Review, 1913).
7. Cited in Donaldson 1962, 166. Also in Searle 1987, 3.
8. Donaldson 1962, 11.
9. Cited in Donaldson 1962, 54.
10. Great Britain, House of Commons, *Debates*, Vol 42, October 11, 1912, 667.
11. Ibid., 739, cited in Donaldson 1962, 39.
12. Ibid., 717–21, cited in Select Committee Report, 1913.
13. Material in the following paragraphs referring to the Select Committee hearings is taken from Select Committee Report, 1913.
14. Minutes of Evidence, January 27, 1913, OX 222.
15. Cited in Donaldson 1962, 96–97.
16. Ibid., 233.
17. Ibid., 124.
18. G. Kemp, Diaries (exp.), March 1, 1913, OX 102.
19. G. Kemp, Diaries (exp.), March 11, 1911, OX 100.
20. G. Kemp to GM, March 6, 1913, OX 193.
21. As late as April 1916, by which time Manchester had given up Kylemore, Kemp was meeting with the estate manager, a bailiff, and local herdsmen to try to establish the property's boundaries (G. Kemp, Diaries (exp.), March 29, 1916, April 3, 1916, OX 105). The Letterfrack station was closed in August 1917. There were still concerns regarding the tenants, and in March 1918 Kemp advised giving them £10 "as a final offer to keep the peace" (G. Kemp, Diaries (exp.), March 5, 1918, OX 107).
22. G. Kemp, Diaries (exp.), March 26, 1913, and April 26, 1913, both in OX 102.
23. Minutes of Evidence, May 7, 1913.
24. Donaldson 1962, 102.

25. J.E. Milholland, Diaries, July 1, 1913, MFP.

26. Cited in Donaldson 1962, 136.

27. Ibid., 159.

28. Advisory Committee on Wireless Telegraphy, "Report of the Committee appointed by the Postmaster-General [H.L. Samuel] to consider and report on the merits of the existing systems of long distance wireless telegraphy, and in particular as to their capacity for continuous communication over the distances required by the Imperial Chain," April 30, 1913, OX 207.

29. Minutes of Evidence, May 1913, 524, OX 222.

30. Ibid., 526.

31. Ibid., 527.

32. Donaldson 1962, 69.

33. Ibid., 171–89.

34. "Agreement between MWTC Ltd, The Rt Hon Herbert Louis Samuel, Postmaster General, and Commendatore Guglielmo Marconi of Marconi House aforesaid Chief Scientific Adviser and Director of the Company of the third part," July 30, 1913, OX 223.

35. *Reports from the Select Committee on MWTC Ltd. Agreement*, cited in Donaldson 1962.

36. Frank Owen, *Tempestuous Journey: Lloyd George, His Life and Times* (London: Hutchinson, 1954), 236, cited in Donaldson 1962, 207.

37. Donaldson 1962, 213.

38. Ibid., 215.

39. Ibid., 217–20.

40. Searle 1987, 172–200.

41. Ibid., 314.

42. GCI testimony in Isaacs v. Hobhouse, July 17–18, 1918, OX 540.

43. Searle 1987, 187.

44. Sir Edward Sassoon was a member of an aristocratic and prominent Anglo-Jewish family. Born in Bombay, he was married to Aline de Rothschild, a daughter of Baron Gustave de Rothschild of Paris. In February 1902, he succeeded Sir Joseph Sebag-Montefiore as president of the London Spanish and Portuguese Congregation, one of Britain's most important Jewish associations (*Jewish Encyclopedia*, 1906 edition, http://www.jewishencyclopedia.com/articles/13218-sassoon#anchor4).

45. Wheatcroft 1996, 111.

46. Julius 2010, 60–61.

47. R. Kipling, "Gehazi" (1915), in *Rudyard Kipling's Verse. Inclusive Edition, 1885–1926*, Garden City, NY: Doubleday, Page & Co., 1927, 278. Nearly fifty years later, Frances Donaldson was refused permission to quote from the poem in her definitive book *The Marconi Scandal* (Donaldson 1962, 237).

48. Donaldson 1962, 48–49.

49. G.K. Chesterton, "The Sign of the World's End: An Open Letter to Lord Reading," Appendix A in Donaldson 1962, 256–59. Also in Ward 1942, 359–62.

50. Chesterton 1936, 205–6.
51. Donaldson 1962, 10.

CHAPTER 22

1. F.H. Betts to GM, May 8, 1913, OX 538. Judge Van Vechten Veeder ruled "that the claims in issue are valid, not anticipated, and infringed" (Marconi Wireless Telegraph Co. of America v. National Electric Signaling Co., District Court, Eastern District of New York, 213 F. 815; 1914 U.S. Dist. LEXIS 1000, March 17, 1914). Marconi's testimony in the NESCO case, which began on June 13, 1913, became the basis for his most extensive personal statement, *Brief Story of My Life* (G. Marconi 1913, OX 54).
2. GM to BOM, May 28, 1913, MB.
3. Lumsden 2004, 91–94 (including note 16 on page 94); *New York Sun*, July 16, 1913; *NYT*, July 18, 1913.
4. J.E. Milholland, Diaries, July 12, 1913, and July 31, 1913, MFP.
5. J.E. Milholland, Diaries, June 13, 1913, MFP.
6. "Eugen Boissevain," http://spartacus-educational.com/Jboissevain.htm.
7. John Tepper Marlin, correspondence with author, July 26, 2012.
8. Correspondence, IM to E. Boissevain, October–November 1913, IMP, folder 2.
9. Dromoland guest book (Grania Weir, correspondence with author, August 23, 2013); D. Marconi 1962, 215.
10. "Dr. Montessori's Reception in Rome," *La Tribuna*, January 2, 1914 (in A.G. Bell's *Beinn Breagh Recorder* 15, 145, AGBMA).
11. Documents of February 24 and 27, 1914, ACS, MRC, Divisione Prima, Segretaria Reale (1914), b.619, f. 690. Marconi, Comm. Guglielmo. *La conferenza di Guglielmo Marconi all'Augusteum* was published in full by the journal *L'Elettricista*, which had previously reported on Marconi's Italian Navy experiments in 1897 and published his speech at Rome's Campidoglio in 1903.
12. Solari 2011, 186–87.
13. Reports, press releases and newspaper cuttings relating to wireless telephony, OX 307 and OX 309. For example, H.J. Round, "Wireless Telegraphy and Telephony Experiments made on board Italian warships at Augusta (Sicily), March 1914," OX 307. Round's pioneering work with direction-finding systems during the 1914–18 war is described in H. Wicker (MWTC), typescript, n.d., OX 418.
14. de Sousa 1919, 373, OX 55.
15. JAF to GM, February 18, 1914, OX 307.
16. "Marconi Triumph," n.d., OX 307. See also Reuters, "Wireless Telephone at Sea," n.d., OX 309.
17. "Transatlantic Telephony Tests," *Wireless World*, June 16, 1926, OX 309.
18. H.B. Tree to GM, June 25, 1914, OX 18/93.
19. JHH to GM, June 16, 1912; Stamfordham to JHH, June 18, 1912; JHH to GM, June 24, 1912; Stamfordham to JHH, June 20, 1914; JHH to GM, June 22, 1914;

JHH to H.H. Asquith, June 24, 1914 (copy); H.H. Asquith to JHH, June 27, 1914 (copy); JHH to GM, June 29, 1914; JHH to GM, July 19, 1914; all in OX 28.

20. MacMillan 2013, chapter 20.

21. W.J. Baker 1970, 158.

22. Ibid., based on Begbie 1919, OX 361.

23. Begbie 1919, chapter 1, 7, OX 361.

24. MacMillan 2013, 627.

25. de Sousa 1919, 364, OX 55.

26. "Note of an intercepted wireless message in German announcing the outbreak of war, copied down by H.J. Round, 1914," OX 358.

27. Headrick 1988, 115; Magder 2011, 13.

28. Albrecht-Carrié 1938, 20.

29. Ministero della Marina to GM, August 11, 1914, OX 20/10.

30. B. Marconi, Memoir, FP; D. Marconi 1962, 218–9; G. Marconi, Diaries and Notebooks, August 14, 1914, OX 5.

31. G. Marconi, Diaries and Notebooks, August 14, 1914, OX 5; French embassy, London, to French civil and military authorities, August 13, 1914, OX 22/57–58; GM to BOM, August 22, 1914, MB; A. di San Giuliano to G. Imperiali, August 21, 1914, DDI, Quinta Serie, I, doc. 382, 209.

32. "Anonymous typescript memorandum discussing conditions under which Italy might enter the war on the side of the Allies, c.1915," OX 21/1–7.

33. G. Marconi, Diaries and Notebooks, August 18, 1914, OX 5; GM to BOM, August 22, 1914; GM to BOM, August 26, 1914; GM to BOM, August 27, 1914; all in MB.

34. GM to BOM, September 2, 1914, MB; H.J. Round to GM, September 10, 1914, OX 307.

35. GCI to GM, September 10, 1914, OX 358; GM to BOM, September 16, 1914, MB.

36. GM to BOM, September 25, 1915, MB.

37. GM to BOM, October 5, 1914 (emphasis in original), MB.

38. GM to W. Churchill, n.d., OX 18/95.

39. Canziani 2006; Falchero 1990.

40. G. Marconi, Diaries and Notebooks, October 19, 1914, OX 5.

41. J. Pender, Home Office, to managing director, MWTC, November 17, 1914, OX 358.

42. Paoloni and Simili 1996, 111.

43. "Earthquake Havoc," LT, January 16, 1915; AM to GM, January 10, 1915 (postscript in AJM to GM, January 10, 1915, RI 80).

44. Albrecht-Carrié 1938, 29.

45. GM to BOM, March 4, 1915, MB.

46. LS to GM, April 9, 1915, OX 212. The agreement, between the Italian Government and Marconi personally, was approved by the Italian parliament on April 15, 1915. See Ufficio Marconi, "Convenzione fra il Governo Italiano e Guglielmo Marconi per l'impiego della Radiotelegrafia e Radiotelefonia in Italia e nelle

colonie," Rome, (April) 1915, OX 51/258–75 (English trans.: "Agreement between the Italian goverment and Gugliemo Marconi for the use of Radiotelegraphy and Radiotelephony in Italy and in the Colonies," 1915, OX 212).

47. GM to BOM, March 20, 1915, MB. Marconi's role as president of the Sconto was meant to be concrete, not merely symbolic. Marconi leant his personal and corporate prestige to the bank's efforts to raise capital for Italian investments, notably in Africa. He rarely participated in the bank's board meetings, but his right-hand man Luigi Solari was a member of the board and acted as a liaison with Marconi and his companies in this as in so many other of Marconi's affairs. See Falchero 1990.

48. GM to BOM, March 11, 1915, MB; Italy, *Atti Parlamentari della Camera del Senato,* Leg. XXIV, Iª s. 1913–18, *Discussioni,* March 15, 1915, 1381–83; "*Una dimostrazione di simpatia a G. Marconi,*" *Messaggero,* March 17, 1915.

49. GM to BOM, March 20, 1915, MB; "Major S. Flood-Page," *LT,* April 10, 1915; "Honour for Mr. Marconi," *LT,* April 13, 1915.

50. Rotscheid and Quäck 1928, 196; "*40 Jahre Groß-Sendestation Nauen,*" *Funk- und Fernseh-Illustrierte* 47 (1956), 18–19. Telefunken Unternehmensarchiv I.4.NL.048 G. Goebel 0300, 18–19.

51. F.H. Betts to GM, January 29, 1915, OX 418; de Sousa 1919, 376, OX 55; Sheffield & Betts to MWTC, December 27, 1916, OX 585.

52. FDR, ASN, Box 2, "Operations Radio"; Box 3, "Personnel: Radio Service"; Box 20, "Radio."

53. Ibid.; Howeth 1963, chapter XIX.

54. Drafts dated May 11 and June 1, 1915, and GCI to GM, September 13, 1915; all in OX 297.

55. Burrows 1924, 29.

56. Howeth 1963, chapter XIX.

57. Larson 2015.

58. Telegram, BOM to GM, May 8, 1915, OX 297; GM to BOM, May 14, 1915, MB.

59. "I have heard with satisfaction from the Minister of Foreign Affairs that Italy has also signed the so-called Treaty of London, and I trust that this will bring about complete co-operation and loyal and reciprocal confidence among the Allies..." (Italy, *Atti Parlamentari della Camera del Senato,* Leg. XXIV, Iª s. 1913–18, *Discussioni,* December 16, 1915, 1910–12. English trans. in "In England circa 1915/18," OX 21 /24–27).

60. Albrecht-Carrié 1938, 30–31.

61. GM to BOM, May 14, 1915, MB.

62. Maclaurin (1949) says he was invited; Archer (1938) says he invited himself.

63. de Sousa 1919, 376, OX 55; Dunlap 1937, 239.

64. E.J. Nally to GM, May 19, 1915, OX 192.

65. "Marconi Returns to Serve Italy," *New York Tribune,* May 23, 1915.

66. Ibid.

67. Inez Milholland Boissevain, "Submarine Pursued St. Paul to Mersey. American Liner Chased by U-Boat in Effort To Get Marconi," *New York Tribune,* June 2, 1915.

68. "St. Paul's Voyage a Nervous Trial," *New York Tribune,* May 31, 1915. This brief, unsigned article, cabled from Liverpool, may also have been dispatched by Inez.

69. Inez Milholland Boissevain, "Warlike France Bustles While England Blusters," *New York Tribune,* July 20 1915.

70. GM to Ministro della Guerra (draft), June 11, 1915, RI 159; Ministro della Guerra to GM, June 17, 1915, OX 20.

71. GM to LOB, June 1, 1915, RI 154; IM to E. Boissevain, June 6, 1915, IMP, folder 3.

72. IM to E. Boissevain, June 6, 1915, IMP, folder 3.

73. E. Boissevain to IM, May 27, 1915, IMP, folder 6.

74. Albrecht-Carrié 1938, 36.

75. Inez Milholland Boissevain, "Italy's Heroic Response Is Blind War Hysteria," *New York Tribune,* July 24, 1915.

76. Italian Ministry of War to GM, June 17 and 19, 1915; both in OX 20/12 and /14; Central News dispatch, n.d., OX 21/8–11, with note attached dated July 26, 1915, asking Senator Marconi to kindly read and return with his approval; J.E. Milholland, Diaries, June 14, 1917, MFP.

77. Andreoli 2004.

78. "Roman Theatre Scene of Battle," *New York Herald,* December 11, 1901, OX 160.

79. See R. Hughes 2011; Hughes-Hallett 2013.

80. Some clues to the affinities are found in D'Annunzio's writings. Although D'Annunzio's signature character—"the child of pleasure" Andrea Sperelli—could in no way be mistaken for Marconi, D'Annunzio's description of young Andrea's first love prefigured Marconi's early relationship with Beatrice: "His love for Conny had been a very delicate affair, for she was a very sweet little creature.... Lively, chattering, never still, lavish of infantile diminutives and silvery peals of laughter, easily moved to sudden caresses and as sudden melancholies and quick bursts of anger, she contributed to her share of love a vast amount of movement, much variety and many caprices. But Conny Landbrooke's melodious twitterings had left no more mark on Andrea's heart than the light musical echo left in one's ear by some gay *ritornella.* More than once in some pensive hour of twilight melancholy, she had said to him with a mist of tears before her eyes—'I know you do not love me.' And in truth he did not love her, she did not by any means satisfy his longings. His ideal was less northern in character..." (*The Child of Pleasure (Il Piacere).* New York/Berlin: Mondial Books, 2006, 17–18, first published in 1889; 1898 translation by Georgina Harding). Marconi told actress Francesca Bertini that *Il Piacere* was his favourite D'Annunzio novel (Bertini 1969, 147).

81. Undated, untitled typescript, on stationery of the Grand Hotel, Rome, IMP, folder 35.

82. On the relationship between D'Annunzio and Marconi, see Andreoli 2004.

83. Solari 2011, 196.

84. GM to BOM, June 20, 1915, MB.

85. A fascinating, previously unpublished letter from D'Annunzio to Marconi came to light in 1977. Dated June 29, 1915, and addressed to "Lieutenant Guglielmo Marconi, Grand Hotel," the letter reads as follows: "My dear great friend, you leave tomorrow and I believe I will leave as well, having arrived today in my automobile. I consider it of the greatest fortune in this prodigious time to have met you and become tied to you in such true friendship. If you are free, let's spend the final evening together. Let's dine together. I shall ask Miss Inez to join us and to indicate to me another or other guests. Thank you for the admirable little book. I hug you, Yours always, Gabriele d'Annunzio, San Pietro, 1915" (Webster and Costa 1977).

86. Hughes-Hallett 2013, 376.

87. L'eroe magico, in various forms. The twenty-two-page manuscript, handwritten in pencil on heavy manila paper, dated July 1915, is in the Marconi Archives in Oxford (OX 40 /5–26). A short version of the story appeared in the Italian-American émigré journal Progresso italo-americano, August 1, 1915; it has been republished many times, most notably in Le vie del mare e dell'aria, 3 (1919), 241–4, and recently as "Il mio pomeriggio con Marconi, il mago che incanta le onde," Corriere della sera, April 11, 2003. Also in Andreoli 2004, 9–12. Translation, GCR, 15/30.

88. GM to BOM, June 20, 1915, MB.

89. Central News dispatch, n.d., OX 21/8–11; Ministry of War to GM, June 21, 1915, OX 212.

90. Jacot and Collier 1935, 150.

91. New York Tribune, July 15 and 21, 1915.

92. Jacot and Collier 1935, 150; W.J. Baker 1970, 172.

93. Author, of course, of the so-called de Sousa manuscript, Marconi's ghostwritten, unpublished 1919 autobiography (OX 55). De Sousa was Marconi's private secretary from July 1915 to July 1922.

94. Jacot and Collier 1935, 153, and LT, July 16 and 27, 1915, and August 19, 1915.

95. Jacot and Collier 1935, 152; GM to BOM, September 21, 1915, MB; NYT, September 30, 1915; B.R. James (War Office) to GM, September 16, 1915, September 27, 1915, October 5, 1915, all in OX 358.

96. GM to BOM, November 7, 1915; GM to BOM, November 22, 1915; both in MB. Atti parlamentari, Senato del Regno, Legislatura XXIV, 1e sessione 1913–18, Discussioni, Tornata del 16 diciembre 1915, 1910–12.

97. English trans. of Senate speech, in "In England circa 1915/18," OX 21 /24–27; LT, December 17, 1915, and January 3 and 11, 1916; Kitchener to GM, December 20, 1915, OX 358.

98. "Papers concerning early experimental work on short wave transmission, 1916–24," OX 232, especially C.S. Franklin to GM, "A Summary of the Experimental Results at Livorno Aug 1st to Oct 22 1916," handwritten report/notes, October 29, 1916; C.S. Franklin to GM, memo and "Report on Short Wave Experiments at Carnarvon. Dec 1916 to Sept 1917," February 1, 1918; and reports of later work done at Poldhu and on Marconi's yacht, *Elettra*. See also "Franklin of Poldhu. His Life and Times," typescript, October 10, 1962, OX 683.

99. Chambers 1989, 42. Except as otherwise noted, Marconi's relationship with Sheridan is based on Chambers 1989.

100. Chambers 1989, 45.

101. B. Marconi, Memoir, FP.

102. M. Sheridan to GM, March 19, 1937, LIN 6/346.

103. G. Marconi, Diaries and Notebooks, March 3, 1916, OX 6.

104. GM to GCI, March 3, 1916, OX 232. The *Times* reported on June 15, 1916, that Marconi had been asked to join the Italian government.

105. GCI to GM, March 16, 1916, OX 232.

106. Carson appeared for the plaintiffs, Herbert Samuel and Rufus Isaacs, in the *Matin* case, for Godfrey Isaacs in his libel suit against Cecil Chesterton, and for the company in its "petition of right" seeking compensation from the government for cancelling the Post Office contract in 1914.

107. Ireland, National Archives, Bureau of Military History 1913–21. Statement by witness Edward Balfe, 28 Shannon Hill, Enniscorthy, Co., Wexford, March 12, 1956 (W.S. 1373; File No. S.2692). "Irish Volunteer activities, Enniscorthy, Co. Wexford, 1913–1921," http://www.bureauofmilitaryhistory.ie/reels/bmh/BMH.WS1373.pdf.

108. Clarke 1995, 23. Except for Clifden and Letterfrack, the company's coastal stations were already taken over by the GPO in 1909. Clifden continued to operate commercially during the war, albeit with a government censor present (Shane Joyce, correspondence with author, September 23, 2014).

109. Articles to be censored—Marconi Wireless Telegraph Co. re Irish riots, LAC RG6-E.

110. Pine 2002, 10–11.

111. McLuhan 1964, 304; Fisher 1978, xv; Pine 2002, 10–11.

112. "Irish Republic declared in Dublin today. Irish troops have captured city and are in full possession. Enemy cannot move in city. The whole country rising." From "1916 Irish Rebellion morse code broadcast," YouTube video, 2:03, posted by "luachnambrog," September 15, 2013, https://www.youtube.com/watch?v=Oym9j_oRscg.

113. Clarke 1995, 23; Pine 2002, 11.

114. M. Gorham, *Forty Years of Irish Broadcasting* (Dublin: Talbot Press, 1967), cited in Sexton 2005, 102.

115. O'Brien made his remarks during a debate on "pirate radio." He added that "like most war propaganda [the rebels' message] was designedly inaccurate

and misleading" (C.C. O'Brien, Symposium on Direct Satellite Broadcasting, Dublin, European Space Agency/European Broadcasting Union, 1977, cited in Pine 2002, 11).

116. Pine 2002, 10.

117. Another noteworthy connection to Marconi was that the British undersecretary for Ireland from 1914 to 1916 was Sir Matthew Nathan, the former permanent secretary of the Post Office who had negotiated the Marconi imperial chain contract for the government. Appointed just after the passage of the Home Rule Act, Nathan was in Dublin when the rising began and was virtually housebound in his offices at Dublin Castle. Criticized for not having anticipated and prevented the rising, and then responsible for coordinating the British response, Nathan resigned on May 3, 1916.

118. Although the text was published in the *New York Times* of December 7, 1921, the Anglo-Irish Treaty ("Treaty between Great Britain and Ireland signed 6th December 1921 at London," http://treaty.nationalarchives.ie/document-gallery/ anglo-irish-treaty-6-december-1921/) officially remained a secret document for ninety years, until published by the National Archives of Ireland on December 6, 2011 ("'Secret' Anglo-Irish Treaty of 6 December 1921 now online," http:// www.thejournal.ie/secret-anglo-irish-treaty-of-6-december-1921-now-online- 298689-Dec2011/). Under the Irish Free State (Agreement) Act, the British government formally transferred power to the Provisional Government of Southern Ireland on April 1, 1922. The Irish Free State Constitution Act, which received royal assent on December 5, 1922, created the Irish Free State.

119. Pine 2002, 45.

120. "Mr. Marconi's Voyage. Wireless Tests at Sea. Cut off by I.R.A," *LT*, July 28, 1922.

121. Clarke 1995, 24; Shane Joyce, correspondence with author, September 23, 2014.

122. Clarke 1995, 24. The IRA attack on Clifden was widely reported in the British and Irish press: *Irish Independent,* July 28, 1922; *Manchester Guardian,* July 28, 1922; *Daily Mail,* July 28, 1922; *Daily Chronicle,* July 29, 1922; *Irish Independent,* July 29, 1922; *Manchester Guardian,* August 2, 1922; all in OX 193.

123. *Manchester Guardian,* August 2, 1922.

124. "Treaty between Great Britain and Ireland signed 6th December 1921 at London," Annex, http://treaty.nationalarchives.ie/document-gallery/anglo- irish-treaty-6-december-1921/.

125. Cited in Pine 2002, 45.

126. Reproduced in Clarke 1995, 25; the original marconigram is in OX 193.

127. "Rule of the I.R.A. Misery and Ruin in West of Ireland," *LT*, November 20, 1922.

128. Cited in Pine 2002, 20.

129. Clarke 1995, 25.

130. "Irish Free State and Marconi Reach Agreement; Court Steps In," *Christian Science Monitor,* June 3, 1925, OX 193.

131. A similar pattern sealed the fate of the Canadian stations; on December 7, 1926, the Paris edition of the *New York Herald* reported that the pioneer stations at Glace Bay and Louisbourg, now obsolete, would soon be closed.

132. The company did obtain the contract to supply transmitter and studio equipment to the Dublin station that began broadcasting in 1926, prefiguring its role as a supplier of equipment and technology, rather than an actual broadcaster (Clarke 1995, 25).

CHAPTER 23

1. Unidentified typescript, OX 1809; A.R. Burrows to K. Shackleton, January 1, 1921, OX 311; "Battle of Jutland. How the Admiralty Knew. A Miracle of Wireless," *Daily Telegraph*, December 21, 1920.

2. de Sousa 1919, 389–99, OX 55.

3. *LT*, May 19, June 14, and July 6, 1916; *Atti parlamentari*, Senato del Regno, Legislatura XXIV, 1ª sessione 1913–18, *Discussioni*, July 4, 1916, 2, 590–94.

4. Ministero della Marina to GM, August 7, 1916, OX 329; GM to BOM, September 5, 1916, FP; C.S. Franklin to GM, October 29, 1916, OX 232.

5. GCI to GM, November 18, 1916, OX 232.

6. "The Marconi Company and Germany," *Electrician*, November 17, 1916; GCI to GM, November 18, 1916, OX 232; *LT*, November 2, 10, and 16, 1916; GM, handwritten fragment on Savoy stationery, n.d., OX 20/16.

7. Lumsden 2004, 164–69.

8. J.E. Milholland, Diaries, July 14, 1917, MFP.

9. Luigi Mascheroni, "*Le lettere segrete dello scienziato in linea con l'amore*," June 10, 2008, http://www.ilgiornale.it/news/lettere-segrete-dello-scienziato-linea-l-amore.html.

10. B. Marconi, Memoir, FP.

11. Albrecht-Carrié 1938, 55–56, note 56. Di Cellere died in October 1919; his papers were published posthumously under a pseudonym in 1920.

12. S. Sonnino to V. Macchi di Cellere, April 30, 1917, DDI, Quinta Serie, VII, doc. 854, 631; "Italy Selects mission. Prince Udine the Leader, Marconi a Member," *NYT*, May 3, 1917.

13. Albrecht-Carrié 1938, 54.

14. "Marconi Brings Anti–U-Boat Plan," *NYT*, May 24, 1917; de Sousa 1919, 380, OX 55; "Italian Mission Greets President," *NYT*, May 25, 1917; "In USA circa 1915/18," n.d. [1918], typescript, OX 21/13; "Italy To Fight on till Liberty Is Safe," *NYT* May 28, 1917; "House Welcomes Italian Mission," *NYT*, June 3, 1917; "Marconi on the War Needs and Ideals of Italy," *NYT*, June 3, 1917, cited in "Marconi—Man of Action," *Wireless Age*, July 1917, GCR 16/2.

15. V. Macchi di Cellere to S. Sonnino, July 2, 1917, DDI, Quinta Serie, VIII, doc. 523, 337–38.

16. de Sousa 1919, 380, OX 55; "Italy's Mission Stirs New York to Ovation," *NYT*, June 22, 1917.

17. "Marconi Tells New War History at Italian Dinner," *NYT*, June 23, 1917.

18. Max Rhoade, "The Italian Commission and Zionism," *American Jewish Chronicle* 3, no. 7 (June 22, 1917): 209–10.

19. Sonnino's father was a Jewish convert to Anglicanism, his mother was a Welsh Protestant, and Sonnino was raised Anglican. A liberal-conservative in the Italian post-Risorgimento tradition, he sat in parliament from 1880 to 1919, including two brief stints as prime minister in 1906 and 1909 to 1910. He served as foreign minister in three different governments between 1914 and 1919 and was the main Italian negotiator in Paris until the resignation of Prime Minister Vittorio Emanuele Orlando. He did not run for re-election in 1919.

20. The interview with Nitti and Marconi is mentioned as an important example of early Italian support for Zionism in de Felice 2001, an authoritative work by Italy's most eminent historian of Fascism (at page 48).

21. V. Macchi di Cellere to S. Sonnino, July 2, 1917, DDI, Quinta Serie, VIII, doc. 523, 337–38.

22. J.E. Milholland, Diaries, July 14, 1917, MFP.

23. GM to P. Boselli, August 18, 1917, ACS, Carte Boselli, b.3, f.190, ins.29, and GM to F. Nitti, September 18, 1917, ACS, FSN, b.98, f.640/C, ins. Marconi; both cited in Paoloni and Simili 1996, 120.

24. "Palestine for the Jews. Official Sympathy," *LT*, November 9, 1917.

25. de Sousa 1919, 388, OX 55.

26. B. Marconi, Memoir, FP.

27. *LT*, January 17, 1918.

28. His sister-in-law Lilah wrote that in Rome Marconi "succumbed to flattery and being chased after. . . . His penchant was very young girls and flappers, who loved to play games with, flirt a little, possibly who would distract him, but not take up his time" (L. O'Brien 1952, FP).

29. Chambers 1989, 57.

30. Hollywood also released a film, *La Tosca,* based on the Victorien Sardou play in 1918, starring Pauline Frederick; not to be confused with the Puccini opera.

31. Bertini 1969, 149.

32. Ibid.

33. G. Marconi, Diaries and Notebooks, May 26 and September 2, 1918, OX 7.

34. Bertini 1969, 191, 149, and 230; "*Francesca, la divina attrice che disse di no a Marconi,*" *Corriere della sera*, March 7, 2004.

35. Bertini 1969, 147.

36. Michael Braga, interview with author, March 29, 2012.

37. V. Macchi di Cellere to S. Sonnino, February 5, 1918, ASMAE, AI, b.296; V. Macchi di Cellere to S. Sonnino, March 6, 1918, ASMAE, AI, b.296, sub-file "Marconi—Inchiesta Tozzi," 1918. Marconi's Senate speech was on March 3, 1918.

38. *LT*, July 3 and 6, 1918.

39. *LT*, July 6, 1918.
40. "Dinner offered by His Majesty's Government at the Royal Gallery of the House of Lords—Remarks by Senator Marconi, 5/7/18," July 5, 1918, OX 21/14–16. Reported in *LT*, July 6, 1918.
41. Untitled typescript, hand-dated (possibly in error) April 1918, OX 21.
42. Just as the parliamentary conference was ending, Marconi received a request from the League of Free Nations Association (LFNA) to allow his name to be included as an honorary member. The LFNA hoped "to establish the closest relationship with kindred societies in Allied countries." As the conference had promoted the idea of economic solidarity among the Allies, the LFNA hoped Marconi would endorse its aims and objects (Invitation from the League of Free Nations Association, July 5, 1918, OX 21 /66).
43. Taylor 1972, 155.
44. W.J. Baker 1970, 183.
45. Begbie 1919, chapter 14, 4, OX 361.
46. W.J. Baker 1970, 173.
47. de Sousa 1919, 321, OX 55.
48. "Marconi Rouses Austria. Warlike Speech by the Inventor Deplored and Resented," *NYT*, December 8, 1906; see also *NYT*, December 16, 1906.

CHAPTER 24

1. "Official American Commentary on the Fourteen Points," in E.M. House, *The Intimate Papers of Colonel House*, Boston: Houghton Mifflin, 1926–28, vol. IV, 197 (cited in Albrecht-Carrié 1938, 63).
2. Albrecht-Carrié 1938, 68–69.
3. Ibid., 79–80, note 66.
4. Ibid., 78–79.
5. MacMillan 2003, 287.
6. See Albrecht-Carrié 1938, 79, for example.
7. House had spent much of the war in Europe and was enormously influential with Wilson. His papers (EMH), including a detailed diary kept at Paris, are the quintessential source on the backroom politics of the conference.
8. Albrecht-Carrié 1938, 80–81 (House quote cited on p. 80).
9. EMH, Series II, Diaries, 1858–1926, Vol. 7, January 7, 1919.
10. Albrecht-Carrié 1938, 89, note 15; MacMillan 2003, 288.
11. G. Imperiali to S. Sonnino, n.d., ASMAE, GI, b.1.
12. MacMillan 2003; Charnovitz 2003; ASMAE, CP.
13. MacMillan 2003, 289.
14. A.J. Balfour to G. Imperiali, April 3, 1919, ASMAE, GI, b.2.
15. Albrecht-Carrié 1938, 131.
16. EMH, Series II, Diaries, 1858–1926, Vol. 7, April 13, 1919.
17. "*Aide-mémoire présenté par le Conseil national, la municipalité et le député de Fiume*," (hand-noted "*Presentato alla Conferenza 15/III 1919*"), OX 21.

18. "Report of the Conference Between President Wilson and Deputy Ossoinack held in the presence of President Orlando, April 14th, 1919, Paris," OX 21/85–93.

19. EMH, Series II, Diaries, 1858–1926, Vol. 7, April 15, 1919.

20. Albrecht-Carrié 1938, 133.

21. EMH, Series II, Diaries, 1858–1926, Vol. 7, April 22, 1919.

22. EMH, Collection Contents, Description, Series II, Diaries, 1858–1926, Vol. 7, January 1 to June 29, 1919.

23. V.E. Orlando to Vittorio Emanuele III, May 23, 1919, DDI, Sesta Serie, III, doc. 582, 597–98.

24. EMH, Series II, Diaries, 1858–1926, Vol. 7, May 21–24, 1919, 213–14.

25. Jacot and Collier 1935, 159 (no source is given for this quote. Jacot and Collier state that Marconi had several meetings with Wilson in Paris); Crespi 1937, 600–601; Solari 1928 (translation in GCR 15/30).

26. V.E. Orlando to Vittorio Emanuele III, May 23, 1919, DDI, Sesta Serie, III, doc. 582, 597–98.

27. Crespi 1937, 600–601.

28. V.E. Orlando to Vittorio Emanuele III, May 23, 1919, DDI, Sesta Serie, III, doc. 582, 597–98.

29. EMH, Series II, Diaries, 1858–1926, Vol. 7, May 25 and 28, 1919; G. Marconi, Diaries and Notebooks, May 28, 1919, OX 8.

30. Colonnini to A. Ossoinack, May 31, 1919, OX 21; "*Proteste di Fiume al Senato,*" June 2, 1919, OX 21.

31. G. Marconi, Diaries and Notebooks, June 8–14, 1919, OX 8.

32. Tittoni was another exemplar of post-Risorgimento liberal politics. Born in Rome in 1855, his father, a large-scale tenant farmer, had fought with Garibaldi. Tittoni was first elected to parliament in 1886 and named a senator in 1902.

33. "Papers relating to Marconi's service as Italian Delegate Plenipotentiary to the Peace Conference, 1919–20," OX 21/67–99; copy of telegram, unaddressed, signed "Tittoni," June 23, 1919, ASMAE, CP, b.14, p.3, c.3/A, sc.3; telegram, Tommasini (Rome) to Italian Embassy (London), June 24, 1919, ASMAE, AL, b.472, f.1; *LT,* June 25, 1919.

34. Documenting a Democracy, "Treaty of Versailles 1919 (including Covenant of the League of Nations)," Museum of Australian Democracy, http://foundingdocs .gov.au/scan-sid-8.html.

35. GM to BOM, June 30, 1919, FP.

36. EMH, Series II, Diaries, 1858–1926, Vol. 8, July 14, 1919; GM to BOM, July 5, 1919, FP; P.J. Baker to Chancellor, Italian Embassy (London), July 5, 1919, ASMAE, AL, b.472, f.1; telegram, GM to P.J. Baker, August 11, 1919, ASMAE, CP, b.70, c.24, sc.38, f.9; British Delegation (Paris) to T. Tittoni, September 13, 1919, ASMAE, AL, b.472, f.3.

37. T. Tittoni to F.S. Nitti, July 15, 1919, ASMAE, CP, b.16, c.3/E, sc.4; [author's name illegible] to G. Imperiali, July 23, 1919, ASMAE, GI, b.3; T. Tittoni to delegates, July 25, 1919, ASMAE, CP, b.14, p.3, c.3/A, sc.2.

38. T. Tittoni to delegates, July 25, 1919, ASMAE, CP, b.14, p.3, c.3/A, sc.2.
39. GM to BOM, July 29, 1919, FP.
40. *LT*, August 1, 1919.
41. GM to BOM, August 18, 1919, FP.
42. GM to F.S. Nitti, August 21, 1919, ACS, FSN, b.48, f.156.
43. MacMillan 2003, 302.
44. S. Crespi to C. Schanzer, August 31, 1919, ASMAE, CP.
45. GM to BOM, September 3, 1919, FP.
46. GM to BOM, September 8, 1919, FP.
47. Albrecht-Carrié 1938, 56; "Count di Cellere, retiring Italian ambassador, dies on operating table in Washington," *NYT*, October 21, 1919.
48. Telegram, GM to BOM, October 21, 1919, FP.

CHAPTER 25

1. All material from Begbie over the following paragraphs is taken from "Incomplete typescript draft of 'Wireless in the War' by Harold Begbie, with related correspondence, 1919–20," OX 361. The critique of the War Office provides some interesting insight into the mindset of British military intelligence in 1919: "Most civilized countries must have realized that we did possess a highly organized Wireless Intelligence and a Cryptographic Bureau, but practically no information has ever been available to them as to the pitch of perfection that these organisations reached, and it is most desirable that this information should still remain unavailable." Naval intelligence vetted the manuscript and sought assurance from the company that the parts they indicated would be omitted from the published text. By all indications, the company tried to comply with the request, but a note in the file indicates succinctly: "Harold Begbie's book started January 1919, turned down by Admiralty, December 1920" (Director of Military Intelligence to Director of Naval Intelligence, June 14, 1920; War Office, "Criticism of Mr. Begbie's book 'Wireless'," n.d.; Naval Staff, Intelligence Division, to GCI, memo marked "Secret & Confidential," June 17, 1920; "Historical Information," note to file, n.d.; all in OX 361). A.R. Burrows, in his 1924 *The Story of Broadcasting*, alludes to the Begbie manuscript: "There exists, I am told, in a strong room less than a mile from Charing Cross, the manuscript of a book which is unlikely to be published in the lifetime of the present generation. It may never see light. Within are set down in considerable detail some of the most romantic stories imaginable. They are on record, and that is all" (Burrows 1924, 32).
2. GM to BOM, April 14, 1919, FP.
3. Cited in Jacot and Collier 1935, 159; B. Marconi, Memoir, FP.
4. Information culled from documents in OX 1865.
5. Jacot and Collier 1935, 160–64; Landini 1960, 82–92; documents in OX 25, OX 90, and OX 1865.

6. Begbie 1919, chapter 21, 2, OX 361.
7. GM to BOM, March 26, 1920, FP.
8. G. Marconi 1923, OX 52.
9. Ibid.
10. Marconi kept an autographed photograph of the poet on display on the yacht. It was inscribed: "To Marconi of the snow-white ship [*la nave bianca*], who navigates in miracles and animates the silences." Later, Marconi also had an autographed photo of Mussolini on board, inscribed: "To Marconi, Magician of Space and Conqueror of the Ether" (Jacot and Collier 1935, 162). According to Landini, photos on display on the *Elettra* included various royalty, Pope Pius XI, Marie Curie, Enrico Caruso, Thomas Edison, Mary Pickford, as well as the one of Popov, dedicated to "Guglielmo Marconi the father of wireless telegraphy," that the Russian physicist presented to Marconi in Kronstadt on July 14, 1902 (Landini 1960, 91–92).
11. W.H. Bullard memos, February 9, 1914, and May 1, 1915, FDR, ASN, Box 3, "Personnel: Radio Service."
12. Lyons 1966, 82.
13. Archer 1938, 129–31.
14. Aitken 1985, 307–8.
15. Howeth 1963, chapter XXVII.
16. Maclaurin 1949, 99.
17. Archer 1938, 151.
18. "Memorandum on Wire and Radio Communications. Paris, 12 February, 1919," submitted to President Wilson by Walter S. Rogers, communications expert of the American Commission to Negotiate Peace, Document 63 in Ray Stannard Baker, *Woodrow Wilson and World Settlement*, Vol. 3 (Garden City, NY: Doubleday, 1922), 428ss.
19. Howeth 1963, chapter XXX. Daniels was later greatly disappointed in Congressional failure to make its war monopoly permanent.
20. Ray Stannard Baker, *Woodrow Wilson and World Settlement*, Vol. 3 (Garden City, NY: Doubleday: 1922), 425.
21. Maclaurin 1949, 100.
22. Howeth 1963, chapter XXX.
23. Douglas 1987, 285.
24. Archer 1938, 163; Maclaurin 1949, 101; W.H. Bullard, in *U.S. Naval Institute Proceedings*, October 1923 (in Maclaurin 1949, 101).
25. Aitken 1985, 338.
26. Howeth 1963, chapter XXX.
27. "Brief Company Histories from the Radio Industry, 1900–1930s," GCR online catalogue.
28. Howeth 1963, chapter XXX.
29. Aitken 1985, 388.
30. Howeth 1963, chapter XXX.

31. O. Young to E. J. Nally, October 23, 1919, GCR, 16/ 36, "E. J. Nally correspondence." American Marconi's corporate culture remained so unchanged after the takeover that as far as the personnel was concerned, the only difference noted "was a different company name on the pay check" (G.H. Clark, "The Formation of the Radio Corporation of America," unpublished manuscript, n.d., 71, GCR; cited in Howeth 1963, chapter XXX).

32. "Agreement between RCA and MWTC Ltd," November 21, 1919, OX 382.

33. Howeth 1963, chapter XXX.

34. Headrick 1991, 181.

35. Aitken 1985, 25.

36. Sobel 1986, 26.

37. In December 1929, Rear Admiral Cary T. Grayson, who had been Wilson's physician in Paris, testified as follows before the Senate Committee on Interstate Commerce: "Senator Pitman:...was there considerable discussion in Paris at that time with regard to radio regulation and control? Admiral Grayson: Yes, sir. Mr. Marconi was there, Senator" (transcript cited in Archer 1938, 154). According to Howeth, Wilson and Bullard were concurrently in Paris between March 16 and 21, 1919 (chapter XXX). I have seen no record of Marconi's whereabouts during those dates, but he was in London in mid-February and mid-April, so a March meeting with the Americans in Paris is plausible.

38. The Italian Socialist Party received 32.3 percent of the vote and gained more than 100 seats (to bring their total to 156); the new Italian Popular Party was second with 20.5 percent and 100 seats. Nitti's Radical Party made small gains but was third with 15.9 percent and 96 seats; Giolitti's new Social Democratic Party was fourth with 10.9 percent (60 seats), and the new Liberal Party was fifth with 8.6 percent (41 seats). So Nitti was able to govern in a Radical-Liberal-SocDem coalition with 35.4 percent and 197 seats, but without a majority.

39. GM to F.S. Nitti, November 20, 1919; F.S. Nitti to GM, November 21, 1919, both in ACS, FSN, b.53, f.183.

40. Romano to T. Tittoni, November 25, 1919, ASMAE, AI, b.288.

41. T. Tittoni to Marconi, November 22, 1919, ASMAE, CP, b.70, c.24, sc.38, f.3. Tittoni resigned as foreign minister a few days later and was promptly named president of the Senate, remaining in that position until January 1929. He supported Mussolini's government and was the first president of the new Royal Academy of Italy until he was succeeded on September 16, 1930, by Marconi. See chapter 31.

42. "Bulgaria Signs," LT, November 28, 1919.

43. Various correspondence, December 7, 8, 9, 10, 12, and 13, 1919, ASMAE, AL, b.472, f.2.

44. Paternô to F.S. Nitti and London Embassy, December 12, 1919, ASMAE, AL, b.472. f.1; G. Imperiali to V. Scialoja and memo, n.d., December 13, 1919, ASMAE, AL, b.472, f.2.

45. F.S. Nitti personal correspondence, January 1920, ACS, FSN, b.53; F.S. Nitti correspondence, ASMAE, CP, b.70; and F.S. Nitti to Italian embassy, Washington, March 3, 1920, ASMAE, AI, b.287.

46. GM to BOM, March 24 and 26, 1920, FP.

47. Bianchi (London) to Delegation (Paris), January 19, 1920, ASMAE, CP, b.14, p.3, c.3/A, sc.2.

48. F.S. Nitti to GM, April 22, 1920, OX 21/99–100.

49. Masini 1995, 278; Pession 1941, 167.

50. Begbie 1919, chapter 19, 2, OX 361.

51. "Political to USA. Occasion and use unknown," [1919], OX 21/57–61.

52. B. Marconi, Memoir, FP.

53. *Wireless World*, June 16, 1920, OX 22.

54. Cemetery records show that Davis purchased the right of burial on November 7, 1913. The right was transferred to Guglielmo Marconi on November 7, 1930, and permission to erect a marker was granted a few months later, on July 7, 1931. A stone monument still standing today reads: "To the never fading memory of our dearly beloved mother Annie Fenwick Jameson Marconi who died 3rd June 1920 in her 81st year. This memorial is erected by her sons Alfonso Marconi and Guglielmo Marconi." Alfonso was buried in the same grave on April 28, 1936 (London Cemetery Company, Friends of Highgate Cemetery, correspondence with author, August 20–22, 2013).

55. Nitti's trajectory provides an interesting counterpoint to Marconi's. Still a member of the Italian parliament, he offered resistance to the nascent power of fascism, and exiled himself in 1924. He returned to Italy and active politics, as an elected senator, after the Second World War. Giolitti's 1920–21 premiership was marked by radical agitation, fears of communist takeover, and the rise of Mussolini. He stepped down after disappointing election results in May 1921 and initially supported Mussolini's government from 1922–24, remaining in parliament until his death in 1928.

56. Albrecht-Carrié 1938, 293.

57. Ibid., 250.

58. R. Hughes 2011, 388.

59. Campbell 2006, 32.

60. Ledeen 1977, x, cited in Campbell 2006, 32.

61. Salaris 2002, 12, cited in Campbell 2006, 33.

62. MacMillan 2003, 302.

63. Biographical statement to Italian court hearing on the Sconto affair, December 11, 1922, ASS, ACG, Carte del processo agli Amministratori della Banca Italiana di Sconto, b.279, c.16bis, f.15.

64. *Il Piccolo* (Trieste), September 16, 1920.

65. "D'Annunzio decorates Signor Marconi," *LT*, September 25, 1920; *La Vedetta d'Italia* (Fiume), September 24, 1920. See also Hughes-Hallett 2013, 552–53. Marconi did, nonetheless, report personally on his mission to Giolitti after-

wards; see GM to Luigi Freddi, October 9, 1920, and November 15, 1920, Bergé auction items (http://www.pba-auctions.com/html/recherche.jsp?lang=en&q uery=marconi&Submit=OK&filterDate=2&npp=20&_exact=&email=&all WordsMatch=true&minEstim=&maxEstim=&ordre=1&heigthMin=&hei gthMax=&widthMin=&widthMax=&depthMin=&depthMax=). See also Campbell 2006, 34.

66. In Solari 1928, translation in GCR 15/30. Campbell (2006) describes this speech as "one of the most important (and overlooked) documents of modern Italian culture." The original full text is Gabriele D'Annunzio, "*Saluto a Guglielmo Marconi in Fiume d'Italia*" ("Greeting to Guglielmo Marconi upon his Arrival in Italian Fiume"), in *La Vedetta d'Italia* (Fiume), September 23, 1920, reprinted in Gabriele D'Annunzio, *La penultima ventura: Scritti e discorsi fiumani, a cura di Renzo de Felice* (Milan: Arnoldo Mondadori Editore, 1974), 354–59. See also press coverage of D'Annunzio's speech in *La Vedetta d'Italia*, September 23 and 24, 1920.

67. Campbell 2006, 33.

68. Webster and Costa 1977; Andreoli 2004.

69. Andreoli 2004, 21–22. See also *La Vedetta d'Italia*, September 24, 1920. The speech establishes that Marconi and D'Annunzio first met in 1915, and had not met since.

70. Albrecht-Carrié 1938, 308.

71. The Socialists also lost support, weakened by a split with the new Communist Party (formed in 1921 and headed by Antonio Gramsci). The results gave the Socialists 24.7 percent, 123 seats; Popular Party, 20.4 percent, 108 seats; National Bloc (including Mussolini), 19.1 percent, 105 seats; Nitti's Radical Party, 10.4 percent, 68 seats; Liberal Party, 7.1 percent, 43 seats; Giolitti's Social Democrat Party, 4.7 percent, 29 seats; Communist Party, 4.6 percent, 15 seats. The new prime minister was Ivanoe Bonomi of the Reform Socialist Party; he was succeeded by Luigi Facta in February 1922.

72. See, for example, Crespi 1937, postscript.

CHAPTER 26

1. Kertzer 2014, 19–20; D. M. Smith 1981.

2. "Interview with Mr. Charles Bertelli, to appear in all the Hearst publications in America. (Given by Senatore Marconi)," October 19, 1920, OX 21/62–65.

3. Falchero 1990; Canziani 2006; "Heavy Withdrawals," *LT*, December 30, 1921.

4. "Bank Directors Suspended," *LT*, January 5, 1922.

5. "The Italian Bank Suspension. An Official Statement," *LT*, January 6, 1922.

6. "Agreements with company etc.," n.d.; "Draft Agreement as to provision for certain out-of-pocket expenses" between MWTC and GM, December 1, 1924; "Proof of His Excellency The Marchese Marconi," 1934; all in OX 22. According to a note in the Archives, Marconi owned nearly no common shares

in MWTC in his own name at this time ("MWTC Ltd. Shares held by Senatore Guglielmo Marconi and Nominees as at the 17th January, 1921," OX 22).

7. GM to BOM, December 30, 1921, FP.

8. GM to BOM, January 11, 1922, FP; Prefetto di Palazzo to BOM, October 11, 1911, OX 29.

9. GM to A. Pogliani, August 30, 1921; A. Pogliani to GM, October 4, 1921; LS, testimony, December 20, 1922; all in ASS, ACG, b.279, c.16bis, f.15.

10. GM to LS, January 11, 1922, ASS, ACG, b.306, c.42, f.32, sf.6.

11. GM to BOM, March 24, 1922, FP.

12. Affidavit, Italian Embassy (London), November 8, 1922, ASS, ACG, b.279, c.16bis, f.15.

13. Correspondence, April 21, 1922, ASS, ACG, b.298, c.34, f.28, sf.6.

14. "Procedimento penale del Senatore Marconi Guglielmo ed altri," typed statement and transcript of court testimony, December 11, 1922, ASS, ACG, b.279, c.16bis, f.15. A copy of this statement is in Mussolini's file at ACS, PCM (1922), f.6/1–1, n.1005, (in Paoloni and Simili 1996, 127–30).

15. LS testimony, December 20, 1922, ASS, ACG, b.279, c.16bis, f.15.

16. "Banca di Sconto Inquiry. Charges against Directors," *LT*, June 2, 1923.

17. "Banca di Sconto's Failure. Signor Marconi Exonerated," *LT*, November 27, 1923.

18. "Banca di Sconto Trial. Acquittal of All the Accused," *LT*, March 3, 1926.

19. One of the leading Italian Marconi scholars, Giovanni Paoloni, has written that Mussolini's support for Marconi in the Sconto case "probably contributed" to convincing him to adhere publicly and formally to fascism (G. Paoloni, "*Un anglosassone amico del Duce*," *La Repubblica*, November 14, 2001).

20. B. Mussolini, declaration to the National Fascist Party congress, Naples, October 24, 1922, cited in Francis L. Carsten, *The Rise of Fascism,* Berkeley: University of California Press, 1982, 62.

21. See Guerin 1973 on the relationship of fascism and the Italian business elite.

22. Kertzer 2014, 451, note 1.

23. Bernabei 1997; Herman 1984; D.M. Smith 1981. Richard Owen, the long-time *Times* correspondent in Rome, cites Bernabei as referring to documents in the Public Records Office (R. Owen, "Marconi Book Broadcasts His Fascist Role," *LT*, March 22. 1997).

24. R. Owen, ibid. The home section of the British Secret Service, known after 1916 as MI5, was established in 1909 to counter German espionage. After the First World War, its focus shifted to the Irish Republican movement and Soviet support for revolutionary activities in Britain. A large part of its efforts concentrated on the shadowing of alien refugees. In 2009, Cambridge historian Peter Martland reported research showing that MI5 had recruited Mussolini in 1917 to agitate for keeping Italy in the war. (T. Kington, "Recruited by MI5: The Name's Mussolini. Benito Mussolini," *Guardian*, October 13, 2009; also Andrew 2009.)

25. Jackson 1964, 78.

26. Bernabei 1997.

27. "The Gathering of the Allies," *LT*, December 9, 1922.

28. Bernabei 1997, 54-6; Baldoli 2003, 118.

29. "Mussolini Faces Struggle in Italy," December 28, 1922; memos of December 28 and 30, 1922, ASMAE, AI, b.287.

30. Bernabei 1997, 121. See also Baldoli 2003.

31. Typed carbon copy sheet (in English): "Telegram received 21st. January 1923 from The Prime Minister of Italy," OX 230; GM to BOM, January 31, 1923 and February 12, 1923, FP. He stayed in Rome longer than expected, at Mussolini's behest, before going on to Paris to see his eye doctor (GM to BOM, February 28, 1923, FP).

32. *LT*, May 18, 1923.

33. Cesarò resigned as minister in February 1924 and soon joined the anti-fascist movement—proving that there were other paths than the one chosen by Marconi. He was suspected of being involved in a 1926 attempt on Mussolini's life (*Enciclopedia italiana*, Treccani.it).

34. Monteleone 1976; Monticone 1978; Balbi 2010a, 2010b.

35. Various sources, including GM to BOM, February 8, 1922, FP, where he tells her the rate of exchange is 100 lire per pound sterling.

36. F. Bonacci to G. Acerbo, November 18, 1922, ACS, PCM (1934–36), f. 12/1, n.2057, sf.1, cited in Monteleone 1976, 235–40.

37. Monticone 1978, 10–14.

38. Solari 1940, 254; LS to BM, January 19, 1923, ACS, PCM (1934–36), f.13/1 n.2057, sf.1, cited in Monteleone 1976, 20; GM to BM, n.d., cited in Monteleone 1976, 21.

39. Monteleone 1976, 23.

40. GM to LS, June 25, 1923, in Solari 1942. Marconi filled out his party application form in London on June 15, 1923, the day after returning from his research cruise. A copy of his application form and a personal letter from the head of the Milan section acknowledging his adherence to the party are in OX 21. Solari (2011) claimed he had been urging Marconi to join the party since its foundation in 1921.

41. GM to BM, ACS, PCM (1934–36), f.12/1, n.2057, sf.1, cited in Monteleone 1976, 242.

42. Solari 1942, 259.

43. Monteleone 1976, 22.

44. Extract from "Italo Radio" contract, August 29, 1923, OX 212.

45. Monteleone 1976, 240–42.

46. Monticone 1978, 7.

47. GM to BOM, October 1 and 12, 1923; both in FP.

48. GM to BM, October 17, 1923, English translation, OX 212. On the same date, in a separate letter, he also communicated his "irrevocable decision" to Cesarò (GM to G. di Cesarò, October 17, 1923, English translation, OX 212).

49. Typescript in Italian, hand-dated November 1923, with hand revisions by GM; possibly published in *Giornale d'Italia*, November 4, 1923, OX 52/163–81.

50. See, for example, "Mussolini—D'Annunzio—Marconi," in *Le Vie del mare e dell'aria* 11 (1923): 165–68.

51. See cuttings from November 15, 1923, J. Reith to GM, November 20, 1923, and broadcast draft of November 28, 1923, all in OX 52.

52. Monticone 1978, 7.

53. Balbi 2010a, 786–808.

54. Monteleone 1976, 26. Commercial provision of news was an ancillary service Marconi offered his marine clients. In 1918, Solari had created the Agenzia Radiotelegrafica Italiana, on behalf of the Marconi group, to provide news bulletins to ships at sea and to Italy's African colonies. In April 1923, the agency was transformed into a new entity known as La Radio Nazionale. Agenzia d'Informazione Radiotelegrafiche (Monticone 1978, 15).

55. *Enciclopedia italiana*, Treccani.it, and Bosworth 2006, 219. See also Moseley 1999, 6.

56. C. Ciano to GM, n.d. [received in London, February 28, 1924], English translation, OX 212.

57. Balbi 2010b, 162.

58. Bosworth 2006, 219.

59. Solari 1939; Solari 2011, 254.

60. Balbi 2010b.

61. For a synthetic history of Marconi's corporate role in the origins of Italian broadcasting, see LS to GM, June 5, 1936, LIN 17/199 (Paoloni and Simili 1996, 182–86). URI began regular broadcasts in October 1924 (Monticone 1978, 1). Ranieri continued lobbying somewhat recklessly; on April 25, 1925, he wrote to Mussolini, attacking Ciano for sanctioning "an ably camouflaged Marconi monopoly" (Balbi 2010b, 162).

62. Historian Mauro Canali has provided the most thorough treatment of the Matteotti affair. Although it has never been proven, Canali suggests that Mussolini probably did order the murder, because of the revelations that Matteotti was about to make (Canali 1997).

CHAPTER 27

1. "Mr. Marconi's Voyage. Wireless Experiments in Atlantic," *LT*, May 25, 1922.

2. *NYT*, May 30, 1922.

3. "No Mars Message yet, Marconi Radios; Ends Yacht Trip 'Listening In' on Planet Today," *NYT*, June 16, 1922.

4. "Developing Secret Wireless Service. Marconi and Franklin Can Defy 'Listening In' on 100-Mile Distance. Seeks a Martian Message. Marconi, Due Here Next Month, Hopes to Hear Mysterious Sounds Reported in 1920," *NYT*, May 6, 1922.

5. "Marconi Here, Hopes to Conquer Static," *NYT*, June 17, 1922. The possibilities afforded by wireless for extraterrestrial communication was a popular

topic; the new weekly *Popular Wireless* featured an article on "Messages from Mars" in its launch issue the week Marconi sailed (*LT*, June 2, 1922). A Marconi interview with the London *Daily Mail* on January 26 and 27, 1920, in which he told of receiving "mysterious undecipherable signals" while doing research on the *Elettra*, had set off a spate of articles in the New York press (*NYT*, January 27, 28, 29, 30, and 31, 1920), in which scientists in the United States and Europe speculated about his investigations, but Marconi was always cautious and prudent on the topic, claiming that his comments about trying to communicate with Mars were only jokes intended to unbalance reporters. The furthest he went was to say the possibility of signals originating from a foreign planet could not be ruled out.

6. *NYT*, June 17, 1922.

7. Ibid.

8. *NYT*, June 21, 1922.

9. G. Marconi, "Radio Telegraphy," *Proceedings of The Institute of Radio Engineers* 10, no. 4, (August 1922): 215–38, 237; also in *Journal of the American Institute of Electrical Engineers* XVI, no. 8 (August 1922): 561–70; and di Benedetto 1974a, 69–90. The typescript version of this speech, in OX 52, contains an archivist's notation: "Radar!"

10. W.J. Baker 1970, 301.

11. H. Gernsback, *Ralph 124c 41+* (Lincoln, NE: University of Nebraska Press: 2000), 207. First published as a twelve-part serial in *Modern Electrics* magazine beginning in April 1911 and then as a novel in 1925.

12. Howeth 1963, chapter XXXVIII.

13. W.J. Baker 1970, 301–2; Howeth 1963, chapter XXXVIII; Joint Board on Scientific Information Policy, *Radar: A Report on Science at War*, Washington, DC: Office of Scientific Research and Development/War Department/Navy Department. Government Printing Office, 1945, 4 and 9.

14. Sarkar et al 2006, 133, and others.

15. "Marconi on Yacht Will Radio to Liner," *NYT*, June 24, 1922.

16. "New Radio Sender Eliminates Noise. Marconi Shows Deep Interest in Tube Being Developed in Schenectady," *NYT*, June 27, 1922.

17. "Marconi Honored by Engineers Here," *NYT*, July 7, 1922; "Radio's Great Tomorrow," *NYT*, July 9, 1922.

18. *NYT*, June 24, 1922; A.C. Doyle to MWTC, August 17, 1921, OX 18/103.

19. Another powerful believer in spiritualism was Marconi's erstwhile nemesis Oliver Lodge, who had also lost a son in the war. Doyle's 1926 book, *The History of Spiritualism*, is dedicated to Sir Oliver Lodge, "a great leader both in physical and in psychic science."

20. "Doyle Says England Needs Baseball," *NYT*, June 19, 1922; "Conan Doyle Comes Out for Prohibition. Returning, Will Advocate It for England. May Revive Sherlock Holmes," *NYT*, June 23, 1922; "Finds No Spirit Flappers. Sir Arthur Conan Doyle Thinks Radio Can Communicate with the Dead,"

NYT, April 29, 1922; "Doyle Says Ghosts May Use the Radio," *NYT*, May 5, 1922.

21. "Doyle Goes in for Radio. Novelist Plans to Use It in Psychic Investigations," *NYT*, June 15, 1922.

22. Brandon 2003, 239.

23. A.C. Doyle 1930; also, Sandford 2011, 134–38. Houdini (1924) mentions the Atlantic City séance, somewhat mockingly, in *A Magician among the Spirits*. Lady Doyle acted as medium at the séance, recording Mother Houdini's poignant message to her son with pencil and paper. Houdini later told his wife how strange it was that his mother, who during her lifetime had communicated with him only in Yiddish, would suddenly have learned English in the after-world (Sandford 2011, 136; see also Rapaport 2010, 170).

24. A.C. Doyle to GM, n.d., OX 18/104.

25. Marconi and Beatrice sent a rare postcard, signed by the two of them, to their daughter Degna from Atlantic City on June 22, 1922 (FP).

26. Sandford 2011, 139.

27. *NYT*, June 24, 1922.

28. For example, Stemman 2005.

29. D. C. Doyle 1937, 19.

30. GM to BOM, July 25, 1922, FP.

31. G. Marconi, "Report on recent visit to the United States," July 26, 1922, OX 36/64.

32. "Note on Franklin's S.W. Experiments leading up to the development of the Beam System in 1922–24," unsigned typescript, n.d., OX 1791.

33. "Proposal for a Network of Wireless Communications to serve the needs of the whole British Empire. Submitted by MWTC Ltd. Feb 1920," OX 230; "Imperial Wireless Telegraphy Committee. 1919–1920. Report," June 1920, OX 230; see also OX 207. This was followed by a more technical report, recapitulating most of the recommendations, "Report of Wireless Telegraphy Commission," December 9, 1921 (printed 1922), OX 207.

34. W.J. Rickard to Secretary of the Wolverhampton Chamber of Commerce, February 19, 1923, OX 230.

35. "Fifty Years Ago," unsigned, n.d., for *Point to Point* [1963], OX 1791.

36. "Prime Minister's statement regarding Imperial wireless communications," March 5, 1923, OX 230 (Great Britain, House of Commons, *Debates*, March 5, 1923, Vol. 161, cc29–30); Winseck and Pike 2007, 307; "Empire Wireless—A Compromise," *LT*, March 6, 1923.

37. G. Marconi, Diaries and Notebooks, April 15–June 14, 1923, OX 10; GM to C.S. Franklin, April 30, 1923; GM to GCI, April 30, 1923; telegram, GM to A. Gray, May 27, 1923; telegram, GM to C.S. Franklin, May 27, 1923, all in OX 232; "Senator Marconi's Speech at Press Conference," MWTC press release, April 9, 1927, OX 231.

38. G. Marconi, "Report on the results of the long distance receiving tests carried out on board the S.Y. 'Elettra' with 100 metres continuous waves, transmitted

by the experimental station at Poldhu," handwritten and typed versions, June 12, 1923, OX 232.

39. Supply Committee debate on Post Office estimates. Sir Laming Worthington-Evans, Postmaster-General (July 24, 1923), OX 230. See also "Statements on Imperial Communications," typed transcripts from meeting of Imperial Economic Conference, October 16, 1923, OX 230.

40. "Statements on Imperial Communications," typed transcripts from meeting of Imperial Economic Conference, October 16, 1923, OX 230.

41. "Imperial Wireless Communications," MWTC press release, October 19, 1923, OX 230.

42. Imperial Economic Conference, resolution passed at final sitting of the Conference, devoted to Imperial Wireless communications, November 9, 1923, OX 230.

43. GM/MWTC letter to the press, November 16, 1923, OX 230.

44. "The British Government and Wireless Communication," internal MWTC document, n.d., OX 230.

45. *Report of the Imperial Wireless Telegraphy Committee, 1924* (chaired by Robert Donald), OX 231. This was the second Donald Committee; the first one reported to Bonar Law in 1922; "Empire Wireless. Case for State Control," *LT,* February 29, 1924.

46. "Imperial Wireless," *LT*, February 29, 1924.

47. "Statement by Mr Godfrey Isaacs," MWTC press release, February 29, 1924, OX 231. See also C.S. Franklin to GCI, January 10, 1924, OX 234, and GCI to Post Office, January 26, 1924 (reported in *LT*, March 1, 1924).

48. GM and GCI, to J. Ramsay MacDonald, March 4, 1924, OX 231. Also reported in *LT*, March 8, 1924. Before writing this letter, Isaacs took care to have the company prepare a list of the GPO's past "obstructions" to the company's development (A. Gray to H.W. Allen, February 7, 1924, OX 197).

49. "Cabinet Committee on Report of Imperial Wireless Service, in Whitehall. Final Copy of Shorthand Notes," April 14, 1924, OX 231.

50. GM, "Private & Confidential," in "Cabinet Committee on Report of Imperial Wireless Service, in Whitehall. Final Copy of Shorthand Notes," April 14, 1924, OX 231.

51. *LT*, April 28, 1924; typescript extract from *Daily Telegraph*, June 16, 1924, OX 234; Rear Admiral in Charge, Gibraltar, to Secretary of the Admiralty, n.d., OX 234.

52. "Empire Wireless. State Control of Stations. Government and Marconi Co.," *LT*, July 24, 1924; Official Treasury Minute, "Agreement between MWTC Ltd and Postmaster General," July 28, 1924, OX 231.

53. GM memo, n.d., OX 234; see also "'Beam' System Wireless. Terms with the Marconi Co.," *LT*, August 1, 1924.

54. *LT*, August 16, 1924.

CHAPTER 28

1. G. Marconi, "Radio Telegraphy," *Proceedings of The Institute of Radio Engineers* 10, no. 4, (August 1922): 215–38, 238.
2. G. Marconi, foreword to Burrows 1924.
3. Cited in Dunlap 1937, 274.
4. Vyvyan 1933, 199.
5. "Proof of His Excellency The Marchese Marconi," 1934, OX 22 /96–108.
6. Briggs and Burke 2009, 143–45. A similar system was also imagined, even earlier, by the American utopian socialist Edward Bellamy in his novel *Looking Backward, 2000–1887* (Boston: Houghton Mifflin, 1888).
7. Aitken 1985, 166; Hong 2001, 165.
8. Briggs and Burke 2009, 149.
9. JAF to GM, n.d., OX 367; J.A. Fleming and MWTC, "Improvements in instruments for detecting and measuring alternating electric currents," UK patent 24850/04, November 16, 1904, copy in OX 412.
10. Hong 2001, 119.
11. Hong 2001, 155 and 192.
12. Briggs and Burke 2009, 152.
13. Douglas 1987, 293 and 303.
14. Ibid., 293.
15. The Radio League replaced Gernsback's earlier Wireless Association of America. Its letterhead listed four "honorary members": Reginald Fessenden, Lee de Forest, Nikola Tesla, and W.H.G. Bullard (*The Electrical Experimenter*, December 1915, 381, in http://earlyradiohistory.us/1915rla.htm).
16. Douglas 1987, 296.
17. "History of Wisconsin AM stations," http://www.qsl.net/k9ez/history.htm.
18. Douglas 1987, 293–95; Briggs and Burke 2009, 152; *NYT*, November 8, 1916.
19. De Forest eventually soured on broadcasting when he saw the effects of advertising; in a 1932 statement to a Canadian parliamentary committee, he stated: "Within the span of a few years we in the United States have seen broadcasting so debased by commercial advertising that many a householder regards it as he does the brazen salesman who tries to thrust his foot in at the door." De Forest called for a ban on direct advertising and pleaded with Canadian parliamentarians "to lead us out of the morass" (Canada, Parliament, House of Commons, Special Committee on Radio Broadcasting, Minutes of Proceedings and Evidence, April 13, 1932, 491–92).
20. Archer 1938, 90.
21. Lyons 1966, 46 and 182.
22. Lyons 1966, 71–72.
23. Archer 1938; Maclaurin 1949.
24. Maclaurin 1949, 111, note 2.

25. The date of Sarnoff's memo to Nally is given variously as 1915 or 1916 by different Sarnoff biographers and other secondary sources. According to Louise M. Benjamin (1993), there is no evidence of a 1915 memo except in Sarnoff's later claims. Most broadcast historians consider Sarnoff's role as an originator of commercial radio to be exaggerated, although there is no doubt of his importance as a corporate broadcasting executive. See http://earlyradiohistory.us/WJY.htm#next.

26. Archer 1938, 200–1.

27. Maclaurin 1949, 139.

28. Douglas 1987, 302–3.

29. Maclaurin 1949, 139.

30. Briggs and Burke 2009, 156.

31. "Broadcasting Dates," undated typescript, OX 312; "Wireless broadcasting in England," typescript c. January 1923, OX 312; "London Remembers," http://www.londonremembers.com/memorials/marconi-wc2.

32. "London Speaks to Geneva by Wireless Telephone," MWTC press release, December 12, 1920, OX 312; "Wireless Telephone to Geneva. A London Experiment," *LT*, December 13, 1920.

33. See also A.R. Burrows to H.W. Allen, January 31, 1922, OX 311.

34. "London Remembers," http://www.londonremembers.com/memorials/marconi-wc2.

35. The Marconi company claimed more than fifty showcase broadcasts between May and November 1922, culminating in an election results broadcast on November 15. Four of the first six UK stations were built by Marconi's (Cardiff, London, Glasgow, and Newcastle); the other two were Metropolitan-Vickers' station at Manchester, and a Western Electric station at Birmingham (MWTC, "The Development of Broadcasting in Great Britain. A General Review of the Lines along which Broadcasting has been Developed," typescript, n.d., OX 323). Marconi himself was rarely involved in broadcasting developments, except as a privileged user. A memo from the company's publicity department on September 28, 1922, provides an example of how programming could be customized to the demands of the boss during these early days: "In addition to the concerts at 5 and 6 o'clock today we have been asked by Mr. Marconi to arrange a third special concert to be transmitted at 9 p.m. to the Italian Embassy where the Crown prince of Italy, Mr. Marconi and Madame Tetrazzini will be listening in" (OX 311).

36. "Arrangements for Broadcasting," *Evening Standard*, May 18, 1922; "Wireless Firms Asked to Produce a Joint Scheme," *Manchester Guardian*, May 19, 1922.

37. Hilmes 2002, 57, based on records in BBC Written Archives Centre, Caversham Park.

38. "The Radio Ban Decision," *Daily News*, July 14, 1922; Great Britain, Broadcasting Committee (chaired by Sir Frederick Sykes), *Report*, London: King's Printer, 1923; "The Development of Broadcasting in Great Britain. A General Review

of the Lines along which Broadcasting has been Developed," undated type-script, OX 323.

39. *Electrician,* August 4, 1922.

40. *Evening News,* August 14, 1922.

41. "Politics and the Money-Power," *Nation and the Atheneum,* December 16, 1922.

42. *Shrewsbury Chronicle,* December 15, 1922.

43. "Mr. F.G. Kellaway, 'Father of British Broadcasting,'" *LT,* April 15, 1933.

44. "Memorandum and Articles of Association, British Broadcasting Company Ltd," December 15, 1922, OX 313.

45. *NYT Magazine,* February 13, 1927.

46. "Memorandum on Wire and Radio Communications. Paris, 12 February, 1919," submitted to President Wilson by Walter S. Rogers, communications expert of the American Commission to Negotiate Peace, Document 63 in Ray Stannard Baker, *Woodrow Wilson and World Settlement,* Vol. 3 (Garden City, NY: Doubleday, 1922), 428ss.

47. Archer 1938, 164–65.

48. Breckinridge Long, FDR's US ambassador to Italy from 1933 to 1936, was controversially favourable to Mussolini. During the Second World War, he was responsible for refugee policy within the US State Department and is remembered today as a leading obstructionist to admitting wartime refugees to the United States.

49. Winseck and Pike 2007, 262.

50. Ibid., 264.

51. Ibid., 270.

52. See, for example, *NYT,* March 24, 1912, in which he explicitly frames voice transmission as wireless telephony.

53. Douglas 1987, 303, citing Barnouw 1966.

54. "Other historical highlights," unsigned, n.d., OX 1865; Dowsett 1951, 454–55, OX 1834.

55. McChesney 1993.

56. Douglas 1987, 292–314.

57. Sykes report, article 6.

58. Ibid., article 7.

59. Great Britain. Broadcasting Committee (chaired by the Earl of Crawford), *Report,* article 3, London: King's Printer, 1926.

60. Crawford report, article 4.

61. Ibid., article 20a.

62. Canadian Marconi Company, *The People Station, CFCF Radio 600* (brochure, 1969), cited in Pagé 2007, 59.

63. Broadcaster Arthur Burrows remembers visiting the Marconi station in Montreal in the autumn of 1920 and finding there "an able and enthusiastic staff bound hand and foot by Government regulations" (Burrows 1924, 54).

64. Pagé 2007, 59. Pagé has done yeoman's work uncovering archival documents that rewrite the company's own self-congratulatory historiography.

65. Ministry of Fisheries annual report for 1919–20, cited in Pagé 2007, 59.

66. Pagé 2007, 41, note 13.
67. The call letters for Marconi's CFCF first appeared in an ad in the *Montreal Herald* on June 24, 1922.
68. Canadian Marconi Company, *The People Station, CFCF Radio 600* (brochure 1969), cited in Pagé 2007, 59; A.R. Burrows, "Reflections at a 'Milestone,'" *World-Radio*, November 11, 1932, OX 322.
69. *Montreal Herald*, May 27, 1922, cited in Pagé 2007, 65.
70. Pagé 2007, 61 and 65.
71. Raboy 1990, 17–47. The quotation is from testimony by Canadian Radio League spokesman Graham Spry in Canada, Parliament, House of Commons, Special Committee on Radio Broadcasting, Minutes of Proceedings and Evidence, March 11, 1932, 45–46 (cited in Raboy 1990, 40).

CHAPTER 29

1. Headrick 1991, 204.
2. Vyvyan 1933, 76–77.
3. Winseck and Pike (2007, 310) see the quick reintegration of Germany in this regard as a display of US foreign policy.
4. Maclaurin 1949, 107.
5. Howeth 1963, chapter XXX.
6. Great Britain, House of Commons, *Debates,* July 24, 1923, Vol. 167, 315.
7. "Cabinet Committee on Report of Imperial Wireless Service, in Whitehall. Final Copy of Shorthand Notes," April 14, 1924, OX 231.
8. "Retirement of Managing Director. Appointment of the Right Honourable F.G. Kellaway, P.C.," OX 684. Isaacs resigned from the boards of MWTC and MIMCC on February 9, 1925 (GCI to GM, February 9, 1925, OX 393).
9. *LT,* April 18, 1925.
10. "Speech by Senator Guglielmo Marconi at Archiginnasio di Bologna at the Commemoration of the thirtieth anniversary of his first patent in wireless telegraphy," June 13, 1926, typescript translation, OX 52 /335–44.
11. "Beam Wireless," two-page typescript, March 17, 1925, OX 238.
12. MWTC, "Canadian beam wireless stations. Construction well advanced," n.d., OX 237; *LT*, April 1 and 3, 1925; GPO agreement with Canadian Marconi Company, May 31, 1926, LAC, MG 28 III 72, vol. 82; "Memorandum submitted on behalf of the [Canadian] Department of Marine and Fisheries respecting certain subjects on the agenda for the Imperial Conference 1926," October 4, 1926, OX 231.
13. *LT*, October 21, 22, 25, and 26, 1926; G. Marconi, Diaries and Notebooks, November 13, 1926, OX 12; "Mussolini and Marconi," *Daily Mail*, November 17, 1926; *LT*, November 22, 1926.
14. "For King and Country," *Daily Mail*, May 11, 1926.
15. *British Gazette*, May 5, 1926.

16. "Great Strike Terminated," *British Worker*, May 13, 1926; "General Strike Off," *British Gazette*, May 13, 1926.

17. H.C. Kenworthy to Chief of Field and Air Division, MWTC, May 16, 1926, OX 370.

18. Ibid.

19. West 2006, 52.

20. Ibid., 252.

21. Wright 1987, 2.

22. MWTC directors' meeting minutes, May 20, 1926, OX 394.

23. For example, " 'Mankind's Debt to Wireless.' Value During a Crisis," *Daily Telegraph,* September 30, 1931.

24. MWTC, Technical and Engineering departments' general correspondence, 1910–28, OX 814.

25. "Proof of His Excellency The Marchese Marconi," 1934, OX 22 /96-108.

26. G. Marconi, "Board meetings attended since appointment as Chairman," n.d., OX 22 /115–22.

27. "Proof of His Excellency The Marchese Marconi," 1934, OX 22 /96-108.

28. The information in this paragraph is taken from several of Marconi's personal records, including: "Company shares and finances," n.d., OX 22/114–44; untitled document, n.d., OX 22/158. Marconi had, of course, other sources of income during this time: director's fees, the Sconto presidency, his Italian business, et cetera.

29. Agreement between GM and MIMCC, January 17, 1922; Agreement between GM and MWTC, January 18, 1922; both in OX 524.

30. MWTC, "Organisation Chart," April 1923, OX 654.

31. "MWTC Ltd. Shares held by Senatore Guglielmo Marconi and Nominees as at the 17th January, 1921," OX 654.

32. Dowsett 1951, 426, OX 1834.

33. "MWTC Ltd. A Study of a Great Organisation," 1932, typescript, OX 1838.

34. Dowsett 1951, 429, OX 1834.

35. MWTC directors' meeting minutes, July 26, 1927, OX 394.

36. ITU, International Radiotelegraph Conference (Washington, 1927), International Radiotelegraph Convention.

37. Anonymous 1933, 11.

38. *NYT*, October 5, 1927; see also Howeth 1963, chapter XLII.

39. "Commercial Rivalry in Radio" (editorial), *Ottawa Citizen*, October 11, 1927.

40. "Says Radio Cannot Replace Newspaper," *Ottawa Citizen*, October 13, 1927.

41. G. Marconi, "Address delivered at a banquet given by the Radio Corporation of America to the delegates to the International Radio Telegraphic Conference: Hotel Plaza, New York City, October 15, 1927," OX 206.

42. G. Marconi, "Radio Communications," lecture to the American Institute of Electrical Engineers and the Institute of Radio Engineers, New York, October 17, 1927, OX 53.

43. For example, the North American Regional Agreement, negotiated in 1929: Dominion of Canada, Treaty Series, 1929, No. 6. "Exchange of Notes (Feb. 26 and 28, 1929, and subsequent dates) Constituting an Agreement between Canada, the United States, Cuba and Newfoundland relative to the assignment of high frequencies to radio stations on the North American continent," Ottawa: King's Printer, 1930.

44. Howeth 1963, chapter XLII; also "Report to the Minister of Marine and Fisheries by the Canadian Delegation to the International Radiotelegraph Conference, Washington, October and November 1927," Ottawa: King's Printer, 1928.

45. GM to DM, December 20, 1927, RI 117.

46. Headrick 1988, 133.

47. Winseck and Pike (2007, 318–29), who also cite as a looming issue the rise of the US-based International Telegraph and Telephone Company (ITT) as the dominant firm in world communication, and the threat that represented to both Marconi and the British cable companies.

48. Dowsett 1951, 435, OX 1834.

49. Headrick 1988, 134.

50. Winseck and Pike 2007, 325. The company name was later changed to Cable and Wireless.

51. Dowsett 1951, 440, OX 1834.

52. Ibid., 444.

53. J. Denison-Pender to FGK, August 2, 1928, OX 524.

54. Service Agreement Between Cables and Wireless Ltd. and Guglielmo Marconi, May 7, 1929, OX 524.

55. FGK to J. Denison-Pender, November 7, 1928, OX 524.

56. Dowsett 1951, 418–44, OX 1834.

57. Jolly 1972, 252–53.

58. See for example, "Marconi Follows the Beam but Not as Merlin Did," *NYT*, April 29, 1928.

CHAPTER 30

1. B. Marconi, Memoir, FP.

2. Bertini 1969, 119; Bosworth 2006, 372.

3. Marconi had invited the Medicis on the *Elettra*'s maiden voyage in 1919, a trip that Beatrice described as "one of the most miserable moments of my life." Beatrice was also supposed to accompany Marconi on his famous trip to Fiume in 1920 but left the boat when she suspected, correctly, that the Medicis were going to join them (B. Marconi, Memoir, FP; *Il Piccolo* [Trieste], September 16, 1920).

4. B. Marconi, Memoir, FP.

5. Solari 2011, 208.

6. Sachs wrote that the pianist found Paola somewhat unpleasant when they first met, in Venice, in 1923, but she fell quite in love and "virtually threw herself at

him," and a five-year on-again-off-again affair began. Rubinstein described Paola in his memoir, *My Many Years*, as exceptionally beautiful, "with a delicate round face and a pale complexion. She parted her shiny black hair on one side and let it fall freely to cover her ears in a curl. Her dark eyes were shadowed by long lashes and heavy eyebrows and they looked at you with a bit of haughtiness but with a touch of humor as well" (Sachs 1995, 222; Rubinstein 1980, 176).

7. B. Marconi, Memoir, FP.
8. GM to BOM, February 8, 1922, FP.
9. GM to BOM, February 21, 1922, FP.
10. GM to BOM, August 17, 1922, FP.
11. Untitled document agreeing that GM and BOM will live separately, et cetera, December 13, 1922, OX 22/71–72.
12. D. Marconi 1962, 241.
13. GM to LOB, February 28, 1923, enclosing letter to BOM, FP. Emphasis in original.
14. GM to BOM, March 6, and 18, 1923, FP; March 29, 1923, MB.
15. GM to BOM, April 11, 1923; GM to BOM, June 6, 1923; both in FP.
16. GM to BOM, October 12 and 19, 1923; both in FP.
17. Masini 1995, 294.
18. Untitled document signed by GM and BOM outlining the terms of their divorce, December 18, 1923, MB.
19. Solari 2011, 250; also GM to LS, March 8, 1924, cited in Solari 2011, 251.
20. The two-page holograph will dated May 15, 1917, in Italian, is in the Marconi Archives in Oxford (OX 22/46rv). It is over-written with a bold "X" and a note at the end stating, *"Annullato Marzo 1924 GM."* At the top of the document he has written *"Tutto annullato Guglielmo Marconi,"* and confirmed the cancellation with another note a year later, stating *"Tutto annullato Guglielmo Marconi. Annullato in seguito a scioglimente di matrimonio con Beatrice O'Brien Marzo 1925. G Marconi."* He did not write a new will until 1935; GM to BOM, March 19, 1924, FP.
21. GM to BOM, June 14, 1924, FP; GM to BOM, November 28, 1924, MB.
22. B. Marconi, Memoir, FP; GM to BOM, January 31, 1925, MB.
23. "Marconi's Love Romance. Engagement to a Cornish Girl Expected. Her 18th Birthday. Wireless Talks from his Yacht," *Daily Express*, April 8, 1925.
24. Marnham 2006, 42.
25. Ibid., 43.
26. GM to Lady Falmouth, March 30, 1925, OX 18/141.
27. "Marconi's Romantic Easter Visit," *Evening Standard*, April 13, 1925; GM to FGK, April 13, 1925, OX 18/139; FGK to GM, April 16, 1925, OX 18/138.
28. Marnham 2006, 41; GM to Avvocato, April 20, 1925, OX 19. On the reporting of the relationship, see also Jolly 1972, 255–57.
29. Jolly 1972, 256.

30. "Tuff" to GM, "Tuesday" [postmarked June 8, 1925], OX 18/143; G. Marconi, Diaries and Notebooks, December 30, 1925, OX 11; March 18, 1926, and September 29, 1926, both in OX 12.
31. GM to Green & Abbott Ltd., December 24, 1924, OX 18/134.
32. G. Marconi, Diaries and Notebooks, 1925 and 1926, OX 11–12.
33. GM to BOM, August 13, 1925, FP; GM to DM, July 6, 1925, RI 112. Emphasis in original.
34. B. Marconi, Memoir, FP.
35. Cited in D. Marconi 1962, 252. From Marconi's correspondence (GM to BOM, September 2, 1925, MB, below), it appears that Beatrice wrote this letter on August 22, 1925.
36. GM to BOM, September 2, 1925, MB.
37. GM to BOM, April 5, 1925, FP.
38. D. Marconi 1962, 259–60.
39. Secretary of State to F. Pacelli, November 29, 1930; unaddressed, unsigned letter re: Count Francesco Bezzi-Scali, exemption from jury duty, December 2, 1930; both in ASV, Segreteria di Stato, Schedario (1930–35), Anno 1930. r. 25. f. 1.
40. D. Colonna, *Via dei Condotti* (Rome: Ello de Rosa editore, 2001), 35–37.
41. M.C. Marconi 1999, 19.
42. G. Marconi, Diaries and Notebooks, October 24 and October 31, 1925, both in OX 11; January 1, 1926 and ff, OX 12; M.C. Marconi 1999, 26
43. GM to MCM, June 27, 1926, cited in M.C. Marconi 1999, 26.
44. "Discussion of Catholic Church's position on divorce—circa 1923 [sic: it must be at least 1927]. Writer unknown," MB.
45. Notes at the back of Marconi's diary for 1926 (OX 12) include details of the key stages of the divorce and annulment proceedings; Beatrice's Church of Ireland baptism information; entries from the Bologna city register for March 29 and April 1, 1924, from the Ministry of Justice in Rome on April 2, 1924, and the Ministry of Foreign Affairs on April 3, 1924; the Fiume civil decree, February 12, 1924, as seen by the Fiume Court of Appeal, March 8, 1926; the Bologna civil notice of annulment of marriage dated April 4, 1924, with copies dated April 7 and 8, 1924; and the notation that these were all sent to Monsignor Surmont, the vicar-general at Westminster, on July 7, 1926. (In their civil divorce proceeding, Marconi and Beatrice both gave their civic addresses as Bologna.)
46. G. Marconi, Diaries and Notebooks, May 30, 1926, OX 12.
47. GM to BOM, June 22, 1926, FP.
48. GM to BOM, July 4, 1926, MB.
49. GM to BOM, July 8, 1926, FP.
50. Ibid.
51. GM to BOM, August 10, 1926, FP; August 24, 1926, September 1 and 16, 1926, FP.
52. GM to BOM, November 10, 1926, FP.
53. Unsigned letter to C.C. Dominioni, November 9, 1926, ASV, Segreteria di Stato Pre-1930, Rubricelle, 1926, Vol. 684; "Marconi, Senatore, Circa causa mat-

rimoniale," November 23, 1926, ASV, Segreteria di Stato Pre-1930, Rubricelle, 1926, Vol. 684.

54. GM to BOM, November 26, 1926, FP. Gasparri, who was Vatican secretary of state from 1914 to 1930, was a personal friend of the Bezzi-Scalis. He was the only cardinal to serve two popes as secretary of state—Benedict XV (1914–22) and Pius XI (1922–30).

55. G. Marconi, Diaries and Notebooks, November 24, 1926, OX 12; GM to BOM, November 26 and 30, 1926; both in FP.

56. *Catholic Herald*, December 11, 1964, and various contemporary press reports.

57. *Spectator*, November 27, 1926.

58. GM to BOM, December 6, 1926, FP.

59. L. Marignoli, Memoir, FP; also in original (Italian) version of Masini (1975, appendix).

60. GM to BOM, n.d. [December 27, 1926], FP. Marignoli claimed that it was at his insistence that Beatrice went ahead and gave testimony that she felt was not fully truthful, in exchange for Marconi's assurance that their children would be properly looked after. According to Marignoli, he and Beatrice met with Marconi a few days later to discuss Marconi's plans to look after the children if the annulment was granted. Marignoli proves to be Marconi's equal at using the system to achieve his ends, arranging an intervention with Cardinal Gasparri to urge him to insist that Marconi settle the situation of his children properly before proceeding to his new marriage. Marconi was furious when he learned of the approach to Gasparri; he called Marignoli and proclaimed, "When I make a promise I keep it." Writing this memoir after Marconi's death, Marignoli stated that he made a grave error in not insisting on a written and signed statement from Marconi about his legacy intentions (L. Marignoli, Memoir, FP).

61. S. Romana Rota, "Acta Tribunalium. Sacra Romana Rota. Westmonasteriensi. Nullitatis Matrimonii (Marconi-O'Brien)." Official summary of the case and the decision, AAS, 19, 1927, 217–27 (www.vatican.va/archive/aas/index_sp.htm).

62. Ibid.

63. See "Discussion of Catholic Church's position on divorce—circa 1923 [sic]. Writer unknown," MB. See also Sheed 1959, which summarizes the Rota judgment (chapter 4, unpaginated). Matrimonial records continue to preoccupy the custodians of Vatican secrets. My request to see the ASV files on Marconi, in March 2012, brought the reply that while some were available, "the documentation related to the annulment of Marconi's marriage is excluded from consultation" (M. Grilli, Secretary of the Prefecture, to author, March 10, 2012).

64. Agent 394, report, May 6, 1927, ACS, PP, b.60/A.

65. Canali 2004, 118.

66. Canali 2009, 227.

67. Bocchini has been described as "cynical, apolitical, pitiless and able" (Bosworth 2006, 5). He was one of the many high Fascist officials who were unhappy about Mussolini's cozying up to Nazi Germany in the late 1930s.

68. D.M. Smith 1981, 129 and 146.

69. "S.E. Guglielmo Marconi," ACS, PP, b.60/A. There are two main files regarding Marconi in the Italian state archives in Rome: one in the Segreteria Particolare del Duce—Carteggio Ordinario (the open or general files)—contains mainly the abundant correspondence between Marconi and Mussolini; the other contains the reports of the fascist political police. (A third file covers Marconi's correspondence with Mussolini regarding the National Research Council.) Mussolini also had files on both of Marconi's wives as well as each of his children. These papers remained in Rome after the fall of the fascist regime and are easy to consult—if one knows how to ask for them—at the Central Archives of the State.

70. Bice Pupeschi was a character straight out of a steamy pulp novel. Described on her arrest in 1945 as "beautiful, shapely, dark and tall," she posed as a writer and poet, had been married and divorced, and had several well-placed lovers, including an industrialist, a baron, and, notably, Bocchini (Canali 2004, 283).

71. All of the detail in this book identifying particular Polpol agents is based on Canali 2004. The reports in Marconi's Polpol file are identified by the agents' numbers only.

72. "*Senatore Marconi circa benedizione suo matrimonio*," May 20, 1927, ASV, Rubricelle, 1927, Vol. 690; slip of paper attached to Pacelli's copy of the original letter.

73. "C.G." to E. Pacelli, May 26, 1927, and note from P. Gasparri, May 28, 1927, in ASV, Rubricelle, 1927, Vol. 690.

74. GM to A. Chiavolini (Mussolini's chief of staff), May 24, 1927; A. Chiavolini to GM, May 26, 1927, in ACS, SPD CO GMF, 197.598/1; telegram, BM to GM, June 14, 1927, ACS, SPD CO GMF, 197.598/1.

75. M.C. Marconi 1999, 22–23; and GM to MCM, May 14, 1927, cited in M.C. Marconi 1999, 34–35; GM to BOM, June 14, 1927, FP.

76. GM to BOM, October 4, 1922, MB; November 9, 1924, FP; March 14, 1925, FP; July 24, 1925, FP; September 2, 1925, FP; April 27, 1927, FP.

77. GM to DM, July 26, 1924, and July 10, 1925, RI 107 and RI 113. Michael Braga, interview with author, March 29, 2012.

78. See, for example, GM to DM, March 30, 1931, RI 137.

79. GM to Clare Armar Lowry-Corry, November 7, 1931, and February 14, 1932, MB.

80. L. O'Brien 1952, FP.

81. GM to BOM, September 28, 1924, FP; November 9, 1924, FP; September 15, 1925, FP; October 17, 1925, FP.

82. GM to DM, July 26, 1924, RI 107.

83. GM to DM, March 1, 1927, RI 116.

84. GM to DM, August 22, 1929, RI 126; DM to GM, August 27, 1929, OX 17.

85. D. Marconi 1962, 276.

86. GM to DM, August 13, 1930, RI 134.

87. GM to BOM, August 10, 1926; August 17, 1926; August 24, 1926; July 20, 1926; December 6, 1926; and December 11, 1926; all in FP.

CHAPTER 31

1. Goethe 1962, 133.
2. Bosworth (2006, xxiv) calls it Rome's "model Fascist suburb."
3. On the role of Italy's business elite in the rise and rule of fascism, see Guerin 1973.
4. Bosworth 2006, 201–2.
5. Foot 2009, 55.
6. He later installed his family in the palatial Villa Torlonia on Via Nomentana north of the city centre, but he rarely slept there.
7. Kertzer 2014, 62.
8. The Lincei was founded in 1603. Its most celebrated member, Galileo Galilei, was inducted in 1611; the academy supported Galileo in his trials with the Roman Inquisition, then disappeared abruptly in 1651. It was revived after unification in 1874 as the Accademia Nazionale Reale dei Lincei (Royal National Lincean Academy), now encompassing the humanities as well as the sciences.
9. Paoloni and Simili 2008, 139. A statute was published in October 1924 (Simili and Paoloni 2001, 5).
10. Ibid., 140.
11. Ibid.
12. Simili 2001b; OX 21/103.
13. Simili 2001b, 142–43.
14. See, for example, the many references to Marconi in Simili and Paoloni 2001.
15. See, for example, G. Marconi 1941.
16. Paoloni and Simili 2008, 141.
17. Turchetti 2006, 157.
18. "Dispositions Regarding Higher Education," Decree, August 28, 1931, published in *Gazzetta Ufficiale*, October 8, 1931, and effective as of November 1, 1931. Among the international interventions against the loyalty oath was one by Albert Einstein, who wrote a letter of protest to Minister of Justice Alfredo Rocco, with absolutely no effect (A. Einstein to A. Rocco, November 16, 1931, Albert Einstein Archives, Jewish National & University Library, Hebrew University of Jerusalem, cited in Paoloni and Simili 2008, note 74). Volterra, like the others, received a letter from his university rector on November 18, 1931, inviting him to come in to sign the oath. He wrote back that he could not do so in good conscience (Simili 2001a, 119) and was dismissed from the university in January 1932.
19. Paoloni and Simili 2008, 140–1.
20. See Gamba and Schiera 2005; also Ianniello 1995.
21. G. Marconi, "*Le radiocomunicazioni a fascio*," lecture, Regia Università Italiana per Stranieri di Perugia, September 8, 1927, BS.
22. "Discorso di G. Marconi," September 19, 1927, in Gamba and Schiera 2005, 219–27 (English typescript, OX 53/15).
23. A photograph of Marconi and Mussolini on this occasion is on display in the museum at the Villa Torlonia.

24. CNR board meeting minutes, September 20, 1927, in Paoloni and Simili 1996, 145, as well as Simili and Paoloni 2001, appendix.

25. "*Il Duce fissa in un messaggio a Guglielmo Marconi le direttive del Consiglio Nazionale delle Ricerche. Un'altra iniziativa da Mussolini per l'incremento economico dell'Italia,*" *Il Popolo di Roma,* January 7–8, 1928, ACS, SPD CO CNR; Simili 2001b, 135; Mussolini message to Marconi, ACS, CNR, b.2, f.12.

26. D.M. Smith 1981, 133–34.

27. *Inventario,* chapter 1, ACS, CNR.

28. Maiocchi 2009, 62.

29. *Inventario,* chapter 1, ACS, CNR.

30. G. Bottai to BM, January 15, 1937, ACS, PCM, Gabinetto 1940–1943, 3/3–8n.1501, sf.1–1; Paoloni and Simili 1996, 152.

31. CNR board meeting minutes, December 20, 1927, cited in Simili 2001b, 128–29.

32. Simili 2001b, 128–29.

33. BM to P. Fedele, April 13, 1928, ACS, SPD CO CNR.

34. A. Giannini to Minister of Foreign Affairs, March 24, 1928, ACS, CNR, b.23, f.415; also in Simili 2001a, 117.

35. G. Magrini, untitled undated handwritten notes, 1928–29, ACS, CNR, b.23, f.420; see also Simili 2001a, 117.

36. Paoloni and Simili 2008.

37. Undated (1929) notes for a draft brief by Giovanni Magrini, ACS, CNR, b.23, f.418; cited in Simili 2001a, 118.

38. February 2, 1929, typescript in OX 53/225; Reports in *L'Impero* (Rome), February 3, 1929; *LT,* February 4, 1929.

39. "Scientific Research in Italy. Organisation urged," *LT,* February 4, 1929. Copy of speech in OX 53.

40. Kertzer 2014, 52.

41. See, for example, GM to BOM, October 19, 1923, FP.

42. Kertzer 2014, 61 and 68; ASV.

43. Kertzer 2014, xxii.

44. Ibid., 19 and 59.

45. Ibid., 59, 48, and 57.

46. Ibid., 51.

47. *Civiltà cattolica* IV (1922): 111–21. The article became a central plank of the emerging ideology of Germany's Nazi party.

48. Kertzer 2001, 263; Kertzer 2014, 196.

49. *The Roman-Vatican Settlement* (translation of the Lateran documents of February 11, 1929. London: The Royal Institute of International Affairs, Chatham House, July 1929).

50. Kertzer 2014, 137.

51. G. Marconi, "*Congratulazione per la conciliazione fra la Santa Sede e l'Italia,*" February 12, 1929, ASV, Rubricelle 1929, Vol. 707, Prot. 77418, f.1.

52. Nadel 1999, 87; see also Campbell 2006.

53. Cristina wrote that she was present at this audience, which she placed incorrectly in 1930 (M.C. Marconi 1999, 126).

54. Carlo Confalonieri, *Pio XI visto da vicino*, Torino: SAIE, 1957, 147–48, cited in Bea and de Carolis 2011, vol. 1, 20. Confalonieri was the pope's private secretary.

55. F. Pacelli to GM, June 14, 1929 (translation), OX 212.

56. B. Marconi, Memoir, FP.

57. MWTC press release, January 5, 1925, OX 316.

58. M.C. Marconi 1999, 22.

59. Kertzer 2014, 149.

60. D. Kertzer, correspondence with author, August 11, 2015.

61. Matelski 1995, xvi; Bea and de Carolis 2011, vol. 1, 19–21.

62. *LT*, June 29, 1929.

63. Agent 37, report, November 19, 1929, ACS, PP, b.60/A.

64. "Inauguration of the wireless station at the Vatican City," MWTC press release, March 24, 1931, OX 242; Matelski 1995, 19.

65. GM to HAW, August 23, 1930, OX 525.

66. GM to HAW, September 6, 1930; GM to FGK, September 6, 1930; both in OX 525; MWTC, "Vatican Wireless Services," April 1, 1932, OX 316.

67. *L'Osservatore Romano*, September 22, 1930; *LT*, September 23, 1930.

68. *NYT*, February 13, 1931; *L'Osservatore Romano*, February 14, 1931.

69. *Globe*, February 13, 1931; "150 Stations Carry Program to Nation," *NYT*, February 13, 1931; "Pope's Voice Clear on New York Radios," *NYT*, February 13, 1931.

70. "Pontiff Heard in Other Countries," *Daily Boston Globe*, February 13, 1931; "Broadcast by the Pope," *LT*, February 23, 1931; "Pope Pius XI's Speech," *NYT*, February 13, 1931; Matelski 1995, xvi and 19.

71. A. Cortesi, "Pope Speaks to World in Greatest Broadcast; Message Stresses Peace. Calls for Justice to All," *NYT*, February 13, 1931.

72. "Inauguration of the wireless station at the Vatican City," MWTC press release, March 24, 1931, OX 242.

73. A. Cortesi, *NYT*, February 13, 1931.

74. "Academy Ceremony Heard on Air Here. Text of Pontiff's Address Conferring Membership on Senator Marconi. Praises Inventive Genius. Pope Congratulates Him for His Leadership and Calls Him 'a Very Great Man,'" *NYT*, February 13, 1931.

75. A. Cortesi, *NYT*, February 13, 1931.

76. Matelski 1995, 20.

77. Agent 40, report, March 5, 1931, ACS, PP, b.60/A.

78. Papal Encyclicals Online, "On Catholic Action in Italy, Non Abbiamo Bisogno, Encyclical Of Pope Pius XI promulgated on June 29, 1931," http://www.papalencyclicals.net/Pius11/P11FAC.HTM.

79. *World-Radio*, November 20, 1936, OX 322.

80. Matelski 1995, xvii. Historian Marilyn Matelski's experience shows that as late as 1993, Vatican Radio was reluctant to allow researchers access to its archives (see Matelski 1995, xi–xii). In 2012, they were still politely denying accessibility. Like Matelski, I had to rely on secondary sources plus a few documents gleaned here and there in the other archives I studied. Some important aspects of Marconi's relationship to the Vatican remain to be fully explored.

81. Kertzer 2014, 121.

82. Masini 1995, 309.

83. GM, correspondence with Mussolini's office, February 20, 21, and 25, 1929, ACS, SPD CO CNR. Marconi and Mussolini met at the Palazzo Venezia on February 27, 1929.

84. GM to BM, September 10, 1929, ACS, SPD CO CNR.

85. Berezin 1997, 110.

86. See, for example, G. Magrini to GM, January 23, 1930, ACS, CNR b.231/1, cited in Simili 2001b, 152; GM, correspondence with Mussolini's office, January 24 and 25, 1930, ACS, SPD CO CNR (Marconi and Mussolini met at the Palazzo Venezia on January 27, 1930); GM to BM, January 28, 1930, with cover letter to A. Giannini, ACS, CNR, b.1, f.9, sf.2; *Il Popolo d'Italia*, February 12, 1930.

87. Paoloni and Simili 1996, 150.

88. Simili 2001b, 128–29. Marconi was forever inviting Mussolini to visit him on the yacht. It was the topic of one of his first letters to him, on September 9, 1926, and most recently on April 17, 1930 (both in ACS, SPD CO GMF, 197.598/1). There is, however, only one recorded visit, on June 7, 1930, discussed in chapter 32.

89. D.M. Smith 1981, 133.

90. Marconi's candidacy to the academy was officially proposed by mathematician Francesco Severi, and after his election he was promptly and unanimously designated president. There is no question but that this was on Mussolini's initiative.

91. BM to A. Mosconi, September 24, 1930, ACS, SPD CO CNR.

92. BM to GM, n.d., ACS, CNR, b.2, f.14–15.

93. Cagiano de Azevedo and Gerardi 2005, introduction; *Inventario*, preface and chapter 4, ACS, CNR.

94. G. Marconi, transcript of simultaneous address to the World Power Conference in Berlin and the National Electric Light Association Conference in San Francisco, broadcast June 18, 1930, OX 53 /265; "Information on World Power Conference for Senator Marconi," LIN 18/2.

95. Telegram, GM to A. Bocchini, September 11, 1931, ACS, PP, b.60/A.

96. Typescript, "The Development of Wireless Telephony," September 19, 1931 (hand-noted "Marconi's broadcast Sept 19/31"), OX 309; Marchese Marconi, "Development of Wireless Communication," *LT*, September 21, 1931.

97. GM to BM, September 29, 1931, ACS, SPD CO CNR.

98. Bodanis 2000, 97.

99. Turchetti 2006.

100. University of Rome, "Orso Mario Corbino (1876–1937)," http://www.phys .uniroma1.it/DipWeb/museo/corbino.html.

101. In 1934, Corbino named Marconi to a newly created chair in electromagnetic physics (O.M. Corbino, note and resolution, May 27, 1934; GM to O.M. Corbino, May 29, 1934; both in LIN 2/95).

102. Ianniello 1995, 31.

103. Ibid., 31–32.

104. Fermi's coalescing role in this conference was a good example of the collaboration between the CNR and the Royal Academy under Marconi's presidency (see RAL, Titolo I: Presidenza, b.1, f.4.3).

105. LT, October 13, 1931; O. Lodge to GM, May 26, 1931; GM to O. Lodge, June 10, 1931; both in LIN 18/2.

CHAPTER 32

1. W.J. Baker 1970, 291; G.A. Mathieu, "Preliminary Report on Micro-Wave Radio Beacon," Rome, December 11, 1933; G. Marconi, "On the Propagation of Microwaves at Long Distances," (translation) August 14, 1933; both in OX 377.

2. G. Marconi, "Radio Communication by Means of Very Short Electric Waves," December 2, 1932, OX 818 and OX 377; Dowsett 1954, OX 1834.

3. "MWTC Ltd. A great wireless concern and its world-wide ramifications," n.d., OX 1838.

4. "Proof of His Excellency The Marchese Marconi," 1934, OX 22/96–108.

5. GM to H.M. Dowsett, n.d. [1933?], LIN 2/121.

6. MWTC personnel file, "G.A. Mathieu," OX 685; "Note on the Present American Broadcasting. From G.A. Mathieu to Senator Marconi July 1922," OX 30/78.

7. For example, G.A. Mathieu, lecture to Dorchester Rotary Club, July 7, 1925, OX 237.

8. See, for example, MWTC press conference, April 9, 1927, OX 231; G. Marconi, "Radio Communications," speech to the American Institutes of Electrical and Radio Engineers, New York, October 17, 1927, OX 53.

9. La Presse (Montreal), October 27, 1926.

10. GM to GAM, October 15, 1926, OX 238.

11. GAM to GM, January 10, 1928, OX 234; MWTC, "Broadcast of the Westminster Abbey Thanksgiving Service. Relayed to Canada and Australia by Beam Telephony," July 9, 1929, OX 237; MWTC, "Inter-Empire Broadcasting by Beam," July 15, 1929, OX 238.

12. GM to GAM, February 19, 1930, Telefunken archive, DTM.

13. GAM to H.M. Dowsett, November 16, 1932, OX 818.

14. MWTC personnel file, "About Gerald Isted," OX 684; GAM reference letter for Isted, August 27, 1935, OX 819; GAM to G. Isted, October 22, 1929, OX 818.

15. "Report from GA Mathieu to The General Management re The Marconi Quasi-Optical System and 1932 Work Programme," January 15, 1932, OX 818.
16. *L'Osservatore Romano*, February 13–14, 1933. Bea and de Carolis, vol. 1, 77–81.
17. Agent 390, report, March 7, 1933, ACS, PP, b.60/A.
18. Lord Inverforth, "New Research Work," speech given at MWTC 35th Annual General Meeting, May 23, 1933, OX 402.
19. Telegram, GM to J. Denison-Pender, August 13, 1933, LIN 20/20.
20. GM to HAW, August 28, 1933, OX 525.
21. Agent 35, report, August 13, 1933, ACS, PP b.60/A; "*Sulla Propagazione di Micro-onde a Notevole Distanza*," typescript, April 14, 1933, ACS, SPD CO GMF, 197.598/1.
22. Agent 297, report, March 27, 1930, ACS, PP, b.60/A.
23. Cristina described the event (although she erroneously dated it in March, not June): "Suddenly we received an urgent message announcing the imminent arrival of Mussolini. He arrived in a naval launch wearing yachting gear—as my husband always did—and came on board with his customary self-assurance. Guglielmo welcomed him in a dignified manner and immediately took him to the radio station. Mussolini observed everything with great interest and attention, complimenting my husband. He was also very gracious toward me" (M.C. Marconi 1999, 193–94).
24. For example, GM to FGK, October 6, 1930, OX 525.
25. Agent 351, report, November 10, 1930, ACS, PP, b.60/A.
26. Agent 304, report, April 16, 1931, ACS, PP, b.60/A.
27. *Daily Herald*, April 7, 1931; *Daily Sketch*, April 8, 1931; both in OX 1504. Agent 304, report, April 16, 1931, ACS, PP, b.60/A.
28. Agent 498, report, July 10, 1934, ACS, PP, b.60/A.
29. For example, "Protest from Senatore Marconi," *New York Herald* (European edition), May 25, 1935.
30. Agent 545, report, August 13, 1935; Agent 484, report, August 16, 1935; both in ACS, PP, b.60/A.
31. *Daily Telegraph*, August 29, 1935.
32. *News Chronicle*, August 29, 1935.
33. *March of Time*, August 28, 1935, http://xroads.virginia.edu/~ma04/wood/mot/html/ethiopia.htm.
34. *NYT,* November 1, 1935.
35. Di Marco 1945, 5, OX 1757 and GCR 30/2.
36. This version of the story was provided by Umberto Di Marco in an unpublished article responding to a sensational report in *Stampa Libera* in February 1945 alleging that Marconi had committed suicide rather than have Mussolini take over his "death ray" (Di Marco 1945, OX 1757 and GCR 30/2). The demonstration on the Rome–Ostia road is described in a memoir by Rachele Mussolini, the dictator's widow, who was present (1980, 10–11). Hers is the only eyewitness claim.
37. N. Sipos, *Gente,* April 1981, LSP 5/26 (translation LSP).

38. Campbell 2006, 155; L. Sergio, "Guglielmo Marconi, The Silent Man Who Made the Ether Speak," typescript, LSP 7/40.

39. Memo, GM to BM, August 1, 1932, ACS, SPD CO CNR.

40. "Guglielmo Marconi, President of the Royal Academy of Italy, addresses the scholars, scientists, and artists of the world," October 15, 1932, LIN 19/11; versions in Italian, English, French, and German in RAL, Titolo I: Presidenza, f.4.5. Also in OX 21.

41. For example, the National Assembly of Intellectuals, October 1, 1932; the Congress of the Italian Society for the Progress of Science, October 9, 1932.

42. Berezin 1997, 109 and 138.

43. G. Marconi address, October 15, 1932, LIN 19/11 and OX 21.

44. Agent 62, report, October 15, 1932, ACS, PP, b.60/A.

45. LT, October 17, 1932.

46. Agent 52/B51, "Radio messaggio Marconi," October 17, 1932; also Agent 62, report, October 15, 1932; both in ACS, PP, b.60/A.

47. G. Marconi, "La ricerca scientifica"; cited in Paoloni and Simili 1996, 10.

48. Marconi 1941, 407–18.

49. File notes regarding memos from GM to BM, March 11 and 31, 1932, ACS, SPD CO CNR.

50. File notes regarding memos from GM to BM, August 1, 1932 and October 14, 1932; GM to BM, November 7, 1932, all in ACS, SPD CO CNR; Il Messaggero, November 20, 1932.

51. "Il Duce—Scientific Dictator" (editorial), NYT, December 4, 1932.

52. ACS, AEI.

53. Although Freddi's mandate was to develop the use of cinema for propaganda purposes—Cinecittà's slogan was "Il cinema è l'arma più forte" (Cinema is the most powerful weapon)—he, interestingly, decided to focus on the Hollywood-style commercial entertainment model rather than, say, Soviet socialist realism.

54. Four letters auctioned by Pierre Bergé, Paris, November 22, 2010, online; Director General for Cinematography to GM, August 1, 1935, LIN 16/188. For more regarding Fascist cinema policy, see LS to GM, June 21, 1935, Paoloni and Simili 1996, 180.

55. G. Steiner, "The Hollow Man" (review of Denis Mack Smith's Mussolini), New Yorker, August 30, 1982, 86–90, 87.

56. Agent 37, report, November 18, 1932, ACS, PP, b.60/A.

57. Agent 40, report, September 19, 1933, ACS, PP, b.60/A.

58. Unidentified agent, report, December 1, 1934; Agent 453, report, February 11, 1934; Agent 329, report, July 23, 1934; all in ACS, PP, b.60/A.

59. Agent 352, report, May 15, 1934, ACS, PP, b.60/A.

60. Agent 498, report, December 4, 1934; Agent 374, report, April 5, 1935; both in ACS PP, b.60/A.

61. GM to DM, September 17, 1933, RI 143; D. Marconi 1962, 293.

62. "Progressive Support for Italian and German Fascism," DiscovertheNetworks .org, http://www.discoverthenetworks.org/viewSubCategory.asp?id=1223.

63. G.H. Clark to Glenn Tucker, September 22, 1933, GCR 15/29.

64. Ostensibly there to provide security, Lettieri answered to Arturo Bocchini, head of the political police, indicating that part of his job may have been to help keep an eye on Marconi. Marconi's police file contains details of Lettieri's assignments and expense reports.

65. NYT, September 22, 1933.

66. M.C. Marconi 1999, 205.

67. NYT, September 29, 1933; Clark 1936, OX 36/112–48.

68. Clark 1936, OX 36/112–48.

69. OED to DS, September 30, 1933, GCR 15/28.

70. Clark 1936, OX 36/112–48.

71. "Report of Banquet in Honor of Marchese Guglielmo Marconi at the Drake October 1, 1933," LIN 19/15.

72. MWTC press release, untitled, October 10, 1933, OX 36/111. Marconi was the second prominent member of the Italian Fascist Party to visit the 1933 World's Fair. In July, Italo Balbo, one of the original architects of the March on Rome and Mussolini's minister for the air force, made a sensational aviation exploit, leading a squadron of two dozen seaplanes on a seven-leg flight from Rome to Chicago. Balbo was invited to the White House where he was presented with the Distinguished Flying Cross. It was the success of Balbo's visit that gave RCA and the fair organizers the idea to invite Marconi ("G. Marconi Visits the Century of Progress at Chicago 1933," n.d., GCR 16/1).

73. October 1–2, 1933, FDR, DBD.

74. Dunlap 1937, 337.

75. October 2, 1933, FDR, DBD.

76. BM to FDR, n.d. (reply to FDR's letter of May 14, 1933), FDR, GTC.

77. NYT, June 5, 1933.

78. Sullivan 1991, 339. The FDR online diary website contains one "world event" for October 1933: the opening, in New York, of the University in Exile, conceived by New School president Alvin Johnson as a haven for European scholars endangered by Hitler's and Mussolini's regimes. It later became the New School for Social Research. The item is dated October 2, 1933, the date of FDR's Chicago meeting with Marconi.

79. Clark 1936, OX 36/112–48

80. October 11, 1933, FDR, DBD.

81. William Baehr (Archives Technician, FDR Presidential Library), correspondence with author, December 21, 2012.

82. Jacot and Collier 1935, 243.

83. Clark 1936, OX 36/112–48.

84. Cited in Clark 1936, OX 36/112–48.

85. Telegram, GM to E. Roosevelt, November 2, 1933, FDR, AER, Box 579.

86. M.C. Marconi 1999, 211 and 213.

87. G. Marconi, Diaries and Notebooks, April 29, 1926, OX 12.

88. M.C. Marconi 1999, 215.

89. *San Francisco News*, October 24, 1933. In the same interview Marconi also "waved off the subject of Bolshevism. 'I am glad of America's forthcoming recognition of Russia as a trade measure, but I do not like Bolshevism,' he smiled."

90. M.C. Marconi 1999, 218–31.

91. "Japanese Honour for Marchese Marconi," MWTC press release, November 21, 1933, OX 36/117; "Japan Pays Tribute to Wireless Genius," *Trans-Pacific* (Tokyo), November 23, 1933.

92. *Peking Chronicle*, December 3, 1933.

93. The *Peking Chronicle* news clipping and Antoniutti's report are in ASV 1934-314-f.1–129110.

94. Italian Legation, Shanghai, to Ministero degli Affari Esteri, Rome, December 13, 1933, Paoloni and Simili 1996, 177.

95. Telegram, GM to BM, January 8, 1934, LIN 13/154.

96. "*L'Opera del Consiglio Nazionale delle Ricerche*," in G. Marconi 1941, 451–66.

97. Mussolini speech to CNR, March 8, 1934, ACS, CNR, b.2, f.12.

98. G. Marconi, "*Per la Ricerca Scientifica*," October 11, 1934, ACS, CNR, b.2, f.20.

99. G. Magrini ("Il Presidente"), CNR, to Ministry of National Education, January 29, 1935, ACS, CNR, b.23, f.382. Maiocchi (1999) argues that "scientific" legitimation of racial supremacy in the 1930s created the backdrop for Italy's antisemitic race laws that were decreed in November 1938.

100. The traces Marconi has left indicate that he was more focused on his British activities during his months convalescing in London. On January 8, 1935, for example, he published a letter in the *Times* congratulating the paper on its 150th anniversary and thanking it for supporting him and his company during their hard times. However, he also handled correspondence regarding his Italian functions, mostly through his private secretary Umberto Di Marco (ACS, AEI), and on January 12, 1935, still from his London hospital bed, he renewed his Fascist Party membership (OX 21/121). The indefatigable Magrini died unexpectedly in 1935.

CHAPTER 33

1. *Guardian,* March 19, 2002.

2. The *Guardian* article was based on Simonetta Fiori, "*Così Marconi escluse gli ebrei*," *Repubblica* (Rome), March 18, 2002. See Capristo 2001 and Capristo 2005.

3. See, for example, Levi 2009.

4. See "*Promemoria per S.E. Marconi, per il colloquio con S.E. Il Capo del Governo*," (brief prepared for Marconi, for his meeting with Mussolini), January 9, 1932, RAL, Titolo I: Presidenza, b.2, f.4.6.

5. Ludwig 1932, cited in Capristo 2001, 1. The interviews took place between March 23 and April 4, 1932.

6. Handwritten notation on a biography of Della Seta, part of a file of biographical notes on the candidates prepared by the academy for the press, n.d. [March 1932] (RAL, Titolo II: Accademici, b.6, f.15).

7. Capristo 2001, 5–6.

8. RAL, Titolo II: Accademici, b.6, f.15-21; b.9.

9. Marconi wrote to D'Annunzio on February 27, 1932, inviting him to be a candidate for membership of the Academy (Andreoli 2004, 37–38).

10. Cited in Capristo 2003, 251.

11. Capristo 2001, 8. Capristo herself (2003) points out that the issue is blurred by the fact that a few days after the appearance of the Ludwig interview, Mussolini named a Jewish minister of finance, Guido Jung.

12. This is the view of Italian historian Michele Sarfatti ("*Il pregiudizio esisteva anche prima 1938*," *Repubblica*, March 18, 2002).

13. Part of the complexity of this question can be grasped through the lens of Mussolini's relationship with Margherita Sarfatti. Like most members of Mussolini's entourage, she was followed closely by the regime's political police, who reported on March 14, 1929, that she was suspected of being a "Jewish internationalist" and very sympathetic to the Zionist movement (ACS, PP, b.88A, cited in Capristo 2001, 14).

14. "Testamento olografo di Guglielmo Marconi e tutto scritto di ma mano," May 15, 1917, OX 22; various correspondence in OX 24 and LIN 4/190 and 4/236.

15. "*Verbale dell'adunanza generale segreta del 13 Aprile 1933–XI*," and GM to BM, April 17, 1933, both in RAL, Titolo II: Accademici, b.6, f.16. The letter is annotated, in Mussolini's hand, as "seen" ("*Visto/Mussolini*").

16. A. Marpicati to GM, September 7, 1936, LIN 15/155. See chapter 34.

17. Nautilus, "Two Guglielmo Marconi's Original Photographs," http://www.thenautilus.it/portfolio/2-guglielmo-marconis-original-photos/. The press agency photo is dated April 21, 1933.

18. C. Camagna, secretary of the Fascio di Londra, to the secretary-general of the Fasci Italiani all'Estero, Rome, November 6, 1933, ASMAE, AL, b.805. f. 2. Marconi was a largely absentee, if enthusiastic, honorary president of the Fascio di Londra. The London branch of the Italian Fascist Party's foreign section was largely a social club cum mutual aid society for émigré party members and sympathizers, but it was an important site for displays of solidarity and propaganda. Marconi made a rare appearance at a Sunday meeting of the group on December 18, 1932, where he gave a long, celebratory speech. In a letter to the party's London paper, *Italia Nostra*, around the same time, he declared his pleasure at renewing contact with the London Fascio, deploring that he had only kept up nominal contact with the group in recent years because of his busy activities (ASMAE, AL, b.805, f.2).

19. Hesse was a great-grandson of Queen Victoria and was married to Princess Mafalda of Savoy, the daughter of Vittorio Emanuele III. In 1929, Philipp and Mafalda visited the Marconis on the *Elettra*, getting the usual visitors' treatment of lunch followed by an audition of wireless broadcasts (M.C. Marconi 1999, 190).

20. Undated [1929] notes for a draft brief by G. Magrini, ACS, CNR, b.23, f.418; cited in Simili 2001, 118. See also RAL, Titolo II: Accademici, b.6, f.17; Capristo 2003; Luconi 2004, 1–17.

21. "Materiale iconografico," LIN (2014), Serie II, Carte di famiglia, b.59.

22. *Italy and Abyssinia,* Società editrice di novissima, Rome, a. XIV [1935], LIN (2014), Serie II, Attività istituzionale, IV, Attività politica, b.34, f.2, sf.5; Pakenham 1991.

23. GM to GIU, March 4, 1896, HWL, copy in OX 1849.

24. *Italy and Abyssinia* [1935], LIN (2014), Serie II, Attività istituzionale, IV, Attività politica, b.34, f.2, sf.5.

25. BM to GM, July 18, 1935; GM to BM, July 27, 1935; BM's personal secretary to GM, n.d. [August 1935]; Military command to personal secretary, September 5, 1935; all in ACS, SPD CO GMF 197.598/1.

26. *Morning Post,* August 28, 1935.

27. *Daily Mail,* August 29, 1935; *Daily Telegraph,* August 29, 1935.

28. Di Marco 1945, 4, OX 1757 and GCR 30/2, and Clark 1936, OX 36/112–48; A. Starace to GM and GM to A. Starace, September 1935, LIN 7/361. There is also a copy of Starace's telegram thanking Marconi for his request, dated September 1, 1935, in RAL (Titolo I: Presidenza, b.2, f.4.8); here someone has written in pencil: "request?"

29. Agent 53, report, September 12, 1935, ACS, PP, b.60/A.

30. "*Viaggio in Brasile con Marconi. Diario di Antonio Bezzi Scali,*" LIN 19/7.

31. GM to BM, September 14, 1935, ACS, SPD CO GMF 197.598/1.

32. "*Viaggio in Brasile con Marconi. Diario di Antonio Bezzi Scali,*" LIN 19/7.

33. It was not unusual for Marconi to be photographed wearing the fascist uniform; see, for example, London's *Daily Express,* November 4, 1935.

34. R. Cantalupo to BM, May 1, 1936, in DDI Ottava Serie, III, doc. 809, 866.

35. "Speech of Baron Aloisi. Head of the Italian Delegation to the League of Nations, before the Assembly on October 9th, 1935," LIN (2014), Serie II, Attività istituzionale, IV, Attività politica, b.34, fasc.2, sf.5.

36. Telegram, GM to DS, October 23, 1935, LIN 6/337; telegram, DS to GM, October 25, 1935, LIN 6/352; telegram, GM to DS, October 30, 1935, LIN 6/337.

37. GM, "Broadcast to the U.S.A.," October 31, 1935, LIN 18/4.

38. F. Suvitch, undersecretary of state for foreign affairs, to GM, October 5, 1935, LIN 20/21.

39. F. Jacomoni (Ministero degli Affari Esteri) to UDM, November 3, 1935; UDM to GM, November 4, 1935; both in LIN 20/21; Private secretary to Mussolini, November 9, 1935, ACS, SPD CO GMF 197.598/1.

40. Typescript, LIN 20/21. The speech was published in a booklet, *The Family of Nations*, New York: Carnegie Endowment, 1935, with November 11, 1935, addresses from New York, Paris, Buenos Aires, Tokyo, London, and Rome.

41. Under the League of Nations sanctions, property of Italian nationals—such as patents—could be seized (Turchetti 2012, 28). Marconi never seemed concerned about this, perhaps since his patents were the property of his company.

42. "'Inhuman crusade.' Marconi Pledges Support for Mussolini," *Star*, November 18, 1935, OX 1504.

43. *Il Giornale d'Italia*, November 19, 1935; "Meeting of the Fascist Grand Council," *LT*, November 19, 1935.

44. W.G. Richards to GM, November 11, 1935; UDM to W.G. Richards, November 16, 1935; both in LIN 10/46.

45. "Dinner offered by His Majesty's Government at the Royal Gallery of the House of Lords, Remarks by Senator Marconi," July 5, 1918, OX 21 14–16.

46. "Marconi Not Allowed to Broadcast," *Daily Mail,* November 20, 1935; "'It Was Not To Be a Hostile Broadcast' says Marconi," *Evening Standard*, November 20, 1935; "Marconi to see Sir J. Reith," *Daily Telegraph*, November 21, 1935.

47. D. Collier to GM, November 23, 1935, LIN 2/85; H. Addison, "'Nation Shall Not Speak unto Nation' Is the Real BBC Motto," *Sunday Pictorial,* November 24, 1935; "No Admission," *Morning Post*, November 26, 1935.

48. "*Circa il rifiuto a Guglielmo Marconi di parlare alla radio di Londra*," ASV, ARRSS (AA.EE.SS.), Vol. IV/7, Italia. 967 P.O. Vol. XII, 1935. Conflitto Italo-Etiopico, Stampa III; Agent 52/B51, report, November 27, 1935, ACS, PP, b.60/A.

49. J. Reith to GM, November 29, 1935; GM to J. Reith, December 2, 1935; both in LIN 10/46; Reith 1949, 236.

50. Stuart 1975, 56–57.

51. Luigi Solari wrote that Marconi said the future King Edward VIII had told him, during this trip, "confidentially" that he was in disagreement with the British government regarding Italy and Germany (Solari 2011, 296).

52. Maiocchi 1999.

53. See chapter 6.

54. D. Collier to GM, January 21, 1937, LIN 2/85.

55. GM to G.S. Clive, December 3, 1935, LIN 2/80.

56. See for example, UDM to A. Godoy, February 3, 1936, in LIN 3/162.

57. Marconi also received letters of support from correspondents in the United States and Britain expressing undeniably antisemitic sentiment; his archives contain no traces of any replies to these letters (LIN 20/21).

58. C. Petrie to GM, February 29, 1936; GM to C. Petrie, March 3, 1936; both in LIN 5/292.

59. *LT*, December 16, 1935.

60. M. Canali, correspondence with author, November 3, 2014.

61. Agent "Valiani," two reports, December 14, 1935, ACS, PP, b.60/A.

62. Note from Mussolini's personal secretary, December 14, 1935, ACS, SPD CO GMF 197.598/1; "Broadcast to Australia," December 14, 1935, LIN 20/22.

63. Prefetto Giovenco, Pisa, to "MIN/INT/GAB e PS," December 17, 1935, ACS, SPD CO GMF 197.598/1.

64. Note regarding Di Marco's phone call, December 18, 1935, ACS, SPD CO GMF 197.598/1.

65. R. Bova Scoppa to BM, December, 18, 1935, DDI, Ottava Serie, II, doc. 877, 864–5, Geneva (League of Nations).

66. See, for example, "Duce's Speech of Defiance. Italian Purpose Unmoved. Claims in Ethiopia," LT, December 19, 1935. Mussolini's government routinely instructed the Italian press on the handling of breaking news stories. On December 18, 1935, the press was told: "If newspapers receive news of a sudden illness of Senator Marconi, nothing is to be published on the subject." It was then announced that Marconi had attended the meeting of the Grand Council. This example of deliberate misinformation was reported in the Paris-based anti-fascist newspaper *Giustizia e Libertà*, and cited by the *Manchester Guardian* in January 1936 as an illustration of Italian propaganda practices ("More Secret Instructions to the Italian Press," *Manchester Guardian*, January 7, 1936).

67. Di Marco 1945, 8, OX 1757 and GCR 30/2.

68. A.H. Ginman, to GAM, May 17, 1935, LIN 5/244; Lord Inverforth to GM, May 17, 1935, LIN 4/188; GM to Lord Inverforth, May 21, 1935, LIN 4/188; GAM to Lord Inverforth, June 18, 1935, LIN 5/244; HAW to GAM, June 27, 1935, LIN 5/244; GM to GAM, June 28, 1935, LIN 5/244.

69. GM to HAW, August 15, 1935, LIN 11/83.

70. Ibid.

71. HAW to GM, August 23, 1935; GM to HAW, August 28, 1935; both in LIN 11/83.

72. HAW to GM, August 30, 1935, LIN 11/83.

73. G. Isted to W. Maconachie, March 23, 1961, OX 1757. Various drafts of Isted's memoir are in OX 3991 and OX 3992.

74. "*L'arbitro societario denunziato di Marconi,*" *La Tribuna,* January 19, 1936.

75. "Marconi Decided To 'Cut' Britain—Until Peace Time," *Daily Mirror*, April 21, 1936.

76. Mussolini used these words in an instruction to General Rodolfo Graziani on July 8, 1936.

77. L. Sergio, English broadcast transcript, May 9, 1936, LSP, 11/75.

78. This was a fine example of Marconi's engaging with fascist revisionist historicism. Liberal Italians considered the 1918 Battle of Vittorio Veneto, which ended the war on the Italian front and helped seal the defeat of the Austro-Hungarian Empire, as the culmination of the Risorgimento and the unification of Italy.

79. Italy, *Atti Parlamentari della Camera del Senato*, Leg. XXIX, 1 as. *Discussioni,* May 16, 1936, 2,143–45.

80. *LT*, May 22, 1936.

81. "Haile Selassie of Ethiopia Dies at 83," *NYT*, August 28, 1975; "Ethiopia: The Lion is Freed," *Time,* September 8, 1975.

82. The Italian-American crisis didn't last long, however. On November 19, 1936, Mussolini wrote to FDR congratulating him on his re-election and expressing satisfaction that relations between the United States and Italy had been re-established (FDR, GTC).

83. GM, broadcast from Bologna to the American people, typescript, June 14, 1936, LIN 18/4.

84. Dunlap 1937, 345; W. Maconachie to Publicity Manager, MWTC, March 21, 1961.

85. A. Rosso to Ministero degli Affari Esteri, Rome, July 9, 1936, LIN 14/154(II).

86. GM, broadcast to the American people, typescript, October 12, 1936, LIN 22/24.

87. L. Jaffe to GM, July 16, 1935; UDM to L. Jaffe, July 31, 1935; both in LIN 9/36.

88. J.J. Thomson to GM, January 26, 1935, LIN 7/368.

89. GM to J.J. Thomson, February 19, 1936, LIN 7/368; GM to Lord Inverforth, February 19, 1936, LIN 4/188.

90. HAW to GM, March 31, 1936, LIN 6/337; GM to J.J. Thomson, April 6, 1936, LIN 7/368.

91. A. Korn to GM, March 5, 1936, LIN 4/203.

92. The most famous case was that of Isaac Shoenberg, an employee of the Russian Marconi company who was stranded in England in 1914. A brilliant mathematician, Shoenberg rose to head the company's patents department, and then became the first research director of a new Marconi partnership with the electronics firm EMI in 1934. He was knighted in 1962 for his contribution to the development of television ("Administrative correspondence relating to the employment of Isaac Shoenberg, 1914–17," OX 360).

93. UDM to A. Korn, March 9, 1936, LIN 4/203.

94. Arendt 1963, 133; "Hans Frank Apologizes," *NYT*, April 19, 1946.

95. *Frankfurter Zeitung,* April 6, 1936.

96. Knox 2000, 141; Strang 1999, 173.

97. Knox 2000, 142.

98. Ibid., 143; D.M. Smith 1981, 206.

99. Knox 2000, 143; *NYT*, April 3, 1936. The following timeline is based on reports in *Il Popolo d'Italia, Corriere delle sera, La Tribuna,* and *Il Messaggero,* April 3–5, 1936. The most extensive report of Frank's speech of April 3 appeared in *Corriere delle sera,* April 4, 1936.

100. *L'Osservatore Romano,* April 5, 1936; *NYT*, April 5, 1936.

101. RAL, Titolo XI: Corrispondenza con gli accademici e sulle loro funzioni, b.15, f.103/50.

102. Frank's speech and reception by Marconi were enthusiastically reported in the German press. The Nazi organ *Völkischer Beobachter* reported on April 4, 1936, that Frank spoke on the guiding principles of German law, stressing similari-

ties between the developments of the state in Italy and Germany, lauding Hitler, and distinguishing national socialism and fascism, on the one hand, from liberalism and parliamentarianism, on the other. Frank elaborated on the Nazi legal system, stressing similarities to classical Roman law, with its Aryan roots, and ending with a call for closer collaboration between German and Italian legal science. In a second report (April 5, 1936), the paper noted the presence of the Nazi swastika and the Italian flag flanking the speakers' podium, and the warm applause offered by the crowd of dignitaries, which included Marconi, in his capacity as president of the Royal Academy. The *Berliner Tageblatt,* also noting Marconi's presence, reported that Frank's speech was interrupted several times by applause and that his comments with regard to the Jewish question, in particular, found a lively interest (April 4, 1936). The *Völkischer Beobachter, Berliner Tageblatt,* and *Frankfurter Zeitung* all reported separately on Frank's visit to the academy and his warm reception there by Marconi. According to the *Völkischer Beobachter* (April 6, 1936), Marconi welcomed Frank as "the representative of a distinguished Führer, the Minister of a great nation and a great populace" and called for closer cultural links between the two countries. Frank thanked Marconi and said he appreciated Marconi's words with regard to closer collaboration between the two countries and shared this position. The article reported that Frank and Marconi had a long conversation. This was the most extensive published report anywhere on the personal contact between the two figures.

103. *"Ministro del reich Dr. Hans Frank. Ricevimento in Accademia,"* April 4, 1936, RAL, Titolo I: Presidenza, b.2, f.8/57–78; Marconi's statement, *"Saluto rivolto da S.E. Marconi a S.E. Hans Frank, Ministro del Reich, il 4 aprile 1936—XIV,"* is b.2, f.8/59–60.
104. *Jewish Criterion* 87, no. 23, April 17, 1936.
105. *Jewish Criterion* 90, no. 11, July 23, 1937; *American Jewish Outlook* 6, no. 8, July 30, 1937.
106. Arendt 1963, 171.

CHAPTER 34

1. G.S. Viereck, "Faith Is Man's Bridge to Infinity," *Liberty,* January 23, 1937.
2. G.S. Viereck, "A Talkie of Marconi," typescript in OX 53 /234–47; distributed by the London-based Anglo American Newspaper Service and published as "'Man Need Not Die!' Declares Marconi," *New York American,* September 29, 1929.
3. G.S. Viereck to GM, February 2, 1935, and January 22, 1937; GM to G.S. Viereck, March 22, 1937; all in LIN 7/385.
4. A. Kirk (US embassy) to GM, 1936, LIN 4/201.
5. LS to UDM, March 8, 1935, LIN 6/352.

6. Correspondence, D.B. Collier and GM, December 27, 1934 to November 19, 1935, LIN 2/85. The book was dedicated to Cristina, "in gratitude for her kindness," while Marconi provided an autographed photograph for the frontispiece. But the outcome was disappointing to Collier. Her publisher, Hutchinson, hired a second writer, B.L. Jacot, to flesh out some sections, and when the book appeared she was shocked to see Jacot's name first.

7. de Sousa 1919; L. de Sousa to MWTC, August 15, 1919; G.G. Hopkins, "MWTC Ltd.," January 1959; W.G. Richards, undated note; all in OX 55; L. de Sousa to UDM, March 7, 1935; L. de Sousa to LS, March 8, 1935; both in LIN 6/355; M. Pettinati, "The Life Story of Marconi. An Authorized Biography," *Sunday Graphic and Sunday News*, March 17–April 28, 1935. De Sousa was still flogging a Marconi manuscript in 1950, according to correspondence in the Marconi Archives in Oxford. Company officials were trying to determine "whether any publication by him at the present time would contravene the agreement made between him and the Company when he handed us the manuscript" (W.G. Richards to E. Blake, January 26, 1950, OX 55). But a letter from de Sousa dated August 15, 1919, in the same file, leaves no doubt that the company owned all rights to the original manuscript.

8. OED to DS, September 30, 1933; DS to OED, October 7, 1933; both in GCR 15/28.

9. GM to DS, March 1, 1935, LIN 12/103; also in GCR 15/28; OED to GM, March 5, 1935; GM to OED, May 27, 1935; both in LIN 3/124; OED to DS, July 26, 1935, GCR 15/28.

10. Oxford University Press to G.H. Clark, January 29, 1935; G.H. Clark to L. de Sousa, February 6, 1935; UDM to G.H. Clark, February 13, 1935; G.H. Clark to UDM, April 8, 1935; UDM to G.H. Clark, April 25, 1935; all in GCR 15/29. Clark did indeed have a huge collection, not only of "marconiana" but of "radioana"—as he called the collection, which was gifted to the Smithsonian Institution in 1959.

11. OED to DS, January 29, 1934, GCR 15/28; OED to GM, November 19, 1935, LIN 3/124; OED to AM, April 17, 1936, LIN 3/124; OED to GM, May 29, 1936, LIN 3/124; F. Mullen to DS, July 29, 1936, GCR 15/28; F. Mullen to OED, July 23, 1936, GCR 15/28; GM to OED, November 16 and 22, 1936, LIN 3/124; OED to GM, November 23, 1936, LIN 3/124.

12. GM to OED, November 29, 1936, LIN 6/337; GM to DS, November 29, 1936, LIN 6/337; F. Mullen to DS, December 14, 1936, GCR 15/28; DS to GM, n.d., LIN 3/124; DS to GM, December 18, 1936, LIN 6/337.

13. GM to OED, January 25, 1937; OED to GM, February 15, 1937; GM to OED, "Enclosure" (list of corrections to manuscript), January 25, 1937; all in LIN 3/124. The Marconi company in London first heard about the book when it appeared, and queried Di Marco asking how much of it had been approved by Marconi; Di Marco replied that it had been a long, complicated process but that the final contents were indeed approved by Marconi (W.G. Richards to UDM, April [?], 1937; UDM to W.G. Richards, April 28, 1937; both in LIN 11/83).

14. G.H. Clark to GM, January 7, 1937 and March 8, 1937; both in LIN 2/77.

15. Dowsett 1951, 682, OX 1834, cited in W.J. Baker 1970, 296.

16. Goethe 1962, 361.

17. L. Sergio, "Marconi: Wizard of the Airwaves," *Rotarian,* September 1974. Sergio visited Marconi at the academy on June 24, 1937, a few days before leaving Italy.

18. Foreign Office to D. Grandi, January 25, 1937; G. Ciano to GM, February 8, 1937; both in LIN 13/154(II).

19. GM to H. Burn, April 5, 1937, LIN 3/146; UDM to W.G. Richards, April 28, 1937, LIN 11/183; Correspondence, L.J. King and UDM, April 1937, LIN 11/81.

20. D. Grandi to GM, May 3, 1937; GM to D. Grandi, May 5, 1937; both in LIN 3/166; G.J. Freshwater to GM, May 28, 1937; UDM to G.J. Freshwater, June 2, 1937; both in LIN 3/146; GM to F. Gaylor, July 15–16, 1937, LIN 4/231; E. Fisk to GM, July 16, 1937, LIN 3/141. Degna Marconi states that her father was still thinking of making a trip to London at the time of his death.

21. M.M. Montessori to GM, received July 19, 1937; invitation, n.d.; both in LIN 13/154(I); M. Sheridan to GM, March 19, 1937, LIN 6/346; L. de Sousa to GM, July 11, 1937, LIN 6/355.

22. *Chicago Tribune* to GM, February 17, 1937, LIN 12/136.

23. G. Marconi, "Broadcast to the *Chicago Tribune* Forum," March 11, 1937, typescript in LIN 22/26.

24. Sarfatti and Tedeschi 2006; Venzo and Migliau, 2003.

25. Giacomelli and Bertocchi 1994.

26. A. Marpicati to GM, September 5, 1936; GM to A. Marpicati, September 6, 1936, both in LIN 15/155.

27. A. Marpicati to GM, September 7, 1936; GM to A. Marpicati, September 14, 1936 [misdated October 14, 1936]; both in LIN 15/155.

28. "*Impressioni minute al Congreso di Norimberga,*" unsigned five-page typescript, n.d., on academy letterhead, LIN 14/154(II).

29. M.C. Marconi 1999, 194.

30. G. Marconi, "*Schema di testo unico delle disposizioni di legge riguardanti il Consiglio nazionale delle ricerche. Promemoria per S.E. Il Capo del Governo,*" September 1936, ACS, CNR, b.1, f. 4.

31. Fascist Federation of Industrialists to BM, February 19, 1937, ACS, CNR, b.1, f.7.

32. Maiocchi 2009.

33. O. Sebastiani to BM, March 23, 1937, ACS, SPD CNR; Simili 2001b.

34. Simili 2001b, 167.

35. Biographical/Historical Note, Lisa Sergio Papers, Georgetown University, Washington, DC. https://repository.library.georgetown.edu/bitstream/handle/10822/558940/GTM.GAMMS172.html?sequence=1#ref2241.

36. The date and circumstances of Sergio's and Marconi's first meeting vary in Sergio's writings. She often wrote that they met at a reception at the Italian

embassy in London, but gives the date variously as 1921, 1923, and 1929; she also wrote that they met in Rome in 1932.

37. Spaulding 2008, 146.

38. L. Sergio, autobiographical fragment, LSP 5/28.

39. One of the first things she did was research an article on Mussolini's interest in archaeology. In 1936 she won an award from the Italian Ministry of Press and Propaganda for an article in the *Washington Post* entitled "Ancient Rome Emerges from Ruins under the Hand of Mussolini," published on October 20, 1935 (*New York Herald Tribune*, June 20, 1936, in LSP 19/1).

40. Documents regarding employment of Lisa Sergio, April 1930–December 1933, ASMAE, MCP, b.642.

41. Agent 431, report, August 30, 1933, ACS, PP, b.1255 ("Lisa Sergio"). The suggestion that Sergio was a member of OVRA was made again in a Polpol report of January 2, 1936 (Agent 571, report, January 2, 1936, ACS, PP, b.1255).

42. Balbi 2010a, 786–808.

43. *Daily Telegraph*, October 10, 1935, cited in Spaulding 2008, 147; J. Flanner, "Paris Letter," *New Yorker*, October 28, 1936, cutting in LSP 19/1.

44. J. Flanner, "Paris Letter."

45. Again, Sergio later used different dates for this event. According to the introduction to Lisa Sergio Papers, Georgetown University Library, Mussolini signed an order for her dismissal on March 10, 1937. Spaulding (2008) cites a document in Sergio's FBI file, copied from Italian documents captured after the war, stating that Sergio resigned from the Ministry of Popular Culture on April 29, 1937.

46. According to research by American scholar Stacy Spaulding, the FBI took an interest in Sergio after her arrival in the United States and eventually built a three-hundred-page file in which it was suggested that her problems in Italy were as much tied to boasting about her romantic affairs with high officials of the regime as to any anti-fascist activity. According to the FBI's informers, Sergio had many high-placed lovers and a tendency to boast about them after she had a few drinks. Leaving aside the flagrantly sexist connotation of such a consequence (can one imagine a prominent male journalist in the 1930s losing his job for boasting about his romantic conquests?), the suggestion is plausible. Sergio's Polpol file also indicates some interest in the topic (ACS, PP, b.1255, September 11, 1935, June 27, 1936).

47. L. Sergio, "Your Eyes Are Like Your Father's," LSP 5/25, cited in Spaulding 2008, 148.

48. GM to DS, June 21, 1937, LIN 6/337; also in LSP 19/2.

49. "Lisa Sergio, Radio Commentator in Italy and New York, Dies at 84," *NYT*, June 26, 1989. The Lincei and Sergio archives contain copies of a telegram sent by Marconi to Sergio on the *Conte di Savoia* on June 28, 1937, wishing her "bon voyage and good luck" (Telegram, GM to L. Sergio, June 28, 1937, LIN 6/337 and LSP 19/2).

50. W.F. Ryan, "Lisa Sergio: The Golden Voice of Rome, The Progressive Complainer in America," *Virginia Country*, August 1990; cited in Spaulding 2008.

51. Spaulding 2008, 151.

52. Sergio wrote that he would address her as "young lady" ("Guglielmo Marconi—A portrait," LSP 7/39); in letters he would address her as *Cara Signorina Sergio*, "Dear Miss Sergio" (GM to L. Sergio, May 22, 1937, LSP 19/1).

53. Lansbury 1938, 157. The interview was reported in the daily and specialized press in the days following Marconi's death; see, for example, *Manchester Guardian,* July 21, 1937, and *Telegraph & Telephone Age*, August 1, 1937.

54. D.M. Smith 1981, 316.

55. Correspondence, F. Capaldo and UDM, February 16–April 17, 1937, LIN 25.

56. MWTC to Società Italiana Marconi, June 10, 1937, LIN 25.

57. GM to G. Stagnaro, July 9, 1937; UDM to G. Stagnaro, July 13, 1937; both in LIN 25.

58. Schueler 1982, 127 and 146, copy in OX 1757.

59. M. Braga, interview with author, March 29, 2012.

60. GM to DM, December 23, 1934, RI 144.

61. GM to Sir John Simon, Secretary of State for Home Affairs, London, requesting renewal of UK work permit for Giulio, December 4, 1935, LIN 6/349; GM to HAW, July 7, 1936, LIN 7/399; LS to GM, September 12, 1936, LIN 6/352.

62. GM to DS, January 12, 1937; DS to GM, February 26, 1937; GM to DS, March 26, 1937; all in LIN 6/337.

63. Giulio Marconi, official biography (provided by MWTC), 1962, OX 1758; GM to GLO, May 17, 1937, RI 149–51.

64. GM to DM, March 11, 1937, RI 145.

65. BOM to O. Sebastiani, January 31, 1934, ACS, SPD CO GMF 197.598/2.

66. "GM," April 7, 1934, annotated as seen by Mussolini on April 29, 1934, ACS, SPD CO GMF 197.598/1.

67. Agent 352, report, May 15, 1934, ACS, PP, b.60/A.

68. "Promemoria," May 1934 (seen by Mussolini on June 2, 1934), ACS, SPD CO GMF 197.598/2.

69. Draft agreement between C&W and GM, cancelling their existing agreements, July 2, 1934, OX 525. Under the agreement signed on November 19, 1934, Marconi would continue to receive £5,000 per annum for research expenses.

70. Correspondence, O. Sebastiani and G. Ciano, November 13–21, 1934, ACS, SPD CO GMF 197.598/2.

71. GM to DM, December 23, 1934, RI 144.

72. "*R. Tribunale civile di Perugia, per il Marchese Marignoli Liborio, Convenuto, contro il Consorzio nazionale per il credito agrario di miglioramento, Attore*," August 9, 1935, ACS, SPD CO GMF 197.598/2.

73. L. Marignoli to O. Sebastiani, May 10, 1935, ACS, SPD CO GMF 197.598/2; DM to O. Sebastiani, September 1, 1935; O. Sebastiani to DM, September 14, 1935; both in ACS, SPD CO GMF 197.598/2. The file contains further correspondence

where Marignoli and Beatrice offer to sell various properties to the government; this continued well after Marconi's death.

74. Various correspondence in LIN 4/232, 1934–35, especially Manager, Westminster Bank to GM, May 28, 1935, and GM to Manager, Westminster Bank, May 31, 1935; for material pertaining to Gioia and Giulio, see, for example, GM to GLO, January 20, 1934, LIN 4/236; GM to unidentified London firm, July 31, 1936, LIN 12/136.

75. GM to D. O'Brien, June 7, 1935, LIN 5/266.

76. D. Marconi 1962, 302.

77. D. Marconi 1962, 303–5.

78. L. O'Brien 1952, FP.

79. F. Paresce, interview with author, May 24, 2012.

80. *Chicago Examiner,* October 3, 1933.

81. E. Marconi, interview with author, May 26, 2014.

82. Draft agreement between MWTC and GM, July 16, 1937; Draft agreement between Società Italiana Marconi and GM, July 16, 1937; both in OX 525.

83. HAW to LS, July 16, 1937 and copy of draft agreement, OX 525.

84. D. Marconi 1962, 303.

85. *Radio Intelligence* 1936, copy in GCR. Also in *LT,* April 27 and 29, 1936.

86. Cited in *Free Press,* August 30, 1952.

87. Di Marco 1945, OX 1757 and GCR 30/2.

88. M. La Stella, "*L'austerità di vita di Guglielmo Marconi, nelle confidenze del suo seg-retario,*" April 27, 1939, ASS.

89. Marpicati 1954.

90. On July 6, 1937, London informed Marconi that they would no longer pay Di Marco's salary or the research expenses of the yacht as of January 1, 1939 (appendix to draft agreement, July 6, 1937, OX 525).

91. Secretary to BM, July 12, 1937, SPD CO GMF 197.598/1.

92. D. Marconi 1962, 307.

93. Di Marco 1945, 9–10, OX 1757 and GCR 30/2.

94. A. Marpicati, "*Le ultime ore di Guglielmo Marconi,*" extract from *Marpicati 1954,* in *La Settimana Incom illustrate* 7, 1954.

95. Di Marco 1945, 10, OX 1757 and GCR 30/2.

96. Solari 2011, 301–3.

97. Di Marco 1945, 10, OX 1757 and GCR 30/2.

98. D. Marconi 1962, 308.

99. Ibid., 309.

CHAPTER 35

1. RAL, Titolo XI: Corrispondenza con gli accademici e sulle loro funzioni, b.13, f.103/1–2. Giacomo Medici del Vascello was, coincidentally, a relation of Luigi Medici del Vascello, the husband of Marconi's one-time mistress Paola.

2. BOM to DM, July 22, 1937, RI 92.

3. Ginevra Lovatelli and Adriano Morabito, *Secret Rome*, Versailles: Éditions Jonglez, 2009, 237–43.
4. Stefani agency reports, July 22, 1937, ASMAE, MCP, b. 455; A. Cortesi, "Rites for Marconi Are Held in Rome," *NYT*, July 22, 1937; "State Funeral in Rome," *LT*, July 22, 1937.
5. A. Cortesi, "Rites," *NYT*, July 22, 1937.
6. "Marconi Funeral Scenes. The Duce Follows on Foot," *Morning Post,* July 22, 1937.
7. A. Cortesi, "Rites," *NYT*, July 22, 1937; also "State Funeral for Marconi. Scenes in Rome. The 'Fascist Rite' Given," *Manchester Guardian,* July 22, 1937.
8. "Silent Tribute. A Wireless Pause," *LT,* July 22, 1937; "Silence Kept for Marconi. Britain's Radio Tribute. All Stations Close Down," *Daily Telegraph,* July 22, 1937.
9. "Marconi," *Manchester Guardian*, July 21, 1937; "Marconi," *Yorkshire Post*, July 21, 1937; "Marconi's Genius," *Observer*, July 25, 1937; AP, "Marconi Is Dead of Heart Attack," *NYT*, July 20, 1937; "Marconi" (editorial), *NYT*, July 21, 1937; "A Modern Magician," *Kilmarnock Standard*, July 24, 1937; "The Great Marconi," *Irish Independent*, July 24, 1937; "Marconi," *The Irish Times*, July 21, 1937.
10. "Marconi," *LT*, July 21, 1937; "Obituary: Marconi. Master of Wireless Development," *LT*, July 21, 1937.
11. *Eastern Daily Press* (Norwich), July 21, 1937, OX 1504.
12. "*Marconi è morto*," *Berliner Börsen-Zeitung*, July 20, 1937; "*Il 'Mago dell'invisibile*,'" *Berliner Tageblatt*, July 21, 1937; translated typescripts, both in ASMAE, MCP, b.455.
13. AP, "Marconi Is Dead of Heart Attack," *NYT*, July 20, 1937; AP, "Marconi Death Premonition Is Told by Vatican," *NYT, Chicago Daily Tribune*, July 20, 1937; see also "Marconi Told the Pope End Was Near," *Daily Express*, July 21, 1937, reporting that the pope told Cardinal Pacelli Marconi felt he was going to die and asked for a special benediction.
14. In an odd coincidence—recorded in Mussolini's press archive, another indication of Mussolini's boundless fascination with celebrity gossip—Marconi's one-time fiancée Betty Paynter married in Cornwall just days after his death. The *Sunday Times* of July 25, 1937, reported that "One of the last letters of the famous wireless inventor was to wish Miss Paynter happiness." The story was also reported in the *Daily Express*, which made it seem that the very public romance had been secret ("Marconi's Love Story Revealed. Cornish Girl Weds Today," *Daily Express*, July 24, 1937). Cuttings of these articles are conserved in ASMAE, MCP, b.455, the Ministry of Foreign Affairs's file of Marconi obituaries.
15. "*La morte cristiana*," *L'Osservatore Romano*, July 21, 1937. The false account was also reported by other papers, including in "*L'attacco del male e la fine*," *Il Messaggero*, July 21, 1937.
16. "*S.E. Guglielmo Marconi, condoglianze per la sua morte, relazione sulla morte*," Vicariato di Roma, August 20, 1937, ASV, Segreteria di Stato, Schedario (1936–39),

1937, Istituti 25 (163417). Rapa swore his affidavit on August 16, 1937, Bengoa on August 18, 1937.

17. Agent 40, report, September 7, 1937, ACS, PP, b.60/A.

18. LS to MWTC, July 22, 1937, OX 1767.

19. "*Il messaggio postumo di Guglielmo Marconi,*" *Il Giornale d'Italia*, July 24, 1937 (as read by Solari on Bologna radio, July 23, 1937). The English version cited here was published on July 23, 1937, in the *Huddersfield Daily Examiner* ("Marconi's last message").

20. Various agents, reports, July 1937, ACS, PP, b.60/A.

21. Agent 40, report, July 22, 1937, ACS, PP, b.60/A.

22. Agent 62, report, July 23, 1937, ACS, PP, b.60/A.

23. Agent 346, report, January 24, 1938, ACS, PP, b.60/A.

24. Agent 40/Puberto, report, July 30, 1937, ACS, PP, b.60/A. Marconi's friends in the UK were also surprised that there were no charitable bequests to British organizations (*Daily Mail*, July 22, 1937).

25. Agent 40, report, July 31, 1937, ACS, PP, b.60/A. A second agent, former journalist and known Nazi sympathizer Fernando Gori, reported essentially the same news ten days later, prompting his police controller to scribble on the report: "This informant is a tremendous moron!" (Agent 668, report, August 10, 1937, ACS, PP, b.60/A).

26. "'Marconi.' Statement by David Sarnoff Made on His Arrival in NY on the Steamer Paris, Sept 25, 1937," *R.C.A. Family Circle*, OX 1767; *NYT*, September 26, 1937; Lyons 1966, 183. The regime was interested in Sarnoff's connection to Marconi as well as in Marconi's finances. On August 25, 1937, Solari addressed a long letter to Mussolini's chief of staff, Osvaldo Sebastiani, briefing him in great detail on Marconi's relationship with "the very able Israelite" David Sarnoff of RCA as well as Marconi's financial arrangements with the British company (LS to O. Sebastiani, August 25, 1937, ACS, SPD CO GM, 197.598/1).

27. *Daily Telegraph*, March 21, 1938.

28. "*Inventario,*" FP.

29. This was confirmed by company secretary L.J. King in a letter to the company's solicitor on February 27, 1952, clarifying the situation in light of a query from Cristina as to whether the estate was entitled to any further royalties: "It is true nevertheless that, notwithstanding this compensation payment, agreements with associated and subsidiary companies continued up to the date of his death, but definitely no payments were due under any of those agreements from the date that he died" (OX 524). Correspondence between attorneys acting for the two sides of Marconi's estate and the company in London reinforces this point: the company held no securities or property in Marconi's name; he had assigned all his patents to the company; and there was no provision for the payment of royalties or any other monies after his death (correspondence, September 1937 to August 1940, OX 525).

30. Solari claimed, in his letter to Sebastiani, that Marconi owned a substantial number of shares in RCA. He also owned stock in the Canadian Marconi Company—twenty shares, worth a grand total of CAD$150 (LAC, RG117-A-3, Vol 2512).

31. L.J. King to Stephenson et al, June 7, 1938, OX 525.

32. "Private. Note for the Secretary" and report attached, November 10, 1938, OX 525.

33. Italy, *Atti Parlamentari della Camera del Senato,* Leg. XXIX, 1ᵃs., *Discussioni,* December 9, 1937, 3,233–35; "Italy and the League," *LT*, December 11, 1937.

34. *Il Messaggero,* January 26, 1938.

35. Unidentified agent, report, September 8, 1938, ACS, PP, b.60/A; agent 650, report, April 10, 1938, ACS, PP, b.60/A.

36. Unidentified agent, report, August 28, 1939, ACS, PP, b.60/A. More than four years after Marconi's death, an agent reported a macabre bit of posthumous iconography—the doctor who had removed Marconi's eye in 1912 was suggesting to officials that it be put on display as an exhibition (Agent 770, report, December 2, 1941, ACS, PP, b.60/A). The final item in Marconi's Polpol file, dated September 1, 1942, refers to Cristina, to Mussolini's great trust in her, and to speculation that she herself was secretly an informer (Agent 650, report, September 1, 1942, ACS, PP, b.60/A).

37. O. Sebastiani to BM, August 4, 1937, ACS, SPD CO GMF 197.598/1.

38. BOM to BM, September 3, 1937, ACS, SPD CO GMF 197.598/2.

39. Ciano 2002, 37.

40. DM to E. Mussolini Ciano, April 28, 1939, ACS, SPD CO GMF 197.598/3.

41. MCM to BM, July 21, 1939, ACS, SPD CO GMF 197.598/1; annotated copy in July 21, 1939, ACS, SPD CO GM; *Il Messaggero,* July 21, 1939.

42. O. Sebastiani to BM, n.d. [December 1939], ACS, SPD CO GMF 197.598/3; O. Sebastiani to G. Ciano, December 4, 1939, ACS, SPD CO GMF 197.598/3.

43. DM to BM, June 25, 1941 and July 11, 1941; GIO to BM, October 25, 1941; all in ACS, SPD CO GMF 197.598/3; note regarding Mussolini receiving Maria Cristina Marconi, November 6, 1941; MCM to BM, November 8, 1941; both in ACS, SPD CO GMF 197.598/1.

44. *Il Messaggero,* November 19, 1941; receipt in file, January 18, 1942, ACS, SPD CO GMF 197.598/1.

45. Agent 40, reports, November 24 and 26, 1941, ACS, PP, b.60/A.

46. Giacomelli and Bertocchi 1994.

47. Press reports in *Il Messaggero,* October 7–8, 1941, cuttings in ACS, SPD CO GM. A photographic record of the event can be accessed on the AIL website, http://www.archivioluce.com/archivio/.

48. The story of the *Elletra*'s fate is taken from the following sources: documents, June–November 1938, OX 1759; Italy, *Atti Parlamentari della Camera del Senato,* Leg. XXIX, 1 as, *Discussioni,* December 15, 1938, 4, 261–62; *Daily Telegraph,* November 20, 1943, and July 13, 1961; "The *Elettra*... ," undated note, OX 1865;

"Panfilo Elettra," July–December 1945, ACS, PCM 44–47, 42059, 13.4; "Elettra Back on Air," Ari Fedenza, June 6–7, 2015, http://arifidenza.it/LaSezione/elettra2015/ElettraMarconi_eng.asp.

49. Stefani agency (NY), "*Les juifs en Italie*," July 20, 1937, ASMAE, MCP, b.455.

50. In *Informazione diplomatica* 14 (February 16, 1938), cited in Venzo and Migliau 2003, 11.

51. Translation http://users.dickinson.edu/~rhyne/232/Nine/RacistScientists.html. See also Maiocchi 1999; "*Troppi Ebrei al Consiglio Nazionale delle Ricerche*," *La Vita Italiana*, July 31, 1938; correspondence, A. Giannini and O. Sebastiani, August 22–September 22, 1938, ACS, SPD CR, b.145.

52. de Felice 2001, 288.

53. According to Renzo de Felice (2011, 345), it is false that Mussolini decided to persecute Italy's Jews under pressure from Germany, although he was increasingly concerned at this time to remain on the good side of Hitler, and the Nazis were beginning to see the Italian Fascists as weak and indecisive.

54. Leone Paserman, preface to Venzo and Migliau 2003, 7.

55. It can be viewed online on the AIL website, http://www.archivioluce.com/archivio/jsp/schede/fotoPlayer.jsp?doc=27751&db=fotograficoLuceCRONO LOGICO&index=1&id=undefined&section=/.

56. *LT*, May 4, 1938.

57. A. Landini, "*Marconi persuase Mussolini a non attaccare la flotta inglese*," *Domenica del Corriere* (Sunday supplement to *Corriere delle Sera*, March 3, 1963). See also Landini 1960.

58. Pier Ugo Calzolari, "*Honoris causa*," in Giorgi and Valotti 2010, 6.

59. Dunlap 1937, 287 and 339.

60. F.D. Roosevelt, "Address of the President on the Fall of Rome," June 5, 1944 (fireside chat no 29), http://www.presidency.ucsb.edu/ws/index.php?pid=16514.

61. Ray Reynolds, "Last Experiments of Marconi," *Sunday Stars and Stripes, Mediterranean Magazine* supplement (Rome), March 11, 1945. The article was based largely on a revisionist interpretation of his own assessment of Marconi that Solari began making in the Italian press and elsewhere as the war in Europe was ending.

62. There is an extensive correspondence on this matter in OX 1757.

63. R.P. Raikes, publicity manager, MWTC, to publicity manager MIMCC, March 17, 1961, OX 1757.

64. W. Maconachie to L. Reade, March 23, 1961, OX 1757.

65. L.J. King to publicity manager, MWTC, March 17, 1961, OX 1757, cited in Reade 1963, 123.

66. E.A. Payne, "Senatore Marchese Marconi," OX 1757, cited in Reade 1963, 123.

67. G.A. Isted to W. Maconachie, March 23, 1961, OX 1757; second portion cited in Reade 1963, 123.

68. L. Reade to W. Maconachie, March 23, 1961, with last part cited in Reade 1963, 122.

69. W. Maconachie to publicity manager, MWTC, March 28, 1961, OX 1757.

70. Reade 1963, 128. When Reade's book appeared in 1963, an unsigned review in the *Times Educational Supplement* said: "Mr. Reade's explanation that Marconi thought that he was acting for the good of his country is at best charitable and certainly unconvincing." Reade, who considered himself a long-time active opponent of fascism, claimed the comment was a slur on his professional integrity and sued for libel. The *Times* replied that its reviewer had made "an unfortunate choice of words"; Reade had actually described the explanation of Marconi's enthusiasm for fascism as "at best charitable" and "an unsatisfying excuse." The *Times* agreed that there was no question of suggesting that Reade had condoned Marconi's fascist sympathies or explained them away; a settlement was reached by which the Times Publishing Company paid Reade a sum for damages and costs (*LT*, February 12, 1964; *Times Educational Supplement*, February 14, 1964).

71. W.J. Baker 1970, 296.

72. Dowsett 1951, 678, OX 1834.

CHAPTER 36

1. The headline on a review in *24domenica* by historian Valerio Castronovo captured the enigma that still shrouds Marconi's memory in Italy today: *"Genio sagace e spregiudicato"*—which can mean either "wise and open-minded genius" or "shrewd and unscrupulous genius"; of course, he was both and neither, politically naive and sophisticated; a saint and a scoundrel (November 24, 2013). A review in *Cultura* declared Marconi to be "The Italian who invented the future" (November 26, 2013).

2. *"Guglielmo Marconi e la propagazione delle onde elettromagnetiche nell'alta atmosfera,"* in *Guglielmo Marconi*, 1938 commemorative journal of SIPS (*Società italiana per il progresso delle scienze*), translation in Masini 1995, 366.

3. Belrose 2006.

4. See G.H. Clark to F. Mullen, memo, August 26, 1938, GCR 15/29, about trying to organize some kind of US memorial for Marconi, and its possible publicity value for RCA.

5. MWTC of America v. United States, 320 US 1 (1943).

6. Ibid.

7. The new title was chosen in 1963 "to avoid the restrictive description of 'wireless telegraphy' as the main occupation of the Company, whose interests include radar, sound and television broadcasting, telecommunications, aeronautical and maritime navigation and communication equipment, and the latest types of data handling displays and air traffic control computing systems" (*Marconi News*, August 21, 1963, OX 1773).

8. "Company History," Catalogue of the Marconi Archives, Bodleian Library, University of Oxford; see also, "Ericsson to Acquire Key Assets of Marconi's Telecommunications Business," October 25, 2005, http://www.ericsson.com/news/1017515. The transaction was completed on January 23, 2006.

9. Christie's, "The Marconi Archive Centenary" (press release), January 29, 1997, OX 1864; "Marconi Daughter Joins Battle to Save Archive," *LT*, February 10, 1997; Lord Asa Briggs, letter, *LT,* February 10, 1997; G. Pattie, chairman GEC-Marconi Ltd., letter, *LT*, February 12, 1997; Science Museum, London, "The Marconi Collection" (press release), March 25, 1997, OX 1864.

10. See http://www.bodley.ox.ac.uk/dept/scwmss/wmss/online/modern/marconi/marconi.html.

11. This tale is told, among others, by Ross King (2000, 42).

12. W.H. Preece, "Signalling through Space Without Wires," Royal Institution lecture, London, June 4, 1897, published in *Electrician* 39 (1897): 216–18 and *Proceedings of the Royal Institution* 15 (1896–97): 467-76; also cited in Hong 2001, 40.

13. C.S. Franklin, "Notes by Mr Franklin," undated typescript, OX 683.

14. Hong 2001, 23.

15. "Journey from the Edge of Space: Father and Son Capture Amazing Video Footage...Using an iPhone and HD Camera," *Daily Mail*, October 11, 2010; "Special Report: Personal Technology," *Economist*, October 8, 2011. This was the same issue of *The Economist* that reported the death of Steve Jobs in a cover story headlined "The Magician."

16. B. Marconi, Memoir, FP; L. O'Brien 1952, FP.

17. Transcript of lecture by Mrs. George Atkinson Braga (Gioia Marconi), London, November 21, 1985, OX 1771.

18. "Sudden Death of Marconi," *LT*, July 21, 1937.

19. D. Marconi 1962, 87.

20. *Manchester Guardian*, June 23, 1904.

21. "The Playboy Interview: Marshall McLuhan," *Playboy*, March 1969.

22. Toscano 2012, xi–xiv.

Bibliography

Listed below are materials consulted during the writing of this book. For ease of access, primary source materials with a single author—such as diaries, memoirs, and unpublished typescripts—appear here. A full list of archival material used during research and writing can be found in Sources and Abbreviations.

Aitken, Hugh G. J. *Syntony and Spark: The Origins of Radio.* New York: John Wiley & Sons, 1976.

Aitken, Hugh G. J. *The Continuous Wave: Technology and American Radio, 1900–1932.* Princeton, New Jersey: Princeton University Press, 1985.

Albrecht-Carrié, René. *Italy at the Paris Peace Conference.* New York: Columbia University Press, 1938.

Andreoli, Annamaria, ed. *Onde d'inchiostro. Marconi, D'Annunzio storia di un'amicizia.* Texts collected by Giorgio Zanetti. Bologna: Abacus, 2004.

Andrew, Christopher. *The Defence of the Realm. The Authorized History of MI5.* London: Allen Lane, 2009.

Andrews, Ian. "Telegraphic Language," 1991. http://ian-andrews.org/texts/telegraphic.pdf.

Anonymous. *Broadcasting and Peace. Studies and Projects in the Matter of International Agreements.* Paris: League of Nations International Institute of Intellectual Co-operation, 1933.

Archer, Gleason L. *History of Radio to 1926.* New York: American Historical Society, 1938.

Arendt, Hannah. *Eichmann in Jerusalem. A Report on the Banality of Evil.* New York: Viking Press, 1963.

Bailey, V.A., and K. Landecker. "On the 'Inventor' of Radio-Communication." *Australian Journal of Science* 9 (February 1947): 126ff.

Baker, Duncan C. "Wireless Telegraphy in South Africa at the Turn of the Twentieth Century." In *History of Wireless,* edited by T.K. Sarkar et al., 421–54. Hoboken, New Jersey: John Wiley & Sons, 2006.

Baker, W.J. *A History of the Marconi Company (1874–1965).* London: Methuen, 1970.

Balbi, Gabriele. "Radio before Radio: Araldo Telefonico and the Invention of Italian Broadcasting." *Technology and Culture* 51, no. 4 (October 2010): 786–808.

Balbi, Gabriele. *La radio prima della radio. L'Araldo Telefonico e l'invenzione del broadcasting in Italia*. Rome: Bulzoni Editore, 2010.

Baldoli, Claudia. *Exporting Fascism. Italian Fascists and Britain's Italians in the 1930s*. Oxford/New York: Berg, 2003.

Baring-Gould, Sabine. *Historic Oddities and Strange Events*. London: Methuen, 1889.

Barlow, John Perry. *A Declaration of the Independence of Cyberspace*, 1996. https://projects.eff.org/~barlow/Declaration-Final.html.

Barnouw, Erik. *A History of Broadcasting in the United States. Volume 1: A Tower in Babel, to 1933*. New York: Oxford University Press, 1966.

Bea, Fernando, and Alessandro de Carolis. *Ottant'anni della Radio del Papa*. Vatican City: Libreria Editrice Vaticana, 2011.

Begbie, Harold. *Wireless in the War*. Typescript, 1919. University of Oxford, Bodleian Library, ms. Marconi 361.

Belrose, John S. "The Development of Wireless Telegraphy and Telephone, and Pioneering Attempts to Achieve Transatlantic Wireless Communications." In *History of Wireless*, edited by T.K. Sarkar et al., 349–420. Hoboken, New Jersey: John Wiley & Sons, 2006.

Benjamin, Louise M. "In Search of the Sarnoff 'Radio Music Box' Memo." *Journal of Broadcasting and Electronic Media* 37, no. 3 (1993): 325–35.

Berezin, Mabel. *Making the Fascist Self: The Political Culture of Interwar Italy*. Ithaca, NY: Cornell University Press, 1997.

Bernabei, Alfio. *Esuli ed emigrati Italiani nel Regno Unito 1920–1940*. Milan: Ugo Mursia Editore, 1997.

Bertini, Francesca. *Il resto non conta*. Pisa: Giardini, 1969.

Bettòlo, G.B. Marini. "Guglielmo Marconi: Personal Memories and Documents." *Rivista di Storia della Scienza* 3 (1986): 447–58.

Bird, Roger, ed. *Documents of Canadian Broadcasting*. Ottawa: Carleton University Press, 1988.

Bodanis, David. *E=mc². A Biography of the World's Most Famous Equation*. Toronto: Anchor Canada, 2000.

Bondyopadhyay, Probir K. "Fleming and Marconi: The Cooperation of the Century." *Radioscientist* 5, no. 2 (June 1994): 52–60.

Bosworth, R.J.B. *Mussolini's Italy. Life under the Dictatorship*. London: Penguin, 2006.

Brandon, Ruth. *The Life and Many Deaths of Harry Houdini*. New York: Random House, 2003. Originally published in 1993.

Briggs, Asa, and Peter Burke. *A Social History of the Media. From Gutenberg to the Internet*. Cambridge (UK): Polity, 2009.

Bruton, Elizabeth, and Graeme Gooday. "Collaboration then Competition. Marconi and the British Post Office, 1896–1906." In *Guglielmo Marconi. Wireless Laureate*, edited by M. Giorgi and B. Valotti, 20–32. Bologna: Bononia University Press, 2010.

Burrows, A.R. *The Story of Broadcasting*. London: Cassell, 1924.

Bussey, Gordon. *Marconi's Atlantic Leap*. Coventry (UK): Marconi Communications, 2000.

Cagiano de Azevedo, Paola, and Elvira Gerardi, eds. *Reale Accademia d'Italia: Inventario dell'archivio*. Rome: Ministero per i Beni Culturali e Ambientali, 2005.

Campbell, Timothy C. *Wireless Writing in the Age of Marconi*. Minneapolis: University of Minnesota, 2006.

Canali, Mauro. *Il delitto Matteotti. Affarismo e politica nel primo governo Mussolini*. Bologna: Il Mulino, 1997.

Canali, Mauro. *Le spie del regime*. Bologna: Il Mulino, 2004.

Canali, Mauro. "Crime and Repression." In *The Oxford Handbook of Fascism*, edited by R.J.B. Bosworth, 221–38. New York: Oxford University Press, 2009.

Candow, James E. *The Lookout: A History of Signal Hill*. St. John's, NL: Creative Publishers, 2011.

Canziani, Arnaldo. "Between Politics and Double Entry. The Crisis of Banca Italiana di Sconto and Banco di Roma and the Italian Banking System, 1914–1924." Università degli Studi di Brescia, Dipartimento di Economia Aziendale, Paper No. 57, December 2006.

Capristo, Annalisa. "*L'esclusione degli ebrei dall'Accademia d'Italia*." *La Rassegna Mensile di Israel* 67, no. 3 (2001): 1–36.

Capristo, Annalisa. "*Tullio Levi-Civita e l'Accademia d'Italia*." *Rassegna Mensile di Israel* 69, no, 1 (2003): 237–54.

Capristo, Annalisa. "The Exclusion of Jews from Italian Academies." In Joshua D. Zimmerman, editor, *Jews in Italy under Fascist and Nazi Rule, 1922–1945*, 81–95. Cambridge (UK): Cambridge University Press, 2005.

Catania, Basilio. "*Alla ricerca della verità su Antonio Meucci e sulla invenzione del telefono*." *L'Elettrotecnica* 77, no. 10 (1990): 937–43.

Catania, Basilio. "*Sulle tracce di Antonio Meucci—Appunti di viaggio*." *L'Elettrotecnica* 79, no. 10 (1992): 973–84.

Catania, Basilio. "The U.S. Government versus Alexander Graham Bell: An Important Acknowledgement for Antonio Meucci." *Bulletin of Science, Technology and Society* 22 (2002): 426–42.

Cecil, Robert. *Life in Edwardian England*. London: B.T. Batsford, 1969.

Chambers, Anne. *La Sheridan, Adorable Diva: Margaret Burke Sheridan, Irish Prima Donna, 1889–1958*. Dublin: Wolfhound Press, 1989.

Charnovitz, Steve. "The Emergence of Democratic Participation in Global Governance (Paris, 1919)." *Indiana Journal of Global Legal Studies* 10, no. 1 (2003): 45–77.

Chesterton, G.K. *The Autobiography of G. K. Chesterton*. New York: Sheed & Ward, 1936.

Chiaberge, Riccardo. *Wireless. Scienza, amori e avventure di Guglielmo Marconi*. Milan: Garzanti, 2013.

Ciano, Galeazzo. *Diary 1937–1943*. London: Phoenix Press, 2002.

Clark, George H. Untitled account of Marconi's 1933 visit to the US, 1936.

Typescript, University of Oxford, Bodleian Library, ms. Marconi 36, folios 112–48.

Clarke, Paddy. "Marconi's Irish Connections Recalled." In *100 Years of Radio*. IEE Conference Publication 411, September 5–7, 1995, 20–25.

Costain, Thomas B. *The Chord of Steel*. New York: Doubleday, 1960.

Crawford, Elisabeth T. *The Beginnings of the Nobel Institution. The Science Prizes, 1901–1915*. Cambridge (UK): Cambridge University Press / Paris: Éditions de la Maison des sciences de l'homme, 1984.

Crawford, Elisabeth T. *The Nobel Population, 1901–1937: A Census of the Nominators and Nominees for the Prizes in Physics and Chemistry*. Berkeley: Office for History of Science and Technology, University of California, Berkeley, 1987.

Crespi, Silvio. *Alla difesa d'Italia in Guerra e a Versailles*. Milan: Mondadori, 1937.

D'Annunzio, Gabriele. *The Child of Pleasure (Il Piacere)*. New York/Berlin: Mondial Books, 2006. Originally published in 1898.

Dalle Donne, Giancarlo, and Barbara Valotti. "Practice vs. Theory. Behind the Scenes of Marconi's Success: A Lively Controversy (1897–1909)." In *Guglielmo Marconi. Wireless Laureate*, edited by M. Giorgi and B. Valotti, 64–74. Bologna: Bononia University Press, 2010.

Dam, H.J.W. "The New Telegraphy. An Interview with Signor Marconi." *Strand Magazine* 13, no. 35 (March 1897): 273–80.

de Felice, Renzo. *The Jews in Fascist Italy. A History*. New York: Enigma Books, 2001.

de Forest, Lee. *Father of Radio: The Autobiography of Lee de Forest*. Chicago: Wilcox & Follett, 1950.

de Sousa, Leon. *Wireless Telegraphy 1895–1919, by Guglielmo Marconi*. Typescript, 1919. University of Oxford, Bodleian Library, ms. Marconi 55.

di Benedetto, Giovanni, ed. *Bibliografia marconiana*. Rome: Consiglio Nazionale delle Ricerche, 1974.

di Benedetto, Giovanni. "Marconi Chronology." In *Bibliografia marconiana*, edited by G. di Benedetto, 135–72. Rome: Consiglio Nazionale delle Ricerche, 1974.

Di Marco, Umberto. Untitled typescript, March 1945. University of Oxford, Bodleian Library, ms. Marconi 1757 and George H. Clark Radioana Collection, Archives Center, National Museum of American History, Smithsonian Institution, Washington, DC, Box 30, folder 2.

Densham, William. Diaries, 1899–1900. University of Oxford, Bodleian Library, ms. Marconi 4007.

Donaldson, Frances. *The Marconi Scandal*. London: Rupert Hart-Davis, 1962.

Douglas, Susan J. *Inventing American Broadcasting, 1899–1922*. Baltimore: Johns Hopkins University Press, 1987.

Dowsett, Henry M. *History of the Marconi Company*, 1951. Typescript, University of Oxford, Bodleian Library, ms. Marconi 1834.

Doyle, Arthur Conan. *The History of Spiritualism*, Volume 1. New York: G.H. Doran, 1926.

Doyle, Arthur Conan. *The Edge of the Unknown*. New York: G.P. Putnam's Sons, 1930.

Doyle, Denis Conan. "My Father—Spiritual Marconi." *Scribner's Commentator* 1, no. 5 (1937): 14–19.

Dragoni, Giorgio. "*Augusto Righi e Guglielmo Marconi tra didattica e ricerca: trace e testimonianze di una collaborazione.*" In *Cento anni di radio da Marconi al futuro delle telecomunicazioni*, edited by G. Paoloni, F. Monteleone and M.G. Ianniello, 11–17. Venice: Marsilio, 1995.

Dunlap, Orrin E., Jr. *Marconi: The Man and His Wireless*. New York: Macmillan, 1937.

Dunlap, Orrin E., Jr. *100 Men of Science. Biographical Narratives of Pathfinders in Electronics and Television*. New York/London: Harper, 1944.

Fahie, J.J. *A History of Wireless Telegraphy 1838–1899*. Edinburgh & London: William Blackwood and Sons, 1899.

Falchero, Anna Maria. *La Banca italiana di sconto 1914–21. Sette anni di guerra*. Milan: Franco Angeli, 1990.

Faleni, Lorenzo, and Amadeo Serafini. *Album Marconi*. Buenos Aires, 1904.

Feldman, Burton. *The Nobel Prize. A History of Genius, Controversy, and Prestige*. New York: Arcade, 2000.

Fessenden, Helen M. *Fessenden: Builder of Tomorrows*. New York: Coward-McCann, 1940.

Fessenden, Reginald A. "The Inventions of Reginald A. Fessenden." *Radio News*, January-November 1925.

Figueira de Faria, Miguel. *Companhia Portuguesa Radio Marconi. 75 anos de Communicaçoes Internacionais*. Lisbon: Companhia Portuguesa Radio Marconi, 2000.

Fisher, Desmond. *Broadcasting in Ireland*. London: Routledge & Kegan Paul, 1978.

Fleming, James Ambrose. "Marconi and the Development of Radiocommunication." *Journal of Royal Society of Arts* XXVI (1937): 42–63.

Foot, John. *Italy's Divided Memory*. New York/Basingstoke: Palgrave Macmillan, 2009.

Friedenwald, Michael. "The Beginnings of Radio Communication in Germany, 1897–1918." *Journal of Radio Studies* 7, no. 2 (2000): 441–63.

Gagné, Wallace, ed. *Nationalism, Technology and the Future of Canada*. Toronto: Macmillan, 1976.

Gamba, Aldo, and Pierangelo Schiera, eds. *Fascismo e scienza: le celebrazioni voltiane e il congresso internazionale dei fisici del 1927*. Bologna: Il Mulino, 2005.

Gervasi, Manfredo. "*Una lettera inedita di Marconi sui primi esperimenti in Inghilterra.*" *Giornale di Fisica* 14, no. 3 (1973): 280–86.

Giacomelli, Alfeo, and Giorgio Bertocchi. "*Guglielmo Marconi: come nasce un genio. Le origini montane e l'ascesa della famiglia.*" *Nuèter—Ricerche* (Bologna/Porretta Terme), no. 2 (1994): 161–92.

Giorgi, Mario, and Barbara Valotti, eds. *Guglielmo Marconi. Wireless Laureate*. Bologna: Bononia University Press, 2010.

Goethe, Johann Wolfgang von. *Italian Journey*. London: William Collins & Sons, 1962. Originally published in 1816.

Gray, Charlotte. *Reluctant Genius: The Passionate Life and Inventive Mind of Alexander Graham Bell*. Toronto: HarperCollins Canada, 2006.

Gray, John. *The Immortalization Commission: Science and the Strange Quest to Cheat Death*. New York: Farrar, Straus and Giroux, 2011.

Guagnini, Anna. "*Il Comitato di radiotelegrafia e gli sviluppi delle radiocomunicazioni*." In *Per una storia del Consiglio Nazionale delle Ricerche*, Vol. 1, edited by R. Simili and G. Paoloni, 365–405. Rome-Bari: Editori Laterza, 2001.

Guagnini, Anna. "Patent Agents, Legal Advisers and Guglielmo Marconi's Breakthrough in Wireless Telegraphy." *History of Technology* 24 (2002): 171–201.

Guagnini, Anna. "*Dall'invenzione all'impresa. Marconi e la Wireless Telegraph Company*." In *Storia, Scienza e Società. Ricerche sulla scienza in Italia nell'età moderna e contemporanea*, edited by Paola Govoni, 175–212. Università di Bologna, Dipartimento di Filosofia, Centro Internazionale per la Storia della Università e della Scienza, 2006.

Guerin, Daniel. *Fascism and Big Business*. New York: Pathfinder, 1973. Originally published in 1939.

Hall, H. Cuthbert. "The Marconi System and the Berlin Conference: The Case for the English Company." *The Empire Review* (July 1906): 3–15.

Hall, H. Cuthbert. "The Flaw in British Government." *The National Review* (August 1915): 874–78.

Hardin, Herschel A. *A Nation Unaware: The Canadian Economic Culture*. North Vancouver: J.J. Douglas, 1974.

Haynes, Renée. *The Society for Psychical Research, 1882–1982. A History*. London & Sydney: Macdonald & Company, 1982.

Headrick, Daniel R. *The Tentacles of Progress: Technology Transfer in the Age of Imperialism, 1850–1940*. New York: Oxford University Press, 1988.

Headrick, Daniel R. *The Invisible Weapon: Telecommunications and International Politics, 1851–1945*. New York: Oxford University Press, 1991.

Heather, Alanna. *Errislannan: Scenes from a Painter's Life*. Dublin: The Lilliput Press, 1993.

Hellemans, Frank. "Towards Techno-Poetics and Beyond." In *The Turn of the Century*, edited by Christian Berg, Frank Durieux, and Geert Lernout, 291–301. Berlin: De Gruyter, 1995.

Herman, Ros. "The Italian Experiment." *New Scientist* 1401 (March 15, 1984): 40–46.

Hilmes, Michele. "Who We Are, Who We Are Not: Battle of the Global Paradigms." In *Planet TV: A Global Television Reader*, edited by Lisa Parks and Shanti Kumar, 53–73. New York: New York University Press, 2002.

Hochschild, Adam. *King Leopold's Ghost: A Story of Greed, Terror, and Heroism in Colonial Africa*. New York: Houghton Mifflin, 1998.

Holman, David Emory. *The Holmans in America*. New York: Grafton Press, 1909.

Holman, Josephine Bowen. Diary, January–April 1902. Guglielmo Marconi correspondence, Huntington Library, San Marino, California, Box 2, folder 38.

Hong, Sungook. *Wireless: From Marconi's Black-Box to the Audion*. Cambridge, MA: MIT Press, 2001.

Houdini, Harry. *A Magician among the Spirits*. New York: Harper & Brothers, 1924.

Howe, Daniel Walker. *What Hath God Wrought: The Transformation of America, 1815–1848*. New York: Oxford University Press, 2007.

Howeth, Linwood S. *History of Communications-Electronics in the United States Navy*. Washington, DC: United States Government Printing Office, 1963.

Hughes, Michael, and Katherine Bosworth. *Titanic Calling. Wireless Communications During the Great Disaster*. Oxford: The Bodleian Library, 2012.

Hughes, Robert. *Rome: A Cultural, Visual and Personal History*. New York: Knopf, 2011.

Hughes, Thomas P. *Networks of Power: Electrification in Western Society, 1880–1930*. Baltimore and London: Johns Hopkins University Press, 1983.

Hughes-Hallett, Lucy. *The Pike: Gabriele d'Annunzio. Poet, Seducer & Preacher of War*. London: Fourth Estate, 2013.

Ianniello, Maria Grace. "*Marconi e la comunità dei fisici italiani (1927–1931)*." In *Cento anni di radio da Marconi al futuro delle telecomunicazioni*, edited by G. Paoloni, F. Monteleone, and M.G. Ianniello, 29–33. Venice: Marsilio, 1995.

Innis, Harold A. *Empire and Communications*. Oxford: Clarendon Press, 1950.

Innis, Harold A. *The Bias of Communication*. Toronto: University of Toronto Press, 1951.

Jackson, Stanley. *The Savoy: The Romance of a Great Hotel*. London: Frederick Muller, 1964.

Jacot, B.L., and D.M.B. Collier. *Marconi—Master of Space: An Authorized Biography of the Marchese Marconi*. London: Hutchinson & Company, 1935.

Jolly, W.P. *Marconi*. New York: Stein & Day, 1972.

Jome, Hiram Leonard. *Economics of the Radio Industry*. Chicago: A.W. Shaw, 1925.

Julius, Anthony. *Trials of the Diaspora. A History of Anti-Semitism in England*. Oxford: Oxford University Press, 2010.

Kemp, George S. Diaries, 1898–1932. University of Oxford, Bodleian Library, mss. Marconi 56–88.

Kemp, George S. Expanded extracts from diaries, 1897–1920. University of Oxford, Bodleian Library, mss. Marconi 89–111.

Kertzer, David I. *The Popes against the Jews. The Vatican's Role in the Rise of Modern Anti-Semitism*. New York: Knopf, 2001.

Kertzer, David I. *The Pope and Mussolini: The Secret History of Pius XI and the Rise of Fascism in Europe*. New York: Random House, 2014.

King, Ross. *Brunelleschi's Dome: How a Renaissance Genius Reinvented Architecture*. New York: Penguin, 2000.

King, Ross. *Defiant Spirits. The Modernist Revolution of the Group of Seven*. Vancouver: Douglas & McIntyre, 2010.

Kittler, Friedrich A. *Grammophon, Film, Typewriter*. Berlin: Brinkmann & Bose, 1986.

Knox, MacGregor. *Common Destiny: Dictatorship, Foreign Policy and War in Fascist Italy and Nazi Germany*. Cambridge (UK): Cambridge University Press, 2000.

Kramer, Rita. *Maria Montessori*. New York: G.P. Putnam's Sons, 1976.

Landini, Adelmo. *Marconi sulle vie dell'etere: La storica impresa narrata dall'Ufficiale Marconista dell'Elettra*, 2nd ed. Turin: Società Editrice Internazionale, 1960.

Lansbury, George. *My Pilgrimage for Peace*. New York: Henry Holt and Co., 1938.

Larson, Erik. *Dead Wake: The Last Crossing of the* Lusitania. New York: Crown, 2015.

Ledeen, Michael A. *D'Annunzio: The First Duce*. Baltimore: Johns Hopkins University Press, 1977.

Levi, Lia. *The Jewish Husband. A Story of Love and Secrets in Fascist Italy*. New York: Europa Editions, 2009.

Lindell, I.V. "Wireless before Marconi." In *History of Wireless*, edited by T.K. Sarkar et al., 247–66. Hoboken, New Jersey: John Wiley & Sons, 2006.

Lodge, Oliver. *Talks About Wireless*. London: Cassell, 1925.

Luconi, Stefano. "Recent Trends in the Study of Italian Anti-Semitism under the Fascist Regime." *Patterns of Prejudice* 38, no. 1 (2004): 1–17.

Ludwig, Emil. *Mussolinis Gespräche mit Emil Ludwig*. Berlin: P. Zsolnay Verglag, 1932.

Lumsden, Linda J. *Inez: The Life and Times of Inez Milholland*. Bloomington: Indiana University Press, 2004.

Lynch, Brendan. *Yesterday We Were in America. Alcock and Brown. First to Fly the Atlantic Non-Stop*. Somerset: Haynes Publishing, 2009.

Lyons, Eugene. *David Sarnoff: A Biography*. New York: Harper & Row, 1966.

Lyttelton, Adrian. "Futurism, Politics, and Society." In *Italian Futurism 1909–1944: Reconstructing the Universe*, edited by Vivien Greene, 58–78. New York: Guggenheim Museum Publications, 2014.

Mackinnon, William Alexander. *On the Rise, Progress and Present State of Public Opinion in Great Britain and Other Parts of the World*. Shannon: Irish University Press, 1971. Originally published in 1828 (London: Saunders and Otley).

Maclaurin, W. Rupert. *Invention and Innovation in the Radio Industry*. New York: Macmillan, 1949.

MacLeod, Mary K. "A Report on Guglielmo Marconi's Two Stations in Glace Bay, 1902–1946." Typescript, June 1982. Guglielmo Marconi Collection, Beaton Institute, University College of Cape Breton, Sydney, Nova Scotia.

MacLeod, Mary K. *Whisper in the Air: Marconi, The Canada Years, 1902–1946*. Hantsport, NS: Lancelot Press, 1992.

MacMillan, Margaret. *Paris 1919*. New York: Random House, 2003.

MacMillan, Margaret. *The War that Ended Peace. The Road to 1914*. Toronto: Allen Lane, 2013.

Magder, Ted. "The Origins of International Agreements and Global Media: The Post, the Telegraph, and Wireless Communication Before World War I." In *The Handbook of Global Media and Communication Policy*, edited by Robin Mansell and Marc Raboy, 23–39. Malden, MA: Wiley-Blackwell, 2011.

Maiocchi, Roberto. *Scienza italiana e razzismo fascista*. Florence: La Nuova Italia, 1999.

Maiocchi, Roberto. "Fascist Autarky and the Italian Scientists." *Journal of History of Science and Technology* 3 (2009): 62–73.

Manchester, Duke of. *My Candid Recollections*. London: Grayson & Grayson, 1932.

Mango, Andrew. *Atatürk*. London: John Murray, 1999.

Marconi, Beatrice. Memoir (untitled, unpublished manuscript) [1950s]. Francesco Paresce Collection, Bologna.

Marconi, Degna. *My Father, Marconi*. New York: McGraw-Hill, 1962.

Marconi, Guglielmo. Untitled typescript on Marconi's work to 1899, attributed to Marconi, 1899. University of Oxford, Bodleian Library, ms. Marconi 48, folios 1–147.

Marconi, Guglielmo. Diaries and Notebooks, 1901–26. University of Oxford, Bodleian Library, mss. Marconi 1–16.

Marconi, Guglielmo. *Brief Story of My Life*, 1913. Printed manuscript, University of Oxford, Bodleian Library, ms. Marconi 54.

Marconi, Guglielmo. "My Laboratories Ashore and Afloat," circa 1923. Typescript, University of Oxford, Bodleian Library, ms. Marconi 52, folios 223–30.

Marconi, Guglielmo. "How I Made my Discovery: Wireless. Marconi" [1937]. Typescript, Archivio Guglielmo Marconi, Accademia Nazionale dei Lincei, Rome, Box 3, folder 124, folios 3181–7.

Marconi, Guglielmo. "Scienza e Fascismo, 1932." In *Scritti di Guglielmo Marconi*, G. Marconi, 407–18. Rome: Reale Accademia d'Italia, 1941.

Marconi, Guglielmo. *Scritti di Guglielmo Marconi*. Rome: Reale Accademia d'Italia, 1941.

Marconi, Maria Cristina. *Marconi My Beloved*. With Elettra Marconi. Boston: Dante University of America Press, 1999.

Marignoli, Liborio. Memoir (unpublished manuscript), n.d. Francesco Paresce Collection, Bologna (also in G. Masini, *Guglielmo Marconi*, Turin: Unione Tipografico-Editrice Torinese, 1975, 452–56).

Marinetti, Filippo Tommaso. "Destruction of Syntax—Wireless Imagination—Words-in-Freedom, 1913." In *Modernism: An Anthology*, edited by Lawrence Rainey et al, 27–34. Hoboken, New Jersey: Wiley-Blackwell, 2005.

Marnham, Patrick. *Wild Mary*. London: Chatto & Windus, 2006.

Marpicati, Arturo. *Questi nostri occhi*. Turin: Società Editrice Internazionale, 1954.

Masini, Giancarlo. *Guglielmo Marconi*. Turin: Unione Tipografico-Editrice Torinese, 1975.

Masini, Giancarlo. *Marconi*. New York: Marsilio Publishers, 1995.

Maskelyne, Jasper. *White Magic: The Story of Maskelynes*. London: S. Paul & Company, 1936.

Massie, Keith, and Stephen D. Perry. "Hugo Gernsback and Radio Magazines: An Influential Intersection in Broadcast History." *Journal of Radio Studies* 9, no. 2 (2002): 264–81.

Matelski, Marilyn J. *Vatican Radio. Propagation by the Airwaves*. Westport, CT: Praeger, 1995.

McChesney, Robert W. *Telecommunications, Mass Media, and Democracy: The Battle for the Control of U.S. Broadcasting, 1928–1935*. New York: Oxford University Press, 1993.

McGrath, P.T. "Marconi and His Transatlantic Signal." *The Century* 63, no. 92 (1902): 769–82.

McLuhan, Marshall. *Understanding Media*. New York: McGraw Hill, 1964.

McMeekin, Sean. *The Berlin–Baghdad Express*. London: Penguin, 2011.

Milholland, John Elmer. Diaries, 1903–17. Milholland Family Papers, Ticonderoga Historical Society, New York.

Monteleone, Franco. *La radio italiana nel periodo fascista. Studio e documenti: 1922–1945*. Venice: Marsilio, 1976.

Monticone, Alberto. *Il Fascismo al microfono. Radio e politica in Italia (1924–1945)*. Rome: Edizioni Studium, 1978.

Morel, Edward D. *The Black Man's Burden: The White Man in Africa from the Fifteenth Century to World War I*. Manchester: National Labour Press, 1920.

Moseley, Ray. *Mussolini's Shadow: The Double Life of Count Galeazzo Ciano*. New Haven, CT: Yale University Press, 1999.

Mumford, Lewis. *The Myth of the Machine: Technics and Human Development*. New York: Harcourt, Brace & World, 1966.

Mussolini, Rachele. *Mussolini privato*. Milan: Rusconi, 1980.

Nadel, Ira B., ed. *The Cambridge Companion to Ezra Pound*. Cambridge (UK): Cambridge University Press, 1999.

O'Brien, Lilah. Memoir (unpublished manuscript), 1952. Francesco Paresce Collection, Bologna.

Pagé, Pierre. *Histoire de la radio au Québec*. Montréal: Fides, 2007.

Pakenham, Thomas. *The Scramble for Africa, 1873–1910*. London: Weidenfeld and Nicolson, 1991.

Paoloni, G., F. Monteleone, and M.G. Ianniello, eds. *Cento anni di radio da Marconi al futuro delle telecomunicazioni*. Venice: Marsilio, 1995.

Paoloni, Giovanni, and Raffaella Simili. *Guglielmo Marconi e l'Italia: Mostra storico-documentaria: Roma 30 marzo–30 aprile 1996*. Rome: Accademia Nazionale dei Lincei, 1996.

Paoloni, Giovanni, and Raffaella Simili. "Vito Volterra and the Making of Research Institutions in Italy and Abroad." In *The Migration of Ideas,* edited by Roberto Scazzieri and Raffaella Simili, 123–50. Sagamore Beach, MA: Watson Publishing International, 2008.

Paresce, Francesco. "Personal Reflections of an 'Italian Adventurer'." Marconi Society website. www.marconisociety.org/aboutus/familyhistory/francescobiography .html.

Partheil, Gustav. *Die drahtlose Telegraphie und Telephonie*, 2nd ed. Berlin: Gerdes & Hödel, 1907.

Partheil, Gustav. *Der gegenwärtige Stand der drahtlosen Telegraphie und Telephonie*. Berlin: Gerdes & Hödel, 1910.

Pession, Giuseppe. *Marconi*. Turin: Unione Tipografico, 1941.

Pine, Richard. *2RN and the Origins of Irish Radio*. Dublin: Four Courts Press, 2002.

Prescott, Daisy. "Reminiscences" (unpublished manuscript), 1910. Francesco Paresce Collection, Bologna.

Raboy, Marc. *Missed Opportunities: The Story of Canada's Broadcasting Policy*. Montreal and Kingston: McGill-Queen's University Press, 1990.

Rapaport, Brooke Kamin. *Houdini: Art and Magic*. New York: The Jewish Museum, 2010.

Reade, Leslie. *Marconi and the Discovery of Wireless*. London: Faber and Faber, 1963.

Reith, J.C.W. *Into the Wind*. London: Hodder & Stoughton, 1949.

Rittenberg, Max. *Every Man His Price*. New York: G.W. Dillingham, 1914.

Rollo, M.F., and M.I. Queiroz. *Marconi em Lisboa*. Lisbon: Fundação Portugal Telecom, 2007.

Roman-Vatican Settlement. Translation of the Lateran documents of February 11, 1929. London: Royal Institute of International Affairs, 1929.

Rotscheid, Emil, and Erich Quäck. "Transradio." In *25 Jahre Telefunken. Festschrift, 1903–1928*, edited by Fritz Schröter, 196–211. Berlin: Telefunken, 1928.

Rowe, Ted. *Connecting the Continents. Heart's Content and the Atlantic Cable*. St. John's, NL: Creative Publishers, 2009.

Rubinstein, Arthur. *My Many Years*. New York: Knopf, 1980.

Sachs, Harvey. *Rubinstein: A Life*. New York: Grove Press, 1995.

Salaris, Claudia. *Alla festa della rivoluzione: Artisti e libertari con D'Annunzio a Fiume*. Bologna: Il Mulino, 2002.

Sandford, Christopher. *Masters of Mystery. The Strange Friendship of Arthur Conan Doyle and Harry Houdini*. New York: Palgrave Macmillan, 2011.

Sarfatti, Michele, and Anne C. Tedeschi. *The Jews in Mussolini's Italy: From Equality to Persecution*. Madison: University of Wisconsin Press, 2006.

Sarkar, T.K., R.J. Mailloux, A.A. Oliner, M. Salazar-Palma, and D.L. Sengupta, eds. *History of Wireless*. Hoboken, New Jersey: John Wiley & Sons, 2006.

Satia, Priya. "War, Wireless, and Empire. Marconi and the British Warfare State, 1896–1903." *Technology and Culture* 51, no. 4 (October 2010): 829–53.

Sato, Gentei, and Motoyuki Sato. "The Antenna Development in Japan: Past and Present." In *History of Wireless*, edited by T.K. Sarkar et al., 455–72. Hoboken, New Jersey: John Wiley & Sons, 2006.

Savorgnan di Brazzà, Francesco. *Da Leonardo a Marconi*. Milan: U. Hoepli Editore, 1939.

Schueler, Donald G. "Inventor Marconi: Brilliant, Dapper, Tough to Live With." *Smithsonian* 12, no. 12 (March 1982): 126–47.

Searle, G.R. *Corruption in British Politics, 1895–1930*. Oxford: Clarendon Press, 1987.

Seifer, Marc J. *Wizard: The Life and Times of Nikola Tesla. Biography of a Genius*. Secaucus, New Jersey: Birch Lane Press, 1996.

Sexton, Michael. *Marconi: The Irish Connection*. Dublin: Four Courts Press, 2005.

Sheed, F.J. *Nullity of Marriage*. New York: Sheed & Ward, 1959.

Simili, Raffaella. "La Presidenza Volterra." In *Per una storia del Consiglio Nazionale delle Ricerche*, Vol. 1, edited by R. Simili and G. Paoloni, 72–127. Rome-Bari: Editori Laterza, 2001.

Simili, Raffaella. "La Presidenza Marconi." In *Per una storia del Consiglio Nazionale delle Ricerche*, Vol. 1, edited by R. Simili and G. Paoloni, 128–72. Rome-Bari: Editori Laterza, 2001.

Simili, Raffaella, and Giovanni Paoloni, eds. *Per una storia del Consiglio Nazionale delle Ricerche*, Vol. 1. Rome-Bari: Editori Laterza, 2001.

Slaby, Adolf. "The New Telegraphy." *Century Magazine* 55 (April 1898): 867–74.

Smith, Denis Mack. *Mussolini*. London: Weidenfeld and Nicolson, 1981.

Smith, William. "How Marconi Came to Canada." Typescript, n.d. Library and Archives Canada, Fonds William Smith, MG30-D18, vol. 4, folder 17.

Smyth, Howard McGaw. *Secrets of the Fascist Era: How Uncle Sam Obtained Some of the Top-Level Documents of Mussolini's Period*. Carbondale: Southern Illinois University Press, 1975.

Sobel, Robert. *RCA*. New York: Stein and Day, 1986.

Solari, Luigi. *Marconi: Dalla Borgata di Pontecchio a Sydney d'Australia*. Naples: A. Morano, 1928.

Solari, Luigi. *Storia della Radio*. Milan: Mondadori, 1939.

Solari, Luigi. *Sui mare e sui continenti con le onde elettriche. Il trionfo di Marconi*. Milan: Fratelli Bocca Editori, 1942.

Solari, Luigi. *Guglielmo Marconi*. Bologna: Odoya, 2011. Originally published as *Marconi: Nell'intimità e nel lavoro* (Milan: Mondadori, 1940).

Spaulding, Stacy. "Totalitarian Refugee or Fascist Mistress? Comparing Lisa Sergio's Autobiography to Her FBI File." *Journalism History* 34, no. 3 (2008): 145–54.

Stemman, Roy. *Spirit Communication*. London: Piatkus, 2005.

Sterling, Christopher H. *History of Telecommunications Technology: An Annotated Bibliography*. Lanham, MD: Scarecrow Press, 2000.

Sterling, Christopher H., ed. *Military Communications: From Ancient Times to the 21st Century*. Santa Barbara, CA: ABC-CLIO, 2007.

Strang, G. Bruce. "War and Peace: Mussolini's Road to Munich." *Diplomacy and Statecraft* 10, no. 2–3 (1999): 160–90.

Stross, Randall E. *The Wizard of Menlo Park. How Thomas Alva Edison Invented the Modern World*. New York: Crown, 2007.

Stuart, Charles, ed. *The Reith Diaries*. London: Collins, 1975.

Sullivan, Brian R. "'A Highly Commendable Action': William J. Donovan's Intelligence Mission for Mussolini and Roosevelt, December 1935–February 1936." *Intelligence and National Security* 6, no. 2 (1991): 334–66.

Sullivan, Brian R. "Introduction." In *My Fault: Mussolini as I Knew Him*, Margherita Sarfatti, edited by Brian R. Sullivan. New York: Enigma Books, 2014.

Symons, A.J.A. *The Quest for Corvo. An Experiment in Biography*. New York: New York Review of Books, 2001. Originally published in 1934 (London: Cassell).

Tabarroni, Giorgio. "Marconi's Formative Years and the Development of His Personality." In *Bibliografia marconiana*, edited by G. di Benedetto, 49–58. Rome: Consiglio Nazionale delle Ricerche, 1974.

Tarrant, D.R. *Marconi's Miracle: The Wireless Bridging of the Atlantic*. St. John's, NL: Flanker Press, 2001.

Taylor, A.J.P. *Beaverbrook: A Biography*. New York: Simon and Schuster, 1972.

Tesla, Nikola. *My Inventions. The Autobiography of Nikola Tesla*. Rockville, MD: Wildside Press, 2005.

Toscano, Aaron A. *Marconi's Wireless and the Rhetoric of a New Technology*. Dordrecht: Springer, 2012.

Turchetti, Simone. "The Invisible Businessman: Nuclear Physics, Patenting Practices, and Trading Activities in the 1930s." *Historical Studies in the Physical and Biological Sciences* 37, no. 1 (2006): 153–72.

Turchetti, Simone. *The Pontecorvo Affair*. Chicago: University of Chicago Press, 2012.

Vaccari, Olimpia, et al. *Breve storia di Livorno*. Pisa: Pacini Editore, 2006.

Valotti, Barbara. *La Formazione di Guglielmo Marconi*. Laurea thesis, University of Bologna, 1994–95.

Valotti, Barbara. "The Roots of Invention: New Sources on Young Marconi." *Universitas* 7, (January 1995): 1–5.

Valotti, Barbara. "*Oltre il mito dell'autodidatta. Le origini e la formazione di Guglielmo Marconi*." In *Guglielmo Marconi: genio, storia e modernità*, edited by Gabriele Falciasecca and Barbara Valotti. Rome: Mondadori, 2003.

Valotti, Barbara. "Beyond the Myth of the Self-taught Inventor: The Learning Process and Formative Years of Young Guglielmo Marconi." *History of Technology* 32 (2014): 249–76.

Venzo, Manola Ida, and Bice Migliau, eds. *The Racial Laws and the Jewish Community of Rome, 1938–1945*. Rome: Gangemi Editore, 2003.

Villiers-Tuthill, Kathleen. *History of Kylemore Castle and Abbey*. Connemara, Galway: Kylemore Abbey Publications, 2002.

Vyvyan, R.N. *Wireless over Thirty Years*. London: Routledge, 1933.

Walsh, Dan. *Family Industry in Enniscorthy*. [1983?], University of Oxford, Bodleian Library, ms. Marconi 1845.

Walsh, Dan. "Enniscorthy: Industry and Trade." In *Enniscorthy: A History*, edited by Colm Tóibín, 191–209. Wexford: Wexford County Council Public Library Service, 2010.

Ward, Maisie. *Gilbert Keith Chesterton*. London: Sheed & Ward, 1942.

Webster, Richard A., and Gustavo Costa. "*Una lettera inedita di D'Annunzio a Marconi*." *Modern Philology* 74, no. 4 (May 1977): 394–96.

Weightman, Gavin. *Signor Marconi's Magic Box: The Most Remarkable Invention of the 19th Century & the Amateur Inventor Whose Genius Sparked a Revolution*. Cambridge, MA: Da Capo Press, 2003.

Weightman, Gavin. *The Industrial Revolutionaries: The Making of the Modern World, 1776–1914*. New York: Grove Press, 2007.

West, Nigel. *Historical Dictionary of International Intelligence*. Lanham, MD: Scarecrow Press/Rowman & Littlefield, 2006.

Whatley, Michael E. *Marconi Wireless on Cape Cod*. South Wellfleet, MA, 1987.

Wheatcroft, Geoffrey. *The Controversy of Zion or How Zionism Tried to Resolve the Jewish Question*. London: Sinclair-Stevenson, 1996.

Whitcomb, Ed. *A Short History of Nova Scotia*. Ottawa: From Sea to Sea Enterprises, 2009.

Williams, Hari. *Marconi and His Wireless Stations in Wales*. Llanrwst, Wales: Gwasg Carreg Gwalch, 1999.

Williams, Raymond. *Television: Technology and Cultural Form*. London and New York: Routledge, 2003. Originally published in 1974.

Winseck, Dwayne, and Robert Pike. *Communication and Empire: Media, Markets, and Globalization, 1860–1930*. Durham, NC: Duke University Press, 2007.

Worrall, Dan M. "David Edward Hughes: Concertinist and Inventor." *Papers of the International Concertina Association*, Volume 4 (2007), edited by Allan Atlas.

Wright, Peter. *Spycatcher: The Candid Autobiography of a Senior Intelligence Officer*. New York: Viking, 1987.

Wu, Tim. *The Master Switch. The Rise and Fall of Information Empires*. New York: Knopf, 2010.

Zuill, W.S. "The Forgotten Father of Radio." *American Heritage of Invention & Technology* 17, no. 1 (Summer 2001): 40–47.

Index